New Challenges in Neutrosophic Theory and Applications

New Challenges in Neutrosophic Theory and Applications

Editors

Stefan Vladutescu
Mihaela Colhon

MDPI • Basel • Beijing • Wuhan • Barcelona • Belgrade • Manchester • Tokyo • Cluj • Tianjin

Editors
Stefan Vladutescu
University of Craiova
Romania

Mihaela Colhon
University of Craiova
Romania

Editorial Office
MDPI
St. Alban-Anlage 66
4052 Basel, Switzerland

This is a reprint of articles from the Special Issue published online in the open access journal *Mathematics* (ISSN 2227-7390) (available at: https://www.mdpi.com/journal/mathematics/special_issues/New_Challenges_Neutrosophic_Theory_Applications).

For citation purposes, cite each article independently as indicated on the article page online and as indicated below:

LastName, A.A.; LastName, B.B.; LastName, C.C. Article Title. *Journal Name* **Year**, *Article Number*, Page Range.

ISBN 978-3-03943-288-2 (Hbk)
ISBN 978-3-03943-289-9 (PDF)

Contents

About the Editors vii

Preface to "New Challenges in Neutrosophic Theory and Applications" ix

Wadei F. Al-Omeri and Saeid Jafari
On Generalized Closed Sets and Generalized Pre-Closed Sets in Neutrosophic Topological Spaces
Reprinted from: *Mathematics* **2018**, 7, 1, doi:10.3390/math7010001 1

Muhammad Zahir Khan, Muhammad Farid Khan, Muhammad Aslam and Abdur Razzaque Mughal
Design of Fuzzy Sampling Plan Using the Birnbaum-Saunders Distribution
Reprinted from: *Mathematics* **2019**, 7, 9, doi:10.3390/math7010009 13

Derya Bakbak, Vakkas Uluçay and Memet Şahin
Neutrosophic Soft Expert Multiset and Their Application to Multiple Criteria Decision Making
Reprinted from: *Mathematics* **2019**, 7, 50, doi:10.3390/math7010050 23

Vakkas Uluçay and Memet Şahin
Neutrosophic Multigroups and Applications
Reprinted from: *Mathematics* **2019**, 7, 95, doi:10.3390/math7010095 41

Changxing Fan, Jun Ye, Sheng Feng, En Fan and Keli Hu
Multi-Criteria Decision-Making Method Using Heronian Mean Operators under a Bipolar Neutrosophic Environment
Reprinted from: *Mathematics* **2019**, 7, 97, doi:10.3390/math7010097 59

Muhammad Gulistan, Majid Khan, Seifedine Kadry and Khaleed Alhazaymeh
Neutrosophic Cubic Einstein Hybrid Geometric Aggregation Operators with Application in Prioritization Using Multiple Attribute Decision-Making Method
Reprinted from: *Mathematics* **2019**, 7, 346, doi:10.3390/math7040346 75

Florentin Smarandache
Refined Neutrosophy and Lattices vs. Pair Structures and YinYang Bipolar Fuzzy Set
Reprinted from: *Mathematics* **2019**, 7, 353, doi:10.3390/math7040353 91

Changxing Fan, Sheng Feng and Keli Hu
Linguistic Neutrosophic Numbers Einstein Operator and Its Application in Decision Making
Reprinted from: *Mathematics* **2019**, 7, 389, doi:10.3390/math7050389 107

Vasantha Kandasamy W.B., Ilanthenral Kandasamy and Florentin Smarandache
Semi-Idempotents in Neutrosophic Rings
Reprinted from: *Mathematics* **2019**, 7, 507, doi:10.3390/math7060507 119

Vasantha Kandasamy W.B., Ilanthenral Kandasamy and Florentin Smarandache
Neutrosophic Triplets in Neutrosophic Rings
Reprinted from: *Mathematics* **2019**, 7, 563, doi:10.3390/math7060563 127

Muhammad Aslam and Mohammed Albassam
Inspection Plan Based on the Process Capability Index Using the Neutrosophic Statistical Method
Reprinted from: *Mathematics* **2019**, 7, 631, doi:10.3390/math7070631 137

About the Editors

Stefan Vladutescu (Professor Dr.) is a Professor of Communication and Information at University of Craiova, Romania. He is a graduate of University of Craiova and University of Bucharest and obtained his doctorate from University of Bucharest. He is a member of International Association of Communication (ICA) (USA), a member of Neutrosophic Science International Association (Gallup, NM, USA), and serves on the board of the Web of Science journal *Neutrosophic Sets and Systems* (USA) and *Polish Journal of Management Studies* (Poland). He is also director of Social Sciences and Education Research Review (Romania) and a member of the editorial board of *European Scientific Journal* (Macedonia). He is author or co-author of 15 books and more than 100 scientific papers (including ISI/Web of Science articles) and proceedings of international seminars and conferences.

Mihaela Colhon (Assoc. Prof. Dr.) is Associate Professor at Department of Computer Science, University of Craiova, Romania. She received her Ph.D. in 2009 in Computer Science for her work at the Department of Computer Science, Faculty of Mathematics and Computer Science, University of Pitești, Romania. Her research field is artificial intelligence, with specialization in knowledge representation, natural language processing (NLP), and human language technologies (HLT) as well as computational statistics and data analysis with applications in NLP.

Preface to "New Challenges in Neutrosophic Theory and Applications"

Neutrosophic theory has representatives on all continents and, therefore, it can be said to be a universal theory. On the other hand, according to the three volumes of "The Encyclopedia of Neutrosophic Researchers" (2016, 2018, 2019), plus numerous others not yet included in Encyclopedia book series, about 1200 researchers from 73 countries have applied both the neutrosophic theory and method.

Neutrosophic theory was founded by Professor Florentin Smarandache in 1998; it constitutes further generalization of fuzzy and intuitionistic fuzzy theories. The key distinction between the neutrosophic set/logic and other types of sets/logics lies in the introduction of the degree of indeterminacy/neutrality (I) as an independent component in the neutrosophic set. Thus, neutrosophic theory involves the degree of membership-truth (T), the degree of indeterminacy (I), and the degree of non-membership-falsehood (F). In recent years, the field of neutrosophic set, logic, measure, probability and statistics, precalculus and calculus, etc., and their applications in multiple fields have been extended and applied in various fields, such as communication, management, and information technology.

We believe that this book serves as useful guidance for learning about the current progress in neutrosophic theories. In total, 22 studies have been presented and reflect the call of the thematic vision. The contents of each study included in the volume are briefly described as follows.

The first contribution, authored by Wadei Al-Omeri and Saeid Jafari, addresses the concept of generalized neutrosophic pre-closed sets and generalized neutrosophic pre-open sets in neutrosophic topological spaces. In the article "Design of Fuzzy Sampling Plan Using the Birnbaum-Saunders Distribution", the authors Muhammad Zahir Khan, Muhammad Farid Khan, Muhammad Aslam, and Abdur Razzaque Mughal discuss the use of probability distribution function of Birnbaum–Saunders distribution as a proportion of defective items and the acceptance probability in a fuzzy environment.

Further, the authors Derya Bakbak, Vakkas Uluçay, and Memet Şahin present the "Neutrosophic Soft Expert Multiset and Their Application to Multiple Criteria Decision Making" together with several operations defined for them and their important algebraic properties.

In "Neutrosophic Multigroups and Applications", Vakkas Uluçay and Memet Şahin propose an algebraic structure on neutrosophic multisets called neutrosophic multigroups, deriving their basic properties and giving some applications to group theory.

Changxing Fan, Jun Ye, Sheng Feng, En Fan, and Keli Hu introduce the "Multi-Criteria Decision-Making Method Using Heronian Mean Operators under a Bipolar Neutrosophic Environment" and test the effectiveness of their new methods.

Another decision-making study upon an everyday life issue which empowered us to organize the key objective of the industry developing is given in "Neutrosophic Cubic Einstein Hybrid Geometric Aggregation Operators with Application in Prioritization Using Multiple Attribute Decision-Making Method" written by Khaleed Alhazaymeh, Muhammad Gulistan, Majid Khan, and Seifedine Kadry.

In "Refined Neutrosophy and Lattices vs. Pair Structures and YinYang Bipolar Fuzzy Set", Florentin Smarandache presents the lattice structures of neutrosophic theories, classifies Zhang-Zhang's YinYang bipolar fuzzy sets, and shows that the number of types of neutralities (sub-indeterminacies) may be any finite or infinite number.

The linguistic neutrosophic environment is treated in the study of Changxing Fan, Sheng Feng, and Keli Hu entitled "Linguistic Neutrosophic Numbers Einstein Operator and Its Application in Decision Making".

Vasantha Kandasamy W.B., Ilanthenral Kandasamy, and Florentin Smarandache propose several properties of "Semi-Idempotents in Neutrosophic Rings" and also suggest some open problems.

This continuation of this study is presented in the next article entitled "Neutrosophic Triplets in Neutrosophic Rings" by the same authors.

An article about neutrosophic statistics applied in a variable sampling plan is proposed by Muhammad Aslam and Mohammed Albassam in "Inspection Plan Based on the Process Capability Index Using the Neutrosophic Statistical Method".

"Measures of Probabilistic Neutrosophic Hesitant Fuzzy Sets and the Application in Reducing Unnecessary Evaluation Processes" are investigated by Songtao Shao and Xiaohong Zhang in their applicability as concerns investment problems.

In the article "Neutrosophic Quadruple Vector Spaces and Their Properties", Vasantha Kandasamy W.B., Ilanthenral Kandasamy, and Florentin Smarandache introduce, for the first time in the literature, the concept of neutrosophic quadruple (NQ) vector spaces and neutrosophic quadruple linear algebras.

In the next study, Muhammad Aslam and Osama Hasan Arif propose the use of "Classification of the State of Manufacturing Process under Indeterminacy" in an uncertainty environment in order to eliminate the non-conforming items and increase the profit of the company.

The neutrosophic statistics under the assumption that the product lifetime follows a Weibull distribution is studied by Muhammad Aslam, P. Jeyadurga, Saminathan Balamurali, and Ali Hussein AL-Marshadi in their article "Time-Truncated Group Plan under a Weibull Distribution based on Neutrosophic Statistics".

Muhammad Aslam, Ali Hussein AL-Marshadi, and Nasrullah Khan propose "A New X-Bar Control Chart for Using Neutrosophic Exponentially Weighted Moving Average" for monitoring data under an uncertainty environment. The modern portfolio theory is addressed by Marcel-Ioan Boloș, Ioana-Alexandra Bradea, and Camelia Delcea in their paper "Neutrosophic Portfolios of Financial Assets. Minimizing the Risk of Neutrosophic Portfolios" using an innovative approach determined by the use of the neutrosophic triangular fuzzy numbers.

Next, Xiaogang An, Xiaohong Zhang, and Yingcang Ma propose the notion of "Generalized Abel-Grassmann's Neutrosophic Extended Triplet Loop" together with its properties.

Based on the theories of AG-groupoid, neutrosophic extended triplet and semigroup, Wangtao Yuan and Xiaohong Zhang present some important results in "Regular CA-Groupoids and Cyclic Associative Neutrosophic Extended Triplet Groupoids (CA-NET-Groupoids) with Green Relations".

In "Multi-Attribute Group Decision Making Based on Multigranulation Probabilistic Models with Interval-Valued Neutrosophic Information", the authors Chao Zhang, Deyu Li, Xiangping Kang, Yudong Liang, Said Broumi, and Arun Kumar Sangaiah present an approach intended to handle MAGDM issues with interval-valued neutrosophic information.

Nguyen Tho Thong, Luong Thi Hong Lan, Shuo-Yan Chou, Le Hoang Son, Do Duc Dong, and Tran Thi Ngan propose "An Extended TOPSIS Method with Unknown Weight Information in Dynamic Neutrosophic Environment" together with a practical example intended to illustrate the feasibility and effectiveness of the proposed method.

The last article included in this volume is dedicated to a popular fuzzy tool used to describe the deviation information in uncertain complex situations. The study "Single-Valued Neutrosophic Linguistic Logarithmic Weighted Distance Measures and Their Application to Supplier Selection of Fresh Aquatic Products", written by Jiefeng Wang, Shouzhen Zeng, and Chonghui Zhang, is based on SVNLS and also presents a case study for testing the performance of the proposed framework.

This book would not have been possible without the skills and efforts of many people: first, the advisory board who guided the editors through the editorial process; second, the contributors who have provided perspectives of their neutrosophic works; and third, the reviewers for their service in critically reviewing book chapters.

Stefan Vladutescu, Mihaela Colhon
Editors

mathematics

MDPI

Article

On Generalized Closed Sets and Generalized Pre-Closed Sets in Neutrosophic Topological Spaces

Wadei Al-Omeri [1,*,†,‡] **and Saeid Jafari** [2,‡]

1 Department of Mathematics, Al-Balqa Applied University, Salt 19117, Jordan
2 Department of Mathematics, College of Vestsjaelland South, Herrestraede 11, 4200 Slagelse, Denmark; jafaripersia@gmail.com
* Correspondence: wadeialomeri@bau.edu.jo; Tel.: +962-77-6690-543
† Current address: Department of Mathematics, Al-Balqa Applied University, Salt 19117, Jordan.
‡ These authors contributed equally to this work.

Received: 17 November 2018; Accepted: 13 December 2018; Published: 20 December 2018

Abstract: In this paper, the concept of generalized neutrosophic pre-closed sets and generalized neutrosophic pre-open sets are introduced. We also study relations and various properties between the other existing neutrosophic open and closed sets. In addition, we discuss some applications of generalized neutrosophic pre-closed sets, namely neutrosophic $pT_{\frac{1}{2}}$ space and neutrosophic $gpT_{\frac{1}{2}}$ space. The concepts of generalized neutrosophic connected spaces, generalized neutrosophic compact spaces and generalized neutrosophic extremally disconnected spaces are established. Some interesting properties are investigated in addition to giving some examples.

Keywords: neutrosophic topology; neutrosophic generalized topology; neutrosophic generalized pre-closed sets; neutrosophic generalized pre-open sets; neutrosophic $pT_{\frac{1}{2}}$ space; neutrosophic $gpT_{\frac{1}{2}}$ space; generalized neutrosophic compact and generalized neutrosophic compact

1. Introduction

Zadeh [1] introduced the notion of fuzzy sets. After that, there have been a number of generalizations of this fundamental concept. The study of fuzzy topological spaces was first initiated by Chang [2,3] in 1968. Atanassov [4] introduced the notion of intuitionistic fuzzy sets (IFs). This notion was extended to intuitionistic L-fuzzy setting by Atanassov and Stoeva [5], which currently has the name "intuitionistic L-topological spaces". Coker [6] introduced the notion of intuitionistic fuzzy topological space by using the notion of (IFs). The concept of generalized fuzzy closed set was introduced by Balasubramanian and Sundaram [7]. In various recent papers, Smarandache generalizes intuitionistic fuzzy sets and different types of sets to neutrosophic sets (NSs). On the non-standard interval, Smarandache, Peide and Lupianez defined the notion of neutrosophic topology [8–10]. In addition, Zhang et al. [11] introduced the notion of an interval neutrosophic set, which is a sample of a neutrosophic set and studied various properties.

Recently, Al-Omeri and Smarandache [12,13] introduced and studied a number of the definitions of neutrosophic closed sets, neutrosophic mapping, and obtained several preservation properties and some characterizations about neutrosophic of connectedness and neutrosophic connectedness continuity.

This paper is arranged as follows. In Section 2, we will recall some notions that will be used throughout this paper. In Section 3, we mention some notions in order to present neutrosophic generalized pre-closed sets and investigate its basic properties. In Sections 4 and 5, we study the neutrosophic generalized pre-open sets and present some of their properties. In addition, we provide an application of neutrosophic generalized pre-open sets. Finally, the concepts of generalized neutrosophic connected space, generalized neutrosophic compact space and generalized neutrosophic extremally disconnected spaces are introduced and established in Section 6 and some of their properties in neutrosophic topological spaces are studied.

This class of sets belongs to the important class of neutrosophic generalized open sets which is very useful not only in the deepening of our understanding of some special features of the already well-known notions of neutrosophic topology but also proves useful in neutrosophic multifunction theory in neutrosophic economy and also in neutrosophic control theory. The applications are vast and the researchers in the field are exploring these realms of research.

2. Preliminaries

Definition 1. *Let $\mathscr{X}$ be a non-empty set. A neutrosophic set (NS for short) $\tilde{S}$ is an object having the form $\tilde{S} = \{\langle k, \mu_{\tilde{S}}(k), \sigma_{\tilde{S}}(k), \gamma_{\tilde{S}}(k)\rangle : k \in \mathscr{X}\}$, where $\gamma_{\tilde{S}}(k)$, $\sigma_{\tilde{S}}(k), \mu_{\tilde{S}}(k)$, and the degree of non-membership (namely $\gamma_{\tilde{S}}(k)$), the degree of indeterminacy (namely $\sigma_{\tilde{S}}(k)$), and the degree of membership function (namely $\mu_{\tilde{S}}(k)$), of each element $k \in \mathscr{X}$ to the set $\tilde{S}$, see [14].*

A neutrosophic set $\tilde{S} = \{\langle k, \mu_{\tilde{S}}(k), \sigma_{\tilde{S}}(k), \gamma_{\tilde{S}}(k)\rangle : k \in \mathscr{X}\}$ can be identified as $\langle \mu_{\tilde{S}}(k), \sigma_{\tilde{S}}(k), \gamma_{\tilde{S}}(k)\rangle$ in $\rfloor 0^-, 1^+ \lfloor$ on $\mathscr{X}$.

Definition 2. *Let $\tilde{S} = \langle \mu_{\tilde{S}}(k), \sigma_{\tilde{S}}(k), \gamma_{\tilde{S}}(k)\rangle$ be an NS on $\mathscr{X}$. [15] The complement of the set $\tilde{S}(C(\tilde{S})$, for short) may be defined as follows:*

(i) $C(\tilde{S}) = \{\langle k, 1 - \mu_{\tilde{S}}(k), 1 - \gamma_{\tilde{S}}(k)\rangle : k \in \mathscr{X}\}$,
(ii) $C(\tilde{S}) = \{\langle k, \gamma_{\tilde{S}}(k), \sigma_{\tilde{S}}(k), \mu_{\tilde{S}}(k)\rangle : k \in \mathscr{X}\}$,
(iii) $C(\tilde{S}) = \{\langle k, \gamma_{\tilde{S}}(k), 1 - \sigma_{\tilde{S}}(k), \mu_{\tilde{S}}(k)\rangle : k \in \mathscr{X}\}$.

Neutrosophic sets (*NSs*) 0_N and 1_N [14] in $\mathscr{X}$ are introduced as follows:
$1 - 0_N$ can be defined as four types:

(i) $0_N = \{\langle k, 0, 0, 1\rangle : k \in \mathscr{X}\}$,
(ii) $0_N = \{\langle k, 0, 1, 1\rangle : k \in \mathscr{X}\}$,
(iii) $0_N = \{\langle k, 0, 1, 0\rangle : k \in \mathscr{X}\}$,
(iv) $0_N = \{\langle k, 0, 0, 0\rangle : k \in \mathscr{X}\}$.

$2 - 1_N$ can be defined as four types:

(i) $1_N = \{\langle k, 1, 0, 0\rangle : k \in \mathscr{X}\}$,
(ii) $1_N = \{\langle k, 1, 0, 1\rangle : k \in \mathscr{X}\}$,
(iii) $1_N = \{\langle k, 1, 1, 0\rangle : k \in \mathscr{X}\}$,
(iv) $1_N = \{\langle k, 1, 1, 1\rangle : k \in \mathscr{X}\}$.

Definition 3. *Let k be a non-empty set, and generalized neutrosophic sets GNSs $\tilde{S}$ and $\tilde{R}$ be in the form $\tilde{S} = \{k, \mu_{\tilde{S}}(k), \sigma_{\tilde{S}}(k), \gamma_{\tilde{S}}(k)\}$, $B = \{k, \mu_{\tilde{R}}(k), \sigma_{\tilde{R}}(k), \gamma_{\tilde{R}}(k)\}$. Then, we may consider two possible definitions for subsets ($\tilde{S} \subseteq \tilde{R}$) [14]:*

(i) $\tilde{S} \subseteq B \Leftrightarrow \mu_{\tilde{S}}(k) \leq \mu_B(k), \sigma_{\tilde{S}}(k) \geq \sigma_B(k)$, *and* $\gamma_{\tilde{S}}(k) \leq \gamma_B(k)$,
(ii) $\tilde{S} \subseteq B \Leftrightarrow \mu_{\tilde{S}}(k) \leq \mu_B(k), \sigma_{\tilde{S}}(k) \geq \sigma_B(k)$, *and* $\gamma_{\tilde{S}}(k) \geq \gamma_B(k)$.

Definition 4. *Let $\{\tilde{S}_j : j \in J\}$ be an arbitrary family of NSs in $\mathscr{X}$. Then,*

(i) $\cap\tilde{S}_j$ *can defined as two types:*
$\cap\tilde{S}_j = \langle k, \underset{j\in J}{\wedge}\mu_{\tilde{S}_j}(k), \underset{j\in J}{\wedge}\sigma_{\tilde{S}_j}(k), \underset{j\in J}{\vee}\gamma_{\tilde{S}_j}(k)\rangle$,
$\cap\tilde{S}_j = \langle k, \underset{j\in J}{\wedge}\mu_{\tilde{S}_j}(k), \underset{j\in J}{\vee}\sigma_{\tilde{S}_j}(k), \underset{j\in J}{\vee}\gamma_{\tilde{S}_j}(k)\rangle$.
(ii) $\cup\tilde{S}_j$ *can defined as two types:*
$\cup\tilde{S}_j = \langle k, \underset{j\in J}{\vee}\mu_{\tilde{S}_j}(k), \underset{j\in J}{\vee}\sigma_{\tilde{S}_j}(k), \underset{j\in J}{\wedge}\gamma_{\tilde{S}_j}(k)\rangle$,
$\cup\tilde{S}_j = \langle k, \underset{j\in J}{\vee}\mu_{\tilde{S}_j}(k), \underset{j\in J}{\wedge}\sigma_{\tilde{S}_j}(k), \underset{j\in J}{\wedge}\gamma_{\tilde{S}_j}(k)\rangle$, *see [14].*

Definition 5. *A neutrosophic topology (NT for short) [16] and a non empty set $\mathscr{X}$ is a family Γ of neutrosophic subsets of $\mathscr{X}$ satisfying the following axioms:*

(i) $0_N, 1_N \in \Gamma$,
(ii) $\tilde{S}_1 \cap \tilde{S}_2 \in \Gamma$ *for any* $\tilde{S}_1, \tilde{S}_2 \in \Gamma$,
(iii) $\cup \tilde{S}_i \in \Gamma, \forall \{\tilde{S}_i | j \in J\} \subseteq \Gamma$.

In this case, the pair $(\mathscr{X}, \Gamma)$ is called a neutrosophic topological space (NTS for short) and any neutrosophic set in Γ is known as neutrosophic open set NOS $\in \mathscr{X}$. The elements of Γ are called neutrosophic open sets. A closed neutrosophic set $\tilde{R}$ if and only if its $C(\tilde{R})$ is neutrosophic open.

Note that, for any NTS $\tilde{S}$ in $(\mathscr{X}, \Gamma)$, we have $NCl(\tilde{S}^c) = [NInt(\tilde{S})]^c$ and $NInt(\tilde{S}^c) = [NCl(\tilde{S})]^c$.

Definition 6. *Let $\tilde{S} = \{\mu_{\tilde{S}}(k), \sigma_{\tilde{S}}(k), \gamma_{\tilde{S}}(k)\}$ be a neutrosophic open set and $B = \{\mu_B(k), \sigma_B(k), \gamma_B(k)\}$ a neutrosophic set on a neutrosophic topological space $(\mathscr{X}, \Gamma)$. Then,*

(i) *$\tilde{S}$ is called neutrosophic regular open [14] iff $\tilde{S} = NInt(NCl(\tilde{S}))$.*
(ii) *If $B \in NCS(\mathscr{X})$, then B is called neutrosophic regular closed [14] iff $\tilde{S} = NCl(NInt(\tilde{S}))$.*

Definition 7. *Let (k, Γ) be NT and $\tilde{S} = \{k, \mu_{\tilde{S}}(k), \sigma_{\tilde{S}}(k), \gamma_{\tilde{S}}(k)\}$ an NS in $\mathscr{X}$. Then,*

(i) $NCL(\tilde{S}) = \cap\{U : U$ *is an NCS in* $\mathscr{X}$, $\tilde{S} \subseteq U\}$,
(ii) $NInt(\tilde{S}) = \cup\{V : V$ *is an NOS in* $\mathscr{X}$, $V \subseteq \tilde{S}\}$, *see [14].*

It can be also shown that $NCl(\tilde{S})$ is an NCS and $NInt(\tilde{S})$ is an NOS in $\mathscr{X}$. We have

(i) *$\tilde{S}$ is in $\mathscr{X}$ iff $NCl(\tilde{S})$.*
(ii) *$\tilde{S}$ is an NCS in $\mathscr{X}$ iff $NInt(\tilde{S}) = \tilde{S}$.*

Definition 8. *Let $\tilde{S}$ be an NS and $(\mathscr{X}, \Gamma)$ an NT. Then,*

(i) *Neutrosophic semiopen set (NSOS) [12] if $\tilde{S} \subseteq NCl(NInt(\tilde{S}))$,*
(ii) *Neutrosophic preopen set (NPOS) [12] if $\tilde{S} \subseteq NInt(NNCl(\tilde{S}))$,*
(iii) *Neutrosophic α-open set (NαOS) [12] if $\tilde{S} \subseteq NInt(NNCl(NInt(\tilde{S})))$,*
(iv) *Neutrosophic β-open set (NβOS) [12] if $\tilde{S} \subseteq NNCl(NInt(NCl(\tilde{S})))$.*

The complement of $\tilde{S}$ is an NSOS, NαOS, NPOS, and NROS, which is called NSCS, NαCS, NPCS, and NRCS, resp.

Definition 9. *Let $\tilde{S} = \{\tilde{S}_1, \tilde{S}_2, \tilde{S}_3\}$ be an NS and $(\mathscr{X}, \Gamma)$ an NT. Then, the $*$-neutrosophic closure of $\tilde{S}$ ($* - NCl(\tilde{S})$ for short [12]) and $*$-neutrosophic interior ($* - NInt(\tilde{S})$ for short [12]) of $\tilde{S}$ are defined by*

(i) $\alpha NCl(\tilde{S}) = \cap\{V : V$ *is an NRC in* $\mathscr{X}$, $\tilde{S} \subseteq V\}$,
(ii) $\alpha NInt(\tilde{S}) = \cup\{U : U$ *is an NRO in* $\mathscr{X}$, $U \subseteq \tilde{S}\}$,
(iii) $pNCl(\tilde{S}) = \cap\{V : V$ *is an NPC in* $\mathscr{X}$, $\tilde{S} \subseteq V\}$,
(iv) $pNInt(\tilde{S}) = \cup\{U : U$ *is an NPO in* $\mathscr{X}$, $U \subseteq \tilde{S}\}$,
(v) $sNCl(\tilde{S}) = \cap\{V : V$ *is an NSC in* $\mathscr{X}$, $\tilde{S} \subseteq V\}$,
(vi) $sNInt(\tilde{S}) = \cup\{U : U$ *is an NSO in* $\mathscr{X}$, $U \subseteq \tilde{S}\}$,
(vii) $\beta NCl(\tilde{S}) = \cap\{V : V$ *is an NCβC in* $\mathscr{X}$, $\tilde{S} \subseteq V\}$,
(viii) $\beta NInt(\tilde{S}) = \cup\{U : U$ *is a NβO in* $\mathscr{X}$, $U \subseteq \tilde{S}\}$,
(ix) $rNCl(\tilde{S}) = \cap\{V : V$ *is an NRC in* $\mathscr{X}$, $\tilde{S} \subseteq V\}$,
(x) $rNInt(\tilde{S}) = \cup\{U : U$ *is an NRO in* $\mathscr{X}$, $U \subseteq \tilde{S}\}$.

Definition 10. *An (NS) $\tilde{S}$ of an NT $(\mathscr{X}, \Gamma)$ is called a generalized neutrosophic closed set [17] (GNC in short) if $NCl(\tilde{S}) \subseteq \tilde{B}$ wherever $\tilde{S} \subset \tilde{B}$ and $\tilde{B}$ is a neutrosophic closed set in $\mathscr{X}$.*

Definition 11. *An NS $\tilde{S}$ in an NT $\mathscr{X}$ is said to be a neutrosophic α generalized closed set (NαgCS [18]) if $N\alpha NCl(\tilde{S}) \subseteq \tilde{B}$ whensoever $\tilde{S} \subseteq \tilde{B}$ and $\tilde{B}$ is an NOS in $\mathscr{X}$. The complement $C(\tilde{S})$ of an NαgCS $\tilde{S}$ is an NαgOS in $\mathscr{X}$.*

3. Neutrosophic Generalized Connected Spaces, Neutrosophic Generalized Compact Spaces and Generalized Neutrosophic Extremally Disconnected Spaces

Definition 12. *Let $(\mathscr{X},\Gamma)$ and $(\mathscr{K},\Gamma_1)$ be any two neutrosophic topological spaces.*

(i) A function $g:(\mathscr{X},\Gamma)\longrightarrow(\mathscr{K},\Gamma_1)$ is called generalized neutrosophic continuous(GN-continuous) g^{-1} of every closed set in $(\mathscr{X},\Gamma_1)$ is GN-closed in $(\mathscr{X},\Gamma)$.
Equivalently, if the inverse image of every open set in $(\mathscr{X},\Gamma_1)$ is GN-open in $(\mathscr{X},\Gamma)$:
(ii) A function $g:(\mathscr{X},\Gamma)\longrightarrow(\mathscr{K},\Gamma_1)$ is called generalized neutrosophic irresolute g^{-1} of every GN-closed set in $(\mathscr{X},\Gamma_1)$ is GN-closed in $(\mathscr{X},\Gamma)$.
Equivalently g^{-1} of every GN-open set in $(\mathscr{X},\Gamma_1)$ is GN-open in $(\mathscr{X},\Gamma)$
(iii) A function $g:(\mathscr{X},\Gamma)\longrightarrow(\mathscr{K},\Gamma_1)$ is said to be strongly neutrosophic continuous if $g^{-1}(\tilde{S})$ is both neutrosophic open and neutrosophic closed in $(\mathscr{X},\Gamma)$ for each neutrosophic set $\tilde{S}$ in $(\mathscr{X},\Gamma_1)$.
(iv) A function $g:(\mathscr{X},\Gamma)\longrightarrow(\mathscr{K},\Gamma_1)$ is said to be strongly GN-continuous if the inverse image of every GN-open set in $(\mathscr{X},\Gamma_1)$ is neutrosophic open in $(\mathscr{X},\Gamma)$, see ([17] for more details).

Definition 13. *An NTS $(\mathscr{X},\Gamma)$ is said to be neutrosophic-$T_{\frac{1}{2}}$ ($NT_{\frac{1}{2}}$ in short) space if every GNC in $\mathscr{X}$ is an NC in $\mathscr{X}$.*

Definition 14. *Let $(\mathscr{X},\Gamma)$ be any neutrosophic topological space. $(\mathscr{X},\Gamma)$ is said to be generalized neutrosophic disconnected (in shortly GN-disconnected) if there exists a generalized neutrosophic open and generalized neutrosophic closed set $\tilde{R}$ such that $\tilde{R}\neq 0_N$ and $\tilde{R}\neq 1_N$.$(\mathscr{X},\Gamma)$ is said to be generalized neutrosophic connected if it is not generalized neutrosophic disconnected.*

Proposition 1. *Every GN-connected space is neutrosophic connected. However, the converse is not true.*

Proof. For a GN-connected $(\mathscr{X},\Gamma)$ space and let $(\mathscr{X},\Gamma)$ not be neutrosophic connected. Hence, there exists a proper neutrosophic set, $\tilde{S}=\langle\mu_{\tilde{S}}(x),\sigma_{\tilde{S}}(x),\gamma_{\tilde{S}}(x)\rangle$ $\tilde{S}\neq 0_N$, $\tilde{S}\neq 1_N$, such that $\tilde{S}$ is both neutrosophic open and neutrosophic closed in $(\mathscr{X},\Gamma)$. Since every neutrosophic open set is GN-open and neutrosophic closed set is GN-closed, $\mathscr{X}$ is not GN-connected. Therefore, $(\mathscr{X},\Gamma)$ is neutrosophic connected. □

Example 1. *Let $\mathscr{X}=\{u,v,w\}$. Define the neutrosophic sets $\tilde{S}$, $\tilde{R}$ and $\mathscr{X}$ in $\mathscr{X}$ as follows: $\tilde{S}=\langle x,(\frac{a}{0.4},\frac{b}{0.5},\frac{c}{0.5}),(\frac{a}{0.4},\frac{b}{0.5},\frac{c}{0.5}),(\frac{a}{0.5},\frac{b}{0.5},\frac{c}{0.5})\rangle$, $\tilde{R}=\langle x,(\frac{a}{0.7},\frac{b}{0.6},\frac{c}{0.5}),(\frac{a}{0.7},\frac{b}{0.6},\frac{c}{0.5}),(\frac{a}{0.3},\frac{b}{0.4},\frac{c}{0.5})\rangle$. Then, the family $\Gamma=\{0_N,1_N,\tilde{S},\tilde{R}\}$ is neutrosophic topology on $\mathscr{X}$. It is obvious that $(\mathscr{X},\Gamma)$ is NTS. Now, $(\mathscr{X},\Gamma)$ is neutrosophic connected. However, it is not a GN-connected for $\tilde{Z}=\langle x,(\frac{a}{0.5},\frac{b}{0.6},\frac{c}{0.5}),(\frac{a}{0.5},\frac{b}{0.6},\frac{c}{0.5}),(\frac{a}{0.5},\frac{b}{0.6},\frac{c}{0.5})\rangle$ is GN open and GN closed in $(\mathscr{X},\Gamma)$.*

Theorem 1. *Let $(\mathscr{X},\Gamma)$ be a neutrosophic $T_{\frac{1}{2}}$ space; then, $(\mathscr{X},\Gamma)$ is neutrosophic connected iff $(\mathscr{X},\Gamma)$ is GN-connected.*

Proof. Suppose that $(\mathscr{X},\Gamma)$ is not GN-connected, and there exists a neutrosophic set $\tilde{S}$ which is both GN-open and GN-closed. Since $(\mathscr{X},\Gamma)$ is neutrosophic $T_{\frac{1}{2}}$, $\tilde{S}$ is both neutrosophic open and neutrosophic closed. Hence, $(\mathscr{X},\Gamma)$ is GN-connected. Conversely, let $(\mathscr{X},\Gamma)$ is GN-connected. Suppose that $(\mathscr{X},\Gamma)$ is not neutrosophic connected, and there exists a neutrosophic set $\tilde{S}$ such that $\tilde{S}$ is both NCs and $NOs\in(\mathscr{X},\Gamma)$. Since the neutrosophic open set is GN-open and the neutrosophic closed set is GN-closed, $(\mathscr{X},\Gamma)$ is not GN-connected. Hence, $(\mathscr{X},\Gamma)$ is neutrosophic connected. □

Proposition 2. *Suppose $(\mathscr{X},\Gamma)$ and $(\mathscr{K},\Gamma_1)$ are any two NTSs. If $g:(\mathscr{X},\Gamma)\longrightarrow(\mathscr{K},\Gamma_1)$ is GN-continuous surjection and $(\mathscr{X},\Gamma)$ is GN-connected, then $(\mathscr{K},\Gamma_1)$ is neutrosophic connected.*

Proof. Suppose that $(\mathscr{K},\Gamma_1)$ is not neutrosophic connected, such that the neutrosophic set $\tilde{S}$ is both neutrosophic open and neutrosophic closed in $(\mathscr{K},\Gamma_1)$. Since g is GN-continuous, $g^{-1}(\tilde{S})$ is GN-open and GN-closed in $((\mathscr{K},\Gamma)$. Thus, $(\mathscr{K},\Gamma)$ is not GN connected. Hence, $(\mathscr{K},\Gamma_1)$ is neutrosophic connected. □

Definition 15. *Let $(\mathscr{K},\Gamma)$ be an NT. If a family $\{\langle k,\mu_{G_i}(k),\sigma_{G_i}(k),\gamma_{G_i}(k):i\in J\rangle\}$ of GN open sets in $(\mathscr{K},\Gamma)$ satisfies the condition $\bigcup\{\langle k,\mu_{G_i}(k),\sigma_{G_i}(k),\gamma_{G_i}(k):i\in J\rangle\}=1_N$, then it is called a GN open cover of $(\mathscr{K},\Gamma)$. A finite subfamily of a GN open cover $\{\langle k,\mu_{G_i}(k),\sigma_{G_i}(k),\gamma_{G_i}(k):i\in J\rangle\}$ of $(\mathscr{L},\Gamma)$, which is also a GN open cover of $(\mathscr{K},\Gamma)$ is called a finite subcover of*

$$\{\langle k,\mu_{G_i}(k),\sigma_{G_i}(k),\gamma_{G_i}(k):i\in J\rangle\}.$$

Definition 16. *An NT $(\mathscr{K},\Gamma)$ is called GN compact iff every GN open cover of $(\mathscr{K},\Gamma)$ has a finite subcover.*

Theorem 2. *Let $(\mathscr{K},\Gamma)$ and $(\mathscr{K},\Gamma_1)$ be any two NTs, and $g:(\mathscr{L},\Gamma)\longrightarrow(\mathscr{K},\Gamma_1)$ be GN continuous surjection. If $(\mathscr{K},\Gamma)$ is GN-compact, hence so is $(\mathscr{K},\Gamma_1)$.*

Proof. Let $G_i=\{\langle y,\mu_{G_i}(x),\sigma_{G_i}(x),\gamma_{G_i}(x):i\in J\rangle\}$ be a neutrosophic open cover in $(\mathscr{K},\Gamma_1)$ with

$$\widetilde{\bigcup}\{\langle y,\mu_{G_i}(x),\sigma_{G_i}(x),\gamma_{G_i}(x):i\in J\rangle\}=\widetilde{\bigcup_{i\in J}}G_i=1_N.$$

Since g is GN continuous, $g^{-1}(G_i)=G_i=\{\langle y,\mu_{g^{-1}(G_i)}(x),\sigma_{g^{-1}(G_i)}(x),\gamma_{g^{-1}(G_i)}(x):i\in J\rangle\}$ is GN open cover of $(\mathscr{K},\Gamma)$. Now,

$$\widetilde{\bigcup_{i\in J}}g^{-1}(G_i)=g^{-1}(\widetilde{\bigcup_{i\in J}}G_i)=1_N.$$

Since $(\mathscr{K},\Gamma)$ is GN compact, there exists a finite subcover $J_0\subset J$, such that

$$\widetilde{\bigcup_{i\in J_0}}g^{-1}(G_i)=1_N.$$

Hence,

$$g\Big(\widetilde{\bigcup_{i\in J_0}}g^{-1}(G_i)=1_N\Big),g^{-1}\Big(\widetilde{\bigcup_{i\in J_0}}(G_i)=1_N\Big).$$

That is,

$$\widetilde{\bigcup_{i\in J_0}}(G_i)=1_N.$$

Therefore, $(\mathscr{K},\Gamma_1)$ is neutrosophic compact. □

Definition 17. *Let $(\mathscr{K},\Gamma)$ be an NT and K be a neutrosophic set in $(\mathscr{L},\Gamma)$. If a family $\{\langle k,\mu_{G_i}(k),\sigma_{G_i}(k),\gamma_{G_i}(k):i\in J\rangle\}$ of GN open sets in $(\mathscr{K},\Gamma)$ satisfies the condition $K\subseteq\bigcup\{\langle k,\mu_{G_i}(k),\sigma_{G_i}(k),\gamma_{G_i}(k):i\in J\rangle\}=1_N$, then it is called a GN open cover of K. A finite subfamily of a GN open cover $\{\langle k,\mu_{G_i}(k),\sigma_{G_i}(k),\gamma_{G_i}(k):i\in J\rangle\}$ of K, which is also a GN open cover of K is called a finite subcover of $\{\langle k,\mu_{G_i}(k),\sigma_{G_i}(k),\gamma_{G_i}(k):i\in J\rangle\}$.*

Definition 18. *An NT $(\mathscr{K},\Gamma)$ is called GN compact iff every GN open cover of K has a finite subcover.*

Theorem 3. *Let $(\mathscr{K},\Gamma)$ and $(\mathscr{K},\Gamma_1)$ be any two NTs, and $g:(\mathscr{L},\Gamma)\longrightarrow(\mathscr{K},\Gamma_1)$ be an GN continuous function. If K is GN-compact, then so is $g(K)$ in $(\mathscr{K},\Gamma_1)$.*

Proof. Let $G_i = \{\langle y, \mu_{G_i}(x), \sigma_{G_i}(x), \gamma_{G_i}(x) : i \in J\rangle\}$ be a neutrosophicopen cover of $g(K)$ in $(\mathscr{K}, \Gamma_1)$. That is,

$$g(K) \subseteq \widetilde{\bigcup_{i\in J}} G_i.$$

Since g is GN continuous, $g^{-1}(G_i) = \{\langle x, \mu_{g^{-1}(G_i)}(x), \sigma_{g^{-1}(G_i)}(x), \gamma_{g^{-1}(G_i)}(x) : i \in J\rangle\}$ is GN open cover of K in $(\mathscr{X}, \Gamma)$. Now,

$$K \subseteq g^{-1}(\widetilde{\bigcup_{i\in J}} G_i) \subseteq \widetilde{\bigcup_{i\in J}} g^{-1}(G_i).$$

Since K is $(\mathscr{X}, \Gamma)$ is GN compact, there exists a finite subcover $J_0 \subset J$, such that

$$K \subseteq \widetilde{\bigcup_{i\in J_0}} g^{-1}(G_i) = 1_N.$$

Hence,

$$g(K) \subseteq g\Big(\widetilde{\bigcup_{i\in J_0}} g^{-1}(G_i)\Big) \widetilde{\bigcup_{i\in J_0}} (G_i).$$

Therefore, $g(K)$ is neutrosophic compact. □

Proposition 3. *Let $(\mathscr{X}, \Gamma)$ be a neutrosophic compact space and suppose that K is a GN-closed set of $(\mathscr{X}, \Gamma)$. Then, K is a neutrosophic compact set.*

Proof. Let $K_j == \{\langle y, \mu_{G_i}(x), \sigma_{G_i}(x), \gamma_{G_i}(x) : i \in J\rangle\}$ be a family of neutrosophic open set in $(\mathscr{X}, \Gamma)$ such that

$$K \subseteq \widetilde{\bigcup_{i\in J}} K_j.$$

Since K is GN-closed, $NCl(K) \subseteq \widetilde{\bigcup}_{i\in J} K_j$. Since $(\mathscr{X}, \Gamma)$ is a neutrosophic compact space, there exists a finite subcover $J_0 \subseteq J$. Now, $NCl(K) \subseteq \widetilde{\bigcup}_{i\in J_0} K_j$. Hence, $K \subseteq NCl(K) \subseteq \widetilde{\bigcup}_{i\in J_0} K_j$. Therefore, K is a neutrosophic compact set. □

Definition 19. *Let $(\mathscr{X}, \Gamma)$ be any neutrosophic topological space. $(\mathscr{X}, \Gamma)$ is said to be GN extremally disconnected if $NCl(K)$ neutrosophic open and K is GN open.*

Proposition 4. *For any neutrosophic topological space $(\mathscr{X}, \Gamma)$, the following are equivalent:*

(i) *$(\mathscr{X}, \Gamma)$ is GN extremally disconnected.*
(ii) *For each GN closed set K, $NGNInt(\tilde{S})$ is a GN closed set.*
(iii) *For each GN open set K, we have $NGNCl(K) + NGNCl(1 - NGNCl(\tilde{S})) = 1$.*
(iv) *For each pair of GN open sets K and M in $(\mathscr{X}, \Gamma)$, $NGNCl(K) + M = 1$, we have $NGNCl(K) + NGNCl(B) = 1$.*

4. Generalized Neutrosophic Pre-Closed Set

Definition 20. *An NS $\tilde{S}$ is said to be a neutrosophic generalized pre-closed set (GNPCS in short) in $(\mathscr{X}, \Gamma)$ if $pNCl(\tilde{S}) \subseteq \tilde{B}$ whensoever $\tilde{S} \subseteq \tilde{B}$ and $\tilde{B}$ is an NO in $\mathscr{X}$. The family of all GNPCSs of an NT $(\mathscr{X}, \Gamma)$ is defined by $GNPC(\mathscr{X})$.*

Example 2. *Let $\mathscr{X} = \{a, b\}$ and $\Gamma = \{0_N, 1_N, T\}$ be a neutrosophic topology on $\mathscr{X}$, where $T = \langle (0.2, 0.3, 0.5), (0.8, 0.7, 0.7)\rangle$. Then, the NS $\tilde{S} = \langle (0.2, 0.2, 0.2), (0.8, 0.7, 0.7)\rangle$ is GNPCs $\in \mathscr{X}$.*

Theorem 4. *Every NC is a GNPC, but the converse is not true.*

Proof. Let $\tilde{S}$ be an NC in $\mathscr{X}$, $\tilde{S} \subseteq \tilde{B}$ and $\tilde{B}$ is NOS in $(\mathscr{X}, \Gamma)$. Since $pNCl(\tilde{S}) \subseteq NCl(\tilde{S})$ and $\tilde{S}$ is NCS in $\mathscr{X}$, $pNCl(\tilde{S}) \subseteq NCl(\tilde{S}) = \tilde{S} \subseteq \tilde{B}$. Therefore, $\tilde{S}$ is $GNPCs \in \mathscr{X}$. □

Example 3. *Let $\mathscr{X} = \{u, v\}$ and $\Gamma = \{0_N, 1_N, H\}$ be a neutrosophic topology on $\mathscr{X}$, where $H = \langle(0.2, 0.3, 0.5), (0.8, 0.7, 0.7)\rangle$. Then, the NS $\tilde{S} = \langle(0.2, 0.2, 0.2), (0.8, 0.7, 0.7)\rangle$ is a GNPC in $\mathscr{X}$ but not an NCS $\in \mathscr{X}$.*

Theorem 5. *Every $N\alpha CS$ is GNPC, but the converse is not true.*

Proof. Let $\tilde{S}$ be an $N\alpha CS$ in $\mathscr{X}$ and let $\tilde{S} \subseteq \tilde{B}$ and $\tilde{B}$ is an NOS in $(\mathscr{X}, \Gamma)$. Now, $NCl(NInt(NCl(\tilde{S}))) \subseteq \tilde{S}$. Since $\tilde{S} \subseteq NCl(\tilde{S})$, $NCl(NInt(\tilde{S})) \subseteq NCl(NInt(NCl(\tilde{S}))) \subseteq \tilde{S}$. Hence, $pNCl(\tilde{S}) \subseteq \tilde{S} \subseteq \tilde{B}$. Therefore, $\tilde{S}$ is $GNPCs \in \mathscr{X}$. □

Example 4. *Let $\mathscr{X} = \{u, v\}$ and let $\Gamma = \{0_N, 1_N, H\}$ is a neutrosophic topology on $\mathscr{X}$, where $H = \langle(0.4, 0.2, 0.5), (0.6, 0.7, 0.6)\rangle$. Then, the NS $\tilde{S} = \langle(0.3, 0.1, 0.4), (0.7, 0.8, 0.7)\rangle$ is a GNPC in $\mathscr{X}$ but not $N\alpha Cs$ in $\mathscr{X}$ since $NCl(NInt(NCl(\tilde{S}))) = \langle(0.5, 0.6, 0.5), (0.5, 0.3, 0.6)\rangle \not\subseteq \tilde{S}$.*

Theorem 6. *Every $GN\alpha C$ is a GNPC, but the converse is not true.*

Proof. Let $\tilde{S}$ be $GN\alpha Cs \in \mathscr{X}$, $\tilde{S} \subseteq \tilde{B}$, $\tilde{B}$ be an NOs in $(\mathscr{X}, \Gamma)$. By Definition 6, $\tilde{S} \cup NCl(NInt(NCl(\tilde{S}))) \subseteq \tilde{B}$. This implies $NCl(NInt(NCl(\tilde{S}))) \subseteq \tilde{B}$ and $NCl(NInt(\tilde{S})) \subseteq \tilde{B}$. Therefore, $pNCl(\tilde{S}) = \tilde{S} \cup NCl(NInt(\tilde{S})) \subseteq \tilde{B}$. Hence, $\tilde{S}$ is $GNPCs \in \mathscr{X}$. □

Example 5. *Let $\mathscr{X} = \{u, v\}$ and $\Gamma = \{0_N, 1_N, H\}$ be a neutrosophic topology on $\mathscr{X}$, where $H = \langle(0.5, 0.6, 0.6), (0.5, 0.4, 0.4)\rangle$. Then, the NS $\tilde{S} = \langle(0.4, 0.5, 0.5), (0.6, 0.5, 0.5)\rangle$ is GNPC in $\mathscr{X}$ but not $GN\alpha C$ in $\mathscr{X}$ since $\alpha NCl(\tilde{S}) = 1_N \not\subseteq H$.*

Definition 21. *An NS $\tilde{S}$ is said to be a neutrosophic generalized pre-closed set (GNSCS) in $(\mathscr{X}, \Gamma)$ if $SNCl(\tilde{S}) \subseteq \tilde{B}$ whensoever $\tilde{S} \subseteq \tilde{B}$ and $\tilde{B}$ is an NO in $\mathscr{X}$. The family of all GNSCSs of an NT $(\mathscr{X}, \Gamma)$ is defined by $GNSC(\mathscr{X})$.*

Proposition 5. *Let $\tilde{S}$, B be a two GNPCs of an NT $(\mathscr{X}, \Gamma)$. NGSC and NGPC are independent.*

Example 6. *Let $\mathscr{X} = \{u, v\}$, $\Gamma = \{0_N, 1_N, H\}$ be a neutrosophic topology on $\mathscr{X}$, where $H = \langle(0.5, 0.4, 0.4), (0.5, 0.6, 0.5)\rangle$. Then, the NS $\tilde{S} = H$ is GNSC but not GNPC in $\mathscr{X}$ since $\tilde{S} \subseteq H$ but $pNCl(\tilde{S}) = \langle(0.5, 0.6, 0.4), (0.5, 0.4, 0.5)\rangle \not\subseteq H$*

Example 7. *Let $\mathscr{X} = \{u, v\}$, $\Gamma = \{0_N, 1_N, H\}$ be a neutrosophic topology on $\mathscr{X}$, where $H = \langle(0.7, 0.9, 0.7), (0.3, 0.1, 0.1)\rangle$. Then, the NS $\tilde{S} = \langle(0.6, 0.7, 0.6), (0.4, 0.3, 0.4)\rangle$ is GNPC but not GNsC in $\mathscr{X}$ since $sNCl(\tilde{S}) = 1_N \subseteq H$.*

Proposition 6. *NSC and GNPC are independent.*

Example 8. *Let $\mathscr{X} = \{a, b\}$, $\Gamma = \{0_N, 1_N, T\}$ be a neutrosophic topology on $\mathscr{X}$, where $T = \langle(0.5, 0.2, 0.3), (0.5, 0.6, 0.5)\rangle$. Then, the NS $\tilde{S} = T$ is an NSC but not GNPC in $\mathscr{X}$ since $\tilde{S} \subseteq T$ but $pNCl(\tilde{S}) = 1\langle(0.5, 0.6, 0.5), (0.5, 0.2, 0.3)\rangle \not\subseteq T$.*

Example 9. *Let $\mathscr{X} = \{u, v\}$, $\Gamma = \{0_N, 1_N, H\}$ be a neutrosophic topology on $\mathscr{X}$, where $H = \langle(0.8, 0.8, 0.8), (0.2, 0.2, 0.2)\rangle$. Then, the NS $\tilde{S} = \langle(0.8, 0.8, 0.8), (0.2, 0.2, 0.2)\rangle$ is GNPC but not an NSC in $\mathscr{X}$ since $NInt(NCl(\tilde{S})) \not\subseteq \tilde{S}$.*

The following Figure 1 shows the implication relations between *GNPC* set and the other existed ones.

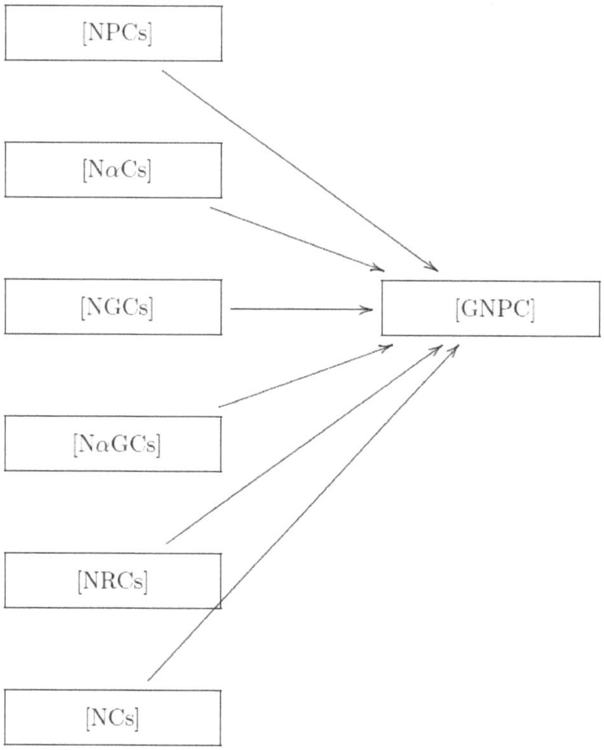

Figure 1. Relation between *GNPC* and others exists set.

Remark 1. *Let $\tilde{S}$, B be a two GNPCs of an NT $(\mathscr{X},\Gamma)$. Then, the union of any two GNPCs is not a GNPC in general—see the following example.*

Example 10. *Let $(\mathscr{X},\Gamma)$ be a neutrosophic topology set on $\mathscr{X}$, where $\mathscr{X} = \{u,v\}$, $T = \langle(0.6,0.8,0.6),(0.4,0.2,0.2)\rangle$. Then, $\Gamma = \{0_N,1_N,T\}$ is neutrosophic topology on $\mathscr{X}$ and the NS $\tilde{S} = \langle(0.2,0.9,0.3),(0.8,0.2,0.6)\rangle$, $B = \langle(0.6,0.7,0.6),(0.4,0.3,0.4)\rangle$ are GNPCSs but $\tilde{S} \cup B$ is not a GNPC in $\mathscr{X}$.*

5. Generalized Neutrosophic Pre-Open Sets

In this section, we present generalized neutrosophic pre-open sets and investigate some of their properties.

Definition 22. *An NS $\tilde{S}$ is said to be a generalized neutrosophic pre-open set (GNPOS) in $(\mathscr{X},\Gamma)$ if the complement $\tilde{S}^c$ is a GNPCS in $\mathscr{X}$. The family of all GNPOSs of NTS $(\mathscr{X},\Gamma)$ is denoted by $GNPO(\mathscr{X})$.*

Example 11. *Let $\mathscr{X} = \{u,v\}$ and $\Gamma = \{0_N,1_N,H\}$ be a neutrosophic topology on $\mathscr{X}$, where $H = \langle(0.8,0.7,0.8),(0.3,0.4,0.3)\rangle$. Then, the NS $\tilde{S} = \langle(0.9,0.8,0.8),(0.3,0.3,0.3)\rangle$ is $GNPO \in \mathscr{X}$.*

Theorem 7. *Let* $(\mathscr{X}, \Gamma)$ *be an* NT. *Then, for every* $\tilde{S} \in GNPO(\mathscr{X})$ *and for every* $\tilde{R} \in NS(\mathscr{X})$, $pNInt(\tilde{S}) \subseteq \tilde{R} \subseteq \tilde{S}$ *implies* $\tilde{R} \in GNPO(\mathscr{X})$.

Proof. By Theorem $\tilde{S}^c \subseteq \tilde{R}^c \subseteq (pNInt(\tilde{S}))^c$. Let $\tilde{R}^c \subseteq \tilde{R}$ and $\tilde{R}$ be NOs. Since $\tilde{S}^c \subseteq B^c$, $\tilde{S}^c \subseteq \tilde{R}$. However, $\tilde{S}^c$ is a $GNPCs$, $pNCl(\tilde{S}^c) \subseteq \tilde{R}$. In addition, $\tilde{R}^c \subseteq (pNInt(\tilde{S}))^c = pNCl(\tilde{S}^c)$ (by theorem). Therefore, $pNCl(\tilde{R}^c) \subseteq pNCl(\tilde{S}^c) \subseteq \tilde{R}$. Hence, B^c is $GNPC$. This implies that $\tilde{R}$ is a $GNPO$ of $\mathscr{X}$. □

Remark 2. *Let* $\tilde{S}$, $\tilde{R}$ *be two* $GNPOs$ *of an* NT $(\mathscr{X}, \Gamma)$. *The intersection of any two* $GNPOSs$ *is not a* $GNPO$ *in general.*

Example 12. *Let* $\mathscr{X} = \{u, v\}$ *and* $\Gamma = \{0_N, 1_N, H\}$ *be a neutrosophic topology on* $\mathscr{X}$, *where* $H = \langle(0.6, 0.8, 0.6), (0.4, 0.2, 0.4)\rangle$. *Then, the NSs,* $\tilde{S} = \langle(0.9, 0.2, 0.1), (0.1, 0.8, 0.2)\rangle$ *and* $\tilde{R} = \langle(0.4, 0.3, 0.4), (0.6, 0.7, 0.6)\rangle$ *is* $GNPO$, *but* $\tilde{S} \cap \tilde{R}$ *is not* $GNPO \in \mathscr{X}$.

Theorem 8. *For any an* NTS $(\mathscr{X}, \Gamma)$, *the following hold:*

(i) Every NO *is* $GNPO$,
(ii) Every NSO *is* $GNPO$,
(iii) Every $N\alpha O$ *is* $GNPO$,
(iv) Every NPO *is* $GNPO$.

Proof. The proof is clear, so it has been omitted. □

The converses are not true in general.

Example 13. *Let* $\mathscr{X} = \{u, v\}$ *and* $H = \langle(0.2, 0.3, 0.2), (0.8, 0.7, 0.7)\rangle$. *Then,* $\Gamma = \{0_N, 1_N, H\}$ *is a neutrosophic topology on* $\mathscr{X}$, *an NS* $\tilde{S} = \langle(0.8, 0.7, 0.7), (0.2, 0.2, 0.2)\rangle$ *is an* NSO *in* $(\mathscr{X}, \Gamma)$ *but not an* $NO \in \mathscr{X}$.

Example 14. *Let* $\mathscr{X} = \{u, v\}$ *and* $\Gamma = \{0_N, 1_N, H\}$ *be neutrosophic topology on* $\mathscr{X}$, *where* $H = \langle(0.6, 0.4, 0.7), (0.7, 0.4, 0.6)\rangle$. *Then, an NS* $\tilde{S} = \langle(0.2, 0.7, 0.7), (0.8, 0.3, 0.8)\rangle$ *is* $GNPO$ *but not an* $NSO \in \mathscr{X}$.

Example 15. *Let* $\mathscr{X} = \{u, v\}$ *and* $\Gamma = \{0_N, 1_N, H\}$ *be a neutrosophic topology on* $\mathscr{X}$, *where* $H = \langle(0.4, 0.2, 0.4), (0.6, 0.7, 0.6)\rangle$. *Then, an NS* $\tilde{S} = \langle(0.8, 0.9, 0.8), (0.4, 0.2, 0.3)\rangle$ *is* $GNPO$ *but not an* $N\alpha O \in \mathscr{X}$.

Example 16. *Let* $\mathscr{X} = \{u, v\}$ *and* $\Gamma = \{0_N, 1_N, H\}$ *be a neutrosophic topology on* $\mathscr{X}$, *where* $H = \langle(0.6, 0.5, 0.6), (0.5, 0.6, 0.5)\rangle$. *Then, an NS* $\tilde{S} = \langle(0.8, 0.7, 0.8), (0.4, 0.5, 0.3)\rangle$ *is* $GNPO$ *but not an* $NPO \in \mathscr{X}$.

Theorem 9. *Let* $(\mathscr{X}, \Gamma)$ *be an* NT. *If* $\tilde{S} \in GNPO(\mathscr{X})$, *then* $\tilde{R} \subseteq NInt(NCl(\tilde{S}))$ *whensoever* $\tilde{R} \subseteq \tilde{S}$ *and* $\tilde{R}$ *is an* NC *in* $\mathscr{X}$.

Proof. Let $\tilde{S} \in GNPO(\mathscr{X})$. Then, $\tilde{S}^c$ is $GnPCS$ in $\mathscr{X}$. Therefore, $pNCl(\tilde{S}^c) \subseteq \tilde{B}$ whensoever $\tilde{S}^c \subseteq \tilde{B}$ and $\tilde{B}$ is an NO in $\mathscr{X}$. That is, $NCl(NInt(\tilde{S}^c)) \subseteq \tilde{B}$. This implies $\tilde{B}^c \subseteq NInt(NCl(\tilde{S}))$ whensoever $\tilde{B}^c \subseteq \tilde{S}$ and $\tilde{B}^c$ is NCs in $\mathscr{X}$. Replacing $\tilde{B}^c$, by $\tilde{R}$, we get $\tilde{R} \subseteq NInt(NCl(\tilde{S}))$ whensoever $\tilde{R} \subseteq \tilde{S}$ and $\tilde{R}$ is an NC in $\mathscr{X}$. □

Theorem 10. *For NS* $\tilde{S}$, $\tilde{S}$ *is an* NO *and* $GNPC$ *in* $\mathscr{X}$ *if and only if* $\tilde{S}$ *is an* NRO *in* $\mathscr{X}$.

Proof. $\Longrightarrow$ Let $\tilde{S}$ be an NO and a $GNPCS$ in $\mathscr{X}$. Then, $pNCl(\tilde{S}) \subseteq \tilde{S}$. This implies $NCl(NInt(\tilde{S})) \subseteq \tilde{S}$. Since $\tilde{S}$ is an NO, it is an NPO. Hence, $\tilde{S} \subseteq NInt(NCl(\tilde{S}))$. Therefore, $\tilde{S} = NInt(NCl(\tilde{S}))$. Hence, $\tilde{S}$ is an NRO in $\mathscr{X}$.

$\Longleftarrow$ Let $\tilde{S}$ be an NRO in $\mathscr{X}$. Therefore, $\tilde{S} = NInt(NCl(\tilde{S}))$. Let $\tilde{S} \subseteq \tilde{B}$ and $\tilde{B}$ be an NO in $\mathscr{X}$. This implies $pNCl(\tilde{S}) \subseteq \tilde{S}$. Hence, $\tilde{S}$ is $GNPC$ in $\mathscr{X}$. □

Theorem 11. *An NS $\tilde{S}$ of an NT $(\mathscr{X},\Gamma)$ is a GNPO iff $H \subseteq pNInt(\tilde{S})$, whensoever H is an NC and $H \subseteq \tilde{S}$.*

Proof. $\Longrightarrow$ Let $\tilde{S}$ be $GNPO$ in $\mathscr{X}$. Let H be an NCs and $H \subseteq \tilde{S}$. Then, H^c is an NOS in $\mathscr{X}$ such that $\tilde{S}^c \subseteq H^c$. Since $\tilde{S}^c$ is $GNPC$, we have $pNCl(\tilde{S}^c) \subseteq H^c$. Hence, $(pNInt(\tilde{S}))^c \subseteq H^c$. Therefore, $H \subseteq pNInt(\tilde{S})$.

$\Longleftarrow$ Suppose $\tilde{S}$ is an NS of $\mathscr{X}$ and let $H \subseteq pNInt(\tilde{S})$ whensoever H is an NC and $H \subseteq \tilde{S}$. Then, $\tilde{S}^c \subseteq H^c$ and H^c is an NO. By assumption, $(pNInt(\tilde{S}))^c \subseteq H^c$, which implies $pNCl(\tilde{S}^c) \subseteq H^c$. Therefore, $\tilde{S}^c$ is $GNPCs$ of $\mathscr{X}$. Hence, $\tilde{S}$ is a $GNPOS$ of $\mathscr{X}$. □

Corollary 1. *An NS $\tilde{S}$ of an NTS $(\mathscr{X},\Gamma)$ is GNPO iff $H \subseteq NInt(NCl(\tilde{S}))$, whensoever H is an NC and $H \subseteq \tilde{S}$.*

Proof. $\Longrightarrow$ Let $\tilde{S}$ is a $GNPOS$ in $\mathscr{X}$. Let H be an NCS and $H \subseteq \tilde{S}$. Then, H^c is an NOS in $\mathscr{X}$ such that $\tilde{S}^c \subseteq H^c$. Since $\tilde{S}^c$ is $GNPC$, we have $pNCl(\tilde{S}^c) \subseteq H^c$. Therefore, $NCl(NInt(\tilde{S}^c)) \subseteq H^c$. Hence, $(NInt(NCl(\tilde{S})))^c \subseteq H^c$. This implies $H \subseteq NInt(NCl(\tilde{S}))$.

$\Longleftarrow$ Suppose $\tilde{S}$ be an NS of $\mathscr{X}$ and $H \subseteq NInt(NCl(\tilde{S}))$, whensoever H is an NC and $H \subseteq \tilde{S}$. Then, $\tilde{S}^c \subseteq H^c$ and H^c is an NO. By assumption, $(NInt(NCl(\tilde{S})))^c \subseteq H^c$. Hence, $NCl(NInt(\tilde{S}^c)) \subseteq H^c$. This implies $pNCl(\tilde{S}^c) \subseteq H^c$. Hence, $\tilde{S}$ is a $GNPOS$ of $\mathscr{X}$. □

6. Applications of Generalized Neutrosophic Pre-Closed Sets

Definition 23. *An NTS $(\mathscr{X},\Gamma)$ is said to be neutrosophic-$pT_{\frac{1}{2}}$ ($NpT_{\frac{1}{2}}$ in short) space if every GNPC in $\mathscr{X}$ is an NCs $\in \mathscr{X}$.*

Definition 24. *An NTS $(\mathscr{X},\Gamma)$ is said to be neutrosophic-$gpT_{\frac{1}{2}}$ ($NgpT_{\frac{1}{2}}$ in short) space if every GNPC in $\mathscr{X}$ is an NPCs $\in \mathscr{X}$.*

Theorem 12. *Every $NpT_{\frac{1}{2}}$ space is an $NgpT_{\frac{1}{2}}$ space.*

Proof. Let $\mathscr{X}$ be an $NpT_{\frac{1}{2}}$ space and $\tilde{S}$ be $GNPC \in \mathscr{X}$. By assumption, $\tilde{S}$ is NCs in $\mathscr{X}$. Since every NC is an NPC, $\tilde{S}$ is an NPC in $\mathscr{X}$. Hence, $\mathscr{X}$ is an $NgpT_{\frac{1}{2}}$ space. □

The converse is not true.

Example 17. *Let $\mathscr{X} = \{u,v\}$, $H = \langle(0.9,0.9,0.9),(0.1,0.1,0.1)\rangle$ and $\Gamma = \{0_N, 1_N, H\}$. Then, $(\mathscr{X},\Gamma)$ is an $NgpT_{\frac{1}{2}}$ space, but it is not $NpT_{\frac{1}{2}}$ since an NS $H = \langle(0.2,0.3,0.3),(0.8,0.7,0.7)\rangle$ is GNPC but not an NCS $\in \mathscr{X}$.*

Theorem 13. *Let $(\mathscr{X},\Gamma)$ be an NT and $\mathscr{X}$ is an $NpT_{\frac{1}{2}}$ space; then,*

(i) the union of GNPCs is GNPC,
(ii) the intersection of GNPOs is GNPO.

Proof. (i) Let $\{\tilde{S}_i\}_{i \in J}$ be a collection of $GNPCs$ in an $NpT_{\frac{1}{2}}$ space $(\mathscr{X},\Gamma)$. Thus, every $GNPCs$ is an NCS. However, the union of an NC is an NCS. Therefore, the Union of $GNPCs$ is $GNPCs$ in $\mathscr{X}$.
(ii) Proved by taking complement in (i). □

Theorem 14. *An NT $\mathscr{X}$ is an $NgpT_{\frac{1}{2}}$ space iff $GNPO(\mathscr{X}) = NPO(\mathscr{X})$.*

Proof. $\Longrightarrow$ Let $\tilde{S}$ be a $GNPOs$ in $\mathscr{X}$; then, $\tilde{S}^c$ is $GNPCs$ in $\mathscr{X}$. By assumption, $\tilde{S}^c$ is an $NPCs$ in $\mathscr{X}$. Thus, $\tilde{S}$ is $NPOs$ in $\mathscr{X}$. Hence, $GNPO(\mathscr{X}) = NPO(\mathscr{X})$.

$\Longleftarrow$ Let $\tilde{S}$ be $GNPC \in \mathscr{X}$. Then, $\tilde{S}^c$ is $GNPO$ in $\mathscr{X}$. By assumption, $\tilde{S}^c$ is an NPO in $\mathscr{X}$. Thus, $\tilde{S}$ is an $NPC \in \mathscr{X}$. Therefore, $\mathscr{X}$ is an $NgpT_{\frac{1}{2}}$ space. □

Theorem 15. *For an NTS* $(\mathscr{X}, \Gamma)$, *the following are equivalent:*

(i) $(\mathscr{X}, \Gamma)$ *is a neutrosophic pre-*$T_{\frac{1}{2}}$ *space.*
(ii) *Every non-empty set of* $\mathscr{X}$ *is either an* $NPCS$ *or* $NPOS$.

Proof. $(i) \Longrightarrow (ii)$. Suppose that $(\mathscr{X}, \Gamma)$ is a neutrosophic pre-$T_{\frac{1}{2}}$ space. Suppose that $\{x\}$ is not an $NPCS$ for some $x \in \mathscr{X}$. Then, $\mathscr{X} - \{x\}$ is not an $NPOS$ and hence $\mathscr{X}$ is the only an $NPOS$ containing $\mathscr{X} - \{x\}$. Hence, $\mathscr{X} - \{x\}$ is an $NPGCS$ in $(\mathscr{X}, \Gamma)$. Since $(\mathscr{X}, \Gamma)$ is a neutrosophic pre-$T_{\frac{1}{2}}$ space, then $\mathscr{X} - \{x\}$ is an $NPCS$ or equivalently $\{x\}$ is an $NPOS$. $(ii) \Longrightarrow (i)$. Let every singleton set of $\mathscr{X}$ be either $NPCS$ or $NPOS$. Let $\tilde{S}$ be an $NPGCS$ of $(\mathscr{X}, \Gamma)$. Let $x \in \mathscr{X}$. We show that $x \in \mathscr{X}$ in two cases.

Case (i): Suppose that $\{x\}$ is $NPCS$. If $x \notin \tilde{S}$, then $x \in pNCl(\tilde{S}) - \tilde{S}$. Now, $pNCl(\tilde{S}) - \tilde{S}$ contains a non—empty $NPCS$. Since $\tilde{S}$ is $NPGCS$, by Theorem 7, we arrived to a contradiction. Hence, $x \in \mathscr{X}$.

Case (ii): Let $\{x\}$ be $NPOS$. Since $x \in pNCl(\tilde{S})$, then $\{x\} \cap \tilde{S} \neq \phi$. Thus, $x \in \mathscr{X}$. Thus, in any case $x \in \mathscr{X}$. Thus, $PNCl(\tilde{S}) \subseteq \tilde{S}$. Hence, $\tilde{S} = pNCl(\tilde{S})$ or equivalently $\tilde{S}$ is an $NPCS$. Thus, every $NPGCS$ is an NCS. Therefore, $(\mathscr{X}, \Gamma)$ is neutrosophic pre-$T_{\frac{1}{2}}$ space. □

7. Conclusions

We have introduced generalized neutrosophic pre-closed sets and generalized neutrosophic pre-open sets over neutrosophic topology space. Many results have been established to show how far topological structures are preserved by these neutrosophic pre-closed. We also have provided examples where such properties fail to be preserved. In this paper, we have studied a few ideas only; it will be necessary to carry out more theoretical research to establish a general framework for decision-making and to define patterns for complex network conceiving and practical application.

Author Contributions: All authors have contributed equally to this paper. The individual responsibilities and contribution of all authors can be described as follows: the idea of this paper was put forward by W.A.-O. W.A.-O. completed the preparatory work of the paper. S.J. analyzed the existing work. The revision and submission of this paper was completed by W.A.-O.

Funding: This research received no external funding.

Conflicts of Interest: The authors declare no conflicts of interest.

References

1. Zadeh, L. Fuzzy sets. *Inform Control* **1965**, *8*, 338–353. [CrossRef]
2. Chang, C.L. Fuzzy topological spaces. *J. Math. Anal. Appl.* **1968**, *24*, 182–190. [CrossRef]
3. Scellato, S.; Fortuna, L.; Frasca, M.; Gómez-Gardenes, J.; Latora, V. Traffic optimization in transport networks based on local routing. *Eur. Phys. J. B* **2010**, *73*, 303–308. [CrossRef]
4. Atanassov, K. Intuitionistic fuzzy sets. In *VII ITKR's Session*; Publishing House: Sofia, Bulgaria, 1983.
5. Atanassov, K.; Stoeva, S. Intuitionistic L-fuzzy sets. In *Cybernetics and System Research*; Trappl, R., Ed.; Elsevier: Amsterdam, The Netherlands, 1984; Volume 2, pp. 539–540.
6. Coker, D. An introduction to intuitionistic fuzzy topological space. *Fuzzy Sets Syst.* **1997**, *88*, 81–89. [CrossRef]
7. Balasubramanin, G.; Sundaram, P. On some generalizations of fuzzy continuous functions. *Fuzzy Sets Syst.* **1997**, *86*, 93–100. [CrossRef]
8. Smarandache, F. *A Unifying Field in Logics: Neutrosophic Logic. Neutrosophy, Neutrosophic Set, Neutrosophic Probability*, 3rd ed.; American Research Press: Rehoboth, NM, USA, 1999.
9. Lupianez, F.G. On neutrosophic sets and topology. *Procedia Comput. Sci.* **2017**, *120*, 975–982. [CrossRef]

10. Liu, P.; Wang, P. Some q-Rung Orthopair Fuzzy Aggregation Operators and Their Applications to Multiple-Attribute Decision Making. *Int. J. Intell. Syst.* **2018**, *33*, 259–280. [CrossRef]
11. Zhou, L. Wu, W. Z. and Zhang, W. X. On characterization of intuitionistic fuzzy rough sets based on intuitionistic fuzzy implicators. *Inf. Sci.* **2009**, *179*, *7*, 883–898. [CrossRef]
12. Al-Omeri, W.; Smarandache, F. New Neutrosophic Sets via Neutrosophic Topological Spaces. In *Neutrosophic Operational Research*; Smarandache, F., Pramanik, S., Eds.; Pons Editions: Brussels, Belgium, 2017; Volume I, pp. 189–209.
13. Al-Omeri, W.F. Neutrosophic crisp sets via neutrosophic crisp topological spaces. *Neutrosophic Sets Syst.* **2016**, *13*, 96–104.
14. Salama, A.A.; Broumi, S.; Alblowi, S.A. Introduction to neutrosophic topological spatial region, possible application to gis topological rules. *Inf. Eng. Electron. Bus.* **2014**, *6*, 15–21. [CrossRef]
15. Salama, A.; Alblowi, S. Generalized neutrosophic set and generalized neutrousophic topological spaces. *J. Comput. Sci. Eng.* **2012**, *2*, 29–32.
16. Salama, A.A.; Smarandache, F.; Kroumov, V. Neutrosophic closed set and neutrosophic continuous functions. *Neutrosophic Sets Syst.* **2014**, *4*, 4–8.
17. Dhavaseelan, R.; Jafari, S. Generalized neutrosophic closed sets. In *New Trends in Neutrosophic Theory and Application*; Smarandache, F., Pramanik, S., Eds.; Pons Editions: Brussels, Belgium., 2018, Volume 2, pp. 261–274.
18. Jayanthi, D. Generalized closed sets in neutrosophic topological spaces. *Int. J. Math. Trends Technol. Spec. Issue* **2018**, *10*, 88–91.

mathematics

MDPI

Article

Design of Fuzzy Sampling Plan Using the Birnbaum-Saunders Distribution

Muhammad Zahir Khan [1,*], Muhammad Farid Khan [1], Muhammad Aslam [2] and Abdur Razzaque Mughal [1]

1 Department of Mathematics and Statistics, Riphah International University Islamabad, Islamabad 45710, Pakistan; muhammad.farid@riphah.edu.pk (M.F.K.); abdur_razzaque@live.com (A.R.M.)
2 Department of Statistics, Faculty of Science, King Abdulaziz University, Jeddah 21551, Saudi Arabia; aslam_ravian@hotmail.com
* Correspondence: zerishkh@gmail.com; Tel.: +92995670137

Received: 18 November 2018; Accepted: 14 December 2018; Published: 21 December 2018

Abstract: Acceptance sampling is one of the essential areas of quality control. In a conventional environment, probability theory is used to study acceptance sampling plans. In some situations, it is not possible to apply conventional techniques due to vagueness in the values emerging from the complexities of processor measurement methods. There are two types of acceptance sampling plans: attribute and variable. One of the important elements in attribute acceptance sampling is the proportion of defective items. In some situations, this proportion is not a precise value, but vague. In this case, it is suitable to apply flexible techniques to study the fuzzy proportion. Fuzzy set theory is used to investigate such concepts. It is observed there is no research available to apply Birnbaum-Saunders distribution in fuzzy acceptance sampling. In this article, it is assumed that the proportion of defective items is fuzzy and follows the Birnbaum-Saunders distribution. A single acceptance sampling plan, based on binomial distribution, is used to design the fuzzy operating characteristic (FOC) curve. Results are illustrated with examples. One real-life example is also presented in the article. The results show the behavior of curves with different combinations of parameters of Birnbaum-Saunders distribution. The novelty of this study is to use the probability distribution function of Birnbaum-Saunders distribution as a proportion of defective items and find the acceptance probability in a fuzzy environment. This is an application of Birnbaum-Saunders distribution in fuzzy acceptance sampling.

Keywords: fuzzy operating characteristic curve; fuzzy OC band; Birnbaum-Sunders distribution; single acceptance sampling plan

1. Introduction

An acceptance sampling plan is used to determine how many units can be selected from a lot, or consignment, and how many defective units are allowed in that sample. If the number of defective units is above the preset number of defective items, the lot is excluded. According to the rule of acceptance sampling, quality can be monitored by checking a few units from the whole lot. The plan that mentions guidelines for sampling and the associated criteria for accepting or rejecting a lot is called the acceptance sampling plan. This acceptance sampling plan can be implemented to check raw material, the material in a process or finished goods. An acceptance sampling plan can be classified as an attribute acceptance sampling plan and a variable acceptance sampling plan. An acceptance sampling plan can be classified with further attributes as a single sampling plan, double sampling plan, multiple sampling plans, and sequential sampling plan. An elementary acceptance sampling plan is a single sampling plan. In a single sampling plan, we select (n) units from the entire lot. This consists of (N) units. After selection of n units they are examined; if the number of damaged units

(d) is more than the specified number of defective items (c), the lot will be disallowed. Otherwise, it will be passed. The performance of any acceptance sampling plan can be judged by its operating characteristic (OC) curve. It determines how well an acceptance sampling plan distinguishes between good and bad lots. This OC curve has two parameters (n, c), where n is sample size and c is acceptance number. In an acceptance sampling plan, two groups are involved: the supplier and buyer. The supplier desires to avoid rejection of a good lot (producer's risk) and the buyer tries to avert acceptance of a bad lot (consumer's risk). In case a bad lot is accepted, it is the responsibility of the consumer. The producer's risk is denoted by α. This is the probability of rejection of the lot having an average quality level (AQL). Similarly, the consumer's risk is denoted with β. This shows the probability of acceptance of the lot, having low quality (LQL) [1]. The proportion of defective items is denoted by p and treated as a precise number. However, in some situations, it is not possible to get the precise numerical value of p. Mostly this value is determined by the expert, based on his judgment. It is used to calculate fuzzy acceptance probability. Further, this fuzzy p value and fuzzy acceptance probability are used to design a fuzzy OC curve [1]. In the study presented in Reference [2], the authors suggested a double acceptance sampling plan, based on assumption that lifetime of the product follows a generalized logistic distribution with known shape parameters, and analyzed the operating characteristic curve to several ratios of the true median life to the specified life. In the study presented in Reference [3], the authors proposed the double sampling plan and specified the design parameters fulfilling both the producer's and consumer's risks at the same time for a stated reliability, in the form of the mean ratio to the specific life. Moreover, double sampling and group sampling plans are constructed using the two-point technique, with the assumption that the lifetime of the product follows the Birnbaum-Saunders distribution. In the study presented in Reference [4], the pioneer of fuzzy set theory gave scientific structure to study imprecise and ambiguous concepts that are based on human judgment; comprising verbal expressions, contentment degree and significance degree, that are often fuzzy. A linguistic variable consists of expressions in a natural language, but not the number. In Reference [5] the authors applied fuzzy set theory to help explain complex and not easy to express linguistic terms, in traditional measurable terms. In Reference [6] the authors proposed a single acceptance sampling plan with a fuzzy parameter and explained the single acceptance sampling plan with fuzzy probability theory. In Reference [7] the authors used the expression for the OC curve and various values to help accept or reject a lot for a particular number of defective items. Proficiency of different acceptance sampling plans can be assessed by using the OC curve. These OC curves are used to determine the producer's risk, as well as the consumer's risk [8]. In Reference [9] the authors suggested using acceptance sampling in the fuzzy environment using Poisson distribution. In Reference [10] the authors explored if N is large, then the defective items will have a fuzzy binomial distribution. In Reference [11] the authors applied parameters of the acceptance sampling plan, sample size n, and acceptance number c, in a fuzzy environment. Acceptance probabilities of two major discrete distributions were also derived. The multiple deferred sampling plans and characteristic curves were proposed—where (p) proportion of defective items was treated as a fuzzy number—and also proposed fuzzy OC curves with different combinations of parameters [12]. Multiple deferred acceptance sampling plans with inspection errors were proposed by the authors of Reference [13]. In the study presented in Reference [14], the authors investigated the inspection errors and their impact on a single acceptance sampling plan, when the proportion of defective items was not known exactly. In Reference [15] the authors proposed an acceptance sampling plan for geospatial data with uncertainty in the proportion of defective items. In Reference [16] the authors investigated a double acceptance sampling plan with the fuzzy parameter. Average outgoing quality (AOQ) and average total inspection (ATI) in a double acceptance sampling plan with the imprecise proportion of defective items were presented [17]. In Reference [18] the authors suggested the fuzzy parameter for quality interval acceptance sampling plan, applying Poisson distribution. The fuzzy double acceptance sampling plan for Poisson distribution was proposed by the authors in Reference [19]. In Reference [20] the authors proposed an application of Weibull distribution in an acceptance sampling plan in the fuzzy environment and calculated fuzzy acceptance probabilities for different sample sizes using real-life data. In Reference [21] the authors proposed truncated life time, based on the Birnbaum-Saunders (BS) distribution. This distribution is used to define the number of stress cycles until failure of the material. In Reference [22] the authors applied the concept of the failure process of materials due to weariness, to design the BS distribution. Estimation of parameters based on crack length data was proposed in Reference [23]. In Reference [24] the authors presented a literature review of the BS distribution and discussed in detail the importance of this distribution and its application in different fields. In this study [25], they developed an acceptance sampling plan using the BS distribution to get the minimum sample size, n.

The aim of this article was to apply a single acceptance sampling plan when data were fuzzy and the proportion of defective items followed the BS distribution. According to the best of our knowledge, there is no work on the fuzzy plan using the BS distribution in the literature. In this paper, we will develop the fuzzy sampling plan using this distribution. The application of the proposed sampling will be given with the aid of a real example.

2. Materials and Methods

Design of Proposed Plan

Probability distribution function (Pdf) of BS distribution

$$F_T(t,\alpha,\lambda) = \Phi\left(\frac{1}{\alpha}\xi\left(\frac{t}{\lambda}\right)\right),\ 0 < t < \infty, \lambda > 0 \tag{1}$$

where α is the shape parameter and λ is the scale parameter, $\Phi(.)$ is the standard normal cumulative function and $\xi(t/\lambda) = \sqrt{\frac{t}{\lambda}} - \sqrt{\frac{\lambda}{t}}$. It can be shown that the median of the BS distribution is equal to the scale parameter and the mean of the BS distribution is

$$\mu = \lambda\left(1 + \alpha^2/2\right) \tag{2}$$

Here we write the assumptions for the BS distribution.

Let $t_0 = a\mu_0; a$ be called the termination ratio. The cumulative distribution function (Cdf) given in Equation (5) can be rewritten as

$$F_T(t_0,\alpha,\lambda) = \Phi\left(\frac{1}{\alpha}\xi\left(\frac{\alpha(1+\alpha^2/2)}{\mu/\mu_0}\right)\right), \tag{3}$$

The acceptance probability

According to [26], the acceptance probability of sampling plans can be obtained by using the binomial distribution. The lot acceptance probability of a lot in a single acceptance sampling plan (SASP) case is given as

$$L(p) = \left[\sum_{i=0}^{n}\binom{n}{i}p^i(1-p)^{n-i}\right] \tag{4}$$

The proportion of defective items in the fuzzy form.

According to the equation proposed by the authors of Reference [27].

$$\widetilde{p}K = (K, b_2 + K, b_3 + K, b_4 + K).p_K \in \widetilde{p}K[\alpha], q_K \in \widetilde{q}K[\alpha], p_K + q_K = 1 \tag{5}$$

$b_i = a_i - a_2$, i = 2, 3, 4 and $K = [0, 1 - b_4]$.

α-cut of $\widetilde{p}K$

$$\widetilde{p}K(\alpha) = (K + (b_2 + K - K)\alpha, b_3 + K + (b_3 - b_4)\alpha) \tag{6}$$

α-cut of $\widetilde{p}K$ at α = 0

$$\widetilde{p}K(0) = (K, b_4 + K)$$

where p is the $F_T(t_0, \alpha, \lambda)$ in Equation (4).

Fuzzy acceptance probability

According to Reference [11], the fuzzy acceptance probability can be calculated as

$$\widetilde{p}_k\ \alpha \left\{ \left(\begin{array}{c} n \\ k \end{array} \right) p^k\ q^{n-k} \middle| p \in p\ \alpha, q \in q\ \alpha \right\},\ 0 \le \alpha \le 1 \tag{7}$$

$$\widetilde{p}_k\ \alpha = [p_{kl}, p_{kr}]$$

$$p_{kl}\ \alpha = min \left\{ \left(\begin{array}{c} n \\ k \end{array} \right) p^k\ q^{n-k} \middle| p \in p\ \alpha, q \in q\ \alpha \right\} \tag{8}$$

$$p_{kr}\ \alpha = max \left\{ \left(\begin{array}{c} n \\ k \end{array} \right) p^k\ q^{n-k} \middle| p \in p\ \alpha, q \in q\ \alpha \right\} \tag{9}$$

The fuzzy acceptance probability when the number of defective items, c = 0, and α = 0

$$\widetilde{p}_K(0)[0] = \left(1 - \widetilde{p}_K{}^U[\alpha]\right)^n, \left(1 - \widetilde{p}_K{}^L[\alpha]\right)^n \tag{10}$$

$$\widetilde{p}_K{}^L = K \widetilde{p}_K{}^U = b_4 + K$$

The fuzzy acceptance probability based when the number of defective items, c = 1 and α = 0

$$\widetilde{p}_K(1)[0] = \left(1 - \widetilde{p}_K{}^U[\alpha]\right)^n + n\left(1 - \widetilde{p}_K{}^U[\alpha]\right)^{n-1}, \left(1 - \widetilde{p}_K{}^L[\alpha]\right)^n + n\left(1 - \widetilde{p}_K{}^L[\alpha]\right)^{n-1} \tag{11}$$

Here $\widetilde{p}_K{}^L[\alpha]$, $\widetilde{p}_K{}^U[\alpha]$ are calculated using CDF of the BS distribution

The design for a single acceptance sampling plan (ASP) to generate a fuzzy operating characteristic curve (FOC)

Step 1. The sample size for a lot is n.
Step 2. Specify the acceptance number (or action limit) c for a sample and the experiment time t_0.
Step 3. Perform the experiment for the sample size n and record the number of failures for a sample.
Step 4. Accept the lot if at most c failures are observed in the sample. Truncate the experiment and reject the lot if more than c failures are observed in the sample.
Step 5. Calculation of fuzzy p (proportion of defective items) using Equation (6).
Step 6. Calculation of fuzzy acceptance probability using Equations (10) and (11).
Step 7. Design of fuzzy OC curve (FOC) includes k and the acceptance probability, where k is the transformation of the fuzzy proportion of defective items.

The advantage of the fuzzy OC curve is that it is flexible and can be applied when the proportion of defective items is fuzzy. Secondly, the width of the fuzzy OC curve indicates the quality. Where the wider the width, the lesser the quality, and vice versa. In this study, the width of the fuzzy OC curve is influenced by the mean ratio. When the mean ratio is higher, the width of the band decreases. When the mean ratio is lower, the width increases. The advantage of this approach is that it can be applied to study any fuzzy data, which follows the BS distribution. This approach is more flexible than conventional p because it considers intermediate values of the fuzzy curve.

The fuzzy proportion of defective item k at α = 0 is denoted by $\widetilde{p}_K[0]$ and the fuzzy acceptance probability as $\widetilde{P}_K(0)[0]$. Values for sample size n = 5 and acceptance number c = 0, will therefore be (0.00, 0.001), and (0.95, 0.96), respectively at k = 0.01. Similarly, values of proportion and acceptance probability for sample size n = 5 and acceptance number c = 0, will be (0.052, 0.054) and (0.77, 0.78), respectively at k = 0.01. Furthermore, the fuzzy proportion of defective item k at $\alpha = 0$ is denoted by $\widetilde{p}_K[0]$ and the fuzzy acceptance probability as $\widetilde{P}_K(1)[0]$. These values for sample size n = 5 and acceptance number c = 1, will be the proportion of defective items (0.01, 0.029), and the acceptance probability (0.99, 0.995) at K = 0.01.

3. Real Life Example

In this section, we will discuss the application of the proposed sampling plan using real data selected from [28] and [29]. As mentioned above, the BS distribution is also known as the fatigue life distribution. It is used extensively in reliability applications to model failure times. The BS distribution is used in circumstances where occurring of events is independent of each other, from one cycle to another cycle, with same random distribution [29]. In this study, the failure life data given in Reference [28] is used. The authors of Reference [28] found that the data follow the BS distribution. The fatigue life data of aluminum coupons having n = 101 observations are shown in Table 1.

Table 1. Fatigue data of aluminum in hours.

70	90	96	97	99	100	103	104	104	105	107	108	108	108	109
109	112	112	113	114	114	114	116	119	120	120	120	121	121	123
124	124	124	124	124	128	128	129	129	130	130	130	131	131	131
131	131	132	132	132	133	134	134	134	134	134	136	136	137	138
138	138	139	139	141	141	142	142	142	142	142	142	144	144	145
146	148	148	149	151	151	152	155	156	157	157	157	157	158	159
162	163	163	164	166	166	168	170	174	196	212.				

We assume that data follows the BS distribution, the proportion of defective items p is fuzzy and the shape parameter is taken as the trapezoidal fuzzy number, $\hat{\alpha}$ = (0.15, 0.16, 0.17, 0.18). When the actual mean is $\mu_0 = 134$, the termination ratio a = 0.5 is then truncated, time will be t = 67 for c = 0, and at $\frac{\mu_0}{\mu_T} = 1$, the proportion of defective items is $\widetilde{p} = 0.0211673$. The acceptance probabilities are calculated by using Equations (9) and (10), for different sample sizes n = (5,25,75,100) and K = (0.0, 0.01,0.02, 0.03, 0.04, 0.05). The fuzzy OC curve is designed using fuzzy p values and fuzzy acceptance probabilities for c = 0. Similarly, when $\mu_0 = 134$, termination ratio a = 0.67 is then truncated, time will be t = 89.7 for c = 1. In this case, acceptance number c = 1, and $\frac{\mu_0}{\mu_T} = 1$, $\widetilde{p} = 0.01923$. The acceptance probabilities are calculated using Equation (11) for different sample sizes n = (5,25,75,100) and K = (0.0, 0.01, 0.02, 0.03, 0.04, 0.05). The fuzzy OC curve is developed using p values and acceptance probabilities for c = 1. Fuzzy acceptance probability for Birnbaum-Saunders distribution is presented in Tables 2 and 3 using real life data and their respective fuzzy OC curves are shown in Figures 1–3. Entire calculations and graphs were completed using R software and codes were given in Appendix A. Acceptance probability is influenced by mean ratio, when mean ratio increases it reduces Uuncertainty and bandwidth of fuzzy OC curve become narrow while decreasing mean ration increases the width of fuzzy OC curve. The fuzzy OC curves show more convexity when sample size n increases. Fuzzy OC curve with c = 0 shows less uncertainty than c = 1. We presented acceptance probabilities and fuzzy OC curves with c = 0, it is almost equal to conventional OC curve. The fuzzy OC curve for c = 0 and c = 1 is more convex at large sample size as compared to small sample size.

Table 2. Fatigue life data of aluminum coupons.

K	$\widetilde{p}_K[0]$	$\widetilde{P}_K(1)[0]$, $n = 5$, $c = 0$, $\frac{\mu}{\mu_0} = 1$	$\widetilde{P}_K(1)[0]$, $n = 25$, $c = 0$, $\frac{\mu}{\mu_0} = 1$	$\widetilde{P}_K(1)[0]$, $n = 75$, $c = 0$, $\frac{\mu}{\mu_0} = 1$	$\widetilde{P}_K(1)[0]$, $n = 100$, $c = 0$, $\frac{\mu}{\mu_0} = 1$
0.00	[0.00, 0.001]	[0.98, 1.00]	[0.99, 1.00]	[0.99, 1.00]	[0.996, 1.00]
0.01	[0.01, 0.011]	[0.95, 0.96]	[0.81, 0.82]	[0.73, 0.76]	[0.604, 0.620]
0.02	[0.02, 0.022]	[0.93, 0.94]	[0.65, 0.66]	[0.53, 0.54]	[0.3640, 0.372]
0.03	[0.03, 0.041]	[0.85, 0.86]	[0.53, 0.54]	[0.40, 0.43]	[0.213, 0.219]
0.04	[0.04, 0.051]	[0.81, 0.83]	[0.41, 0.43]	[0.29, 0.31]	[0.125, 0.129]
0.05	[0.052, 0.054]	[0.77, 0.78]	[0.35, 0.36]	[0.21, 0.23]	[0.076, 0.077]

Table 3. The fuzzy acceptance probability of Birnbaum-Saunders distribution for c = 0.

K	$\widetilde{p}_K[0]$	$\widetilde{P}_K(1)[0]$, $n = 5$, $c = 1$, $\frac{\mu}{\mu_0} = 1$	$\widetilde{P}_K(1)[0]$, $n = 25$, $c = 1$, $\frac{\mu}{\mu_0} = 1$	$\widetilde{P}_K(1)[0]$, $n = 75$, $c = 1$, $\frac{\mu}{\mu_0} = 1$	$\widetilde{P}_K(1)[0]$, $n = 100$, $c = 1$, $\frac{\mu}{\mu_0} = 1$
0.00	[0.00, 0.019]	[1.000, 1.00]	[1.0000, 1.00]	[1.00, 1.00]	[0.99, 1.00]
0.01	[0.01, 0.029]	[0.99, 0.995]	[0.979, 0.97]	[0.827, 0.82]	[0.82, 0.82]
0.02	[0.02, 0.039]	[0.94, 0.96]	[0.99, 0.949]	[0.55, 0.556]	[0.65, 0.660]
0.03	[0.03, 0.077]	[0.93, 0.95]	[0.82, 0.826]	[0.338, 0.338]	[0.63, 0.54]
0.04	[0.04, 0.050]	[0.91, 0.92]	[0.73, 0.730]	[0.190, 0.190]	0.44, 0.42]
0.05	[0.05, 0.067]	[0.89, 0.900]	[0.720, 0.7202]	[0.160, 0.160]	[0.35, 0.37]

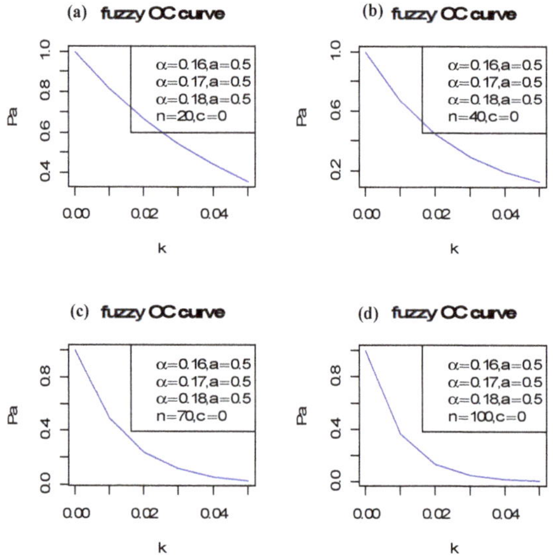

Figure 1. The fuzzy operating characteristic (OC) Curve of the Birnbaum-Saunders distribution at c = 0 (**a**) n = 20, (**b**) n = 40, (**c**) n = 70, and (**d**) n = 100.

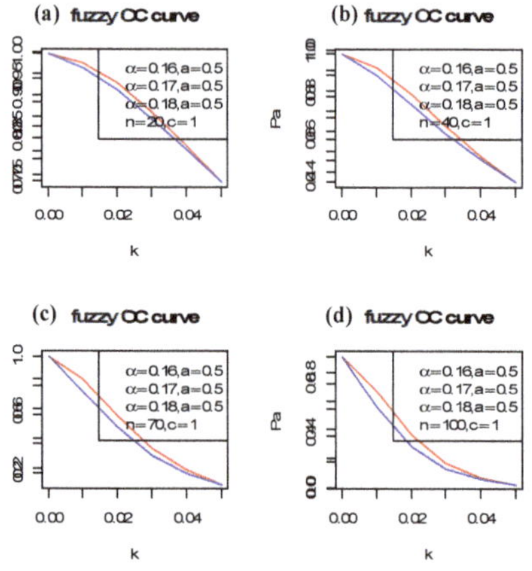

Figure 2. The fuzzy operating characteristic (OC) curve of Birnbaum-Saunders distribution at c = 1, mean ratio = 0.5, (**a**) n = 20, (**b**) n = 40, (**c**) n = 70, and (**d**) n = 100.

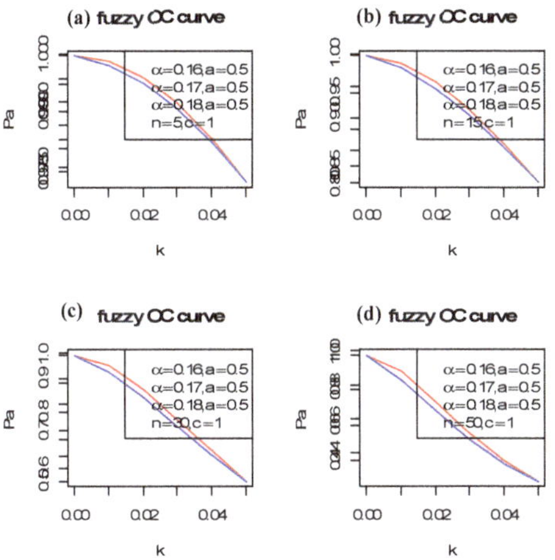

Figure 3. The fuzzy operating characteristic (OC) curve of Birnbaum-Saunders distribution at c = 1, mean ratio = 0.5, (**a**) n = 5, (**b**) n = 15, (**c**) n = 30, and (**d**) n = 50.

4. Conclusions

Acceptance sampling is one of the important aspects of statistical quality control. When the data follow the Birnbaum-Saunders distribution and the proportion of defective items is fuzzy, acceptance probability and the OC curve can be presented in a fuzzy form. In this article, the fuzzy OC curve of the Birnbaum-Saunders distribution is presented in a single acceptance sampling plan, using the binomial distribution. The fuzzy OC curve has a band with two bounds, lower and upper. The width of the band depends upon the uncertainty in the proportion of defective items in the fuzzy environment. Less uncertainty will give a narrow width. The fuzzy OC curves also show more convexity at a large sample size. The mean ratio in the Birnbaum-Saunders distribution is another important factor in quality. Here, a lower value of the mean ratio causes the width of the band of the fuzzy OC curve to increase. This indicates more uncertainty. The advantage of this approach is that it can be used to calculate the proportion of defective items when fuzzy data follows a Birnbaum-Saunders distribution, because mostly we assume the value of the proportion of defective items without using any distribution. Secondly, the fuzzy acceptance probability based on the Birnbaum-Saunders distribution is calculated. The fuzzy OC curve of the Birnbaum-Saunders distribution is constructed based on fuzzy p and the acceptance probability. The OC curve is more convex at large sample sizes, as compared to small sample sizes. It was concluded that when data followed the Birnbaum-Saunders distribution, this proposed approach was suitable to calculate the proportion (p), the acceptance probability, and the OC curve in both conventional and fuzzy form. In the future, we will apply the same concept to group acceptance sampling and chain acceptance sampling in a fuzzy environment.

Author Contributions: Conceptualization, M.F.K. and M.Z.K.; methodology, M.Z.K.; software, M.A.; validation, A.R.M.

Funding: This research received no funding.

Acknowledgments: The authors are deeply thankful to the reviewers and editor for their valuable suggestions to improve the quality of the paper.

Conflicts of Interest: The authors declare that there is no conflict of interest regarding the publication of this paper.

Abbreviations

OC	operating characteristic
OC curve	operating characteristic curve
(FOC) curve	fuzzy operating characteristic curve
AQL	average quality level
LQL	low quality level
pdf	probability density function.
cdf	Cumulative distribution function
SASP	single acceptance sampling plan
BS	Birnbaum-Saunders

Appendix A

R codes

```
#When n = 5, 20, 30, 30, c = 0
rm(list=ls ())
windows ()
par(mfrow=c(2,2))
a = 0.5 #For c = 0 at t = 67
alpha = c(0.15, 0.16, 0.17, 0.18)
K = seq(0, 0.05, 0.01)
x = a*((1 + alpha^2)/2) #Here b is teated as Alpha
y = 0.5# 1, 2, 3, 4, 5, 6 values of ratio of
X = c((1/alpha)*(sqrt(x/y)−sqrt(y/x)))
FX = pnorm(X, mean = 0, sd = 1, lower.tail = TRUE, log.p = FALSE)
FX
p = FX
p = 2.470246e-16
K = seq(0,0.05,0.01)
W = p + K
p = 0.226627400
W = p + K
#p = 0.226627400
p = 0.019
W = p + K
K = seq(0,0.05,0.01)
c = o, a = 0.5) and (c = 1 when a = 0.67) for a = 0.5, p = 2.470246e-16, for a = 0.67 p = 0.01923
B = dbinom(0,10,K) # B = (1−K)^5 When n = 5, c = 0 (1)
A = dbinom(0,10, W)# A = (1−(K+p))^5#When n = 5, c = 0 (2)
data.frame(A,B)
data.frame(K,W,A,B)
#B = (1-K)^5 # THIS WILL GIVE US UPPER BAND HIGHER PROBABILITY
#A = (1-(K+p))^5#THIS WILL GIVE US LOWER BAND LOWER PROBABILITY
plot(K,A,type = "l", col = "red", xlab = "K", ylab = "Pa ", main = "fuzzy OC curve")
par(new = TRUE)
plot(W,B,type = "l", col = "blue", xlab = "k", ylab = " ", main = "")
legend("topright",c(expression(paste(alpha==0.15,",",a==0.67)),expression(paste(alpha==0.16,",",a==
0.67)),expression(paste(alpha==0.17,",",a==0.1)),expression(paste(alpha==0.18,",",a==0.67)),expression
(paste(n==5,",",c==0))))
```

References

1. Kahraman, C.; Kaya, İ. Fuzzy acceptance sampling plans. In *Production Engineering and Management under Fuzziness*; Springer: Heidelberg, Germany, 2010; pp. 457–481.
2. Aslam, M.; Jun, C.-H. A double acceptance sampling plan for generalized log-logistic distributions with known shape parameters. *J. Appl. Stat.* **2010**, *37*, 405–414. [CrossRef]
3. Aslam, M.; Jun, C.-H.; Ahmad, M. New acceptance sampling plans based on life tests for Birnbaum–Saunders distributions. *J. Stat. Comput. Simul.* **2011**, *81*, 461–470. [CrossRef]

4. Zadeh, L.A. Information and control. *Fuzzy Sets* **1965**, *8*, 338–353.
5. Zimmermann, H.J. *Fuzzy Set Theory and Its Applications*; Kluwer Academic Publishers: Dordrecht, The Netherlands, 1991.
6. Sadeghpour Gildeh, B.; Jamkhaneh, E.B.; Yari, G. Acceptance single sampling plan with fuzzy parameter. *Iran. J. Fuzzy Syst.* **2011**, *8*, 47–55.
7. Grzegorzewski, P. Acceptance sampling plans by attributes with fuzzy risks and quality levels. In *Frontiers in Statistical Quality Control*; Springer: Heidelberg, Germany, 2001; pp. 36–46.
8. Montgomery, D.C. *Ihntroduction to Statistical Quality Control*; John Wiley & Sons: New York, NY, USA, 2009.
9. Jamkhaneh, E.B.; Sadeghpour-Gildeh, B.; Yari, G. Acceptance single sampling plan with fuzzy parameter with the using of poisson distribution. *World Acad. Sci. Eng. Technol.* **2009**, *3*, 1–20.
10. Buckley, J.J. *Fuzzy Probability and Statistics*; Springer: Heidelberg, Germany, 2006.
11. Turanoğlu, E.; Kaya, İ.; Kahraman, C. Fuzzy acceptance sampling and characteristic curves. *Int. J. Comput. Intell. Syst.* **2012**, *5*, 13–29. [CrossRef]
12. Afshari, R.; Gildeh, B.S.; Sarmad, M. Multiple Deferred State Sampling Plan with Fuzzy Parameter. *Int. J. Fuzzy Syst.* **2017**, *20*, 549–557. [CrossRef]
13. Afshari, R.; Gildeh, B.S.; Sarmad, M. Fuzzy multiple deferred state attribute sampling plan in the presence of inspection errors. *J. Intell. Fuzzy Syst.* **2017**, *33*, 503–514. [CrossRef]
14. Jamkhaneh, E.B.; Sadeghpour-Gildeh, B.; Yari, G. Inspection error and its effects on single sampling plans with fuzzy parameters. *Struct. Multidiscip. Optim.* **2011**, *43*, 555–560. [CrossRef]
15. Tong, X.; Wang, Z. Fuzzy acceptance sampling plans for inspection of geospatial data with ambiguity in quality characteristics. *Comput. Geosci.* **2012**, *48*, 256–266. [CrossRef]
16. Sadeghpour-Gildeh, B.; Yari, G.; Jamkhaneh, E.B. Acceptance double sampling plan with fuzzy parameter. Available online: http://citeseerx.ist.psu.edu/viewdoc/download?doi=10.1.1.864.8328&rep=rep1&type=pdf (accessed on 1 November 2018).
17. Jamkhaneh, E.B.; Gildeh, B.S. Notice of Retraction AOQ and ATI for Double Sampling Plan with Using Fuzzy Binomial Distribution. In Proceedings of the International Conference on Intelligent Computing and Cognitive Informatics (ICICCI), Kuala Lumpur, Malaysia, 22–23 June 2010.
18. Divya, P. Quality interval acceptance single sampling plan with fuzzy parameter using poisson distribution. *Int. J. Adv. Res. Technol.* **2012**, *1*, 115–125.
19. Jamkhaneh, E.B.; Gildeh, B.S. Acceptance Double Sampling Plan using Fuzzy Poisson Distribution. *World Appl. Sci. J.* **2012**, *15*, 1692–1702.
20. Venkatesh, A.; Subramani, G. Acceptance sampling for the secretion of Gastrin using crisp and fuzzy Weibull distribution. *Int. J. Eng. Res. Appl.* **2014**, *4*, 564–569.
21. Wu, C.-J.; Tsai, T.-R. Acceptance sampling plans for Birnbaum-Saunders distribution under truncated life tests. *Int. J. Reliab. Qual. Saf. Eng.* **2005**, *12*, 507–519. [CrossRef]
22. Birnbaum, Z.W.; Saunders, S.C. A new family of life distributions. *J. Appl. Probab.* **1969**, *6*, 319–327. [CrossRef]
23. Birnbaum, Z.W.; Saunders, S.C. Estimation for a family of life distributions with applications to fatigue. *J. Appl. Probab.* **1969**, *6*, 328–347. [CrossRef]
24. Balakrishnan, N.; Kundu, D. Birnbaum-Saunders Distribution: A Review of Models, Analysis and Applications. *arXiv* 2018, arXiv:1805.06730.
25. Balakrishnan, N.; Leiva, V.; Lopez, J. Acceptance sampling plans from truncated life tests based on the generalized Birnbaum–Saunders distribution. *Commun. Stat. Simul. Comput.* **2007**, *36*, 643–656. [CrossRef]
26. Stephens, K.S. *The Handbook of Applied Acceptance Sampling: Plans, Principles, and Procedures*; Asq Press: Milwaukee, WI, USA, 2001.
27. Jamkhaneh, E.B.; Gildeh, B.S. Chain sampling plan using Fuzzy probability theory. *J. Appl. Sci.* **2011**, *11*, 3830–3838. [CrossRef]
28. From, S.G.; Li, L. Estimation of the parameters of the Birnbaum–Saunders distribution. *Commun. Stat. Theory Methods* **2006**, *35*, 2157–2169. [CrossRef]
29. Croarkin, C.; Tobias, P.; Zey, C. *Engineering Statistics Handbook*; NIST: Gaithersburg, MD, USA, 2002.

mathematics

Article

Neutrosophic Soft Expert Multiset and Their Application to Multiple Criteria Decision Making

Derya Bakbak [1], Vakkas Uluçay [2,*] and Memet Şahin [2]

1 TBMM Public Relations Building 2nd Floor, B206 room Ministries, 06543 Ankara, Turkey; derya.bakbak@tbmm.gov.tr

2 Department of Mathematics, Gaziantep University, 27310 Gaziantep, Turkey; mesahin@gantep.edu.tr

* Correspondence: vulucay27@gmail.com; Tel.: +90-0537-643-5034

Received: 11 December 2018; Accepted: 2 January 2019; Published: 6 January 2019

Abstract: In this paper, we have investigated neutrosophic soft expert multisets (NSEMs) in detail. The concept of NSEMs is introduced. Several operations have been defined for them and their important algebraic properties are studied. Finally, we define a NSEMs aggregation operator to construct an algorithm for a NSEM decision-making method that allows for a more efficient decision-making process.

Keywords: aggregation operator; decision making; neutrosophic soft expert sets; neutrosophic soft expert multiset

1. Introduction

Multiple criteria decision making (MCDM) is an important part of modern decision science and relates to many complex factors, such as economics, psychological behavior, ideology, military and so on. For a proper description of objects in an uncertain and ambiguous environment, indeterminate and incomplete information has to be properly handled. Intuitionistic fuzzy sets were introduced by Atanassov [1], followed by Molodtsov [2] on soft set and neutrosophy logic [3] and neutrosophic sets [4] by Smarandache. The term neutrosophy means knowledge of neutral thought and this neutral represents the main distinction between fuzzy and intuitionistic fuzzy logic and set. Presently, work on soft set theory is progressing rapidly. Various operations and applications of soft sets were developed rapidly, including multi-adjoint t-concept lattices [5], signatures, definitions, operators and applications to fuzzy modelling [6], fuzzy inference system optimized by genetic algorithm for robust face and pose detection [7], fuzzy multi-objective modeling of effectiveness and user experience in online advertising [8], possibility fuzzy soft set [9], soft multiset theory [10], multiparameterized soft set [11], soft intuitionistic fuzzy sets [12], Q-fuzzy soft sets [13–15], and multi Q-fuzzy sets [16–18], thereby opening avenues to many applications [19,20]. Later, Maji [21] introduced a more generalized concept, which is a combination of neutrosophic sets and soft sets and studied its properties. Alkhazaleh and Salleh [22] defined the concept of fuzzy soft expert sets, which were later extended to vague soft expert set theory [23], generalized vague soft expert set [24], and multi Q-fuzzy soft expert set [25]. Şahin et al. [26] introduced neutrosophic soft expert sets, while Hassan et al. [27] extended it further to Q-neutrosophic soft expert sets. Broumi et al. [28] defined neutrosophic parametrized soft set theory and its decision making. Deli [29] introduced refined neutrosophic sets and refined neutrosophic soft sets.

Since membership values are inadequate for providing complete information in some real problems which has different membership values for each element, different generalizations of fuzzy sets, intuitionistic fuzzy sets and neutrosophic sets have been introduced called the multi fuzzy set [30], intuitionistic fuzzy multiset [31] and neutrosophic multiset [32,33], respectively. In the multisets, an element of a universe can be constructed more than once with possibly the same or different membership values. Some work on the multi fuzzy set [34,35], on the intuitionistic fuzzy multiset [36–39] and on the neutrosophic multiset [40–43] have been studied. The above set theories have been applied to many different areas including real decision-making problems [44–47]. The aim of this paper is allow the neutrosophic set to handle problems involving incomplete, indeterminacy and awareness of inconsistency knowledge, and this is further developed to neutrosohic soft expert sets.

The initial contributions of this paper involve the introduction of various new set-theoretic operators on neutrosophic soft expert multisets (NSEMs) and their properties. Later, we intend to extend the discussion further by proposing the concept of NSEMs and its basic operations, namely complement, union, intersection AND and OR, along with a definition of a NSEMs-aggregation operator to construct an algorithm of a NSEMs decision method. Finally we provide an application of the constructed algorithm to solve a decision-making problem.

2. Preliminaries

In this section we review the basic definitions of a neutrosophic set, neutrosophic soft set, soft expert sets, neutrosophic soft expert sets, and NP-aggregation operator required as preliminaries.

Definition 1 ([4]). *A neutrosophic set A on the universe of discourse $\mathcal{U}$ is defined as $A = \{\langle u, (\mu_A(u), v_A(u), w_A(u))\rangle : u \in U, \mu_A(u), v_A(u), w_A(u) \in [0,1]\}$. There is no restriction on the sum of $\mu_A(u)$; $v_A(u)$ and $w_A(u)$, so $0^- \le \mu_A(u) + v_A(u) + w_{\mathcal{A}}(u) \le 3^+$.*

Definition 2 ([21]). *Let $\mathcal{U}$ be an initial universe set and E be a set of parameters. Consider $A \subseteq E$. Let $NS(\mathcal{U})$ denotes the set of all neutrosophic sets of $\mathcal{U}$. The collection (F, A) is termed to be the neutrosophic soft set over $\mathcal{U}$, where F is a mapping given by $F : A \to NS(\mathcal{U})$.*

Definition 3 ([22]). *$\mathcal{U}$ is an initial universe, E is a set of parameters X is a set of experts (agents), and $O = \{agree = 1, disagree = 0\}$ a set of opinions. Let $Z = E \times X \times O$ and $A \subseteq Z$. A pair (F, A) is called a soft expert set over $\mathcal{U}$, where F is mapping given by $F : A \to P(\mathcal{U})$ where $P(\mathcal{U})$ denote the power set of $\mathcal{U}$.*

Definition 4 ([26]). *A pair (F, A) is called a neutrosophic soft expert set over $\mathcal{U}$, where F is mapping given by*

$$F : A \to P(\mathcal{U}) \tag{1}$$

where $P(\mathcal{U})$ denotes the power neutrosophic set of U.

Definition 5 ([26]). *The complement of a neutrosophic soft expert set (F, A) denoted by $(F, A)^c$ and is defined as $(F, A)^c = (F^c, A)$ where $F^c = \neg A \to P(\mathcal{U})$ is mapping given by $F^c(x)$ = neutrosophic soft expert complement with $\mu_{F^c(x)} = w_{F(x)}$, $v_{F^c(x)} = v_{F(x)}$, $w_{F^c(x)} = \mu_{F(x)}$.*

Definition 6 ([26]). *The agree-neutrosophic soft expert set $(F, A)_1$ over $\mathcal{U}$ is a neutrosophic soft expert subset of (F, A) is defined as*

$$(F, A)_1 = \{F_1(m) : m \in E \times X \times \{1\}\}. \tag{2}$$

Definition 7 ([26]). *The disagree-neutrosophic soft expert set $(F, A)_0$ over $\mathcal{U}$ is a neutrosophic soft expert subset of (F, A) is defined as*

$$(F, A)_0 = \{F_0(m) : m \in E \times X \times \{0\}\}. \tag{3}$$

Definition 8 ([26]). *Let (H, A) and (G, B) be two NSESs over the common universe U. Then the union of (H, A) and (G, B) is denoted by "$(H, A) \tilde{\cup} (G, B)$" and is defined by $(H, A) \tilde{\cup} (G, B) = (K, C)$, where*

$C = A \cup B$ *and the truth-membership, indeterminacy-membership and falsity-membership of* (K, C) *are as follows:*

$$\mu_{K(e)}(m) = \begin{cases} \mu_{H(e)}(m), & if\ e \in A - B, \\ \mu_{G(e)}(m), & if\ e \in B - A, \\ \max\left(\mu_{H(e)}(m), \mu_{G(e)}(m)\right), & if\ e \in AB. \end{cases}$$
$$v_{K(e)}(m) = \begin{cases} v_{H(e)}(m), & if\ e \in A - B, \\ v_{G(e)}(m), & if\ e \in B - A, \\ \frac{v_{H(e)}(m) + v_{G(e)}(m)}{2}, & if\ e \in AB. \end{cases} \quad (4)$$
$$w_{K(e)}(m) = \begin{cases} w_{H(e)}(m), & if\ e \in A - B, \\ w_{G(e)}(m), & if\ e \in B - A, \\ \min\left(w_{H(e)}(m), w_{G(e)}(m)\right), & if\ e \in AB. \end{cases}$$

Definition 9 ([26]). *Let* (H, A) *and* (G, B) *be two NSESs over the common universe* U. *Then the intersection of* (H, A) *and* (G, B) *is denoted by "*$(H, A) \tilde{\cap} (G, B)$*" and is defined by* $(H, A) \tilde{\cap} (G, B) = (K, C)$, *where* $C = A \cap B$ *and the truth-membership, indeterminacy-membership and falsity-membership of* (K, C) *are as follows:*

$$c\mu_{K(e)}(m) = \min\left(\mu_{H(e)}(m), \mu_{G(e)}(m)\right),$$
$$v_{K(e)}(m) = \frac{v_{H(e)}(m) + v_{G(e)}(m)}{2}, \quad (5)$$
$$w_{K(e)}(m) = \max\left(w_{H(e)}(m), w_{G(e)}(m)\right),$$

$if\ e \in AB.$

Definition 10 ([29]). *Let* $\mathcal{U}$ *be a universe. A neutrosophic multiset set (Nms)* A *on* $\mathcal{U}$ *can be defined as follows:*

$$A = \left\{ \prec u, \left(\mu_A^1(u), \mu_A^2(u), \ldots, \mu_A^p(u)\right), \left(v_A^1(u), v_A^2(u), \ldots, v_A^p(u)\right), \left(w_A^1(u), w_A^2(u), \ldots, w_A^p(u)\right) \succ : u \in \mathcal{U} \right\}$$

where,

$$c\,\mu_A^1(u), \mu_A^2(u), \ldots, \mu_A^p(u) : \mathcal{U} \to [0, 1],$$
$$v_A^1(u), v_A^2(u), \ldots, v_A^p(u) : \mathcal{U} \to [0, 1],$$

and

$$w_A^1(u), w_A^2(u), \ldots, w_A^p(u) : \mathcal{U} \to [0, 1],$$

such that

$$0 \le sup\mu^i{}_A(u) + supv^i{}_A(u) + supw^i{}_A(u) \le 3$$

$(i = 1, 2, \ldots, P)$ *and*

$$\left(\mu_A^1(u), \mu_A^2(u), \ldots, \mu_A^p(u)\right), \left(v_A^1(u), v_A^2(u), \ldots, v_A^p(u)\right) \text{ and } \left(w_A^1(u), w_A^2(u), \ldots, w_A^p(u)\right)$$

This is the truth-membership sequence, indeterminacy-membership sequence and falsity-membership sequence of the element u, respectively. Also, P is called the dimension (cardinality) of Nms A, denoted $d(A)$. We arrange the truth-membership sequence in decreasing order but the corresponding indeterminacy-membership and falsity-membership sequence may not be in decreasing or increasing order.

The set of all neutrosophic multisets on $\mathcal{U}$ is denoted by NMS($\mathcal{U}$).

Definition 11 ([28]). *Let $\Psi_K \in$ NP-soft set. Then an NP-aggregation operator of Ψ_K, denoted by Ψ_K^{agg} is defined by*

$$\Psi_K^{agg} = \left\{ \left(\langle u, T_K^{agg}, I_K^{agg}, F_K^{agg} \rangle \right) : u \in U \right\},$$

which is a neutrosophic set over U,

$$\begin{array}{ll} T_K^{agg} : U \to [0,1] & T_K^{agg}(u) = \frac{1}{|U|} \sum\limits_{\substack{e \in E \\ u \in U}} T_K(u).\lambda f_{K(x)}(u), \\ I_K^{agg} : U \to [0,1] & I_K^{agg}(u) = \frac{1}{|U|} \sum\limits_{\substack{e \in E \\ u \in U}} I_K(u).\lambda f_{K(x)}(u), \\ F_K^{agg} : U \to [0,1] & F_K^{agg} = \frac{1}{|U|} \sum\limits_{\substack{e \in E \\ u \in U}} F_K(u).\lambda f_{K(x)}(u) \end{array} \tag{6}$$

and where,

$$\lambda f_{K(x)}(u) = \begin{cases} 1, & x \in f_{K(x)}(u), \\ 0, & \text{otherwise.} \end{cases} \tag{7}$$

$|U|$ is the cardinality of U.

3. Neutrosophic Soft Expert Multiset (NSEM) Sets

This section introduces neutrosophic soft expert multiset as a generalization of neutrosophic soft expert set. Throughout this paper, V is an initial universe, E is a set of parameters X is a set of experts (agents), and $O = \{\text{agree} = 1, \text{disagree} = 0\}$ a set of opinions. Let $Z = E \times X \times O$ and $G \subseteq Z$ and u is a membership function of G; that is, $\Omega : G \to= [0,1]$.

Definition 12. *A pair (F^{Ω}, G) is called a neutrosophic soft expert multiset over V, where F^{Ω} is mapping given by*

$$F^{\Omega} : G \to \mathcal{N}(V)\times, \tag{8}$$

where $\mathcal{N}(V)$ be the set of all neutrosophic soft expert subsets of U. For any parameter $e \in G$, $F(e)$ is referred as the neutrosophic value set of parameter e, i.e.,

$$F(e) = \left\{ \langle \frac{v}{\left(D^1_{F(e)}(v), \ldots, D^n_{F(e)}\right), \left(I^1_{F(e)}(v), \ldots, I^n_{F(e)}\right), \left(Y^1_{F(e)}(v), \ldots, Y^n_{F(e)}\right)} \rangle \right\}, \tag{9}$$

where $D^i,^i, Y^i : U \to [0,1]$ are the membership sequence of truth, indeterminacy and falsity respectively of the element $v \in V$. For any $v \in V$, $e \in G$ and $i = 1, 2, \ldots, n$.

$$0 \leq D^i{}_{F(e)}(v) +^i {}_{F(e)}(v) + Y^i{}_{F(e)}(v) \leq 3$$

In fact F^{Ω} is a parameterized family of neutrosophic soft expert multisets on V, which has the degree of possibility of the approximate value set which is prepresented by $\Omega(e)$ for each parameter e. So we can write it as follows:

$$F^{\Omega}(e) = \left\{ \left(\frac{v_1}{F(e)(v_1)}, \frac{v_2}{F(e)(v_2)}, \frac{v_3}{F(e)(v_3)}, \cdots, \frac{v_n}{F(e)(v_n)} \right), \Omega(e) \right\}. \tag{10}$$

Example 1. *Suppose that $V = \{v_1\}$ is a set of computers and $E = \{e_1, e_2\}$ is a set of decision parameters. Let $X = \{p, r\}$ be set of experts. Suppose that*

$$cF^{\Omega}(e_1, p, 1) = \left\{\left(\frac{v_1}{(0.4, 0.3, \ldots, 0.2), (0.5, 0.7, \ldots, 0.2), (0.6, 0.1, \ldots, 0.3)}\right), 0.4\right\}$$
$$F^{\Omega}(e_1, r, 1) = \left\{\left(\frac{v_1}{(0.3, 0.2, \ldots, 0.5), (0.8, 0.1, \ldots, 0.4), (0.5, 0.6, \ldots, 0.2)}\right), 0.8\right\}$$
$$F^{\Omega}(e_2, p, 1) = \left\{\left(\frac{v_1}{(0.7, 0.3, \ldots, 0.6), (0.3, 0.2, \ldots, 0.6), (0.8, 0.2, \ldots, 0.1)}\right), 0.5\right\}$$
$$F^{\Omega}(e_2, r, 1) = \left\{\left(\frac{v_1}{(0.8, 0.3, \ldots, 0.4), (0.3, 0.1, \ldots, 0.5), (0.2, 0.3, \ldots, 0.4)}\right), 0.4\right\}$$
$$F^{\Omega}(e_1, p, 0) = \left\{\left(\frac{v_1}{(0.5, 0.1, \ldots, 0.2), (0.6, 0.3, \ldots, 0.4), (0.7, 0.2, \ldots, 0.6)}\right), 0.1\right\}$$
$$F^{\Omega}(e_1, r, 0) = \left\{\left(\frac{v_1}{(0.4, 0.2, \ldots, 0.1), (0.6, 0.1, \ldots, 0.3), (0.7, 0.2, \ldots, 0.4)}\right), 0.4\right\}$$
$$F^{\Omega}(e_2, p, 0) = \left\{\left(\frac{v_1}{(0.8, 0.1, \ldots, 0.5), (0.2, 0.1, \ldots, 0.4), (0.6, 0.3, \ldots, 0.1)}\right), 0.6\right\}$$
$$F^{\Omega}(e_2, r, 0) = \left\{\left(\frac{v_1}{(0.7, 0.2, \ldots, 0.3), (0.4, 0.1, \ldots, 0.6), (0.3, 0.2, \ldots, 0.1)}\right), 0.2\right\}$$

The neutrosophic soft expert multiset (F, Z) is a parameterized family $\{F(e_i),\ i = 1, 2, \ldots\}$ of all neutrosophic multisets of V and describes a collection of approximation of an object.

Definition 13. *For two neutrosophic soft expert multisets (NSEMs) (F^{Ω}, G) and (H^{η}, R) over U, (F^{Ω}, G) is called a neutrosophic soft expert subset of (H^{η}, R) if*

i. $R \subseteq G$,
ii. *for all $\varepsilon \in H$, $H^{\eta}(\varepsilon)$ is neutrosophic soft expert subset $F^{\Omega}(\varepsilon)$.*

Example 2. *Consider Example 1. Suppose that G and R are as follows.*

$$cG = \{(e_1, p, 1), (e_2, p, 1), (e_2, p, 0), (e_2, r, 1)\}$$
$$R = \{(e_1, p, 1), (e_2, r, 1)\}$$

Since R is a neutrosophic soft expert subset of G, clearly $R \subset G$. Let (H^{η}, R) and (F^{Ω}, G) be defined as follows:

$$c\left(F^{\Omega}, G\right) = \left\{\left[(e_1, p, 1), \left(\frac{v_1}{(0.4, 0.3, \ldots, 0.2), (0.5, 0.7, \ldots, 0.2), (0.6, 0.1, \ldots, 0.3)}\right), 0.4\right],\right.$$
$$\left[(e_2, p, 1), \left(\frac{v_1}{(0.7, 0.3, \ldots, 0.6), (0.3, 0.2, \ldots, 0.6), (0.8, 0.2, \ldots, 0.1)}\right), 0.5\right],$$
$$\left[(e_2, p, 0), \left(\frac{v_1}{(0.8, 0.1, \ldots, 0.5), (0.2, 0.1, \ldots, 0.4), (0.6, 0.3, \ldots, 0.1)}\right), 0.6\right],$$
$$\left.\left[(e_2, r, 1), \left(\frac{v_1}{(0.8, 0.3, \ldots, 0.4), (0.3, 0.1, \ldots, 0.5), (0.2, 0.3, \ldots, 0.4)}\right), 0.4\right]\right\}.$$
$$(H^{\eta}, R) = \left\{\left[(e_1, p, 1), \left(\frac{v_1}{(0.4, 0.3, \ldots, 0.2), (0.5, 0.7, \ldots, 0.2), (0.6, 0.1, \ldots, 0.3)}\right), 0.4\right],\right.$$
$$\left.\left[(e_2, r, 1), \left(\frac{v_1}{(0.8, 0.3, \ldots, 0.4), (0.3, 0.1, \ldots, 0.5), (0.2, 0.3, \ldots, 0.4)}\right), 0.4\right]\right\}.$$

Therefore $(H^{\eta}, R) \subseteq (F^{\Omega}, G)$.

Definition 14. *Two NSEMs (F^{Ω}, G) and (G^{η}, B) over V are said to be equal if (F^{Ω}, G) is a NSEM subset of (H^{η}, R) and (H^{η}, R) is a NSEM subset of (F^{Ω}, G).*

Definition 15. *Agree-NSEMs $(F^{\Omega}, G)_1$ over V is a NSEM subset of (F^{Ω}, G) defined as follows.*

$$\left(F^{\Omega}, G\right)_1 = \{F_1(\Delta) : \Delta \in E \times X \times \{1\}\}. \tag{11}$$

Example 3. *Consider Example 1. The agree- neutrosophic soft expert multisets $(F^{\Omega}, Z)_1$ over V is*

$$\begin{aligned} c(F^{\Omega}, Z)_1 = \Bigg\{ &\left[(e_1, p, 1), \left(\frac{v_1}{(0.4, 0.3, \ldots, 0.2), (0.5, 0.7, \ldots, 0.2), (0.6, 0.1, \ldots, 0.3)}\right), 0.4\right], \\ &\left[(e_1, r, 1), \left(\frac{v_1}{(0.3, 0.2, \ldots, 0.5), (0.8, 0.1, \ldots, 0.4), (0.5, 0.6, \ldots, 0.2)}\right), 0.8\right], \\ &\left[(e_2, p, 1), \left(\frac{v_1}{(0.7, 0.3, \ldots, 0.6), (0.3, 0.2, \ldots, 0.6), (0.8, 0.2, \ldots, 0.1)}\right), 0.5\right], \\ &\left[(e_2, r, 1), \left(\frac{v_1}{(0.8, 0.3, \ldots, 0.4), (0.3, 0.1, \ldots, 0.5), (0.2, 0.3, \ldots, 0.4)}\right), 0.4\right] \Bigg\}. \end{aligned}$$

Definition 16. *A disagree-NSEMs $(F^{\Omega}, G)_0$ over V is a NSES subset of (F^{Ω}, G) is defined as follows:*

$$(F^{\Omega}, A)_0 = \{F_0(\Delta) : \Delta \in E \times X \times \{0\}\}. \tag{12}$$

Example 4. *Consider Example 1. The disagree- neutrosophic soft expert multisets $(F^{\Omega}, Z)_0$ over V are*

$$\begin{aligned} (F^{\Omega}, Z)_0 = \Bigg\{ &\left[(e_1, p, 0), \left(\frac{v_1}{(0.5, 0.1, \ldots, 0.2), (0.6, 0.3, \ldots, 0.4), (0.7, 0.2, \ldots, 0.6)}\right), 0.1\right], \\ &\left[(e_1, r, 0), \left(\frac{v_1}{(0.4, 0.2, \ldots, 0.1), (0.6, 0.1, \ldots, 0.3), (0.7, 0.2, \ldots, 0.4)}\right), 0.4\right], \\ &\left[(e_2, p, 0), \left(\frac{v_1}{(0.8, 0.1, \ldots, 0.5), (0.2, 0.1, \ldots, 0.4), (0.6, 0.3, \ldots, 0.1)}\right), 0.6\right], \\ &\left[(e_2, r, 0), \left(\frac{v_1}{(0.7, 0.2, \ldots, 0.3), (0.4, 0.1, \ldots, 0.6), (0.3, 0.2, \ldots, 0.1)}\right), 0.2\right] \Bigg\}. \end{aligned}$$

4. Basic Operations on NSEMs

Definition 17. *The complement of a neutrosophic soft expert multiset (F^{Ω}, G) is denoted by $(F^{\Omega}, G)^c$ and is defined by $(F^{\Omega}, G)^c = \left(F^{\Omega(c)}, \neg G\right)$ where $F^{u(c)} : \neg G \to \mathcal{N}(V)\times$ is mapping given by*

$$F^{\Omega(c)}(\Delta) = \left\{ {D^i}_{F(\Delta)^{(c)}} = {Y^i}_{F(\Delta)}, \ {I^i}_{F(\Delta)^{(c)}} = \overline{1} - {I^i}_{F(\Delta)}, \ {Y^i}_{F(\Delta)^{(c)}} = {D^i}_{F(\Delta)} \text{ and } \Omega^c(\Delta) = \overline{1} - \Omega(\Delta) \right\} \tag{13}$$

for each $\Delta \in E$.

Example 5. *Consider Example 1. The complement of the neutrosophic soft expert multiset* F^{Ω} *denoted by* $F^{\Omega(c)}$ *is given by as follows:*

$$c(F^{\Omega(c)}, Z) = \Big\{ \Big[(\neg e_1, p, 1), \Big(\frac{v_1}{(0.2, 0.7, \ldots, 0.4), (0.2, 0.3, \ldots, 0.5), (0.3, 0.9, \ldots, 0.6)} \Big), 0.6 \Big],$$
$$\Big[(\neg e_1, r, 1), \Big(\frac{v_1}{(0.5, 0.8, \ldots, 0.3), (0.4, 0.9, \ldots, 0.8), (0.2, 0.4, \ldots, 0.5)} \Big), 0.2 \Big],$$
$$\Big[(\neg e_2, p, 1), \Big(\frac{v_1}{(0.6, 0.7, \ldots, 0.7), (0.6, 0.8, \ldots, 0.3), (0.1, 0.8, \ldots, 0.8)} \Big), 0.5 \Big],$$
$$\Big[(\neg e_2, r, 1), \Big(\frac{v_1}{(0.4, 0.7, \ldots, 0.8), (0.5, 0.9, \ldots, 0.3), (0.4, 0.7, \ldots, 0.2)} \Big), 0.6 \Big],$$
$$\Big[(\neg e_1, p, 0), \Big(\frac{v_1}{(0.2, 0.9, \ldots, 0.5), (0.4, 0.7, \ldots, 0.6), (0.6, 0.8, \ldots, 0.7)} \Big), 0.9 \Big],$$
$$\Big[(\neg e_1, r, 0), \Big(\frac{v_1}{(0.1, 0.8, \ldots, 0.4), (0.3, 0.9, \ldots, 0.6), (0.4, 0.8, \ldots, 0.7)} \Big), 0.6 \Big],$$
$$\Big[(\neg e_2, p, 0), \Big(\frac{v_1}{(0.5, 0.9, \ldots, 0.8), (0.4, 0.9, \ldots, 0.2), (0.1, 0.7, \ldots, 0.6)} \Big), 0.4 \Big],$$
$$\Big[(\neg e_2, r, 0), \Big(\frac{v_1}{(0.3, 0.8, \ldots, 0.7), (0.6, 0.9, \ldots, 0.4), (0.1, 0.8, \ldots, 0.3)} \Big), 0.8 \Big] \Big\}.$$

Proposition 1. *If* (F^{Ω}, G) *is a neutrosophic soft expert multiset over* V*, then*

1. $((F^{\Omega}, G)^c)^c = (F^{\Omega}, G)$
2. $((F^{\Omega}, G)_1)^c = (F^{\Omega}, G)_0$
3. $((F^{\Omega}, G)_0)^c = (F^{\Omega}, G)_1$

Proof. (1) From Definition 17, we have $(F^{\Omega}, G)^c = \left(F^{\Omega(c)}, \neg G\right)$ where $F^{\Omega(c)}(\Delta) = D^i_{F(\Delta)^{(c)}} = Y^i_{F(\Delta)}$, $I^i_{F(\Delta)^{(c)}} = \bar{1} - I^i_{F(\Delta)}$, $Y^i_{F(\Delta)^{(c)}} = D^i_{F(\Delta)}$ and $\Omega^c(\Delta) = \bar{1} - \Omega(\Delta)$ for each $\Delta \in E$. Now $((F^{\Omega}, G)^c)^c = \left(\left(F^{\Omega(c)}\right)^c, G\right)$ where

$$\begin{aligned} \left(F^{\Omega(c)}\right)^c(\Delta) &= [D^i_{F(\Delta)^{(c)}} = Y^i_{F(\Delta)}, & I^i_{F(\Delta)^{(c)}} &= \bar{1} - I^i_{F(\Delta)}, & Y^i_{F(\Delta)^{(c)}} &= D^i_{F(\Delta)}, & (\Omega^i)^c(\Delta) &= \bar{1} - \Omega^i(\Delta)]^c \\ &= D^i_{F(\Delta)} = Y^i_{F(\Delta)^{(c)}}, & I^i_{F(\Delta)} &= \bar{1} - I^i_{F(\Delta)^{(c)}}, & Y^i_{F(\Delta)} &= D^i_{F(\Delta)^{(c)}}, & \Omega^i(\Delta) &= \bar{1} - (\Omega^i)(\Delta) \\ & & &= \bar{1} - (\bar{1} - I^i_{F(\Delta)}) & & & &= \bar{1} - (\bar{1} - \Omega^i(\Delta)) \\ & & &= I^i_{F(\Delta)} & & & &= \Omega^i(\Delta) \end{aligned}$$

Thus $\left((F^{\Omega}, G)^c\right)^c = \left(\left(F^{\Omega(c)}\right)^c, G\right) = (F^{\Omega}, G)$, for all $\Delta \in E$.
The Proofs (2) and (3) can proved similarly. □

Definition 18. *The union of two NSEMs* (F^{Ω}, G) *and* (K^{ρ}, L) *over* V*, denoted by* $(F^{\Omega}, G) \tilde{\cup} (K^{\rho}, L)$ *is a NSEMs* (H^{σ}, C) *where* $C = G \cup L$ *and* $\forall\, e \in C$*,*

$$(H^{\sigma}, C) = \begin{cases} max\left(D^i{}_{(F^{\Omega}(e)}(m), D^i{}_{(K^{\rho}(e)}(m)\right) & if\ \Delta \in G \cap L \\ min\left(I^i{}_{(F^{\Omega}(e)}(m), I^i{}_{(K^{\rho}(e)}(m)\right) & if\ \Delta \in G \cap L \\ min\left(Y^i{}_{(F^{\Omega}(e)}(m), Y^i{}_{(K^{\rho}(e)}(m)\right) & if\ \Delta \in G \cap L \end{cases} \tag{14}$$

where $\sigma(m) = max\left(\Omega_{(e)}(m), \rho_{(e)}(m)\right)$.

Example 6. *Suppose that* (F^{Ω}, G) *and* (K^{ρ}, L) *are two NSEMs over* V*, such that*

$$c(F^{\Omega}, G) = \left\{ \left[(e_1, p, 1), \left(\frac{v_1}{(0.7, 0.3, \ldots, 0.6), (0.5, 0.2, \ldots, 0.4), (0.7, 0.6, \ldots, 0.3)} \right), 0.3 \right], \right.$$
$$\left[(e_2, q, 1), \left(\frac{v_1}{(0.4, 0.3, \ldots, 0.6), (0.8, 0.2, \ldots, 0.4), (0.5, 0.1, \ldots, 0.7)} \right), 0.6 \right],$$
$$\left. \left[(e_3, r, 1), \left(\frac{v_1}{(0.8, 0.2, \ldots, 0.3), (0.6, 0.3, \ldots, 0.7), (0.4, 0.2, \ldots, 0.8)} \right), 0.5 \right] \right\}.$$
$$(K^{\rho}, L) = \left\{ \left[(e_1, p, 1), \left(\frac{v_1}{(0.4, 0.3, \ldots, 0.1), (0.7, 0.2, \ldots, 0.3), (0.5, 0.4, \ldots, 0.7)} \right), 0.6 \right], \right.$$
$$\left. \left[(e_3, r, 1), \left(\frac{v_1}{(0.8, 0.3, \ldots, 0.2), (0.6, 0.1, \ldots, 0.2), (0.3, 0.5, \ldots, 0.3)} \right), 0.7 \right] \right\}$$

Then $(F^{\Omega}, G) \,\widetilde{\cup}\, (K^{\rho}, L) = (H^{\sigma}, C)$ *where*

$$c(H^{\sigma}, C) = \left\{ \left[(e_1, p, 1), \left(\frac{v_1}{(0.7, 0.3, \ldots, 0.1), (0.7, 0.2, \ldots, 0.3), (0.7, 0.4, \ldots, 0.3)} \right), 0.6 \right], \right.$$
$$\left[(e_2, q, 1), \left(\frac{v_1}{(0.4, 0.3, \ldots, 0.6), (0.8, 0.2, \ldots, 0.4), (0.5, 0.1, \ldots, 0.7)} \right), 0.6 \right],$$
$$\left. \left[(e_3, r, 1), \left(\frac{v_1}{(0.8, 0.2, \ldots, 0.2), (0.6, 0.1, \ldots, 0.2), (0.4, 0.2, \ldots, 0.3)} \right), 0.7 \right] \right\}.$$

Proposition 2. *If* (F^{Ω}, G)*,* (K^{ρ}, L) *and* (H^{Ω}, C) *are three NSEMs over* V*, then*

1. $\left((F^{\Omega}, G) \,\widetilde{\cup}\, (K^{\rho}, L) \right) \widetilde{\cup}\, (H^{\sigma}, C) = (F^{\Omega}, G) \,\widetilde{\cup}\, \left((K^{\rho}, L) \,\widetilde{\cup}\, (H^{\sigma}, C) \right)$
2. $(F^{\Omega}, G)(F^{\Omega}, G) \subseteq (F^{\Omega}, G)$.

Proof. (1) We want to prove that

$$\left((F^{\Omega}, G) \,\widetilde{\cup}\, (K^{\rho}, L) \right) \widetilde{\cup}\, (H^{\sigma}, C) = (F^{\Omega}, G) \,\widetilde{\cup}\, \left((K^{\rho}, L) \,\widetilde{\cup}\, (H^{\sigma}, C) \right)$$

by using Definition 18, we consider the case when if $e \in G \cap L$ as other cases are trivial. We will have

$$\begin{aligned} &(F^{\Omega}, G) \,\widetilde{\cup}\, (K^{\rho}, L) \\ &= \Big\{ \Big(v / max\big(D^i{}_{F^{\Omega}(e)}(m), D^i{}_{G^{\rho}(e)}(m) \big), min\big(I^i{}_{F^{\Omega}(e)}(m), I^i{}_{G^{\rho}(e)}(m) \big), min\big(Y^i{}_{F^{\Omega}(e)}(m), Y^i{}_{G^{\rho}(e)}(m) \big) \Big), \\ &\qquad max\big(\Omega_{(e)}(m), \rho_{(e)}(m) \big), v \in V \Big\} \end{aligned}$$

Also consider the case when $e \in H$ as the other cases are trivial. We will have

$$\begin{aligned} &\left((F^{u}, A) \,\widetilde{\cup}\, (G^{\eta}, B) \right) \widetilde{\cup}\, (H^{\Omega}, C) \\ &= \Big\{ \Big(v / max\big(D^i{}_{F^{\Omega}(e)}(m), D^i{}_{G^{\rho}(e)}(m) \big), min\big(I^i{}_{F^{\Omega}(e)}(m), I^i{}_{G^{\rho}(e)}(m) \big), min\big(Y^i{}_{F^{\Omega}(e)}(m), Y^i{}_{G^{\rho}(e)}(m) \big) \Big), \\ &\Big(v / D^i{}_{H^{\Omega}(e)}(m), I^i{}_{H^{\Omega}(e)}(m), Y^i{}_{H^{\Omega}(e)}(m) \Big), max\big(u_{(e)}(m), \eta_{(e)}(m), \Omega(m) \big), v \in V \Big\} \\ &= \left\{ \begin{array}{c} \Big(v / D^i{}_{F^{\Omega}(e)}(m), I^i{}_{F^{\Omega}(e)}(m), Y^i{}_{F^{\Omega}(e)}(m) \Big), \\ \Big(v / max\big(D^i{}_{G^{u}(e)}(m), D^i{}_{H^{\eta}(e)}(m) \big), min\big(I^i{}_{G^{u}(e)}(m), I^i(m) \big), \; min\big(Y^i{}_{G^{u}(e)}(m), Y^i(m) \big) \Big) \end{array} \right. \\ &max\big(\Omega_{(e)}(m), \rho_{(e)}(m), \sigma(m) \big), v \in V \Big\} \\ &= (F^{\Omega}, G) \,\widetilde{\cup}\, \left((K^{\rho}, L) \,\widetilde{\cup}\, (H^{\sigma}, C) \right). \end{aligned}$$

(2) The proof is straightforward. □

Definition 19. *The intersection of two NSEMs* (F^{Ω}, G) *and* (K^{ρ}, L) *over* V*, denoted by* $(F^{\Omega}, G) \tilde{\cap} (K^{\rho}, L) = (P^{\delta}, C)$ *where* $C = G \cap L$ *and* $\forall\, e \in C$,

$$\left(P^{\delta}, C\right) = \begin{cases} min\left(D^{i}{}_{(F^{\Omega}(e)}(m), D^{i}{}_{(K^{\rho}(e)}(m)\right) & if\ e \in G \cap L \\ max\left(I^{i}{}_{(F^{\Omega}(e)}(m), I^{i}{}_{(K^{\rho}(e)}(m)\right) & if\ e \in G \cap L \\ max\left(Y^{i}{}_{(F^{\Omega}(e)}(m), Y^{i}{}_{(K^{\rho}(e)}(m)\right) & if\ e \in G \cap L \end{cases} \tag{15}$$

where $\delta(m) = min\left(\Omega_{(e)}(m), \rho_{(e)}(m)\right)$.

Example 7. *Suppose that* (F^{Ω}, G) *and* (K^{ρ}, L) *are two NSEMs over* V*, such that*

$$c(F^{\Omega}, G) = \left\{ \left[(e_3, r, 1), \left(\frac{v_1}{(0.8, 0.3, \ldots, 0.2), (0.6, 0.1, \ldots, 0.2), (0.3, 0.5, \ldots, 0.3)}\right), 0.4\right], \right.$$
$$\left[(e_1, q, 1), \left(\frac{v_1}{(0.8, 0.2, \ldots, 0.2), (0.7, 0.3, \ldots, 0.2), (0.4, 0.2, \ldots, 0.3)}\right), 0.7\right],$$
$$\left. \left[(e_3, q, 0), \left(\frac{v_1}{(0.4, 0.3, \ldots, 0.6), (0.8, 0.2, \ldots, 0.4), (0.5, 0.1, \ldots, 0.7)}\right), 0.6\right] \right\}.$$
$$(K^{\rho}, L) = \left\{ \left[(e_1, p, 1), \left(\frac{v_1}{(0.7, 0.3, \ldots, 0.1), (0.7, 0.2, \ldots, 0.3), (0.7, 0.4, \ldots, 0.3)}\right), 0.3\right], \right.$$
$$\left. \left[(e_3, r, 1), \left(\frac{v_1}{(0.4, 0.7, \ldots, 0.8), (0.5, 0.9, \ldots, 0.3), (0.4, 0.7, \ldots, 0.2)}\right), 0.8\right] \right\}$$

Then $(F^{\Omega}, G) \tilde{\cap} (K^{\rho}, L) = (P^{\delta}, C)$ *where*

$$\left(P^{\delta}, C\right) = \left\{ \left[(e_3, r, 1), \left(\frac{v_1}{(0.4, 0.3, \ldots, 0.2), (0.6, 0.9, \ldots, 0.3), (0.4, 0.7, \ldots, 0.3)}\right), 0.4\right] \right\}.$$

Proposition 3. *If* (F^{Ω}, G)*,* (K^{ρ}, L) *and* (H^{Ω}, C) *are three NSEMs over* V*, then*

1. $\left((F^{\Omega}, G) \tilde{\cap} (K^{\rho}, L)\right) \tilde{\cap} (H^{\sigma}, C) = (F^{\Omega}, G) \tilde{\cap} \left((K^{\rho}, L) \tilde{\cap} (H^{\sigma}, C)\right)$
2. $(F^{\Omega}, G) \tilde{\cap} (F^{\Omega}, G) \subseteq (F^{\Omega}, G)$.

Proof. (1) We want to prove that

$$\left((F^{\Omega}, G) \tilde{\cap} (K^{\rho}, L)\right) \tilde{\cap} (H^{\sigma}, C) = (F^{\Omega}, G) \tilde{\cap} \left((K^{\rho}, L) \tilde{\cap} (H^{\sigma}, C)\right)$$

by using Definition 19, we consider the case when if $e \in G \cap L$ as other cases are trivial. We will have

$$(F^{\Omega}, G) \tilde{\cap} (K^{\rho}, L)$$
$$= \left\{ \left(v/min\left(D^{i}{}_{F^{\Omega}(e)}(m), D^{i}{}_{G^{\rho}(e)}(m)\right), max\left(I^{i}{}_{F^{\Omega}(e)}(m), I^{i}{}_{G^{\rho}(e)}(m)\right), max\left(Y^{i}{}_{F^{\Omega}(e)}(m), Y^{i}{}_{G^{\rho}(e)}(m)\right)\right), \right.$$
$$\left. min\left(\Omega_{(e)}(m), \rho_{(e)}(m)\right), v \in V \right\}$$

Also consider the case when $\Delta \in H$ as the other cases are trivial. We will have

$$\left((F^u, A) \tilde{\cap} (G^\eta, B)\right) \tilde{\cap} (H^\Omega, C)$$
$$= \Big\{ \Big(v/max\left(D^i{}_{F^{\Omega}(e)}(m), D^i{}_{G^{\rho}(e)}(m)\right), min\left(I^i{}_{F^{\Omega}(e)}(m), I^i{}_{G^{\rho}(e)}(m)\right), min\left(Y^i{}_{F^{\Omega}(e)}(m), Y^i{}_{G^{\rho}(e)}(m)\right)\Big),$$
$$\left(v/D^i{}_{H^{\Omega}(e)}(m), I^i{}_{H^{\Omega}(e)}(m), Y^i{}_{H^{\Omega}(e)}(m)\right), min\left(u_{(e)}(m), \eta_{(e)}(m), \Omega(m)\right), v \in V \Big\}$$
$$= \left\{ \begin{array}{c} \left(v/D^i{}_{F^{\Omega}(e)}(m), I^i{}_{F^{\Omega}(e)}(m), Y^i{}_{F^{\Omega}(e)}(m)\right), \\ \left(v/min\left(D^i{}_{G^u(e)}(m), D^i{}_{H^\eta(e)}(m)\right), max\left(I^i{}_{G^u(e)}(m), I^i(m)\right), max\left(Y^i{}_{G^u(e)}(m), Y^i(m)\right)\right) \end{array} \right.$$
$$min\left(\Omega_{(e)}(m), \rho_{(e)}(m), \sigma(m)\right), v \in V \Big\}$$
$$= (F^\Omega, G) \tilde{\cap} \left((K^\rho, L) \tilde{\cap} (H^\sigma, C)\right).$$

(2) The proof is straightforward. □

Proposition 4. *If (F^Ω, G), (K^ρ, L) and (H^Ω, C) are three NSEMs over V. Then*

1. $\left((F^\Omega, G) \tilde{\cup} (K^\rho, L)\right) \tilde{\cap} (H^\sigma, C) = \left((F^\Omega, G) \tilde{\cap} (H^\sigma, C)\right) \tilde{\cup} \left((K^\rho, L) \tilde{\cap} (H^\sigma, C)\right)$.
2. $\left((F^\Omega, G) \tilde{\cap} (K^\rho, L)\right) \tilde{\cup} (H^\sigma, C) = \left((F^\Omega, G) \tilde{\cup} (H^\sigma, C)\right) \tilde{\cap} \left((K^\rho, L) \tilde{\cup} (H^\sigma, C)\right)$.

Proof. The proofs can be easily obtained from Definitions 18 and 19. □

5. AND and OR Operations

Definition 20. *Let (F^Ω, G) and (K^ρ, L) be any two NSEMs over V, then $(F^\Omega, G) AND (K^\rho, L)$" denoted $(F^\Omega, G) \wedge (K^\rho, L)$ is defined by*

$$(F^\Omega, G) \wedge (K^\rho, L) = (H^\sigma, G \times L) \tag{16}$$

where $(H^\sigma, G \times L) = H^\sigma(\alpha, \beta)$ such that $H^\sigma(\alpha, \beta) = F^\Omega(\alpha) \cap K^\rho(\beta)$ for all $(\alpha, \beta) \in G \times L$ where $\cap$ represent the basic intersection.

Example 8. *Suppose that (F^Ω, G) and (K^ρ, L) are two NSEMs over V, such that*

$$c(F^\Omega, G) = \left\{ \left[(e_1, p, 1), \left(\frac{v_1}{(0.2, 0.3, \ldots, 0.6), (0.2, 0.1, \ldots, 0.8), (0.3, 0.2, \ldots, 0.6)}\right), 0.1\right], \right.$$
$$\left. \left[(e_2, r, 0), \left(\frac{v_1}{(0.5, 0.3, \ldots, 0.4), (0.6, 0.5, \ldots, 0.4), (0.2, 0.4, \ldots, 0.3)}\right), 0.5\right] \right\}.$$
$$(K^\rho, L) = \left\{ \left[(e_1, p, 1), \left(\frac{v_1}{(0.3, 0.2, \ldots, 0.1), (0.5, 0.2, \ldots, 0.3), (0.8, 0.3, \ldots, 0.4)}\right), 0.2\right], \right.$$
$$\left. \left[(e_2, q, 0), \left(\frac{v_1}{(0.6, 0.4, \ldots, 0.7), (0.3, 0.4, \ldots, 0.2), (0.6, 0.1, \ldots, 0.5)}\right), 0.6\right] \right\}.$$

Then $(F^\Omega, G) \wedge (K^\rho, L) = (H^\sigma, G \times L)$ where

$$c(H^\sigma, G \times L) = \left\{ \left[(e_1, p, 1), (e_1, p, 1)\left(\frac{v_1}{(0.2, 0.2, \ldots, 0.1), (0.5, 0.2, \ldots, 0.8), (0.8, 0.3, \ldots, 0.6)}\right), 0.1\right], \right.$$
$$\left[(e_1, p, 1), (e_2, q, 0), \left(\frac{v_1}{(0.2, 0.3, \ldots, 0.6), (0.3, 0.4, \ldots, 0.8), (0.6, 0.2, \ldots, 0.6)}\right), 0.1\right],$$
$$\left[(e_2, r, 0), (e_1, p, 1), \left(\frac{v_1}{(0.3, 0.2, \ldots, 0.1), (0.6, 0.5, \ldots, 0.4), (0.8, 0.4, \ldots, 0.4)}\right), 0.2\right],$$
$$\left. \left[(e_2, r, 0), (e_2, q, 0), \left(\frac{v_1}{(0.5, 0.3, \ldots, 0.4), (0.6, 0.5, \ldots, 0.4), (0.6, 0.4, \ldots, 0.5)}\right), 0.5\right] \right\}.$$

Definition 21. *Let (F^{Ω}, G) and (K^{ρ}, L) be any two NSEMs over V, then $(F^{\Omega}, G) OR (K^{\rho}, L)$" denoted $(F^{\Omega}, G) \vee (K^{\rho}, L)$ is defined by*

$$(F^{\Omega}, G) \vee (K^{\rho}, L) = (H^{\sigma}, G \times L) \tag{17}$$

where $(H^{\sigma}, G \times L) = H^{\sigma}(\alpha, \beta)$ such that $H^{\sigma}(\alpha, \beta) = F^{\Omega}(\alpha) \cup K^{\rho}(\beta)$ for all $(\alpha, \beta) \in G \times L$ where $\cup$ represent the basic union.

Example 9. *Suppose that (F^{Ω}, G) and (K^{ρ}, L) are two NSEMs over V, such that*

$$c(F^{\Omega}, G) = \left\{ \left[(e_1, p, 1), \left(\frac{v_1}{(0.2, 0.3, \ldots, 0.6), (0.2, 0.1, \ldots, 0.8), (0.3, 0.2, \ldots, 0.6)} \right), 0.1 \right], \right.$$
$$\left. \left[(e_2, r, 0), \left(\frac{v_1}{(0.5, 0.3, \ldots, 0.4), (0.6, 0.5, \ldots, 0.4), (0.2, 0.4, \ldots, 0.3)} \right), 0.5 \right] \right\}.$$
$$(K^{\rho}, L) = \left\{ \left[(e_1, p, 1), \left(\frac{v_1}{(0.3, 0.2, \ldots, 0.1), (0.5, 0.2, \ldots, 0.3), (0.8, 0.3, \ldots, 0.4)} \right), 0.2 \right], \right.$$
$$\left. \left[(e_2, q, 0), \left(\frac{v_1}{(0.6, 0.4, \ldots, 0.7), (0.3, 0.4, \ldots, 0.2), (0.6, 0.1, \ldots, 0.5)} \right), 0.6 \right] \right\}.$$

Then $(F^{\Omega}, G) \vee (K^{\rho}, L) = (H^{\sigma}, G \times L)$ where

$$c(H^{\sigma}, G \times L) = \left\{ \left[(e_1, p, 1), (e_1, p, 1) \left(\frac{v_1}{(0.3, 0.3, \ldots, 0.6), (0.2, 0.1, \ldots, 0.3), (0.3, 0.2, \ldots, 0.4)} \right), 0.2 \right], \right.$$
$$\left[(e_1, p, 1), (e_2, q, 0), \left(\frac{v_1}{(0.6, 0.4, \ldots, 0.7), (0.2, 0.1, \ldots, 0.2), (0.3, 0.1, \ldots, 0.5)} \right), 0.6 \right],$$
$$\left[(e_2, r, 0), (e_1, p, 1), \left(\frac{v_1}{(0.5, 0.3, \ldots, 0.4), (0.5, 0.2, \ldots, 0.3), (0.2, 0.3, \ldots, 0.3)} \right), 0.2 \right],$$
$$\left. \left[(e_2, r, 0), (e_2, q, 0), \left(\frac{v_1}{(0.6, 0.4, \ldots, 0.7), (0.3, 0.4, \ldots, 0.2), (0.2, 0.1, \ldots, 0.3)} \right), 0.6 \right] \right\}.$$

Proposition 5. *Let (F^{Ω}, G) and (K^{ρ}, L) be NSEMs over V. Then*

1. $((F^{\Omega}, G) \wedge (K^{\rho}, L))^c = (F^u, A)^c \vee (G^{\eta}, B)^c$
2. $((F^{\Omega}, G) \vee (K^{\rho}, L))^c = (F^u, A)^c \wedge (G^{\eta}, B)^c$

Proof. (1) Suppose that (F^{Ω}, G) and (K^{ρ}, L) be NSEMs over V defined as:

$$\begin{aligned}
(F^{\Omega}, G) \wedge (K^{\rho}, L) &= (F^{\Omega}(\alpha) \wedge K^{\rho}(\beta))^c \\
&= (F^{\Omega}(\alpha) \cap K^{\rho}(\beta))^c \\
&= (F^{\Omega}(\alpha) \cap K^{\rho}(\beta))^c \\
&= \left(F^{\Omega(c)}(\alpha) \cup K^{\rho(c)}(\beta) \right) \\
&= \left(F^{\Omega(c)}(\alpha) \vee K^{\rho(c)}(\beta) \right) \\
&= (F^u, A)^c \vee (G^{\eta}, B)^c
\end{aligned}$$

(2) The proofs can be easily obtained from Definitions 20 and 21. □

6. NSEMs-Aggregation Operator

In this section, we define a NSEMs-aggregation operator of NSEMs to construct a decision method by which approximate functions of a soft expert set are combined to produce a neutrosophic set that can be used to evaluate each alternative.

Definition 22. *Let $\Gamma_G \in NSEMs$. Then NSEMs-aggregation operator of Γ_G, denoted by Γ_G^{agg}, is defined by*

$$\Gamma_G^{agg} = \left\{ \left(\langle v, \left(D^i\right)_G^{agg}(v), \left(I^i\right)_G^{agg}(v), \left(Y^i\right)_G^{agg}(v) \rangle \right) : v \in V \right\},$$

which are NSEMs over V,

$$\begin{aligned}
\left(D^i\right)_G^{agg} : V \to [0,1] \quad & \left(D^i\right)_G^{agg}(v) = \left(\frac{1}{|V|} \sum_{\substack{e \in E \\ v \in V}} D^i{}_G(v) \right).\Omega, \\
\left(Y^i\right)_G^{agg} : V \to [0,1] \quad & \left(Y^i\right)_G^{agg}(v) = \left(\frac{1}{|V|} \sum_{\substack{e \in E \\ v \in V}} Y^i{}_G(v) \right).\Omega, \\
\left(I^i\right)_G^{agg} : V \to [0,1] \quad & \left(I^i\right)_G^{agg}(v) = \left(\frac{1}{|V|} \sum_{\substack{e \in E \\ v \in V}} I^i{}_G(v) \right).\Omega
\end{aligned} \tag{18}$$

where $|V|$ is the cardinality of V and Ω^i is defined below

$$\Omega = \frac{1}{n}.\sum_{i=1}^{n} \Omega(e_i). \quad (e_i,\ i = 1, 2, 3, \ldots, n) \tag{19}$$

Definition 23. *Let $\Gamma_G \in NSEMs$, Γ_G^{agg} be NSEMs. Then a reduced fuzzy set of Γ_G^{agg} is a fuzzy set over is denoted by*

$$\Gamma_G^{agg} = \left\{ \frac{\lambda\Gamma_G^{agg}(v)}{v} : v \in V \right\}, \tag{20}$$

where $\lambda\Gamma_G^{agg}(v) : V \to [0,1]$ and $v_i = \left| \left(D^i\right)_{G_i}^{agg} - \left(Y^i\right)_{G_i}^{agg} - \left(I^i\right)_{G_i}^{agg} \right|$.

7. An Application of NSEMs

In this section, we present an application of NSEMs theory in a decision-making problem. Based on Definitions 22 and 23, we construct an algorithm for the NSEMs decision-making method as follows:

Step 1-Choose a feasible subset of the set of parameters.
Step 2-Construct the NSEMs for each opinion (agree, disagree) of expert.
Step 3-Compute the aggregation NSEMS Γ_G^{agg} of Γ_G and the reduced fuzzy set $\left(D^i\right)_{G_i}^{agg}, \left(Y^i\right)_{G_i}^{agg}, \left(I^i\right)_{G_i}^{agg}$ of Γ_G^{agg}
Step 4-Score $(v_1) = (\max - agree(v_i)) - (\min - disagree(v_i))$
Step 5-Choose the element of v_i that has maximum membership. This will be the optimal solution.

Example 10. *In the architectural design process, let us assume that the design outputs used in the design of moving structures are taken by a few experts at certain time intervals. So, let us take the samples at three different timings in a day (in 08:30, 14:30 and 20:30) The design of moving structures consists of the architectural design, the design of the mechanism and the design of the surface covering membrane. Architectural design will be evaluated from these designs.,* $V = \{v_1, v_2, v_3\}$. *Suppose there are three parameters* $E = \{e_1, e_2, e_3\}$ *where the parameters* e_i $(i = 1, 2, 3)$ *stand for "time", "temperature" and "spatial needs" respectively. Let* $X = \{p, q\}$ *be a set of experts. After a serious discussion, the experts construct the following NSEMs.*

Step 1-Choose a feasible subset of the set of parameters:

$$
\begin{aligned}
&(F^{\Omega}, Z) = \\
&\left\{\left[(e_1,p,1), \left(\frac{v_1}{(0.3,0.1,0.4),(0.2,0.1,0.5),(0.5,0.2,0.6)}\right), \left(\frac{v_2}{(0.4,0.2,0.3),(0.7,0.1,0.6),(0.3,0.2,0.6)}\right), \left(\frac{v_3}{(0.5,0.3,0.4),(0.2,0.1,0.8),(0.4,0.2,0.3)}\right), 0.7\right],\right. \\
&\left[(e_1,q,1), \left(\frac{v_1}{(0.4,0.2,0.5),(0.3,0.1,0.2),(0.6,0.3,0.4)}\right), \left(\frac{v_2}{(0.5,0.3,0.2),(0.8,0.2,0.4),(0.5,0.3,0.2)}\right), \left(\frac{v_3}{(0.6,0.3,0.8),(0.3,0.2,0.1),(0.5,0.4,0.3)}\right), 0.6\right], \\
&\left[(e_2,p,1), \left(\frac{v_1}{(0.6,0.4,0.2),(0.3,0.1,0.4),(0.8,0.2,0.5)}\right), \left(\frac{v_2}{(0.8,0.3,0.4),(0.2,0.1,0.5),(0.4,0.3,0.5)}\right), \left(\frac{v_3}{(0.8,0.3,0.2),(0.3,0.1,0.4),(0.2,0.1,0.4)}\right), 0.8\right], \\
&\left[(e_2,q,1), \left(\frac{v_1}{(0.5,0.2,0.4),(0.3,0.2,0.5),(0.6,0.1,0.3)}\right), \left(\frac{v_2}{(0.6,0.4,0.7),(0.5,0.3,0.2),(0.6,0.2,0.4)}\right), \left(\frac{v_3}{(0.6,0.5,0.4),(0.1,0.3,0.2),(0.6,0.2,0.3)}\right), 0.4\right], \\
&\left[(e_3,p,1), \left(\frac{v_1}{(0.8,0.1,0.5),(0.2,0.3,0.4),(0.5,0.2,0.3)}\right), \left(\frac{v_2}{(0.7,0.2,0.5),(0.1,0.2,0.3),(0.3,0.2,0.1)}\right), \left(\frac{v_3}{(0.4,0.3,0.7),(0.3,0.1,0.4),(0.5,0.3,0.2)}\right), 0.3\right], \\
&\left[(e_3,q,1), \left(\frac{v_1}{(0.7,0.2,0.4),(0.3,0.1,0.5),(0.6,0.2,0.1)}\right), \left(\frac{v_2}{(0.9,0.4,0.5),(0.2,0.4,0.5),(0.1,0.2,0.3)}\right), \left(\frac{v_3}{(0.6,0.8,0.9),(0.2,0.1,0.6),(0.3,0.1,0.4)}\right), 0.4\right], \\
&\left[(e_1,q,0), \left(\frac{v_1}{(0.7,0.1,0.4),(0.3,0.2,0.1),(0.4,0.2,0.5)}\right), \left(\frac{v_2}{(0.6,0.5,0.4),(0.4,0.2,0.1),(0.8,0.2,0.6)}\right), \left(\frac{v_3}{(0.9,0.4,0.5),(0.2,0.1,0.3),(0.6,0.2,0.3)}\right), 0.7\right], \\
&\left[(e_2,p,0), \left(\frac{v_1}{(0.6,0.5,0.7),(0.3,0.5,0.4),(0.6,0.3,0.4)}\right), \left(\frac{v_2}{(0.5,0.2,0.3),(0.2,0.1,0.3),(0.4,0.3,0.5)}\right), \left(\frac{v_3}{(0.6,0.3,0.4),(0.1,0.2,0.4),(0.5,0.3,0.2)}\right), 0.8\right], \\
&\left[(e_2,q,0), \left(\frac{v_1}{(0.3,0.1,0.2),(0.4,0.1,0.3),(0.5,0.2,0.6)}\right), \left(\frac{v_2}{(0.7,0.2,0.4),(0.4,0.3,0.6),(0.5,0.1,0.6)}\right), \left(\frac{v_3}{(0.7,0.3,0.5),(0.2,0.4,0.3),(0.5,0.2,0.3)}\right), 0.4\right], \\
&\left[(e_3,p,0), \left(\frac{v_1}{(0.8,0.5,0.4),(0.2,0.4,0.3),(0.6,0.3,0.4)}\right), \left(\frac{v_2}{(0.5,0.2,0.3),(0.4,0.1,0.2),(0.2,0.1,0.4)}\right), \left(\frac{v_3}{(0.4,0.3,0.2),(0.2,0.1,0.6),(0.7,0.3,0.2)}\right), 0.5\right], \\
&\left.\left[(e_3,q,0), \left(\frac{v_1}{(0.6,0.1,0.4),(0.2,0.1,0.5),(0.4,0.2,0.3)}\right), \left(\frac{v_2}{(0.7,0.2,0.5),(0.4,0.3,0.2),(0.1,0.2,0.3)}\right), \left(\frac{v_3}{(0.5,0.3,0.4),(0.3,0.2,0.4),(0.4,0.2,0.3)}\right), 0.1\right]\right\}.
\end{aligned}
$$

Step 2-Construct the neutrosophic soft expert tables for each opinion (agree, disagree) of expert.

Step 3-Now calculate the score of agree (v_i) by using the data in Table 1 to obtain values in Table 2.

$$
\begin{aligned}
\left(\mathrm{D}^1\right)^{agg}_{G_1} &= \left(\frac{D^1_{G_1}+D^1_{G_2}+D^1_{G_3}}{3}\right).\left(\frac{\Omega_1+\Omega_2+\Omega_3}{3}\right)\\
&= \left(\frac{0.3+0.6+0.8}{3}\right).\left(\frac{0.7+0.8+0.3}{3}\right)\\
&= 0.34\\
\left(\mathrm{D}^2\right)^{agg}_{G_1} &= \left(\frac{D^2_{G_1}+D^2_{G_2}+D^2_{G_3}}{3}\right).\left(\frac{\Omega_1+\Omega_2+\Omega_3}{3}\right)\\
&= \left(\frac{0.1+0.4+0.1}{3}\right).\left(\frac{0.7+0.8+0.3}{3}\right)\\
&= 0.12\\
\left(\mathrm{D}^3\right)^{agg}_{G_1} &= \left(\frac{D^3_{G_1}+D^3_{G_2}+D^3_{G_3}}{3}\right).\left(\frac{\Omega_1+\Omega_2+\Omega_3}{3}\right)\\
&= \left(\frac{0.4+0.2+0.5}{3}\right).\left(\frac{0.7+0.8+0.3}{3}\right)\\
&= 0.22\\
(\mathrm{D})^{agg}_{G_1}(p,v_1) &= \frac{0.34+0.12+0.22}{3} = 0.2267\\
\left(\mathrm{I}^1\right)^{agg}_{G_1} &= \left(\frac{I^1_{G_1}+I^1_{G_2}+I^1_{G_3}}{3}\right).\left(\frac{\Omega_1+\Omega_2+\Omega_3}{3}\right)\\
&= \left(\frac{0.2+0.3+0.2}{3}\right).\left(\frac{0.7+0.8+0.3}{3}\right)\\
&= 0.1404\\
\left(\mathrm{I}^2\right)^{agg}_{G_1} &= \left(\frac{I^2_{G_1}+I^2_{G_2}+I^2_{G_3}}{3}\right).\left(\frac{\Omega_1+\Omega_2+\Omega_3}{3}\right)\\
&= \left(\frac{0.1+0.1+0.3}{3}\right).\left(\frac{0.7+0.8+0.3}{3}\right)\\
&= 0.1002\\
\left(\mathrm{I}^3\right)^{agg}_{G_1} &= \left(\frac{I^3_{G_1}+I^3_{G_2}+I^3_{G_3}}{3}\right).\left(\frac{\Omega_1+\Omega_2+\Omega_3}{3}\right)\\
&= \left(\frac{0.5+0.4+0.4}{3}\right).\left(\frac{0.7+0.8+0.3}{3}\right)\\
&= 0.2604\\
(\mathrm{I})^{agg}_{G_1}(p,v_1) &= \frac{0.1404+0.1002+0.2604}{3} = 0.167\\
\left(\mathrm{Y}^1\right)^{agg}_{G_1} &= \left(\frac{Y^1_{G_1}+Y^1_{G_2}+Y^1_{G_3}}{3}\right).\left(\frac{\Omega_1+\Omega_2+\Omega_3}{3}\right)\\
&= \left(\frac{0.5+0.8+0.5}{3}\right).\left(\frac{0.7+0.8+0.3}{3}\right)\\
&= 0.36\\
\left(\mathrm{Y}^2\right)^{agg}_{G_1} &= \left(\frac{Y^2_{G_1}+Y^2_{G_2}+Y^2_{G_3}}{3}\right).\left(\frac{\Omega_1+\Omega_2+\Omega_3}{3}\right)\\
&= \left(\frac{0.2+0.2+0.2}{3}\right).\left(\frac{0.7+0.8+0.3}{3}\right)\\
&= 0.12\\
\left(\mathrm{Y}^3\right)^{agg}_{G_1} &= \left(\frac{Y^3_{G_1}+Y^3_{G_2}+Y^3_{G_3}}{3}\right).\left(\frac{\Omega_1+\Omega_2+\Omega_3}{3}\right)\\
&= \left(\frac{0.6+0.5+0.3}{3}\right).\left(\frac{0.7+0.8+0.3}{3}\right)\\
&= 0.2802\\
(\mathrm{Y})^{agg}_{G_1}(p,v_1) &= \frac{0.36+0.12+0.2802}{3} = 0.2534\\
v_1 &= \left|(\mathrm{D})^{agg}_{G_1} - (\mathrm{I})^{agg}_{G_1} - (\mathrm{Y})^{agg}_{G_1}\right| = |0.2267 - 0.167 - 0.2534| = 0.1937
\end{aligned}
$$

Table 1. Agree-neutrosophic soft expert multiset.

	v_1	v_2	v_3	Ω
(e_1,p)	$\langle(0.3,0.1,0.4),(0.2,0.1,0.5),(0.5,0.2,0.6)\rangle$	$\langle(0.4,0.2,0.3),(0.7,0.1,0.6),(0.3,0.2,0.6)\rangle$	$\langle(0.5,0.3,0.4),(0.2,0.1,0.8),(0.4,0.2,0.3)\rangle$	0.7
(e_2,p)	$\langle(0.6,0.4,0.2),(0.3,0.1,0.4),(0.8,0.2,0.5)\rangle$	$\langle(0.8,0.3,0.4),(0.2,0.1,0.5),(0.4,0.3,0.5)\rangle$	$\langle(0.8,0.3,0.2),(0.3,0.1,0.4),(0.2,0.1,0.4)\rangle$	0.8
(e_3,p)	$\langle(0.8,0.1,0.5),(0.2,0.3,0.4),(0.5,0.2,0.3)\rangle$	$\langle(0.7,0.2,0.5),(0.1,0.2,0.3),(0.3,0.2,0.1)\rangle$	$\langle(0.4,0.3,0.7),(0.3,0.1,0.4),(0.5,0.3,0.2)\rangle$	0.3
(e_1,q)	$\langle(0.4,0.2,0.5),(0.3,0.1,0.2),(0.6,0.3,0.4)\rangle$	$\langle(0.5,0.3,0.2),(0.8,0.2,0.4),(0.5,0.3,0.2)\rangle$	$\langle(0.6,0.3,0.8),(0.3,0.2,0.1),(0.5,0.4,0.3)\rangle$	0.6
(e_2,q)	$\langle(0.5,0.2,0.4),(0.3,0.2,0.5),(0.6,0.1,0.3)\rangle$	$\langle(0.6,0.4,0.7),(0.5,0.3,0.2),(0.6,0.2,0.4)\rangle$	$\langle(0.6,0.5,0.4),(0.1,0.3,0.2),(0.6,0.2,0.3)\rangle$	0.4
(e_3,q)	$\langle(0.7,0.2,0.4),(0.3,0.1,0.5),(0.6,0.2,0.1)\rangle$	$\langle(0.9,0.4,0.5),(0.2,0.4,0.5),(0.1,0.2,0.3)\rangle$	$\langle(0.6,0.8,0.9),(0.2,0.1,0.6),(0.3,0.1,0.4)\rangle$	0.4

Table 2. Degree table of agree- neutrosophic soft expert multiset.

	v_1	v_2	v_3
p	0.1136	0.1267	0.093
q	0.1142	0.0933	0.015

Now calculate the score of disagree (v_i) by using the data in Table 3 to obtain values in Table 4.

Table 3. Disagree-neutrosophic soft expert multiset.

	v_1	v_2	v_3	Ω
(e_1, p)	$\langle(0.5,0.1,0.7),(0.4,0.2,0.3),(0.5,0.4,0.1)\rangle$	$\langle(0.8,0.2,0.3),(0.2,0.1,0.4),(0.3,0.4,0.5)\rangle$	$\langle(0.5,0.2,0.6),(0.3,0.4,0.1),(0.2,0.3,0.1)\rangle$	0.9
(e_2, p)	$\langle(0.6,0.5,0.7),(0.3,0.5,0.4),(0.6,0.3,0.4)\rangle$	$\langle(0.5,0.2,0.3),(0.2,0.1,0.3),(0.4,0.3,0.5)\rangle$	$\langle(0.6,0.3,0.4),(0.1,0.2,0.4),(0.5,0.3,0.2)\rangle$	0.8
(e_3, p)	$\langle(0.8,0.5,0.4),(0.2,0.4,0.3),(0.6,0.3,0.4)\rangle$	$\langle(0.5,0.2,0.3),(0.4,0.1,0.2),(0.2,0.1,0.4)\rangle$	$\langle(0.4,0.3,0.2),(0.2,0.1,0.6),(0.7,0.3,0.2)\rangle$	0.5
(e_1, q)	$\langle(0.7,0.1,0.4),(0.3,0.2,0.1),(0.4,0.2,0.5)\rangle$	$\langle(0.6,0.5,0.4),(0.4,0.2,0.1),(0.8,0.2,0.6)\rangle$	$\langle(0.9,0.4,0.5),(0.2,0.1,0.3),(0.6,0.2,0.3)\rangle$	0.7
(e_2, q)	$\langle(0.3,0.1,0.2),(0.4,0.1,0.3),(0.5,0.2,0.6)\rangle$	$\langle(0.7,0.2,0.4),(0.4,0.3,0.6),(0.5,0.1,0.6)\rangle$	$\langle(0.7,0.3,0.5),(0.2,0.4,0.3),(0.5,0.2,0.3)\rangle$	0.4
(e_3, q)	$\langle(0.6,0.1,0.4),(0.2,0.1,0.5),(0.4,0.2,0.3)\rangle$	$\langle(0.7,0.2,0.5),(0.4,0.3,0.2),(0.1,0.2,0.3)\rangle$	$\langle(0.5,0.3,0.4),(0.3,0.2,0.4),(0.4,0.2,0.3)\rangle$	0.1

Table 4. Degree table of disagree-neutrosophic soft expert multiset.

	v_1	v_2	v_3
p	0.1631	0.1468	0.1386
q	0.1155	0.0933	0.04

Step 4-The final score of v_i is computed as follows:

$$\text{Score}(v_1) = 0.1142 - 0.1155 = -0.0013,$$
$$\text{Score}(v_2) = 0.1267 - 0.0933 = 0.0334,$$
$$\text{Score}(v_3) = 0.093 - 0.04 = 0.053.$$

Step 5-Clearly, the maximum score is the score 0.053, shown in the above for the v_3. Hence the best decision for the experts is to select worker v_2 as the company's employee.

8. Comparison Analysis

The NSEMs model give more precision, flexibility and compatibility compared to the classical, fuzzy and/or neutrosophic models.

In order to verify the feasibility and effectiveness of the proposed decision-making approach, a comparison analysis using neutrosophic soft expert decision method, with those methods used by Alkhazaleh and Salleh [18], Maji [17], Sahin et al. [22], Hassan et al. [23] and Ulucay et al. [40] are given in Table 5, based on the same illustrative example as in An Application of NSEMs. Clearly, the ranking order results are consistent with those in [17,18,22,23,40].

Table 5. Comparison of fuzzy soft set and its extensive set theory.

	Fuzzy Soft Expert	**Neutrosophic Soft Set**	**Neutrosophic Soft Expert**	**Q-Neutrosophic Soft Expert**	**Generalized Neutrosophic Soft Expert**	**NSEMs**
Methods	Alkhazaleh and Salleh [22]	Maji [21]	Sahin et al. [26]	Hassan et al. [27]	Ulucay et al. [48]	Proposed Method in this paper
Domain	Universe of discourse	Universe of discourse	Universe of discourse	Universe of discourse	Universe of discourse	Universe of discourse
True	Yes	Yes	Yes	Yes	Yes	Yes
Falsity	No	Yes	Yes	Yes	No	No
Indeterminacy	No	Yes	Yes	Yes	No	No
Expert	Yes	No	Yes	Yes	Yes	No
Q	No	No	No	Yes	Yes	Yes
Ranking	$v_1 > v_3 > v_2$	$v_1 > v_3 > v_2$	$v_1 > v_2 > v_3$	$v_1 > v_3 > v_2$	$v_1 > v_3 > v_2$	$v_3 > v_2 > v_1$
Membershipvalued	Membership-valued	Single-valued	single-valued	Single-valued	Single-valued	Multi-valued

9. Conclusions

In this paper, we reviewed the basic concepts of neutrosophic set, neutrosophic soft set, soft expert sets, neutrosophic soft expert sets and NP-aggregation operator before establishing the concept of neutrosophic soft expert multiset (NSEM). The basic operations of NSEMs, namely complement, union, intersection AND and OR were defined. Subsequently a definition of NSEM-aggregation operator is proposed to construct an algorithm of a NSEM decision method. Finally an application of the constructed algorithm to solve a decision-making problem is provided. This new extension will provide a significant addition to existing theories for handling indeterminacy, and spurs more developments of further research and pertinent applications.

Author Contributions: All authors contributed equally.

Conflicts of Interest: The authors declare no conflict of interest.

References

1. Atanassov, K. Intuitionistic fuzzy sets. *Fuzzy Set Syst.* **1986**, *20*, 87–96. [CrossRef]
2. Molodtsov, D. Soft set theory-first results. *Comput. Math. Appl.* **1999**, *37*, 19–31. [CrossRef]
3. Smarandache, F. Neutrosophic set—A generalization of the intuitionistic fuzzy sets. *Int. J. Pure Appl. Math.* **2005**, *24*, 287–297.
4. Smarandache, F. *Neutrosophy: Neutrosophic Probability, Set, and Logic*; American Research Press: Rehoboth, IL, USA, 1998.
5. Medina, J.; Ojeda-Aciego, M. Multi-adjoint t-concept lattices. *Inf. Sci.* **2010**, *180*, 712–725. [CrossRef]
6. Pozna, C.; Minculete, N.; Precup, R.E.; Kóczy, L.T.; Ballagi, Á. Signatures: Definitions, operators and applications to fuzzy modelling. *Fuzzy Sets Syst.* **2012**, *201*, 86–104. [CrossRef]
7. Moallem, P.; Mousavi, B.S.; Naghibzadeh, S.S. Fuzzy inference system optimized by genetic algorithm for robust face and pose detection. *Int. J. Artif. Intell.* **2015**, *13*, 73–88.
8. Jankowski, J.; Kazienko, P.; Wątróbski, J.; Lewandowska, A.; Ziemba, P.; Zioło, M. Fuzzy multi-objective modeling of effectiveness and user experience in online advertising. *Expert Syst. Appl.* **2016**, *65*, 315–331. [CrossRef]
9. Alkhazaleh, S.; Salleh, A.R.; Hassan, N. Possibility fuzzy soft set. *Adv. Decis. Sci.* **2011**, *2011*, 479756. [CrossRef]
10. Alkhazaleh, S.; Salleh, A.R.; Hassan, N. Soft multisets theory. *Appl. Math. Sci.* **2011**, *5*, 3561–3573.
11. Salleh, A.R.; Alkhazaleh, S.; Hassan, N.; Ahmad, A.G. Multiparameterized soft set. *J. Math. Stat.* **2012**, *8*, 92–97.
12. Alhazaymeh, K.; Halim, S.A.; Salleh, A.R.; Hassan, N. Soft intuitionistic fuzzy sets. *Appl. Math. Sci.* **2012**, *6*, 2669–2680.
13. Adam, F.; Hassan, N. Q-fuzzy soft matrix and its application. *AIP Conf. Proc.* **2014**, *1602*, 772–778.
14. Adam, F.; Hassan, N. Q-fuzzy soft set. *Appl. Math. Sci.* **2014**, *8*, 8689–8695. [CrossRef]
15. Adam, F.; Hassan, N. Operations on Q-fuzzy soft set. *Appl. Math. Sci.* **2014**, *8*, 8697–8701. [CrossRef]
16. Adam, F.; Hassan, N. Multi Q-fuzzy parameterized soft set and its application. *J. Intell. Fuzzy Syst.* **2014**, *27*, 419–424.
17. Adam, F.; Hassan, N. Properties on the multi Q-fuzzy soft matrix. *AIP Conf. Proc.* **2014**, *1614*, 834–839.
18. Adam, F.; Hassan, N. Multi Q-fuzzy soft set and its application. *Far East J. Math. Sci.* **2015**, *97*, 871–881. [CrossRef]
19. Varnamkhasti, M.; Hassan, N. A hybrid of adaptive neurofuzzy inference system and genetic algorithm. *J. Intell. Fuzzy Syst.* **2013**, *25*, 793–796.
20. Varnamkhasti, M.; Hassan, N. Neurogenetic algorithm for solving combinatorial engineering problems. *J. Appl. Math.* **2012**, *2012*, 253714. [CrossRef]
21. Maji, P.K. Neutrosophic soft set. *Ann. Fuzzy Math. Inform.* **2013**, *5*, 157–168.
22. Alkhazaleh, S.; Salleh, A.R. Fuzzy soft expert set and its application. *Appl. Math.* **2014**, *5*, 1349. [CrossRef]
23. Hassan, N.; Alhazaymeh, K. Vague soft expert set theory. *AIP Conf. Proc.* **2013**, *1522*, 953–958.
24. Alhazaymeh, K.; Hassan, N. Mapping on generalized vague soft expert set. *Int. J. Pure Appl. Math.* **2014**, *93*, 369–376. [CrossRef]

25. Adam, F.; Hassan, N. Multi Q-Fuzzy soft expert set and its applications. *J. Intell. Fuzzy Syst.* **2016**, *30*, 943–950. [CrossRef]
26. Sahin, M.; Alkhazaleh, S.; Ulucay, V. Neutrosophic soft expert sets. *Appl. Math.* **2015**, *6*, 116–127. [CrossRef]
27. Hassan, N.; Uluçay, V.; Şahin, M. Q-neutrosophic soft expert set and its application in decision making. *Int. J. Fuzzy Syst. Appl.* **2018**, *7*, 37–61. [CrossRef]
28. Broumi, S.; Deli, I.; Smarandache, F. Neutrosophic parametrized soft set theory and its decision making. *Int. Front. Sci. Lett.* **2014**, *1*, 1–11. [CrossRef]
29. Deli, I. Refined Neutrosophic Sets and Refined Neutrosophic Soft Sets: Theory and Applications. In *Handbook of Research on Generalized and Hybrid Set Structures and Applications for Soft Computing*; IGI Global: Hershey, PA, USA, 2016; p. 321.
30. Syropoulos, A. On generalized fuzzy multisets and their use in computation. *Iran. J. Fuzzy Syst.* **2012**, *9*, 113–125.
31. Shinoj, T.K.; John, S.J. Intuitionistic fuzzy multisets and its application in mexdical diagnosis. *World Acad. Sci. Eng. Technol.* **2012**, *6*, 1–28.
32. Wang, H.; Smarandache, F.; Zhang, Y.Q.; Sunderraman, R. Single valued neutrosophic sets. *Multispace Multistruct.* **2010**, *4*, 410–413.
33. Ye, S.; Ye, J. Dice similarity measure between single valued neutrosophic multisets and its application in medical diagnosis. *Neutrosophic Sets Syst.* **2014**, *6*, 48–52.
34. Riesgo, Á.; Alonsoa, P.; Díazb, I.; Montesc, S. Basic operations for fuzzy multisets. *Int. J. Approx. Reason.* **2018**, *101*, 107–118. [CrossRef]
35. Sebastian, S.; John, R. Multi-fuzzy sets and their correspondence to other sets. *Ann. Fuzzy Math. Inform.* **2016**, *11*, 341–348.
36. Uluçay, V.; Deli, I.; Şahin, M. Intuitionistic trapezoidal fuzzy multi-numbers and its application to multi-criteria decision-making problems. In *Complex & Intelligent Systems*; Springer: Berlin, Germany, 2018; pp. 1–14.
37. Kunnambath, S.T.; John, S.J. Compactness ın ıntuıtıonıstıc fuzzy multıset topology. *J. New Theory* **2017**, *16*, 92–101.
38. Dhivya, J.; Sridevi, B. A New Similarity Measure between Intuitionistic Fuzzy Multisets based on their Cardinality with Applications to Pattern Recognition and Medical Diagnosis. *Asian J. Res. Soc. Sci. Humanit.* **2017**, *7*, 764–782. [CrossRef]
39. Ejegwa, P.A. On Intuitionistic fuzzy multisets theory and its application in diagnostic medicine. *MAYFEB J. Math.* **2017**, *4*, 13–22.
40. Fan, C.; Fan, E.; Ye, J. The Cosine Measure of Single-Valued Neutrosophic Multisets for Multiple Attribute Decision-Making. *Symmetry* **2018**, *10*, 154. [CrossRef]
41. Şahin, M.; Deli, I.; Ulucay, V. Extension principle based on neutrosophic multi-sets and algebraic operations. *J. Math. Ext.* **2018**, *12*, 69–90.
42. Al-Quran, A.; Hassan, N. Neutrosophic vague soft multiset for decision under uncertainty. *Songklanakarin J. Sci. Technol.* **2018**, *40*, 290–305.
43. Hu, Q.; Zhang, X. New Similarity Measures of Single-Valued Neutrosophic Multisets Based on the Decomposition Theorem and Its Application in Medical Diagnosis. *Symmetry* **2018**, *10*, 466. [CrossRef]
44. Şahin, M.; Uluçay, V.; Olgun, N.; Kilicman, A. On neutrosophic soft lattices. *Afr. Mat.* **2017**, *28*, 379–388.
45. Şahin, M.; Olgun, N.; Kargın, A.; Uluçay, V. Isomorphism theorems for soft G-modules. *Afr. Mat.* **2018**, *29*, 1237–1244. [CrossRef]
46. Ulucay, V.; Şahin, M.; Olgun, N. Time-Neutrosophic Soft Expert Sets and Its Decision Making Problem. *Matematika* **2018**, *34*, 246–260. [CrossRef]
47. Uluçay, V.; Kiliç, A.; Yildiz, I.; Sahin, M. A new approach for multi-attribute decision-making problems in bipolar neutrosophic sets. *Neutrosophic Sets Syst.* **2018**, *23*, 142–159.
48. Uluçay, V.; Şahin, M.; Hassan, N. Generalized neutrosophic soft expert set for multiple-criteria decision-making. *Symmetry* **2018**, *10*, 437. [CrossRef]

mathematics

MDPI

Article

Neutrosophic Multigroups and Applications

Vakkas Uluçay * and Memet Şahin

Department of Mathematics, Gaziantep University, 27310 Gaziantep, Turkey; mesahin@gantep.edu.tr
* Correspondence: vulucay27@gmail.com; Tel.: +90-537-643-5034

Received: 12 December 2018; Accepted: 14 January 2019; Published: 17 January 2019

Abstract: In recent years, fuzzy multisets and neutrosophic sets have become a subject of great interest for researchers and have been widely applied to algebraic structures include groups, rings, fields and lattices. Neutrosophic multiset is a generalization of multisets and neutrosophic sets. In this paper, we proposed a algebraic structure on neutrosophic multisets is called neutrosophic multigroups which allow the truth-membership, indeterminacy-membership and falsity-membership sequence have a set of real values between zero and one. This new notation of group as a bridge among neutrosophic multiset theory, set theory and group theory and also shows the effect of neutrosophic multisets on a group structure. We finally derive the basic properties of neutrosophic multigroups and give its applications to group theory.

Keywords: neutrosophic sets; neutrosophic multisets; neutrosophic multigroups; neutrosophic multisubgroups

1. Introduction

In the real world, there are much uncertainty information which cannot be handled by crisp values. The fuzzy set theory [1] has been an age old and effective tool to tackle uncertainty information by introduced Zadeh but it can be applied only on random process. Therefore, on the basis of fuzzy set theory, Sebastian and Ramakrishnan [2] introduced Multi-Fuzzy Sets, Atanassov [3] proposed intuitionistic fuzzy set theory, Shinoj and John [4] initiated intuitionistic fuzzy multisets. Recently, the above theories have developed in many directions and found its applications in a wide variety of fields including algebraic structures. For example, on fuzzy sets [5–7], on fuzzy multi sets [8–10], on intuitionistic fuzzy sets [11–19], on intuitionistic fuzzy multi sets [20] are some of the selected works.

But these theories cannot manage the all types of uncertainties, such as indeterminate and inconsistent information some decision-making problems. For instance, "when we ask the opinion of an expert about certain statement, he or she may that the possibility that the statement is true is 0.5 and the statement is false is 0.6 and the degree that he or she is not sure is 0.2" [21]. In order to overcome this shortage, Smarandache [22] introduced neutrosophic set theory to makes the theory of Atanassov [3] very convenient and easily applicable in practice. Then, Wang et al. [21] gave the some operations and results of single valued neutrosophic set theory. In order to establish the algebraic structures of neutrosophic sets, some authors gave definition of neutrosophic groups [23–26] that is actually a example of a group. To develop the neutrosophic set theory, the concept of neutrosophic multi sets was initiated by Deli et al. [27] and Ye [28,29] for modeling vagueness and uncertainty. Using their definitions, in this paper, we define a new type of neutrosophic group on a neutrosophic multi set, which we call neutrosophic multi set group. Since this new concept a brings the neutrosophic multi set theory, set theory and the group theory together, it is very functional in the sense of improving the neutrosophic multi set theory with respect to group structure. Rosenfeld [30] extended the classical group theory to fuzzy set. By using the definitions and results on fuzzy sets in [6,30] and on intuitionistic fuzzy multiset in [20], we applied the definitions and results to neutrosophic multi set theory.The above set theories have been applied to many different areas including neutrosophic environments have been studied by many researchers in [31–39]. In this paper the notion of neutrosophic multigroup along with some related properties have been introduced by follow the results of intuitionistic fuzzy group theory. This concept will bring a new opportunity in research and development of neutrosophic sets theory.

The paper is organized as follows. In Section 2, we briefly review some preliminary concepts that will be used in the paper. In Section 3, we introduce the concept of neutrosophic multi group and give several basic properties and operations. In Section 4, we give some applications to the group theory with respect to neutrosophic multi groups. In Section 5, we make some concluding remarks and suggest.

2. Preliminary

In this section, we present basic definitions of fuzzy set theory, multi fuzzy set theory, intuitionistic fuzzy set theory, intuitionistic fuzzy multi set theory, neutrosophic set theory and neutrosophic multi set theory. For more detailed explanations related to this section, we refer to the earlier studies [1,2,4,6,20,22,27,30].

Definition 1 ([1]). *Let E be a universe.*

Then, a fuzzy set X over E is defined by

$$X = \{(\mu_X(x)/x) : x \in E\}, \tag{1}$$

where μ_X is called membership function of X and defined by $\mu_X : E \to [0,1]$. For each $x \in E$, the value $\mu_X(x)$ represents the degree of x belonging to the fuzzy set X.

Definition 2 ([2]). *Let X be a non-empty set. A multi-fuzzy set A on X is defined as:*

$$A = \{< x, \mu_1(x), \mu_2(x), \mu_3(x), ..., \mu_i... : x \in E \}, \tag{2}$$

where $\mu_i : X \to [0,1]$ for all $i \in \{1,2,...,p\}$ and $x \in E$.

Definition 3 ([4]). *Let X be a nonempty set. An Intuitionistic Fuzzy Multi-set A denoted by IFMS drawn from X is characterized by two functions: 'count membership' of $A(CM_A)$ and 'count non membership' of $A(CN_A)$ given respectively by $A(CM_A) : X \to Q$ and $A(CN_A) : X \to Q$ where Q is the set of all crisp multi-sets drawn from the unit interval $[0,1]$ such that, for each $x \in X$, the membership sequence is defined as a decreasingly ordered sequence of elements in $CM_A(x)$, which is denoted by $(\mu_A^1(x), \mu_A^2(x), ..., \mu_A^P(x))$ where $\mu_A^1(x) \geq \mu_A^2(x) \geq ... \geq \mu_A^P(x)$ and the corresponding non membership sequence will be denoted by $(\nu_A^1(x), \nu_A^2(x), ..., \nu_A^P(x))$ such that $0 \leq \mu_A^i(x) + \nu_A^i(x) \leq 1$ for every $x \in X$ and $i = (1,2,3,...,p)$. An IFMS A is denoted by*

$$A = \{\langle x : (\mu_A^1(x), \mu_A^2(x), ..., \mu_A^P(x)), (\nu_A^1(x), \nu_A^2(x), ..., \nu_A^P(x))\rangle : x \in X \}. \tag{3}$$

Definition 4 ([4]). *Length of an element x in an IFMS. A defined as the Cardinality of $CM_A(x)$ or $CN_A(x)$ for which $0 \leq \mu_A^j(x) + \nu_A^j(x) \leq 1$ and it is denoted by $L(x : A)$. That is,*

$$L(x : A) = |CM_A(x)| = |CN_A(x)|. \tag{4}$$

Proposition 1 ([20]). *Let $A, B, A_i \in IFMS(X)$; then, the following results hold:*

1. $[A^{-1}]^{-1} = A$.
2. $A \subseteq B \Rightarrow A^{-1} \subseteq B^{-1}$.
3. $[\cup_{i=1}^{n} A_i]^{-1} = \cup_{i=1}^{n}[A_i^{-1}]$.

4. $[\bigcap_{i=1}^{n} A_i]^{-1} = \bigcap_{i=1}^{n}[A_i^{-1}]$.
5. $(A \circ B)^{-1} = B^{-1} \circ A^{-1}$.
6. $$\begin{aligned} CM_{A\circ B}(x) &= \vee_{y\in X}\{CM_A(y) \wedge CM_B(y^{-1}x)\} \ \forall\, x \in X \\ &= \vee_{y\in X}\{CM_A(xy^{-1}) \wedge CM_B(y)\} \ \forall\, x \in X. \end{aligned}$$
$$\begin{aligned} CN_{A\circ B}(x) &= \wedge_{y\in X}\{CN_A(y) \vee CN_B(y^{-1}x)\} \ \forall\, x \in X \\ &= \wedge_{y\in X}\{CN_A(xy^{-1}) \vee CN_B(y)\} \ \forall\, x \in X. \end{aligned}$$

Definition 5 ([20]). *Let X be a group. An intuitionistic fuzzy multiset G over X is an intuitionistic fuzzy multi group $(IFMG)$ over X if the counts(count membership and non membership) of G satisfies the following four conditions:*

1. $CM_G(xy) \geq CM_G(x) \wedge CM_G(y) \ \forall\, x,y \in X$.
2. $CM_G(x^{-1}) \geq CT_G(x) \ \forall\, x \in X$.
3. $CN_G(xy) \leq CN_G(x) \wedge CI_G(y) \ \forall\, x,y \in X$.
4. $CN_G(x^{-1}) \leq CN_G(x) \ \forall\, x \in X$.

Definition 6 ([22]). *Let X be a space of points (objects), with a generic element in X denoted by x. A neutrosophic set(N-set) A in X is characterized by a truth-membership function T_A, a indeterminacy-membership function I_A and a falsity-membership function F_A. $T_A(x)$, $I_A(x)$ and $F_A(x)$ are real standard or nonstandard subsets of $[^-0, 1^+]$.*

It can be written as

$$A = \{< x, (T_A(x), I_A(x), F_A(x)) >: x \in X,\ T_A(x), I_A(x), F_A(x) \in [0,1]\}. \tag{5}$$

There is no restriction on the sum of $T_A(x)$; $I_A(x)$ and $F_A(x)$, so $^-0 \leq supT_A(x) + supI_A(x) + supF_A(x) \leq 3^+$.

Here, 1^+ = 1+ε, where 1 is its standard part and ε its non-standard part. Similarly, $^-0$ = 1+ε, where 0 is its standard part and ε its non-standard part.

Definition 7 ([27]). *Let E be a universe. A neutrosophic multiset set(Nms) A on E can be defined as follows:*

$$\begin{aligned} A &= \{< x, (T_A^1(x), T_A^2(x), ..., T_A^P(x)), (I_A^1(x), I_A^2(x), ..., I_A^P(x)), \\ &\qquad (F_A^1(x), F_A^2(x), ..., F_A^P(x)) >:\ x \in E\}, \end{aligned} \tag{6}$$

where

$$T_A^1(x), T_A^2(x), ..., T_A^P(x) : E \to [0,1],$$

$$I_A^1(x), I_A^2(x), ..., I_A^P(x) : E \to [0,1],$$

and

$$F_A^1(x), F_A^2(x), ..., F_A^P(x) : E \to [0,1]$$

such that

$$0 \leq supT_A^i(x) + supI_A^i(x) + supF_A^i(x) \leq 3$$

$(i = 1, 2, ..., P)$ and

$$T_A^1(x) \leq T_A^2(x) \leq ... \leq T_A^P(x)$$

for any $x \in E$.

$(T_A^1(x), T_A^2(x), ..., T_A^P(x))$, $(I_A^1(x), I_A^2(x), ..., I_A^P(x))$ and $(F_A^1(x), F_A^2(x), ..., F_A^P(x))$ is the truth-membership sequence, indeterminacy-membership sequence and falsity-membership sequence of the element x, respectively. In addition, P is called the dimension(cardinality) of Nms A, denoted d(A).

We arrange the truth-membership sequence in decreasing order, but the corresponding indeterminacy-membership and falsity-membership sequence may not be in decreasing or increasing order.

Definition 8 ([27,28]). *Let A, B be two Nms. Then,*

1. *A is said to be Nm-subset of B is denoted by* $A\tilde{\subseteq}B$ *if* $T^i_A(x) \leq T^i_B(x)$, $I^i_A(x) \geq I^i_B(x)$, $F^i_A(x) \geq F^i_B(x)$, $\forall x \in E$ *and* $i = 1, 2, ..., P$.
2. *A is said to be neutrosophic equal of B is denoted by* $A = B$ *if* $T^i_A(x) = T^i_B(x)$, $I^i_A(x) = I^i_B(x)$, $F^i_A(x) = F^i_B(x)$, $\forall x \in E$ *and* $i = 1, 2, ..., P$.
3. *The union of A and B is denoted by* $A\tilde{\cup}B = C$ *and is defined by*

$$C = \{< x, (T^1_C(x), T^2_C(x), ..., T^P_C(x)), (I^1_C(x), I^2_C(x), ..., I^P_C(x)), (F^1_C(x), F^2_C(x), ..., F^P_C(x)) >: x \in E\},$$

where $T^i_C = T^i_A(x) \vee T^i_B(x)$, $I^i_C = I^i_A(x) \wedge I^i_B(x)$, $F^i_C = F^i_A(x) \wedge F^i_B(x)$, $\forall x \in E$ *and* $i = 1, 2, ..., P$.

4. *The intersection of A and B is denoted by* $A\tilde{\cap}B = D$ *and is defined by*

$$D = \{< x, (T^1_D(x), T^2_D(x), ..., T^P_D(x)), (I^1_D(x), I^2_D(x), ..., I^P_D(x)), (F^1_D(x), F^2_D(x), ..., F^P_D(x)) >: x \in E\},$$

where $T^i_D = T^i_A(x) \wedge T^i_B(x)$, $I^i_D = I^i_A(x) \vee I^i_B(x)$, $F^i_D = F^i_A(x) \vee F^i_B(x)$, $\forall x \in E$ *and* $i = 1, 2, ..., P$.

3. Neutrosophic Multigroups

In this section, we introduce neutrosophic multigroups and investigate their basic properties. Throughout this section,

1. Let X be a group with a binary operation and the identity element is e.
2. $NMS(X)$ denotes the set of all neutrosophic multisets over the X.
3. $NMG(X)$ denotes the set of all neutrosophic multi groups NMG over the group X.

Definition 9. *Let X be a group* $A \in NMS(X)$. *Then,* A^{-1} *is defined as*

$$\begin{aligned} A^{-1} &= \{< x, (T^{1\,-1}_A(x), T^{2\,-1}_A(x), ..., T^P_A(x))^{-1}, (I^1_A(x)^{-1}, I^{2\,-1}_A(x), ..., I^{P\,-1}_A(x)), \\ &\qquad (F^{1\,-1}_A(x), F^{2\,-1}_A(x), ..., F^{P-1}_A(x)) >: x \in E\}, \end{aligned} \tag{7}$$

where $T'^{-1}_A(x) = T'_A(x^{-1})$, $I'^{-1}_A(x) = I'_A(x^{-1})$ *and* $F'^{-1}_A(x) = F'_A(x^{-1})$ *for all* $i = 1, 2, ..., P$.

Definition 10. *Let X be a classical group* $A \in NMS(X)$. *Then, A is called a neutrosophic multi groupoid over X if*

1. $T^i{}_G(xy) \geq T^i{}_G(x) \wedge T^i{}_G(y)$,
2. $I^i{}_G(xy) \leq I^i{}_G(x) \vee I^i{}_G(y)$,
3. $F^i{}_G(xy) \leq F^i{}_G(x) \vee F^i{}_G(y)$,

for all $x, y \in X$ *and* $i = 1, 2, ..., P$.

A is called a neutrosophic multi group(NM-group) over X if the neutrosophic multi groupoid satisfies

1. $T^i{}_G(x^{-1}) \geq T^i{}_G(x)$,
2. $I^i{}_G(x^{-1}) \leq I^i{}_G(x)$,
3. $F^i{}_G(x^{-1}) \leq F^i{}_G(x)$,

for all $x \in X$ *and* $i = 1, 2, ..., P$.

Example 1. *Assume that* $(Z_3, +)$ *is a classical group. Then,*

$$\begin{aligned} A = &\{\langle 0; (0.8, 0.7, 0.6, 0.4), (0.1, 0.1, 0.2, 0.3), (0.2, 0.3, 0.4, 0.5)\rangle, \langle 1; (0.7, 0.6, 0.4, 0.3), \\ &(0.2, 0.3, 0.2, 0.3), (0.3, 0.4, 0.5, 0.6)\rangle, \langle 2; (0.8, 0.6, 0.6, 0.4), (0.1, 0.2, 0.2, 0.3), (0.2, 0.4, 0.4, 0.5)\rangle\} \end{aligned}$$

is a NM-group. However,

$$B = \{\langle 0;(0.8,0.7,0.6,0.4),(0.1,0.1,0.2,0.3),(0.2,0.3,0.4,0.5)\rangle,\langle 1;(0.9,0.5,0.4,0.3),(0.2,0.1,0.2,0.3),(0.3,0.3,0.5,0.4)\rangle,\langle 2;(0.8,0.7,0.6,0.4),(0.1,0.3,0.2,0.3),(0.2,0.4,0.4,0.6)\rangle\}$$

is not a NM-group because ${T^i}_B(1^{-1})$ *is not greater than or equal to* ${T^i}_B(1)$.

From the Definition 10 and Example 1, it is clear that a NM-group is a generalized case of fuzzy group and intuitionistic fuzzy multi group.

Proposition 2. *Let X be a classical group and* $A \in NMS(X)$. *If* $A \in NMG(X)$; *then,*

1. ${T^i}_A(e) \geq {T^i}_A(x)\ \forall\, x \in X,$
2. ${I^i}_A(e) \leq {I^i}_A(x)\ \forall\, x \in X,$
3. ${F^i}_A(e) \leq {F^i}_A(x)\ \forall\, x \in X,$

for all $x \in X$ *and* $i = 1,2,...,P$.

Proof. Since A an $NM-group$ over X, then

1.
$$\begin{aligned} {T^i}_A(e) &= {T^i}_A(x.x^{-1}) \\ &\geq {T^i}_A(x) \wedge {T^i}_A(x^{-1}) \\ &\geq {T^i}_A(x) \wedge {T^i}_A(x) \\ &= {T^i}_A(x) \end{aligned}$$
for all $x \in X$ and $i = 1,2,...,P$.

2.
$$\begin{aligned} {I^i}_A(e) &= {I^i}_A(x.x^{-1}) \\ &\leq {I^i}_A(x) \vee {I^i}_A(x^{-1}) \\ &\leq {I^i}_A(x) \vee {I^i}_A(x) \\ &= {I^i}_A(x) \end{aligned}$$
for all $x \in X$ and $i = 1,2,...,P$.

3.
$$\begin{aligned} {F^i}_A(e) &= {F^i}_A(x.x^{-1}) \\ &\leq {F^i}_A(x) \vee {F^i}_A(x^{-1}) \\ &\leq {F^i}_A(x) \vee {F^i}_A(x) \\ &= {F^i}_A(x) \end{aligned}$$
for all $x \in X$ and $i = 1,2,...,P$.

□

Proposition 3. *Let X be a classical group and* $A \in NMS(X)$. *If* $A \in NMG(X)$, *then*

1. ${T^i}_A(x^n) \geq {T^i}_A(x)\ \forall\, x \in X,$
2. ${I^i}_A(x^n) \leq {I^i}_A(x)\ \forall\, x \in X,$
3. ${F^i}_A(x^n) \leq {F^i}_A(x)\ \forall\, x \in X,$

for all $x \in X$ *and* $i = 1,2,...,P$.

Proof. Since A an $NM-group$ over X, then

1.
$$\begin{aligned} T^i{}_A(x^n) &\geq T^i{}_A(x) \wedge T^i{}_A(x^{n-1}) \\ &\geq T^i{}_A(x) \wedge T^i{}_A(x) \wedge ... \wedge T^i{}_A(x) \\ &= T^i{}_A(x) \end{aligned}$$

for all $x \in X$ and $i = 1, 2, ..., P$.

2.
$$\begin{aligned} I^i{}_A(x^n) &\leq I^i{}_A(x) \vee I^i{}_A(x^{n-1}) \\ &\leq I^i{}_A(x) \vee I^i{}_A(x) \vee ... \vee I^i{}_A(x) \\ &= I^i{}_A(x) \end{aligned}$$

for all $x \in X$ and $i = 1, 2, ..., P$.

3.
$$\begin{aligned} F^i{}_A(x^n) &\leq F^i{}_A(x) \vee F^i{}_A(x^{n-1}) \\ &\leq F^i{}_A(x) \vee F^i{}_A(x) \vee ... \vee F^i{}_A(x) \\ &= F^i{}_A(x) \end{aligned}$$

for all $x \in X$ and $i = 1, 2, ..., P$.

□

Definition 11. *Let Y be a subgroup of X, $B \in NMG(Y)$, $B \tilde{\subseteq} A$ and $A \in NMG(X)$. If $B \in NMG(Y)$, then B is called a neutrosophic multi subgroup of A over X and denoted by $B \tilde{\leq} A$.*

Example 2. *Assume that $(Z_3, +)$ is a classical group. We define A and B neutrosophic multi group over $(Z_3, +)$ by*

$$\begin{aligned} A &= \{\langle 0; (0.4, 0.3, 0.3, 0.2), (0.1, 0.1, 0.2, 0.3), (0.2, 0.3, 0.4, 0.6)\rangle, \\ &\quad \langle 1; (0.6, 0.5, 0.3, 0.2), (0.2, 0.4, 0.2, 0.3), (0.3, 0.2, 0.5, 0.6)\rangle, \\ &\quad \langle 2; (0.8, 0.7, 0.5, 0.4), (0.1, 0.3, 0.2, 0.3), (0.2, 0.1, 0.4, 0.5)\rangle\}. \end{aligned}$$

$$\begin{aligned} B &= \{\langle 0; (0.4, 0.3, 0.3, 0.2), (0.1, 0.1, 0.2, 0.3), (0.2, 0.3, 0.4, 0.6)\rangle, \\ &\quad \langle 1; (0.6, 0.5, 0.3, 0.2), (0.2, 0.4, 0.2, 0.3), (0.3, 0.2, 0.5, 0.6)\rangle\}. \end{aligned}$$

Then, B is a neutrosophic multi subgroup of A over $(Z_3, +)$ and denoted by $B \tilde{\leq} A$.

Theorem 1. *Let X be a group $A \in NMS(X)$. Then, A is an $NM - group$ if and only if $T^i{}_A(xy^{-1}) \geq T^i{}_A(x) \wedge T^i{}_A(y)$, $I^i{}_A(xy^{-1}) \leq I^i{}_A(x) \vee I^i{}_A(y)$ and $F^i{}_A(xy^{-1}) \leq F^i{}_A(x) \vee F^i{}_A(y)$ for all $x, y \in X$.*

Proof. Assume that A is an $NM - group$ over X. Then,

$$\begin{aligned} T^i{}_A(xy^{-1}) &\geq T^i{}_A(x) \wedge T^i{}_A(y^{-1}) \\ &\geq T^i{}_A(x) \wedge T^i{}_A(y) \end{aligned}$$

for all $x, y \in X$ and $i = 1, 2, ..., P$.

$$\begin{aligned} I^i{}_A(xy^{-1}) &\leq I^i{}_A(x) \vee I^i{}_A(y^{-1}) \\ &\leq I^i{}_A(x) \vee I^i{}_A(y) \end{aligned}$$

for all $x, y \in X$ and $i = 1, 2, ..., P$.

$$\begin{aligned} F^i{}_A(xy^{-1}) &\leq F^i{}_A(x) \vee F^i{}_A(y^{-1}) \\ &\leq F^i{}_A(x) \vee F^i{}_A(y) \end{aligned}$$

for all $x, y \in X$ and $i = 1, 2, ..., P$.

Conversely, the given condition be satisfied. Firstly,

$$\begin{aligned} T^i{}_A(x^{-1}) &= T^i{}_A(e.x^{-1}) \\ &\geq T^i{}_A(e) \wedge T^i{}_A(x) \\ &= T^i{}_A(x) \end{aligned}$$

$$\begin{aligned} T^i{}_A(xy) &\geq T^i{}_A(x) \wedge T^i{}_A(y^{-1}) \\ &\geq T^i{}_A(x) \wedge T^i{}_A(y). \end{aligned}$$

Secondly,

$$\begin{aligned} I^i{}_A(x^{-1}) &= I^i{}_A(e.x^{-1}) \\ &\leq I^i{}_A(e) \vee I^i{}_A(x) \\ &= I^i{}_A(x) \end{aligned}$$

$$\begin{aligned} I^i{}_A(xy) &\leq I^i{}_A(x) \vee I^i{}_A(y^{-1}) \\ &\leq I^i{}_A(x) \vee I^i{}_A(y). \end{aligned}$$

Thirdly,

$$\begin{aligned} F^i{}_A(x^{-1}) &= F^i{}_A(e.x^{-1}) \\ &\leq F^i{}_A(e) \vee F^i{}_A(x) \\ &= F^i{}_A(x) \end{aligned}$$

$$\begin{aligned} F^i{}_A(xy) &\leq F^i{}_A(x) \vee F^i{}_A(y^{-1}) \\ &\leq F^i{}_A(x) \vee F^i{}_A(y) \end{aligned}$$

so the proof is complete. □

Definition 12. *Let $A, B \in NMS(X)$. Then, their "AND" operation is denoted by $A \barwedge B$ and is defined by*

$$A \barwedge B = \{(x,y),\ T^i{}_{A \barwedge B}(x,y),\ I^i{}_{A \barwedge B}(x,y),\ F^i{}_{A \barwedge B}(x,y) : (x,y) \in X \times X\}, \tag{8}$$

where $T^i{}_{A \barwedge B}(x,y) = T^i{}_A(x) \wedge T^i{}_B(y)$, $I^i{}_{A \barwedge B}(x,y) = I^i{}_A(x) \vee I^i{}_B(y)$, $F^i{}_{A \barwedge B}(x,y) = F^i{}_A(x) \vee F^i{}_B(y)$.

Theorem 2. *Let $A, B \in NMG(X)$. Then, $A \barwedge B$ is a neutrosophic multi group over X.*

Proof. Let $(x_1, y_1), (x_2, y_2) \in X \times X$. Then,

$$\begin{aligned} T^i{}_{A \barwedge B}((x_1,y_1),(x_2,y_2)^{-1}) &= T^i{}_{A \barwedge B}(x_1x_2^{-1}, y_1y_2^{-1}) \\ &= T^i{}_A(x_1x_2^{-1}) \wedge T^i{}_B(y_1y_2^{-1}) \\ &\geq (T^i{}_A(x_1) \wedge T^i{}_A(x_2)) \wedge (T^i{}_B(y_1) \wedge T^i{}_B(y_2)) \\ &= (T^i{}_A(x_1) \wedge T^i{}_B(y_1)) \wedge (T^i{}_A(x_2) \wedge T^i{}_B(y_2)) \\ &= T^i{}_{A \barwedge B}(x_1,y_1) \wedge T^i{}_{A \barwedge B}(x_2,y_2) \end{aligned}$$

$$\begin{aligned} I^i{}_{A \barwedge B}((x_1,y_1),(x_2,y_2)^{-1}) &= I^i{}_{A \barwedge B}(x_1x_2^{-1}, y_1y_2^{-1}) \\ &= I^i{}_A(x_1x_2^{-1}) \vee I^i{}_B(y_1y_2^{-1}) \\ &\leq (I^i{}_A(x_1) \vee I^i{}_A(x_2)) \vee (I^i{}_B(y_1) \vee I^i{}_B(y_2)) \\ &= (I^i{}_A(x_1) \vee I^i{}_B(y_1)) \vee (I^i{}_A(x_2) \vee I^i{}_B(y_2)) \\ &= I^i{}_{A \barwedge B}(x_1,y_1) \vee I^i{}_{A \barwedge B}(x_2,y_2) \end{aligned}$$

and

$$\begin{aligned} F^i{}_{A \barwedge B}((x_1,y_1),(x_2,y_2)^{-1}) &= F^i{}_{A \barwedge B}(x_1x_2^{-1}, y_1y_2^{-1}) \\ &= F^i{}_A(x_1x_2^{-1}) \vee F^i{}_B(y_1y_2^{-1}) \\ &\leq (F^i{}_A(x_1) \vee F^i{}_A(x_2)) \vee (F^i{}_B(y_1) \vee F^i{}_B(y_2)) \\ &= (F^i{}_A(x_1) \vee F^i{}_B(y_1)) \vee (F^i{}_A(x_2) \vee F^i{}_B(y_2)) \\ &= F^i{}_{A \barwedge B}(x_1,y_1) \vee F^i{}_{A \barwedge B}(x_2,y_2) \end{aligned}$$

for all $(x_1, y_1), (x_2, y_2) \in X$ and $i = 1, 2, ..., P$. Therefore, $A \tilde{\wedge} B$ is a neutrosophic multi group over X, hence the proof.

□

Example 3. *Let us take into consideration the classical group* $(Z_3, +)$*. Define the neutrosophic multiset* A, B *on* $(Z_3, +)$ *as follows:*

$$\begin{aligned}
A &= \{\langle 0; (0.5, 0.3, 0.2, 0.1), (0.1, 0.1, 0.2, 0.3), (0.2, 0.3, 0.4, 0.5)\rangle, \\
&\langle 1; (0.6, 0.4, 0.3, 0.2), (0.1, 0.3, 0.3, 0.4), (0.1, 0.2, 0.4, 0.6)\rangle, \\
&\langle 2; (0.7, 0.5, 0.3, 0.2), (0.2, 0.2, 0.3, 0.4), (0.3, 0.3, 0.4, 0.6)\rangle, \\
B &= \{\langle 0; (0.6, 0.5, 0.4, 0.2), (0.2, 0.2, 0.3, 0.4), (0.3, 0.3, 0.4, 0.6)\rangle, \\
&\langle 1; (0.8, 0.6, 0.4, 0.3), (0.1, 0.1, 0.2, 0.3), (0.2, 0.2, 0.3, 0.4)\rangle\} \\
&are\ NM - groups. \\
A \tilde{\wedge} B &= \{\langle (0,0); (0.5, 0.3, 0.2, 0.1), (0.2, 0.2, 0.2, 0.4), (0.2, 0.3, 0.4, 0.6)\rangle, \\
&\langle (0,1); (0.5, 0.3, 0.2, 0.1), (0.1, 0.1, 0.2, 0.3), (0.2, 0.2, 0.3, 0.5)\rangle, \\
&\langle (1,0); (0.6, 0.4, 0.3, 0.2), (0.2, 0.2, 0.3, 0.4), (0.3, 0.3, 0.4, 0.6)\rangle, \\
&\langle (1,1); (0.6, 0.4, 0.3, 0.2), (0.1, 0.3, 0.3, 0.4), (0.2, 0.2, 0.4, 0.6)\rangle, \\
&\langle (2,0); (0.6, 0.5, 0.3, 0.2), (0.2, 0.2, 0.3, 0.4), (0.3, 0.3, 0.4, 0.6)\rangle, \\
&\langle (2,1); (0.7, 0.5, 0.3, 0.2), (0.2, 0.2, 0.3, 0.4), (0.3, 0.3, 0.4, 0.6)\rangle\}.
\end{aligned}$$

Then, $A \tilde{\wedge} B \in NMG(X)$.

Definition 13. *Let* X *be a classical group and* $A, B \in NMS(X)$*. Then, their "OR" operation is denoted by* $A \tilde{\vee} B$ *and is defined by*

$$A \tilde{\vee} B = \{(x,y), T^i{}_{A\tilde{\vee}B}(x,y), I^i{}_{A\tilde{\vee}B}(x,y), F^i{}_{A\tilde{\vee}B}(x,y) : (x,y) \in X \times X\} \tag{9}$$

where $T^i{}_{A\tilde{\vee}B}(x,y) = T^i{}_A(x) \vee T^i{}_B(y)$, $I^i{}_{A\tilde{\vee}B}(x,y) = I^i{}_A(x) \wedge I^i{}_B(y)$, $F^i{}_{A\tilde{\vee}B}(x,y) = F^i{}_A(x) \wedge F^i{}_B(y)$.

Proposition 4. *Let* $A, B \in NMG(X)$*. Then,* $T^i{}_{A\tilde{\vee}B}(x) \leq T^i{}_{A\tilde{\vee}B}(x^{-1})$, $I^i{}_{A\tilde{\vee}B}(x) \geq I^i{}_{A\tilde{\vee}B}(x^{-1})$, $F^i{}_{A\tilde{\vee}B}(x) \geq F^i{}_{A\tilde{\vee}B}(x^{-1})$.

Proof. Let $(x_1, y_1), (x_2, y_2) \in X \times X$. Then,

$$\begin{aligned}
T^i{}_{A\tilde{\vee}B}((x_1,y_1),(x_2,y_2)^{-1}) &= T^i{}_{A\tilde{\vee}B}(x_1x_2^{-1}, y_1y_2^{-1}) \\
&= T^i{}_A(x_1x_2^{-1}) \vee T^i{}_B(y_1y_2^{-1}) \\
&\leq (T^i{}_A(x_1) \vee T^i{}_A(x_2)) \vee (T^i{}_B(y_1) \vee T^i{}_B(y_2)) \\
&= (T^i{}_A(x_1) \vee T^i{}_B(y_1)) \vee (T^i{}_A(x_2) \vee T^i{}_B(y_2)) \\
&= T^i{}_{A\tilde{\vee}B}(x_1,y_1) \vee T^i{}_{A\tilde{\vee}B}(x_2,y_2)
\end{aligned}$$

$$\begin{aligned}
I^i{}_{A\tilde{\vee}B}((x_1,y_1),(x_2,y_2)^{-1}) &= I^i{}_{A\tilde{\vee}B}(x_1x_2^{-1}, y_1y_2^{-1}) \\
&= I^i{}_A(x_1x_2^{-1}) \wedge I^i{}_B(y_1y_2^{-1}) \\
&\geq (I^i{}_A(x_1) \wedge I^i{}_A(x_2)) \wedge (I^i{}_B(y_1) \wedge I^i{}_B(y_2)) \\
&= (I^i{}_A(x_1) \wedge I^i{}_B(y_1)) \wedge (I^i{}_A(x_2) \wedge I^i{}_B(y_2)) \\
&= I^i{}_{A\tilde{\vee}B}(x_1,y_1) \wedge I^i{}_{A\tilde{\vee}B}(x_2,y_2)
\end{aligned}$$

and

$$\begin{aligned}
F^i{}_{A\tilde{\vee}B}((x_1,y_1),(x_2,y_2)^{-1}) &= F^i{}_{A\tilde{\vee}B}(x_1x_2^{-1}, y_1y_2^{-1}) \\
&= F^i{}_A(x_1x_2^{-1}) \wedge F^i{}_B(y_1y_2^{-1}) \\
&\geq (F^i{}_A(x_1) \wedge F^i{}_A(x_2)) \wedge (F^i{}_B(y_1) \wedge F^i{}_B(y_2)) \\
&= (F^i{}_A(x_1) \wedge F^i{}_B(y_1)) \wedge (F^i{}_A(x_2) \wedge F^i{}_B(y_2)) \\
&= F^i{}_{A\tilde{\vee}B}(x_1,y_1) \wedge F^i{}_{A\tilde{\vee}B}(x_2,y_2)
\end{aligned}$$

for all $(x_1, y_1), (x_2, y_2) \in X$ and $i = 1, 2, ..., P$—hence the proof.

From this, it is clear that, if $A, B \in NMG(X)$, then $A\tilde{\vee}B \in NMG(X)$ iff $T^i{}_{A\tilde{\vee}B}(x,y) \geq T^i{}_{A\tilde{\vee}B}(x) \wedge T^i{}_{A\tilde{\vee}B}(y), I^i{}_{A\tilde{\vee}B}(x,y) \leq I^i{}_{A\tilde{\vee}B}(x) \vee I^i{}_{A\tilde{\vee}B}(y), F^i{}_{A\tilde{\vee}B}(x,y) \leq F^i{}_{A\tilde{\vee}B}(x) \vee F^i{}_{A\tilde{\vee}B}(y)$. □

Corollary 1. *Let $A, B \in NMG(X)$. Then, $A\tilde{\vee}B$ need not be an element of $NMG(X)$.*

Example 4. *Let us take into consideration the classical group $(Z_4, +)$. Define the neutrosophic multiset A, B on $(Z_4, +)$ as follows:*

$$\begin{aligned}
A &= \{\langle 0; (0.5, 0.3, 0.2, 0.1), (0.1, 0.1, 0.2, 0.3), (0.2, 0.3, 0.4, 0.5)\rangle, \\
&\quad \langle 1; (0.6, 0.4, 0.3, 0.2), (0.1, 0.3, 0.3, 0.4), (0.1, 0.2, 0.4, 0.6)\rangle, \\
&\quad \langle 2; (0.7, 0.5, 0.3, 0.2), (0.2, 0.2, 0.3, 0.4), (0.3, 0.3, 0.4, 0.6)\rangle, \\
&\quad \langle 3; (0.7, 0.6, 0.4, 0.3), (0.2, 0.1, 0.2, 0.3), (0.3, 0.2, 0.1, 0.3)\rangle\} \\
B &= \{\langle 0; (0.6, 0.5, 0.4, 0.2), (0.2, 0.2, 0.3, 0.4), (0.2, 0.3, 0.4, 0.6)\rangle, \\
&\quad \langle 1; (0.8, 0.6, 0.4, 0.3), (0.1, 0.1, 0.2, 0.3), (0.2, 0.2, 0.3, 0.4)\rangle\} \\
&\quad \text{are } NM-groups. \\
A\tilde{\vee}B &= \{\langle (0,0); (0.6, 0.5, 0.4, 0.2), (0.1, 0.1, 0.2, 0.3), (0.2, 0.3, 0.4, 0.5)\rangle, \\
&\quad \langle (0,1); (0.8, 0.6, 0.4, 0.3), (0.1, 0.1, 0.2, 0.3), (0.2, 0.2, 0.3, 0.4)\rangle, \\
&\quad \langle (1,0); (0.6, 0.5, 0.4, 0.2), (0.1, 0.2, 0.3, 0.4), (0.1, 0.2, 0.4, 0.6)\rangle, \\
&\quad \langle (1,1); (0.8, 0.6, 0.4, 0.3), (0.1, 0.1, 0.2, 0.3), (0.1, 0.2, 0.3, 0.4)\rangle, \\
&\quad \langle (2,0); (0.7, 0.5, 0.4, 0.2), (0.1, 0.2, 0.3, 0.4), (0.2, 0.3, 0.4, 0.6)\rangle, \\
&\quad \langle (2,1); (0.8, 0.6, 0.4, 0.3), (0.1, 0.1, 0.2, 0.3), (0.2, 0.2, 0.3, 0.4)\rangle, \\
&\quad \langle (3,0); (0.7, 0.6, 0.4, 0.3), (0.2, 0.1, 0.2, 0.3), (0.2, 0.2, 0.1, 0.3)\rangle, \\
&\quad \langle (3,1); (0.8, 0.6, 0.4, 0.3), (0.1, 0.1, 0.2, 0.3), (0.2, 0.2, 0.1, 0.3)\rangle\}.
\end{aligned}$$

However, $T^i{}_{A\tilde{\vee}B}(3,0) \geq T^i{}_{A\tilde{\vee}B}(1,0)$. Then, $A\tilde{\vee}B \notin NMG(X)$.

Theorem 3. *Let X be a classical group and $A \in NMG(X)$. Then, the followings are equivalent:*

1. $T^i_A(yx) = T^i_A(xy), I^i_A(yx) = I^i_A(xy)$ *and* $F^i_A(yx) = F^i_A(xy)$ *for all* $x, y \in X$.
2. $T^i_A(xyx^{-1}) = T^i_A(y), I^i_A(xyx^{-1}) = I^i_A(y)$ *and* $F^i_A(xyx^{-1}) = F^i_A(y)$ *for all* $x, y \in X$.
3. $T^i_A(xyx^{-1}) \geq T^i_A(y), I^i_A(xyx^{-1}) \leq I^i_A(y)$ *and* $F^i_A(xyx^{-1}) \leq F^i_A(y)$ *for all* $x, y \in X$.
4. $T^i_A(xyx^{-1}) \leq T^i_A(y), I^i_A(xyx^{-1}) \geq I^i_A(y)$ *and* $F^i_A(xyx^{-1}) \geq F^i_A(y)$ *for all* $x, y \in X$.

Proof.
1. (1) ⇒ (2): Let $x, y \in X$. Then,

$$\begin{aligned}
T^i_A(xyx^{-1}) &= T^i_A(x^{-1}xy) = T^i_A(y), \\
I^i_A(xyx^{-1}) &= I^i_A(x^{-1}xy) = I^i_A(y), \\
F^i_A(xyx^{-1}) &= F^i_A(x^{-1}xy) = F^i_A(y).
\end{aligned}$$

2. (2) ⇒ (3): Immediate.
3. (3) ⇒ (4)

$$\begin{aligned}
T^i_A(xyx^{-1}) &\leq T^i_A(x^{-1}xy(x^{-1})^{-1}) = T^i_A(y), \\
I^i_A(xyx^{-1}) &\geq I^i_A(x^{-1}xy(x^{-1})^{-1}) = I^i_A(y), \\
F^i_A(xyx^{-1}) &\geq F^i_A(x^{-1}xy(x^{-1})^{-1}) = F^i_A(y).
\end{aligned}$$

4. (4) ⇒ (1): Let $x, y \in X$. Then,

$$\begin{aligned}
T^i_A(xy) &= T^i_A(x.yx.x^{-1}) \\
&\leq T^i_A(yx) \\
&= T^i_A(y.xy.y^{-1}) \\
&\leq T^i_A(xy),
\end{aligned}$$

$$\begin{aligned} I^i_A(xy) &= I^i_A(x.yx.x^{-1}) \\ &\leq I^i_A(yx) \\ &= I^i_A(y.xy.y^{-1}) \\ &\leq I^i_A(xy), \\ F^i_A(xy) &= F^i_A(x.yx.x^{-1}) \\ &\leq F^i_A(yx) \\ &= F^i_A(y.xy.y^{-1}) \\ &\leq F^i_A(xy) \end{aligned}$$

Hence, $T^i_A(yx) = T^i_A(xy), I^i_A(yx) = I^i_A(xy), F^i_A(yx) = F^i_A(xy)$.

□

Definition 14. *Let X be a group, $A \in NMS(X)$ and B is a nonempty neutrosophic multi subset of A over X. Then, B is called an abelian neutrosophic multi subset of A if $T^i_A(yx) = T^i_A(xy), I^i_A(yx) = I^i_A(xy)$ and $F^i_A(yx) = F^i_A(xy)$ for all $x, y \in X$.*

Example 5. *1_X and 1_e are normal neutrosophic multi subgroup of X. If X is a commutative group, every neutrosophic multi subgroup of X is normal.*

Definition 15. *Let X be a group, $A \in NMG(X)$ and B is a neutrosophic multi subgroup of A over X. Then, B is called an a normal neutrosophic multi subgroup of A, denoted by $B\tilde{\triangledown}A$ if it is an abelian neutrosophic multi subset of A over X.*

Example 6. *Assume that $(Z_3, +)$ is a classiccal group. Define the neutrosophic multisets A and B on $(Z_3, +)$ as follows:*

$$\begin{aligned} A &= \{\langle 0; (0.6, 0.5, 0.4, 0.2), (0.1, 0.1, 0.2, 0.3), (0.2, 0.3, 0.4, 0.6)\rangle, \\ &\langle 1; (0.5, 0.4, 0.4, 0.3), (0.2, 0.1, 0.2, 0.3), (0.3, 0.4, 0.5, 0.6)\rangle, \\ &\langle 2; (0.9, 0.7, 0.6, 0.5), (0.1, 0.1, 0.2, 0.3), (0.2, 0.2, 0.3, 0.5)\rangle\} \end{aligned}$$

is a NM-group. If

$$\begin{aligned} B &= \{\langle 0; (0.6, 0.5, 0.4, 0.2), (0.1, 0.1, 0.2, 0.3), (0.2, 0.3, 0.4, 0.6)\rangle, \\ &\langle 1; (0.5, 0.4, 0.4, 0.3), (0.2, 0.1, 0.2, 0.3), (0.3, 0.3, 0.5, 0.4)\rangle\}, \end{aligned}$$

then B is a neutrosophic multi subgroup of A over $(Z_3, +)$ and denoted by $B\tilde{\leq}A$. Therefore, $B\tilde{\triangledown}A$.

Corollary 2. *Let $A \in NMG(X)$ and B be a neutrosophic multi subgroup of A over X. If X is an abelian group, then B is a normal neutrosophic multi subgroup of A over X.*

4. Applications of Neutrosophic Multi Groups

In this section, we give some applications to the group theory with respect to neutrosophic multi groups.

Definition 16. *Let A be a neutrosophic multiset on X and $\alpha \in [0, 1]$. Define the α-level sets of A as follows:*

$$\begin{aligned} (T^i{}_A)_\alpha &= \{x \in X : T^i{}_A(x) \geq \alpha\}, \\ (I^i{}_A)^\alpha &= \{x \in X : I^i{}_A(x) \leq \alpha\}, \\ (F^i{}_A)^\alpha &= \{x \in X : F^i{}_A(x) \leq \alpha\}. \end{aligned}$$

It is easy to verify that

(1) $If\, A \tilde{\subseteq} B$ *and* $\alpha \in [0,1]$, *then*
$(T^i{}_A)_\alpha \subseteq (T^i{}_B)_\alpha, (I^i{}_A)^\alpha \supseteq (I^i{}_B)^\alpha$ *and* $(F^i{}_A)^\alpha \supseteq (F^i{}_B)^\alpha$.

(2) $\alpha \leq \beta$ *implies* $(T^i{}_A)_\alpha \supseteq (T^i{}_A)_\beta, (I^i{}_A)^\alpha \subseteq (I^i{}_A)^\beta$ *and* $(F^i{}_A)^\alpha \subseteq (F^i{}_A)^\beta$.

Proposition 5. *A is a neutrosophic multi group of a classical group X if and only if for all* $\alpha \in [0,1]$, *α-level sets of A,* $(T^i{}_A)_\alpha, (I^i{}_A)_\alpha$ *and* $(F^i{}_A)^\alpha$ *are classical subgroups of X.*

Proof. Let A be a neutrosophic multi subgroup of X, $\alpha \in [0,1]$ and $x.y \in (T^i{}_A)_\alpha$ (similarly $x.y \in (I^i{}_A)_\alpha, (F^i{}_A)^\alpha$). By the assumption, $T^i{}_A(xy^{-1}) \geq T^i{}_A(x) \wedge T^i{}_A(y) \geq \alpha \wedge \alpha = \alpha$ (and similarly, $I^i{}_A(xy^{-1}) \leq \alpha$ and $F^i{}_A(xy^{-1}) \leq \alpha$). Hence, $xy^{-1} \in (T^i{}_A)_\alpha$ (and similarly $xy^{-1} \in (I^i{}_A)^\alpha, (F^i{}_A)^\alpha$) for each $\alpha \in [0,1]$. This means that $(T^i{}_A)_\alpha$ (and similarly $(I^i{}_A)^\alpha, (F^i{}_A)^\alpha$) is a classical subgroup of X for each $\alpha \in [0,1]$.

Conversely, let $(T^i{}_A)_\alpha$ be a classical subgroup of X, for each $\alpha \in [0,1]$. Let $x, y \in X, \alpha = T^i{}_A(x) \wedge T^i{}_A(y)$ and $\beta = T^i{}_A(x)$. Since $(T^i{}_A)_\alpha$ and $(T^i{}_A)_\beta$ are classical subgroups of X, $x.y \in (T^i{}_A)_\alpha$ and $x^{-1} \in (T^i{}_A)_\beta$. Thus, $T^i{}_A(xy^{-1}) \geq \alpha = T^i{}_A(x) \wedge T^i{}_A(y)$ and $T^i{}_A(x^{-1}) \geq \beta = T^i{}_A(x)$. Similarly, $I^i{}_A(xy^{-1}) \leq I^i{}_A(x) \vee I^i{}_A(y)$ and $F^i{}_A(xy^{-1}) \leq F^i{}_A(x) \vee F^i{}_A(y)$. □

Theorem 4. *Let* X_1, X_2 *be the classical groups and* $g : X_1 \to X_2$ *be a group homomorphism. If A is a neutrosophic multi subgroup of* X_1, *then the image of A,* $g(A)$ *is a neutrosophic multi subgroup of* X_2.

Proof. Let $A \in NMS(X_1)$ and $y_1, y_2 \in X_2$. If $g^{-1}(y_1) = \emptyset$ or $g^{-1}(y_2) = \emptyset$, then it is clear that $g(A) \in NMS(X_2)$. Let us assume that there exists $x_1, x_2 \in X_1$ such that $g(x_1) = y_1$ and $g(x_2) = y_2$. Since g is a group homomorphism,

$$
\begin{aligned}
g(T^i{}_A)(y_1y_2^{-1}) &= \vee_{y_1y_2^{-1}=g(x)} T^i{}_A(x) \geq T^i{}_A(x_1x_2^{-1}), \\
g(I^i{}_A)(y_1y_2^{-1}) &= \wedge_{y_1y_2^{-1}=g(x)} I^i{}_A(x) \leq I^i{}_A(x_1x_2^{-1}), \\
g(F^i{}_A)(y_1y_2^{-1}) &= \wedge_{y_1y_2^{-1}=g(x)} F^i{}_A(x) \leq F^i{}_A(x_1x_2^{-1}).
\end{aligned}
$$

By using the above inequalities, let us prove that $g(A)(y_1y_2^{-1}) \geq g(A)(y_1) \wedge g(A)(y_2)$:

$$
\begin{aligned}
g(A)(y_1y_2^{-1}) &= g(T^i{}_A)(y_1y_2^{-1}), g(I^i{}_A)(y_1y_2^{-1}), g(F^i{}_A)(y_1y_2^{-1}) \\
&= \vee_{y_1y_2^{-1}=g(x)} T^i{}_A(x), \wedge_{y_1y_2^{-1}=g(x)} I^i{}_A(x), \wedge_{y_1y_2^{-1}=g(x)} F^i{}_A(x) \\
&\geq (T^i{}_A(x_1x_2^{-1}), I^i{}_A(x_1x_2^{-1}), F^i{}_A(x_1x_2^{-1})) \\
&\geq (T^i{}_A(x_1) \wedge T^i{}_A(x_2), I^i{}_A(x_1) \vee I^i{}_A(x_2), F^i{}_A(x_1) \vee F^i{}_A(x_2)) \\
&= (T^i{}_A(x_1), I^i{}_A(x_1), F^i{}_A(x_1)) \wedge (T^i{}_A(x_2), I^i{}_A(x_2), F^i{}_A(x_2)).
\end{aligned}
$$

This is satisfied for each $x_1, x_2 \in X_1$ with $g(x_1) = y_1$ and $g(x_2) = y_2$, then it is obvious that

$$
\begin{aligned}
g(A)(y_1y_2^{-1}) &\geq (\vee_{y_1=g(x_1)} T^i{}_A(x_1), \wedge_{y_1=g(x_1)} I^i{}_A(x_1), \wedge_{y_1=g(x_1)} F^i{}_A(x_1)) \\
&\wedge (\vee_{y_2=g(x_2)} T^i{}_A(x_2), \wedge_{y_2=g(x_2)} I^i{}_A(x_2), \wedge_{y_2=g(x_2)} F^i{}_A(x_2)) \\
&= (g(T^i{}_A)(y_1), g(I^i{}_A)(y_1), g(F^i{}_A)(y_1)) \wedge (g(T^i{}_A)(y_2), g(I^i{}_A)(y_2), g(F^i{}_A)(y_2)) \\
&= g(A)(y_1) \wedge g(A)(y_2).
\end{aligned}
$$

Hence, the image of a neutrosophic multi subgroup is also a neutrosophic multi subgroup. □

Theorem 5. *Let* X_1, X_2 *be the classical groups and* $g : X_1 \to X_2$ *be a group homomorphism. If B is a neutrosophic multi subgroup of* X_2, *then the preimage* $g^{-1}(B)$ *is a neutrosophic multi subgroup of* X_1.

Proof. Let $B \in NMS(X_2)$ and $x_1, x_2 \in X_1$. Since g is a group homomorphism, the following inequality is obtained:

$$\begin{aligned} g^{-1}(B)(x_1x_2^{-1}) \quad & (T^i{}_B(g(x_1x_2^{-1})), I^i{}_B(g(x_1x_2^{-1})), F^i{}_B(g(x_1x_2^{-1}))) \\ &= (T^i{}_B(g(x_1)g(x_2)^{-1}), I^i{}_B(g(x_1)g(x_2)^{-1}), F^i{}_B(g(x_1)g(x_2)^{-1})) \\ &\geq (T^i{}_B(g(x_1)) \wedge T^i{}_B(g(x_2)), I^i{}_B(g(x_1)) \vee I^i{}_B(g(x_2)), F^i{}_B(g(x_1)) \vee F^i{}_B(g(x_2))) \\ &= (T^i{}_B(g(x_1)), I^i{}_B(g(x_1)), F^i{}_B(g(x_1))) \wedge (T^i{}_B(g(x_2)), I^i{}_B(g(x_2)), F^i{}_B(g(x_2))) \\ &= g^{-1}(B)(x_1) \wedge g^{-1}(B)(x_2). \end{aligned}$$

Therefore, $g^{-1}(B) \in NMS(X_1)$. □

Definition 17. *Let X be a classical group. $A \in NMG(X)$; then, the compound function of A and A is defined as*

$$A\check{\circ}A(z) = \{z, T^i{}_{A\check{\circ}A}(z), I^i{}_{A\check{\circ}A}(z), F^i{}_{A\check{\circ}A}(z) : \forall z \in X\}, \tag{10}$$

where $T^i{}_{A\check{\circ}A}(z) = (\vee_{xy=z} T^i{}_A(y) \wedge T^i{}_A(zy^{-1}))$, $I^i{}_{A\check{\circ}A}(z) = (\wedge_{xy=z} I^i{}_A(y) \vee I^i{}_A(zy^{-1}))$ and $F^i{}_{A\check{\circ}A}(z) = (\wedge_{xy=z} F^i{}_A(y) \vee F^i{}_A(zy^{-1}))$.

Theorem 6. *Let $A \in NMS(X)$. Then, $A \in NMG(X)$ iff $A\check{\circ}A \tilde{\subseteq} A$ and $A \tilde{\subseteq} A^{-1}$.*

Proof. Let $A \in NMS(X)$ and $x, y, z \in X$.

$$\begin{aligned} &\Rightarrow T^i{}_A(xy) \geq T^i{}_A(x) \wedge T^i{}_A(y) \\ &\Rightarrow T^i{}_A(z) \geq \vee\{T^i{}_A(x) \wedge T^i{}_A(y); xy = z\} \\ &= T^i{}_{A\check{\circ}A}(z) \end{aligned}$$

$$\begin{aligned} &\Rightarrow I^i{}_A(xy) \leq T^i{}_A(x) \vee I^i{}_A(y) \\ &\Rightarrow I^i{}_A(z) \leq \wedge\{I^i{}_A(x) \vee I^i{}_A(y); xy = z\} \\ &= I^i{}_{A\check{\circ}A}(z) \end{aligned}$$

$$\begin{aligned} &\Rightarrow F^i{}_A(xy) \leq F^i{}_A(x) \vee T^i{}_A(y) \\ &\Rightarrow F^i{}_A(z) \leq \wedge\{F^i{}_A(x) \vee F^i{}_A(y); xy = z\} \\ &= F^i{}_{A\check{\circ}A}(z) \end{aligned}$$

$\Rightarrow A\check{\circ}A \tilde{\subseteq} A$.

Now, by Proposition 2, we get the conditions. Conversely, suppose $A\check{\circ}A \tilde{\subseteq} A$ and $A \tilde{\subseteq} A^{-1}$

$$\begin{aligned} &\Rightarrow T^{i}{}^{-1}_A(x) \geq T^i{}_A(x) \text{ but } T^{i}{}^{-1}_A(x) = T^i{}_A(x^{-1}) \Rightarrow T^i{}_A(x^{-1}) \geq T^i{}_A(x) \\ &\Rightarrow I^{i}{}^{-1}_A(x) \leq T^i{}_A(x) \text{ but } I^{i}{}^{-1}_A(x) = I^i{}_A(x^{-1}) \Rightarrow I^i{}_A(x^{-1}) \leq I^i{}_A(x) \\ &\Rightarrow F^{i}{}^{-1}_A(x) \leq F^i{}_A(x) \text{ but } F^{i}{}^{-1}_A(x) = F^i{}_A(x^{-1}) \Rightarrow F^i{}_A(x^{-1}) \leq F^i{}_A(x) \end{aligned}$$

since $A \in NMS(X)$; then, to prove $A \in NMG(X)$, it enough to prove that $T^i{}_A(xy^{-1}) \geq T^i{}_A(x) \wedge T^i{}_A(y)$, $I^i{}_A(xy^{-1}) \leq I^i{}_A(x) \vee I^i{}_A(y)$ and $F^i{}_A(xy^{-1}) \leq F^i{}_A(x) \vee F^i{}_A(y)$ $\forall\, x, y \in X$.

Now,

$$\begin{aligned} T^i{}_A(xy^{-1}) \quad &\geq T^i{}_{A\check{\circ}A}(xy^{-1}) \\ &= \vee_{z \in X}\{T^i{}_A(z) \wedge T^i{}_A(z^{-1}xy^{-1})\} \\ &\geq \{T^i{}_A(x) \wedge T^i{}_A(y^{-1}); z = x\} \\ &\geq T^i{}_A(x) \wedge T^i{}_A(y) \end{aligned}$$

$$\begin{aligned} I^i{}_A(xy^{-1}) \quad &\leq I^i{}_{A\check{\circ}A}(xy^{-1}) \\ &= \wedge_{z \in X}\{I^i{}_A(z) \vee I^i{}_A(z^{-1}xy^{-1})\} \\ &\leq \{I^i{}_A(x) \vee I^i{}_A(y^{-1}); z = x\} \\ &\leq I^i{}_A(x) \vee I^i{}_A(y) \end{aligned}$$

$$\begin{aligned} F^i{}_A(xy^{-1}) &\leq F^i{}_{A\tilde{\circ}A}(xy^{-1}) \\ &= \wedge_{z\in X}\{F^i{}_A(z) \vee F^i{}_A(z^{-1}xy^{-1})\} \\ &\leq \{F^i{}_A(x) \vee F^i{}_A(y^{-1}); z = x\} \\ &\leq F^i{}_A(x) \vee F^i{}_A(y), \end{aligned}$$

hence the proof. □

Corollary 3. *Let $A \in NMS(X)$. Then, $A \in NMG(X)$ iff $A\tilde{\circ}A = A$ and $A\tilde{\subseteq}A^{-1}$.*

Proof. Let $A \in NMG(X)$. Then,

$$\begin{aligned} T^i{}_{A\tilde{\circ}A}(x) &= \vee\{T^i{}_A(y) \wedge T^i{}_A(z); y, z \in X \text{ and } yz = x\} \\ &\geq \{T^i{}_A(e) \wedge T^i{}_A(e^{-1}x)\} \\ &= T^i{}_A(x) \end{aligned}$$

$$\begin{aligned} I^i{}_{A\tilde{\circ}A}(x) &= \wedge\{I^i{}_A(y) \vee I^i{}_A(z); y, z \in X \text{ and } yz = x\} \\ &\leq \{I^i{}_A(e) \vee I^i{}_A(e^{-1}x)\} \\ &= I^i{}_A(x) \end{aligned}$$

$$\begin{aligned} F^i{}_{A\tilde{\circ}A}(x) &= \wedge\{F^i{}_A(y) \vee F^i{}_A(z); y, z \in X \text{ and } yz = x\} \\ &\leq \{F^i{}_A(e) \vee F^i{}_A(e^{-1}x)\} \\ &= F^i{}_A(x). \end{aligned}$$

Therefore, $A\tilde{\subseteq}A\tilde{\circ}A$.

Hence, by the above theorem, the proof is complete. □

Theorem 7. *Let X be a classical group and $A, B \in NMS(X)$. If $A, B \in NMG(X)$, then $A\tilde{\cap}B \in NMG(X)$.*

Proof. Let $x, y \in X$ be arbitrary:

$$\Rightarrow T^i{}_A(xy^{-1}) \geq T^i{}_A(x) \wedge T^i{}_A(y^{-1}), T^i{}_B(xy^{-1}) \geq T^i{}_B(x) \wedge T^i{}_B(y^{-1})$$

$$I^i{}_A(xy^{-1}) \leq I^i{}_A(x) \vee I^i{}_A(y^{-1}), I^i{}_B(xy^{-1}) \leq I^i{}_B(x) \vee T^i{}_B(y^{-1})$$

$$F^i{}_A(xy^{-1}) \leq F^i{}_A(x) \vee F^i{}_A(y^{-1}), F^i{}_B(xy^{-1}) \leq F^i{}_B(x) \vee F^i{}_B(y^{-1}).$$

Now,

$$\begin{aligned} T^i{}_{A\tilde{\cap}B}(xy^{-1}) &= T^i{}_{A\tilde{\cap}B}(x) \wedge T^i{}_{A\tilde{\cap}B}(y^{-1}) \text{ by definition intersection} \\ &\geq [T^i{}_A(x) \wedge T^i{}_A(y^{-1})] \wedge [T^i{}_B(x) \wedge T^i{}_B(y^{-1})] \\ &= [T^i{}_A(x) \wedge T^i{}_B(x)] \wedge [T^i{}_A(y^{-1}) \wedge T^i{}_B(y^{-1})] \text{ by commutative property of minimum} \\ &\geq [T^i{}_A(x) \wedge T^i{}_B(x)] \wedge [T^i{}_A(y) \wedge T^i{}_B(y)] \text{ since } A,\ B \in NMG(X) \\ &= T^i{}_{A\tilde{\cap}B}(x) \wedge T^i{}_{A\tilde{\cap}B}(y) \text{ by definition intersection} \\ &\Rightarrow T^i{}_{A\tilde{\cap}B}(xy^{-1}) \geq T^i{}_{A\tilde{\cap}B}(x) \wedge T^i{}_{A\tilde{\cap}B}(y) \ \ (1) \end{aligned}$$

$$\begin{aligned} I^i{}_{A\tilde{\cap}B}(xy^{-1}) &= I^i{}_{A\tilde{\cap}B}(x) \vee I^i{}_{A\tilde{\cap}B}(y^{-1}) \text{ by definition intersection} \\ &\leq [I^i{}_A(x) \vee I^i{}_A(y^{-1})] \vee [I^i{}_B(x) \vee I^i{}_B(y^{-1})] \\ &= [I^i{}_A(x) \vee I^i{}_B(x)] \vee [I^i{}_A(y^{-1}) \vee I^i{}_B(y^{-1})] \text{ by commutative property of maximum} \\ &\leq [I^i{}_A(x) \vee I^i{}_B(x)] \vee [I^i{}_A(y) \vee I^i{}_B(y)] \text{ since } A,\ B \in NMG(X) \\ &= I^i{}_{A\tilde{\cap}B}(x) \vee I^i{}_{A\tilde{\cap}B}(y) \text{ by definition intersection} \\ &\Rightarrow I^i{}_{A\tilde{\cap}B}(xy^{-1}) \leq I^i{}_{A\tilde{\cap}B}(x) \wedge I^i{}_{A\tilde{\cap}B}(y) \ \ (2) \end{aligned}$$

$$\begin{aligned} F^i{}_{A\tilde{\cap}B}(xy^{-1}) &= F^i{}_{A\tilde{\cap}B}(x) \vee F^i{}_{A\tilde{\cap}B}(y^{-1}) \text{ by definition intersection} \\ &\leq [F^i{}_A(x) \vee F^i{}_A(y^{-1})] \vee [F^i{}_B(x) \vee F^i{}_B(y^{-1})] \\ &= [F^i{}_A(x) \vee F^i{}_B(x)] \vee [F^i{}_A(y^{-1}) \vee F^i{}_B(y^{-1})] \text{ by commutative property of maximum} \\ &\leq [F^i{}_A(x) \vee F^i{}_B(x)] \vee [F^i{}_A(y) \vee F^i{}_B(y)] \text{ since } A,\ B \in NMG(X) \\ &= F^i{}_{A\tilde{\cap}B}(x) \vee F^i{}_{A\tilde{\cap}B}(y) \text{ by definition intersection} \\ &\Rightarrow F^i{}_{A\tilde{\cap}B}(xy^{-1}) \leq F^i{}_{A\tilde{\cap}B}(x) \wedge F^i{}_{A\tilde{\cap}B}(y) \quad (3) \end{aligned}$$

From (1), (2) and (3), $A\tilde{\cap}B \in NMG(X)$, hence the proof. □

Remark 1. *Let X be a classical group and* $\{A_i; i \in I\}$ *be neutrosophic multiset on X. If* $\{A_i; i \in I\}$ *is a family of* $NMG(X)$ *over X, then their intersection* $\tilde{\bigcap}_{i\in I} A_i$ *is also a* $NMG(X)$ *over X.*

Proposition 6. *Let* $A, B \in NMG(X)$*. Then,* $T^i{}_{A\tilde{\cup}B}(x) \leq T^i{}_{A\tilde{\cup}B}(x^{-1}), I^i{}_{A\tilde{\cup}B}(x) \geq I^i{}_{A\tilde{\cup}B}(x^{-1}),\ F^i{}_{A\tilde{\cup}B}(x) \geq F^i{}_{A\tilde{\cup}B}(x^{-1})$.

Proof. Let $x, y \in X$. Now,

$$\begin{aligned} T^i{}_{A\tilde{\cup}B}(x^{-1}) &= \vee\{T^i{}_A(x^{-1}), T^i{}_B(x^{-1})\} \\ &\geq \vee\{T^i{}_A(x), T^i{}_B(x)\} \textit{ since } A,\ B \in NMG(X) \\ &= T^i{}_{A\tilde{\cup}B}(x) \end{aligned}$$

$$\begin{aligned} I^i{}_{A\tilde{\cup}B}(x^{-1}) &= \wedge\{I^i{}_A(x^{-1}), I^i{}_B(x^{-1})\} \\ &\leq \wedge\{I^i{}_A(x), I^i{}_B(x)\} \textit{ since } A,\ B \in NMG(X) \\ &= I^i{}_{A\tilde{\cup}B}(x) \end{aligned}$$

$$\begin{aligned} F^i{}_{A\tilde{\cup}B}(x^{-1}) &= \wedge\{F^i{}_A(x^{-1}), F^i{}_B(x^{-1})\} \\ &\leq \wedge\{F^i{}_A(x), F^i{}_B(x)\} \textit{ since } A,\ B \in NMG(X) \\ &= F^i{}_{A\tilde{\cup}B}(x), \end{aligned}$$

hence the proof.

From this, it is clear that, if $A, B \in NMG(X)$, then $A\tilde{\cup}B \in NMG(X)$ iff $T^i{}_{A\tilde{\cup}B}(xy) \geq T^i{}_{A\tilde{\cup}B}(x) \wedge T^i{}_{A\tilde{\cup}B}(y), I^i{}_{A\tilde{\cup}B}(xy) \leq T^i{}_{A\tilde{\cup}B}(x) \vee I^i{}_{A\tilde{\cup}B}(y), F^i{}_{A\tilde{\cup}B}(xy) \leq F^i{}_{A\tilde{\cup}B}(x) \vee F^i{}_{A\tilde{\cup}B}(y)$. □

Corollary 4. *Let* $A, B \in NMG(X)$*. Then,* $A\tilde{\cup}B$ *need not be an element of* $NMG(X)$*.*

Example 7. *Assume that* $X = \{1, -1, i, -i\}$ *is a classical group. Then,*

$$\begin{aligned} A &= \{\langle 1; (0.5, 0.3, 0.2, 0.1), (0.1, 0.1, 0.2, 0.3), (0.2, 0.3, 0.4, 0.5)\rangle, \\ &\langle -1; (0.7, 0.6, 0.4, 0.3), (0.1, 0.2, 0.2, 0.4), (0.2, 0.5, 0.4, 0.3)\rangle\}, \\ &\langle i; (0.6, 0.4, 0.3, 0.2), (0.1, 0.3, 0.3, 0.4), (0.1, 0.2, 0.4, 0.6)\rangle, \\ &\langle -i; (0.6, 0.4, 0.3, 0.2), (0.1, 0.3, 0.3, 0.4), (0.1, 0.2, 0.4, 0.6)\rangle\}, \\ B &= \{\langle 1; (0.5, 0.6, 0.6, 0.4), (0.1, 0.2, 0.2, 0.3), (0.2, 0.4, 0.4, 0.5)\rangle, \\ &\langle -1; (0.7, 0.6, 0.4, 0.3), (0.2, 0.1, 0.2, 0.3), (0.3, 0.4, 0.5, 0.3)\rangle\} \\ &\textit{are } NM - \textit{groups}. \\ A \cup B &= \{\langle 1; (0.5, 0.6, 0.6, 0.4), (0.1, 0.1, 0.2, 0.3), (0.2, 0.3, 0.4, 0.5)\rangle, \\ &\langle -1; (0.7, 0.6, 0.4, 0.3), (0.1, 0.2, 0.2, 0.3), (0.2, 0.4, 0.4, 0.3)\rangle, \\ &\langle i; (0.6, 0.4, 0.3, 0.2), (0.1, 0.3, 0.3, 0.4), (0.1, 0.2, 0.4, 0.6)\rangle, \\ &\langle -i; (0.6, 0.4, 0.3, 0.2), (0.1, 0.3, 0.3, 0.4), (0.1, 0.2, 0.4, 0.6)\rangle\}. \end{aligned}$$

However, $T^i{}_{A\tilde{\cup}B}(1) \geq T^i{}_{A\tilde{\cup}B}(i) \wedge T^i{}_{A\tilde{\cup}B}(-i)$ *as* $i.(-i) = 1$*. Then,* $A\tilde{\cup}B \notin NMG(X)$*.*

Proposition 7. *If* $A \in NMG(X)$ *and* X_1 *is a subgroup of X, then* $A_{|X_1}$ *(i.e., A restricted to* X_1*)* $\in NM-group(X_1)$ *and is a neutrosophic multi subgroup of A.*

Proof. Let $x, y \in X_1$. Then, $xy^{-1} \in X_1$. Now,

$$T^i{}_{A_{|X_1}}(xy^{-1}) = T^i{}_A(xy^{-1}) \geq T^i{}_A(x) \wedge T^i{}_A(y) = T^i{}_{A_{|X_1}}(x) \wedge T^i{}_{A_{|X_1}}(y),$$

$$I^i{}_{A_{|X_1}}(xy^{-1}) = I^i{}_A(xy^{-1}) \leq I^i{}_A(x) \vee I^i{}_A(y) = I^i{}_{A_{|X_1}}(x) \vee I^i{}_{A_{|X_1}}(y),$$

$$F^i{}_{A_{|X_1}}(xy^{-1}) = F^i{}_A(xy^{-1}) \leq F^i{}_A(x) \vee F^i{}_A(y) = F^i{}_{A_{|X_1}}(x) \vee F^i{}_{A_{|X_1}}(y).$$

The second part is trivial. □

Definition 18. *Let $A \in NMG(X)$ and $B \in NMG(Y)$ be two neutrosophic multi groups over the groups X and Y, respectively. Then, the Cartesian product of A and B is defined as $(A \tilde{\times} B)(x, y) = A(x) \tilde{\times} B(y)$ where*

$$A \tilde{\times} B = \{(x, y), T^i{}_{A \tilde{\times} B}(x, y), I^i{}_{A \tilde{\times} B}(x, y), F^i{}_{A \tilde{\times} B}(x, y) : (x, y) \in X \times Y\}, \tag{11}$$

where $T^i{}_{A \tilde{\times} B}(x, y) = T^i{}_A(x) \vee T^i{}_B(y)$, $I^i{}_{A \tilde{\times} B}(x, y) = I^i{}_A(x) \wedge I^i{}_B(y)$, $F^i{}_{A \tilde{\times} B}(x, y) = F^i{}_A(x) \wedge F^i{}_B(y)$.

Example 8. *Assume that $(Z_2, +)$ and $(Z_3, +)$ are classiccal groups. Define the neutrosophic multi group A on $(Z_2, +)$ and B on $(Z_3, +)$ as follows:*

$$\begin{aligned} A &= \{\langle 0; (0.6, 0.5, 0.4, 0.2), (0.2, 0.2, 0.3, 0.4), (0.2, 0.3, 0.4, 0.6)\rangle, \\ &\quad \langle 1; (0.6, 0.5, 0.4, 0.3), (0.2, 0.1, 0.2, 0.3), (0.1, 0.2, 0.3, 0.4)\rangle\} \end{aligned}$$

$$\begin{aligned} B &= \{\langle 0; (0.7, 0.6, 0.5, 0.4), (0.2, 0.2, 0.3, 0.4), (0.3, 0.4, 0.5, 0.5)\rangle, \\ &\quad \langle 1; (0.7, 0.6, 0.5, 0.3), (0.3, 0.4, 0.3, 0.4), (0.3, 0.2, 0.5, 0.5)\rangle, \\ &\quad \langle 2; (0.8, 0.7, 0.5, 0.4), (0.1, 0.3, 0.2, 0.3), (0.2, 0.1, 0.3, 0.5)\rangle\}. \end{aligned}$$

$$\begin{aligned} A \tilde{\times} B &= \{\langle (0, 0); (0.6, 0.5, 0.4, 0.2), (0.2, 0.2, 0.3, 0.4), (0.2, 0.3, 0.4, 0.6)\rangle, \\ &\quad \langle (0, 1); (0.6, 0.5, 0.4, 0.2), (0.3, 0.4, 0.3, 0.4), (0.3, 0.3, 0.5, 0.6)\rangle, \\ &\quad \langle (0, 2); (0.6, 0.5, 0.4, 0.2), (0.2, 0.3, 0.3, 0.4), (0.2, 0.3, 0.4, 0.6)\rangle, \\ &\quad \langle (1, 0); (0.6, 0.5, 0.4, 0.3), (0.2, 0.2, 0.3, 0.4), (0.3, 0.4, 0.5, 0.5)\rangle, \\ &\quad \langle (1, 1); (0.6, 0.5, 0.4, 0.3), (0.3, 0.4, 0.3, 0.4), (0.3, 0.2, 0.5, 0.5)\rangle, \\ &\quad \langle (1, 2); (0.6, 0.5, 0.4, 0.3), (0.2, 0.3, 0.2, 0.3), (0.2, 0.2, 0.3, 0.5)\rangle\}. \end{aligned}$$

Then, $A \tilde{\times} B$ is a neutrosophic multi group.

Theorem 8. *Let $A, B \in NMG(X)$. The cartesian product of A and B is denoted by $A \tilde{\times} B \in NMG(X)$.*

Proof. From the Theorem 1, it is clear that a $NMG(X)$ is a neutrosophic multi group:

$$\begin{aligned} T^i{}_{A \tilde{\times} B}((x_1, y_1), (x_2, y_2)^{-1}) &= T^i{}_{A \tilde{\times} B}(x_1 x_2^{-1}, y_1 y_2^{-1}) \\ &= T^i{}_A(x_1 x_2^{-1}) \wedge T^i{}_B(y_1 y_2^{-1}) \\ &\geq (T^i{}_A(x_1) \wedge T^i{}_A(x_2)) \wedge (T^i{}_B(y_1) \wedge T^i{}_B(y_2)) \\ &= (T^i{}_A(x_1) \wedge T^i{}_B(y_1)) \wedge (T^i{}_A(x_2) \wedge T^i{}_B(y_2)) \\ &= T^i{}_{A \tilde{\times} B}(x_1, y_1) \wedge T^i{}_{A \tilde{\times} B}(x_2, y_2) \end{aligned}$$

$$\begin{aligned} I^i{}_{A \tilde{\times} B}((x_1, y_1), (x_2, y_2)^{-1}) &= I^i{}_{A \tilde{\times} B}(x_1 x_2^{-1}, y_1 y_2^{-1}) \\ &= I^i{}_A(x_1 x_2^{-1}) \vee I^i{}_B(y_1 y_2^{-1}) \\ &\leq (I^i{}_A(x_1) \vee I^i{}_A(x_2)) \vee (I^i{}_B(y_1) \vee I^i{}_B(y_2)) \\ &= (I^i{}_A(x_1) \vee I^i{}_B(y_1)) \vee (I^i{}_A(x_2) \vee I^i{}_B(y_2)) \\ &= I^i{}_{A \tilde{\times} B}(x_1, y_1) \vee I^i{}_{A \tilde{\times} B}(x_2, y_2) \end{aligned}$$

and

$$\begin{aligned} F^i{}_{A\tilde{\times}B}((x_1,y_1),(x_2,y_2)^{-1}) &= F^i{}_{A\tilde{\times}B}(x_1x_2^{-1},y_1y_2^{-1}) \\ &= F^i{}_A(x_1x_2^{-1}) \vee F^i{}_B(y_1y_2^{-1}) \\ &\leq (F^i{}_A(x_1) \vee F^i{}_A(x_2)) \vee (F^i{}_B(y_1) \vee F^i{}_B(y_2)) \\ &= (F^i{}_A(x_1) \vee F^i{}_B(y_1)) \vee (F^i{}_A(x_2) \vee F^i{}_B(y_2)) \\ &= F^i{}_{A\tilde{\times}B}(x_1,y_1) \vee F^i{}_{A\tilde{\times}B}(x_2,y_2) \end{aligned}$$

for all $x, y \in X$ and $i = 1, 2, ..., P$—hence the proof. □

5. Conclusions

The concept of a group is of fundamental importance in the study of algebra. In this paper, the algebraic structure of neutrosophic multiset is introduced as a neutrosophic multigroup. The neutrosophic multigroup is a generalized case of intuitionistic fuzzy multigroup and fuzzy multigroup. The various basic operations, definitions and theorems related to neutrosophic multigroup have been discussed. The foundations which we made through this paper can be used to get an insight into the higher order structures of group theory.

Author Contributions: All authors contributed equally.

Funding: This research received no external funding.

Conflicts of Interest: The authors declare no conflict of interest.

Abbreviations

The following abbreviations are used in this manuscript:

NMG	Neutrosophic Multigroup
NMS	Neutrosophic Multiset
IFMS	Intuitionistic Fuzzy Multiset

References

1. Zadeh, L.A. Fuzzy Sets. *Inf. Control* **1965**, *8*, 338–353. [CrossRef]
2. Sebastian, S.; Ramakrishnan, T.T. Multi-Fuzzy Sets. *Int. Math. Forum* **2010**, *5*, 2471–2476.
3. Atanassov, K. Intuitionistic fuzzy sets. *Fuzzy Set Syst.* **1986**, *20*, 87–96. [CrossRef]
4. Shinoj, T.K.; John, S.S. Intuitionistic fuzzy multisets and its application in medical diagnosis. *World Acad. Sci. Eng. Technol.* **2012**, *6*, 1418–1421.
5. Abdullah, S.; Naeem, M. A new type of interval valued fuzzy normal subgroups of groups. *New Trends Math. Sci.* **2015**, *3*, 62–77.
6. Mordeson, J.N.; Bhutani, K.R.; Rosenfeld, A. *Fuzzy Group Theory*; Springer: Berlin/Heidelberg, Germany, 2005.
7. Liu, Y.L. Quotient groups induced by fuzzy subgroups. *Quasigroups Related Syst.* **2004**, *11*, 71–78.
8. Baby, A.; Shinoj, T.K.; John, S.J. On Abelian Fuzzy Multi Groups and Orders of Fuzzy Multi Groups. *J. New Theory* **2015**, *5*, 80–93.
9. Muthuraj, R.; Balamurugan, S. Multi-Fuzzy Group and its Level Subgroups. *Gen* **2013**, *17*, 74–81.
10. Tella, Y. On Algebraic Properties of Fuzzy Membership Sequenced Multisets. *Br. J. Math. Comput. Sci.* **2015**, *6*, 146–164. [CrossRef]
11. Fathi, M.; Salleh, A.R. Intuitionistic fuzzy groups. *Asian J. Algebra* **2009**, *2*, 1–10. [CrossRef]
12. Li, X.P.; Wang, G.J. (l, a)-Homomorphisms of Intuitionistic Fuzzy Groups. *Hacettepe J. Math. Stat.* **2011**, *40*, 663–672.
13. Palaniappan, N.; Naganathan, S.; Arjunan, K. A study on Intuitionistic L-fuzzy Subgroups. *Appl. Math. Sci.* **2009**, *3*, 2619–2624.
14. Sharma, P.K. Homomorphism of Intuitionistic fuzzy groups. *Int. Math. Forum* **2011**, *6*, 3169–3178.
15. Sharma, P.K. On the direct product of Intuitionistic fuzzy subgroups. *Int. Math. Forum* **2012**, *7*, 523–530.
16. Sharma, P.K. On intuitionistic fuzzy abelian subgroups. *Adv. Fuzzy Sets Syst.* **2012**, *12*, 1–16.

17. Xu, C.Y. Homomorphism of Intuitionistic Fuzzy Groups. In Proceedings of the 2007 International Conference on Machine Learning and Cybernetics, Hong Kong, China, 19–22 August 2007.
18. Li, X. Homomorphism and isomorphism of the intuitionistic fuzzy normal subgroups. *BUSEFAL* **2001**, *85*, 162–175.
19. Yuan, X.H.; Li, H.X.; Lee, E.S. On the definition of the intuitionistic fuzzy subgroups. *Comput. Math. Appl.* **2010**, *59*, 3117–3129. [CrossRef]
20. Shinoj, T.K.; John, S.J. Intuitionistic fuzzy multigroups. *Ann. Pure Appl. Math.* **2015**, *9*, 131–143.
21. Wang, H.; Smarandache, F.Y.; Zhang, Q.; Sunderraman, R. Single valued neutrosophic sets. *Multispace Multistruct.* **2010**, *4*, 410–413.
22. Smarandache, F. *A Unifying Field in Logics*; Infinite Study: Conshohocken, PA, USA, 1998.
23. Agboola, A.; Akwu, A.D.; Oyebo, Y.T. Neutrosophic groups and subgroups. *Math. Comb.* **2012**, *3*, 1–9.
24. Ali, M.; Smarandache, F. Neutrosophic Soluble Groups, Neutrosophic Nilpotent Groups and Their Properties. In Proceedings of the Annual Symposium of the Institute of Solid Mechanics (SISOM 2015), Bucharest, Romania, 21–22 May 2015.
25. Cetkin, V.; Aygün, H. An approach to neutrosophic subgroup and its fundamental properties. *J. Intell. Fuzzy Syst.* **2015**, *29.5*, 1941–1947. [CrossRef]
26. Shabir, M.; Ali, M.; Naz, M.; Smarandache, F. Soft neutrosophic group. *Neutrosophic Sets Syst.* **2013**, *1*, 13–25.
27. Deli, I.; Broumi, S.; Smarandache, F. On neutrosophic multisets and its application in medical diagnosis. *J. New Theory* **2015**, *6*, 88–98.
28. Ye, S.; Ye, J. Dice Similarity Measure between Single Valued Neutrosophic Multisets and Its Application in Medical Diagnosis. *Neutrosophic Sets Syst.* **2014**, *6*, 49–54.
29. Ye, S.; Fu, J.; Ye, J. Medical Diagnosis Using Distance-Based Similarity Measures of Single Valued Neutrosophic Multisets. *Neutrosophic Sets Syst.* **2015**, *7*, 47–52.
30. Rosenfeld, A. Fuzzy groups. *J. Math. Anal. Appl.* **1971**, *35*, 512–517. [CrossRef]
31. Şahin, M.; Uluçay, V.; Olgun, N.; Kilicman, A. On neutrosophic soft lattices. *Afr. Matematika* **2017**, 28, 379–388.
32. Şahin, M.; Olgun, N.; Kargın, A.; Uluçay, V. Isomorphism theorems for soft G-modules. *Afr. Matematika* **2018**, *29*, 1–8. [CrossRef]
33. Uluçay, V.; Kiliç, A.; Yildiz, I.; Şahin, M. A new approach for multi-attribute decision-making problems in bipolar neutrosophic sets. *Neutrosophic Sets Syst.* **2018**, *23*, 142–159.
34. Şahin, M.; Deli, I.; Ulucay, V. Extension principle based on neutrosophic multi-sets and algebraic operations. *J. Math. Ext.* **2018**, *12*, 69–90.
35. Hu, Q.; Zhang, X. New Similarity Measures of Single-Valued Neutrosophic Multisets Based on the Decomposition Theorem and Its Application in Medical Diagnosis. *Symmetry* **2018** , *10*, 466. [CrossRef]
36. Smarandache, F.; Şahin, M.; Kargın, A. Neutrosophic Triplet G-Module. *Mathematics* **2018**, *6*, 53. [CrossRef]
37. Şahin, M.; Olgun, N.; Uluçay, V. Normed Quotient Rings. *New Trends Math. Sci.* **2018**, *6*, 52–58.
38. Uluçay, V.; Şahin, M.; Olgun, N. Soft Normed Rings. *SpringerPlus* **2016**, *5*, 1950. [CrossRef] [PubMed]
39. Sharma, P.K. Modal Operator F, ß in Intuitionistic Fuzzy Groups. *Ann. Pure Appl. Math.* **2014**, *7*, 19–28.

mathematics

MDPI

Article

Multi-Criteria Decision-Making Method Using Heronian Mean Operators under a Bipolar Neutrosophic Environment

Changxing Fan [1,*], Jun Ye [2], Sheng Feng [1], En Fan [1] and Keli Hu [1]

1 Department of Computer Science, Shaoxing University, 508 Huancheng West Road, Shaoxing 312000, China; fengsheng_13@aliyun.com (S.F.); efan@usx.edu.cn (E.F.); ancimoon@gmail.com (K.H.)
2 Department of Electrical and Information Engineering, Shaoxing University, 508 Huancheng West Road, Shaoxing 312000, China; yehjun@aliyun.com
* Correspondence: fcxjszj@usx.edu.cn; Tel.: +86-575-8820-2669

Received: 23 November 2018; Accepted: 11 January 2019; Published: 17 January 2019

Abstract: In real applications, most decisions are fuzzy decisions, and the decision results mainly depend on the choice of aggregation operators. In order to aggregate information more scientifically and reasonably, the Heronian mean operator was studied in this paper. Considering the advantages and limitations of the Heronian mean (HM) operator, four Heronian mean operators for bipolar neutrosophic number (BNN) are proposed: the BNN generalized weighted HM (BNNGWHM) operator, the BNN improved generalized weighted HM (BNNIGWHM) operator, the BNN generalized weighted geometry HM (BNNGWGHM) operator, and the BNN improved generalized weighted geometry HM (BNNIGWGHM) operator. Then, their propositions were examined. Furthermore, two multi-criteria decision methods based on the proposed BNNIGWHM and BNNIGWGHM operator are introduced under a BNN environment. Lastly, the effectiveness of the new methods was verified with an example.

Keywords: bipolar neutrosophic number (BNN); BNN improved generalized weighted HM (BNNIGWHM) operator; BNN improved generalized weighted geometry HM (BNNIGWGHM) operator; decision-making

1. Introduction

In the real world, there is lots of uncertain information in science, technology, daily life, and so on. Particularly under the background of big data, the uncertainty of information is more complex and diverse. Now, how to make use of mathematical tools to deal with the uncertain information is an urgent problem for researchers. In order to describe uncertain information, Zadeh [1] put forward the concept of fuzzy sets. Considering the complexities and changes of uncertainty in the real environment, there was a certain limit on fuzzy sets to describe complex uncertainty; then, some extension theories [2–4] were put forward. Afterword, the neutrosophic set (NS) containing three neutrosophic components and the single-valued neutrosophic set were proposed by Smarandache [5], and the single-valued neutrosophic set was also mentioned by Wang and Smarandache [6]. Wang and Zhang [7] put forward an interval neutrosophic set (INS) theory. Furthermore, an n-value neutrosophic set [8] theory was proposed by Smarandache. The fuzzy set theory changed the binary view of people, but ignored the bipolarity of things. Under the background of big data, the confliction between data became more and more obvious. Traditional fuzzy sets could not do well in analyzing and handing uncertain information with incompatible bipolarity; this phenomenon was identified in 1994. For the first time, Zhang [9] introduced incompatible bipolarity into the fuzzy set theory, and put forward the bipolar fuzzy set (BFS). The founder of the fuzzy set theory, Zadeh, also affirmed that the bipolar fuzzy set theory was a breakthrough in traditional fuzzy set theory [10]. Then, Zemankova et al. [11] discussed a more generalized multipolar fuzzy problem, and pointed out that the multipolar fuzzy problem can be divided into multiple bipolar fuzzy problems. Chen et al. [12] studied m-polar fuzzy sets. Bosc and Pivert [13] introduced a study on fuzzy bipolar relational algebra. Manemaran and Chellappa [14] gave some applications of bipolar fuzzy groups. Zhou and Li [15] introduced some applications of bipolar fuzzy sets in semiring. Deli et al. [16] put forward a bipolar neutrosophic set (BNS), which can describe bipolar information. Later, some studies about BNS were put forward [17–20]. In this paper, we propose four Heronian mean operators for bipolar neutrosophic number (BNN). Compared with the literature [17–19], the HM operator can embody the interaction between attributes to avoid unreasonable situations in information aggregation. Compared with the literature [20], the Bonferroni mean (BM) aggregation operator not only neglects the relationship between each attribute and itself, but also considers the relationship between each attribute and other attributes repeatedly. However, the BM aggregation operator has large computational complexity, but the Heronian mean (HM) can overcome these two shortcomings.

The remaining sections are organized as follows: some related concepts are reviewed in Section 2. The four operators are defined and their properties are investigated in Section 3; these four operators are BNN generalized weighted HM (BNNGWHM), BNN improved generalized weighted HM (BNNIGWHM), BNN generalized weighted geometry HM (BNNGWGHM), and BNN improved generalized weighted geometry HM (BNNIGWGHM). Multi-criteria decision-making (MCDM) methods based on the BNNIGWHM and BNNIGWGHM operators are established in Section 4. A numerical example is provided and the effects of parameters p and q are analyzed in Section 5. The conclusion of this paper is given in Section 6.

2. Some Basic Concepts

2.1. BNN and Its Operational Laws

Definition 1 [16]. *Let $U = \{u_1, u_2, \ldots, u_n\}$ be a universe; a BNS Γ in U is defined as follows:*

$$\Gamma = \{\langle u, \alpha_\Gamma^+(u), \beta_\Gamma^+(u), \gamma_\Gamma^+(u), \alpha_\Gamma^-(u), \beta_\Gamma^-(u), \gamma_\Gamma^-(u)\rangle | u \in U\},$$

in which $\alpha_\Gamma^+(u) : U \to [0,1]$ means a truth-membership function, $\gamma_\Gamma^+(u) : U \to [0,1]$ means a falsity-membership function and $\beta_\Gamma^+(u) : U \to [0,1]$ means an indeterminacy-membership function, corresponding to a BNS Γ and $\alpha_\Gamma^-(u), \gamma_\Gamma^-(u), \beta_\Gamma^-(u) : U \to [-1,0]$ mean, respectively, the truth membership, false membership, and indeterminate membership to some implicit counter-property corresponding to a BNS Γ.

Definition 2 [16]. *Let U be a universe, and Γ_1 and Γ_2 be two BNSs.*

$$\Gamma_1 = \{\langle u, \alpha_{\Gamma_1}^+(u), \beta_{\Gamma_1}^+(u), \gamma_{\Gamma_1}^+(u), \alpha_{\Gamma_1}^-(u), \beta_{\Gamma_1}^-(u), \gamma_{\Gamma_1}^-(u)\rangle | u \in U\},$$

$$\Gamma_2 = \{\langle u, \alpha_{\Gamma_2}^+(u), \beta_{\Gamma_2}^+(u), \gamma_{\Gamma_2}^+(u), \alpha_{\Gamma_2}^-(u), \beta_{\Gamma_2}^-(u), \gamma_{\Gamma_2}^-(u)\rangle | u \in U\}.$$

Then, the operations of Γ_1 and Γ_2 are defined as follows [16]:

① $\Gamma_1 \subseteq \Gamma_2$, *if and only if $\alpha_{\Gamma_1}^+(u) \le \alpha_{\Gamma_2}^+(u)$, $\beta_{\Gamma_1}^+(u) \ge \beta_{\Gamma_2}^+(u), \gamma_{\Gamma_1}^+(u) \ge \gamma_{\Gamma_2}^+(u)$, and $\alpha_{\Gamma_1}^-(u) \ge \alpha_{\Gamma_2}^-(u), \beta_{\Gamma_1}^-(u) \le \beta_{\Gamma_2}^-(u), \gamma_{\Gamma_1}^-(u) \le \gamma_{\Gamma_2}^-(u)$;*

② $\Gamma_1 = \Gamma_2$, *if and only if $\alpha_{\Gamma_1}^+(u) = \alpha_{\Gamma_2}^+(u)$, $\beta_{\Gamma_1}^+(u) = \beta_{\Gamma_2}^+(u), \gamma_{\Gamma_1}^+(u) = \gamma_{\Gamma_2}^+(u)$, and $\alpha_{\Gamma_1}^-(u) = \alpha_{\Gamma_2}^-(u), \beta_{\Gamma_1}^-(u) = \beta_{\Gamma_2}^-(u), \gamma_{\Gamma_1}^-(u) = \gamma_{\Gamma_2}^-(u)$;*

$$③\ \Gamma_1 \cup \Gamma_2 = \{\left\langle \begin{array}{c} u, \max\left(\alpha_{\Gamma_1}^+(u), \alpha_{\Gamma_2}^+(u)\right), \frac{\beta_{\Gamma_1}^+(u)+\beta_{\Gamma_2}^+(u)}{2}, \min\left(\gamma_{\Gamma_1}^+(u), \gamma_{\Gamma_2}^+(u)\right), \\ \min\left(\alpha_{\Gamma_1}^-(u), \alpha_{\Gamma_2}^-(u)\right), \frac{\beta_{\Gamma_1}^-(u)+\beta_{\Gamma_2}^-(u)}{2}, max\left(\gamma_{\Gamma_1}^-(u), \gamma_{\Gamma_2}^-(u)\right) \end{array} \right\rangle | u \in U\};$$

$$④\ \Gamma_1 \cap \Gamma_2 = \{\langle \begin{matrix} u, \min\left(\alpha^+_{\Gamma_1}(u), \alpha^+_{\Gamma_2}(u)\right), \frac{\beta^+_{\Gamma_1}(u)+\beta^+_{\Gamma_2}(u)}{2}, \max\left(\gamma^+_{\Gamma_1}(u), \gamma^+_{\Gamma_2}(u)\right), \\ \max\left(\alpha^-_{\Gamma_1}(u), \alpha^-_{\Gamma_2}(u)\right), \frac{\beta^-_{\Gamma_1}(u)+\beta^-_{\Gamma_2}(u)}{2}, \min\left(\gamma^-_{\Gamma_1}(u), \gamma^-_{\Gamma_2}(u)\right) \end{matrix} \rangle | u \in U\};$$

For convenience, we denote a bipolar neutrosophic number (BNN) by $\tau = \langle \alpha^+_\tau, \beta^+_\tau, \gamma^+_\tau, \alpha^-_\tau, \beta^-_\tau, \gamma^-_\tau \rangle$.

Definition 3 [16]. *Let* τ_1 *and* τ_2 *be two BNNs,* $\tau_1 = \langle \alpha^+_{\tau_1}, \beta^+_{\tau_1}, \gamma^+_{\tau_1}, \alpha^-_{\tau_1}, \beta^-_{\tau_1}, \gamma^-_{\tau_1} \rangle$ *and* $\tau_2\langle= \alpha^+_{\tau_2}, \beta^+_{\tau_2}, \gamma^+_{\tau_2}, \alpha^-_{\tau_2}, \beta^-_{\tau_2}, \gamma^-_{\tau_2} \rangle$*, and* $\delta > 0$*; then, the operations for BNNs are defined as follows [16]:*

$$\tau_1 \oplus \tau_2 = \langle \alpha^+_{\tau_1} + \alpha^+_{\tau_2} - \alpha^+_{\tau_1}\alpha^+_{\tau_2}, \beta^+_{\tau_1}\beta^+_{\tau_2}, \gamma^+_{\tau_1}\gamma^+_{\tau_2}, -\alpha^-_{\tau_1}\alpha^-_{\tau_2}, -(-\beta^-_{\tau_1} - \beta^-_{\tau_2} - \beta^-_{\tau_1}\beta^-_{\tau_2}), -(-\gamma^-_{\tau_1} - \gamma^-_{\tau_2} - \gamma^-_{\tau_1}\gamma^-_{\tau_2}) \rangle; \quad (1)$$

$$\tau_1 \otimes \tau_2 = \langle \alpha^+_{\tau_1}\alpha^+_{\tau_2}, \beta^+_{\tau_1} + \beta^+_{\tau_2} - \beta^+_{\tau_1}\beta^+_{\tau_2}, \gamma^+_{\tau_1} + \gamma^+_{\tau_2} - \gamma^+_{\tau_1}\gamma^+_{\tau_2}, -(-\alpha^-_{\tau_1} - \alpha^-_{\tau_2} - \alpha^-_{\tau_1}\alpha^-_{\tau_2}), -\beta^-_{\tau_1}\beta^-_{\tau_2}, -\gamma^-_{\tau_1}\gamma^-_{\tau_2} \rangle; \quad (2)$$

$$\delta\tau_1 = \langle 1 - (1 - \alpha^+_{\tau_1})^\delta, (\beta^+_{\tau_1})^\delta, (\gamma^+_{\tau_1})^\delta, -(-\alpha^-_{\tau_1})^\delta, -\left(1 - (1 - (-\beta^-_{\tau_1}))^\delta\right), -\left(1 - (1 - (-\gamma^-_{\tau_1}))^\delta\right) \rangle; \quad (3)$$

$$\tau_1{}^\delta = \langle (\alpha^+_{\tau_1})^\delta, 1 - (1 - \beta^+_{\tau_1})^\delta, 1 - (1 - \gamma^+_{\tau_1})^\delta, -\left(1 - (1 - (-\alpha^-_{\tau_1}))^\delta\right), -(-\beta^-_{\tau_1})^\delta, -(-\gamma^-_{\tau_1})^\delta \rangle. \quad (4)$$

Definition 4 [16]. *Let* $\tau = \langle \alpha^+_\tau, \beta^+_\tau, \gamma^+_\tau, \alpha^-_\tau, \beta^-_\tau, \gamma^-_\tau \rangle$ *be a BNN; then, we define* $s(\tau)$*,* $a(\tau)$*, and* $c(\tau)$ *as the score, accuracy, and certain functions, respectively; they are as follows:*

$$s(\tau) = \frac{1}{6}(\alpha^+_\tau + 1 - \beta^+_\tau + 1 - \gamma^+_\tau + 1 + \alpha^-_\tau - \beta^-_\tau - \gamma^-_\tau); \quad (5)$$

$$a(\tau) = \alpha^+_\tau - \gamma^+_\tau + \alpha^-_\tau - \gamma^-_\tau; \quad (6)$$

$$c(\tau) = \alpha^+_\tau - \gamma^+_\tau. \quad (7)$$

Definition 5 [16]. *Let* τ_1 *and* τ_2 *be two BNNs,* $\tau_1\langle= \alpha^+_{\tau_1}, \beta^+_{\tau_1}, \gamma^+_{\tau_1}, \alpha^-_{\tau_1}, \beta^-_{\tau_1}, \gamma^-_{\tau_1} \rangle$ *and* $\tau_2 = \langle \alpha^+_{\tau_2}, \beta^+_{\tau_2}, \gamma^+_{\tau_2}, \alpha^-_{\tau_2}, \beta^-_{\tau_2}, \gamma^-_{\tau_2} \rangle$*; then, we can get Figure 1.*

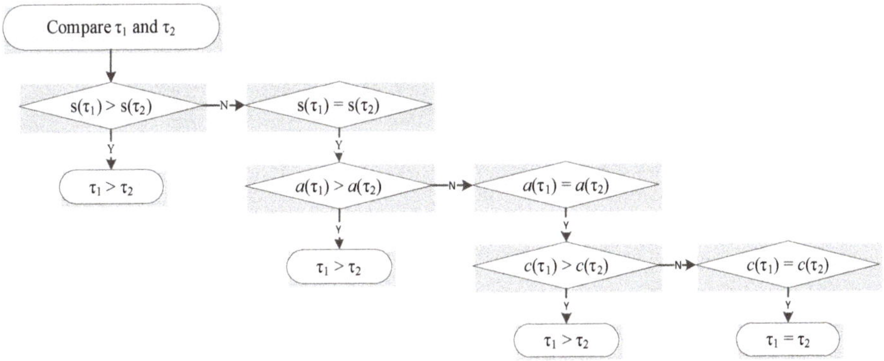

Figure 1. The relationship between τ_1 and τ_2.

2.2. Generalized Weighted HM (GWHM), Improved Generalized Weighted HM (IGWHM), Generalized Weighted Geometry HM (GWGHM), and Improved Generalized Weighted Geometry HM (IGWGHM) Operators

Definition 6 [21]. *Let* $\varepsilon = (\varepsilon_1, \varepsilon_2, \cdots, \varepsilon_k)$ *be the weight vector of a collection of non-negative real numbers* $(\tau_1, \tau_2, \ldots, \tau_k)$, $\sum_{j=1}^{k} \varepsilon_j = 1$ *and* $\varepsilon_j \in [0,1]$, *and* $t, s \geq 0$. *Then,*

$$GWHM^{t,s}(\tau_1, \tau_2, \ldots, \tau_k) = \left(\frac{2}{k(k+1)} \sum_{j=1}^{k} \sum_{i=j}^{k} (\varepsilon_j \tau_j)^t (\varepsilon_i \tau_i)^s \right)^{\frac{1}{t+s}}, \tag{8}$$

which is called a GWHM operator.

Definition 7 [22]. *Let* $\varepsilon = (\varepsilon_1, \varepsilon_2, \cdots, \varepsilon_k)$ *be the weight vector of a collection of non-negative real numbers* $(\tau_1, \tau_2, \ldots, \tau_k)$, $\sum_{j=1}^{k} \varepsilon_j = 1$ *and* $\varepsilon_j \in [0,1]$, *and* $t, s \geq 0$. *Then,*

$$GWHM^{t,s}(\tau_1, \tau_2, \ldots, \tau_k) = \left(\frac{1}{\lambda} \mathop{\oplus}_{j=1}^{k} \mathop{\oplus}_{i=j}^{k} ({\varepsilon_j}^t {\varepsilon_i}^s {\tau_j}^t \otimes {\tau_i}^s) \right)^{\frac{1}{t+s}}, \tag{9}$$

where $\lambda = \sum_{j=1}^{k} \sum_{i=j}^{k} {\varepsilon_j}^t {\varepsilon_i}^s$ *is called an IGWHM operator.*

Definition 8 [21]. *Let* $\varepsilon = (\varepsilon_1, \varepsilon_2, \cdots, \varepsilon_k)$ *be the weight vector of a collection of non-negative real numbers* $(\tau_1, \tau_2, \ldots, \tau_k)$, $\sum_{j=1}^{k} \varepsilon_j = 1$ *and* $\varepsilon_j \in [0,1]$, *and* $t, s \geq 0$. *Then,*

$$GWGHM^{t,s}(\tau_1, \tau_2, \ldots, \tau_k) = \frac{1}{t+s} \mathop{\otimes}_{j=1}^{k} \mathop{\otimes}_{i=j}^{k} \left((t\tau_j)^{\varepsilon_j} \oplus (s\tau_i)^{\varepsilon_i} \right)^{\frac{2}{k(k+1)}}, \tag{10}$$

which is called a GWGHM operator.

Definition 9 [22]. *Let* $\varepsilon = (\varepsilon_1, \varepsilon_2, \cdots, \varepsilon_k)$ *be the weight vector of a collection of non-negative real numbers* $(\tau_1, \tau_2, \ldots, \tau_k)$, $\sum_{j=1}^{k} \varepsilon_j = 1$ *and* $\varepsilon_j \in [0,1]$, *and* $t, s \geq 0$. *Then,*

$$IGWGHM^{t,s}(\tau_1, \tau_2, \ldots, \tau_k) = \frac{1}{t+s} \left(\mathop{\otimes}_{j=1}^{k} \mathop{\otimes}_{i=j}^{k} (t\tau_j \oplus s\tau_i)^{\frac{2(k+1-j)}{k(k+1)} \frac{\varepsilon_i}{\sum_{m=j}^{k} \varepsilon_m}} \right), \tag{11}$$

which is called an IGWGHM operator.

3. Some BNN Aggregation Operators

3.1. GWHM Operators for BNNs

Definition 10. *Let* $t, s \geq 0$, *and* $t + s \neq 0$, *a collection* $\tau_j = \langle \alpha^+_{\tau_j}, \beta^+_{\tau_j}, \gamma^+_{\tau_j}, \alpha^-_{\tau_j}, \beta^-_{\tau_j}, \gamma^-_{\tau_j} \rangle$ $(j = 1, 2, \cdots, k)$ *of BNN; then, we define the BNNGWHM operator as follows:*

$$BNNGWHM^{t,s}(\tau_1, \tau_2, \ldots, \tau_k) = \left(\frac{2}{k(k+1)} \sum_{j=1}^{k} \sum_{i=j}^{k} (\varepsilon_j \tau_j)^t (\varepsilon_i \tau_i)^s \right)^{\frac{1}{t+s}}, \tag{12}$$

where $\sum_{j=1}^{k} \varepsilon_j = 1$ *and* $\varepsilon_j \in [0,1]$.

According to Definitions 3 and 10, the following theorem can be attained:

Theorem 1. *Set a collection* $\tau_j = \langle \alpha^+_{\tau_j}, \beta^+_{\tau_j}, \gamma^+_{\tau_j}, \alpha^-_{\tau_j}, \beta^-_{\tau_j}, \gamma^-_{\tau_j} \rangle$ $(j = 1, 2, \cdots, k)$ *of BNNs, using the BNNGWHM operator; then, the aggregation result is still a BNN, which is given by the following form:*

$$BNNGWHM^{t,s}(\tau_1, \tau_2, \ldots, \tau_k) = \left(\frac{2}{k(k+1)} \sum_{j=1}^{k} \sum_{i=j}^{k} (\varepsilon_j \tau_j)^t (\varepsilon_i \tau_i)^s \right)^{\frac{1}{t+s}} = \tag{13}$$
$$\left\langle \begin{array}{c} \left(1-\prod_{j=1}^{k}\prod_{i=j}^{k}\left(1-\left(1-\left(1-\alpha^+_{\tau_j}\right)^{\varepsilon_j}\right)^t\left(1-\left(1-\alpha^+_{\tau_i}\right)^{\varepsilon_i}\right)^s\right)^{\frac{1}{\lambda}}\right)^{\frac{1}{t+s}}, 1-\left(1-\prod_{j=1}^{k}\prod_{i=j}^{k}\left(1-\left(1-\left(\beta^+_{\tau_j}\right)^{\varepsilon_j}\right)^t\left(1-\left(\beta^+_{\tau_i}\right)^{\varepsilon_i}\right)^s\right)^{\frac{1}{\lambda}}\right)^{\frac{1}{t+s}}, \\ 1-\left(1-\prod_{j=1}^{k}\prod_{i=j}^{k}\left(1-\left(1-\left(\gamma^+_{\tau_j}\right)^{\varepsilon_j}\right)^t\left(1-\left(\gamma^+_{\tau_i}\right)^{\varepsilon_i}\right)^s\right)^{\frac{1}{\lambda}}\right)^{\frac{1}{t+s}}, -\left(1-\left(1-\prod_{j=1}^{k}\prod_{i=j}^{k}\left(1-\left(1-\left(-\alpha^-_{\tau_j}\right)^{\varepsilon_j}\right)^t\left(1-\left(-\alpha^-_{\tau_i}\right)^{\varepsilon_i}\right)^s\right)^{\frac{1}{\lambda}}\right)^{\frac{1}{t+s}}\right), \\ -\left(1-\prod_{j=1}^{k}\prod_{i=j}^{k}\left(1-\left(1-\left(1-\left(-\beta^-_{\tau_j}\right)\right)^{\varepsilon_j}\right)^t\left(1-\left(1-\left(-\beta^-_{\tau_i}\right)\right)^{\varepsilon_i}\right)^s\right)^{\frac{1}{\lambda}}\right)^{\frac{1}{t+s}}, -\left(1-\prod_{j=1}^{k}\prod_{i=j}^{k}\left(1-\left(1-\left(1-\left(-\gamma^-_{\tau_j}\right)\right)^{\varepsilon_j}\right)^t\left(1-\left(1-\left(-\gamma^-_{\tau_i}\right)\right)^{\varepsilon_i}\right)^s\right)^{\frac{1}{\lambda}}\right)^{\frac{1}{t+s}} \end{array} \right\rangle$$

where $\frac{1}{\lambda} = \frac{2}{k(k+1)}, \sum_{j=1}^{k} \varepsilon_j = 1$ *and* $\varepsilon_j \in [0, 1]$.

Proof.

$$(1)\ \varepsilon_j \tau_j = \langle 1-\left(1-\alpha^+_{\tau_j}\right)^{\varepsilon_j}, \left(\beta^+_{\tau_j}\right)^{\varepsilon_j}, \left(\gamma^+_{\tau_j}\right)^{\varepsilon_j}, -\left(-\alpha^-_{\tau_j}\right)^{\varepsilon_j}, -\left(1-\left(1-\left(-\beta^-_{\tau_j}\right)\right)^{\varepsilon_j}\right), -\left(1-\left(1-\left(-\gamma^-_{\tau_j}\right)\right)^{\varepsilon_j}\right) \rangle;$$

$$(2)\ \varepsilon_i \tau_i = \langle 1-\left(1-\alpha^+_{\tau_i}\right)^{\varepsilon_i}, \left(\beta^+_{\tau_i}\right)^{\varepsilon_i}, \left(\gamma^+_{\tau_i}\right)^{\varepsilon_i}, -\left(-\alpha^-_{\tau_i}\right)^{\varepsilon_i}, -\left(1-\left(1-\left(-\beta^-_{\tau_i}\right)\right)^{\varepsilon_i}\right), -\left(1-\left(1-\left(-\gamma^-_{\tau_i}\right)\right)^{\varepsilon_i}\right) \rangle;$$

$$(3) (\varepsilon_j \tau_j)^t = \langle \left(1-\left(1-\alpha^+_{\tau_j}\right)^{\varepsilon_j}\right)^t, 1-\left(1-\left(\beta^+_{\tau_j}\right)^{\varepsilon_j}\right)^t, 1-\left(1-\left(\gamma^+_{\tau_j}\right)^{\varepsilon_j}\right)^t,$$
$$-\left(1-\left(1-\left(-\alpha^-_{\tau_j}\right)^{\varepsilon_j}\right)^t\right), -\left(1-\left(1-\left(-\beta^-_{\tau_j}\right)\right)^{\varepsilon_j}\right)^t, -\left(1-\left(1-\left(-\gamma^-_{\tau_j}\right)\right)^{\varepsilon_j}\right)^t \rangle;$$

$$(4)\ (\varepsilon_i \tau_i)^s = \langle \left(1-\left(1-\alpha^+_{\tau_i}\right)^{\varepsilon_i}\right)^s, 1-\left(1-\left(\beta^+_{\tau_i}\right)^{\varepsilon_i}\right)^s, 1-\left(1-\left(\gamma^+_{\tau_i}\right)^{\varepsilon_i}\right)^s,$$
$$-\left(1-\left(1-\left(-\alpha^-_{\tau_i}\right)^{\varepsilon_i}\right)^s\right), -\left(1-\left(1-\left(-\beta^-_{\tau_i}\right)\right)^{\varepsilon_i}\right)^s, -\left(1-\left(1-\left(-\gamma^-_{\tau_i}\right)\right)^{\varepsilon_i}\right)^s \rangle;$$

$$(5)\ (\varepsilon_j \tau_j)^t (\varepsilon_i \tau_i)^s = \langle \left(1-\left(1-\alpha^+_{\tau_j}\right)^{\varepsilon_j}\right)^t \left(1-\left(1-\alpha^+_{\tau_i}\right)^{\varepsilon_i}\right)^s,$$
$$1-\left(1-\left(\beta^+_{\tau_j}\right)^{\varepsilon_j}\right)^t + 1-\left(1-\left(\beta^+_{\tau_i}\right)^{\varepsilon_i}\right)^s - \left(1-\left(1-\left(\beta^+_{\tau_j}\right)^{\varepsilon_j}\right)^t\right)\left(1-\left(1-\left(\beta^+_{\tau_i}\right)^{\varepsilon_i}\right)^s\right),$$
$$1-\left(1-\left(\gamma^+_{\tau_j}\right)^{\varepsilon_j}\right)^t + 1-\left(1-\left(\gamma^+_{\tau_i}\right)^{\varepsilon_i}\right)^s - \left(1-\left(1-\left(\gamma^+_{\tau_j}\right)^{\varepsilon_j}\right)^t\right)\left(1-\left(1-\left(\gamma^+_{\tau_i}\right)^{\varepsilon_i}\right)^s\right),$$
$$-\left(\left(1-\left(1-\left(-\alpha^-_{\tau_j}\right)^{\varepsilon_j}\right)^t\right) + \left(1-\left(1-\left(-\alpha^-_{\tau_i}\right)^{\varepsilon_i}\right)^s\right) - \left(1-\left(1-\left(-\alpha^-_{\tau_j}\right)^{\varepsilon_j}\right)^t\right)\left(1-\left(1-\left(-\alpha^-_{\tau_i}\right)^{\varepsilon_i}\right)^s\right)\right),$$
$$-\left(1-\left(1-\left(-\beta^-_{\tau_j}\right)\right)^{\varepsilon_j}\right)^t\left(1-\left(1-\left(-\beta^-_{\tau_i}\right)\right)^{\varepsilon_i}\right)^s, -\left(1-\left(1-\left(-\gamma^-_{\tau_j}\right)\right)^{\varepsilon_j}\right)^t\left(1-\left(1-\left(-\gamma^-_{\tau_i}\right)\right)^{\varepsilon_i}\right)^s \rangle$$

$$(6)\ \sum_{j=1}^{k} \sum_{i=j}^{k} (\varepsilon_j \tau_j)^t (\varepsilon_i \tau_i)^s = \langle 1-\prod_{j=1}^{k}\prod_{i=j}^{k}\left(1-\left(1-\left(1-\alpha^+_{\tau_j}\right)^{\varepsilon_j}\right)^t\left(1-\left(1-\alpha^+_{\tau_i}\right)^{\varepsilon_i}\right)^s\right),$$
$$\prod_{j=1}^{k}\prod_{i=j}^{k}\left(1-\left(1-\left(\beta^+_{\tau_j}\right)^{\varepsilon_j}\right)^t\left(1-\left(\beta^+_{\tau_i}\right)^{\varepsilon_i}\right)^s\right), \prod_{j=1}^{n}\prod_{i=j}^{n}\left(1-\left(1-\left(\gamma^+_{\tau_j}\right)^{\varepsilon_j}\right)^t\left(1-\left(\gamma^+_{\tau_i}\right)^{\varepsilon_i}\right)^s\right),$$
$$-\prod_{j=1}^{k}\prod_{i=j}^{k}\left(1-\left(1-\left(-\alpha^-_{\tau_j}\right)^{\varepsilon_j}\right)^t\left(1-\left(-\alpha^-_{\tau_i}\right)^{\varepsilon_i}\right)^s\right),$$
$$-\left(1-\prod_{j=1}^{k}\prod_{i=j}^{k}\left(1-\left(1-\left(1-\left(-\beta^-_{\tau_j}\right)\right)^{\varepsilon_j}\right)^t\left(1-\left(1-\left(-\beta^-_{\tau_i}\right)\right)^{\varepsilon_i}\right)^s\right)\right),$$
$$-\left(1-\prod_{j=1}^{k}\prod_{i=j}^{k}\left(1-\left(1-\left(1-\left(-\gamma^-_{\tau_j}\right)\right)^{\varepsilon_j}\right)^t\left(1-\left(1-\left(-\gamma^-_{\tau_i}\right)\right)^{\varepsilon_i}\right)^s\right)\right)\rangle;$$

$$(7)\ \frac{2}{k(k+1)}\sum_{j=1}^{k}\sum_{i=j}^{k}(\varepsilon_j\tau_j)^t(\varepsilon_i\tau_i)^s = \frac{1}{\lambda}\sum_{j=1}^{k}\sum_{i=j}^{k}(\varepsilon_j\tau_j)^t(\varepsilon_i\tau_i)^s =$$

$$\left\langle \begin{array}{c} 1-\prod_{j=1}^{k}\prod_{i=j}^{k}\left(1-\left(1-\left(1-\alpha_{\tau_j}^{+}\right)^{\varepsilon_j}\right)^t\left(1-\left(1-\alpha_{\tau_i}^{+}\right)^{\varepsilon_i}\right)^s\right)^{\frac{1}{\lambda}}, \\ \prod_{j=1}^{k}\prod_{i=j}^{k}\left(1-\left(1-\left(\beta_{\tau_j}^{+}\right)^{\varepsilon_j}\right)^t\left(1-\left(\beta_{\tau_i}^{+}\right)^{\varepsilon_i}\right)^s\right)^{\frac{1}{\lambda}}, \\ \prod_{j=1}^{k}\prod_{i=j}^{k}\left(1-\left(1-\left(\gamma_{\tau_j}^{+}\right)^{\varepsilon_j}\right)^t\left(1-\left(\gamma_{\tau_i}^{+}\right)^{\varepsilon_i}\right)^s\right)^{\frac{1}{\lambda}}, \\ -\prod_{j=1}^{k}\prod_{i=j}^{k}\left(1-\left(1-\left(-\alpha_{\tau_j}^{-}\right)^{\varepsilon_j}\right)^t\left(1-\left(-\alpha_{\tau_i}^{-}\right)^{\varepsilon_i}\right)^s\right)^{\frac{1}{\lambda}}, \\ -\left(1-\prod_{j=1}^{k}\prod_{i=j}^{k}\left(1-\left(1-\left(1-\left(-\beta_{\tau_j}^{-}\right)\right)^{\varepsilon_j}\right)^t\left(1-\left(1-\left(-\beta_{\tau_i}^{-}\right)\right)^{\varepsilon_i}\right)^s\right)^{\frac{1}{\lambda}}\right), \\ -\left(1-\prod_{j=1}^{k}\prod_{i=j}^{k}\left(1-\left(1-\left(1-\left(-\gamma_{\tau_j}^{-}\right)\right)^{\varepsilon_j}\right)^t\left(1-\left(1-\left(-\gamma_{\tau_i}^{-}\right)\right)^{\varepsilon_i}\right)^s\right)^{\frac{1}{\lambda}}\right) \end{array} \right\rangle;$$

$$(8)\ \left(\frac{1}{\lambda}\sum_{j=1}^{k}\sum_{i=j}^{k}(\varepsilon_j\tau_j)^t(\varepsilon_i\tau_i)^s\right)^{\frac{1}{t+s}} =$$

$$\left\langle \begin{array}{c} \left(1-\prod_{j=1}^{k}\prod_{i=j}^{k}\left(1-\left(1-\left(1-\alpha_{\tau_j}^{+}\right)^{\varepsilon_j}\right)^t\left(1-\left(1-\alpha_{\tau_i}^{+}\right)^{\varepsilon_i}\right)^s\right)^{\frac{1}{\lambda}}\right)^{\frac{1}{t+s}}, \\ 1-\left(1-\prod_{j=1}^{k}\prod_{i=j}^{k}\left(1-\left(1-\left(\beta_{\tau_j}^{+}\right)^{\varepsilon_j}\right)^t\left(1-\left(\beta_{\tau_i}^{+}\right)^{\varepsilon_i}\right)^s\right)^{\frac{1}{\lambda}}\right)^{\frac{1}{t+s}}, \\ 1-\left(1-\prod_{j=1}^{k}\prod_{i=j}^{k}\left(1-\left(1-\left(\gamma_{\tau_j}^{+}\right)^{\varepsilon_j}\right)^t\left(1-\left(\gamma_{\tau_i}^{+}\right)^{\varepsilon_i}\right)^s\right)^{\frac{1}{\lambda}}\right)^{\frac{1}{t+s}}, \\ -\left(1-\left(1-\prod_{j=1}^{k}\prod_{i=j}^{k}\left(1-\left(1-\left(-\alpha_{\tau_j}^{-}\right)^{\varepsilon_j}\right)^t\left(1-\left(-\alpha_{\tau_i}^{-}\right)^{\varepsilon_i}\right)^s\right)^{\frac{1}{\lambda}}\right)^{\frac{1}{t+s}}\right), \\ -\left(1-\prod_{j=1}^{k}\prod_{i=j}^{k}\left(1-\left(1-\left(1-\left(-\beta_{\tau_j}^{-}\right)\right)^{\varepsilon_j}\right)^t\left(1-\left(1-\left(-\beta_{\tau_i}^{-}\right)\right)^{\varepsilon_i}\right)^s\right)^{\frac{1}{\lambda}}\right)^{\frac{1}{t+s}}, \\ -\left(1-\prod_{j=1}^{k}\prod_{i=j}^{k}\left(1-\left(1-\left(1-\left(-\gamma_{\tau_j}^{-}\right)\right)^{\varepsilon_j}\right)^t\left(1-\left(1-\left(-\gamma_{\tau_i}^{-}\right)\right)^{\varepsilon_i}\right)^s\right)^{\frac{1}{\lambda}}\right)^{\frac{1}{t+s}} \end{array} \right\rangle.$$

This proves Theorem 1. □

Theorem 2. *(Monotonicity). Set* $\tau_j = \langle \alpha_{\tau_j}^{+}, \beta_{\tau_j}^{+}, \gamma_{\tau_j}^{+}, \alpha_{\tau_j}^{-}, \beta_{\tau_j}^{-}, \gamma_{\tau_j}^{-}\rangle$ $(j = 1, 2, \cdots, k)$ *and* $\sigma_j = \langle \alpha_{\sigma_j}^{+}, \beta_{\sigma_j}^{+}, \gamma_{\sigma_j}^{+}, \alpha_{\sigma_j}^{-}, \beta_{\sigma_j}^{-}, \gamma_{\sigma_j}^{-}\rangle$ $(j = 1, 2, \cdots, k)$ *as two collections of BNNs; if* $\alpha_{\tau_j}^{+} \le \alpha_{\sigma_j}^{+}, \beta_{\tau_j}^{+} \ge \beta_{\sigma_j}^{+}, \gamma_{\tau_j}^{+} \ge \gamma_{\sigma_j}^{+}$ *and* $\alpha_{\tau_j}^{-} \ge \alpha_{\sigma_j}^{-}, \beta_{\tau_j}^{-} \le \beta_{\sigma_j}^{-}, \gamma_{\tau_j}^{-} \le \gamma_{\sigma_j}^{-}$*, then*

$$BNNGWHM^{t,s}(\tau_1, \tau_2, \ldots, \tau_k) \le BNNGWHM^{t,s}(\sigma_1, \sigma_2, \ldots, \sigma_k).$$

Proof. For $\alpha_{\tau_j}^{+} \le \alpha_{\sigma_j}^{+}, \beta_{\tau_j}^{+} \ge \beta_{\sigma_j}^{+}, \gamma_{\tau_j}^{+} \ge \gamma_{\sigma_j}^{+}$ and $\alpha_{\tau_j}^{-} \ge \alpha_{\sigma_j}^{-}, \beta_{\tau_j}^{-} \le \beta_{\sigma_j}^{-}, \gamma_{\tau_j}^{-} \le \gamma_{\sigma_j}^{-}$, it is obvious that

$$\left(1-\left(1-\alpha_{\tau_j}^{+}\right)^{\varepsilon_j}\right)^t\left(1-\left(1-\alpha_{\tau_i}^{+}\right)^{\varepsilon_i}\right)^s \le \left(1-\left(1-\alpha_{\sigma_j}^{+}\right)^{\varepsilon_j}\right)^t\left(1-\left(1-\alpha_{\sigma_i}^{+}\right)^{\varepsilon_i}\right)^s,$$

$$\left(1-\prod_{j=1}^{k}\prod_{i=j}^{k}\left(1-\left(1-\left(1-\alpha_{\tau_j}^{+}\right)^{\varepsilon_j}\right)^{t}\left(1-\left(1-\alpha_{\tau_i}^{+}\right)^{\varepsilon_i}\right)^{s}\right)^{\frac{1}{\lambda}}\right)^{\frac{1}{t+s}}$$
$$\leq\left(1-\prod_{j=1}^{k}\prod_{i=j}^{k}\left(1-\left(1-\left(1-\alpha_{\sigma_j}^{+}\right)^{\varepsilon_j}\right)^{t}\left(1-\left(1-\alpha_{\sigma_i}^{+}\right)^{\varepsilon_i}\right)^{s}\right)^{\frac{1}{\lambda}}\right)^{\frac{1}{t+s}}.$$

Similarly

$$1-\left(1-\prod_{j=1}^{k}\prod_{i=j}^{k}\left(1-\left(1-\left(\beta_{\tau_j}^{+}\right)^{\varepsilon_j}\right)^{t}\left(1-\left(\beta_{\tau_i}^{+}\right)^{\varepsilon_i}\right)^{s}\right)^{\frac{1}{\lambda}}\right)^{\frac{1}{t+s}}\geq$$
$$1-\left(1-\prod_{j=1}^{k}\prod_{i=j}^{k}\left(1-\left(1-\left(\beta_{\sigma_j}^{+}\right)^{\varepsilon_j}\right)^{t}\left(1-\left(\beta_{\sigma_i}^{+}\right)^{\varepsilon_i}\right)^{s}\right)^{\frac{1}{\lambda}}\right)^{\frac{1}{t+s}},$$
$$1-\left(1-\prod_{j=1}^{k}\prod_{i=j}^{k}\left(1-\left(1-\left(\gamma_{\tau_j}^{+}\right)^{\varepsilon_j}\right)^{t}\left(1-\left(\gamma_{\tau_i}^{+}\right)^{\varepsilon_i}\right)^{s}\right)^{\frac{1}{\lambda}}\right)^{\frac{1}{t+s}}$$
$$\geq 1-\left(1-\prod_{j=1}^{k}\prod_{i=j}^{k}\left(1-\left(1-\left(\gamma_{\sigma_j}^{+}\right)^{\varepsilon_j}\right)^{t}\left(1-\left(\gamma_{\sigma_i}^{+}\right)^{\varepsilon_i}\right)^{s}\right)^{\frac{1}{\lambda}}\right)^{\frac{1}{t+s}},$$
$$-\left(1-\left(1-\prod_{j=1}^{k}\prod_{i=j}^{k}\left(1-\left(1-\left(-\alpha_{\tau_j}^{-}\right)^{\varepsilon_j}\right)^{t}\left(1-\left(-\alpha_{\tau_i}^{-}\right)^{\varepsilon_i}\right)^{s}\right)^{\frac{1}{\lambda}}\right)^{\frac{1}{t+s}}\right)$$
$$\geq-\left(1-\left(1-\prod_{j=1}^{k}\prod_{i=j}^{k}\left(1-\left(1-\left(-\alpha_{\sigma_j}^{-}\right)^{\varepsilon_j}\right)^{t}\left(1-\left(-\alpha_{\sigma_i}^{-}\right)^{\varepsilon_i}\right)^{s}\right)^{\frac{1}{\lambda}}\right)^{\frac{1}{t+s}}\right),$$
$$-\left(1-\prod_{j=1}^{k}\prod_{i=j}^{k}\left(1-\left(1-\left(1-\left(-\beta_{\tau_j}^{-}\right)\right)^{\varepsilon_j}\right)^{t}\left(1-\left(1-\left(-\beta_{\tau_i}^{-}\right)\right)^{\varepsilon_i}\right)^{s}\right)^{\frac{1}{\lambda}}\right)^{\frac{1}{t+s}}$$
$$\leq-\left(1-\prod_{j=1}^{k}\prod_{i=j}^{k}\left(1-\left(1-\left(1-\left(-\beta_{\sigma_j}^{-}\right)\right)^{\varepsilon_j}\right)^{t}\left(1-\left(1-\left(-\beta_{\sigma_i}^{-}\right)\right)^{\varepsilon_i}\right)^{s}\right)^{\frac{1}{\lambda}}\right)^{\frac{1}{t+s}},$$

and

$$-\left(1-\prod_{j=1}^{k}\prod_{i=j}^{k}\left(1-\left(1-\left(1-\left(-\gamma_{\tau_j}^{-}\right)\right)^{\varepsilon_j}\right)^{t}\left(1-\left(1-\left(-\gamma_{\tau_i}^{-}\right)\right)^{\varepsilon_i}\right)^{s}\right)^{\frac{1}{\lambda}}\right)^{\frac{1}{s+t}}\leq$$
$$-\left(1-\prod_{j=1}^{k}\prod_{i=j}^{k}\left(1-\left(1-\left(1-\left(-\gamma_{\sigma_j}^{-}\right)\right)^{\varepsilon_j}\right)^{t}\left(1-\left(1-\left(-\gamma_{\sigma_i}^{-}\right)\right)^{\varepsilon_i}\right)^{s}\right)^{\frac{1}{\lambda}}\right)^{\frac{1}{s+t}}.$$

Thus, $BNNGWHM^{t,\,s}(\tau_1,\tau_2,\ldots,\tau_k)\leq BNNGWHM^{t,\,s}(\sigma_1,\sigma_2,\ldots,\sigma_k)$; this proves Theorem 2. □

3.2. Improved Generalized Weighted HM Operators for BNNs

Definition 11. *Let $t,s\geq 0$, and $t+s\neq 0$, a collection $\tau_j\langle=\alpha_{\tau_j}^{+},\beta_{\tau_j}^{+},\gamma_{\tau_j}^{+},\alpha_{\tau_j}^{-},\beta_{\tau_j}^{-},\gamma_{\tau_j}^{-}\ (j=1,2,\cdots,k)\rangle$ of BNN; then, we define the BNNIGWHM operator as follows:*

$$BNNIGWHM^{\,t,\,s}(\tau_1,\ \tau_2,\ldots,\ \tau_k)=\left(\frac{1}{\sum_{j=1}^{k}\sum_{i=j}^{k}\varepsilon_j\varepsilon_i}\bigoplus_{j=1}^{k}\bigoplus_{i=j}^{k}\left(\varepsilon_j\varepsilon_i{\tau_j}^{t}\otimes{\tau_i}^{s}\right)\right)^{\frac{1}{t+s}},\tag{14}$$

where $\sum_{j=1}^{k}\varepsilon_j=1$ and $\varepsilon_j\in[0,1]$.

According to Definitions 3 and 11, the following theorem can be attained:

Theorem 3. *Set a collection* $\tau_j\langle= \alpha^+_{\tau_j}, \beta^+_{\tau_j}, \gamma^+_{\tau_j}, \alpha^-_{\tau_j}, \beta^-_{\tau_j}, \gamma^-_{\tau_j}\ (j = 1, 2, \cdots, k)\rangle$ *of BNNs, using BNNIGWHM operator; then, the aggregation result is still a BNN, which is given by the following form:*

$$BNNIGWHM^{t,\,s}(\tau_1,\ \tau_2,\ldots,\ \tau_k) = \left(\frac{1}{\sum_{j=1}^{k}\sum_{i=j}^{k}\varepsilon_j\varepsilon_i}\bigoplus_{j=1}^{k}\bigoplus_{i=j}^{k}\left(\varepsilon_j\varepsilon_i\tau_j{}^t\otimes\tau_i{}^s\right)\right)^{\frac{1}{t+s}} =$$

$$\left\langle\begin{array}{c}
\left(1-\left(\prod_{j=1}^{k}\prod_{i=j}^{k}\left(1-\left(\alpha^+_{\tau_j}\right)^t\left(\alpha^+_{\tau_i}\right)^s\right)^{\varepsilon_j\varepsilon_i}\right)^{\frac{1}{\lambda}}\right)^{\frac{1}{t+s}}\\
1-\left(1-\left(\prod_{j=1}^{k}\prod_{i=j}^{k}\left(1-\left(1-\beta^+_{\tau_j}\right)^t\left(1-\beta^+_{\tau_i}\right)^s\right)^{\varepsilon_j\varepsilon_i}\right)^{\frac{1}{\lambda}}\right)^{\frac{1}{t+s}},\\
1-\left(1-\left(\prod_{j=1}^{k}\prod_{i=j}^{k}\left(1-\left(1-\gamma^+_{\tau_j}\right)^t\left(1-\gamma^+_{\tau_i}\right)^s\right)^{\varepsilon_j\varepsilon_i}\right)^{\frac{1}{\lambda}}\right)^{\frac{1}{t+s}},\\
-\left(1-\left(1-\left(\prod_{j=1}^{k}\prod_{i=j}^{k}\left(1-\left(1-\left(-\alpha^-_{\tau_j}\right)\right)^t\left(1-\left(-\alpha^-_{\tau_i}\right)\right)^s\right)^{\varepsilon_j\varepsilon_i}\right)^{\frac{1}{\lambda}}\right)^{\frac{1}{t+s}}\right),\\
-\left(1-\left(\prod_{j=1}^{k}\prod_{i=j}^{k}\left(1-\left(-\beta^-_{\tau_j}\right)^t\left(-\beta^-_{\tau_i}\right)^s\right)^{\varepsilon_j\varepsilon_i}\right)^{\frac{1}{\lambda}}\right)^{\frac{1}{t+s}},\\
-\left(1-\left(\prod_{j=1}^{k}\prod_{i=j}^{k}\left(1-\left(-\gamma^-_{\tau_j}\right)^t\left(-\gamma^-_{\tau_i}\right)^s\right)^{\varepsilon_j\varepsilon_i}\right)^{\frac{1}{\lambda}}\right)^{\frac{1}{t+s}}
\end{array}\right\rangle, \tag{15}$$

where $\lambda = \sum_{j=1}^{k}\sum_{i=j}^{k}\varepsilon_j\varepsilon_i$, $\sum_{j=1}^{k}\varepsilon_j = 1$ *and* $\varepsilon_j \in [0, 1]$.

The proof of Theorem 3 can be achieved according to the proof of Theorem 1; thus, we omit it here.

Theorem 4. *(Idempotency). Set a collection* $\tau_j = \langle\alpha^+_{\tau_j}, \beta^+_{\tau_j}, \gamma^+_{\tau_j}, \alpha^-_{\tau_j}, \beta^-_{\tau_j}, \gamma^-_{\tau_j}\rangle$ $(j = 1, 2, \cdots, k)$ *of BNNs; if* $\tau_j = \tau$, *then*

$$BNNIGWHM^{\,t,\,s}(\tau_1, \tau_2, \ldots, \tau_k) = BNNIGWHM^{\,t,\,s}(\tau, \tau, \ldots\tau) = \tau.$$

Proof. For $\tau_j = \tau(j = 1, 2, \ldots, k)$, the following result can be easily attained:

$$BNNIGWHM^{t,s}(\tau_1, \tau_2, \ldots, \tau_k) = BNNIGWHM^{t,s}(\tau, \tau, \ldots \tau) =$$

$$= \left\langle \begin{array}{c} \left(1 - \left(\prod_{j=1}^{k}\prod_{i=j}^{k}\left(1-(\alpha_\tau^+)^t(\alpha_\tau^+)^s\right)^{\varepsilon_j\varepsilon_i}\right)^{\frac{1}{\lambda}}\right)^{\frac{1}{t+s}} \\ 1-\left(1 - \left(\prod_{j=1}^{k}\prod_{i=j}^{k}\left(1-(1-\beta_\tau^+)^t(1-\beta_\tau^+)^s\right)^{\varepsilon_j\varepsilon_i}\right)^{\frac{1}{\lambda}}\right)^{\frac{1}{t+s}}, \\ 1-\left(1 - \left(\prod_{j=1}^{k}\prod_{i=j}^{k}\left(1-(1-\gamma_\tau^+)^t(1-\gamma_\tau^+)^s\right)^{\varepsilon_j\varepsilon_i}\right)^{\frac{1}{\lambda}}\right)^{\frac{1}{t+s}}, \\ -\left(1-\left(1 - \left(\prod_{j=1}^{k}\prod_{i=j}^{k}\left(1-(1-(-\alpha_\tau^-))^t(1-(-\alpha_\tau^-))^s\right)^{\varepsilon_j\varepsilon_i}\right)^{\frac{1}{\lambda}}\right)^{\frac{1}{t+s}}\right), \\ -\left(1 - \left(\prod_{j=1}^{k}\prod_{i=j}^{k}\left(1-(-\beta_\tau^-)^t(-\beta_\tau^-)^s\right)^{\varepsilon_j\varepsilon_i}\right)^{\frac{1}{\lambda}}\right)^{\frac{1}{t+s}}, \\ -\left(1 - \left(\prod_{j=1}^{k}\prod_{i=j}^{k}\left(1-(-\gamma_\tau^-)^t(-\gamma_\tau^-)^s\right)^{\varepsilon_j\varepsilon_i}\right)^{\frac{1}{\lambda}}\right)^{\frac{1}{t+s}} \end{array} \right\rangle.$$

$$= \left\langle \begin{array}{c} \left((\alpha_\tau^+)^{t+s}\right)^{\frac{1}{t+s}}, 1-\left((1-\beta_\tau^+)^{t+s}\right)^{\frac{1}{t+s}}, \\ 1-\left((1-\gamma_\tau^+)^{t+s}\right)^{\frac{1}{t+s}}, -\left(1-\left((1-(-\alpha_\tau^-))^{t+s}\right)^{\frac{1}{t+s}}\right), \\ -\left((-\beta_\tau^-)^{t+s}\right)^{\frac{1}{t+s}}, -\left((-\gamma_\tau^-)^{t+s}\right)^{\frac{1}{t+s}} \end{array} \right\rangle = \langle \alpha_\tau^+, \beta_\tau^+, \gamma_\tau^+, \alpha_\tau^-, \beta_\tau^-, \gamma_\tau^- \rangle = \tau$$

This proves Theorem 4. □

Theorem 5. *(Monotonicity). Set* $\tau_j = \langle \alpha_{\tau_j}^+, \beta_{\tau_j}^+, \gamma_{\tau_j}^+, \alpha_{\tau_j}^-, \beta_{\tau_j}^-, \gamma_{\tau_j}^- \rangle$ $(j = 1, 2, \cdots, k)$ *and* $\sigma_j = \langle \alpha_{\sigma_j}^+, \beta_{\sigma_j}^+, \gamma_{\sigma_j}^+, \alpha_{\sigma_j}^-, \beta_{\sigma_j}^-, \gamma_{\sigma_j}^- \rangle$ $(j = 1, 2, \cdots, k)$ *as two collections of BNNs; if* $\alpha_{\tau_j}^+ \le \alpha_{\sigma_j}^+, \beta_{\tau_j}^+ \ge \beta_{\sigma_j}^+, \gamma_{\tau_j}^+ \ge \gamma_{\sigma_j}^+$ *and* $\alpha_{\tau_j}^- \ge \alpha_{\sigma_j}^-, \beta_{\tau_j}^- \le \beta_{\sigma_j}^-, \gamma_{\tau_j}^- \le \gamma_{\sigma_j}^-$, *then,*

$$BNNIGWHM^{t,s}(\tau_1, \tau_2, \ldots, \tau_k) \le BNNIGWHM^{t,s}(\sigma_1, \sigma_2, \ldots, \sigma_k).$$

The proof of Theorem 5 is similar to Theorem 2; thus, we omit it.

Theorem 6. *(Boundedness). Set a collection* $\tau_j = \langle \alpha_{\tau_j}^+, \beta_{\tau_j}^+, \gamma_{\tau_j}^+, \alpha_{\tau_j}^-, \beta_{\tau_j}^-, \gamma_{\tau_j}^- \rangle$ $(j = 1, 2, \cdots, k)$ *of BNNs, and let* $\tau^- = \left\langle \begin{array}{c} \min\left(\alpha_{\tau_j}^+\right), \max\left(\beta_{\tau_j}^+\right), \max\left(\gamma_{\tau_j}^+\right), \\ \max\left(\alpha_{\tau_j}^-\right), \min\left(\beta_{\tau_j}^-\right), \min\left(\gamma_{\tau_j}^-\right) \end{array} \right\rangle$ *and* $\tau^+ = \left\langle \begin{array}{c} \max\left(\alpha_{\tau_j}^+\right), \min\left(\beta_{\tau_j}^+\right), \min\left(\gamma_{\tau_j}^+\right), \\ \min\left(\alpha_{\tau_j}^-\right), \max\left(\beta_{\tau_j}^-\right), \max\left(\gamma_{\tau_j}^-\right) \end{array} \right\rangle$*; then,*

$$\tau^- \le BNNIGWHM^{t,s}(\tau_1, \tau_2, \ldots, \tau_k) \le \tau^+.$$

Based on Theorems 4 and 5, the following can be obtained:

$$\tau^- = BNNIGWHM^{t,s}(\tau^-, \tau^-, \ldots, \tau^-) \text{and } \tau^+ = BNNIGWHM^{t,s}(\tau^+, \tau^+, \ldots, \tau^+).$$

$$BNNIGWHM^{t,s}(\tau^-, \tau^-, \ldots, \tau^-) \le BNNIGWHM^{t,s}(\tau_1, \tau_2, \ldots, \tau_k) \le BNNIGWHM^{t,s}(\tau^+, \tau^+, \ldots, \tau^+).$$

$$\text{Then}, \tau^{-} \leq BNNIGWHM^{t,\,s}\ (\tau_1,\ \tau_2, \ldots,\ \tau_k) \leq \tau^{+}.$$

This proves Theorem 6.

3.3. GWGHM Operators of BNNs

Definition 12. *Let $t, s \geq 0$, $t + s \neq 0$, a collection $\tau_j = \langle \alpha^{+}_{\tau_j}, \beta^{+}_{\tau_j}, \gamma^{+}_{\tau_j}, \alpha^{-}_{\tau_j}, \beta^{-}_{\tau_j}, \gamma^{-}_{\tau_j} \rangle$ $(j = 1, 2, \cdots, k)$ of BNNs; then, we define the BNNGWGHM operator as follows:*

$$BNNGWGHM^{t,\,s}(\tau_1,\ \tau_2, \ldots,\ \tau_k) = \frac{1}{t+s} \overset{k}{\underset{j=1}{\otimes}} \overset{k}{\underset{i=j}{\otimes}} \left((t\tau_j)^{\varepsilon_j} \oplus (s\tau_i)^{\varepsilon_i} \right)^{\frac{2}{k(k+1)}}, \tag{16}$$

where $\sum_{j=1}^{k} \varepsilon_j = 1$ and $\varepsilon_j \in [0,1]$.

According to Definitions 3 and 12, the following theorem can be attained:

Theorem 7. *Set a collection $\tau_j = \langle \alpha^{+}_{\tau_j}, \beta^{+}_{\tau_j}, \gamma^{+}_{\tau_j}, \alpha^{-}_{\tau_j}, \beta^{-}_{\tau_j}, \gamma^{-}_{\tau_j} \rangle$ $(j = 1, 2, \cdots, k)$ of BNNs, using the BNNGWGHM operator; then, the aggregation result is still a BNN, which is given by the following form:*

$$BNNGWGHM^{t,\,s}(\tau_1,\ \tau_2, \ldots,\ \tau_k) = \tfrac{1}{t+s} \overset{k}{\underset{j=1}{\otimes}} \overset{k}{\underset{i=j}{\otimes}} \left((t\tau_j)^{\varepsilon_j} \oplus (s\tau_i)^{\varepsilon_i} \right)^{\frac{2}{k(k+1)}} =$$

$$\left\langle \begin{array}{c} 1 - \left(1 - \prod_{j=1}^{k}\prod_{i=j}^{k}\left(1 - \left(1 - \left(1 - \left(1 - \left(\alpha^{+}_{\tau_j}\right)\right)^{t}\right)^{\varepsilon_j}\right)\left(1 - \left(1 - \left(1 - \left(\alpha^{+}_{\tau_i}\right)\right)^{s}\right)^{\varepsilon_i}\right)\right)^{\frac{1}{\lambda}}\right)^{\frac{1}{t+s}}, \\ \left(1 - \prod_{j=1}^{k}\prod_{i=j}^{k}\left(1 - \left(1 - \left(1 - \left(\beta^{+}_{\tau_j}\right)^{t}\right)^{\varepsilon_j}\right)\left(1 - \left(1 - \left(\beta^{+}_{\tau_i}\right)^{s}\right)^{\varepsilon_i}\right)\right)^{\frac{1}{\lambda}}\right)^{\frac{1}{t+s}}, \\ \left(1 - \prod_{j=1}^{k}\prod_{i=j}^{k}\left(1 - \left(1 - \left(1 - \left(\gamma^{+}_{\tau_j}\right)^{t}\right)^{\varepsilon_j}\right)\left(1 - \left(1 - \left(\gamma^{+}_{\tau_i}\right)^{s}\right)^{\varepsilon_i}\right)\right)^{\frac{1}{\lambda}}\right)^{\frac{1}{t+s}}, \\ -\left(1 - \prod_{j=1}^{k}\prod_{i=j}^{k}\left(1 - \left(1 - \left(1 - \left(-\alpha^{-}_{\tau_j}\right)^{t}\right)^{\varepsilon_j}\right)\left(1 - \left(1 - \left(-\alpha^{-}_{\tau_i}\right)^{s}\right)^{\varepsilon_i}\right)\right)^{\frac{1}{\lambda}}\right)^{\frac{1}{t+s}}, \\ -\left(1 - \left(1 - \prod_{j=1}^{k}\prod_{i=j}^{k}\left(1 - \left(1 - \left(1 - \left(1 - \left(-\beta^{-}_{\tau_j}\right)\right)^{t}\right)^{\varepsilon_j}\right)\left(1 - \left(1 - \left(1 - \left(-\beta^{-}_{\tau_i}\right)\right)^{s}\right)^{\varepsilon_i}\right)\right)^{\frac{1}{\lambda}}\right)^{\frac{1}{t+s}}\right), \\ -\left(1 - \left(1 - \prod_{j=1}^{k}\prod_{i=j}^{k}\left(1 - \left(1 - \left(1 - \left(1 - \left(-\gamma^{-}_{\tau_j}\right)\right)^{t}\right)^{\varepsilon_j}\right)\left(1 - \left(1 - \left(1 - \left(-\gamma^{-}_{\tau_i}\right)\right)^{s}\right)^{\varepsilon_i}\right)\right)^{\frac{1}{\lambda}}\right)^{\frac{1}{t+s}}\right) \end{array} \right\rangle, \tag{17}$$

where $\frac{1}{\lambda} = \frac{2}{k(k+1)}$, $\sum_{j=1}^{k} \varepsilon_j = 1$ and $\varepsilon_j \in [0,1]$.

Theorem 8. *(Monotonicity). Set $\tau_j = \langle \alpha^{+}_{\tau_j}, \beta^{+}_{\tau_j}, \gamma^{+}_{\tau_j}, \alpha^{-}_{\tau_j}, \beta^{-}_{\tau_j}, \gamma^{-}_{\tau_j} \rangle$ $(j = 1, 2, \cdots, k)$ and $\sigma_j = \langle \alpha^{+}_{\sigma_j}, \beta^{+}_{\sigma_j}, \gamma^{+}_{\sigma_j}, \alpha^{-}_{\sigma_j}, \beta^{-}_{\sigma_j}, \gamma^{-}_{\sigma_j} \rangle$ $(j = 1, 2, \cdots, k)$ as two collections of BNNs; if $\alpha^{+}_{\tau_j} \leq \alpha^{+}_{\sigma_j}, \beta^{+}_{\tau_j} \geq \beta^{+}_{\sigma_j}, \gamma^{+}_{\tau_j} \geq \gamma^{+}_{\sigma_j}$ and $\alpha^{-}_{\tau_j} \geq \alpha^{-}_{\sigma_j}, \beta^{-}_{\tau_j} \leq \beta^{-}_{\sigma_j}, \gamma^{-}_{\tau_j} \leq \gamma^{-}_{\sigma_j}$, then,*

$$BNNGWGHM^{t,\,s}(\tau_1,\ \tau_2, \ldots,\ \tau_k) \leq BNNGWGHM^{t,\,s}(\sigma_1,\ \sigma_2, \ldots,\ \sigma_k).$$

The proofs of theorems about BNNGWGHM are similar to those about BNNGWHM; thus, we omit them.

3.4. IGWGHM Operators of BNNs

Definition 13. *Let t, s* ≥ 0*, and* $t+s \neq 0$*, a collection* $\tau_j = \langle \alpha^+_{\tau_j}, \beta^+_{\tau_j}, \gamma^+_{\tau_j}, \alpha^-_{\tau_j}, \beta^-_{\tau_j}, \gamma^-_{\tau_j} \rangle$ $(j = 1, 2, \cdots, k)$ *of BNNs; then, we define the BNNIGWGHM operator as follows:*

$$BNNIGWGHM^{t,\,s}(\tau_1,\ \tau_2, \ldots,\ \tau_k) = \frac{1}{t+s}\left(\mathop{\oplus}_{j=1}^{k} \mathop{\otimes}_{i=j}^{k} \left(t\tau_j \oplus s\tau_i\right)^{\frac{2(k+1-j)}{k(k+1)} \frac{\varepsilon_j}{\sum_{m=j}^{k} \varepsilon_m}} \right), \quad (18)$$

where $\sum_{j=1}^{k} \varepsilon_j = 1$ *and* $\varepsilon_j \in [0,1]$.

According to Definitions 3 and 13, the following theorem can be attained:

Theorem 9. *Set a collection* $\tau_j = \langle \alpha^+_{\tau_j}, \beta^+_{\tau_j}, \gamma^+_{\tau_j}, \alpha^-_{\tau_j}, \beta^-_{\tau_j}, \gamma^-_{\tau_j} \rangle$ $(j = 1, 2, \cdots, k)$ *of BNNs, using the BNNIGWGHM operator; then, the aggregation result is still a BNN, which is given by the following form:*

$$\begin{array}{c} BNNIGWGHM^{t,\,s}(\tau_1,\ \tau_2, \ldots,\ \tau_k) = \frac{1}{t+s}\left(\mathop{\oplus}_{j=1}^{k} \mathop{\otimes}_{i=j}^{k} \left(t\tau_j \oplus s\tau_i\right)^{\frac{2(k+1-j)}{k(k+1)} \frac{\varepsilon_j}{\sum_{m=j}^{k} \varepsilon_m}} \right) = \\ \left\langle \begin{array}{c} 1 - \left(1 - \prod_{j=1}^{k}\prod_{i=j}^{k}\left(1 - \left(1-\left(\alpha^+_{\tau_j}\right)\right)^t \left(1-\left(\alpha^+_{\tau_i}\right)\right)^s\right)^{\frac{1}{\lambda}}\right)^{\frac{1}{t+s}}, \\ \left(1 - \prod_{j=1}^{k}\prod_{i=j}^{k}\left(1 - \left(\beta^+_{\tau_j}\right)^t \left(\beta^+_{\tau_i}\right)^s\right)^{\frac{1}{\lambda}}\right)^{\frac{1}{t+s}}, \\ \left(1 - \prod_{j=1}^{k}\prod_{i=j}^{k}\left(1 - \left(\gamma^+_{\tau_j}\right)^t \left(\gamma^+_{\tau_i}\right)^s\right)^{\frac{1}{\lambda}}\right)^{\frac{1}{t+s}} \\ -\left(1 - \prod_{j=1}^{k}\prod_{i=j}^{k}\left(1 - \left(-\alpha^-_{\tau_j}\right)^t \left(-\alpha^-_{\tau_i}\right)^s\right)^{\frac{1}{\lambda}}\right)^{\frac{1}{t+s}}, \\ -\left(1 - \left(1 - \prod_{j=1}^{k}\prod_{i=j}^{k}\left(1 - \left(1-\left(-\beta^-_{\tau_j}\right)\right)^t \left(1-\left(-\beta^-_{\tau_i}\right)\right)^s\right)^{\frac{1}{\lambda}}\right)^{\frac{1}{t+s}}\right), \\ -\left(1 - \left(1 - \prod_{j=1}^{k}\prod_{i=j}^{k}\left(1 - \left(1-\left(-\gamma^-_{\tau_j}\right)\right)^t \left(1-\left(-\gamma^-_{\tau_i}\right)\right)^s\right)^{\frac{1}{\lambda}}\right)^{\frac{1}{t+s}}\right) \end{array} \right\rangle, \end{array} \quad (19)$$

where $\frac{1}{\lambda} = \frac{2(k+1-j)}{k(k+1)} \frac{\varepsilon_j}{\sum_{m=j}^{k} \varepsilon_m}$, $\sum_{j=1}^{k} \varepsilon_j = 1$ *and* $\varepsilon_j \in [0,1]$.

Theorem 10. *(Monotonicity). Set* $\tau_j = \langle \alpha^+_{\tau_j}, \beta^+_{\tau_j}, \gamma^+_{\tau_j}, \alpha^-_{\tau_j}, \beta^-_{\tau_j}, \gamma^-_{\tau_j} \rangle$ $(j = 1, 2, \cdots, k)$ *and* $\sigma_j = \alpha^+_{\sigma_j}, \beta^+_{\sigma_j}, \gamma^+_{\sigma_j}, \alpha^-_{\sigma_j}, \beta^-_{\sigma_j}, \gamma^-_{\sigma_j}$ $(j = 1, 2, \cdots, k)$ *as two collections of BNNs; if* $\alpha^+_{\tau_j} \leq \alpha^+_{\sigma_j}, \beta^+_{\tau_j} \geq \beta^+_{\sigma_j}, \gamma^+_{\tau_j} \geq \gamma^+_{\sigma_j}$ *and* $\alpha^-_{\tau_j} \geq \alpha^-_{\sigma_j}, \beta^-_{\tau_j} \leq \beta^-_{\sigma_j}, \gamma^-_{\tau_j} \leq \gamma^-_{\sigma_j}$*, then,*

$$BNNIGWGHM^{t,\,s}(\tau_1,\ \tau_2, \ldots,\ \tau_k) \leq BNNIGWGHM^{t,\,s}(\sigma_1,\ \sigma_2, \ldots,\ \sigma_k).$$

Theorem 11. *(Idempotency). Set a collection* $\tau_j = \langle \alpha^+_{\tau_j}, \beta^+_{\tau_j}, \gamma^+_{\tau_j}, \alpha^-_{\tau_j}, \beta^-_{\tau_j}, \gamma^-_{\tau_j} \rangle$ $(j = 1, 2, \cdots, k)$ *of BNNs; if* $\tau_j = \tau$*, then,*

$$BNNIGWGHM^{t,\,s}(\tau_1,\ \tau_2, \ldots,\ \tau_k) = BNNIGWGHM^{t,\,s}(\tau, \tau, \ldots \tau) = \tau.$$

Theorem 12. *(Boundedness). Set a collection* $\tau_j = \langle \alpha^+_{\tau_j}, \beta^+_{\tau_j}, \gamma^+_{\tau_j}, \alpha^-_{\tau_j}, \beta^-_{\tau_j}, \gamma^-_{\tau_j} \rangle$ $(j = 1, 2, \cdots, k)$ *of BNN, and let* $\tau^- = \langle \min\left(\alpha^+_{\tau_j}\right), \max\left(\beta^+_{\tau_j}\right), \max\left(\gamma^+_{\tau_j}\right), max\left(\alpha^-_{\tau_j}\right), min\left(\beta^-_{\tau_j}\right), min\left(\gamma^-_{\tau_j}\right) \rangle$ *, and* $\tau^+ = \langle max\left(\alpha^+_{\tau_j}\right), min\left(\beta^+_{\tau_j}\right), min(\gamma^+_{\tau_j}), min\left(\alpha^-_{\tau_j}\right), max\left(\beta^-_{\tau_j}\right), max\left(\gamma^-_{\tau_j}\right) \rangle$*; then,*

$$\tau^- \leq BNNIGWHM^{t,\,s}(\tau_1,\ \tau_2, \ldots,\ \tau_k) \leq \tau^+.$$

The proofs of theorems about BNNIGWGHM are similar to those about BNNIGWHM; thus, we omit them.

4. MCDM Methods Based on the BNNIGWHM and BNNIGWGHM Operator

We applied the BNNIGWHM and BNNIGWGHM operator to manage MCDM problems within BNN information in this section.

Suppose that a set $\Gamma = \{\Gamma_1, \Gamma_2, \ldots, \Gamma_n\}$ of alternatives and a set $\Phi = \{\Phi_1, \Phi_2, \ldots, \Phi_m\}$ of attributes, with the weight vector $\varepsilon = (\varepsilon_1, \varepsilon_2, \ldots, \varepsilon_m)$ of $\Phi_j (j = 1, 2, \ldots, m)$, in which $\sum_{j=1}^{n} \varepsilon_j = 1$ and $\varepsilon_j \in [0, 1]$. Decision-makers use BNNs to evaluate the alternatives. The evaluation values τ_{ij} for Γ_i associated with the attribute Φ_j are represented by the form of BNNs. Assume that $(\tau_{ij})_{n\times m} = \left(\langle \alpha^+_{\tau_{ij}}, \beta^+_{\tau_{ij}}, \gamma^+_{\tau_{ij}}, \alpha^-_{\tau_{ij}}, \beta^-_{\tau_{ij}}, \gamma^-_{\tau_{ij}} \rangle\right)_{n\times m}$ is the BNN decision matrix.

Now, based on the BNNIGWHM and BNNIGWGHM operator, we can develop some decision algorithms:

Step 1: Construct the decision matrix:

$$(\tau_{ij})_{n\times m} = \left(\langle \alpha^+_{\tau_{ij}}, \beta^+_{\tau_{ij}}, \gamma^+_{\tau_{ij}}, \alpha^-_{\tau_{ij}}, \beta^-_{\tau_{ij}}, \gamma^-_{\tau_{ij}} \rangle\right)_{n\times m}.$$

Step 2: According to Definition 11 or Definition 13, calculate τ_i.

Step 3: According to the Equation (5), calculate the score value of $s(\tau_i)$ for $\tau_i (i = 1, 2, \ldots, n)$.

Step 4: According to Definition 5, rank all the alternatives corresponding to the values of $s(\tau_i)$.

5. Illustrative Example

In this section, we used a numerical example adapted from the literature [16]. A woman wants to buy a car. Now, four kinds of cars $\Gamma_1, \Gamma_2, \Gamma_3$, and Γ_4 are taken into account according to gasoline consumption (Φ_1), aerodynamics (Φ_2), comfort (Φ_3), and safety performances (Φ_4). The importance of these four attributes is given as $\varepsilon = (0.5, 0.25, 0.125, 0.125)^T$. Then, she evaluates four alternatives under the above four attributes in the form of BNNs.

5.1. The Decision-Making Process Based on the BNNIGWHM Operator or BNNIGWGHM Operator

Step 1: Establish the BNN decision matrix $(\tau_{ij})_{4\times 4}$ provided by customer, as shown in Table 1.

Table 1. The decision matrix $(\tau_{ij})_{4\times 4}$.

	Φ_1	Φ_2	Φ_3	Φ_4
Γ_1	$\langle 0.5, 0.7, 0.2, -0.7, -0.3, -0.6 \rangle$	$\langle 0.4, 0.4, 0.5, -0.7, -0.8, -0.4 \rangle$	$\langle 0.7, 0.7.0.5, -0.8, -0.7, -0.6 \rangle$	$\langle 0.1, 0.5, 0.7, -0.5, -0.2, -0.8 \rangle$
Γ_2	$\langle 0.9, 0.7, 0.5, -0.7, -0.7, -0.1 \rangle$	$\langle 0.7, 0.6, 0.8, -0.7, -0.5, -0.1 \rangle$	$\langle 0.9, 0.4, 0.6, -0.1, -0.7, -0.5 \rangle$	$\langle 0.5, 0.2, 0.7, -0.5, -0.1, -0.9 \rangle$
Γ_3	$\langle 0.3, 0.4, 0.2, -0.6, -0.3, -0.7 \rangle$	$\langle 0.2, 0.2, 0.2, -0.4, -0.7, -0.4 \rangle$	$\langle 0.9, 0.5, 0.5, -0.6, -0.5, -0.2 \rangle$	$\langle 0.7, 0.5, 0.3, -0.4, -0.2, -0.2 \rangle$
Γ_4	$\langle 0.9, 0.7, 0.2, -0.8, -0.6, -0.1 \rangle$	$\langle 0.3, 0.5, 0.2, -0.5, -0.5, -0.2 \rangle$	$\langle 0.5, 0.4, 0.5, -0.1, -0.7, -0.2 \rangle$	$\langle 0.4, 0.2, 0.8, -0.5, -0.5, -0.6 \rangle$

Step 2: According to Definition 11 (suppose $p = q = 1$) and ε of attributes, calculate $\tau_i (i = 1, 2, 3, 4)$:

$$\tau_1 = \langle 0.4656, 0.5984, 0.3248, -0.6874, -0.4906, -0.5832 \rangle,$$

$$\tau_2 = \langle 0.8362, 0.5751, 0.5918, -0.5868, -0.6108, -0.2872 \rangle,$$

$$\tau_3 = \langle 0.4212, 0.3684, 0.2341, -0.5268, -0.4254, -0.5540\rangle,$$

$$\tau_4 = \langle 0.7456, 0.5504, 0.2669, -0.5838, -0.5793, -0.2006\rangle.$$

Step 3: According to Equation (5), calculate thscore value of $s(\tau_i)$ for $\tau_i (i = 1,\ 2,\ 3,\ 4)$:

$$s(\tau_1) = 0.4881; s(\tau_2) = 0.4968;\ s(\tau_3) = 0.5458;\ s(\tau_4) = 0.5207.$$

Step 4: According to Definition 5, rank $\Gamma_3 \succ \Gamma_4 \succ \Gamma_2 \succ \Gamma_1$ corresponding to $s(\tau_i)$; thus, Γ_3 is the best choice among all the alternatives.

Now, we use the BNNIGWGHM operator (set p =1, q = 1) to deal with this problem.

Step 1′: Just as described in step 1.

Step 2′: According to Definition 13 (suppose $p = q = 1$) and ε of attributes, calculate $\tau_i (i = 1,\ 2,\ 3,\ 4)$:

$$\tau_1 = \langle 0.3834, 0.5909, 0.4846, -0.6881, -0.4467, -0.5722\rangle,$$

$$\tau_2 = \langle 0.7371, 0.5369, 0.6627, -0.5747, -0.4484, -0.2381\rangle,$$

$$\tau_3 = \langle 0.4112, 0.3994, 0.2991, -0.5106, -0.3982, -0.3551\rangle,$$

$$\tau_4 = \langle 0.4922, 0.5086, 0.4579, -0.5674, -0.5684, -0.2139\rangle.$$

Step 3′: According to Equation (5), calculate the score value of $s(\tau_i)$. for $\tau_i (i = 1,\ 2,\ 3,\ 4)$:

$$s(\tau_1) = 0.4398;\ s(\tau_2) = 0.4416;\ s(\tau_3) = 0.4926;\ s(\tau_4) = 0.4568.$$

Step 4′: According to Definition 5, rank $\Gamma_3 \succ \Gamma_4 \succ \Gamma_2 \succ \Gamma_1$ corresponding to $s(\tau_i)$; thus, Γ_3 is the best choice among all the alternatives.

5.2. *Analyzing the Effects of the Parameters p and q*

In this section, we took different parameters p and q for calculating $\tau_i (i = 1,\ 2,\ 3,\ 4)$ for the alternative Γ_i, and then we analyzed the influence of the parameters p and q for the ranking result. Tables 2 and 3 show the values of $s(\tau_1)$ to $s(\tau_4)$ and the ranking results.

Table 2. Ranking results with different values of p and q based on bipolar neutrosophic number improved generalized weighted Heronian mean (BNNIGWHM) operator.

No.	p, q	BNNIGWHM	Ranking
1	$p = 1, q = 0$	$s(\tau_1) = 0.4915, s(\tau_2) = 0.4782, s(\tau_3) = 0.5471, s(\tau_4) = 0.5116$	$\Gamma_3 \succ \Gamma_4 \succ \Gamma_1 \succ \Gamma_2$
2	$p = 1, q = 0.5$	$s(\tau_1) = 0.4823, s(\tau_2) = 0.4809, s(\tau_3) = 0.5392, s(\tau_4) = 0.5083$	$\Gamma_3 \succ \Gamma_4 \succ \Gamma_1 \succ \Gamma_2$
3	$p = 1, q = 2$	$s(\tau_1) = 0.5059, s(\tau_2) = 0.5316, s(\tau_3) = 0.5658, s(\tau_4) = 0.5495$	$\Gamma_3 \succ \Gamma_4 \succ \Gamma_2 \succ \Gamma_1$
4	$p = 0, q = 1$	$s(\tau_1) = 0.5021, s(\tau_2) = 0.5433, s(\tau_3) = 0.5659, s(\tau_4) = 0.5517$	$\Gamma_3 \succ \Gamma_4 \succ \Gamma_2 \succ \Gamma_1$
5	$p = 0.5, q = 1$	$s(\tau_1) = 0.4871, s(\tau_2) = 0.4966, s(\tau_3) = 0.5445, s(\tau_4) = 0.5215$	$\Gamma_3 \succ \Gamma_4 \succ \Gamma_2 \succ \Gamma_1$
6	$p = 2, q = 1$	$s(\tau_1) = 0.4981, s(\tau_2) = 0.5161, s(\tau_3) = 0.5589, s(\tau_4) = 0.5346$	$\Gamma_3 \succ \Gamma_4 \succ \Gamma_2 \succ \Gamma_1$
7	$p = 2, q = 2$	$s(\tau_1) = 0.5105, s(\tau_2) = 0.5425, s(\tau_3) = 0.5730, s(\tau_4) = 0.5567$	$\Gamma_3 \succ \Gamma_4 \succ \Gamma_2 \succ \Gamma_1$

Table 3. The ranking with different p and q based on BNN improved generalized weighted geometry HM (BNNIGWGHM) operator.

No.	p, q	BNNIGWGHM	Ranking
1	$p = 1, q = 0$	$s(\tau_1) = 0.5228, s(\tau_2) = 0.5768, s(\tau_3) = 0.5967, s(\tau_4) = 0.5955$	$\Gamma_3 \succ \Gamma_4 \succ \Gamma_2 \succ \Gamma_1$
2	$p = 1, q = 0.5$	$s(\tau_1) = 0.4831, s(\tau_2) = 0.4893, s(\tau_3) = 0.5358, s(\tau_4) = 0.5039$	$\Gamma_3 \succ \Gamma_4 \succ \Gamma_2 \succ \Gamma_1$
3	$p = 1, q = 2$	$s(\tau_1) = 0.3834, s(\tau_2) = 0.3990, s(\tau_3) = 0.4504, s(\tau_4) = 0.4160$	$\Gamma_3 \succ \Gamma_4 \succ \Gamma_2 \succ \Gamma_1$
4	$p = 0, q = 1$	$s(\tau_1) = 0.4190, s(\tau_2) = 0.4376, s(\tau_3) = 0.4841, s(\tau_4) = 0.4584$	$\Gamma_3 \succ \Gamma_4 \succ \Gamma_2 \succ \Gamma_1$
5	$p = 0.5, q = 1$	$s(\tau_1) = 0.4411, s(\tau_2) = 0.4492, s(\tau_3) = 0.4957, s(\tau_4) = 0.4664$	$\Gamma_3 \succ \Gamma_4 \succ \Gamma_2 \succ \Gamma_1$
6	$p = 2, q = 1$	$s(\tau_1) = 0.4275, s(\tau_2) = 0.4211, s(\tau_3) = 0.4791, s(\tau_4) = 0.4341$	$\Gamma_3 \succ \Gamma_4 \succ \Gamma_1 \succ \Gamma_2$
7	$p = 2, q = 2$	$s(\tau_1) = 0.3873, s(\tau_2) = 0.3913, s(\tau_3) = 0.4496, s(\tau_4) = 0.4057$	$\Gamma_3 \succ \Gamma_4 \succ \Gamma_2 \succ \Gamma_1$

From the decision results based on BNNIGWHM in Table 2, we can see that all the ranking orders are $\Gamma_3 \succ \Gamma_4 \succ \Gamma_1 \succ \Gamma_2$ in No. 1–2 and all the ranking orders are $\Gamma_3 \succ \Gamma_4 \succ \Gamma_2 \succ \Gamma_1$ in No. 3–7; thus, the best choice is Γ_3. From the decision results based on BNNIGWGHM in Table 3, we can see that the ranking order is $\Gamma_3 \succ \Gamma_4 \succ \Gamma_1 \succ \Gamma_2$ in No. 6 and the others are $\Gamma_3 \succ \Gamma_4 \succ \Gamma_2 \succ \Gamma_1$; thus, the best choice is also Γ_3.

IGWHM and IGWGHM aggregation operators can take into account the correlation between attribute values and can better reflect the preferences of decision-makers and make the decision results more reasonable and reliable. A BNS has two fully independent parts, one part has three independent positive membership functions and the other has three independent negative membership functions, which can deal with uncertain information containing incompatible polarity. Here, we used the BNNIGWHM and BNNIGWGHM operators to solve real problems and analyze the influences of parameters p and q on the results of decisions, using different parameter values for sorting and comparing the corresponding results. Then, it could be found that the influences of parameters p and q on the results of decisions were small in these both methods. Comparing the results of the two methods, it can be found that their results were consistent; therefore, the proposed methods in this paper have feasibility and generality.

5.3. Comparison with Related Methods

In this section, we compared the methods proposed in this paper with other related methods proposed in the literature [16,19]. Table 4 lists the ranking results.

Table 4. Decision results based on four aggregation operators.

Aggregation Operator	Score Value	Ranking
The bipolar neutrosophic weighted average operator (Aw) and bipolar neutrosophic weighted geometric operator (Gw) proposed in Reference [16]	$\sigma(\tau_1) = 0.50$, $\sigma(\tau_2) = 0.52$, $\sigma(\tau_3) = 0.56$, $\sigma(\tau_4) = 0.54$	$\Gamma_3 \succ \Gamma_4 \succ \Gamma_2 \succ \Gamma_1$
The Similarity measures of bipolar neutrosophic sets proposed in Reference [19] with the following variables:		
$\lambda = 0.25$	$\sigma(\tau_1) = 0.24683$, $\sigma(\tau_2) = 0.11778$, $\sigma(\tau_3) = 0.27833$, $\sigma(\tau_4) = 0.21136$	$\Gamma_3 \succ \Gamma_1 \succ \Gamma_4 \succ \Gamma_2$
$\lambda = 0.3$	$\sigma(\tau_1) = 0.27063$, $\sigma(\tau_2) = 0.19497$, $\sigma(\tau_3) = 0.30222$, $\sigma(\tau_4) = 0.22904$	$\Gamma_3 \succ \Gamma_1 \succ \Gamma_4 \succ \Gamma_2$
$\lambda = 0.6$	$\sigma(\tau_1) = 0.41342$, $\sigma(\tau_2) = 0.29803$, $\sigma(\tau_3) = 0.44555$, $\sigma(\tau_4) = 0.33510$	$\Gamma_3 \succ \Gamma_1 \succ \Gamma_4 \succ \Gamma_2$
$\lambda = 0.9$	$\sigma(\tau_1) = 0.55620$, $\sigma(\tau_2) = 0.40109$, $\sigma(\tau_3) = 0.54313$, $\sigma(\tau_4) = 0.44116$	$\Gamma_1 \succ \Gamma_3 \succ \Gamma_4 \succ \Gamma_2$

In Table 4, we can see that the ranking results were different; Γ_3 was obtained as the optimal alternative except the method in Reference [19] with $\lambda = 0.9$. Compared with these related methods, the BNNIGWHM and BNNIGWGHM operators considered the correlation between attribute values and could better reflect the preferences of decision-makers and make the decision results more reasonable and reliable while dealing with uncertain information containing incompatible polarity. Thus, we think the proposed methods in this paper are more suitable to handle these decision-making problems.

6. Conclusions

This paper firstly proposed the BNNGWHM, BNNIGWHM, BNNGWGHM, and BNNIGWGHM operators for BNNs and discussed the related properties of these four operators. Furthermore, we developed two methods of MCDM in a BNN environment based on the BNNIGWHM and BNNIGWGHM operators. Finally, these two methods were used for a numerical example to establish their effectiveness and application. Dealing with the calculation, we took different values for p and

q to observe the sorting results and found that both parameters had little influence on the decision results. Furthermore, we compared the proposed methods with related methods and discovered that the selection result using the proposed methods was the same as the majority of existing methods. In the future, we will make further research bipolar neutrosophic sets, using, e.g., the technique for order preference by similarity to an ideal solution (TOPSIS) and VIKOR (VIseKriterijumska Optimizacija I Kompromisno Resenje, that means: multicriteria optimization and compromise solution, with pronunciation: vikor) methods with BNS [23], the weighted aggregated sum product assessment (WASPAS) method with BNS [24], the Multi-Attribute Market Value Assessment (MAMVA) method with BNS [25], and so on [26–28].

Author Contributions: C.F. proposed the BNNIGWHM and BNNIGWGHM operators and investigated their properties, C.F., S.F. and K.H. presented the organization and decision making method of this paper, J.Y. and E.F. provided the calculation and analysis of the illustrative example; All authors wrote the paper together.

Funding: This research was funded by [the National Natural Science Foundation of China] grant number [61703280, 61603258], [the Science and Technology Planning Project of Shaoxing City of China] grant number [2017B70056, 2018C10013], [the General Research Project of Zhejiang Provincial Department of Education] grant number [Y201839944], and [the Public Welfare Technology Research Project of Zhejiang Province] grant number [LGG19F020007].

Conflicts of Interest: The author declares no conflict of interest.

References

1. Zadeh, L.A. Fuzzy sets. *Inf. Control* **1965**, *8*, 338–356. [CrossRef]
2. Atanassov, K.T. Intuitionistic fuzzy sets. *Fuzzy Sets Syst.* **1986**, *20*, 87–96. [CrossRef]
3. Chen, T.Y.; Han, Y.; Liao, S.Z. The largest sub-algebra and its generalized tautology in disturbing fuzzy propositional logic. *Chin. J. Eng. Math.* **2003**, *20*, 118–120.
4. Zadeh, L.A. The concept of a linguistic variable and its application to approximate reasoning-I. *Inf. Sci.* **1975**, *8*, 199–249. [CrossRef]
5. Smarandache, F. *Neutrosophy: Neutrosophic Probability, Set, and Logic*; American Research Press: Rehoboth, MA, USA, 1998.
6. Wang, H.; Smarandache, F.; Zhang, Y.Q.; Sunderraman, R. Single valued neutrosophic sets. *Rev. Air Acad.* **2010**, *17*, 1–10.
7. Wang, H.; Smarandache, F.; Zhang, Y.Q.; Sunderraman, R. Interval neutrosophic sets and logic: Theory and applications in computing. *Int. Conf. Comput. Sci.* **2006**, *3991*, 920–923.
8. Smarandache, F. n-Valued Refined Neutrosophic Logic and Its Applications in Physics. *Prog. Phys.* **2013**, *4*, 143–146.
9. Zhang, W.R. Bipolar fuzzy sets and relations: A computational framework for cognitive modeling and multiagent decision analysis. In Proceedings of the The Workshop on Fuzzy Information Processing Society Bi Conference, San Antonio, TX, USA, 18–21 December 1994; pp. 305–309.
10. Zhang, W.R. *Yin Yang Bipolar Relativity: A Unifying Theory of Nature, Agents and Causality with Applications in Quantum Computing, Cognitive Informatics and Life Sciences*; IGI Global: Hershey, PA, USA, 2011.
11. Mesiarova-Zemankova, A.; Ahmad, K. Extended multi-polarity and multi-polar-valued fuzzy sets. *Fuzzy Sets Syst.* **2014**, *234*, 61–78. [CrossRef]
12. Chen, J.J.; Li, S.G.; Ma, S.Q.; Wang, X.P. m-polar fuzzy sets: An extension of bipolar fuzzy sets. *Sci. World J.* **2014**, *3*, 1–8.
13. Bosc, P.; Pivert, O. On a fuzzy bipolar relational algebra. *Inf. Sci.* **2013**, *219*, 1–16. [CrossRef]
14. Manemaran, S.V.; Chellappa, B. Structures on bipolar fuzzy groups and bipolar fuzzy ideals under (T, S) Norms. *Int. J. Comput. Appl.* **2012**, *9*, 24–28. [CrossRef]
15. Zhou, M.; Li, S. Application of bipolar fuzzy sets in semirings. *J. Math. Res. Appl.* **2014**, *34*, 61–72.
16. Deli, I.; Ali, M.; Smarandache, F. Bipolar neutrosophic sets and their application based on multi-criteria decision making problems. In Proceedings of the International Conference on Advanced Mechatronic Systems, Beijing, China, 22–24 August 2015; pp. 249–254.
17. Pramanik, S.; Dey, P.P.; Giri, B.C.; Smarandache, F. Bipolar neutrosophic projection based models for solving multi-attribute decision making problems. *Neutrosophic Sets Syst.* **2017**, *15*, 70–79.

18. Dey, P.P.; Pramanik, S.; Giri, B.C. *TOPSIS for Solving Multi-Attribute Decision Making Problems under Bi-Polar Neutrosophic Environment. New Trends in Neutrosophic Theory and Application, Pons asbl*; European Union: Brussels, Belgium, 2016; pp. 65–77.
19. Uluçay, V.; Deli, I.; Şahin, M. Similarity measures of bipolar neutrosophic sets and their application to multiple criteria decision making. *Neural Comput. Appl.* **2016**, *29*, 739–748. [CrossRef]
20. Wang, L.; Zhang, H.Y.; Wang, J.Q. Frank Choquet Bonferroni mean operators of bipolar neutrosophic sets and their application to multi-criteria decision-making problems. *Int. J. Fuzzy Syst.* **2017**, *3*, 1–16. [CrossRef]
21. Yu, D.J.; Wu, Y.Y. Interval-valued intuitionistic fuzzy Heronian mean operators and their application in multi-criteria decision making. *Afr. J. Bus. Manag.* **2012**, *6*, 4158–4168. [CrossRef]
22. Liu, P.D. *The Evaluation Method and Application Research of Enterprise Information Level Based on Fuzzy Multi-Attribute Decision-Making*; Beijing Jiaotong University: Bejing, China, 2010; pp. 113–129.
23. Pouresmaeil, H.; Shivanian, E.; Khorram, E.; Fathabadi, H.S. An Extended Method Using Topsis And Vikor For Multiple Attribute Decision Making with Multiple Decision Makers And Single Valued Neutrosophic Numbers. *Adv. Appl. Stat.* **2017**, *50*, 261–292. [CrossRef]
24. Zavadskas, E.K.; Bausys, R.; Lazauskas, M. Sustainable assessment of alternative sites for the construction of a waste incineration plant by applying WASPAS method with single-valued neutrosophic set. *Sustainability* **2015**, *7*, 15923–15936. [CrossRef]
25. Zavadskas, E.K.; Bausys, R.; Kaklauskas, A.; Ubartė, I.; Kuzminskė, A.; Gudienė, N. Sustainable market valuation of buildings by the single-valued neutrosophic MAMVA method. *Appl. Soft Comput.* **2017**, *57*, 74–87. [CrossRef]
26. Stanujkic, D.; Zavadskas, E.K.; Smarandache, F.; Brauers, W.K.M.; Karabasevic, D. A neutrosophic extension of the MULTIMOORA method. *Informatica* **2017**, *28*, 181–192. [CrossRef]
27. Li, Y.; Liu, P.; Chen, Y. Some Single Valued Neutrosophic Number Heronian Mean Operators and Their Application in Multiple Attribute Group Decision Making. *Informatica* **2016**, *27*, 85–110. [CrossRef]
28. Zavadskas, E.K.; Bausys, R.; Juodagalviene, B.; Garnyte-Sapranaviciene, I. Model for residential house element and material selection by neutrosophic MULTIMOORA method. *Eng. Appl. Artif. Intell.* **2017**, *64*, 315–324. [CrossRef]

mathematics

MDPI

Article

Neutrosophic Cubic Einstein Hybrid Geometric Aggregation Operators with Application in Prioritization Using Multiple Attribute Decision-Making Method

Khaleed Alhazaymeh [1], Muhammad Gulistan [2,*], Majid Khan [2] and Seifedine Kadry [3]

[1] Department of Basic Science and Mathematics, Faculty of Sciences, Philadelphia University, Amman 19392, Jordan; khazaymeh@philadelphia.edu.jo
[2] Department of Mathematics and Statistics, Hazara University Mansehra, Dhodial 21130, Pakistan; majid_swati@yahoo.com
[3] Department of Mathematics and Computer Science, Faculty of Science, Beirut Arab University, P.O. Box 11-5020, Beirut 11072809, Lebanon; s.kadry@bau.edu.lb
* Correspondence: gulistanmath@hu.edu.pk or gulistanm21@yahoo.com; Tel.: +92-997-414164

Received: 5 January 2019; Accepted: 18 February 2019; Published: 10 April 2019

Abstract: Viable collection is one of the imperative instruments of decision-making hypothesis. Collection operators are not simply the operators that normalize the value; they represent progressively broad values that can underline the entire information. Geometric weighted operators weight the values only, and the ordered weighted geometric operators weight the ordering position only. Both of these operators tend to the value that relates to the biggest weight segment. Hybrid collection operators beat these impediments of weighted total and request total operators. Hybrid collection operators weight the incentive as well as the requesting position. Neutrosophic cubic sets (NCs) are a classification of interim neutrosophic set and neutrosophic set. This distinguishing of neutrosophic cubic set empowers the decision-maker to manage ambiguous and conflicting data even more productively. In this paper, we characterized neutrosophic cubic hybrid geometric accumulation operator (NCHG) and neutrosophic cubic Einstein hybrid geometric collection operator (NCEHG). At that point, we outfitted these operators upon an everyday life issue which empoweredus to organize the key objective to develop the industry.

Keywords: neutrosophic cubic set; neutrosophic cubic hybrid geometric operator; neutrosophic cubic Einstein hybrid geometric operator; multiattributedecision-making (MADM)

1. Introduction

Life is loaded with indeterminacy and vagueness, which makes it hard to get adequate and exact information. This uncertain and obscure information can be tended to by fuzzy set [1], interim-valued fuzzy set (IVFS) [2,3], intuitionistic fuzzy set (IFS) [4], interim-valued intuitionistic fuzzy set (IVIFS) [5], cubic sets [6], neutrosophic set (Ns) [7], single-valued neutrosophic set (SVNs) [8], interim neutrosophic set (INs) [9], and neutrosophic cubic set [10]. Smarandache first investigated the hypothesis of neutrosophic sets [7].

Not long after thisinvestigation, it became a vital tool to manage obscure and conflicting information. The neutrosophic set comprises of three segments: truth enrollment, indeterminant participation, and deception enrollment. These segments can, likewise, be alluded to as participation, aversion, andnon-membership, and these segments range from $]0^-, 1^+[$. For science and designing issues, Wang et al. [8] proposed the idea of a single-valued neutrosophic set, which is a class of neutrosophic set, where the parts of single-valued neutrosophic set are in [0,1]. Wang et al. stretched it outto the interim neutrosophic set [9]. Jun et al. [10] consolidated both of these structures to frame the neutrosophic cubic set, which is the speculation of single-valued neutrosophic set and interim neutrosophic set. These structures drew scientistsinto apply it to various fields of sciences, building day-by-day life issues.

Decision-making is a basic instrument of everyday life issues. Analysts connected distinctive collection operators to neutrosophic sets and its augmentations. Zhan et. al. [11] took a shot at multicriteria decision-making on neutrosophic cubic sets. Banerjee et al. [12] utilized GRA(Grey Rational Analysis) for multicriteria decision-making on neutrosophic cubic sets. Lu and Ye [13] characterized cosine measure inneutrosophic cubic sets. Pramanik et al. [14] utilized a likeness measure to neutrosophic cubic sets. Shi and Ye [15] characterized Dombi total operators on neutrosophic cubic sets. Baolin et al. [16] connected Einstein accumulations to neutrosophic sets. Majid et al. [17] proposed neutrosophic cubic geometric and Einstein geometric collection operators. Different applied aspects of different types of fuzzy sets can be seen in [18–27].

A compelling accumulation is one of the imperative instruments of decision-making. Collection operators are not simply the operators that normalize the value, theyrepresent progressively broad values that can underline the entire data. The geometric weighted operator weights the values just where the requested weighted geometric collection operators weight the requesting position of values. In any case, the issue emerges when the load segments of weight vectors are so that one segment is a lot bigger than the other in parts of the weight vector. Motivated by such a circumstance, the thought of neutrosophic cubic crossbreed geometric and neutrosophic cubic Einstein hybrid geometric total operators are proposed. That is the reason we present the idea of neutrosophic cubic hybrid geometric and neutrosophic cubic Einstein hybrid geometric (NCEHG) collection operators. More often than not, the decision-making strategies are produced to pick one fitting option among the given. Be that as it may, frequently, in certain circumstances, we instead organize the option to pick a suitable one. Roused by such a circumstance, a technique is being created toprioritize the options. A numerical model is outfitted upon these operators to organize the vital objective to develop the industry.

2. Preliminaries

This section consists of some predefined definitions and results. We recommend the reader to see [1–3,6–10,16].

Definition 1. [1] *Mapping ψ: $U \to [0,1]$ is called fuzzy set, $\psi(u)$ is called membership function. Simply denoted by ψ.*

Definition 2. [2,3] *Mapping $\widetilde{\Psi} : U \to D[0,1]$, $D[0,1]$ has interval value of $[0,1]$, and is called interval-valued fuzzy set(IVF). For all $u \in U$ $\widetilde{\Psi}(u) = \left\{\left[\psi^L(u), \psi^U(u)\right] \middle| \psi^L(u), \psi^U(u) \in [0,1] \text{ and } \psi^L(u) \le \psi^U(u)\right\}$ is membership degree of u in $\widetilde{\Psi}$. Simply denoted by $\widetilde{\Psi} = \left[\Psi^L, \Psi^U\right]$.*

Definition 3. [6] *A structure $C = \left\{\left(u, \widetilde{\Psi}(u), \Psi(u)\right) \middle| u \in U\right\}$ is cubic set in U, in which $\widetilde{\Psi}(u)$ is IVF in U, i.e., $\widetilde{\Psi} = \left[\Psi^L, \Psi^U\right]$, and Ψ is fuzzy set in U. Simply denoted by $C = \left(\widetilde{\Psi}, \Psi\right)$. C^U denotes collection of cubic sets in U.*

Definition 4. [7] *A structure $N = \{(T_N(u), I_N(u), F_N(u)) | u \in U\}$ is neutrosophic set (Ns), where $\{T_N(u), I_N(u), F_N(u) \in]0^-, 1^+[\}$ and $T_N(u), I_N(u), F_N(u)$ are truth, indeterminacy, andfalsity function.*

Definition 5. [8] *A structure $N = \{(T_N(u), I_N(u), F_N(u)) | u \in U\}$ is single value neutrosophic set (SVNs), where $\{T_N(u), I_N(u), F_N(u) \in [0,1]\}$ are called truth, indeterminacy, and falsity functions respectively. Simply denoted by $N = (T_N, I_N, F_N)$.*

Definition 6. [9] *An interval neutrosophic set (INs) in U is a structure* $N = \left\{\left(\widetilde{T}_N(u), \widetilde{I}_N(u), \widetilde{F}_N(u)\right) \middle| u \in U\right\}$ *where* $\left\{\widetilde{T}_N(u), \widetilde{I}_N(u), \widetilde{F}_N(u) \in D[0,1]\right\}$ *respectively called truth, indeterminacy, and falsity function in U. Simply denoted by* $N = \left(\widetilde{T}_N, \widetilde{I}_N, \widetilde{F}_N\right)$. *For convenience, we denote* $N = \left(\widetilde{T}_N, \widetilde{I}_N, \widetilde{F}_N\right)$ *by* $N = \left(\widetilde{T}_N = \left[T_N^L, T_N^U\right], \widetilde{I}_N = \left[I_N^L, I_N^U\right], \widetilde{F}_N = \left[F_N^L, F_N^U\right]\right)$.

Definition 7. [10] *A structure* $N = \left\{\left(u, \widetilde{T}_N(u), \widetilde{I}_N(u), \widetilde{F}_N(u), T_N(u), I_N(u), F_N(u)\right) \middle| u \in U\right\}$ *is neutrosophic cubic set in U, in which* $\left(\widetilde{T}_N = \left[T_N^L, T_N^U\right], \widetilde{I}_N = \left[I_N^L, I_N^U\right], \widetilde{F}_N = \left[F_N^L, F_N^U\right]\right)$ *is an interval neutrosophic set and* (T_N, I_N, F_N) *is neutrosophic set in U. Simply denoted by* $N = \left(\widetilde{T}_N, \widetilde{I}_N, \widetilde{F}_N, T_N, I_N, F_N\right)$, $[0,0] \leq \widetilde{T}_N + \widetilde{I}_N + \widetilde{F}_N \leq [3,3]$, *and* $0 \leq T_N + I_N + F_N \leq 3$. N^U *denotes the collection of neutrosophic cubic sets in U. Simply denoted by* $N = \left(\widetilde{T}_N, \widetilde{I}_N, \widetilde{F}_N, T_N, I_N, F_N\right)$.

Definition 8. [16] *The t-operators are basically Union and Intersection operators in the theory of fuzzy sets which are denoted by t-conorm* (Γ^*) *and t-norm* (Γ), *respectively. The role of t-operators is very important in fuzzy theory and its applications.*

Definition 9. [16] $\Gamma^* : [0,1] \times [0,1] \rightarrow [0,1]$ *is called t-conorm if it satisfies the following axioms.*
Axiom 1 $\Gamma^*(1,u) = 1$ *and* $\Gamma^*(0,u) = 0$
Axiom 2 $\Gamma^*(u,v) = \Gamma^*(v,u)$ *for all a and b.*
Axiom 3 $\Gamma^*(u, \Gamma^*(v,w)) = \Gamma^*(\Gamma^*(u,v), w)$ *for all a, b, and c.*
Axiom 4 *If* $u \leq u'$ *and* $v \leq v'$, *then* $\Gamma^*(u,v) \leq \Gamma^*(u',v')$

Definition 10. [16] $\Gamma : [0,1] \times [0,1] \rightarrow [0,1]$ *is called t-norm if it satisfies the following axioms.*
Axiom 1 $\Gamma(1,u) = u$ *and* $\Gamma(0,u) = 0$
Axiom 2 $\Gamma(u,v) = \Gamma(v,u)$ *for all a and b.*
Axiom 3 $\Gamma(u, \Gamma(v,w)) = \Gamma(\Gamma(u,v), w)$ *for all a, b, and c.*
Axiom 4 *If* $u \leq u'$ *and* $v \leq v'$, *then* $\Gamma(u,v) \leq \Gamma(u',v')$

The t-conorms and t-norms families have a vast range, which correspond to unions and intersections, among these, Einstein sum and Einstein product are good choices since they give smooth approximations, like algebraic sum and algebraic product, respectively. Einstein sums $\oplus_E$ and Einstein products $\otimes_E$ are respectively the examples of t-conorm and t-norm:

$$\Gamma_E^*(u,v) = \frac{u+v}{1+uv},$$

$$\Gamma_E(u,v) = \frac{uv}{1+(1-u)(1-v)}.$$

Definition 11. [17] *The sum of two neutrosophic cubic sets,* $A = \left(\widetilde{T}_A, \widetilde{I}_A, \widetilde{F}_A, T_A, I_A, F_A\right)$, *where* $\widetilde{T}_A = \left[T_A^L, T_A^U\right], \widetilde{I}_A = \left[I_A^L, I_A^U\right], \widetilde{F}_A = \left[F_A^L, F_A^U\right]$, *and* $B = \left(\widetilde{T}_B, \widetilde{I}_B, \widetilde{F}_B, T_B, I_B, F_B\right)$, *where* $\widetilde{T}_B = \left[T_B^L, T_B^U\right], \widetilde{I}_B = \left[I_B^L, I_B^U\right], \widetilde{F}_B = \left[F_B^L, F_B^U\right]$ *is defined as*

$$A \oplus B = \begin{pmatrix} \left[T_A^L + T_B^L - T_A^L T_B^L, T_A^U + T_B^U - T_A^U T_B^U\right], \\ \left[I_A^L + I_B^L - I_A^L I_B^L, I_A^U + I_B^U - I_A^U I_B^U\right], \\ \left[F_A^L F_B^L, F_A^U F_B^U\right], \\ T_A T_B, I_A I_B, F_A + F_B - F_A F_B \end{pmatrix}$$

.

Definition 12. [17] *The product between two neutrosophic cubic sets,* $A = \left(\widetilde{T}_A, \widetilde{I}_A, \widetilde{F}_A, T_A, I_A, F_A\right)$, *where* $\widetilde{T}_A = \left[T_A^L, T_A^U\right], \widetilde{I}_A = \left[I_A^L, I_A^U\right], \widetilde{F}_A = \left[F_A^L, F_A^U\right]$, *and* $B = \left(\widetilde{T}_B, \widetilde{I}_B, \widetilde{F}_B, T_B, I_B, F_B\right)$, *where* $\widetilde{T}_B = \left[T_B^L, T_B^U\right], \widetilde{I}_B = \left[I_B^L, I_B^U\right], \widetilde{F}_B = \left[F_B^L, F_B^U\right]$ *is defined as*

$$A \otimes B = \begin{pmatrix} \left[T_A^L T_B^L, T_A^U T_B^U\right], \\ \left[I_A^L I_B^L, I_A^U I_B^U\right], \\ \left[F_A^L + F_B^L - F_A^L F_B^L, F_A^U + F_B^U - F_A^U F_B^U\right], \\ T_A + T_B - T_A T_B, I_A + I_B - I_A I_B, F_A F_B \end{pmatrix}$$

.

Definition 13. [17] *The scalar multiplication on a neutrosophic cubic set* $A = \left(\widetilde{T}_A, \widetilde{I}_A, \widetilde{F}_A, T_A, I_A, F_A\right)$, *where* $\widetilde{T}_A = \left[T_A^L, T_A^U\right], \widetilde{I}_A = \left[I_A^L, I_A^U\right], \widetilde{F}_A = \left[F_A^L, F_A^U\right]$, *and a scalar k is defined.*

$$kA = \begin{pmatrix} \left[1 - (1 - T_A^L)^k, 1 - (1 - T_A^U)^k\right], \\ \left[1 - (1 - I_A^L)^k, 1 - (1 - I_A^U)^k\right], \\ \left[\left(F_A^L\right)^k, \left(F_A^U\right)^k\right], \\ (T_A)^k, (I_A)^k, 1 - (1 - F_A)^k \end{pmatrix}$$

The exponential multiplication is followed by the following result.

Theorem 1. [17] *Let* $A = \left(\widetilde{T}_A, \widetilde{I}_A, \widetilde{F}_A, T_A, I_A, F_A\right)$, *where* $\widetilde{T}_A = \left[T_A^L, T_A^U\right], \widetilde{I}_A = \left[I_A^L, I_A^U\right], \widetilde{F}_A = \left[F_A^L, F_A^U\right]$, *is a neutrosophic cubic value, then, the exponential operation defined by*

$$A^k = \begin{pmatrix} \left[(T_A^L)^k, (T_A^U)^k\right], \\ \left[(I_A^L)^k, (I_A^U)^k\right], \\ \left[1 - \left(1 - F_A^L\right)^k, 1 - \left(1 - F_A^U\right)^k\right], \\ 1 - (1 - T_A)^k, 1 - (1 - I_A)^k, (F_A)^k \end{pmatrix},$$

where $A^k = A \otimes A \otimes, \ldots. \otimes A(k - times)$, *moreover,* A^k *is a neutrosophic cubic value for every positive value of k.*

Definition 14. [17] *The Einstein sum between two neutrosophic cubic sets* $A = \left(\widetilde{T}_A, \widetilde{I}_A, \widetilde{F}_A, T_A, I_A, F_A\right)$, *where* $\widetilde{T}_A = \left[T_A^L, T_A^U\right], \widetilde{I}_A = \left[I_A^L, I_A^U\right], \widetilde{F}_A = \left[F_A^L, F_A^U\right]$, *and* $B = \left(\widetilde{T}_B, \widetilde{I}_B, \widetilde{F}_B, T_B, I_B, F_B\right)$, *where* $\widetilde{T}_B = \left[T_B^L, T_B^U\right], \widetilde{I}_B = \left[I_B^L, I_B^U\right], \widetilde{F}_B = \left[F_B^L, F_B^U\right]$ *is defined as*

$$A \oplus_E B = \begin{pmatrix} \left[\frac{T_A^L + T_B^L}{1 + T_A^L T_B^L}, \frac{T_A^U + T_B^U}{1 + T_A^U T_B^U}\right], \\ \left[\frac{I_A^L + I_B^L}{1 + I_A^L I_B^L}, \frac{I_A^U + I_B^U}{1 + I_A^U I_B^U}\right], \\ \left[\frac{F_A^L F_B^L}{1 + (1 - F_A^L)(1 - F_B^L)}, \frac{F_A^U F_B^U}{1 + (1 - F_A^U)(1 - F_B^U)}\right] \\ \frac{T_A T_B}{1 + (1 - T_A)(1 - T_B)}, \frac{I_A I_B}{1 + (1 - I_A)(1 - I_B)}, \frac{F_A + F_B}{1 + F_A F_B} \end{pmatrix}.$$

Definition 15. [17] *The Einstein product between two neutrosophic cubic sets,* $A = \left(\widetilde{T}_A, \widetilde{I}_A, \widetilde{F}_A, T_A, I_A, F_A\right)$, *where* $\widetilde{T}_A = \left[T_A^L, T_A^U\right], \widetilde{I}_A = \left[I_A^L, I_A^U\right], \widetilde{F}_A = \left[F_A^L, F_A^U\right]$, *and* $B = \left(\widetilde{T}_B, \widetilde{I}_B, \widetilde{F}_B, T_B, I_B, F_B\right)$, *where* $\widetilde{T}_B = \left[T_B^L, T_B^U\right], \widetilde{I}_B = \left[I_B^L, I_B^U\right], \widetilde{F}_B = \left[F_B^L, F_B^U\right]$ *is defined as*

$$A \otimes_E B = \begin{pmatrix} \left[\frac{T_A^L T_B^L}{1+(1-T_A^L)(1-T_B^L)}, \frac{T_A^U T_B^U}{1+(1-T_A^U)(1-T_B^U)}\right], \\ \left[\frac{I_A^L I_B^L}{1+(1-I_A^L)(1-I_B^L)}, \frac{I_A^U I_B^U}{1+(1-I_A^U)(1-I_B^U)}\right], \\ \left[\frac{F_A^L+F_B^L}{1+F_A^L F_B^L}, \frac{F_A^U+F_B^U}{1+F_A^U F_B^U}\right] \\ \frac{T_A+T_B}{1+T_A T_B}, \frac{I_A+I_B}{1+I_A I_B}, \frac{F_A F_B}{1+(1-F_A)(1-F_B)} \end{pmatrix}.$$

Definition 16. [17] *The scalar multiplication on a neutrosophic cubic set,* $A = \left(\widetilde{T}_A, \widetilde{I}_A, \widetilde{F}_A, T_A, I_A, F_A\right)$, *where* $\widetilde{T}_A = \left[T_A^L, T_A^U\right], \widetilde{I}_A = \left[I_A^L, I_A^U\right], \widetilde{F}_A = \left[F_A^L, F_A^U\right]$, *and a scalar k is defined*

$$k_E A = \begin{pmatrix} \left[\frac{(1+T_A^L)^k-(1-T_A^L)^k}{(1+T_A^L)^k+(1-T_A^L)^k}, \frac{(1+T_A^U)^k-(1-T_A^U)^k}{(1+T_A^U)^k+(1-T_A^U)^k}\right], \\ \left[\frac{(1+I_A^L)^k-(1-I_A^L)^k}{(1+I_A^L)^k+(1-I_A^L)^k}, \frac{(1+I_A^U)^k-(1-I_A^U)^k}{(1+I_A^U)^k+(1-I_A^U)^k}\right], \\ \left[\frac{2(F_A^L)^k}{(2-F_A^L)^k+(F_A^L)^k}, \frac{2(F_A^U)^k}{(2-F_A^U)^k+(F_A^U)^k}\right], \\ \frac{2(T_A)^k}{(2-T_A)^k+(T_A)^k}, \frac{2(I_A)^k}{(2-I_A)^k+(I_A)^k}, \frac{(1+F_A)^k-(1-F_A)^k}{(1+F_A)^k+(1-F_A)^k} \end{pmatrix}.$$

The Einstein exponential multiplication is followed by the following result.

Theorem 2. [17] *Let* $A = \left(\widetilde{T}_A, \widetilde{I}_A, \widetilde{F}_A, T_A, I_A, F_A\right)$, *where* $\widetilde{T}_A = \left[T_A^L, T_A^U\right], \widetilde{I}_A = \left[I_A^L, I_A^U\right], \widetilde{F}_A = \left[F_A^L, F_A^U\right]$, *is a neutrosophic cubic value, then, the exponential operation defined by*

$$A^{E^k} = \begin{pmatrix} \left[\frac{2(T_A^L)^k}{(2-T_A^L)^k+(T_A^L)^k}, \frac{2(T_A^U)^k}{(2-T_A^U)^k+(T_A^U)^k}\right], \\ \left[\frac{2(I_A^L)^k}{(2-I_A^L)^k+(I_A^L)^k}, \frac{2(I_A^U)^k}{(2-I_A^U)^k+(I_A^U)^k}\right], \\ \left[\frac{(1+F_A^L)^k-(1-F_A^L)^k}{(1+F_A^L)^k+(1-F_A^L)^k}, \frac{(1+F_A^U)^k-(1-F_A^U)^k}{(1+F_A^U)^k+(1-F_A^U)^k}\right], \\ \frac{(1+T_A)^k-(1-T_A)^k}{(1+T_A)^k+(1-T_A)^k}, \frac{(1+I_A)^k-(1-I_A)^k}{(1+I_A)^k+(1-I_A)^k}, \frac{2(F_A)^k}{(2-F_A)^k+(F_A)^k} \end{pmatrix},$$

where $A^{E^k} = A \otimes_E A \otimes_E \ldots \otimes_E A(k - times)$, *moreover,* A^{E^k} *is a neutrosophic cubic value for every positive value of k.*

To compare two neutrosophic cubic values the score function is defined.

Definition 17. [17] *Let* $N = \left(\widetilde{T}_N, \widetilde{I}_N, \widetilde{F}_N, T_N, I_N, F_N\right)$, *where* $\widetilde{T}_N = \left[T_N^L, T_N^U\right], \widetilde{I}_N = \left[I_N^L, I_N^U\right], \widetilde{F}_N = \left[F_N^L, F_N^U\right]$ *is a neutrosophic cubic value, and the score function is defined as*

$$S(N) = \left[T_N^L - F_N^L + T_N^U - F_N^U + T_N - F_N\right].$$

If the score function of two values are equal, the accuracy function is used.

Definition 18. [17] *Let* $N = \left(\widetilde{T}_N, \widetilde{I}_N, \widetilde{F}_N, T_N, I_N, F_N\right)$, *where* $\widetilde{T}_N = \left[T_N^L, T_N^U\right], \widetilde{I}_N = \left[I_N^L, I_N^U\right], \widetilde{F}_N = \left[F_N^L, F_N^U\right]$ *is a neutrosophic cubic value, and theaccuracy function is defined as*

$$H(u) = \frac{1}{9}\left\{T_N^L + I_N^L + F_N^L + T_N^U + I_N^U + F_N^U + T_N + I_N + F_N\right\}.$$

The following definition describes the comparison relation between two neutrosophic cubic values.

Definition 19. [17] *Let* N_1, N_2 *be two neutrosophic cubic values, with core functions* S_{N_1}, S_{N_2}, *and accuracy function* H_{N_1}, H_{N_2}. *Then,*

(1) $S_{N_1} > S_{N_2} \Rightarrow N_1 > N_2$
(2) *If* $S_{N_1} = S_{N_2}$

   (i) $H_{N_1} > H_{N_2} \Rightarrow N_1 > N_2$
   (ii) $H_{N_1} = H_{N_2} \Rightarrow N_1 = N_2$

Definition 20. [17] *The neutrosophic cubic weighted geometric operator (NCWG) is defined as*

$$NCWG : R^m \to R \; defined \; by \; NCWG_w(N_1, N_2, \ldots, N_m) = \underset{j=1}{\overset{m}{\otimes}} N_j^{w_j},$$

where the weight $W = (w_1, w_2, \ldots, w_m)^T$ *of* $N_j (j = 1, 2, 3, \ldots, m)$, *such that* $w_j \in [0, 1]$ *and* $\sum_{j=1}^{m} = 1$.

Definition 21. [17] *The neutrosophic cubic ordered weighted geometric operator(NCOWG) is defined as*

$$NCOWG : R^m \to R \; defined \; by \; NCOWG_w(N_1, N_2, \ldots, N_m) = \underset{j=1}{\overset{m}{\otimes}} N_{(\gamma)_j}^{w_j},$$

where $N_{(\gamma)_j}$ *is the descending ordered neutrosophic cubic values,* $W = (w_1, w_2, \ldots, w_m)^T$ *of* $N_j (j = 1, 2, 3, \ldots, m)$, *such that* $w_j \in [0, 1]$ *and* $\sum_{j=1}^{m} = 1$.

Definition 22. [17] *The neutrosophic cubic Einstein weighted geometric operator(NCEWG) is defined as*

$$NCEWG : R^m \to R \; defined \; by \; NCEWG_w(N_1, N_2, \ldots, N_m) = \underset{j=1}{\overset{m}{\otimes}} \left(N_j\right)^{E^{w_j}},$$

where $W = (w_1, w_2, \ldots, w_m)^T$ *is weight of* $N_j (j = 1, 2, 3, \ldots, m)$, *such that* $w_j \in [0, 1]$ *and* $\sum_{j=1}^{m} = 1$.

Definition 23. [17] *Order neutrosophic cubic Einstein weighted geometric operator(NCEOWG) is defined as*

$$NCEOWG : R^m \to R \; by \; NCEOWG_w(N_1, N_2, \ldots, N_m) = \underset{j=1}{\overset{m}{\otimes}} \left(B_j\right)^{E^{w_j}},$$

where B_j *is the* j*th largest neutrosophic cubic value, and* $W = (w_1, w_2, \ldots, w_m)^T$ *is weight of* $N_j (j = 1, 2, 3, \ldots, m)$, *such that* $w_j \in [0, 1]$ *and* $\sum_{j=1}^{m} = 1$.

The neutrosophic cubic geometric aggregation operators weight only the neutrosophic cubic values, whereas neutrosophic cubic order geometric aggregation operators weight the orders of the values first then weight them. In the two cases, the amassed values that focused on the value relate to the biggest weight. The accompanying precedents represent the impediments of the NCWG and NCEWG.

Let $W = (0.7, 0.2, 0.1)$ be the weight corresponding to the neutrosophic cubic values

$$\begin{array}{l} N_1 = ([0.2, 0.7], [0.2, 0.4], [0.2, 0.5], 0.8, 0.5, 0.8) \\ N_2 = ([0.4, 0.6], [0.4, 0.7], [0.1, 0.3], 0.2, 0.6, 0.5) \\ N_3 = ([0.5, 0.8], [0.3, 0.6], [0.4, 0.9], 0.5, 0.8, 0.9) \end{array},$$

Then $S(N_1) = 0.2$, $S(N_2) = 0.3$, and $S(N_3) = -0.4$.

Therefore, $NCWG = ([0.251, 0.688], [0.239, 0.465], [0.204, 0.524], 0.711, 0.563, 0.737)$ and $NCOWG = ([0.356, 0.636], [0.338, 0.616], [0.155, 0.544], 0.421, 0.609, 0.737)$.

We observe that the higher the weight component, the aggregated value will tend to the corresponding neutrosophic cubic value of that vector. In NCWG, the value tendsto N_1, as the weight that corresponds to N_1 is highest, and in NCOWG, the highest component of weight corresponds to N_2. This situation often arises in aggregation problems. Motivated by such a situation, the idea of neutrosophic cubic hybrid geometric and neutrosophic cubic Einstein hybrid geometric operators are proposed.

3. Neutrosophic Cubic Hybrid Geometric and Neutrosophic Cubic Einstein Geometric Operators

This segment comprises of the following subsections. In Section 3.1 neutrosophic cubic crossbreed, the geometric operator is characterized. In Section 3.2 neutrosophic cubic Einstein crossbreed, the geometric operator is characterized. In Section 3.3, a calculation is characterized to organize the neutrosophic cubic values utilizing these tasks. In Section 3.4, a numerical model is outfitted upon Section 3.3.

3.1. Neutrosophic Cubic Hybrid Geometric Operator

NCWG operator weights only the neutrosophic cubic values, where NCOWG weights only the ordering positions. The idea of neutrosophic cubic hybrid geometric aggregation operators is developed to overcome these limitations. NCHG weights both the neutrosophic cubic values and its order positioning as well.

Definition 24. *NCHG* $: \Omega^m \to \Omega$ *is a mapping from* **m***-dimenion, which has associated weight* $W = (w_1, w_2, \ldots, w_m)^T$*,such that* $w_j \in [0,1]$ *and* $\sum_{j=1}^{m} w_j = 1$*, such that*

$$NCEOWG: R^m \to R \text{ by } NCEOWG_w(N_1, N_2, \ldots, N_m) = \overset{m}{\underset{j=1}{\otimes}} (B_j)^{E^{w_j}},$$

where $N_j^{\sim}$ *jth largest of the weighted neutrosophic cubic values* $\left\{N^{\sim}_{(j)}\left(N^{\sim}_{(j)} = N_j^{mw_j}\right), j = 1, 2, 3, , , m), W = (w_1, w_2, \ldots, w_m)^T\right\}$*, such that* $w_j \in [0,1]$ *and* $\sum_{j=1}^{m} w_j = 1$*, and m is the balancing coefficient.*

Theorem 3. *Let* $N_j = \left(\widetilde{T}_{N_j}, \widetilde{I}_{N_j}, \widetilde{F}_{N_j}, T_{N_j}, I_{N_j}, F_{N_j}\right)$, *where* $\widetilde{T}_{N_j} = \left[T^L_{N_j}, T^U_{N_j}\right], \widetilde{I}_{N_j} = \left[I^L_{N_j}, I^U_{N_j}\right], \widetilde{F}_{N_j} = \left[F^L_{N_j}, F^U_{N_j}\right]$, $(j = 1, 2, \ldots, m)$ *be collection of neutrosophic cubic values, then the aggregated value (NCHWG) is also a cubic value and*

$$NCHG(N_j) = \begin{pmatrix} \left[\overset{m}{\underset{j=1}{\otimes}}\left(T^{\sim L}_{\sigma(j)}\right)^{w_j}, \overset{m}{\underset{j=1}{\otimes}}\left(T^{\sim U}_{\sigma(j)}\right)^{w_j}\right], \left[\overset{m}{\underset{j=1}{\otimes}}\left(I^{\sim L}_{\sigma(j)}\right)^{w_j}, \left(I^{\sim U}_{\sigma(j)}\right)^{w_j}\right], \\ \left[1 - \overset{m}{\underset{j=1}{\otimes}}\left(1 - F^{\sim L}_{\sigma(j)}\right)^{w_j}, 1 - \overset{m}{\underset{j=1}{\otimes}}\left(1 - F^{\sim U}_{\sigma(j)}\right)^{w_j}\right], \\ 1 - \overset{m}{\underset{j=1}{\otimes}}\left(1 - T^{\sim}_{\sigma(j)}\right)^{w_j}, 1 - \overset{m}{\underset{j=1}{\otimes}}\left(1 - I^{\sim}_{\sigma(j)}\right)^{w_j}, \overset{m}{\underset{j=1}{\otimes}}\left(F^{\sim}_{\sigma(j)}\right)^{w_j} \end{pmatrix},$$

the weight $W = (w_1, w_2, \ldots, w_m)^T$, *such that* $w_j \in [0, 1]$ *and* $\sum_{j=1}^{m} = 1$.

Proof. By mathematical induction for $m = 2$, using

$$\overset{2}{\underset{j=1}{\otimes}} N_j^{w_j} = N_1^{w_1} \otimes N_2^{w_2}.$$

$$= \begin{pmatrix} \left[(T^L_{N\sigma(j)})^{w_1}, (T^U_{N_{\sigma(j)}})^{w_1}\right], \\ \left[(I^L_{N_{\sigma(j)}})^{w_1}, (I^U_{N_{\sigma(j)}})^{w_1}\right], \\ \left[1 - \left(1 - F^L_{N_{\sigma(j)}}\right)^{w_1}, 1 - \left(1 - F^U_{N_{\sigma(j)}}\right)^{w_1}\right], \\ 1 - \left(1 - \left(T_{N_{\sigma(j)}}\right)\right)^{w_1}, 1 - \left(1 - \left(I_{N\sigma(j)}\right)\right)^{w_1}, \left(F_{N_{\sigma(j)}}\right)^{w_1} \end{pmatrix} \otimes \begin{pmatrix} \left[(T^L_{N_{\sigma(j)}})^{w_2}, (T^U_{N_{\sigma(j)}})^{w_2}\right], \\ \left[(I^L_{N_{\sigma(j)}})^{w_2}, (I^U_{N_{\sigma(j)}})^{w_2}\right], \\ \left[1 - \left(1 - F^L_{N_{\sigma(j)}}\right)^{w_2}, 1 - \left(1 - F^U_{N_{\sigma(j)}}\right)^{w_2}\right], \\ 1 - \left(1 - \left(T_{N_{\sigma(j)}}\right)\right)^{w_2}, 1 - \left(1 - \left(I_{N_{\sigma(j)}}\right)\right)^{w_2}, \left(F_{N_{\sigma(j)}}\right)^{w_2} \end{pmatrix}$$

$$= \begin{pmatrix} \left[\overset{2}{\underset{j=1}{\otimes}}(T^L_{N_{\sigma(j)}})^{w_j}, \overset{2}{\underset{j=1}{\otimes}}(T^U_{N_{\sigma(j)}})^{w_j}\right], \left[\overset{2}{\underset{j=1}{\otimes}}(I^L_{N_{\sigma(j)}})^{w_j}, \overset{2}{\underset{j=1}{\otimes}}(I^U_{N_{\sigma(j)}})^{w_j}\right], \\ \left[1 - \overset{2}{\underset{j=1}{\otimes}}\left(1 - F^L_{N_{\sigma(j)}}\right)^{w_j}, 1 - \overset{2}{\underset{j=1}{\otimes}}\left(1 - F^U_{N_{\sigma(j)}}\right)^{w_j}\right], \\ 1 - \overset{2}{\underset{j=1}{\otimes}}\left(1 - \left(T_{N_{\sigma(j)}}\right)\right)^{w_j}, 1 - \overset{2}{\underset{j=1}{\otimes}}\left(1 - \left(I_{N_{\sigma(j)}}\right)\right)^{w_j}, \overset{2}{\underset{j=1}{\otimes}}\left(F_{N_{\sigma(j)}}\right)^{w_j} \end{pmatrix}.$$

Let the results hold for m.

$$\overset{m}{\underset{j=1}{\otimes}} N_j^{w_j} = \begin{pmatrix} \left[\overset{m}{\underset{j=1}{\otimes}}(T^L_{N_{\sigma(j)}})^{w_j}, \overset{m}{\underset{j=1}{\otimes}}(T^U_{N_{\sigma(j)}})^{w_j}\right], \left[\overset{m}{\underset{j=1}{\otimes}}(I^L_{N_{\sigma(j)}})^{w_j}, \overset{m}{\underset{j=1}{\otimes}}(I^U_{N_{\sigma(j)}})^{w_j}\right], \\ \left[1 - \overset{m}{\underset{j=1}{\otimes}}\left(1 - F^L_{N_{\sigma(j)}}\right)^{w_j}, 1 - \overset{m}{\underset{j=1}{\otimes}}\left(1 - F^U_{N_{\sigma(j)}}\right)^{w_j}\right], \\ 1 - \overset{m}{\underset{j=1}{\otimes}}\left(1 - T_{N_{\sigma(j)}}\right)^{w_j}, 1 - \overset{m}{\underset{j=1}{\otimes}}\left(1 - I_{N_{\sigma(j)}}\right)^{w_j}, \overset{m}{\underset{j=1}{\otimes}}\left(F_{N_{\sigma(j)}}\right)^{w_j} \end{pmatrix}$$

We prove the result for $m + 1$,

$$\text{as}\left(N_{j+1}\right)^{w_{j+1}} = \begin{pmatrix} \left[(T^L_{N_{j+1}})^{w_{j+1}}, (T^U_{N_{j+1}})^{w_{j+1}}\right], \left[(I^L_{N_{j+1}})^{w_{j+1}}, (I^U_{N_{j+1}})^{w_{j+1}}\right], \\ \left[1 - \left(1 - F^L_{N_{j+1}}\right)^{w_{j+1}}, 1 - \left(1 - F^U_{N_{j+1}}\right)^{w_{j+1}}\right], \\ 1 - \left(1 - T_{N_{j+1}}\right)^{w_{j+1}}, 1 - \left(1 - I_{N_{j+1}}\right)^{w_{j+1}}, \left(F_{N_{j+1}}\right)^{w_{j+1}} \end{pmatrix}$$

$$\overset{m}{\underset{j=1}{\otimes}} N_j^{w_j} \oplus N_{j+1}^{w_{j+1}} = \begin{pmatrix} \left[\overset{m}{\underset{j=1}{\otimes}}(T^L_{N_{\sigma(j)}})^{w_j}, \overset{m}{\underset{j=1}{\otimes}}(T^U_{N_{\sigma(j)}})^{w_j}\right], \left[\overset{m}{\underset{j=1}{\otimes}}(I^L_{N_{\sigma(j)}})^{w_j}, \overset{m}{\underset{j=1}{\otimes}}(I^U_{N_{\sigma(j)}})^{w_j}\right], \\ \left[1 - \overset{m}{\underset{j=1}{\otimes}}\left(1 - F^L_{N_{\sigma(j)}}\right)^{w_j}, 1 - \overset{m}{\underset{j=1}{\otimes}}\left(1 - F^U_{N_{\sigma(j)}}\right)^{w_j}\right], \\ 1 - \overset{m}{\underset{j=1}{\otimes}}\left(1 - T_{N_{\sigma(j)}}\right)^{w_j}, 1 - \overset{m}{\underset{j=1}{\otimes}}\left(1 - I_{N_{\sigma(j)}}\right)^{w_j}, \overset{m}{\underset{j=1}{\otimes}}\left(F_{N_{\sigma(j)}}\right)^{w_j} \end{pmatrix} \oplus \begin{pmatrix} \left[(T^L_{N_{j+1}})^{w_{j+1}}, (T^U_{N_{j+1}})^{w_{j+1}}\right], \\ \left[(I^L_{N_{j+1}})^{w_{j+1}}, (I^U_{N_{j+1}})^{w_{j+1}}\right], \\ \left[1 - \left(1 - F^L_{N_{j+1}}\right)^{w_{j+1}}, 1 - \left(1 - F^U_{N_{j+1}}\right)^{w_{j+1}}\right], \\ 1 - \left(1 - T_{N_{j+1}}\right)^{w_{j+1}}, 1 - \left(1 - I_{N_{j+1}}\right)_{1}^{w_{j+1}}, \left(F_{N_{j+1}}\right)^{w_{j+1}} \end{pmatrix}$$

$$\overset{m+1}{\underset{j=1}{\otimes}} N_j^{w_j} = \begin{pmatrix} \left[\overset{m}{\underset{j=1}{\otimes}}\left(T^L_{N_{\sigma(j)}}\right)^{w_j}\left(T^L_{N_{m+1}}\right)^{w_{m+1}}, \overset{m}{\underset{j=1}{\otimes}}\left(T^U_{N_{\sigma(j)}}\right)^{w_j}\left(T^U_{N_{m+1}}\right)^{w_{m+1}}\right], \left[\overset{m}{\underset{j=1}{\otimes}}\left(I^L_{N_{\sigma(j)}}\right)^{w_j}\left(I^L_{N_{m+1}}\right)^{w_{m+1}}, \overset{m}{\underset{j=1}{\otimes}}\left(I^U_{N_{\sigma(j)}}\right)^{w_j}\left(I^U_{N_{m+1}}\right)^{w_{m+1}}\right], \\ \begin{bmatrix} 1-\overset{m}{\underset{j=1}{\otimes}}(1-F^L_{N_{\sigma(j)}})^{w_j}+1-(1-F^L_{N_{m+1}})^{w_{m+1}}-\left(1-\overset{m}{\underset{j=1}{\otimes}}(1-F^L_{N_{\sigma(j)}})^{w_j}\right)\left(1-(1-F^L_{N_{m+1}})^{w_{m+1}}\right), \\ 1-\overset{m}{\underset{j=1}{\otimes}}(1-F^U_{N_{\sigma(j)}})^{w_j}+1-(1-F^U_{N_{m+1}})^{w_{m+1}}-\left(1-\overset{m}{\underset{j=1}{\otimes}}(1-F^U_{N_{\sigma(j)}})^{w_j}\right)\left(1-(1-F^U_{N_{m+1}})^{w_{m+1}}\right), \end{bmatrix}, \\ 1-\overset{m}{\underset{j=1}{\otimes}}(1-(T_{N_{\sigma(j)}}))^{w_j}+1-(1-(T_{N_{m+1}}))^{w_{m+1}}-\left(1-\overset{m}{\underset{j=1}{\otimes}}(1-(T_{N_{\sigma(j)}}))^{w_j}\right)\left(1-(1-(T_{N_{m+1}}))^{w_{m+1}}\right) \\ 1-\overset{m}{\underset{j=1}{\otimes}}(1-(I_{N_{\sigma(j)}}))^{w_j}+1-(1-(I_{N_{m+1}}))^{w_{m+1}}-\left(1-\overset{m}{\underset{j=1}{\otimes}}(1-(I_{N_{\sigma(j)}}))^{w_j}\right)\left(1-(1-(I_{N_{m+1}}))^{w_{m+1}}\right) \\ \overset{m}{\underset{j=1}{\otimes}}\left(F_{N_{\sigma(j)}}\right)^{w_j}\left(F_{N_{m+1}}\right)^{w_{m+1}} \end{pmatrix}$$

$$= \begin{pmatrix} \left[\overset{m+1}{\underset{j=1}{\otimes}}\left(T^L_{N_{\sigma(j)}}\right)^{w_j}, \overset{m+1}{\underset{j=1}{\otimes}}\left(T^U_{N_{\sigma(j)}}\right)^{w_j}\right], \left[\overset{m+1}{\underset{j=1}{\otimes}}\left(I^L_{N_{\sigma(j)}}\right)^{w_j}, \overset{m+1}{\underset{j=1}{\otimes}}\left(I^U_{N_{\sigma(j)}}\right)^{w_j}\right], \\ \begin{bmatrix} 2-\overset{m+1}{\underset{j=1}{\otimes}}(1-F^L_{N_{\sigma(j)}})^{w_j}-1+\overset{m+1}{\underset{j=1}{\otimes}}(1-F^L_{N_{\sigma(j)}})^{w_j}+(1-F^L_{N_{m+1}})^{w_{m+1}}-\left(\overset{m+1}{\underset{j=1}{\otimes}}(1-F^L_{N_{\sigma(j)}})^{w_j}\right)(1-F^L_{N_{m+1}})^{w_{m+1}}, \\ 2-\overset{m+1}{\underset{j=1}{\otimes}}(1-F^U_{N_{\sigma(j)}})^{w_j}-1+\overset{m+1}{\underset{j=1}{\otimes}}(1-F^U_{N_{\sigma(j)}})^{w_j}+(1-F^U_{N_{m+1}})^{w_{m+1}}-\left(\overset{m+1}{\underset{j=1}{\otimes}}(1-F^U_{N_{\sigma(j)}})^{w_j}\right)(1-F^U_{N_{m+1}})^{w_{m+1}} \end{bmatrix}, \\ 2-\overset{m+1}{\underset{j=1}{\otimes}}(1-T_{N_{\sigma(j)}})^{w_j}-1+\overset{m+1}{\underset{j=1}{\otimes}}(1-T_{N_{\sigma(j)}})^{w_j}+(1-T_{N_{m+1}})^{w_{m+1}}-\left(\overset{m+1}{\underset{j=1}{\otimes}}(1-T^{w_j}_{N_{\sigma(j)}})\right)(1-T_{N_{m+1}})^{w_{m+1}}, \\ 2-\overset{m+1}{\underset{j=1}{\otimes}}(1-I_{N_{\sigma(j)}})^{w_j}-1+\overset{m+1}{\underset{j=1}{\otimes}}(1-I_{N_{\sigma(j)}})^{w_j}+(1-I_{N_{m+1}})^{w_{m+1}}-\left(\overset{m+1}{\underset{j=1}{\otimes}}(1-I^{w_j}_{N_{\sigma(j)}})\right)(1-I_{N_{m+1}})^{w_{m+1}} \\ ,\overset{m+1}{\underset{j=1}{\otimes}}\left(F_{N_j}\right)^{w_j} \end{pmatrix}$$

$$= \begin{pmatrix} \left[\overset{m+1}{\underset{j=1}{\otimes}}\left(T^L_{N_{\sigma(j)}}\right)^{w_j}, \overset{m+1}{\underset{j=1}{\otimes}}\left(T^U_{N_{\sigma(j)}}\right)^{w_j}\right], \left[\overset{m+1}{\underset{j=1}{\otimes}}\left(I^L_{N_{\sigma(j)}}\right)^{w_j}, \overset{m+1}{\underset{j=1}{\otimes}}\left(I^U_{N_{\sigma(j)}}\right)^{w_j}\right], \\ \left[1-\overset{m+1}{\underset{j=1}{\otimes}}(1-F^L_{N_{\sigma(j)}})^{w_j}, 1-\overset{m+1}{\underset{j=1}{\otimes}}(1-F^U_{N_{\sigma(j)}})^{w_j}\right], \\ 1-\overset{m+1}{\underset{j=1}{\otimes}}\left(1-T_{N_{\sigma(j)}}\right)^{w_j}, 1-\overset{m+1}{\underset{j=1}{\otimes}}\left(1-I_{N_{\sigma(j)}}\right)^{w_j}, \overset{m+1}{\underset{j=1}{\otimes}}\left(F_{N_{\sigma(j)}}\right)^{w_j} \end{pmatrix}$$

which completes the proof. □

Theorem 4. *The NCWG is a special case of NCHG operator.*

Proof. Let $W = (\frac{1}{m}, \frac{1}{m}, \ldots, \frac{1}{m})^T$. Then,

$NCHG(N_1, N_2, \ldots, N_m)^w = \left(N^{\sim}_{\sigma(1)}\right)^{w_1} \otimes \left(N^{\sim}_{\sigma(2)}\right)^{w_2} \otimes, \ldots, \otimes \left(N^{\sim}_{\sigma(m)}\right)^{w_m}$

$= \left(N^{\sim}_{\sigma(1)}\right)^{\frac{1}{m}} \otimes \left(N^{\sim}_{\sigma(2)}\right)^{\frac{1}{m}} \otimes, \ldots, \otimes \left(N^{\sim}_{\sigma(m)}\right)^{\frac{1}{m}}$

$= (N_1, N_2, \ldots, N_m)^{\frac{1}{m}}$

$= (N_1)^{w_1}, (N_2)^{w_2}, \ldots, (N_m)^{w_m}$

$= NCWG(N_1, N_2, \ldots, N_m)$. □

Theorem 5. *The NCOWG is a special case of NCHG.*

Proof. Let $W = (\frac{1}{m}, \frac{1}{m}, \ldots, \frac{1}{m})^T$. Then,

$NCHG(N_1, N_2, \ldots, N_m)^w = \left(N^{\sim}_{\sigma(1)}\right)^{w_1} \otimes \left(N^{\sim}_{\sigma(2)}\right)^{w_2} \otimes, \ldots, \otimes \left(N^{\sim}_{\sigma(m)}\right)^{w_m}$

$= \left(N^{\sim}_{\sigma(1)}\right)^{\frac{1}{m}} \otimes \left(N^{\sim}_{\sigma(2)}\right)^{\frac{1}{m}} \otimes, \ldots, \otimes \left(N^{\sim}_{\sigma(m)}\right)^{\frac{1}{m}}$

$= (N_1, N_2, \ldots, N_m)^{\frac{1}{m}}$

$$= (N_1)^{w_1}, (N_2)^{w_2}, ..., (N_m)^{w_m}$$
$$= NCOWG(N_1, N_2, ..., N_m). \quad \square$$

3.2. Neutrosophic Cubic Einstein Hybrid Geometric Operator

NCEWG operator weights only the neutrosophic cubic values, where NCEOWGA weights only the ordering positions. The idea of neutrosophic cubic Einstein hybrid aggregation operators (NCEHG) is developed to overcome these limitations, which weights both the given neutrosophic cubic value and its order position as well.

Definition 25. *NCEHG* $: \Omega^m \to \Omega$ *is a map from m-dimension which has an associated vector* $W = (w_1, w_2, ..., w_m)^T$*, where* $w_j \in [0,1]$ *and* $\sum_{j=1}^{m} = 1$*, such that*

$$NCEHG_w(N_1, N_2, ..., N_m) = \left(N^{\sim}_{\sigma(1)}\right)^{w_1} \otimes_E \left(N^{\sim}_{\sigma(2)}\right)^{w_2} \otimes_E, ..., \otimes_E \left(N^{\sim}_{\sigma(m)}\right)^{w_m},$$

where $N^{\sim}_j$ *is the jth largest of the weighted neutrosophic cubic values* $\left\{N^{\sim}_{(j)}\left(N^{\sim}_{(j)} = N_j^{mw_j}\right), j = 1,2,3,,,m), W = (w_1, w_2, ..., w_m)^T\right\}$*, with* $w_j \in [0,1]$ *and* $\sum_{j=1}^{m} = 1$*, and m is the balancing coefficient.*

Theorem 6. *Let* $N_j = \left(\widetilde{T}_{N_j}, \widetilde{I}_{N_j}, \widetilde{F}_{N_j}, T_{N_j}, I_{N_j}, F_{N_j}\right)$*, where* $\widetilde{T}_{N_j} = \left[T^L_{N_j}, T^U_{N_j}\right], \widetilde{I}_{N_j} = \left[I^L_{N_j}, I^U_{N_j}\right], \widetilde{F}_{N_j} = \left[F^L_{N_j}, F^U_{N_j}\right]$*,* $N = \left(\widetilde{T}_{N_j}, \widetilde{I}_{N_j}, \widetilde{F}_{N_j}, T_{N_j}, I_{N_j}, F_{N_j}\right)$*, where* $\widetilde{T}_{N_j} = \left[T^L_{N_j}, T^U_{N_j}\right], \widetilde{I}_{N_j} = \left[I^L_{N_j}, I^U_{N_j}\right], \widetilde{F}_{N_j} = \left[F^L_{N_j}, F^U_{N_j}\right]$ $(j = 1, 2, ..., m)$ *is acollection of neutrosophic cubic values, then, their aggregated value by NCEWG operator is also a cubic value and*

$$NCEHG(N_j) = \left(\begin{array}{c} \left[\dfrac{2\bigotimes_{j=1}^{m}\left(T^L_{N_{\sigma(j)}}\right)^{w_j}}{\bigotimes_{j=1}^{m}\left(2-T^L_{N_{\sigma(j)}}\right)^{w_j} + \bigotimes_{j=1}^{m}\left(T^L_{N_{\sigma(j)}}\right)^{w_j}}, \dfrac{2\bigotimes_{j=1}^{m}\left(T^U_{N_{\sigma(j)}}\right)^{w_j}}{\bigotimes_{j=1}^{m}\left(2-T^U_{N_{\sigma(j)}}\right)^{w_j} + \bigotimes_{j=1}^{m}\left(T^U_{N_{\sigma(j)}}\right)^{w_j}} \right], \\ \left[\dfrac{2\bigotimes_{j=1}^{m}\left(I^L_{N_{\sigma(j)}}\right)^{w_j}}{\bigotimes_{j=1}^{m}\left(2-I^L_{N_{\sigma(j)}}\right)^{w_j} + \bigotimes_{j=1}^{m}\left(I^L_{N_{\sigma(j)}}\right)^{w_j}}, \dfrac{2\bigotimes_{j=1}^{m}\left(I^U_{N_{\sigma(j)}}\right)^{w_j}}{\bigotimes_{j=1}^{m}\left(2-I^U_{N_{\sigma(j)}}\right)^{w_j} + \bigotimes_{j=1}^{m}\left(I^U_{N_{\sigma(j)}}\right)^{w_j}} \right], \\ \left[\dfrac{\bigotimes_{j=1}^{m}(1+F^L_{N_{\sigma(j)}})^{w_j} - \bigotimes_{j=1}^{m}(1-F^L_{N_{\sigma(j)}})^{w_j}}{\bigotimes_{j=1}^{m}(1+F^L_{N_{\sigma(j)}})^{w_j} + \bigotimes_{j=1}^{m}(1-F^L_{N_{\sigma(j)}})^{w_j}}, \dfrac{\bigotimes_{j=1}^{m}(1+F^U_{N_{\sigma(j)}})^{w_j} - \bigotimes_{j=1}^{m}(1-F^U_{N_{\sigma(j)}})^{w_j}}{\bigotimes_{j=1}^{m}(1+F^U_{N_{\sigma(j)}})^{w_j} + \bigotimes_{j=1}^{m}(1-F^U_{N_{\sigma(j)}})^{w_j}} \right], \\ \dfrac{\bigotimes_{j=1}^{m}(1+T_{N_{\sigma(j)}})^{w_j} - \bigotimes_{j=1}^{m}(1-T_{N_{\sigma(j)}})^{w_j}}{\bigotimes_{j=1}^{m}(1+T_{N_{\sigma(j)}})^{w_j} + \bigotimes_{j=1}^{m}(1-T_{N_{\sigma(j)}})^{w_j}}, \dfrac{\bigotimes_{j=1}^{m}(1+I_{N_{\sigma(j)}})^{w_j} - \bigotimes_{j=1}^{m}(1-I_{N_{\sigma(j)}})^{w_j}}{\bigotimes_{j=1}^{m}(1+I_{N_{\sigma(j)}})^{w_j} + \bigotimes_{j=1}^{m}(1-I_{N_{\sigma(j)}})^{w_j}}, \\ \dfrac{2\bigotimes_{j=1}^{m}\left(F_{N_{\sigma(j)}}\right)^{w_j}}{\bigotimes_{j=1}^{m}\left(2-F_{N_{\sigma(j)}}\right)^{w_j} + \bigotimes_{j=1}^{m}\left(F_{N_{\sigma(j)}}\right)^{w_j}} \end{array} \right),$$

where $W = (w_1, w_2, ..., w_m)^T$ *is weight of* $N_j (j = 1, 2, 3, ..., m)$*, with* $w_j \in [0,1]$ *and* $\sum_{j=1}^{m} = 1$*.*

Proof. We use mathematical induction to prove this result for $k = 2$, using definition

$$\left(N_1^E\right)^{w_1} = \left(\begin{array}{c} \left[\frac{2\left(T^L_{N_1}\right)^{w_1}}{\left(2-T^L_{N_1}\right)^{w_1}+T^L_{N_1}}, \frac{2\left(T^U_{N_1}\right)^{w_1}}{\left(2-T^U_{N_1}\right)^{w_1}+T^U_{N_1}} \right], \left[\frac{2\left(I^L_{N_1}\right)^{w_1}}{\left(2-I^L_{N_1}\right)^{w_1}+I^L_{N_1}}, \frac{2\left(I^U_{N_1}\right)^{w_1}}{\left(2-I^U_{N_1}\right)^{w_1}+I^U_{N_1}} \right], \\ \left[\frac{(1+F^L_{N_1})^{w_1}-(1-F^L_{N_1})^{w_1}}{(1+F^L_{N_1})^{w_1}+(1-F^L_{N_1})^{w_1}}, \frac{(1+F^U_{N_1})^{w_1}-(1-F^U_{N_1})^{w_1}}{(1+F^U_{N_1})^{w_1}+(1-F^U_{N_1})^{w_1}} \right], \\ \frac{(1+T_{N_1})^{w_1}-(1-T_{N_1})^{w_1}}{(1+T_{N_1})^{w_1}+(1-T_{N_1})^{w_1}}, \frac{(1+I_{N_1})^{w_1}-(1-I_{N_1})^{w_1}}{(1+I_{N_1})^{w_1}+(1-I_{N_1})^{w_1}}, \frac{2\left(F_{N_1}\right)^{w_1}}{\left(2-F_{N_1}\right)^{w_1}+F_{N_1}} \end{array} \right)$$

$$\left(N_2^E\right)^{w_2} = \left(\begin{array}{c} \left[\frac{2\left(T^L_{N_2}\right)^{w_2}}{\left(2-T^L_{N_2}\right)^{w_2}+T^L_{N_2}}, \frac{2\left(T^U_{N_2}\right)^{w_2}}{\left(2-T^U_{N_2}\right)^{w_2}+T^U_{N_2}} \right], \left[\frac{2\left(I^L_{N_2}\right)^{w_2}}{\left(2-I^L_{N_2}\right)^{w_2}+I^L_{N_2}}, \frac{2\left(I^U_{N_2}\right)^{w_2}}{\left(2-I^U_{N_2}\right)^{w_2}+I^U_{N_2}} \right], \\ \left[\frac{(1+F^L_{N_2})^{w_2}-(1-F^L_{N_2})^{w_2}}{(1+F^L_{N_2})^{w_2}+(1-F^L_{N_2})^{w_2}}, \frac{(1+F^U_{N_2})^{w_2}-(1-F^U_{N_2})^{w_2}}{(1+F^U_{N_2})^{w_2}+(1-F^U_{N_2})^{w_2}} \right], \\ \frac{(1+T_{N_2})^{w_2}-(1-T_{N_2})^{w_2}}{(1+T_{N_2})^{w_2}+(1-T_{N_2})^{w_2}}, \frac{(1+I_{N_2})^{w_2}-(1-I_{N_2})^{w_2}}{(1+I_{N_2})^{w_2}+(1-I_{N_2})^{w_2}}, \frac{2\left(F_{N_2}\right)^{w_2}}{\left(2-F_{N_2}\right)^{w_2}+F_{N_2}} \end{array} \right)$$

$$\bigotimes_{j=1}^{2}\left(N_j^E\right)^{w_j} = \left(\begin{array}{c} \left[\frac{2\bigotimes_{j=1}^{2}\left(T^L_{N_{\sigma(j)}}\right)^{w_j}}{\bigotimes_{j=1}^{2}\left(2-T^L_{N_{\sigma(j)}}\right)^{w_j}+\bigotimes_{j=1}^{2}\left(T^L_{N_{\sigma(j)}}\right)^{w_j}}, \frac{2\bigotimes_{j=1}^{2}\left(T^U_{N_{\sigma(j)}}\right)^{w_j}}{\bigotimes_{j=1}^{2}\left(2-T^U_{N_{\sigma(j)}}\right)^{w_j}+\bigotimes_{j=1}^{2}\left(T^U_{N_{\sigma(j)}}\right)^{w_j}} \right], \\ \left[\frac{2\bigotimes_{j=1}^{2}\left(I^L_{N_{\sigma(j)}}\right)^{w_j}}{\bigotimes_{j=1}^{2}\left(2-I^L_{N_{\sigma(j)}}\right)^{w_j}+\bigotimes_{j=1}^{2}\left(I^L_{N_{\sigma(j)}}\right)^{w_j}}, \frac{2\bigotimes_{j=1}^{2}\left(I^U_{N_{\sigma(j)}}\right)^{w_j}}{\bigotimes_{j=1}^{2}\left(2-I^U_{N_{\sigma(j)}}\right)^{w_j}+\bigotimes_{j=1}^{2}\left(I^U_{N_{\sigma(j)}}\right)^{w_j}} \right], \\ \left[\frac{\bigotimes_{j=1}^{2}(1+F^L_{N_{\sigma(j)}})^{w_j}-\bigotimes_{j=1}^{2}(1-F^L_{N_{\sigma(j)}})^{w_j}}{\bigotimes_{j=1}^{2}(1+F^L_{N_{\sigma(j)}})^{w_j}+\bigotimes_{j=1}^{2}(1-F^L_{N_{\sigma(j)}})^{w_j}}, \frac{\bigotimes_{j=1}^{2}(1+F^U_{N_{\sigma(j)}})^{w_j}-\bigotimes_{j=1}^{2}(1-F^U_{N_{\sigma(j)}})^{w_j}}{\bigotimes_{j=1}^{2}(1+F^U_{N_{\sigma(j)}})^{w_j}+\bigotimes_{j=1}^{2}(1-F^U_{N_{\sigma(j)}})^{w_j}} \right], \\ \frac{\bigotimes_{j=1}^{2}(1+T_{N_{\sigma(j)}})^{w_j}-\bigotimes_{j=1}^{2}(1-T_{N_{\sigma(j)}})^{w_j}}{\bigotimes_{j=1}^{2}(1+T_{N_{\sigma(j)}})^{w_j}+\bigotimes_{j=1}^{2}(1-T_{N_{\sigma(j)}})^{w_j}}, \frac{\bigotimes_{j=1}^{2}(1+I_{N_{\sigma(j)}})^{w_j}-\bigotimes_{j=1}^{2}(1-I_{N_{\sigma(j)}})^{w_j}}{\bigotimes_{j=1}^{2}(1+I_{N_{\sigma(j)}})^{w_j}+\bigotimes_{j=1}^{2}(1-I_{N_{\sigma(j)}})^{w_j}}, \\ \frac{2\bigotimes_{j=1}^{2}\left(F_{N_{\sigma(j)}}\right)^{w_j}}{\bigotimes_{j=1}^{2}\left(2-F_{N_{\sigma(j)}}\right)^{w_j}+\bigotimes_{j=1}^{2}\left(F_{N_{\sigma(j)}}\right)^{w_j}} \end{array} \right).$$

Let the result holds for m.

$$\bigotimes_{j=1}^{m}\left(N_j^E\right)^{w_j} = \left(\begin{array}{c} \left[\frac{2\bigotimes_{j=1}^{m}\left(T^L_{N_{\sigma(j)}}\right)^{w_j}}{\bigotimes_{j=1}^{m}\left(2-T^L_{N_{\sigma(j)}}\right)^{w_j}+\bigotimes_{j=1}^{m}\left(T^L_{N_{\sigma(j)}}\right)^{w_j}}, \frac{2\bigotimes_{j=1}^{m}\left(T^U_{N_{\sigma(j)}}\right)^{w_j}}{\bigotimes_{j=1}^{m}\left(2-T^U_{N_{\sigma(j)}}\right)^{w_j}+\bigotimes_{j=1}^{m}\left(T^U_{N_{\sigma(j)}}\right)^{w_j}} \right], \\ \left[\frac{2\bigotimes_{j=1}^{m}\left(I^L_{N_{\sigma(j)}}\right)^{w_j}}{\bigotimes_{j=1}^{m}\left(2-I^L_{N_{\sigma(j)}}\right)^{w_j}+\bigotimes_{j=1}^{m}\left(I^L_{N_{\sigma(j)}}\right)^{w_j}}, \frac{2\bigotimes_{j=1}^{m}\left(I^U_{N_{\sigma(j)}}\right)^{w_j}}{\bigotimes_{j=1}^{m}\left(2-I^U_{N_{\sigma(j)}}\right)^{w_j}+\bigotimes_{j=1}^{m}\left(I^U_{N_{\sigma(j)}}\right)^{w_j}} \right], \\ \left[\frac{\bigotimes_{j=1}^{m}(1+F^L_{N_{\sigma(j)}})^{w_j}-\bigotimes_{j=1}^{m}(1-F^L_{N_{\sigma(j)}})^{w_j}}{\bigotimes_{j=1}^{m}(1+F^L_{N_{\sigma(j)}})^{w_j}+\bigotimes_{j=1}^{m}(1-F^L_{N_{\sigma(j)}})^{w_j}}, \frac{\bigotimes_{j=1}^{m}(1+F^U_{N_{\sigma(j)}})^{w_j}-\bigotimes_{j=1}^{m}(1-F^U_{N_{\sigma(j)}})^{w_j}}{\bigotimes_{j=1}^{m}(1+F^U_{N_{\sigma(j)}})^{w_j}+\bigotimes_{j=1}^{m}(1-F^U_{N_{\sigma(j)}})^{w_j}} \right], \\ \frac{\bigotimes_{j=1}^{m}(1+T_{N_{\sigma(j)}})^{w_j}-\bigotimes_{j=1}^{m}(1-T_{N_{\sigma(j)}})^{w_j}}{\bigotimes_{j=1}^{m}(1+T_{N_{\sigma(j)}})^{w_j}+\bigotimes_{j=1}^{m}(1-T_{N_{\sigma(j)}})^{w_j}}, \frac{\bigotimes_{j=1}^{m}(1+I_{N_{\sigma(j)}})^{w_j}-\bigotimes_{j=1}^{m}(1-I_{N_{\sigma(j)}})^{w_j}}{\bigotimes_{j=1}^{m}(1+I_{N_{\sigma(j)}})^{w_j}+\bigotimes_{j=1}^{m}(1-I_{N_{\sigma(j)}})^{w_j}}, \\ \frac{2\bigotimes_{j=1}^{m}\left(F_{N_{\sigma(j)}}\right)^{w_j}}{\bigotimes_{j=1}^{m}\left(2-F_{N_{\sigma(j)}}\right)^{w_j}+\bigotimes_{j=1}^{m}\left(F_{N_{\sigma(j)}}\right)^{w_j}} \end{array} \right).$$

We prove the result holds for $m+1$.

$$as\left(N_{m+1}^{E}\right)^{w_{m+1}} = \left(\begin{array}{c} \left[\frac{2\left(T^L_{N_{m+1}}\right)^{w_{m+1}}}{\left(2-T^L_{N_{m+1}}\right)^{w_{m+1}}+\left(T^L_N\right)^{w_{m+1}}}, \frac{2\left(T^U_{N_{m+1}}\right)^{w_{m+1}}}{\left(2-T^U_N\right)^{w_{m+1}}+\left(T^U_N\right)^{w_{m+1}}}\right], \\ \left[\frac{2\left(I^L_{N_{m+1}}\right)^{w_{m+1}}}{\left(2-I^L_{N_{m+1}}\right)^{w_{m+1}}+\left(I^L_N\right)^{w_{m+1}}}, \frac{2\left(I^U_{N_{m+1}}\right)^{w_{m+1}}}{\left(2-I^U_{N_{m+1}}\right)^{w_{m+1}}+\left(I^U_N\right)^{w_{m+1}}}\right], \\ \left[\frac{(1+F^L_{N_{m+1}})^{w_{m+1}}-(1-F^L_{N_{m+1}})^{w_{m+1}}}{(1+F^L_{N_{m+1}})^{w_{m+1}}+(1-F^L_{N_{m+1}})^{w_{m+1}}}, \frac{(1+F^U_{N_{m+1}})^{w_{m+1}}-(1-F^U_{N_{m+1}})^{w_{m+1}}}{(1+F^U_{N_{m+1}})^{w_{m+1}}+(1-F^U_{N_{m+1}})^{w_{m+1}}}\right], \\ \frac{(1+T_{N_{m+1}})^{w_{m+1}}-(1-T_{N_{m+1}})^{w_{m+1}}}{(1+T_{N_{m+1}})^{w_{m+1}}+(1-T_{N_{m+1}})^{w_{m+1}}}, \frac{(1+I_{N_{m+1}})^{w_{m+1}}-(1-I_{N_{m+1}})^{w_{m+1}}}{(1+I_{N_{m+1}})^{w_{m+1}}+(1-I_{N_{m+1}})^{w_{m+1}}}, \frac{2\left(F_{N_{m+1}}\right)^{w_{m+1}}}{\left(2-F_{N_{m+1}}\right)^{w_{m+1}}+\left(F_N\right)^{w_{m+1}}} \end{array}\right)$$

$$so \underset{j=1}{\overset{m}{\otimes}}\left(N_j^E\right)^{w_j} \otimes_E \left(N_{m+1}^E\right)^{w_{m+1}} =$$

$$\left(\begin{array}{c} \left[\frac{2\underset{j=1}{\overset{m}{\otimes}}\left(T^L_{N_{\sigma(j)}}\right)^{w_j}}{\underset{j=1}{\overset{m}{\otimes}}\left(2-T^L_{N_{\sigma(j)}}\right)^{w_j}+\underset{j=1}{\overset{m}{\otimes}}\left(T^L_{N_{\sigma(j)}}\right)^{w_j}}, \frac{2\underset{j=1}{\overset{m}{\otimes}}\left(T^U_{N_{\sigma(j)}}\right)^{w_j}}{\underset{j=1}{\overset{m}{\otimes}}\left(2-T^U_{N_{\sigma(j)}}\right)^{w_j}+\underset{j=1}{\overset{m}{\otimes}}\left(T^U_{N_{\sigma(j)}}\right)^{w_j}}\right], \\ \left[\frac{2\underset{j=1}{\overset{m}{\otimes}}\left(I^L_{N_{\sigma(j)}}\right)^{w_j}}{\underset{j=1}{\overset{m}{\otimes}}\left(2-I^L_{N_{\sigma(j)}}\right)^{w_j}+\underset{j=1}{\overset{m}{\otimes}}\left(I^L_{N_{\sigma(j)}}\right)^{w_j}}, \frac{2\underset{j=1}{\overset{m}{\otimes}}\left(I^U_{N_{\sigma(j)}}\right)^{w_j}}{\underset{j=1}{\overset{m}{\otimes}}\left(2-I^U_{N_{\sigma(j)}}\right)^{w_j}+\underset{j=1}{\overset{m}{\otimes}}\left(I^U_{N_{\sigma(j)}}\right)^{w_j}}\right], \\ \left[\frac{\underset{j=1}{\overset{m}{\otimes}}(1+F^L_{N_{\sigma(j)}})^{w_j}-\underset{j=1}{\overset{m}{\otimes}}(1-F^L_{N_{\sigma(j)}})^{w_j}}{\underset{j=1}{\overset{m}{\otimes}}(1+F^L_{N_{\sigma(j)}})^{w_j}+\underset{j=1}{\overset{m}{\otimes}}(1-F^L_{N_{\sigma(j)}})^{w_j}}, \frac{\underset{j=1}{\overset{m}{\otimes}}(1+F^U_{N_{\sigma(j)}})^{w_j}-\underset{j=1}{\overset{m}{\otimes}}(1-F^U_{N_{\sigma(j)}})^{w_j}}{\underset{j=1}{\overset{m}{\otimes}}(1+F^U_{N_{\sigma(j)}})^{w_j}+\underset{j=1}{\overset{m}{\otimes}}(1-F^U_{N_{\sigma(j)}})^{w_j}}\right], \\ \frac{\underset{j=1}{\overset{m}{\otimes}}(1+T_{N_{\sigma(j)}})^{w_j}-\underset{j=1}{\overset{m}{\otimes}}(1-T_{N_{\sigma(j)}})^{w_j}}{\underset{j=1}{\overset{m}{\otimes}}(1+T_{N_{\sigma(j)}})^{w_j}+\underset{j=1}{\overset{m}{\otimes}}(1-T_{N_{\sigma(j)}})^{w_j}}, \frac{\underset{j=1}{\overset{m}{\otimes}}(1+I_{N_{\sigma(j)}})^{w_j}-\underset{j=1}{\overset{m}{\otimes}}(1-I_{N_{\sigma(j)}})^{w_j}}{\underset{j=1}{\overset{m}{\otimes}}(1+I_{N_{\sigma(j)}})^{w_j}+\underset{j=1}{\overset{m}{\otimes}}(1-I_{N_{\sigma(j)}})^{w_j}}, \\ \frac{2\underset{j=1}{\overset{m}{\otimes}}\left(F_{N_{\sigma(j)}}\right)^{w_j}}{\underset{j=1}{\overset{m}{\otimes}}\left(2-F_{N_{\sigma(j)}}\right)^{w_j}+\underset{j=1}{\overset{m}{\otimes}}\left(F_{N_{\sigma(j)}}\right)^{w_j}} \end{array}\right) \oplus_E$$

$$\left(\begin{array}{c} \left[\frac{2\left(T^L_{N_{m+1}}\right)^{w_{m+1}}}{\left(2-T^L_{N_{m+1}}\right)^{w_{m+1}}+\left(T^L_N\right)^{w_{m+1}}}, \frac{2\left(T^U_{N_{m+1}}\right)^{w_{m+1}}}{\left(2-T^U_N\right)^{w_{m+1}}+\left(T^U_N\right)^{w_{m+1}}}\right], \\ \left[\frac{2\left(I^L_{N_{m+1}}\right)^{w_{m+1}}}{\left(2-I^L_{N_{m+1}}\right)^{w_{m+1}}+\left(I^L_N\right)^{w_{m+1}}}, \frac{2\left(I^U_{N_{m+1}}\right)^{w_{m+1}}}{\left(2-I^U_{N_{m+1}}\right)^{w_{m+1}}+\left(I^U_N\right)^{w_{m+1}}}\right], \\ \left[\frac{(1+F^L_{N_{m+1}})^{w_{m+1}}-(1-F^L_{N_{m+1}})^{w_{m+1}}}{(1+F^L_{N_{m+1}})^{w_{m+1}}+(1-F^L_{N_{m+1}})^{w_{m+1}}}, \frac{(1+F^U_{N_{m+1}})^{w_{m+1}}-(1-F^U_{N_{m+1}})^{w_{m+1}}}{(1+F^U_{N_{m+1}})^{w_{m+1}}+(1-F^U_{N_{m+1}})^{w_{m+1}}}\right], \\ \frac{(1+T_{N_{m+1}})^{w_{m+1}}-(1-T_{N_{m+1}})^{w_{m+1}}}{(1+T_{N_{m+1}})^{w_{m+1}}+(1-T_{N_{m+1}})^{w_{m+1}}}, \frac{(1+I_{N_{m+1}})^{w_{m+1}}-(1-I_{N_{m+1}})^{w_{m+1}}}{(1+I_{N_{m+1}})^{w_{m+1}}+(1-I_{N_{m+1}})^{w_{m+1}}}, \\ \frac{2\left(F_{N_{m+1}}\right)^{w_{m+1}}}{\left(2-F_{N_{m+1}}\right)^{w_{m+1}}+\left(F_N\right)^{w_{m+1}}} \end{array}\right)$$

$$\overset{m+1}{\underset{j=1}{\otimes}}\left(N_j^E\right)^{w_j}=\left(\begin{array}{c}\left[\dfrac{2\overset{m+1}{\underset{j=1}{\otimes}}\left(T^L_{N_{\sigma(j)}}\right)^{w_j}}{\overset{m+1}{\underset{j=1}{\otimes}}\left(2-T^L_{N_{\sigma(j)}}\right)^{w_j}+\overset{m+1}{\underset{j=1}{\otimes}}\left(T^L_{N_{\sigma(j)}}\right)^{w_j}},\dfrac{2\overset{m+1}{\underset{j=1}{\otimes}}\left(T^U_{N_{\sigma(j)}}\right)^{w_j}}{\overset{m+1}{\underset{j=1}{\otimes}}\left(2-T^U_{N_{\sigma(j)}}\right)^{w_j}+\overset{m+1}{\underset{j=1}{\otimes}}\left(T^U_{N_{\sigma(j)}}\right)^{w_j}}\right],\\ \left[\dfrac{2\overset{m+1}{\underset{j=1}{\otimes}}\left(I^L_{N_{\sigma(j)}}\right)^{w_j}}{\overset{m+1}{\underset{j=1}{\otimes}}\left(2-I^L_{N_{\sigma(j)}}\right)^{w_j}+\overset{m+1}{\underset{j=1}{\otimes}}\left(I^L_{N_{\sigma(j)}}\right)^{w_j}},\dfrac{2\overset{m+1}{\underset{j=1}{\otimes}}\left(I^U_{N_{\sigma(j)}}\right)^{w_j}}{\overset{m+1}{\underset{j=1}{\otimes}}\left(2-I^U_{N_{\sigma(j)}}\right)^{w_j}+\overset{m+1}{\underset{j=1}{\otimes}}\left(I^U_{N_{\sigma(j)}}\right)^{w_j}}\right],\\ \left[\dfrac{\overset{m+1}{\underset{j=1}{\otimes}}(1+F^L_{N_{\sigma(j)}})^{w_j}-\overset{m+1}{\underset{j=1}{\otimes}}(1-F^L_{N_{\sigma(j)}})^{w_j}}{\overset{m+1}{\underset{j=1}{\otimes}}(1+F^L_{N_{\sigma(j)}})^{w_j}+\overset{m+1}{\underset{j=1}{\otimes}}(1-F^L_{N_{\sigma(j)}})^{w_j}},\dfrac{\overset{m+1}{\underset{j=1}{\otimes}}(1+F^U_{N_{\sigma(j)}})^{w_j}-\overset{m+1}{\underset{j=1}{\otimes}}(1-F^U_{N_{\sigma(j)}})^{w_j}}{\overset{m+1}{\underset{j=1}{\otimes}}(1+F^U_{N_{\sigma(j)}})^{w_j}+\overset{m+1}{\underset{j=1}{\otimes}}(1-F^U_{N_{\sigma(j)}})^{w_j}}\right],\\ \dfrac{\overset{m+1}{\underset{j=1}{\otimes}}(1+T_{N_{\sigma(j)}})^{w_j}-\overset{m+1}{\underset{j=1}{\otimes}}(1-T_{N_{\sigma(j)}})^{w_j}}{\overset{m+1}{\underset{j=1}{\otimes}}(1+T_{N_{\sigma(j)}})^{w_j}+\overset{m+1}{\underset{j=1}{\otimes}}(1-T_{N_{\sigma(j)}})^{w_j}},\dfrac{\overset{m+1}{\underset{j=1}{\otimes}}(1+I_{N_{\sigma(j)}})^{w_j}-\overset{m+1}{\underset{j=1}{\otimes}}(1-I_{N_{\sigma(j)}})^{w_j}}{\overset{m+1}{\underset{j=1}{\otimes}}(1+I_{N_{\sigma(j)}})^{w_j}+\overset{m+1}{\underset{j=1}{\otimes}}(1-I_{N_{\sigma(j)}})^{w_j}},\\ \dfrac{2\overset{m+1}{\underset{j=1}{\otimes}}\left(F_{N_{\sigma(j)}}\right)^{w_j}}{\overset{m+1}{\underset{j=1}{\otimes}}\left(2-F_{N_{\sigma(j)}}\right)^{w_j}+\overset{m+1}{\underset{j=1}{\otimes}}\left(F_{N_{\sigma(j)}}\right)^{w_j}}\end{array}\right),$$

so the result holds for all values of m. □

Theorem 7. *The NCEWG is special case of the NCEHG operator.*

Proof. Followed by Theorem 4. □

Theorem 8. *The NCOWG is a special case of NCEHG.*

Proof. Followed by Theorem 5. □

3.3. An Application of Neutrosophic Cubic Hybrid Geometric and Einstein Hybrid Geometric Aggregation Operator to Group Decision-Making Problems

In this section, we develop an algorithm for group decision-making problems using the neutrosophic cubichybrid geometric and Einstein hybrid geometric aggregation (NCHWG and NCEHWG).

Algorithm 1. Let $F = \{F_1, F_2, ..., F_n\}$ be the set of n alternatives, $H = \{H_1, H_2, ..., H_m\}$ be the m attributes subject to their corresponding weight $W = \{w_1, w_2, ..., w_m\}$, such that $w_j \in [0,1]$ and $\sum_{j=1}^{m} = 1$. The method has the following steps.

Step 1: First of all, we construct neutrosophic cubic decision matrix $D = \left[N_{ij}\right]_{n\times m}$.
Step 2: The attributes $H = \{H_1, H_2, ..., H_m\}$ are weighted to their corresponding weight $W = \{w_1, w_2, ..., w_m\}$, and these values multipliedby the balancing coefficient m.
Step 3: The new weights are calculated using [18] so that we get new weights $V = \{v_1, v_2, ..., v_m\}$.
Step 4: By using aggregation operators like (NCHG, NCEHG), the decision matrix is aggregated by the new weightsassigned to the m attributes.
Step 5: The n alternatives are ranked according to their scores and arranged in descending order to select the alternative with highest score.

3.4. NumericalApplication

A steering committee is interested in prioritizingthe set of information for improvement of the project using a multiple attribute decision-making method. The committee must prioritize the development and implementation of a set of six information technology improvement projects A_j $(j = 1, 2, ..., 6)$. The three factors, B_1 productivity, to increase the effectiveness and efficiency,

B_2 differentiation, from products and services of competitors, and B_3 management, to assist the management in improving their planning, are considered to assess the potential contribution of each project. The list of proposed information systems are A_1 Quality Assurance, (2) A_2 Budget Analysis, (3) A_3 Itemization, (4) A_4 Employee Skills Tracking, (5) A_5 Customer Returns and Complaints, and (6) A_6 Materials Acquisition. Suppose the weight $W = (0.5, 0.3, 0.2)$ corresponds to the B_j, $(j = 1, 2, 3,)$ factors and characteristics of projects A_i $(i = 1, 2, ..., 10)$ by the neutrosophic cubic value N_{ij}.
Step 1: Construction of neutrosophic cubic decision matrix $D = \left[N_{ij}\right]_{6\times 3}$

$$D = \begin{pmatrix} & B_1 & B_2 & B_3 \\ A_1 & \begin{pmatrix} [0.5,0.6],[0.2,0.5], \\ [0.4,0.8],0.7,0.8,0.4 \end{pmatrix} & \begin{pmatrix} [0.3,0.7],[0.1,0.6], \\ [0.3,0.7],0.4,0.7,0.2 \end{pmatrix} & \begin{pmatrix} [0.4,0.8],[0.3,0.6], \\ [0.5,0.7],0.4,0.2,0.6 \end{pmatrix} \\ A_2 & \begin{pmatrix} [0.2,0.5],[0.7,0.9], \\ [0.3,0.7],0.8,0.5,0.3 \end{pmatrix} & \begin{pmatrix} [0.1,0.6],[0.4,0.7], \\ [0.2,0.5],0.6,0.4,0.7 \end{pmatrix} & \begin{pmatrix} [0.2,0.6],[0.1,0.7], \\ [0.3,0.7],0.6,0.4,0.8 \end{pmatrix} \\ A_3 & \begin{pmatrix} [0.4,0.7],[0.2,0.5], \\ [0.5,0.7],0.3,0.6,0.2 \end{pmatrix} & \begin{pmatrix} [0.3,0.8],[0.1,0.4], \\ [0.6,0.7],0.6,0.2,0.6 \end{pmatrix} & \begin{pmatrix} [0.3,0.5],[0.5,0.9], \\ [0.2,0.7],0.7,0.5,0.6 \end{pmatrix} \\ A_4 & \begin{pmatrix} [0.3,0.6],[0.4,0.7], \\ [0.2,0.5],0.6,0.4,0.7 \end{pmatrix} & \begin{pmatrix} [0.5,0.8],[0.1,0.5], \\ [0.3,0.8],0.4,0.8,0.6 \end{pmatrix} & \begin{pmatrix} [0.1,0.7],[0.2,0.6], \\ [0.4,0.7],0.5,0.6,0.8 \end{pmatrix} \\ A_5 & \begin{pmatrix} [0.2,0.5],[0.3,0.7], \\ [0.2,0.6],0.5,0.3,0.8 \end{pmatrix} & \begin{pmatrix} [0.4,0.8],[0.3,0.7], \\ [0.1,0.6],0.4,0.6,0.7 \end{pmatrix} & \begin{pmatrix} [0.6,0.8],[0.5,0.9], \\ [0.4,0.9],0.6,0.8,0.3 \end{pmatrix} \\ A_6 & \begin{pmatrix} [0.1,0.6],[0.3,0.6], \\ [0.4,0.8],0.6,0.9,0.4 \end{pmatrix} & \begin{pmatrix} [0.2,0.7],[0.6,0.9], \\ [0.3,0.6],0.4,0.8,0.3 \end{pmatrix} & \begin{pmatrix} [0.4,0.7],[0.3,0.5], \\ [0.1,0.6],0.4,0.6,0.7 \end{pmatrix} \end{pmatrix}$$

Step 2: The attributes are weighted $W = (0.5, 0.3, 0.2)$ and multiplied by balancing coefficient 3.

$$D = \begin{pmatrix} & B_1 & B_2 & B_3 \\ A_1 & \begin{pmatrix} [0.3535,0.4647], \\ [0.0894,0.3535], \\ [0.5352,0.9105], \\ 0.8356,0.9105,0.2529 \end{pmatrix} & \begin{pmatrix} [0.3383,0.7254], \\ [0.1258,0.6314], \\ [0.27453,0.6616], \\ 0.3685,0.6616,0.2349 \end{pmatrix} & \begin{pmatrix} [0.5770,0.8746], \\ [0.4855,0.7360], \\ [0.4229,0.5144], \\ 0.2639,0.1253,0.7360 \end{pmatrix} \\ A_2 & \begin{pmatrix} [0.0894,0.3535], \\ [0.5856,0.8538], \\ [0.4143,0.8356], \\ 0.9105,0.6464,0.1643 \end{pmatrix} & \begin{pmatrix} [0.1258,0.6314], \\ [0.4383,0.7254], \\ [0.1819,0.4641], \\ 0.5616,0.3685,0.7254 \end{pmatrix} & \begin{pmatrix} [0.3807,0.7360], \\ [0.2511,0.8073], \\ [0.1926,0.5144], \\ 0.4229,0.2639,0.8073 \end{pmatrix} \\ A_3 & \begin{pmatrix} [0.2529,0.5856], \\ [0.0894,0.3535], \\ [0.6464,0.8356], \\ 0.4143,0.7470,0.0894 \end{pmatrix} & \begin{pmatrix} [0.3383,0.8180], \\ [0.1258,0.4383], \\ [0.5616,0.6616], \\ 0.5616,0.1819,0.6314 \end{pmatrix} & \begin{pmatrix} [0.4855,0.6597], \\ [0.6597,0.9387], \\ [0.1253,0.5144], \\ 0.5144,0.3402,0.7360 \end{pmatrix} \\ A_4 & \begin{pmatrix} [0.1643,0.4647], \\ [0.2529,0.5856], \\ [0.2844,0.6464], \\ 0.7470,0.5352,0.5856 \end{pmatrix} & \begin{pmatrix} [0.5358,0.8180], \\ [0.1258,0.5358], \\ [0.2745,0.7650], \\ 0.3685,0.7650,0.6314 \end{pmatrix} & \begin{pmatrix} [0.2511,0.8073], \\ [0.3807,0.7360], \\ [0.2639,0.5144], \\ 0.6402,0.4229,0.8073 \end{pmatrix} \\ A_5 & \begin{pmatrix} [0.0894,0.5353], \\ [0.1643,0.5856], \\ [0.2844,0.7470], \\ 0.6464,0.4143,0.7155 \end{pmatrix} & \begin{pmatrix} [0.4383,0.8180], \\ [0.3383,0.7254], \\ [0.0904,0.5616], \\ 0.3685,0.5616,0.7254 \end{pmatrix} & \begin{pmatrix} [0.7360,0.8073], \\ [0.6597,0.9387], \\ [0.2639,0.7488], \\ 0.4229,0.6192,0.4855 \end{pmatrix} \\ A_6 & \begin{pmatrix} [0.0316,0.4647], \\ [0.1643,0.4647], \\ [0.5352,0.9105], \\ 0.7470,0.9683,0.2529 \end{pmatrix} & \begin{pmatrix} [0.2349,0.7254], \\ [0.6314,0.9035], \\ [0.2745,0.5616], \\ 0.3685,0.7650,0.3383 \end{pmatrix} & \begin{pmatrix} [0.5770,0.8073], \\ [0.4855,0.6597], \\ [0.0612,0.4229], \\ 0.2639,0.4229,0.8073 \end{pmatrix} \end{pmatrix}$$

Step 3: The new weights are calculated using the normal distribution method. Let $W = (0.2429, 0.5142, 0.2429)$ be its weighting vector derived by the normal distribution-based method [18].
Step 4: By neutrosophic cubic weighted geometric aggregation operator (NCWG), the decision matrix is aggregated by the new weights assigned to the m attributes.

$$D = \begin{pmatrix} A_1 & \begin{pmatrix} [0.3892, 0.6812], [0.1606, 0.5692], \\ [0.3840, 0.7325], 0.5273, 0.6914, 0.3270 \end{pmatrix} \\ A_2 & \begin{pmatrix} [0.1852, 0.5692], [0.4107, 0.7745], \\ [0.2480, 0.4946], 0.6813, 0.4306, 0.5190 \end{pmatrix} \\ A_3 & \begin{pmatrix} [0.3441, 0.7158], [0.1731, 0.5005], \\ [0.7078, 0.6899], 0.5178, 0.4161, 0.4076 \end{pmatrix} \\ A_4 & \begin{pmatrix} [0.3344, 0.7107], [0.1950, 0.5913], \\ [0.2743, 0.6904], 0.7010, 0.6550, 0.6580 \end{pmatrix} \\ A_5 & \begin{pmatrix} [0.3378, 0.7375], [0.3338, 0.7331], \\ [0.1848, 0.6649], 0.4633, 0.5454, 0.6557 \end{pmatrix} \\ A_6 & \begin{pmatrix} [0.1795, 0.6681], [0.4271, 0.7122], \\ [0.3068, 0.6813], 0.4751, 0.8937, 0.3893 \end{pmatrix} \end{pmatrix}$$

Step 5: The scores are

$$S(A_1) = 0.1542, S(A_2) = 0.1741, S(A_3) = -0.2276, S(A_4) = 0.1297, S(A_5) = 0.0332, S(A_6) = -0.0547,$$

$$S(A_2) > S(A_1) > S(A_4) > S(A_5) > S(A_6) > S(A_3).$$

List of priorities are as follows.

$$A_2 > A_1 > A_4 > A_5 > A_6 > A_3$$

Hence, the project A_1 has the highest potential contribution to the firm's strategic goal of gaining competitive advantage in the industry.

4. Conclusions

This paper was influenced by the impediment of neutrosophic cubic geometric and Einstein geometric collection operators as preliminarily discussed, that is, we observed that the higher the weight component, the aggregated value tended to the corresponding neutrosophic cubic value of that vector. Consequent upon such circumstances, we characterized neutrosophic cubic hybrid and neutrosophic cubic Einstein hybrid aggregation operators. At that point, these operators are outfitted upon a day-by-day life precedent structure industry to organize the potential contributions that serve to achieve the strategic objective of getting favorable circumstances in industry.

Author Contributions: All authors contributed equally.

Acknowledgments: The authors gratefully acknowledged the financial from the Deanship of Scientific research and Graduate Studies at Philadelphia University in Jordan.

Conflicts of Interest: There is no conflict of interest.

References

1. Zadeh, L.A. Fuzzy Sets. *Inf. Control* **1965**, *8*, 338–353. [CrossRef]
2. Turksen, I.B. Interval valued strict preferences with Zadeh triplet. *Fuzzy Sets Syst.* **1996**, *78*, 183–195. [CrossRef]
3. Zadeh, L.A. Outlines of new approach to the analysis of complex system and decisionprocesses interval valued fuzzy sets. *IEEE Trans.* **1968**. SMC-3, NO.1. [CrossRef]
4. Atanassov, K.T. Intuitionistic fuzzy sets. *Fuzzy Sets Syst.* **1986**, *20*, 87–96. [CrossRef]

5. Atanassov, K.T.; Gargov, G. Interval intuitionistic fuzzy sets. *Fuzzy Sets Syst.* **1989**, *31*, 343–349. [CrossRef]
6. Jun, Y.B.; Kim, C.S.; Yang, K.O. Cubic sets. *Ann. Fuzzy Math. Inf.* **2012**, *1*, 83–98.
7. Smarandache, F. *A Unifying Field in Logics, Neutrosophic Logic, Neutrosophy, Neutrosophic Set and Neutrosophic Probabilty*, 4th ed.; American Research Press: Rehoboth, DE, USA, 1999.
8. Wang, H.; Smarandache, F.; Zhang, Y.; Sunderraman, R. Single valued neutrosophic sets. *Tech. Sci. Appl. Math.* **2010**, *1*. Available online: https://www.researchgate.net/publication/262034557_Single_Valued_Neutrosophic_Sets (accessed on 28 February 2019).
9. Wang, H.; Smarandache, F.; Zhang, Y.Q.; Sunderraman, R. Interval neutrosophic sets and loics. In *Theory and Application in Computing*; Hexis: Phoenox, AZ, USA, 2005.
10. Jun, Y.B.; Smarandache, F.; Kim, C.S. Neutrosophic cubic sets. *New Math. Nat. Comput.* **2015**, *13*, 41–54. [CrossRef]
11. Zhan, J.; Khan, M.; Gulistan, M.; Ali, A. Applications of neutrosophic cubic sets in multi-criteria decision making. *Int. J. Uncertain. Quabtif.* **2017**, *7*, 377–394. [CrossRef]
12. Banerjee, D.; Giri, B.C.; Pramanik, S.; Smarandache, F. GRA for multi attribute decision making in neutrosophic cubic set environment. *Neutrosophic Sets Syst.* **2017**, *15*, 64–73.
13. Lu, Z.; Ye, J. Cosine measure for neutrosophic cubic sets for multiple attribte decision making. *Symmetry* **2017**, *9*, 121. [CrossRef]
14. Pramanik, S.; Dalapati, S.; Alam, S.; Roy, S.; Smarandache, F. Neutrosophic cubic MCGDM method based on similarity measure. *Neutrosophic Sets Syst.* **2017**, *16*, 44–56.
15. Shi, L.; Ye, J. Dombi aggregation operators of neutrosophic cubic set for multiple attribute deicision making algorithms. *Algorithms* **2018**, *11*, 29. [CrossRef]
16. Li, B.; Wang, J.; Yang, L.; Li, X. A Novel Generalized Simplified Neutrosophic Number Einstein Aggregation Operator. *Int. J. Appl. Math.* **2018**, *48*, 67–72.
17. Khan, M.; Gulistan, M.; Yaqoob, N.; Khan, M.; Smarandache, F. Neutrosophic cubic Einstein geometric aggregation operatorswith application to multi-creiteria decision making method. *Symmetry* **2019**, *112*, 47. [CrossRef]
18. Xu, Z. An overview of methods for determining OWA weights. *Int. J. Intell. Syst.* **2005**, *20*, 843–865. [CrossRef]
19. Xu, Z. Intuitionistic Fuzzy Aggregation Operators. *IEEE Trans. Fuzzy syst.* **2007**, *15*, 1179–1187.
20. Yaqoob, N.; Gulistan, M.; Leoreanu-Fotea, V.; Hila, K. Cubic hyperideals in LA-semihypergroups. *J. Int. Fuzzy Syst.* **2018**, *34*, 2707–2721. [CrossRef]
21. Gulistan, M.; Yaqoob, N.; Vougiouklis, T.; Wahab, H.A. Extensions of cubic ideals in weak left almost semihypergroups. *J. Int. Fuzzy Syst.* **2018**, *34*, 4161–4172. [CrossRef]
22. Gulistan, M.; Wahab, H.A.; Smarandache, F.; Khan, S.; Shah, S.I.A. Some Linguistic Neutrosophic Cubic Mean Operators and Entropy with Applications in a Corporation to Choose an Area Supervisor. *Symmetry* **2018**, *10*, 428. [CrossRef]
23. Hashim, R.M.; Gulistan, M.; Smrandache, F. Applications of Neutrosophic Bipolar Fuzzy Sets in HOPE Foundation for Planning to Build a Children Hospital with Different Types of Similarity Measures. *Symmetry* **2018**, *10*, 331. [CrossRef]
24. Gulistan, M.; Nawaz, S.; Hassan, N. Neutrosophic Triplet Non-Associative Semihypergroups with Application. *Symmetry* **2018**, *10*, 613. [CrossRef]
25. Khan, M.; Gulistan, M.; Yaqoob, N.; Shabir, M. Neutrosophic cubic (α, β)-ideals in semigroups with application. *J. Int. Fuzzy Syst.* **2018**, *35*, 2469–2483. [CrossRef]
26. Khaleed, A.; Hassan, N. Possibility vague soft set and its application in decision making. *Int. J. Pure Appl. Math.* **2012**, *77*, 549–563.
27. Varnamkhasti, J.; Hassan, N. Neurogenetic algorithm for solving combinatorial engineering problems. *J. Appl. Math.* **2012**. [CrossRef]

mathematics

MDPI

Article

Refined Neutrosophy and Lattices vs. Pair Structures and YinYang Bipolar Fuzzy Set

Florentin Smarandache

Mathematics Department, University of New Mexico, 705 Gurley Ave., Gallup, NM 87301, USA; smarand@unm.edu

Received: 4 March 2019; Accepted: 1 April 2019; Published: 16 April 2019

Abstract: In this paper, we present the lattice structures of neutrosophic theories. We prove that Zhang-Zhang's YinYang bipolar fuzzy set is a subclass of the Single-Valued bipolar neutrosophic set. Then we show that the pair structure is a particular case of refined neutrosophy, and the number of types of neutralities (sub-indeterminacies) may be any finite or infinite number.

Keywords: neutrosophic set; Zhang-Zhang's YinYang bipolar fuzzy set; single-valued bipolar neutrosophic set; bipolar fuzzy set; YinYang bipolar fuzzy set

---

1. Introduction

First, we prove that Klement Dand Mesiar's lattices [1] do not fit the general definition of neutrosophic set, and we construct the appropriate nonstandard neutrosophic lattices of the first type (as neutrosophically ordered set) [2], and of the second type (as neutrosophic algebraic structure, endowed with two binary neutrosophic laws, $\inf_N$ and $\sup_N$) [2].

We also present the novelties that neutrosophy, neutrosophic logic, set, and probability and statistics, with respect to the previous classical and multi-valued logics and sets, and with the classical and imprecise probability and statistics, respectively.

Second, we prove that Zhang-Zhang's YinYang bipolar fuzzy set [3,4] is not equivalent with but a subclass of the Single-Valued bipolar neutrosophic set.

Third, we show that Montero, Bustince, Franco, Rodríguez, Gómez, Pagola, Fernández, and Barrenechea's paired structure of the knowledge representation model [5] is a particular case of Refined Neutrosophy (a branch of philosophy that generalized dialectics) and of the Refined Neutrosophic Set [6]. We disprove again the claim that the bipolar fuzzy set (renamed as YinYang bipolar fuzzy set) is the same of neutrosophic set as asserted by Montero et al [5].

About the three types of neutralities presented by Montero et al., we show, by examples and formally, that there may be any finite number or an infinite number of types of neutralities n, or that indeterminacy (I), as neutrosophic component, can be refined (split) into $1 \leq n \leq \infty$ number of sub-indeterminacies (not only 3 as Montero et al. said) as needed to each application to solve.

Also, we show, besides numerous neutrosophic applications, many innovatory contributions to science were brought on by the neutrosophic theories, such as: generalization of Yin Yang Chinese philosophy and dialectics to neutrosophy [7], a new branch of philosophy that is based on the dynamics of opposites and their neutralities, the sum of the neutrosophic components T, I, F up to 3, the degrees of dependence/independence between the neutrosophic components [8,9]; the distinction between absolute truth and relative truth in the neutrosophic logic [10], the introduction of nonstandard neutrosophic logic, set, and probability after we have extended the nonstandard analysis [11,12], the refinement of neutrosophic components into subcomponents [6]; the ability to express incomplete information, complete information, paraconsistent (conflicting) information [13,14]; and the extension of the middle principle to the multiple-included middle principle [15], introduction of neutrosophic crisp set and topology [16], and so on.

2. Answers to Erich Peter Klement and Radko Mesiar

2.1. Oversimplification of the Neutrosophic Set

At [1], page 10 (Section 3.3) in their paper, related to neutrosophic sets, they wrote:

"As a straightforward generalization of the product lattice $\left(\mathbb{I}\times\mathbb{I}, \leq_{comp}\right)$*, for each n ∈ N, the n-dimensional unit cube* $\left(\mathbb{I}^n, \leq_{comp}\right)$*, i.e., the n-dimensional product of the lattice* $(\mathbb{I}, \leq_{comp})$*, can be defined by means of (1) and (2).*

The so-called "neutrosophic" sets introduced by F. Smarandache [93] (see also [94–97], which are based on the bounded lattices $\left(\mathbb{I}^3, \leq_{\mathbb{I}^3}\right)$ *and* $\left(\mathbb{I}^3, \leq^{\mathbb{I}^3}\right)$*, where the orders* $\leq_{I^3}$ *and* $\leq_{I^3}$ *on the unit cube* I^3 *are defined by the Equations below.*

$$(x_1, x_2, x_3) \leq_{I^3} (y_1, y_2, y_3) \Leftrightarrow x_1 \leq y_1 \ AND\ x_2 \leq y_2 \ AND\ x_3 \geq y_3 \quad (-13)$$

$$(x_1, x_2, x_3) \leq^{I^3} (y_1, y_2, y_3) \Leftrightarrow x_1 \leq y_1 \ AND\ x_2 \geq y_2 \ AND\ x_3 \geq y_3 \quad (-14)$$

The authors have defined Equations *(1)* and *(2)* as follows:

$$\left(\prod_{i=1}^{n} L_i, \leq_{comp}\right), where\ \left(L_i, \leq_{L_i}\right)\ are\ fuzzy\ lattices,\ for\ all\ 1 \leq i \leq n \quad (1)$$

$$(x_1, x_2, \ldots, x_n) \leq_{comp} (y_1, y_2, \ldots, y_n) \Leftrightarrow x_1 \leq y_1 \ AND\ x_2 \leq y_2 \ AND\ \ldots\ AND\ x_n \leq y_n \quad (2)$$

The authors did not specify what type of lattices they employ: of the first type (lattice, as a partially ordered set), or the second type (lattice, as an algebraic structure). Since their lattices are endowed with some inequality (referring to the neutrosophic case), we assume it is as the first type.

The authors have used the notations:

$\mathbb{I} = [0,\ 1]$,
$\mathbb{I}^2 = [0,\ 1]^2$,
$\mathbb{I}^3 = [0,\ 1]^3$.

The order relationship $\leq_{\text{comp}}$ on $\mathbb{I}^3$ can be defined as:

$$(x_1, x_2, x_3) \leq_{\text{comp}} (y_1, y_2, y_3) \Leftrightarrow x_1 \leq y_1 \text{ and } x_2 \leq y_2 \text{ and } x_3 \leq y_3$$

The three lattices they constructed are denoted by KL_1, KL_2, KL_3, respectively.

$$KL_1 = (\mathbb{I}^3, \leq_{\text{comp}}),\ KL_2 = (\mathbb{I}^3, \leq_{I^3}),\ KL_3 = (\mathbb{I}^3, \leq^{\mathbb{I}^3})$$

Contain only the very particular case of standard single-valued neutrosophic set, i.e., when the neutrosophic components T (truth-membership), I (indeterminacy-membership), and F (false-membership) of the generic element $x(T,\ I,\ F)$, of a neutrosophic set N are single-valued (crisp) numbers from the unit interval $[0, 1]$.

The authors have *oversimplified* the neutrosophic set. Neutrosophic is much more complex. Their lattices do not characterize the *initial definition of the neutrosophic set* ([10], 1998): a set whose elements have the degrees of appurtenance *T*, *I*, *F*, where *T*, *I*, *F* are standard or nonstandard subsets of the nonstandard unit interval: $]^-0,\ 1^+[$, where $]^-0,\ 1^+[$ overpasses the classical real unit interval $[0,\ 1]$ to the left and to the right.

2.2. Neutrosophic Cube vs. Unit Cube

Clearly, their $\mathbb{I}^3 = [0,1]^3 \subsetneq]^-0,\ 1^+[^3$ that is our neutrosophic cube (Figure 1), where $]^-0 = \mu(^-0)$ is the left nonstandard monad of number 0, and $1^+ = \mu(1^+)$ is the right nonstandard monad of number 1.

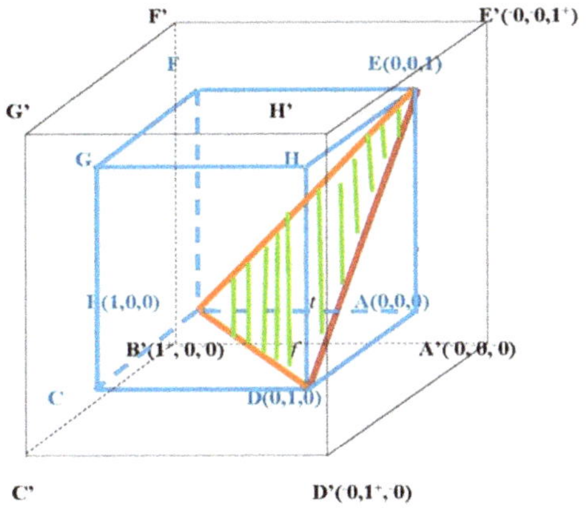

Figure 1. Neutrosophic cube.

The unit cube $\mathbb{I}^3$ used by the authors does not equal the above neutrosophic cube. The neutrosophic cube $A'B'C'D'E'F'G'H'$ was introduced by Dezert [17] in 2002.

2.3. The Most General Neutrosophic Lattices

The authors' lattices are far from catching the *most general definition of the neutrosophic set*.

Let $\mathcal{U}$ be a universe of discourse, and $M \subset \mathcal{U}$ be a set. Then an element $x(T(x), I(x), F(x)) \in M$, where $T(x), I(x), F(x)$ are standard or nonstandard subsets of nonstandard interval: $]^-\Omega,\ \Psi^+[$, where $\Omega \le 0 < 1 \le \Psi$, with $\Omega,\ \Psi \in \mathbb{R}$, whose values Ω and Ψ depend on each application, and

$$]^-\Omega,\ \Psi^+[=_N \left\{\varepsilon, a, a^-, a^{-0},\ a^+,\ a^{+0},\ a^{\mp},\ a^{-0+} \middle|\ \varepsilon, a \in [\Omega,\ \Psi],\ \varepsilon \text{ is infinitesimal}\right\},$$

where $\overset{m}{a}$, $m \in \left\{^-, ^{-0}, ^+, ^{+0}, ^{-+}, ^{-0+}\right\}$ are monads or binads [12].

It follows that the nonstandard neutrosophic mobinad real offsets lattices $\left(]^-\Omega,\ \Psi^+[, \le_N^{nonS}\right)$ and $\left(]^-\Omega,\ \Psi^+[,\ \inf_N,\ \sup_N, ^-\Omega,\ \Psi^+\right)$ of the first type and, respectively, of the second type are the most general (non-refined) neutrosophic lattices.

While the most general refined neutrosophic lattices of the first type is: $\left(]^-\Omega,\ \Psi^+[, \le_{nN}^{nonS}\right)$, where $\le_{nN}^{nonS}$ is the n-tuple nonstandard neutrosophic inequality dealing with nonstandard subsets, defined as:

$$(T_1(x),\ T_2(x),\ \ldots,\ T_p(x);\ I_1(x),\ I_2(x),\ \ldots,\ I_r(x);\ F_1(x),\ F_2(x),\ \ldots,\ F_s(x)\ ) \le_{nN}^{nonS} (T_1(y),\ T_2(y),\ \ldots,\ T_p(y);\ I_1(y),\ I_2(y),\ \ldots,\ I_r(y);\ F_1(y),\ F_2(y),\ \ldots,\ F_s(y)) \text{ iff}$$

$$T_1(x) \le_{nN}^{nonS} T_1(y),\ T_2(x) \le_{nN}^{nonS} T_2(y),\ \ldots,\ T_p(x) \le_{nN}^{nonS} T_p(y)$$
$$I_1(x) \ge_{nN}^{nonS} I_1(y),\ I_2(x) \ge_{nN}^{nonS} I_2(y),\ \ldots,\ I_r(x) \ge_{nN}^{nonS} I_r(y)$$
$$F_1(x) \ge_{nN}^{nonS} F_1(y),\ F_2(x) \ge_{nN}^{nonS} F_2(y),\ \ldots,\ F_s(x) \ge_{nN}^{nonS} F_s(y)$$

2.4. Distinction between Absolute Truth and Relative Truth

The authors' lattices are incapable of making distinctions between absolute truth (when $T = 1^+ >_N 1$) and relative truth (when $T = 1$) in the sense of Leibniz, which is the essence of nonstandard neutrosophic logic.

2.5. Neutrosophic Standard Subset Lattices

Their three lattices are not even able to deal with *standard subsets* [including intervals [8], and hesitant (discrete finite) subsets] $T,\ I,\ F \subseteq [0,\ 1]$, since they have defined the 3D-inequalities with respect to single-valued (crisp) numbers: $x_1, x_2, x_3 \in [0,\ 1]$ and $y_1, y_2, y_3 \in [0,\ 1]$.

In order to deal with standard subsets, they should use *inf/sup*, i.e.,

$$\begin{aligned}&(T_1, I_1, F_1) \leq (T_2, I_2, F_2) \Leftrightarrow \\ &\inf T_1 \leq \inf T_2 \text{ and } \sup T_1 \leq \sup T_2, \\ &\inf I_1 \geq \inf I_2 \text{ and } \sup I_1 \geq \sup I_2, \\ &\text{and } \inf F_1 \geq \inf F_2 \text{ and } \sup F_1 \geq \sup F_2\end{aligned}$$

[I have displayed the most used 3D-inequality by the neutrosophic community.]

2.6. Nonstandard and Standard Refined Neutrosophic Lattices

The *Nonstandard Refined Neutrosophic Set* [2,6,12], defined on $]^-0, 1^+[^n$, strictly includes their n-dimensional unit cube ($\mathbb{I}^n$), and we use a nonstandard neutrosophic inequality, not the classical inequalities, to deal with inequalities of monads and binads, such as $\leq_{nN}^{nonS}$ and $\leq_N^{nonS}$.

Not even the Standard Refined Single-Valued Neutrosophic Set [6] (2013) may be characterized with KL_1, KL_2, and KL_3 nor with $\left(\mathbb{I}^n,\ \leq_{\text{comp}}\right)$, since the n-D neutrosophic inequality is different from n-D $\leq_{\text{comp}}$, and from n-D extensions of $\leq_{I_3}$ or $\leq^{I_3}$ respectively, as follows:

Let T be refined into $T_1, T_2, \ldots, T_p$;

I be refined into $I_1, I_2, \ldots, I_r$;

and F be refined into $F_1, F_2, \ldots, F_s$;

with $p,\ r,\ s\ \geq\ 1$ are integers, and $p + r + s\ =\ n\ \geq\ 4$, produced the following n-D neutrosophic inequality.

Let $x\left(T_1^x, T_2^x, \ldots, T_p^x;\ I_1^x, I_2^x, \ldots, I_r^x;\ F_1^x, F_2^x, \ldots, F_s^x\right)$, and $y\left(T_1^y, T_2^y, \ldots, T_p^y;\ I_1^y, I_2^y, \ldots, I_r^y;\ F_1^y, F_2^y, \ldots, F_s^y\right)$.

Then:

$$x \leq_N y \Leftrightarrow \begin{pmatrix} T_1^x \leq T_1^y, T_2^x \leq T_2^y, \ldots, T_p^x \leq T_p^y; \\ I_1^x \geq I_1^y, I_2^x \geq I_2^y, \ldots, I_r^x \geq I_r^y; \\ F_1^x \geq F_1^y, F_2^x \geq F, \ldots, F_s^x \geq F_s^y. \end{pmatrix}$$

2.7. Neutrosophic Standard Overset/Underset/Offset Lattice

Their three lattices KL_1, KL_2 and KL_3 are no match for neutrosophic overset (when the neutrosophic components $T,\ I,\ F > 1$), nor for neutrosophic underset (when the neutrosophic components $T,\ I,\ F < 0$), and, in general, no match for the neutrosophic offset (when the neutrosophic components $T,\ I,\ F$ take values outside the unit interval $[0,\ 1]$ as needed in real life applications [13,14,18–20] (2006–2018): $[\Omega,\ \Psi]$ with $\Omega \leq 0 < 1 \leq \Psi$.)

Therefore, a lattice may similarly be built on the *non-unitary neutrosophic cube* $[\varphi,\ \psi]^3$.

2.8. Sum of Neutrosophic Components up to 3

The authors do not mention the novelty of neutrosophic theories regarding the sum of single-valued neutrosophic components $T + I + F \leq 3$, extended up to 3, and, similarly, the corresponding inequality when T, I, F are subsets of $[0,\ 1]$: $\sup T + \sup I + \sup F \leq 3$, for neutrosophic set, neutrosophic logic, and neutrosophic probability never done before in the previous classic logic and multiple-valued logics and set theories, nor in the classical or imprecise probabilities.

This makes a big difference, since, for a single-valued neutrosophic set S, all unit cubes $[0,\ 1]^3$ are fulfilled with points, each point $P(a,\ b,\ c)$ into the unit cube may represent the neutrosophic coordinates $(a,\ b,\ c)$ of an element $x(a,\ b,\ c) \in S$, which was not the case for previous logics, sets, and probabilities.

This is not the case for the Picture Fuzzy Set (Cuong [21], 2013) whose domain is $\frac{1}{6}$ of the unit cube (a cube corner):

$$\mathbb{D}^* = \left\{ (x_1, x_2, x_3) \in \mathbb{I}^3 \middle| x_1 + x_2 + x_3 \le 1 \right\}$$

For Intuitionistic Fuzzy Set (Atanassov [22], 1986), the following is true.

$$\mathbb{D}_A = \left\{ (x_1, x_2, x_3) \in \mathbb{I}^3 \middle| x_1 + x_2 + x_3 = 1 \right\}$$

where x_1 = membership degree, x_2 = hesitant degree, and x_3 = nonmembership degree, whose domain is the main cubic diagonal triangle that connects the vertices: (1, 0, 0), (0, 1, 0), and (0, 0,1), i.e., triangle BDE (its sides and its interior) in Figure 1.

2.9. Etymology of Neutrosophy and Neutrosophic

The authors [1] write ironically twice, in between quotations, "neutrosophic" because they did not read the etymology [10] of the word published into my first book (1998), etymology, which also appears into Denis Howe's 1999 *The Free Online Dictionary of Computing* [23], and, afterwards, repeated by many researchers from the neutrosophic community in their published papers:

Neutrosophy [23]: *<philosophy> (From Latin "neuter"—neutral, Greek "sophia"—skill/wisdom). A branch of philosophy, introduced by Florentin Smarandache in 1980, which studies the origin, nature, and scope of neutralities, as well as their interactions with different ideational spectra. Neutrosophy considers a proposition, theory, event, concept, or entity, "A"in relation to its opposite, "Anti-A" and that which is not A, "Non-A", and that which is neither "A" nor "Anti-A", denoted by "Neut-A". Neutrosophy is the basis of neutrosophic logic, neutrosophic probability, neutrosophic set, and neutrosophic statistics.*

While **neutrosophic** means what is derived/resulted from *neutrosophy*.

Unlike the "intuitionistic|" and "picture fuzzy" notions, the notion of *neutrosophic* was carefully and meaningfully chosen, coming from **neutral** (or indeterminate, denoted by <neutA>) between two opposites, $\langle A \rangle$ and $\langle \text{anti}A \rangle$, which made the main distinction between neutrosophic logic/set/probability, and the previous fuzzy, intuitionistic fuzzy logics and sets, i.e.,

- For neutrosophic logic neither true nor false, but neutral (or indeterminate) in between them;
- Similarly for neutrosophic set: neither membership nor non-membership, but in between (neutral, or indeterminate);
- And analogously for neutrosophic probability: chance that an event *E* occurs, chance that the event *E* does not occur, and indeterminate (neutral) chance of the event *E* of occurring or not occuring.

Their irony is malicious and ungrounded.

2.10. Neutrosophy as Extension of Dialectics

Let $\langle A \rangle$ be a concept, notion, idea, or theory.

Then $\langle \text{anti}A \rangle$ is the opposite of $\langle A \rangle$, while $\langle \text{neut}A \rangle$ is the neutral (or indeterminate) part between them.

While in philosophy, Dialectics is the dynamics of opposites ($\langle A \rangle$ and $\langle \text{anti}A \rangle$), Neutrosophy is an extension of dialectics. In other words, neutrosophy is the dynamics of opposites and their neutrals ($\langle A \rangle$, $\langle \text{anti}A \rangle$, $\langle \text{neut}A \rangle$), because the neutrals play an important role in our world, interfering in one side or the other of the opposites.

Refined Neutrosophy is an extension of Neutrosophy, and it is the dynamics of the refined-items <A_1>, <A_2>, … , <A_n>, their refined-opposites <$antiA_1$>, <$antiA_2$>, … , <$antiA_n$>, and their refined-neutrals <$neutA_1$>, <$neutA_2$>, … , <$neutA_n$>.

As an extension of Refined Neutrosophy one has the Plithogeny [24–27].

2.11. Refined Neutrosophic Set and Lattice

At page 11, Klement and Mesiar ([1], 2018) assert that: Considering, for $n > 3$, lattices which are isomorphic to $\left(L_n(\mathbb{I}),\ \leq_{comp}\right)$, further generalizations of "neutrosophic" sets can be introduced.

The authors are uninformed so that a generalization was done in 2013 when we have published a paper [6] that introduced, for the first time, the refined neutrosophic set/logic/probability, where T, I, F were refined into n neutrosophic subcomponents:

$T_1, T_2, \ldots, T_p; I_1, I_2, \ldots, I_r; F_1, F_2, \ldots, F_s,$

With $p,\ r,\ s \geq 1$ are integers and $p + r + s = n \geq 4$.

But in our lattice $(\mathbb{I}^n,\ \leq_{nN})$, the neutrosophic inequality is adjusted to the categories of sub-truths, sub-indeterminacies, and sub-falsehood, respectively.

$$(T_1(x),\ T_2(x),\ \ldots,\ T_p(x);\ I_1(x),\ I_2(x),\ \ldots,\ I_r(x);\ F_1(x),\ F_2(x),\ \ldots,\ F_s(x)) \leq_{nN} (T_1(y),\ T_2(y),\ \ldots,\ T_p(y);\ I_1(y),\ I_2(y),\ \ldots,\ I_r(y);\ F_1(y),\ F_2(y),\ \ldots,\ F_s(y)) \text{ if and only if}$$

$$T_1(x) \leq T_1(y),\ T_2(x) \leq T_2(y),\ \ldots,\ T_p(x) \leq T_p(y)$$

$$I_1(x) \geq I_1(y),\ I_2(x) \geq I_2(y),\ \ldots,\ I_r(x) \geq I_r(y)$$

$$F_1(x) \geq F_1(y),\ F_2(x) \geq F_2(y),\ \ldots,\ F_s(x) \geq F_s(y)$$

Therefore, $\leq_{nN}$ is different from the n-D inequalities $\leq_{\text{comp}}$, and from $\leq_{\mathbb{I}^n}$ and $\leq^{\mathbb{I}^n}$ (extending from authors inequalities $\leq_{\mathbb{I}^3}$ and $\leq^{\mathbb{I}^3}$, respectively).

2.12. Nonstandard Refined Neutrosophic Set and Lattice

Even more, Nonstandard Refined Neutrosophic Set/Logic/Probability (which include infinitesimals, monads, and closed monads, binads and closed binads) has no connection and no isomorphism whatsoever with any of the authors' lattices or extensions of their lattices for *2D* and *3D* to *nD*.

2.13. Nonstandard Neutrosophic Mobinad Real Lattice

We have built ([2], 2018) a more complex Nonstandard Neutrosophic Mobinad Real Lattice, on the nonstandard mobinad unit interval $]^-0,\ 1^+[$ defined as:

$$]^-0,\ 1^+[= \left\{\varepsilon, a, a^-, a^{-0},\ a^+,\ a^{+0},\ a^{-+},\ a^{-0+} \middle|\ \text{with } 0 \leq a \leq 1,\ a \in \mathbb{R}, \text{ and } \varepsilon > 0,\ \varepsilon \text{ infinitesimal},\ \varepsilon \in \mathbb{R}^*\right\}$$

which is both **nonstandard neutrosophic lattice of the first type** (as partially ordered set, under neutrosophic inequality $\leq_N$) and lattice of the second type (as algebraic structure, endowed with two binary nonstandard neutrosophic laws: $\inf_N$ and $\sup_N$).

Now, $]^-0,\ 1^+[^3$ is a nonstandard unit cube, with much higher density than $[0,\ 1]^3$ and which comprise not only real numbers $a \in [0,\ 1]$ but also infinitesimals $\varepsilon > 0$ and monads and binads neutrosophically included in $]^-0,\ 1^+[$.

2.14. New Ideas Brought by the Neutrosophic Theories and Never Done Before

- The sum of the neutrosophic components is up to 3 (previously the sum was up to 1);
- Degree of independence and dependence between the neutrosophic components T, I, F, making their sum $T + I + F$ vary between 0 and 3.

For example, when T, I, and F are totally dependent with each other, then $T + I + F \leq 1$. Therefore, we obtain the particular cases of intuitionistic fuzzy set (when $T + I + F = 1$) and picture set when $T + I + F \leq 1$.

- Nonstandard analysis used in order to distinguish between absolute and relative (truth, membership, chance).

— Refinement of the components into sub-components:

$$\left(T_1, T_2, \ldots, T_p; I_1, I_2, \ldots, I_r; F_1, F_2, \ldots, F_s\right)$$

with the newly introduced Refined Neutrosophic Logic/Set/Probability.

— Ability to express *incomplete information* ($T + I + F < 1$) and paraconsistent (conflicting) and subjective information ($T + I + F > 1$).

— Law of Included Middle explicitly/independently expressed as $\langle \text{neut}A \rangle$ (indeterminacy, neutral).

— Law of Included Middle expanded to the Law of Included Multiple-Middles within the refined neutrosophic set as well as logic and probability.

— A large array of applications [28–30] in a variety of fields, after two decades from their foundation ([10], 1998), such as: Artificial Intelligence, Information Systems, Computer Science, Cybernetics, Theory Methods, Mathematical Algebraic Structures, Applied Mathematics, Automation, Control Systems, Communication, Big Data, Engineering, Electrical, Electronic, Philosophy, Social Science, Psychology, Biology, Biomedical, Engineering, Medical Informatics, Operational Research, Management Science, Imaging Science, Photographic Technology, Instruments, Instrumentation, Physics, Optics, Economics, Mechanics, Neurosciences, Radiology Nuclear, Medicine, Medical Imaging, Interdisciplinary Applications, Multidisciplinary Sciences, and more [30].

Klement's and Mesiar's claim that the neutrosophic set (I do not talk herein about intuitionistic fuzzy set, picture fuzzy set, and Pythagorean fuzzy set that they criticized) is not a new result is far from the truth.

3. Neutrosophy vs. Yin Yang Philosophy

Ying Han, Zhengu Lu, Zhenguang Du, Gi Luo, and Sheng Chen [3] have defined the "YinYang bipolar fuzzy set" (2018).

However, the "YinYang bipolar" is already a pleonasm, because, in Taoist Chinese philosophy, from the 6th century BC, Yin and Yang was already a bipolarity, between negative (Yin)/positive (Yang), or feminine (Yin)/masculine (Yang).

Dialectics was derived, much later in time, from Yin Yang.

Neutrosophy, as the dynamicity and harmony between opposites (Yin <A> and Yang (antiA>) together with their neutralities (things which are neither Yin nor Yang, or things which are blends of both: <neutA>) is an extension of Yin Yang Chinese philosophy. Neutrosophy came naturally since, into the dynamicity, conflict, cooperation, and even ignorance between opposites, the neutrals are attracted and play an important role.

3.1. YinYang Bipolar Fuzzy Set Is the Bipolar Fuzzy Set

The authors sincerely recognize that: "*In the existing papers, YinYang bipolar fuzzy set also was called bipolar fuzzy set [5] and bipolar-valued fuzzy set [13,16].*"

These papers are cited as References [31–33].

We prove that the YinYang bipolar fuzzy set is not equivalent with the neutrosophic set, but a particular case of the bipolar neutrosophic set.

The authors [3] say that: "Denote $I^P = [0, 1]$ and $I^N = [-1, 0]$, and $L = \left\{\tilde{\alpha} = (\tilde{\alpha}^P, \tilde{\alpha}^N) \middle| \tilde{\alpha}^P \in I^P, \tilde{\alpha}^N \in I^N\right\}$, then $\tilde{\alpha}$ is called the YinYang bipolar fuzzy number. (YinYang bipolar fuzzy set) $X = \{x_1, \cdots, x_n\}$ represents the finite discourse. YinYang bipolar fuzzy set in X is defined by the mapping below.

$$\tilde{A} : X \to L, x \to \left(\tilde{A}^P(x), \tilde{A}^N(x)\right), \forall x \in X.$$

where the functions $\tilde{A}^{P}: X \to I^P, x \to \tilde{A}^{P}(x) \in I^P$ and $\tilde{A}^{N}: X \to I^N, x \to \tilde{A}^{N}(x) \in I^N$ define the satisfaction degree of the element $x \in X$ to the property, and the implicit counter-property to the YinYang bipolar fuzzy set $\tilde{A}$ in X, respectively (see [3], page 2).

With simpler notations, the above set L is equivalent to:

$L = \{(a, b)$, *with* $a \in [0, 1]$, $b \in [-1, 0]\}$, and the authors denote (a, b) as the YinYang bipolar fuzzy number.

Further on, again with simpler notations, the so-called *YinYang bipolar fuzzy set* in

$X = \{x_1, \ldots, x_n\}$ is equivalent to:

$X = \{x_1(a_1, b_1), \ldots, x_n(a_n, b_n)\}$, where all $a_1, \ldots, a_n \in [0, 1]$, and all $b_1, \ldots, b_n \in [-1, 0]\}$. Clearly, this is the bipolar fuzzy set and there is no need to call it the "YinYang bipolar fuzzy set." The authors added that: "Montero et al. pointed out that the neutrosophic set is equivalent to the YinYang bipolar fuzzy set in syntax." However, the bipolar fuzzy set is not equivalent to the neutrosophic set at all. The bipolar fuzzy set is actually a particular case of the bipolar neutrosophic set, defined as (keeping the previous notations):

$$X = \{x_1(\,(a_1, b_1), (c_1, d_1), (e_1, f_1)\,), \ldots, x_n(\,(a_n, b_n), (c_n, d_n), (e_n, f_n)\,)\}$$

where

all $a_1, \ldots, a_n, c_1, \ldots, c_n, e_1, \ldots, e_n \in [0, 1]$, and all $b_1, \ldots, b_n, d_1, \ldots, d_n, f_1, \ldots, f_n \in [-1, 0]\}$;

for a generic $x_j((a_j, b_j), (c_j, d_j), (e_j, f_j)) \in X$, $1 \le j \le n$,

a_i = *positive membership degree of* x_i, and b_i = *negative membership degree of* x_i;

c_i = *positive indeterminate-membership degree of* x_i, and d_i = *negative indeterminate membership degree of* x_i;

e_i = *positive non-membership degree of* x_i, and f_i = *negative non-membership degree of* x_i.

Using notations adequate to the neutrosophic environment, one found the following.

Let $\mathcal{U}$ be a universe of discourse, and $M \subset \mathcal{U}$ be a set. M is a **single-valued bipolar fuzzy set** (that authors call *YinYang bipolar fuzzy set*) if, for any element, $x(T^+_{(x)}, T^-_{(x)}) \in M$, $T^+_{(x)} \in [0, 1]$, and $T^-_{(x)} \in [-1, 0]$, where $T^+_{(x)}$ is the positive membership of x, and $T^-_{(x)}$ is the negative membership of x. (BFS).

The authors write that: "*Montero et al. pointed that the neutrosophic set [22] is equivalent to the YinYang bipolar fuzzy set in syntax [17]*".

Montero et al.'s paper is cited below as Reference [5].

If somebody says something, it does not mean it is true. They have to verify. Actually, it is *untrue*, since the neutrosophic set is totally different from the so-called YinYang bipolar fuzzy set.

Let $\mathcal{U}$ be a universe of discourse, and $M \subset \mathcal{U}$ be a set, if for any element.

$$x(T(x), I(x), F(x)) \in M$$

$T(x), I(x), F(x)$ are standard or nonstandard real subsets of the nonstandard real subsets of the nonstandard real unit interval $]^-0, 1^+[$. (NS).

Clearly, the definitions (BFS) and (NS) are totally different. In the so-called YinYang bipolar fuzzy set, there is no indeterminacy $I(x)$, no nonstandard analysis involved, and the neutrosophic components may be subsets as well.

3.2. Single-Valued Bipolar Fuzzy Set as a Particular Case of the Single-Valued Bipolar Neutrosophic Set

The Single-Valued bipolar fuzzy set (alias YinYang bipolar fuzzy set) is a particular case of the Single-Valued bipolar neutrosophic set, employed by the neutrosophic community, and defined as follows:

Let $\mathcal{U}$ be a universe of discourse, and $M \subset \mathcal{U}$ be a set. M is a single-valued bipolar neutrosophic set, if for any element:

$$x(T^{+}_{(x)}, T^{-}_{(x)}; I^{+}_{(x)}, I^{-}_{(x)}; F^{+}_{(x)}, F^{-}_{(x)}) \in M$$

$$T^{+}_{(x)}, I^{+}_{(x)}, F^{+}_{(x)} \in [0, 1]$$

$$T^{-}_{(x)}, I^{-}_{(x)}, F^{-}_{(x)} \in [-1, 0]$$

3.3. Dependent Indeterminacy vs. Independent Indeterminacy

The authors say: *"Attanassov's intuitionistic fuzzy set [4] perfectly reflects indeterminacy but not bipolarity."*

We disagree, since Atanassov's intuitionistic fuzzy set [22] perfectly reflects **hesitancy** between membership and non-membership not **indeterminacy**, since **hesitancy is dependent** on membership and non-membership: $H = 1 - T - F$, where H = hesitancy, T = membership, and F = non-membership.

It is the single-valued neutrosophic set that "perfectly reflects indeterminacy" since indeterminacy (I) in the neutrosophic set is **independent** from membership (T) and from nonmembership (F).

On the other hand, the neutrosophic set perfectly reflects the bipolarity membership/non-membership as well, since the membership (T) and nonmembership (F) are independent of each other.

3.4. Dependent Bipolarity vs. Independent Bipolarity

The *bipolarity* in the single-valued fuzzy set and intuitionistic fuzzy set is **dependent** (restrictive) in the sense that, if the truth-membership is T, then it involves the falsehood-nonmembership $F \leq 1 - T$ while the *bipolarity* in a single-valued neutrosophic set is independent (nonrestrictive): if the truth-membership $T \in [0, 1]$, the falsehood-nonmebership is not influenced at all, then $F \in [0, 1]$.

3.5. Equilibriums and Neutralities

Again: "While, in semantics, the YinYang bipolar fuzzy set suggests *equilibrium*, and neutrosophic set suggests a general *neutrality*. While the neutrosophic set has been successfully applied to a medical diagnosis [9,27], from the above analysis and the conclusion in [31], we see that the YinYang bipolar fuzzy set is clearly the suitable model to a bipolar disorder diagnosis and will be adopted in this paper."

I'd like to add that the single-valued bipolar neutrosophic set suggests:

— three types of equilibrium, between: $T^{+}_{(x)}$ and $T^{-}_{(x)}$, $I^{+}_{(x)}$ and $I^{-}_{(x)}$, and $F^{+}_{(x)}$ and $F^{-}_{(x)}$;

— and two types of neutralities (indeterminacies) between $T^{+}_{(x)}$ and $F^{+}_{(x)}$, and between $T^{-}_{(x)}$ and $F^{-}_{(x)}$.

Therefore, the single-valued bipolar neutrosophic set is $3 \times 2 = 6$ times more complex and more flexible than the YinYang bipolar fuzzy set. Due to higher complexity, flexibility, and capability of catching more details (such as falsehood-nonmembership, and indeterminacy), the single-valued bipolar neutrosophic set is more suitable than the YinYang bipolar fuzzy set to be used in a bipolar disorder diagnosis.

3.6. Zhang-Zhang's Bipolar Model is not Equivalent with the Neutrosophic Set

Montero et al. [5] wrote: *"Zhang-Zhang's bipolar model is, therefore, equivalent to the neutrosophic sets proposed by Smarandache [70]" (p. 56).*

This sentence is false and we proved previously that what Zhang & Zhang proposed in 2004 is a subclass of the single-valued bipolar neutrosophic set.

3.7. *Tripolar and Multipolar Neutrosophic Sets*

Not talking about the fact that, in 2016, we have extended our bipolar neutrosophic set to tripolar and even multipolar neutrosophic sets [18], the sets have become more general than the bipolar fuzzy model.

3.8. *Neutrosophic Overset/Underset/Offset*

Not talking that the unit interval [0, 1] was extended in 2006 below 0 and above 1 into the neutrosophic overset/underset/offset: $[\Omega, \Psi]$ with $\Omega \leq 0 < 1 \leq \Psi$ (as explained above).

3.9. *Neutrosophic Algebraic Structures*

The Montero et al. [5] continue: "*Notice that none of these two equivalent models include any formal structure, as claimed in [48]*".

First, we have proved that these two models (Zhang-Zhang's bipolar fuzzy set, and neutrosophic logic) are not equivalent at all. Zhang-Zhang's bipolar fuzzy set is a subclass of a particular type of neutrosophic set, called the single-valued bipolar neutrosophic set.

Second, since 2013, Kandasamy and Smarandache have developed various algebraic structures (such as neutrosophic semigroup, neutrosophic group, neutrosophic ring, neutrosophic field, neutrosophic vector space, etc.) [28] on the set of neutrosophic numbers:

$S_R = \left\{a + bI|, \text{ where } a,\ b \in \mathbb{R}, \text{ and } I = \text{indeterminacy},\ I^2 = I\right\}$, where $\mathbb{R}$ is the set of real numbers.

And extended on:

$S_C = \left\{a + bI|, \text{ where } a,\ b \in C, \text{ and } I = \text{indeterminacy},\ I^2 = I\right\}$, where C is the set of complex numbers.

However, until 2016 [year of Montero et al.'s published paper], I did not develop a formal structure on the neutrosophic set. Montero et al. are right.

Yet, in 2018, and, consequently at the beginning of 2019, we [2] developed, then generalized, and proved that the neutrosophic set has a structure of the lattice of the first type (as the neutrosophically partially ordered set): $(\,]^{-}0,\ 1^{+}[, \leq_N)$, where $]^{-}0,\ 1^{+}[$ is the nonstandard neutrosophic mobinad (monads and binads) real unit interval, and $\leq_N$ is the nonstandard neutrosophic inequality. Moreover, $\left(\,]^{-}0,\ 1^{+}[, \inf_N, \sup_N, {}^{-}0,\ 1^{+}\right)$ has the structure of the bound lattice of the second type (as algebraic structure), under two binary laws $\inf_N$ (nonstandard neutrosophic infimum) and $\sup_N$ (nontandard neutrosophic supremum).

3.10. *Neutrality (<neutA>)*

Montero et al. [5] continue: "... *the selected denominations within each model might suggest different underlying structures: while the model proposed by Zhang and Zhang suggests conflict between categories (a specific type of neutrality different from Atanassov's indeterminacy), Smarandache suggests a general neutrality that should, perhaps jointly, cover some of the specific types of neutrality considered in our paired approach.*"

In neutrosophy and neutrosophic set/logic/probability, the neutrality <neutA> means everything in between <A> and <antiA>, everything which is neither <A> nor <antiA>, or everything which is a blending of <A> and <antiA>.

Further on, in Refined Neutrosophy and Refined Neutrosophic Set/Logic/Probability [9], the neutrality <neutA> was split (refined) in 2013 into sub-neutralities (or sub-indeterminacies), such as: <neutA$_1$>, <neutA$_2$>, ... , <neutA$_n$> whose number could be finite or infinite depending on each application that needs to be solved.

Thus, the paired structure becomes a particular case of refined neutrosophy (see next).

4. The Pair Structure as a Particular Case of Refined Neutrosophy

Montero et al. [5] in 2016 have defined a **paired structure**: "*composed by a pair of opposite concepts and three types of neutrality as primary valuations: L = {concept, opposite, indeterminacy, ambivalence, conflict}.*"

Therefore, each element $x \in X$, where X is a universe of discourse, is characterized by a degree function, with respect to each attribute value from L:

$$\mu : X \to [0, 1]^5$$

$$\mu(x) = (\mu_1(x), \mu_2(x), \mu_3(x), \mu_4(x), \mu_5(x))$$

where $\mu_1(x)$ represents the degree of x with respect to the concept;

$\mu_2(x)$ represents the degree of x with respect to the opposite (of the concept);

$\mu_3(x)$ represents the degree of x with respect to 'indeterminacy';

$\mu_4(x)$ represents the degree of x with respect to 'ambivalence';

$\mu_5(x)$ represents the degree of x with respect to 'conflict'.

However, this paired structure is a particular case of Refined Neutrosophy.

4.1. Antonym vs. Negation

First, Dialectics is the dynamics of opposites. Denote them by $\langle A \rangle$ and $\langle \text{anti}A \rangle$, where $\langle A \rangle$ may be an item, a concept, attribute, idea, theory, and so on while $\langle \text{anti}A \rangle$ is the opposite of $\langle A \rangle$.

Secondly, Neutrosophy ([10], 1998), as a generalization of Dialectics, and a new branch of philosophy, is the dynamics of opposites and their neutralities (denoted by $\langle \text{neut}A \rangle$). Therefore, Neutrosophy is the dynamics of $\langle A \rangle$, $\langle \text{anti}A \rangle$, and $\langle \text{neut}A \rangle$.

$\langle \text{neut}A \rangle$ means everything, which is neither $\langle A \rangle$ nor $\langle \text{anti}A \rangle$, or which is a mixture of them, or which is indeterminate, vague, or unknown.

The **antonym** of $\langle A \rangle$ is $\langle \text{anti}A \rangle$.

The **negation** of $\langle A \rangle$ (which we denote by $\langle \text{non}A \rangle$) is what is not $\langle A \rangle$, therefore:

$$\neg_N \langle A \rangle = \langle nonA \rangle =_N \langle neutA \rangle \cup_N \langle antiA \rangle$$

We preferred to use the lower index $_N$ (neutrosophic) because we deal with items, concepts, attributes, ideas, and theories such as $\langle A \rangle$ and, in consequence, its derivates $\langle \text{anti}A \rangle$, $\langle \text{neut}A \rangle$, and $\langle \text{non}A \rangle$, whose borders are ambiguous, vague, and not clearly delimited.

4.2. Refined Neutrosophy as an Extension of Neutrosophy

Thirdly, Refined Neutrosophy ([6], 2013), as an extension of Neutrosophy, and a refined branch of philosophy, is the dynamics of refined opposites: $\langle A_1 \rangle, \langle A_2 \rangle, \ldots, \langle A_p \rangle$ with $\langle \text{anti}A_1 \rangle, \langle \text{anti}A_2 \rangle, \ldots, \langle \text{anti}A_s \rangle$, and their refined neutralities: $\langle \text{neut}A_1 \rangle, \langle \text{neut}A_2 \rangle, \ldots, \langle \text{neut}A_r \rangle$, for integers p, r, $s \geq 1$, and $p + r + s = n \geq 4$. Therefore, the item $\langle A \rangle$ has been split into sub-items $\langle A_j \rangle$, $1 \leq j \leq p$, the $\langle \text{anti}A \rangle$ into sub-(anti-items) $\langle \text{anti}A_k \rangle$, $1 \leq l \leq s$, and the $\langle \text{neut}A \rangle$ into sub-(neutral-items) $\langle \text{neut}A_l \rangle$, $1 \leq k \leq r$.

4.3. Qualitative Scale as a Particular Case of Refined Neutrosophy

Montero et al.'s qualitative scale [5] is a particular case of Refined Neutrosophy where the neutralities are split into three parts.

$$L = \{concept, opposite, indeterminacy, ambivalence, conflict\} = \{<A>, <antiA>, <neutA_1>, <neutA_2>, <neutA_3>\}$$

where: $<A>$ = *concept*, $<antiA>$ = *opposite*, $<neutA1>$ = *indeterminacy*, $<neutA2>$ = *ambivalence*, $<neutA3>$ = *conflict*.

Yin Yang, Dialectics, Neutrosophy, and Refined Neutrosophy (the last one having only $\langle \text{neut}A \rangle$ as refined component), are bipolar: $\langle A \rangle$ and $\langle \text{anti}A \rangle$ are the poles.

Montero et al.'s qualitative scale is bipolar ('concept', and its 'opposite').

4.4. Multi-Subpolar Refined Neutrosophy

However, the Refined Neutrosophy, whose at least one of ⟨A⟩ or ⟨antiA⟩ is refined, is *multi-subpolar*.

4.5. Multidimensional Fuzzy Set as a Particular Case of the Refined Neutrosophic Set

Montero et al. [5] defined the *Multidimensional Fuzzy Set* A_L as: $A_t = \left\{< x; (\mu_s(x))_{s\in L} > \middle| x \in X\right\}$, where X is the universe of discourse, L = the previous qualitative scale, and $\mu_s(x) \in S$, where S is a valuation scale (in most cases $S = [0, 1]$), $\mu_s(x)$ is the degree of x with respect to $s \in L$.

A *Single-Valued Neutrosophic Set* is defined as follows. Let $\mathcal{U}$ be a universe of discourse, and $M \subset \mathcal{U}$ a set. For each element $x(T(x), I(x), F(x)) \in M$, $T(x) \in [0, 1]$ is the degree of truth-membership of element x with respect to the set M, $I(x) \in [0, 1]$ is the degree of indeterminacy-membership of element x with respect to the set M, and $F(x) \in [0, 1]$ is the degree of falsehood-nonmembership of element x with respect to the set M.

Let's refine $I(x)$ as $I_1(x), I_2(x)$, and $I_3(x) \in [0, 1]$ sub-indeterminacies. Then we get a single-valued refined neutrosophic set.

$\mu_{concept}(x) = T(x)$ (truth-membership);
$\mu_{opposite}(x) = F(x)$ (falsehood-non-membership);
$\mu_{\text{indeterminacy}}(x) = I_1(x)$ (first sub-indeterminacy);
$\mu_{\text{ambivalence}}(x) = I_2(x)$ (second sub-indeterminacy);
$\mu_{\text{conflict}}(x) = I_3(x)$ (third sub-indeterminacy).

The *Single-Valued Refined Neutrosophic Set* is defined as follows. Let $\mathcal{U}$ be a universe of discourse, and $M \subset \mathcal{U}$ a set. For each element:

$$x\big(T_1(x), T_2(x), \ldots, T_p(x);\ I_1(x), I_2(x), \ldots, I_r(x);\ F_1(x), F_2(x), \ldots, F_s(x)\big) \in M$$

$T_j(x), 1 \le j \le p$, are degrees of subtruth-submembership of element x with respect to the set M.

$I_k(x)$, $1 \le k \le r$, are degrees of subindeterminacy-membership of element x with respect to the set M.

Lastly, $F_l(x)$, $1 \le l \le s$, are degrees of sub-falsehood-sub-non-membership of element x with respect to the set M, where integers p, r, $s \ge 1$, and $p + r + s = n \ge 4$.

Therefore, Montero et al.'s **multidimensional fuzzy set** is a particular case of the **refined neutrosophic set**, when $p = 1$, $r = 3$, and $s = 1$, where $n = 1 + 3 + 1 = 5$.

4.6. Plithogeny and Plithogenic Set

Fourthly, in 2017 and in 2018 [24–27], the Neutrosophy was extended to Plithogeny, which is multipolar, being the dynamics and hermeneutics [methodological study and interpretation] of many opposites and/or their neutrals, together with non-opposites.

⟨A⟩, ⟨neutA⟩, ⟨antiA⟩;
⟨B⟩, ⟨neutB⟩, ⟨antiB⟩; etc.
⟨C⟩, ⟨D⟩, etc.

In addition, the **Plithogenic Set** was introduced, as a generalization of **Crisp**, **Fuzzy**, **Intuitionistic Fuzzy**, and **Neutrosophic Sets**.

Unlike previous sets defined, whose elements were characterized by the attribute 'appurtenance' (to the set), which has only one (membership), or two (membership, nonmembership), or three (membership, nonmembership, indeterminacy) attribute values, respectively. For the Plithogenic Set, each element may be characterized by a multi-attribute, with any number of attribute values.

4.7. Refined Neutrosophic Set as a Unifying View of Opposite Concepts

Montero et al.'s statement [5] from their paper Abstract: *"we propose a consistent and unifying view to all those basic knowledge representation models that are based on the existence of two somehow opposite fuzzy concepts."*

With respect to the "unifying" claim, their statement is not true, since, as we proved before, their **paired structure** together with three types on neutralities (**indeterminacy**, **ambivalence**, and **conflict**) is a simple, particular case of the refined neutrosophic set.

The real unifying view currently is the **Refined Neutrosophic Set**.

{I was notified about this paired structure article [5] by Dr. Said Broumi, who forwarded it to me.}

4.8. Counter-Example to the Paired Structure

As a counter example to the paired structure [5], it cannot catch a simple voting scenario.

The election for the United States President from 2016: Donald Trump vs. Hillary Clinton. USA has 50 states and since, in the country, there is an **Electoral vote**, not a **Popular vote**, it is required to know the winner of each state.

There were two opposite candidates.

The candidate that receives more votes than the other candidate in a state gets all the points of that state.

As in the neutrosophic set, there are three possibilities:

T = percentage of USA people voting for Mr. Trump;

I = percentage of USA people not voting, or voting but giving either a blank vote (not selecting any candidate) or a black vote (cutting all candidates);

F = percentage of USA people voting against Mr. Trump.

The opposite concepts, using Montero et al.'s knowledge representation, are T (voting for, or truth-membership) and F (voting against, or false-membership). However, $T > F$, or $T = F$, or $T < F$, that the Paired Structure can catch, mean only the Popular vote, which does not count in the United States.

Actually, it happened that $T < F$ in the US 2016 presidential election, or Mr. Trump lost the Popular vote, but he won the Presidency using the Electoral vote.

The paired structure is not capable of refining the opposite concepts (T and F), while the indeterminate (I) could be refined by the paired structure only in three parts.

Therefore, the paired structure is not a unifying view of all basic knowledge that uses opposite fuzzy concepts. However, the refined neutrosophic set/logic/probability do.

Using the refined neutrosophic set and logic, and splits (refines) T, I, and F as:

T_j = percentage of American state S_j people voting for Mr. Trump;

I_j = percentage of American state S_j people not voting, or casting a blank vote or a black vote;

F_j = percentage of American state S_j people voting against Mr. Trump, with $T_j, I_j, F_j \in [0, 1]$ and $T_j + I_j + F_j = 1$, for all $j \in \{1, 2, \ldots, 50\}$.

Therefore, one has:

$(T_1, T_2, \ldots, T_{50}; I_1, I_2, \ldots, I_{50}; F_1, F_2, \ldots, F_{50})$.

On the other hand, due to the fact that the sub-indeterminacies $I_1, I_2, \ldots, I_{50}$ did not count towards the winner or looser (only for indeterminate voting statistics), it is not mandatory to refine I. We could simply refine it as:

$(T_1, T_2, \ldots, T_{50}; I; F_1, F_2, \ldots, F_{50})$.

4.9. Finite Number and Infinite Number of Neutralities

Montero et al. [5]: *"(...) we emphasize the key role of certain neutralities in our knowledge representation models, as pointed out by Atanassov [4], Smarandache [70], and others. However, we notice that our notion of neutrality should not be confused with the neutral value in a traditional sense (see [22–24,36,54], among others).*

Instead, we will stress the existence of different kinds of neutrality that emerge (in the sense of Reference [11]) from the semantic relation between two opposite concepts (and notice that we refer to a neutral category that does not entail linearity between opposites)."

In neutrosophy, and, consequently, in the neutrosophic set, logic, and probability, between the opposite items (concepts, attributes, ideas, etc.) $\langle A\rangle$ and $\langle \text{anti}A\rangle$, there may be a large number of neutralities/indeterminacies (all together denoted by $\langle \text{neut}A\rangle$ even an infinite spectrum—depending on the application to solve.

We agree with different kinds of neutralities and indeterminacies (vague, ambiguous, unknown, incomplete, contradictory, linear and non-linear information, and so on), but the authors display only three neutralities.

In our everyday life and in practical applications, there are more neutralities and indeterminacies.

In another example (besides the previous one about Electoral voting), there may be any number of sub indeterminacies/sub neutralities.

The opposite concepts attributes are: $\langle A\rangle$ = white, $\langle \text{anti}A\rangle$ = black, while neutral concepts in between may be: $\langle \text{neut}A_1\rangle$ = yellow, $\langle \text{neut}A_2\rangle$ = orange, $\langle \text{neut}A_3\rangle$ = red, $\langle \text{neut}A_4\rangle$ = violet, $\langle \text{neut}A_5\rangle$ = green, and $\langle \text{neut}A_6\rangle$ = blue. Therefore, we have six neutralities. Example with infinitely many neutralities:

— The opposite concepts: $\langle A\rangle$ = white, $\langle \text{anti}A\rangle$ = black;
— The neutralities: $\langle \text{neut}A_{1,\,2,\,\ldots,\,\infty}\rangle$ = the whole light spectrum between white and black, measured in nanometers (*nn*) [a nanometer is a billionth part of a meter].

5. Conclusions

The neutrosophic community thank the authors for their criticism and interest in the neutrosophic environment, and we wait for new comments and criticism, since, as Winston Churchill had said, *the eagles fly higher against the wind.*

Funding: The author received no external funding.

Conflicts of Interest: The author declares no conflict of interest.

Notations

$\le_{nN}^{nonS}$	means nonstandard n-tuple neutrosophic inequality;
$\le_{nN}$	means standard (real) *n*-tuple inequality;
$\le_{N}^{nonS}$	means nonstandard unary neutrosophic inequality;
$\le_{N}$	mean standard (real) unary neutrosophic inequality;
$=_N$	means neutrosophic equality;
$\neg_N$	means neutrosophic negation;
$\cup_N$	means neutrosophic union;
$=$	means classical equality;
$<, >, \le, \ge$	mean classical inequalities.

References

1. Klement, E.P.; Mesiar, R. L-Fuzzy Sets and Isomorphic Lattices: Are All the "New" Results Really New? *Mathematics* **2018**, *6*, 146. [CrossRef]
2. Smarandache, F. Extended Nonstandard Neutrosophic Logic, Set, and Probability based on Extended Nonstandard Analysis. *arXiv*, 2019; arXiv:1903.04558.
3. Zhang, W.; Zhang, L. YinYang bipolar logic and bipolar fuzzy logic. *Inf. Sci.* **2004**, *165*, 265–287. [CrossRef]
4. Han, Y.; Lu, Z.; Du, Z.; Luo, Q.; Chen, S. A YinYang bipolar fuzzy cognitive TOPSIS method to bipolar disorder diagnosis. *Comput. Methods Programs Biomed.* **2018**, *158*, 1–10. [CrossRef] [PubMed]
5. Montero, J.; Bustince, H.; Franco, C.; Rodríguez, J.T.; Gómez, D.; Pagola, M.; Fernández, J.; Barrenechea, E. Paired structures in knowledge representation. *Knowl. Base D Syst.* **2016**, *100*, 50–58. [CrossRef]

6. Smarandache, F. N-Valued Refined Neutrosophic Logic and Its Applications in Physics. *Prog. Phys.* **2013**, *4*, 143–146.
7. Smarandache, F. Neutrosophy, A New Branch of Philosophy. *Mult. Valued Log. Int. J.* **2002**, *8*, 297–384.
8. Smarandache, F. Degree of Dependence and Independence of the (Sub)Components of Fuzzy Set and Neutrosophic Set. *Neutrosophic Sets Syst.* **2016**, *11*, 95–97.
9. Smarandache, F. Degree of Dependence and Independence of Neutrosophic Logic Components Applied in Physics. In Proceedings of the 2016 Annual Spring Meeting of the APS Ohio-Region Section, Dayton, OH, USA, 8–9 April 2016.
10. Smarandache, F. A Unifying Field in Logics: Neutrosophic Logic. Neutrosophy, Neutrosophic Set, Neutrosophic Probability and Statistics. 1998. Available online: http://fs.unm.edu/eBook-Neutrosophics6.pdf (accessed on 3 April 2019).
11. Smarandache, F. About Nonstandard Neutrosophic Logic (Answers to Imamura's 'Note on the Definition of Neutrosophic Logic'). Available online: https://arxiv.org/ftp/arxiv/papers/1812/1812.02534.pdf (accessed on 3 April 2019).
12. Smarandache, F. Extended Nonstandard Neutrosophic Logic, Set, and Probability based on Extended Nonstandard Analysis. *arXiv*, 2019; arXiv:1903.04558v1.
13. Smarandache, F. *Neutrosophic Overset, Neutrosophic Underset, and Neutrosophic Offset. Similarly for Neutrosophic Over-/Under-/Off-Logic, Probability, and Statistics*; Pons Editions: Bruxelles, Belgique, 2016; 168p. Available online: https://arxiv.org/ftp/arxiv/papers/1607/1607.00234.pdf (accessed on 3 April 2019).
14. Smarandache, F. Operadores con conjunto neutrosóficos de valor único Oversets, Undersets y Off-set. *Neutrosophic Comput. Mach. Learn.* **2018**, *4*, 3–7. [CrossRef]
15. Smarandache, F. *Law of Included Multiple-Middle & Principle of Dynamic Neutrosophic Opposition*; EuropaNova asbl: Brussels, Belgium; The Educational Publisher Inc.: Columbus, OH, USA, 2014; 136 p.
16. Salama, A.A.; Smarandache, F. *Neutrosophic Crisp Set Theory*; Educational Publisher: Columbus, OH, USA, 2015.
17. Dezert, J. Open Questions to Neutrosophic Inferences. *Mult. Valued Log. Int. J.* **2002**, *8*, 439–472.
18. Smarandache, F. Interval-Valued Neutrosophic Oversets, Neutrosophic Understes, and Neutrosophic Offsets. *Int. J. Sci. Eng. Investig.* **2016**, *5*, 1–4.
19. Smarandache, F. Operators on Single-Valued Neutrosophic Oversets, Neutrosophic Undersets, and Neutrosophic Offsets. *J. Math. Inform.* **2016**, *5*, 63–67. [CrossRef]
20. Smarandache, F. *Applications of Neutrosophic Sets in Image Identification, Medical Diagnosis, Fingerprints and Face Recognition and Neutrosophic verset/Underset/Offset*; COMSATS Institute of Information Technology: Abbottabad, Pakistan, 26 December 2017.
21. Cuong, B.C.; Kreinovich, V. Picture fuzzy sets—A new concept for computational intelligence problems. In Proceedings of the Third World Congress on Information and Communication Technologies (WICT 2013), Hanoi, Vietnam, 15–18 December 2013.
22. Atanassov, K.T. Intuitionistic fuzzy sets. *Fuzzy Sets Syst.* **1986**, *20*, 87–96. [CrossRef]
23. Howe, D. The Free Online Dictionary of Computing. Available online: http://foldoc.org/ (accessed on 3 April 2019).
24. Smarandache, F. *Plithogeny, Plithogenic Set, Logic, Probability, and Statistics*; Cornell University, Computer Science—Artificial Intelligence; Pons Publishing House: Brussels, Belgium, 2014; 141p.
25. Smarandache, F. Extension of Soft Set to Hypersoft Set, and then to Plithogenic Hypersoft Set. *Neutrosophic Sets Syst.* **2018**, *22*, 168–170. [CrossRef]
26. Smarandache, F. Plithogenic Set, an Extension of Crisp, Fuzzy, Intuitionistic Fuzzy, and Neutrosophic Sets—Revisited. *Neutrosophic Sets Syst.* **2018**, *21*, 153–166. [CrossRef]
27. Smarandache, F. Physical Plithogenic Set. In Proceedings of the 71st Annual Gaseous Electronics Conference, American Physical Society (APS), Session LW1, Oregon Convention Center Room, Portland, OR, USA, 5–9 November 2018.
28. Kandasamy, W.B.V.; Smarandache, F. Fuzzy Cognitive Maps and Neutrosophic Cognitive Maps. Xiquan, Phoenix, 2003. Available online: http://fs.unm.edu/NCMs.pdf (accessed on 4 April 2019).
29. Vladutescu, S.; Smarandache, F.; Gifu, D.; Tenescu, A. (Eds.) *Topical Communication Uncertainties*; Sitech Publishing House: Craiova, Romania; Zip Publishing: Columbus, OH, USA, 2014; 300p.

30. Peng, X.; Dai, J. A bibliometric analysis of neutrosophic set: Two decades review from 1998 to 2017. *Artif. Intell. Rev.* **2018**. [CrossRef]
31. Bloch, I. Geometry of spatial bipolar fuzzy sets based on bipolar fuzzy numbers and mathematical morphology, fuzzy logic and applications. *Lect. Notes Comput. Sci.* **2009**, *5571*, 237–245.
32. Han, Y.; Shi, P.; Chen, S. Bipolar-valued rough fuzzy set and its applications to decision information system. *IEEE Trans. Fuzzy Syst.* **2015**, *23*, 2358–2370. [CrossRef]
33. Lee, K.M. Bipolar-valued fuzzy sets and their basic operations. In Proceedings of the International Conference, Bangkok, Thailand, 23–25 April 2010; pp. 307–317.

Σ mathematics

MDPI

Article

Linguistic Neutrosophic Numbers Einstein Operator and Its Application in Decision Making

Changxing Fan, Sheng Feng and Keli Hu *

Department of Computer Science, Shaoxing University, Shaoxing 312000, China; fcxjszj@usx.edu.cn (C.F.); fengsheng_13@aliyun.com (S.F.)
* Correspondence: ancimoon@gmail.com

Received: 28 March 2019; Accepted: 25 April 2019; Published: 28 April 2019

Abstract: Linguistic neutrosophic numbers (LNNs) include single-value neutrosophic numbers and linguistic variable numbers, which have been proposed by Fang and Ye. In this paper, we define the linguistic neutrosophic number Einstein sum, linguistic neutrosophic number Einstein product, and linguistic neutrosophic number Einstein exponentiation operations based on the Einstein operation. Then, we analyze some of the relationships between these operations. For LNN aggregation problems, we put forward two kinds of LNN aggregation operators, one is the LNN Einstein weighted average operator and the other is the LNN Einstein geometry (LNNEWG) operator. Then we present a method for solving decision-making problems based on LNNEWA and LNNEWG operators in the linguistic neutrosophic environment. Finally, we apply an example to verify the feasibility of these two methods.

Keywords: multiple attribute group decision making (MAGDM); Linguistic neutrosophic; LNN Einstein weighted-average operator; LNN Einstein weighted-geometry (LNNEWG) operator

1. Introduction

Smarandache [1] proposed the neutrosophic set (NS) in 1998. Compared with the intuitionistic fuzzy sets (IFSs), the NS increases the uncertainty measurement, from which decision makers can use the truth, uncertainty and falsity degrees to describe evaluation, respectively. In the NS, the degree of uncertainty is quantified, and these three degrees are completely independent of each other, so, the NS is a generalization set with more capacity to express and deal with the fuzzy data. At present, the study of NS theory has been a part of research that mainly includes the research of the basic theory of NS, the fuzzy decision of NS, and the extension of NS, etc. [2–14]. Recently, Fang and Ye [15] presented the linguistic neutrosophic number (LNN). Soon afterwards, many research topics about LNN were proposed [16–18].

Information aggregation operators have become an important research topic and obtained a wide range of research results. Yager [19] put forward the ordered weighted average (OWA) operator considering the data sorting position. Xu [20] presented the arithmetic aggregation (AA) of IFS. Xu and Yager [21] presented the geometry aggregation (GA) operator of IFS. Zhao [22] proposed generalized aggregation operators based on IFS and proved that AA and GA were special cases of generalized aggregation operator. The operators mentioned above are established based on the algebraic sum and the algebraic product of number sets. They are respectively referred to as a special case of Archimedes t-conorm and t-norm to establish union or intersection operation of the number set. The union and intersection of Einstein operation is a kind of Archimedes t-conorm and t-norm with good smooth characteristics [23]. Wang and Liu [24] built some IF Einstein aggregation operators and proved that the Einstein aggregation operator has better smoothness than the arithmetic aggregation operator. Zhao and Wei [25] put forward the IF Einstein hybrid-average (IFEHA) operator and IF

Einstein hybrid-geometry (IFEHG) operator. Further, Guo etc. [26] applied the Einstein operation to a hesitate fuzzy set. Lihua Yang etc. [27] put forward novel power aggregation operators based on Einstein operations for interval neutrosophic linguistic sets. However, neutrosophic linguistic sets are different from linguistic neutrosophic sets. The former still use two values to describe the evaluation value, while the latter can use a pure language value to describe the evaluation value. As far as we know, this is the first work on Einstein aggregation operators for LNN. It must be noticed that the aggregation operators in References [15–18] are almost based on the most commonly used algebraic product and algebraic sum of LNNs for carrying the combination process, which is not the only operation law that can be chosen to model the intersection and union on LNNs. Thus, we establish the operation rules of LNN based on Einstein operation and put forward the LNN Einstein weighted-average (LNNEWA) operator and LNN Einstein weighted-geometry (LNNEWG) operator. These operators are finally utilized to solve some relevant problems.

The other organizations: in Section 2, concepts of LNN and Einstein are described, operational laws of LNNs based on Einstein operation are defined, and their performance is analyzed. In Section 3, LNNEWA and LNNEWG operators are proposed. In Section 4, multiple attribute group decision making (MAGDM) methods are built based on LNNEWA and LNNEWG operators. In Section 5, an instance is given. In Section 6, conclusions and future research are given.

2. Basic Theories

2.1. LNN and Its Operational Laws

Definition 1. [15] *Set a finite language set* $\Psi = \{\psi_t | t \in [0,k]|\}$, *where* ψ_t *is a linguistic variable,* $k+1$ *is the cardinality of* Ψ. *Then, we define* $u = \langle \psi_\beta, \psi_\gamma, \psi_\delta \rangle$, *in which* $\psi_\beta, \psi_\gamma, \psi_\delta \in \Psi$ *and* $\beta, \gamma, \delta \in [0,k]$, ψ_β, ψ_δ *and* ψ_γ *expresse truth, falsity and indeterminacy degree, respectively, we call u an LNN.*

Definition 2. [15] *Set three LNNs* $u = \langle \psi_\beta, \psi_\gamma, \psi_\delta \rangle$, $u_1 = \langle \psi_{\beta_1}, \psi_{\gamma_1}, \psi_{\delta_1} \rangle$ *and* $u_2 = \langle \psi_{\beta_2}, \psi_{\gamma_2}, \psi_{\delta_2} \rangle$ *in* Ψ *and* $\lambda \geq 0$, *then, the operational rules are as following:*

$$\oplus u_2 = \langle \psi_{\beta_1}, \psi_{\gamma_1}, \psi_{\delta_1} \rangle \oplus \langle \psi_{\beta_2}, \psi_{\gamma_2}, \psi_{\delta_2} \rangle = \langle \psi_{\beta_1+\beta_2-\frac{\beta_1\beta_2}{k}}, \psi_{\frac{\gamma_1\gamma_2}{k}}, \psi_{\frac{\delta_1\delta_2}{k}} \rangle; \tag{1}$$

$$u_1 \otimes u_2 = \langle \psi_{\beta_1}, \psi_{\gamma_1}, \psi_{\delta_1} \rangle \otimes \langle \psi_{\beta_2}, \psi_{\gamma_2}, \psi_{\delta_2} \rangle = \langle \psi_{\frac{\beta_1\beta_2}{k}}, \psi_{\gamma_1+\gamma_2-\frac{\gamma_1\gamma_2}{k}}, \psi_{\delta_1+\delta_2-\frac{\delta_1\delta_2}{k}} \rangle; \tag{2}$$

$$\lambda u = \lambda \langle \psi_{\beta_1}, \psi_{\gamma_1}, \psi_{\delta_1} \rangle = \langle \psi_{k-k(1-\frac{\beta}{k})^\lambda}, \psi_{k(\frac{\gamma}{k})^\lambda}, \psi_{k(\frac{\delta}{k})^\lambda} \rangle; \tag{3}$$

$$u^\lambda = \langle \psi_{\beta_1}, \psi_{\gamma_1}, \psi_{\delta_1} \rangle^\lambda = \langle \psi_{k(\frac{\beta}{k})^\lambda}, \psi_{k-k(1-\frac{\gamma}{k})^\lambda}, \psi_{k-k(1-\frac{\delta}{k})^\lambda} \rangle. \tag{4}$$

Definition 3. [15] *Set an LNN* $u = \langle \psi_\beta, \psi_\gamma, \psi_\delta \rangle$ *in* Ψ, *we define* $\zeta(u)$ *as the expectation and* $\eta(u)$ *as the accuracy:*

$$\zeta(u) = (2k + \beta - \gamma - \delta)/3k \tag{5}$$

$$\eta(u) = (\beta - \delta)/k \tag{6}$$

Definition 4. [15]: *Set two LNNs* $u_1 = \langle \psi_{\beta_1}, \psi_{\gamma_1}, \psi_{\delta_1} \rangle$ *and* $u_2 = \langle \psi_{\beta_2}, \psi_{\gamma_2}, \psi_{\delta_2} \rangle$ *in* Ψ, *then*

If $\zeta(u_1) > \zeta(u_2)$, *then* $u_1 > u_2$;
If $\zeta(u_1) = \zeta(u_2)$ *then*
If $\eta(u_1) > \eta(u_2)$, *then* $u_1 > u_2$;
If $\eta(u_1) = \eta(u_2)$, *then* $u_1 \sim u_2$.

2.2. Einstein Operation

Definition 5. [28,29] *For any two real Numbers a, b∈ [0,1], Einstein $\oplus_e$ is an Archimedes t-conorms, Einstein $\otimes_e$ is an Archimedes t-norms, then*

$$\mathrm{a}\oplus_e\mathrm{b} = \frac{a+b}{1+ab}, \quad \mathrm{a}\otimes_e\mathrm{b} = \frac{ab}{1+(1-a)(1-b)}. \tag{7}$$

2.3. Einstein Operation Under the Linguistic Neutrosophic Number

Definition 6. *Set $u = \langle \psi_\beta, \psi_\gamma, \psi_\delta \rangle$, $u_1 = \langle \psi_{\beta_1}, \psi_{\gamma_1}, \psi_{\delta_1} \rangle$ and $u_2 = \langle \psi_{\beta_2}, \psi_{\gamma_2}, \psi_{\delta_2} \rangle$ as three LNNs in Ψ, $\lambda \geq 0$, the operation of Einstein $\oplus_e$ and Einstein $\otimes_e$ under the linguistic neutrosophic number are defined as follows:*

$$u_1 \oplus_e u_2 = \langle \psi_{\frac{k^2(\beta_1+\beta_2)}{k^2+\beta_1\beta_2}}, \psi_{\frac{k\gamma_1\gamma_2}{k^2+(k-\gamma_1)(k-\gamma_2)}}, \psi_{\frac{k\delta_1\delta_2}{k^2+(k-\delta_1)(k-\delta_2)}} \rangle; \tag{8}$$

$$u_1 \otimes_e u_2 = \langle \psi_{\frac{k\beta_1\beta_2}{k^2+(k-\beta_1)(k-\beta_2)}}, \psi_{\frac{k^2(\gamma_1+\gamma_2)}{k^2+\gamma_1\gamma_2}}, \psi_{\frac{k^2(\delta_1+\delta_2)}{k^2+\delta_1\delta_2}} \rangle; \tag{9}$$

$$\lambda u = \langle \psi_{k*\frac{(k+\beta)^\lambda-(k-\beta)^\lambda}{(k+\beta)^\lambda+(k-\beta)^\lambda}}, \psi_{k*\frac{2\gamma^\lambda}{(2k-\gamma)^\lambda+\gamma^\lambda}}, \psi_{k*\frac{2\delta^\lambda}{(2k-\delta)^\lambda+\delta^\lambda}} \rangle; \tag{10}$$

$$u^\lambda = \langle \psi_{k*\frac{2\beta^\lambda}{(2k-\beta)^\lambda+\beta^\lambda}}, \psi_{k*\frac{(k+\gamma)^\lambda-(k-\gamma)^\lambda}{(k+\gamma)^\lambda+(k-\gamma)^\lambda}}, \psi_{k*\frac{(k+\delta)^\lambda-(k-\delta)^\lambda}{(k+\delta)^\lambda+(k-\delta)^\lambda}} \rangle. \tag{11}$$

Theorem 1. *Set $u \langle = \psi_\beta, \psi_\gamma, \psi_\delta \rangle$, $u_1 = \langle \psi_{\beta_1}, \psi_{\gamma_1}, \psi_{\delta_1} \rangle$ and $u_2 = \langle \psi_{\beta_2}, \psi_{\gamma_2}, \psi_{\delta_2} \rangle$ as three LNNs in Ψ, $\lambda \geq 0$, then, the operation of Einstein $\oplus_e$ and Einstein $\otimes_e$ have the following performance:*

$$u_1 \oplus_e u_2 = u_2 \oplus_e u_1; \tag{12}$$

$$u_1 \otimes_e u_2 = u_2 \otimes_e u_1; \tag{13}$$

$$\lambda(u_1 \oplus_e u_2) = \lambda u_1 \oplus_e \lambda u_2; \tag{14}$$

$$(u_1 \otimes_e u_2)^\lambda = u_1{}^\lambda \otimes_e u_2{}^\lambda; \tag{15}$$

Proof. Performance (1) and (2) are easy to be obtained, so we omit it; Now we prove the performance (3): According to Definition 6, we can get

$$① \quad u_1 \oplus_e u_2 = \langle \psi_{\frac{k^2(\beta_1+\beta_2)}{k^2+\beta_1\beta_2}}, \psi_{\frac{k\gamma_1\gamma_2}{k^2+(k-\gamma_1)(k-\gamma_2)}}, \psi_{\frac{k\delta_1\delta_2}{k^2+(k-\delta_1)(k-\delta_2)}} \rangle;$$

② $\lambda(u_1 \oplus_e u_2)$

$$= \langle \psi_{k*\frac{(k+\frac{k^2(\beta_1+\beta_2)}{k^2+\beta_1\beta_2})^\lambda-(k-\frac{k^2(\beta_1+\beta_2)}{k^2+\beta_1\beta_2})^\lambda}{(k+\frac{k^2(\beta_1+\beta_2)}{k^2+\beta_1\beta_2})^\lambda+(k-\frac{k^2(\beta_1+\beta_2)}{k^2+\beta_1\beta_2})^\lambda}}, \psi_{k*\frac{2(\frac{k\gamma_1\gamma_2}{k^2+(k-\gamma_1)(k-\gamma_2)})^\lambda}{(2k-\frac{k\gamma_1\gamma_2}{k^2+(k-\gamma_1)(k-\gamma_2)})^\lambda+(\frac{k\gamma_1\gamma_2}{k^2+(k-\gamma_1)(k-\gamma_2)})^\lambda}}, \psi_{k*\frac{2(\frac{k\delta_1\delta_2}{k^2+(k-\delta_1)(k-\delta_2)})^\lambda}{(2k-\frac{k\delta_1\delta_2}{k^2+(k-\delta_1)(k-\delta_2)})^\lambda+(\frac{k\delta_1\delta_2}{k^2+(k-\delta_1)(k-\delta_2)})^\lambda}} \rangle$$

$$= \langle \psi_{k*\frac{(k+\beta_1)^\lambda(k+\beta_2)^\lambda-(k-\beta_1)^\lambda(k-\beta_2)^\lambda}{(k+\beta_1)^\lambda(k+\beta_2)^\lambda+(k-\beta_1)^\lambda(k-\beta_2)^\lambda}}, \psi_{k*\frac{2(\gamma_1\gamma_2)^\lambda}{((2k-\gamma_1)^\lambda(2k-\gamma_2)^\lambda)+(\gamma_1\gamma_2)^\lambda}}, \psi_{k*\frac{2(\delta_1\delta_2)^\lambda}{((2k-\delta_1)^\lambda(2k-\delta_2)^\lambda)+(\delta_1\delta_2)^\lambda}} \rangle;$$

$$③ \quad \lambda u_1 = \langle \psi_{k*\frac{(k+\beta_1)^\lambda-(k-\beta_1)^\lambda}{(k+\beta_1)^\lambda+(k-\beta_1)^\lambda}}, \psi_{k*\frac{2\gamma_1{}^\lambda}{(2k-\gamma_1)^\lambda+\gamma_1{}^\lambda}}, \psi_{k*\frac{2\delta_1{}^\lambda}{(2k-\delta_1)^\lambda+\delta_1{}^\lambda}} \rangle;$$

$$④ \quad \lambda u_2 = \langle \psi_{k*\frac{(k+\beta_2)^\lambda-(k-\beta_2)^\lambda}{(k+\beta_2)^\lambda+(k-\beta_2)^\lambda}}, \psi_{k*\frac{2\gamma_2{}^\lambda}{(2k-\gamma_2)^\lambda+\gamma_2{}^\lambda}}, \psi_{k*\frac{2\delta_2{}^\lambda}{(2k-\delta_2)^\lambda+\delta_2{}^\lambda}} \rangle;$$

⑤ $\lambda u_1 \oplus_e \lambda u_2$

$$= \langle \psi_{\frac{k^2 (k* \frac{(k+\beta_1)^\lambda - (k-\beta_1)^\lambda}{(k+\beta_1)^\lambda + (k-\beta_1)^\lambda} + k* \frac{(k+\beta_2)^\lambda - (k-\beta_2)^\lambda}{(k+\beta_2)^\lambda + (k-\beta_2)^\lambda})}{k^2 + ((k* \frac{(k+\beta_1)^\lambda - (k-\beta_1)^\lambda}{(k+\beta_1)^\lambda + (k-\beta_1)^\lambda})(k* \frac{(k+\beta_2)^\lambda - (k-\beta_2)^\lambda}{(k+\beta_2)^\lambda + (k-\beta_2)^\lambda}))}}, \psi_{\frac{k(k* \frac{2\gamma_1{}^\lambda}{(2k-\gamma_1)^\lambda + \gamma_1{}^\lambda})(k* \frac{2\gamma_2{}^\lambda}{(2k-\gamma_2)^\lambda + \gamma_2{}^\lambda})}{k^2 + (k-(k* \frac{2\gamma_1{}^\lambda}{(2k-\gamma_1)^\lambda + \gamma_1{}^\lambda}))(k-(k* \frac{2\gamma_2{}^\lambda}{(2k-\gamma_2)^\lambda + \gamma_2{}^\lambda}))}}, \psi_{\frac{k(k* \frac{2\delta_1{}^\lambda}{(2k-\delta_1)^\lambda + \delta_1{}^\lambda})(k* \frac{2\delta_2{}^\lambda}{(2k-\delta_2)^\lambda + \delta_2{}^\lambda})}{k^2 + (k-(k* \frac{2\delta_1{}^\lambda}{(2k-\delta_1)^\lambda + \delta_1{}^\lambda}))(k-(k* \frac{2\delta_2{}^\lambda}{(2k-\delta_2)^\lambda + \delta_2{}^\lambda}))}} \rangle$$

$$= \langle \psi_{k* \frac{(k+\beta_1)^\lambda (k+\beta_2)^\lambda - (k-\beta_1)^\lambda (k-\beta_2)^\lambda}{(k+\beta_1)^\lambda (k+\beta_2)^\lambda + (k-\beta_1)^\lambda (k-\beta_2)^\lambda}}, \psi_{k* \frac{2(\gamma_1\gamma_2)^\lambda}{((2k-\gamma_1)^\lambda (2k-\gamma_2)^\lambda) + (\gamma_1\gamma_2)^\lambda}}, \psi_{k* \frac{2(\gamma_1\delta_2)^\lambda}{((2k-\delta_1)^\lambda (2k-\delta_2)^\lambda) + (\delta_1\delta_2)^\lambda}} \rangle$$

So, we can get $\lambda(u_1 \oplus_e u_2) = \lambda u_1 \oplus_e \lambda u_2$.

Now, we prove the performance (4):

$$① \; u_1{}^\lambda = \langle \psi_{k* \frac{2\beta_1{}^\lambda}{(2k-\beta_1)^\lambda + \beta_1{}^\lambda}}, \psi_{k* \frac{(k+\gamma_1)^\lambda - (k-\gamma_1)^\lambda}{(k+\gamma_1)^\lambda + (k-\gamma_1)^\lambda}}, \psi_{k* \frac{(k+\delta_1)^\lambda - (k-\delta_1)^\lambda}{(k+\delta_1)^\lambda + (k-\delta_1)^\lambda}} \rangle;$$

$$② \; u_2{}^\lambda = \langle \psi_{k* \frac{2\beta_2{}^\lambda}{(2k-\beta_2)^\lambda + \beta_2{}^\lambda}}, \psi_{k* \frac{(k+\gamma_2)^\lambda - (k-\gamma_2)^\lambda}{(k+\gamma_2)^\lambda + (k-\gamma_2)^\lambda}}, \psi_{k* \frac{(k+\delta_2)^\lambda - (k-\delta_2)^\lambda}{(k+\delta_2)^\lambda + (k-\delta_2)^\lambda}} \rangle;$$

$$③ \; u_1{}^\lambda \oplus_e u_2{}^\lambda = \langle \begin{matrix} \psi_{\frac{k(k* \frac{2\beta_1{}^\lambda}{(2k-\beta_1)^\lambda + \beta_1{}^\lambda})(k* \frac{2\beta_2{}^\lambda}{(2k-\beta_2)^\lambda + \beta_2{}^\lambda})}{k^2 + (k-(k* \frac{2\beta_1{}^\lambda}{(2k-\beta_1)^\lambda + \beta_1{}^\lambda}))(k-(k* \frac{2\beta_2{}^\lambda}{(2k-\beta_2)^\lambda + \beta_2{}^\lambda}))}}, \\ \psi_{\frac{k^2((k* \frac{(k+\gamma_1)^\lambda - (k-\gamma_1)^\lambda}{(k+\gamma_1)^\lambda + (k-\gamma_1)^\lambda}) + (k* \frac{2\beta_2{}^\lambda}{(2k-\beta_2)^\lambda + \beta_2{}^\lambda}))}{k^2 + (k* \frac{(k+\gamma_1)^\lambda - (k-\gamma_1)^\lambda}{(k+\gamma_1)^\lambda + (k-\gamma_1)^\lambda})(k* \frac{2\beta_2{}^\lambda}{(2k-\beta_2)^\lambda + \beta_2{}^\lambda})}}, \\ \psi_{\frac{k^2((k* \frac{(k+\delta_1)^\lambda - (k-\delta_1)^\lambda}{(k+\delta_1)^\lambda + (k-\delta_1)^\lambda}) + (k* \frac{(k+\delta_2)^\lambda - (k-\delta_2)^\lambda}{(k+\delta_2)^\lambda + (k-\delta_2)^\lambda}))}{k^2 + (k* \frac{(k+\delta_1)^\lambda - (k-\delta_1)^\lambda}{(k+\delta_1)^\lambda + (k-\delta_1)^\lambda})(k* \frac{(k+\delta_2)^\lambda - (k-\delta_2)^\lambda}{(k+\delta_2)^\lambda + (k-\delta_2)^\lambda})}} \end{matrix} \rangle$$

$$= \langle \psi_{k* \frac{2(\beta_1\beta_2)^\lambda}{((2k-\beta_1)^\lambda (2k-\beta_2)^\lambda) + (\beta_1\beta_2)^\lambda}}, \psi_{k* \frac{(k+\gamma_1)^\lambda (k+\gamma_2)^\lambda - (k-\gamma_1)^\lambda (k-\gamma_2)^\lambda}{(k+\gamma_1)^\lambda (k+\gamma_2)^\lambda + (k-\gamma_1)^\lambda (k-\gamma_2)^\lambda}}, \psi_{k* \frac{(k+\delta_1)^\lambda (k+\delta_2)^\lambda - (k-\delta_1)^\lambda (k-\delta_2)^\lambda}{(k+\delta_1)^\lambda (k+\delta_2)^\lambda + (k-\delta_1)^\lambda (k-\delta_2)^\lambda}} \rangle;$$

$$④ \; u_1 \otimes_e u_2 = \langle \psi_{\frac{k\beta_1\beta_2}{k^2 + (k-\beta_1)(k-\beta_2)}}, \psi_{\frac{k^2(\gamma_1+\gamma_2)}{k^2+\gamma_1\gamma_2}}, \psi_{\frac{k^2(\delta_1+\delta_2)}{k^2+\delta_1\delta_2}} \rangle;$$

$$⑤ \; (u_1 \otimes_e u_2)^\lambda = \langle \psi_{k* \frac{2(\frac{k\beta_1\beta_2}{k^2+(k-\beta_1)(k-\beta_2)})^\lambda}{(2k-\frac{k\beta_1\beta_2}{k^2+(k-\beta_1)(k-\beta_2)})^\lambda + (\frac{k\beta_1\beta_2}{k^2+(k-\beta_1)(k-\beta_2)})^\lambda}}, \psi_{k* \frac{(k+\frac{k^2(\gamma_1+\gamma_2)}{k^2+\gamma_1\gamma_2})^\lambda - (k-\frac{k^2(\gamma_1+\gamma_2)}{k^2+\gamma_1\gamma_2})^\lambda}{(k+\frac{k^2(\gamma_1+\gamma_2)}{k^2+\gamma_1\gamma_2})^\lambda + (k-\frac{k^2(\gamma_1+\gamma_2)}{k+\gamma_1\gamma_2})^\lambda}}, \psi_{k* \frac{(k+\frac{k^2(\delta_1+\delta_2)}{k^2+\delta_1\delta_2})^\lambda - (k-\frac{k^2(\delta_1+\delta_2)}{k^2+\delta_1\delta_2})^\lambda}{(k+\frac{k^2(\delta_1+\delta_2)}{k^2+\delta_1\delta_2})^\lambda + (k-\frac{k^2(\delta_1+\delta_2)}{k^2+\delta_1\delta_2})^\lambda}} \rangle$$

$$= \langle \psi_{k* \frac{2(\beta_1\beta_2)^\lambda}{((2k-\beta_1)^\lambda (2k-\beta_2)^\lambda) + (\beta_1\beta_2)^\lambda}}, \psi_{k* \frac{(k+\gamma_1)^\lambda (k+\gamma_2)^\lambda - (k-\gamma_1)^\lambda (k-\gamma_2)^\lambda}{(k+\gamma_1)^\lambda (k+\gamma_2)^\lambda + (k-\gamma_1)^\lambda (k-\gamma_2)^\lambda}}, \psi_{k* \frac{(k+\delta_1)^\lambda (k+\delta_2)^\lambda - (k-\delta_1)^\lambda (k-\delta_2)^\lambda}{(k+\delta_1)^\lambda (k+\delta_2)^\lambda + (k-\delta_1)^\lambda (k-\delta_2)^\lambda}} \rangle;$$

So, we can get $(u_1 \oplus_e u_2)^\lambda = u_1{}^\lambda \oplus_e u_2{}^\lambda$. □

3. Einstein Aggregation Operators

3.1. LNNEWA Operator

Definition 7. *Set a LNN* $u_i = \langle \psi_{\beta_i}, \psi_{\gamma_i}, \psi_{\delta_i} \rangle$ *in* Ψ, *for i = 1,2, ... , z, we define the LNNEWA operator:*

$$LNNEWA(u_1, u_2, \ldots u_z) = \underset{i=1}{\overset{z}{\oplus_e}} \epsilon_i u_i, \tag{16}$$

with the relative weight vector $\epsilon = (\epsilon_1, \epsilon_2, \ldots, \epsilon_z)^T$, $\sum_{i=1}^{z} \epsilon_i = 1$ *and* $\epsilon_i \in [0, 1]$.

Theorem 2. *Set a collection* $u_i = \langle \psi_{\beta_i}, \psi_{\gamma_i}, \psi_{\delta_i} \rangle$ *in* Ψ, *for i = 1,2, ... ,z, then according to the LNNEWA aggregation operator, we can get the following result:*

$$\begin{aligned} LNNEWA(u_1, u_2, \ldots u_z) &= \underset{i=1}{\overset{z}{\oplus_e}} \epsilon_i u_i \\ &= \langle \psi_{k* \frac{\prod_{i=1}^{z} (k+\beta_i)^{\epsilon_i} - \prod_{i=1}^{z} (k-\beta_i)^{\epsilon_i}}{\prod_{i=1}^{z} (k+\beta_i)^{\epsilon_i} + \prod_{i=1}^{z} (k-\beta_i)^{\epsilon_i}}}, \psi_{k* \frac{2 \prod_{i=1}^{z} \gamma_i{}^{\epsilon_i}}{\prod_{i=1}^{z} (2k-\gamma_i)^{\epsilon_i} + \prod_{i=1}^{z} \gamma_i{}^{\epsilon_i}}}, \psi_{k* \frac{2 \prod_{i=1}^{z} \delta_i{}^{\epsilon_i}}{\prod_{i=1}^{z} (2k-\delta_i)^{\epsilon_i} + \prod_{i=1}^{z} \delta_i{}^{\epsilon_i}}} \rangle \end{aligned} \tag{17}$$

with the relative weight vector $\epsilon = (\epsilon_1, \epsilon_2, \ldots, \epsilon_z)^T$, $\sum_{i=1}^{z} \epsilon_i = 1$ *and* $\epsilon_i \in [0,1]$.

Proof.

$$① \ \epsilon_i u_i = \langle \psi_{k*\frac{(k+\beta_i)^{\epsilon_i}-(k-\beta_i)^{\epsilon_i}}{(k+\beta_i)^{\epsilon_i}+(k-\beta_i)^{\epsilon_i}}}, \psi_{k*\frac{2\gamma_i^{\epsilon_i}}{(2k-\gamma_i)^{\epsilon_i}+\gamma_i^{\epsilon_i}}}, \psi_{k*\frac{2\delta_i^{\epsilon_i}}{(2k-\delta_i)^{\epsilon_i}+\delta_i^{\epsilon_i}}} \rangle;$$

$$② \ z = 2, LNNEWA(u_1, u_2) = \underset{i=1}{\overset{2}{\oplus_e}} \epsilon_i u_i$$

$$= \langle \psi_{\frac{k^2(k*\frac{(k+\beta_1)^{\epsilon_1}-(k-\beta_1)^{\epsilon_1}}{(k+\beta_1)^{\epsilon_1}+(k-\beta_1)^{\epsilon_1}}+k*\frac{(k+\beta_2)^{\epsilon_2}-(k-\beta_2)^{\epsilon_2}}{(k+\beta_2)^{\epsilon_2}+(k-\beta_2)^{\epsilon_2}})}{k^2+((k*\frac{(k+\beta_1)^{\epsilon_1}-(k-\beta_1)^{\epsilon_1}}{(k+\beta_1)^{\epsilon_1}+(k-\beta_1)^{\epsilon_1}})(k*\frac{(k+\beta_2)^{\epsilon_2}-(k-\beta_2)^{\epsilon_2}}{(k+\beta_2)^{\epsilon_2}+(k-\beta_2)^{\epsilon_2}})}}, \psi_{\frac{k(k*\frac{2\gamma_1^{\epsilon_1}}{(2k-\gamma_1)^{\epsilon_1}+\gamma_1^{\epsilon_1}})(k*\frac{2\gamma_2^{\epsilon_2}}{(2k-\gamma_2)^{\epsilon_2}+\gamma_2^{\epsilon_2}})}{k^2+(k-(k*\frac{2\gamma_1^{\epsilon_1}}{(2k-\gamma_1)^{\epsilon_1}+\gamma_1^{\epsilon_1}}))(k-(k*\frac{2\gamma_2^{\epsilon_2}}{(2k-\gamma_2)^{\epsilon_2}+\gamma_2^{\epsilon_2}}))}}, \psi_{\frac{k(k*\frac{2\delta_1^{\epsilon_1}}{(2k-\delta_1)^{\epsilon_1}+\delta_1^{\epsilon_1}})(k*\frac{2\delta_2^{\epsilon_2}}{(2k-\delta_2)^{\epsilon_2}+\delta_2^{\epsilon_2}})}{k^2+(k-(k*\frac{2\delta_1^{\epsilon_1}}{(2k-\delta_1)^{\epsilon_1}+\delta_1^{\epsilon_1}}))(k-(k*\frac{2\delta_2^{\epsilon_2}}{(2k-\delta_2)^{\epsilon_2}+\delta_2^{\epsilon_2}}))}} \rangle$$

$$= \langle \psi_{k*\frac{(k+\beta_1)^{\epsilon_1}(k+\beta_2)^{\epsilon_2}-(k-\beta_1)^{\epsilon_1}(k-\beta_2)^{\epsilon_2}}{(k+\beta_1)^{\epsilon_1}(k+\beta_2)^{\epsilon_2}+(k-\beta_1)^{\epsilon_1}(k-\beta_2)^{\epsilon_2}}}, \psi_{k*\frac{2\gamma_1^{\epsilon_1}\gamma_2^{\epsilon_2}}{(2k-\gamma_1)^{\epsilon_1}(2k-\gamma_2)^{\epsilon_2}+\gamma_1^{\epsilon_1}\gamma_2^{\epsilon_2}}}, \psi_{k*\frac{2\delta_1^{\epsilon_1}\delta_2^{\epsilon_2}}{(2k-\delta_1)^{\epsilon_1}(2k-\delta_2)^{\epsilon_2}+\delta_1^{\epsilon_1}\delta_2^{\epsilon_2}}} \rangle$$

$$= \langle \psi_{k*\frac{\prod_{i=1}^{2}(k+\beta_i)^{\epsilon_i}-\prod_{i=1}^{2}(k-\beta_i)^{\epsilon_i}}{\prod_{i=1}^{2}(k+\beta_i)^{\epsilon_i}+\prod_{i=1}^{2}(k-\beta_i)^{\epsilon_i}}}, \psi_{k*\frac{2\prod_{i=1}^{2}\gamma_i^{\epsilon_i}}{\prod_{i=1}^{2}(2k-\gamma_i)^{\epsilon_i}+\prod_{i=1}^{2}\gamma_i^{\epsilon_i}}}, \psi_{k*\frac{2\prod_{i=1}^{2}\delta_i^{\epsilon_i}}{\prod_{i=1}^{2}(2k-\delta_i)^{\epsilon_i}+\prod_{i=1}^{2}\delta_i^{\epsilon_i}}} \rangle.$$

Suppose $z = m$, according t formula (17), we can get

$$\begin{aligned} LNNEWA(u_1, u_2, \ldots u_m) &= \underset{i=1}{\overset{m}{\oplus_e}} \epsilon_i u_i \\ &= \langle \psi_{k*\frac{\prod_{i=1}^{m}(k+\beta_i)^{\epsilon_i}-\prod_{i=1}^{m}(k-\beta_i)^{\epsilon_i}}{\prod_{i=1}^{m}(k+\beta_i)^{\epsilon_i}+\prod_{i=1}^{m}(k-\beta_i)^{\epsilon_i}}}, \psi_{k*\frac{2\prod_{i=1}^{m}\gamma_i^{\epsilon_i}}{\prod_{i=1}^{m}(2k-\gamma_i)^{\epsilon_i}+\prod_{i=1}^{m}\gamma_i^{\epsilon_i}}}, \psi_{k*\frac{2\prod_{i=1}^{m}\delta_i^{\epsilon_i}}{\prod_{i=1}^{m}(2k-\delta_i)^{\epsilon_i}+\prod_{i=1}^{m}\delta_i^{\epsilon_i}}} \rangle, \end{aligned} \tag{18}$$

Then z = m + 1, the following can be found:

$$LNNEWA(u_1, u_2, \ldots \ u_m, u_{m+1}) = (\underset{i=1}{\overset{m}{\oplus_e}} \epsilon_i u_i) \oplus_e \epsilon_{m+1} u_{m+1}$$

$$= \langle \begin{matrix} \psi_{k*\frac{\prod_{i=1}^{m}(k+\beta_i)^{\epsilon_i}-\prod_{i=1}^{m}(k-\beta_i)^{\epsilon_i}}{\prod_{i=1}^{m}(k+\beta_i)^{\epsilon_i}+\prod_{i=1}^{m}(k-\beta_i)^{\epsilon_i}}}, \\ \psi_{k*\frac{2\prod_{i=1}^{m}\gamma_i^{\epsilon_i}}{\prod_{i=1}^{m}(2k-\gamma_i)^{\epsilon_i}+\prod_{i=1}^{m}\gamma_i^{\epsilon_i}}}, \\ \psi_{k*\frac{2\prod_{i=1}^{m}\delta_i^{\epsilon_i}}{\prod_{i=1}^{m}(2k-\delta_i)^{\epsilon_i}+\prod_{i=1}^{m}\delta_i^{\epsilon_i}}} \end{matrix} \rangle \oplus_e \langle \begin{matrix} \psi_{k*\frac{(k+\beta_{m+1})^{\epsilon_{m+1}}-(k-\beta_{m+1})^{\epsilon_{m+1}}}{(k+\beta_{m+1})^{\epsilon_{m+1}}+(k-\beta_{m+1})^{\epsilon_{m+1}}}}, \\ \psi_{k*\frac{2\gamma_{m+1}^{\epsilon_{m+1}}}{(2k-\gamma_{m+1})^{\epsilon_{m+1}}+\gamma_{m+1}^{\epsilon_{m+1}}}}, \\ \psi_{k*\frac{2\delta_{m+1}^{\epsilon_{m+1}}}{(2k-\delta_{m+1})^{\epsilon_{m+1}}+\delta_{m+1}^{\epsilon_{m+1}}}} \end{matrix} \rangle$$

$$= \langle \begin{matrix} \psi_{\frac{k^2((k*\frac{\prod_{i=1}^{m}(k+\beta_i)^{\epsilon_i}-\prod_{i=1}^{m}(k-\beta_i)^{\epsilon_i}}{\prod_{i=1}^{k}(k+\beta_i)^{\epsilon_i}+\prod_{i=1}^{k}(k-\beta_i)^{\epsilon_i}})+(k*\frac{(k+\beta_{m+1})^{\epsilon_{m+1}}-(k-\beta_{m+1})^{\epsilon_{m+1}}}{(k+\beta_{m+1})^{\epsilon_{m+1}}+(k-\beta_{m+1})^{\epsilon_{m+1}}}))}{k^2+(k*\frac{\prod_{i=1}^{m}(k+\beta_i)^{\epsilon_i}-\prod_{i=1}^{m}(k-\beta_i)^{\epsilon_i}}{\prod_{i=1}^{m}(k+\beta_i)^{\epsilon_i}+\prod_{i=1}^{m}(k-\beta_i)^{\epsilon_i}})(k*\frac{(k+\beta_{m+1})^{\epsilon_{m+1}}-(k-\beta_{m+1})^{\epsilon_{m+1}}}{(k+\beta_{m+1})^{\epsilon_{m+1}}+(k-\beta_{m+1})^{\epsilon_{m+1}}})}}, \\ \psi_{\frac{k(k*\frac{2\prod_{i=1}^{m}\gamma_i^{\epsilon_i}}{\prod_{i=1}^{m}(2k-\gamma_i)^{\epsilon_i}+\prod_{i=1}^{m}\gamma_i^{\epsilon_i}})(k*\frac{2\gamma_{m+1}^{\epsilon_{m+1}}}{(2k-\gamma_{m+1})^{\epsilon_{m+1}}+\gamma_{m+1}^{\epsilon_{m+1}}})}{k^2+(k-(k*\frac{2\prod_{i=1}^{m}\gamma_i^{\epsilon_i}}{\prod_{i=1}^{m}(2k-\gamma_i)^{\epsilon_i}+\prod_{i=1}^{m}\gamma_i^{\epsilon_i}}))(k-(k*\frac{2\gamma_{m+1}^{\epsilon_{m+1}}}{(2k-\gamma_{m+1})^{\epsilon_{m+1}}+\gamma_{m+1}^{\epsilon_{m+1}}}))}}, \\ \psi_{\frac{k(k*\frac{2\prod_{i=1}^{m}\delta_i^{\epsilon_i}}{\prod_{i=1}^{m}(2k-\delta_i)^{\epsilon_i}+\prod_{i=1}^{m}\delta_i^{\epsilon_i}})(k*\frac{2\delta_{m+1}^{\epsilon_{m+1}}}{(2k-\delta_{m+1})^{\epsilon_{m+1}}+\delta_{m+1}^{\epsilon_{m+1}}})}{k^2+(k-(k*\frac{2\prod_{i=1}^{m}\delta_i^{\epsilon_i}}{\prod_{i=1}^{m}(2k-\delta_i)^{\epsilon_i}+\prod_{i=1}^{m}\delta_i^{\epsilon_i}}))(k-(k*\frac{2\delta_{m+1}^{\epsilon_{m+1}}}{(2k-\delta_{m+1})^{\epsilon_{m+1}}+\delta_{m+1}^{\epsilon_{m+1}}}))}} \end{matrix} \rangle$$

$$= \langle \psi_{k*\frac{\prod_{i=1}^{m+1}(k+\beta_i)^{\epsilon_i}-\prod_{i=1}^{m+1}(k-\beta_i)^{\epsilon_i}}{\prod_{i=1}^{m+1}(k+\beta_i)^{\epsilon_i}+\prod_{i=1}^{m+1}(k-\beta_i)^{\epsilon_i}}}, \psi_{k*\frac{2\prod_{i=1}^{m+1}\gamma_i^{\epsilon_i}}{\prod_{i=1}^{m+1}(2k-\gamma_i)^{\epsilon_i}+\prod_{i=1}^{m+1}\gamma_i^{\epsilon_i}}}, \psi_{k*\frac{2\prod_{i=1}^{m+1}\delta_i^{\epsilon_i}}{\prod_{i=1}^{m+1}(2k-\delta_i)^{\epsilon_i}+\prod_{i=1}^{m+1}\delta_i^{\epsilon_i}}} \rangle.$$

So, Equation (17) is satisfied for any z according to the above results.

This proves Theorem 1. □

Theorem 3. *(Idempotency). Set an LNN* $u = \langle \psi_\beta, \psi_\gamma, \psi_\delta \rangle$ *in* Ψ, *for every* u_i *in* u *is equal to* u, *we can get:*

$$LNNEWA(u_1, u_2, \ldots u_z) = LNNEWA(u, u \ldots u) = u.$$

Proof. For $u_i = u$, then $\beta_i = \beta; \gamma_i = \gamma; \delta_i = \delta = (i = 1, 2, \ldots, z)$, the following result can be found:

$$LNNEWA(u_1, u_2, \ldots u_z) = LNNEWA\ (u, u \ldots u) = (\overset{z}{\underset{i=1}{\oplus_e}} \epsilon_i u)$$
$$= \langle \psi_{k*\frac{\prod_{i=1}^{z}(k+\beta)^{\epsilon_i} - \prod_{i=1}^{z}(k-\beta)^{\epsilon_i}}{\prod_{i=1}^{z}(k+\beta)^{\epsilon_i} + \prod_{i=1}^{z}(k-\beta)^{\epsilon_i}}}, \psi_{k*\frac{2\prod_{i=1}^{z}\gamma^{\epsilon_i}}{\prod_{i=1}^{z}(2k-\gamma)^{\epsilon_i} + \prod_{i=1}^{z}\gamma^{\epsilon_i}}}, \psi_{k*\frac{2\prod_{i=1}^{z}\delta^{\epsilon_i}}{\prod_{i=1}^{z}(2k-\delta)^{\epsilon_i} + \prod_{i=1}^{z}\delta^{\epsilon_i}}} \rangle$$
$$= \langle \psi_{k*\frac{(k+\beta)-(k-\beta)}{(k+\beta)+(k-\beta)}}, \psi_{k*\frac{2\gamma}{(2k-\gamma)+\gamma}}, \psi_{k*\frac{2\delta}{(2k-\delta)+\delta}} \rangle$$
$$= \langle \psi_\beta, \psi_\gamma, \psi_\delta \rangle = u$$

Theorem 4. *(Monotonicity) set two collections of LNNs* $u_i = \langle \psi_{\beta_i}, \psi_{\gamma_i}, \psi_{\delta_i} \rangle$ *and* $u_i' = \langle \psi_{\beta_i'}, \psi_{\gamma_i'}, \psi_{\delta_i'} \rangle$ *(i = 1, 2, … , z) in* Ψ*, if* $u_i \leq u_i'$ *then*

$$LNNEWA(u_1, u_2, \ldots u_z) \leq LNNEWA(u_1', u_2', \ldots u_z').$$

Proof. For $u_i \leq u_i'$, then $\epsilon_i u_i \leq \epsilon_i u_i'$

So, we can easily obtain:

$$\overset{z}{\underset{i=1}{\oplus_e}} \epsilon_i u_i \leq \overset{z}{\underset{i=1}{\oplus_e}} \epsilon_i u_i'$$

For $LNNEWA(u_1, u_2, \ldots u_z) = \overset{z}{\underset{i=1}{\oplus_e}} \epsilon_i u_i$ and $LNNEWA(u_1', u_2', \ldots u_z') = \overset{z}{\underset{i=1}{\oplus_e}} \epsilon_i u_i'$, then we can get: $LNNEWA(u_1, u_2, \ldots u_z) \leq LNNEWA(u_1', u_2', \ldots u_z')$. □

Theorem 5. *(Boundedness) Let a collection* $u_i = \langle \psi_{\beta_i}, \psi_{\gamma_i}, \psi_{\delta_i} \rangle$ *in* Ψ, $u^- = \langle min(\psi_{\beta_i}), max(\psi_{\gamma_i}), max(\psi_{\delta_i}) \rangle$ *and* $u^+ = \langle max(\psi_{\beta_i}), min(\psi_{\gamma_i}), min(\psi_{\delta_i}) \rangle$*, we can get:*

$$u^- \leq LNNEWA(u_1, u_2, \ldots u_z) \leq u^+.$$

Proof. The following can be obtained by using Theorem 3:

$$u^- = LNNEWA(u^-, u^- \ldots u^-),\ u^+ = LNNEWA(u^+, u^+ \ldots u^+).$$

The following can be obtained by using Theorem 4:

$$LNNEWA\ (u^-, u^- \ldots u^-) \leq LNNEWA(u_1, u_2, \ldots u_z) \leq LNNEWA(u^+, u^+ \ldots u^+).$$

Above all, we can get:

$$u^- \leq LNNEWA(u_1, u_2, \ldots u_z) \leq u^+.$$

□

3.2. LNNEWG Operators

Definition 8. *Set a collection* $u_i = \langle \psi_{\beta_i}, \psi_{\gamma_i}, \psi_{\delta_i} \rangle$ *in* Ψ*, for i = 1, 2, … , z, we define the LNNEWG operator:*

$$LNNEWG(u_1, u_2, \ldots u_z) = \overset{z}{\underset{i=1}{\otimes_e}} (u_i)^{\epsilon_i}, \tag{19}$$

with the relative weight vector $\epsilon = (\epsilon_1, \epsilon_2, \ldots, \epsilon_z)^T$, $\sum_{i=1}^{z} \epsilon_i = 1$ *and* $\epsilon_i \in [0, 1]$.

Theorem 6. *Set a collection* $u_i = \langle \psi_{\beta_i}, \psi_{\gamma_i}, \psi_{\delta_i} \rangle$ *in* Ψ, *for i = 1,2, ... ,z, then according to the LNNEWG aggregation operator, we can get the following result:*

$$\begin{aligned} LNNEWG(u_1, u_2, \ldots u_z) &= \underset{i=1}{\overset{z}{\otimes_e}} (u_i)^{\epsilon_i} \\ &= \langle \psi_{k * \frac{2\prod_{i=1}^{z} \beta_i^{\epsilon_i}}{\prod_{i=1}^{z} (2k-\beta_i)^{\epsilon_i} + \prod_{i=1}^{z} \beta_i^{\epsilon_i}}}, \psi_{k * \frac{\prod_{i=1}^{z} (k+\gamma_i)^{\epsilon_i} - \prod_{i=1}^{z} (k-\gamma_i)^{\epsilon_i}}{\prod_{i=1}^{z} (k+\gamma_i)^{\epsilon_i} + \prod_{i=1}^{z} (k-\gamma_i)^{\epsilon_i}}}, \psi_{k * \frac{\prod_{i=1}^{z} (k+\delta_i)^{\epsilon_i} - \prod_{i=1}^{z} (k-\delta_i)^{\epsilon_i}}{\prod_{i=1}^{z} (k+\delta_i)^{\epsilon_i} + \prod_{i=1}^{z} (k-\delta_i)^{\epsilon_i}}} \rangle \end{aligned} \quad (20)$$

with the relative weight vector $\epsilon = (\epsilon_1, \epsilon_2, \ldots, \epsilon_z)^T$, $\sum_{i=1}^{z} \epsilon_i = 1$ *and* $\epsilon_i \in [0,1]$.

Theorem 7. *(Idempotency) Set a collection* $u_i = \langle \psi_{\beta_i}, \psi_{\gamma_i}, \psi_{\delta_i} \rangle$ *in* Ψ, *for i = 1,2, ... ,z, for every* u_i *in u is equal to u, we can get*

$$LNNEWG(u_1, u_2, \ldots u_z) = LNNEWG(u, u \ldots u) = u.$$

Theorem 8. *(Monotonicity). Set two collections of LNNs* $u_i = \langle \psi_{\beta_i}, \psi_{\gamma_i}, \psi_{\delta_i} \rangle$ *and* $u_i{}' = \langle \psi_{\beta_i{}'}, \psi_{\gamma_i{}'}, \psi_{\delta_i{}'} \rangle$ $(i = 1, 2, \ldots, z)$ *in* Ψ, *if* $u_i \le u_i{}'$ *then*

$$LNNEWG(u_1, u_2, \ldots u_z) \le LNNEWG(u_1{}', u_2{}', \ldots u_z{}').$$

Theorem 9. *(Boundedness) Let a collection* $u_i = \langle \psi_{\beta_i}, \psi_{\gamma_i}, \psi_{\delta_i} \rangle$ *in* Ψ, $u^- = \langle min(\psi_{\beta_i}), max(\psi_{\gamma_i}), max(\psi_{\delta_i}) \rangle$ *and* $u^+ = \langle max(\psi_{\beta_i}), min(\psi_{\gamma_i}), min(\psi_{\delta_i}) \rangle$, *we can get:*

$$u^- \le LNNEWG(u_1, u_2, \ldots u_z) \le u^+$$

We omit the proof here because it is similar to Theorems 2–5.

4. Methods with LNNEWA or LNNEWG Operator

We introduce two MAGDM methods with the LNNEWA or LNNEWG operator in LNN information.

Now, we suppose that a collection of alternatives is expressed $\Theta = \{\Theta_1, \Theta_2, \ldots, \Theta_m\}$ and a collection of attributes is expressed $E = \{E_1, E_2, \ldots, E_n\}$. Then, $\epsilon = (\epsilon_1, \epsilon_2, \ldots, \epsilon_n)^T$ with $\sum_{i=1}^{n} \epsilon_i = 1$ *and* $\epsilon_i \in [0,1]$ is the weight vector of $E_i (i = 1, 2, \ldots, n)$. Establishing a set of experts $D = \{D_1, D_2, \ldots, D_t\}$, $\mu = (\mu_1, \mu_2, \ldots, \mu_t)^T$ with $1 \ge \mu_j \ge 0$ and $\sum_{j=1}^{t} \mu_j = 1$ is the weight vector of $D_i (i = 1, 2, \ldots, t)$. Assuming that the expert $D_y (y = 1, 2, \ldots, t)$ uses the LNNs to give out the assessed value $\theta_{ij}^{(y)}$ for alternative Θ_i with the attribute E_j, the value $\theta_{ij}^{(y)}$ can be written as $\theta_{ij}^{(y)} = \langle \psi_{\beta_{ij}}^{y}, \psi_{\gamma_{ij}}^{y}, \psi_{\delta_{ij}}^{y} \rangle (y = 1, 2, \ldots, t; i = 1, 2, \ldots, m; j = 1, 2, \ldots, n), \psi_{\beta_{ij}}^{y}, \psi_{\gamma_{ij}}^{y}, \psi_{\delta_{ij}}^{y} \in \Psi$. Then, the decision evaluation matrix can be found. Table 1 is the decision matrix.

Table 1. The decision matrix using linguistic neutrosophic numbers (LNN).

	E_1	...	E_n
Θ_1	$\langle \psi_{\beta_{11}}^{y}, \psi_{\gamma_{11}}^{y}, \psi_{\delta_{11}}^{y} \rangle$	...	$\langle \psi_{\beta_{1n}}^{y}, \psi_{\gamma_{1n}}^{y}, \psi_{\delta_{1n}}^{y} \rangle$
Θ_2	$\langle \psi_{\beta_{21}}^{y}, \psi_{\gamma_{21}}^{y}, \psi_{\delta_{21}}^{y} \rangle$	...	$\langle \psi_{\beta_{2n}}^{y}, \psi_{\gamma_{2n}}^{y}, \psi_{\delta_{2n}}^{y} \rangle$
...	...	...	...
Θ_m	$\langle \psi_{\beta_{m1}}^{y}, \psi_{\gamma_{m1}}^{y}, \psi_{\delta_{m1}}^{y} \rangle$	...	$\langle \psi_{\beta_{mn}}^{y}, \psi_{\gamma_{mn}}^{y}, \psi_{\delta_{mn}}^{y} \rangle$

The decision steps are described as follows:

Step 1: the integrated matrix can be obtained by the *LNNEWA* operator:

$$\begin{aligned} \theta_{ij} &= \langle \psi_{\beta_{ij}}, \psi_{\gamma_{ij}}, \psi_{\delta_{ij}} \rangle = LNNEWA(\theta_{ij}^1, \theta_{ij}^2, \ldots, \theta_{ij}^t) = \overset{t}{\underset{l=1}{\oplus_e}} \theta_l \theta_{ij}^l \\ &= \langle \psi_{k*\frac{\prod_{l=1}^t (k+\beta_{ij}^l)^{\mu_l} - \prod_{l=1}^t (k-\beta_{ij}^l)^{\mu_l}}{\prod_{l=1}^t (k+\beta_{ij}^l)^{\mu_l} + \prod_{l=1}^t (k-\beta_{ij}^l)^{\mu_l}}}, \psi_{k*\frac{2\prod_{l=1}^t \gamma_{ij}^{l\ \mu_l}}{\prod_{l=1}^t (2k-\gamma_{ij}^l)^{\mu_l} + \prod_{l=1}^t \gamma_{ij}^{l\ \mu_l}}}, \psi_{k*\frac{2\prod_{l=1}^t \delta_{ij}^{l\ \mu_l}}{\prod_{l=1}^t (2k-\delta_{ij}^l)^{\mu_l} + \prod_{l=1}^t \delta_{ij}^{l\ \mu_l}}} \end{aligned} \tag{21}$$

Step 2: the total collective LNN θ_i $(i = 1, 2, \ldots, m)$ can be obtained by the *LNNWEA* or *LNNEWG* operator.

$$\begin{aligned} \theta_i &= LNNEWA(\theta_{i1}, \theta_{i2}, \ldots, \theta_{in}) = \overset{n}{\underset{j=1}{\oplus_e}} \epsilon_{ij} \theta_{ij} \\ &= \langle \psi_{k*\frac{\prod_{j=1}^n (k+\beta_{ij})^{\epsilon_{ij}} - \prod_{j=1}^n (k-\beta_{ij})^{\epsilon_{ij}}}{\prod_{j=1}^n (k+\beta_{ij})^{\epsilon_{ij}} + \prod_{j=1}^n (k-\beta_{ij})^{\epsilon_{ij}}}}, \psi_{k*\frac{2\prod_{j=1}^n \gamma_{ij}^{\epsilon_{ij}}}{\prod_{j=1}^n (2k-\gamma_{ij})^{\epsilon_{ij}} + \prod_{j=1}^n \gamma_{ij}^{\epsilon_{ij}}}}, \psi_{k*\frac{2\prod_{j=1}^n \delta_{ij}^{\epsilon_{ij}}}{\prod_{j=1}^n (2k-\delta_{ij})^{\epsilon_{ij}} + \prod_{j=1}^n \delta_{ij}^{\epsilon_{ij}}}} \rangle \end{aligned} \tag{22}$$

Or

$$\begin{aligned} \theta_i &= LNNEWG(\theta_{i1}, \theta_{i2}, \ldots, \theta_{in}) = \overset{n}{\underset{j=1}{\otimes_e}} (\theta_{ij})^{\epsilon_{ij}} \\ &= \langle \psi_{k*\frac{2\prod_{j=1}^n \beta_{ij}^{\epsilon_{ij}}}{\prod_{j=1}^n (2k-\beta_{ij})^{\epsilon_{ij}} + \prod_{j=1}^n \beta_{ij}^{\epsilon_{ij}}}}, \psi_{k*\frac{\prod_{j=1}^n (k+\gamma_{ij})^{\epsilon_{ij}} - \prod_{j=1}^n (k-\gamma_{ij})^{\epsilon_{ij}}}{\prod_{j=1}^n (k+\gamma_{ij})^{\epsilon_{ij}} + \prod_{j=1}^n (k-\gamma_{ij})^{\epsilon_{ij}}}}, \psi_{k*\frac{\prod_{j=1}^n (k+\delta_{ij})^{\epsilon_{ij}} - \prod_{j=1}^n (k-\delta_{ij})^{\epsilon_{ij}}}{\prod_{j=1}^n (k+\delta_{ij})^{\epsilon_{ij}} + \prod_{j=1}^n (k-\delta_{ij})^{\epsilon_{ij}}}} \rangle \end{aligned} \tag{23}$$

Step 3: according to Definition 3, we can calculate $\zeta(\theta_i)$ and $\eta(\theta_i)$ of every LNN $\Theta_i (i = 1, 2, \ldots, m)$.
Step 4: According to $\zeta(\theta_i)$, then we can rank the alternatives and the best one can be chosen out.
Step 5: End.

5. Illustrative Examples

5.1. Numerical Example

Now, we adopt illustrative examples of the MAGDM problems to verify the proposed decision methods. An investment company wants to find a company to invest. Now, there are four companies $\Theta = \{\Theta_1, \Theta_2, \Theta_3, \Theta_4\}$ to be considered as candidates, the first is for selling cars (Θ_1), the second is for selling food (Θ_2), the third is for selling computers (Θ_3), and the last is for selling arms (Θ_4). Next, three experts $D = \{D_1, D_2, D_3\}$ are invited to evaluate these companies, their weight vector is $\mu = (0.37, 0.33, 0.3)^T$. The experts make evaluations of the alternatives according to three attributes $E = \{E_1, E_2, E_3\}$, E_1 is the ability of risk, E_2 is the ability of growth, and E_3 is the ability of environmental impact, the weight vector of them is $\epsilon = (0.35, 0.25, 0.4)^T$. Then, the experts use LNNs to make the evaluation values with a linguistic set $\Psi = \{\psi_0 =$ extremely poor, $\psi_1 =$ very poor, $\psi_2 =$ poor, $\psi_3 =$ slightly poor, $\psi_4 =$ medium , $\psi_5 =$ slightlygood, $\psi_6 =$ good, $\psi_7 =$ very good, $\psi_8 =$ extremely good$\}$.

Then, the decision evaluation matrix can be established, Tables 2–4 show them.

Table 2. The decision matrix based on the data of D_1.

	E_1	E_2	E_3
Θ_1	$\langle \psi_6^1, \psi_1^1, \psi_2^1 \rangle$	$\langle \psi_7^1, \psi_2^1, \psi_1^1 \rangle$	$\langle \psi_6^1, \psi_2^1, \psi_2^1 \rangle$
Θ_2	$\langle \psi_7^1, \psi_1^1, \psi_1^1 \rangle$	$\langle \psi_7^1, \psi_3^1, \psi_2^1 \rangle$	$\langle \psi_7^1, \psi_2^1, \psi_1^1 \rangle$
Θ_3	$\langle \psi_6^1, \psi_2^1, \psi_2^1 \rangle$	$\langle \psi_7^1, \psi_1^1, \psi_1^1 \rangle$	$\langle \psi_6^1, \psi_2^1, \psi_2^1 \rangle$
Θ_4	$\langle \psi_7^1, \psi_1^1, \psi_2^1 \rangle$	$\langle \psi_7^1, \psi_2^1, \psi_3^1 \rangle$	$\langle \psi_7^1, \psi_2^1, \psi_1^1 \rangle$

Table 3. The decision matrix based on the data of D_2.

	E_1	E_2	E_3
Θ_1	$\langle \psi_6^2, \psi_1^2, \psi_2^2 \rangle$	$\langle \psi_6^2, \psi_1^2, \psi_1^2 \rangle$	$\langle \psi_4^2, \psi_2^2, \psi_3^2 \rangle$
Θ_2	$\langle \psi_7^2, \psi_2^2, \psi_3^2 \rangle$	$\langle \psi_6^2, \psi_1^2, \psi_1^2 \rangle$	$\langle \psi_4^2, \psi_2^2, \psi_3^2 \rangle$
Θ_3	$\langle \psi_5^2, \psi_1^2, \psi_2^2 \rangle$	$\langle \psi_5^2, \psi_1^2, \psi_2^2 \rangle$	$\langle \psi_5^2, \psi_4^2, \psi_2^2 \rangle$
Θ_4	$\langle \psi_6^2, \psi_1^2, \psi_1^2 \rangle$	$\langle \psi_5^2, \psi_1^2, \psi_1^2 \rangle$	$\langle \psi_5^2, \psi_2^2, \psi_3^2 \rangle$

Table 4. The decision matrix based on the data of D_3.

.	E_1	E_2	E_3
Θ_1	$\langle \psi_7^3, \psi_3^3, \psi_4^3 \rangle$	$\langle \psi_7^3, \psi_3^3, \psi_3^3 \rangle$	$\langle \psi_5^3, \psi_2^3, \psi_5^3 \rangle$
Θ_2	$\langle \psi_6^3, \psi_3^3, \psi_4^3 \rangle$	$\langle \psi_5^3, \psi_1^3, \psi_2^3 \rangle$	$\langle \psi_6^3, \psi_2^3, \psi_3^3 \rangle$
Θ_3	$\langle \psi_7^3, \psi_2^3, \psi_4^3 \rangle$	$\langle \psi_6^3, \psi_1^3, \psi_2^3 \rangle$	$\langle \psi_7^3, \psi_2^3, \psi_4^3 \rangle$
Θ_4	$\langle \psi_7^3, \psi_2^3, \psi_3^3 \rangle$	$\langle \psi_5^3, \psi_2^3, \psi_1^3 \rangle$	$\langle \psi_6^3, \psi_1^3, \psi_1^3 \rangle$

Now, the proposed method is applied to manage this MAGDM problem and the computational procedures are as follows:

Step 1: the overall decision matrix can be obtained by the *LNNEWA* operator in Table 5.

Table 5. The overall decision matrix.

	E_1	E_2	E_3
Θ_1	$\langle \psi_{6.3671}, \psi_{1.4116}, \psi_{2.4888} \rangle$	$\langle \psi_{6.7366}, \psi_{1.8191}, \psi_{1.4116} \rangle$	$\langle \psi_{5.1343}, \psi_{2.000}, \psi_{3.0637} \rangle$
Θ_2	$\langle \psi_{6.7630}, \psi_{1.7705}, \psi_{2.2397} \rangle$	$\langle \psi_{6.2295}, \psi_{1.5275}, \psi_{1.5997} \rangle$	$\langle \psi_{6.0042}, \psi_{2.000}, \psi_{2.0355} \rangle$
Θ_3	$\langle \psi_{6.1200}, \psi_{1.5997}, \psi_{2.4888} \rangle$	$\langle \psi_{6.2067}, \psi_{1.000}, \psi_{1.5564} \rangle$	$\langle \psi_{6.1200}, \psi_{2.5427}, \psi_{2.4888} \rangle$
Θ_4	$\langle \psi_{6.7366}, \psi_{1.2370}, \psi_{1.8191} \rangle$	$\langle \psi_{5.9645}, \psi_{1.5997}, \psi_{1.5275} \rangle$	$\langle \psi_{6.2067}, \psi_{1.6329}, \psi_{1.4602} \rangle$

Step 2: the total collective LNN $\theta_i (i = 1, 2, \ldots, m)$ can be obtained by the *LNNWEA* operator:

$$\theta_1 = \langle \psi_{6.0661}, \psi_{1.7313}, \psi_{2.3644} \rangle, \theta_2 = \langle \psi_{6.0961}, \psi_{1.7929}, \psi_{1.9840} \rangle,$$
$$\theta_3 = \langle \psi_{5.7523}, \psi_{1.7260}, \psi_{2.2199} \rangle, \text{ and } \theta_4 = \langle \psi_{6.4198}, \psi_{1.4753}, \psi_{1.5957} \rangle.$$

Step 3: according to Definition 3, the expected values of $\zeta(\theta_i)$ for $\theta_i (i = 1, 2, 3, 4)$ can be calculated:

$$\zeta(\theta_1) = 0.7488, \zeta(\theta_2) = 0.7633, \zeta(\theta_3) = 0.7419, \text{ and } \zeta(\theta_4) = 0.8062.$$

Based on the expected values, four alternatives can be ranked $\Theta_4 > \Theta_2 > \Theta_1 > \Theta_3$, thus, company Θ_4 is the optimal choice.

Now, the *LNNEWG* operator was used to manage this MAGDM problem:

Step 1′: the overall decision matrix can be obtained by the *LNNEWA* operator;

Step 2′: the total collective LNN θ_i $(i = 1, 2, \ldots, m)$ can be obtained by the *LNNEWG* operator, which are as following:

$$\theta_1 = \langle \psi_{5.9491}, \psi_{1.7507}, \psi_{2.4660} \rangle, \theta_2 = \langle \psi_{6.5864}, \psi_{1.8026}, \psi_{2.0000} \rangle, \theta_3 = \langle \psi_{6.8354}, \psi_{1.8390}, \psi_{2.2614} \rangle,$$
$$\text{and } \theta_4 = \langle \psi_{6.3950}, \psi_{1.4868}, \psi_{1.6033} \rangle.$$

Step 3′: according to Definition 3, the expected values of $\zeta(\theta_i)$ for $\theta_i (i = 1, 2, 3, 4)$ can be calculated:

$$\zeta(\theta_1) = 0.7389, \zeta(\theta_2) = 0.7827, (\theta_3) = 0.7806, \text{ and } \zeta(\theta_4) = 0.8043.$$

Based on the expected values, four alternatives can be ranked $\Theta_4 > \Theta_2 > \Theta_3 > \Theta_1$, thus, company Θ_4 is still the optimal choice.

Clearly, there exists a small difference in sorting between these two kinds of methods. However, we can get the same optimal choice by using the LNNEWA and LNNEWG operators. The proposed methods are effective ranking methods for the MCDM problem.

5.2. Comparative Analysis

Now, we do some comparisons with other related methods for LNN, all the results are shown in Table 6.

Table 6. The ranking orders by utilizing three different methods.

Method	Result	Ranking Order	The Best Alternative
Method 1 based on arithmetic averaging in [15]	$\zeta(\theta_1) = 0.7528$, $\zeta(\theta_2) = 0.7777$, $\zeta(\theta_3) = 0.7613$, $\zeta(\theta_4) = 0.8060$.	$\theta_4 > \theta_2 > \theta_3 > \theta_1$	θ_4
Method 2 based on geometric averaging in [15]	$\zeta(\theta_1) = 0.7397$, $\zeta(\theta_2) = 0.7747$, $\zeta(\theta_3) = 0.7531$, $\zeta(\theta_4) = 0.8035$.	$\theta_4 > \theta_2 > \theta_3 > \theta_1$	θ_4
Method 3 based on Bonferroni Mean in [16] (p = q = 1)	$\zeta(\theta_1) = 0.7298$, $\zeta(\theta_2) = 0.7508$, $\zeta(\theta_3) = 0.7424$ $\zeta(\theta_4) = 0.7864$.	$\theta_4 > \theta_2 > \theta_3 > \theta_1$	θ_4
The proposed method	$\zeta(\theta_1) = 0.7488$, $\zeta(\theta_2) = 0.7633$, $\zeta(\theta_3) = 0.7419$ $\zeta(\theta_4) = 0.8062$.	$\theta_4 > \theta_2 > \theta_1 > \theta_3$	θ_4

As shown in Table 6, we can see that company θ_4 is the best choice for investing by using four methods. Many methods such as arithmetic averaging, geometric averaging, and Bonferroni mean can all be used in LNN to handle the multiple attribute decision-making problems and can get similar results. Additionally, The Einstein aggregation operator is smoother than the algebra aggregation operator, which is used in the literature [15,16]. Compared to the existing literature [2–14], LNNs can express and manage pure linguistic evaluation values, while other literature [2–14] cannot do that. In this paper, a new MAGDM method was presented by using the LNNEWA or LNNEWG operator based on LNN environment.

6. Conclusions

A new approach for solving MAGDM problems was proposed in this paper. First, we applied the Einstein operation to a linguistic neutrosophic set and established the new operation rules of this linguistic neutrosophic set based on the Einstein operator. Second, we combined some aggregation operators with the linguistic neutrosophic set and defined the linguistic neutrosophic number Einstein weight average operator and the linguistic neutrosophic number Einstein weight geometric (LNNEWG) operator according the new operation rules. Finally, by using the LNNEWA and LNNEWG operator, two methods for handling MADGM problem were presented. In addition, these two methods were introduced into a concrete example to show the practicality and advantages of the proposed approach. In future, we will further study the Einstein operation in other neutrosophic environment just like the refined neutrosophic set [30]. At the same time, we will use these aggregation operators in many actual fields, such as campaign management, decision making and clustering analysis and so on [31–33].

Author Contributions: C.F. originally proposed the LNNEWA and LNNEWG operators and their properties; C.F., S.F. and K.H. wrote the paper together.

Acknowledgments: This research was funded by the National Natural Science Foundation of China grant number [61603258], [61703280]; General Research Project of Zhejiang Provincial Department of Education grant number [Y201839944]; Public Welfare Technology Research Project of Zhejiang Province grant number [LGG19F020007]; Public Welfare Technology Application Research Project of Shaoxing City grant number [2018C10013].

Conflicts of Interest: The authors declare no conflict of interest.

References

1. Smarandache, F. *Neutrosophy: Neutrosophic Probability, Set, and Logic, ProQuest Information & Learning*; Infolearnquest: Ann Arbor, MI, USA, 1998; p. 105.
2. Wang, H.; Smarandache, F.; Zhang, Y.Q.; Sunderraman, R. Single valued neutrosophic sets. *Multisp. Multi Struct.* **2010**, *4*, 410–413.

3. Wang, H.; Smarandache, F.; Zhang, Y.Q.; Sunderraman, R. *Interval Neutrosophic Sets and Logic: Theory and Applications in Computing*; Hexis: Phoenix, AZ, USA, 2005.
4. Ye, S.; Ye, J. Dice similarity measure between single valued neutrosophic multisets and its application in medical diagnosis. *Neutrosophic Sets Syst.* **2014**, *6*, 49–54.
5. Ye, J. Improved cosine similarity measures of simplified neutrosophic sets for medical diagnoses. *Artif. Intell. Med.* **2015**, *63*, 171–179. [CrossRef] [PubMed]
6. Ye, J. An extended TOPSIS method for multiple attribute group decision making based on single valued neutrosophic linguistic numbers. *J. Intell. Fuzzy Syst.* **2015**, *28*, 247–255.
7. Ye, J.; Florentin, S. Similarity Measure of Refined Single-Valued Neutrosophic Sets and Its Multicriteria Decision Making Method. *Neutrosophic Sets Syst.* **2016**, *12*, 41–44.
8. Fan, C.X.; Fan, E.; Hu, K. New form of single valued neutrosophic uncertain linguistic variables aggregation operators for decision-making. *Cogn. Syst. Res.* **2018**, *52*, 1045–1055. [CrossRef]
9. Fan, C.X.; Ye, J. Heronian Mean Operator of Linguistic Neutrosophic Cubic Numbers and Their Multiple Attribute Decision-Making Methods. *Math. Probl. Eng.* **2018**, *2018*, 4158264. [CrossRef]
10. Fan, C.; Ye, J. The cosine measure of refined-single valued neutrosophic sets and refined-interval neutrosophic sets for multiple attribute decision-making. *J. Intell. Fuzzy Syst.* **2017**, *33*, 2281–2289. [CrossRef]
11. Ye, J. Multiple attribute group decision making based on interval neutrosophic uncertain linguistic variables. *Int. J. Mach. Learn. Cybern.* **2017**, *8*, 837–848. [CrossRef]
12. Liu, P.D.; Shi, L.L. Some neutrosophic uncertain linguistic number Heronian mean operators and their application to multi-attribute group decision making. *Neural Comput. Appl.* **2017**, *28*, 1079–1093. [CrossRef]
13. Jun, Y.; Shigui, D. Some distances, similarity and entropy measures for interval-valued neutrosophic sets and their relationship. *Int. J. Mach. Learn. Cybern.* **2019**, *10*, 347–355. [CrossRef]
14. Fan, C.X.; Fan, E.; Ye, J. The Cosine Measure of Single-Valued Neutrosophic Multisets for Multiple Attribute Decision-Making. *Symmetry-Basel* **2018**, *10*, 154. [CrossRef]
15. Fang, Z.B.; Ye, J. Multiple Attribute Group Decision-Making Method Based on Linguistic Neutrosophic Numbers. *Symmetry* **2017**, *9*, 111. [CrossRef]
16. Fan, C.; Ye, J.; Hu, K.; Fan, E. Bonferroni Mean Operators of Linguistic Neutrosophic Numbers and Their Multiple Attribute Group Decision-Making Methods. *Information* **2017**, *8*, 107. [CrossRef]
17. Li, Y.Y.; Zhang, H.Y.; Wang, J.Q. Linguistic Neutrosophic Sets and Their Application in Multicriteria Decision-Making Problems. *Int. J. Uncertain. Quantif.* **2017**, *7*, 135–154. [CrossRef]
18. Shi, L.; Ye, J. Cosine Measures of Linguistic Neutrosophic Numbers and Their Application in Multiple Attribute Group Decision-Making. *Information* **2017**, *8*, 10.
19. Yager, R.R. On ordered weighted averaging aggregation operators in multicriteria decision making. *IEEE Trans. Syst. Man Cybern.* **1988**, *18*, 183–190. [CrossRef]
20. Xu, Z.S. Intuitionistic fuzzy aggregation operators. *IEEE Trans. Fuzzy Syst.* **2007**, *15*, 1179–1187.
21. Xu, Z.S.; Yager, R.R. Some geometric aggregation operators based on intuitionistic fuzzy sets. *Int. J. Gen. Syst.* **2006**, *35*, 417–433. [CrossRef]
22. Zhao, H.; Xu, Z.S.; Ni, M.F.; Liu, S.S. Generalized Aggregation Operators for Intuitionistic Fuzzy Sets. *Int. J. Intell. Syst.* **2010**, *25*, 1–30. [CrossRef]
23. Klement, E.P.; Mesiar, R.; Pap, E. Triangular norms. Position paper I: Basic analytical and algebraic properties. *Fuzzy Sets Syst.* **2004**, *143*, 5–26. [CrossRef]
24. Wang, W.Z.; Liu, X.W. Intuitionistic fuzzy geometric aggregation operators based on Einstein operations. *Int. J. Intell. Syst.* **2011**, *26*, 1049–1075. [CrossRef]
25. Zhao, X.F.; Wei, G.W. Some intuitionistic fuzzy Einstein hybrid aggregation operators and their application to multiple attribute decision making. *Knowl. Based Syst.* **2013**, *37*, 472–479. [CrossRef]
26. Guo, S.; Jin, F.F.; Chen, Y.H. Application of hesitate fuzzy Einstein geometry operator. *Comput. Eng. Appl.* **2013**. [CrossRef]
27. Yang, L.; Li, B.; Xu, H. Novel Power Aggregation Operators Based on Einstein Operations for Interval Neutrosophic Linguistic Sets. *IAENG Int. J. Appl. Math.* **2018**, *48*, 4.
28. Xia, M.M.; Xu, Z.S.; Zhu, B. Some issues on intuitionistic fuzzy aggregation operators based on Archimedean t-conorm and t-norm. *Knowl. Based Syst.* **2012**, *31*, 78–88. [CrossRef]
29. Wang, W.Z.; Liu, X.W. Intuitionistic fuzzy information aggregation using Einstein operations. *IEEE Trans. Fuzzy Systs.* **2012**, *20*, 923–938. [CrossRef]

30. Smarandache, F. N-Valued Refined Neutrosophic Logic and Its Applications in Physics. *Prog. Phys.* **2013**, *4*, 143–146.
31. Morente-Molinera, J.A.; Kou, G.; González-Crespo, R.; Corchado, J.M. Solving multi-criteria group decision making problems under environments with a high number of alternatives using fuzzy ontologies and multi-granular linguistic modelling methods. *Knowl. Based Syst.* **2017**, *137*, 54–64. [CrossRef]
32. Carrasco, R.A.; Blasco, M.F.; García-Madariaga, J.; Herrera-Viedma, E. A Fuzzy Linguistic RFM Model Applied to Campaign Management. *Int. J. Interact. Multimed. Artif. Intell.* **2019**, *5*, 21–27. [CrossRef]
33. Khiat, S.; Djamila, H. A Temporal Distributed Group Decision Support System Based on Multi-Criteria Analysis. *Int. J. Interact. Multimedia Artif. Intell.* **2019**, 1–15, In Press. [CrossRef]

mathematics

MDPI

Article

Semi-Idempotents in Neutrosophic Rings

Vasantha Kandasamy W.B. [1], **Ilanthenral Kandasamy** [1,*] and **Florentin Smarandache** [2]

[1] School of Computer Science and Engineering, VIT, Vellore 632014, India; vasantha.wb@vit.ac.in
[2] Department of Mathematics, University of New Mexico, 705 Gurley Avenue, Gallup, NM 87301, USA; smarand@unm.edu
* Correspondence: ilanthenral.k@vit.ac.in

Received: 13 April 2019; Accepted: 27 May 2019; Published: 3 June 2019

Abstract: In complex rings or complex fields, the notion of imaginary element i with $i^2 = -1$ or the complex number i is included, while, in the neutrosophic rings, the indeterminate element I where $I^2 = I$ is included. The neutrosophic ring $\langle R \cup I \rangle$ is also a ring generated by R and I under the operations of R. In this paper we obtain a characterization theorem for a semi-idempotent to be in $\langle Z_p \cup I \rangle$, the neutrosophic ring of modulo integers, where p a prime. Here, we discuss only about neutrosophic semi-idempotents in these neutrosophic rings. Several interesting properties about them are also derived and some open problems are suggested.

Keywords: semi-idempotent; neutrosophic rings; modulo neutrosophic rings; neutrosophic semi-idempotent

MSC: 16-XX; 17C27

---

1. Introduction

According to Gray [1], an element $\alpha \neq 0$ of a ring R is called a semi-idempotent if and only if α is not in the proper two-sided ideal of R generated by $\alpha^2 - \alpha$, that is $\alpha \notin R(\alpha^2 - \alpha)R$ or $R = R(\alpha^2 - \alpha)R$. Here, 0 is a semi-idempotent, which we may term as trivial semi-idempotent. Semi-idempotents have been studied for group rings, semigroup rings and near rings [2–9].

An element I was defined by Smarandache [10] as an indeterminate element. Neutrosophic rings were defined by Vasantha and Smarandache [11]. The neutrosophic ring $\langle R \cup I \rangle$ is also a ring generated by R and the indeterminate element I ($I^2 = I$) under the operations of R [11]. The concept of neutrosophic rings is further developed and studied in [12–16]. As the newly introduced notions of neutrosophic triplet groups [17,18] and neutrosophic triplet rings [19], neutrosophic triplets in neutrosophic rings [20] and their relations to neutrosophic refined sets [21,22] depend on idempotents, thus the relative study of semi-idempotents will be an innovative research for any researcher interested in these fields. Finding idempotents is discussed in [18,23–25]. One can also characterize and study neutrosophic idempotents in these situations as basically neutrosophic idempotents are trivial neutrosophic semi-idempotents. A new angle to this research can be made by studying quaternion valued functions [26].

We call a semi-idempotents x in $\langle R \cup I \rangle$ as neutrosophic semi-idempotents if $x = a + bI$ and $b \neq 0; a, b \in \langle R \cup I \rangle$. Several interesting results about semi-idempotents are derived for neutrosophic rings in this paper. As the study pivots on idempotents it has much significance for the recent studies on neutrosophic triplets, duplets and refined sets.

Here, the notion of semi-idempotents in the case of neutrosophic rings is introduced and several interesting properties associated with them are analyzed. We discuss only about neutrosophic semi-idempotents in these neutrosophic rings. This paper is organized into three sections. Section 1 is introductory in nature. In Section 2, the notion of semi-idempotents in the case of

$$\langle Z_n \cup I \rangle = \{a + bI | a, b \in Z_n; n < \infty; I^2 = I\}$$

is considered. Section 3 gives conclusions and proposes some conjectures based on our study.

2. Semi-Idempotents in the Modulo Neutrosophic Rings $\langle Z_n \cup I \rangle$

Throughout this paper, $\langle Z_n \cup I \rangle = \{a + bI/a, b \in Z_n, 2 \leq n < \infty; I^2 = I\}$ denotes the neutrosophic ring of modulo integers. We illustrate some semi-idempotents of $\langle Z_n \cup I \rangle$ by examples and derive some interesting results related with them.

Example 1. *Let $S = \langle Z_2 \cup I \rangle = \{a + bI/a, b \in Z_2, I^2 = I\}$ be the neutrosophic ring of modulo integers. Clearly, $I^2 = I$ and $(1 + I)^2 = 1 + I$ are the two non-trivial idempotents of S. Here, 0 and 1 are trivial idempotents of S. Thus, S has no non-trivial semi-idempotents as all idempotents are trivial semi-idempotents of S.*

Example 2. *Let*

$$R = \langle Z_3 \cup I \rangle = \{a + bI | a, b \in Z^3, I^2 = I\} = \{0, 1, 2, I, 2I, 1 + I, 2 + I, 1 + 2I, 2 + 2I\}$$

be the neutrosophic ring of modulo integers. The trivial idempotents of R are 0 and 1. The non-trivial neutrosophic idempotents are I and $1 + 2I$. Thus, the idempotents I and $1 + 2I$ are trivial neutrosophic semi-idempotents of R. Clearly, 2 and $2 + 2I$ are units of R as $2 \times 2 = 1(mod\ 3)$ and $2 + 2I \times 2 + 2I = 1(mod\ 3)$. $1 + I \in R$ is such that

$$(1 + I)^2 - (1 + I) = 1 + 2I + I - (1 + I) = 1 + 2 + 2I = 2I.$$

Thus, $1 + I$ is a semi-idempotent as the ideal generated by $1 + I$ is $\langle (1 + I)^2 - (1 + I) \rangle = \langle 2I \rangle$ is such that $1 + I \notin R$. However, it is important to note that $(1 + I) \in R$ is a unit as $(1 + I)^2 = 1 + 2I + I = 1$, hence $1 + I$ is a unit in R but it is also a non-trivial semi-idempotent of R. $2 + I$ is not a semi-idempotent as

$$(2 + I)^2 - (2 + I) = 1 + 4I + I - (2 + I) = 2 + I;$$

hence the claim. $2 + 2I \in R$ is a unit, now $(2 + 2I)^2 = 4 + 8I + 4I^2 = 1$, thus $2 + 2I$ is a unit. However, $(2 + 2I)^2 - (2 + 2I) = 1 + 1 + I = 2 + I$.

Now, the ideal generated by $\langle 2 + I \rangle$ does not contain $2 + 2I$ as $\langle 2 + I \rangle = \{0, 2 + I, 1 + 2I\}$, thus $2 + 2I$ is also a non-trivial semi-idempotent even though $2 + 2I$ is a unit of R. Thus, it is important to note that units in modulo neutrosophic rings contribute to non-trivial semi-idempotents. Let $P = \{0, 2 + 2I, 2 + I, 1 + 2I, I, 1 + I, 1\}$ be the collection of trivial and non-trivial semi-idempotents. 2I is not a semi-idempotent as $(2I)^2 - 2I = I + I = 2I$, hence the claim. Thus, P is not closed under sum or product.

Theorem 1. *Let $S = \{\langle Z_p \cup I \rangle, +, \times\}$ be the ring of neutrosophic modulo integers where p is a prime. x is semi-idempotent if and only if $x \in \langle Z_p \cup I \rangle \setminus \{Z_p I, 0, 1, a + bI \text{ with } a + b = 0\}$.*

Proof. The elements $x = a + bI \in S$ with $b = 0$ are such that $x^2 - x$ generates the ideal, which is S, thus x is a semi-idempotent. Let $y = a + bI$; if $a = 0$, the ideal generated by y is $Z_p I$, thus $y \in Z_p I \subset S$, hence $y \in Z_p I$, therefore y is not a semi-idempotent.

Consider $z = a + bI \in S$ with $a + b = 0(mod\ p)$; then, $z^2 - z$ generates an ideal M of S such that every element $x = d + cI$ in M is such that $d + c \equiv 0(mod\ p)$, thus z is not a semi-idempotent of S. Let $x = a + bI \in S(a \neq 0, b \neq 0 \text{ and } a + b \neq 0)$.

$$x^2 - x = \begin{cases} m & m \in Z_p \text{ or} \\ nI & n \in Z_p \text{ or} \\ n + mI & m + n \neq 0 \end{cases}$$

If $x^2 - x = m$, then the ideal generated by $x^2 - x$ is S, thus x is a semi-idempotent. If $x^2 - x = nI$, then the ideal generated by nI is Z_pI, thus $x \notin Z_pI$, hence again x is a semi-idempotent. If $x^2 - x = n + mI(m + n \neq 0)$, then the ideal generated by $n + mI$ is S, thus x is a semi-idempotent by using properties of Z_p, p a prime. Hence, the theorem is proved. □

If we take $S = \{\langle Z_n \cup I\rangle, +, \times\}$ as a neutrosophic ring where n is not a prime, it is difficult to find all semi-idempotents.

Example 3. *Let $S = \{\langle Z_{15} \cup I\rangle, +, \times\}$ be the neutrosophic ring. How can the non-trivial semi-idempotents of S be found? Some of the neutrosophic idempotents of S are $\{1+9I,\ 6+4I,\ 1+5I,\ 1+14I, 6+5I, 6+9I, I, 6I, 10I, 10, 6, 6+10I, 10+11I, 10+6I, 10+5I\}$.*

The semi-idempotents are $\{1+I, 1+2I, 1+3I, 1+4I, 1+6I, 1+7I, 1+8I, 1+10I, 1+11I, 1+12I, 1+13I, 6+I, 6+2I, 6+3I, 6+6I, 6+7I, 6+8I, 6+11I, 6+12I, 6+13I, 6+14I, 10+I, 10+2I, 10+3I, 10+4I, 10+7I, 10+8I, 10+9I, 10+10I, 10+12I, 10+13I, 10+14I\}$.

Are there more non-trivial neutrosophic idempotents and semi-idempotents?

However, we are able to find all idempotents and semi-idempotents of S other than the once given. In view of all these, we have the following theorem.

Theorem 2. *Let $S = \{\langle Z_{pq} \cup I\rangle; \times, +\}$ where p and q are two distinct primes:*

1. *There are two idempotents in Z_{pq} say r and s.*
2. *$\{r, s, rI, sI, I, r+tI, s+tI | t \in \{Z_{pq} \setminus 0\}\}$ such that $r + t = s$, 1 or 0 and $s + t = 0, 1$ or r is the partial collection of idempotents and semi-idempotents of S.*

Proof. Given $S = \{\langle Z_{pq} \cup I\rangle, +, \times\}$ is a neutrosophic ring where p and q are primes, we know from [12,17,18,20,23–25] that Z_{pq} has two idempotents r and s to prove $A = \{r, s, rIsI, I, r+tI$ and $s + tI/t \in Z_{pq} \setminus \{0\}\}$ are idempotents or semi-idempotents of $S.\{r, s, rI, sI, I\}$ are non-trivial idempotents of S. Now, $r + tI \in A$ and $(r+tI)^2 - (r+tI) = mI$ as$r^2 = r$, thus the ideal generated by mI does not contain r_tI. Therefore, r_tI is a non-trivial semi-idempotent. Similarly, $s + tI$ is a non-trivial semi-idempotent. Hence, the theorem is proved. □

We in addition to this theorem propose the following problem.

Problem 1. *Let $S = \{\langle Z_{pq} \cup I\rangle, I, \times\}$, where p and q are two distinct primes, be the neutrosophic ring. Can S have non-trivial idempotents and non-trivial semi-idempotents other than the ones mentioned in (b) of the above theorem?*

Problem 2. *Can the collection of all trivial and non-trivial semi-idempotents have any algebraic structure defined on them?*

We give an example of Z_{pqr}, where p, q and r are three distinct primes, for which we find all the neutrosophic idempotents.

Example 4. *Let $S = \{\langle Z_{30} \cup I\rangle, +, \times\}$, be the neutrosophic ring. The idempotents of Z_{30} are 6, 10, 15, 16, 21 and 25. The non-trivial semi-idempotents of S are $\{1+I, 1+2I, 1+3I, 1+4I, 1+6I, 1+7I, 1+8I, 1+10I, 1+11I, 1+13I, 1+12I, 1+16I, 1+17I, 1+18I, 1+19I, 1+21I, 1+22I, 1+23I, 1+25I, 1+26I, 1+27I, 1+28I\}$.*

$P_1 = \{1+5I, 1+9I, 1+14I, 1+15I, 1+20I, 1+24I, 1+29I\}$ are non-trivial idempotents of S. $J_2 = \{6+I, 6+2I, 6+3I, 6+5I, 6+6I, 6+7I, 6+8I, 6+11I, 6+12I, 6+13I, 6+14I, 6+16I, 6+17I, 6+18I, 6+20I, 6+21I, 6+22I, 6+23I, 6+26I, 6+27I, 6+28I, 6+29I\}$ are non-trivial neutrosophic semi-idempotents of S. $P_2 = \{6+4I, 6+9I, 6+10I, 6+15I, 6+24I, 6+19I, 6+25I\}$ are non-idempotents of S.

Now, we list the non-trivial semi-idempotents associated with 10 of Z_{30}. $J_3 = \{10+I, 10+2I, 10+3I, 10+4I, 10+7I, 10+8I, 10+9I, 10+10I, 10+11I, 10+12I, 10+13I, 10+14I, 10+16I, 10+17I, 10+18I, 10+19I, 10+22I, 10+23I, 10+24I, 10+25I, 10+27I, 10+28I, 10+29I\}$

$P_3 = \{10+5, 10+6I, 10+15I, 10+20I, 10+21I, 10+26I, 10+11I\}$ are the collection of non-trivial idempotent related with the idempotents. Now, we find the non-trivial idempotents associated with 15: $J_4 = \{15+2I, 15+3I, 15+4I, 15+7I, 15+8I, 15+9I, 15+11I, 15+12I, 15+13I, 15+14I, 15+17I, 15+18I, 15+19I, 15+20I, 15+22I, 15+23I, 15+24I, 15+25I, 15+26I, 15+27I, 15+28I, 15+29I\}$.

$P_4 = \{15+I, 15+5I, 15+6I, 15+10I, 15+15I, 15+16I, 15+21I\}$ are the non-trivial idempotents associated with 15. The collection of non-trivial semi-idempotents associated with 16 are: $J_5 = \{16+I, 16+2I, 16+3I, 16+4I, 16+6I, 16+7I, 16+8I, 16+10I, 16+19I, 16+27I, 16+21I, 16+22I, 16+23I, 16+25I, 16+11I, 16+12I, 16+13I, 16+17I, 16+18I, 16+28I$. $P_5 = \{16+14I, 16+15I, 16+20I, 16+24I, 16+29I, 16+5I, 16+9I\}$ are the set of non-trivial idempotents related with the idempotent. We find the non-trivial semi-idempotents associated with the idempotent 21: $J_6 = \{21+I, 21+2I, 21+3I, 21+5I, 21+6I, 21+7I, 21+8I, 21+12I, 21+11I, 21+13I, 21+14I, 21+16I, 21+17I, 21+18I, 21+20I, 21+21I, 21+22I, 21+23I, 21+26I, 21+27I, 21+28I, 21+29I\}$. $P_6 = \{21+4I, 21+9I, 21+10I, 21+15I, 21+19I, 21+24I, 21+25I\}$ is the collection of non-trivial idempotents related with the real idempotent 21. The collection of all non-trivial semi-idempotents associated with the idempotent 25. $J_7 = \{25+I, 25+2I, 25+3I, 25+4I, 25+7I, 25+8I, 25+9I, 25+10I, 25+12I, 25+13I, 25+14I, 25+16I, 25+24I, 25+17I, 25+18I, 25+19I, 25+22I, 25+23I, 25+27I, 25+28I, 25+29I\}$ $P_7 = \{25+5I, 25+6I, 25+11I, 25+15I, 25+20I, 25+21I, 25+26I\}$ are the non-trivial collection of neutrosophic semi-idempotents related with the idempotent 25.

We tabulate the neutrosophic idempotents associated with the real idempotents in Table 1. Based on that table, we propose some open problems.

Table 1. Idempotents.

S.No	Real	Neutrosophic	Sum	Missing
1	1	$1+5I$ $1+9I$ $1+14I$ $1+15I$ $1+20I$ $1+24I$ $1+29I$	1 + 5 = 6 1 + 9 = 10 1 + 14 = 15 1 + 15 = 16 1 + 20 = 21 1 + 24 = 25 1 + 29 = 0	1
2	6	$6+4I$ $6+9I$ $6+10I$ $6+15I$ $6+24I$ $6+19I$ $6+25I$	6 + 4 = 10 6 + 9 = 15 6 + 10 = 16 6 + 15 = 1 6 + 24 = 0 6 + 19 = 25 6 + 25 ≡ 1	6

Table 1. *Cont.*

S.No	Real	Neutrosophic	Sum	Missing
3	10	$10+5I$ $10+6I$ $10+15I$ $10+20I$ $10+21I$ $10+26I$ $10+11I$	$10+5=15$ $10+6=16$ $10+15=25$ $10+20\equiv 0$ $10+21\equiv 1$ $10+26\equiv 6$ $10+11=21$	10
4	15	$15+I$ $15+5I$ $15+6I$ $15+10I$ $15+15I$ $15+16I$ $15+21I$	$15+1=16$ $15+5=20$ $15+6=21$ $15+10=25$ $15+15\equiv 0$ $15+16\equiv 1$ $15+21\equiv 6$	15
5	16	$16+14I$ $16+15I$ $16+20I$ $16+24I$ $16+29I$ $16+5I$ $16+9I$	$16+14\equiv 0$ $16+15\equiv 1$ $16+20\equiv 6$ $16+24\equiv 10$ $16+29\equiv 15$ $16+5=21$ $16+9=25$	16
6	21	$21+4I$ $21+9I$ $21+10I$ $21+15I$ $21+19I$ $21+24I$ $21+25I$	$21+4=25$ $21+9\equiv 0$ $21+10\equiv 1$ $21+15\equiv 6$ $21+19\equiv 10$ $21+24\equiv 15$ $21+25\equiv 16$	21
7	25	$25+I$ $25+5I$ $25+6I$ $25+10I$ $25+16I$ $25+21I$ $25+26I$	$25+5\equiv 0$ $25+6\equiv 1$ $25+11\equiv 6$ $25+15\equiv 10$ $25+20\equiv 15$ $25+21\equiv 16$ $25+26\equiv 21$	25

We see there are eight idempotents including 0 and 1. It is obvious that using 0 we get only idempotents or trivial semi-idempotents.

In view of all these, we conjecture the following.

Conjecture 1. *Let* $S = \{\langle Z_n \cup I\rangle, +, \times\}$ *be the neutrosophic ring* $n = pqr$*, where* p, q *and* r *are three distinct primes.*

1. $Z_n = Z_{pqr}$ *has only six non-trivial idempotents associated with it.*
2. *If* m_1, m_2, m_3, m_4, m_5 *and* m_6 *are the idempotents, then, associated with each real idempotent* m_i*, we have seven non-trivial neutrosophic idempotents associated with it, i.e.* $\{m_i + n_j I, j = 1,2,\ldots,7\}$*, such that* $m_i + n_j \equiv t$*, where* t_j *takes the seven distinct values from the set* $\{0,1,m_k, k \neq i; k = 1,2,3,\ldots 6\}$*.* $i = 1,2,\ldots,6$*.*

This has been verified for large values of p, q and r, where p, q and r are three distinct primes.

3. Conjectures, Discussion and Conclusions

We have characterized the neutrosophic semi-idempotents in $\langle Z_p \cup I \rangle$, with p a prime. However, it is interesting to find neutrosophic semi-idempotents of $\langle Z_n \cup I \rangle$, with n a non-prime composite number. Here, we propose a few new open conjectures about idempotents in Z_n and semi-idempotents in $\langle Z_n \cup I \rangle$.

Conjecture 2. *Given $\langle Z_n \cup I \rangle$, where $n = p_1, p_2, \ldots p_t; t > 2$ and p_is are all distinct primes, find:*

1. *the number of idempotents in Z_n;*
2. *the number of idempotents in $\langle Z_n \cup I \rangle \setminus Z_n$;*
3. *the number of non-trivial semi-idempotents in Z_n; and*
4. *the number of non-trivial semi-idempotents in $\langle Z_n \cup I \rangle \setminus Z_n$.*

Conjecture 3. *Prove if $\langle Z_n \cup I \rangle$ and $\langle Z_m \cup I \rangle$ are two neutrosophic rings where $n > m$ and $n = p^t q$ ($t > 2$, and p and q two distinct primes) and $m = p_1 p_2 \ldots p_s$ where p_is are distinct primes. $1 \leq i \leq s$, then*

1. *prove Z_n has more number of idempotents than Z_m; and*
2. *prove $\langle Z_m \cup I \rangle$ has more number of idempotents and semi-idempotents than $\langle Z_n \cup I \rangle$.*

Finding idempotents in the case of Z_n has been discussed and problems are proposed in [18,23,24]. Further, the neutrosophic triplets in Z_n are contributed by Z_n. In the case of neutrosophic duplets, we see units in Z_n contribute to them. Both units and idempotents contribute in general to semi-idempotents.

Author Contributions: The contributions of the authors are roughly equal.

Funding: This research received no external funding.

Acknowledgments: The authors would like to thank the reviewers for their reading of the manuscript and many insightful comments and suggestions.

Conflicts of Interest: The authors declare no conflict of interest.

References

1. Gray, M. *A Radical Approach to Algebra*; Addison Wesley: Boston, MA, USA, 1970.
2. Jinnah, M.I.; Kannan, B. On semi-idempotents in rings. *Proc. Jpn. Acad. Ser. A Math. Sci.* **1986**, *62*, 211–212. [CrossRef]
3. Vasantha, W.B. On semi-idempotents in group rings. *Proc. Jpn. Acad. Ser. A Math. Sci.* **1985**, *61*, 107–108. [CrossRef]
4. Vasantha, W.B. On semi-idempotents in semi group rings. *J. Guizhou Inst. Technol.* **1989**, *18*, 105–106. [CrossRef]
5. Vasantha, W.B. Idempotents and semi idempotents in near rings. *J. Sichuan Univ.* **1996**, *33*, 330–332.
6. Vasantha, W.B. Semi idempotents in group rings of a cyclic group over the field of rationals. *Kyungpook Math. J.* **1990**, *301*, 243–251.
7. Vasantha, W.B. A note on semi-idempotents in group rings. *Ultra Sci. Phy. Sci.* **1992**, *4*, 77–81.
8. Vasantha, W.B. A note on units and semi idempotent elements in commutative group rings. *Ganita* **1991**, *42*, 33–34.
9. Vasantha, W.B. *Smarandache Ring*; American Research Press: Santa Fe, NM, USA, 2002.
10. Smarandache, F. Neutrosophy, A New Branch of Philosophy. *Multiple Valued Logic.* **2002**, *8*, 297–384.
11. Vasantha, W.B.; Smaradache, F. *Neutrosophic Rings*; Hexis: Phoenix, AZ, USA, 2006.
12. Agboola, A.A.D.; Akinola, A.D.; Oyebola, O.Y. Neutrosophic Rings I. *Int. J. Math. Comb.* **2011**, *4*, 115.
13. Ali, M.; Smarandache, F.; Shabir, M.; Naz, M. Soft Neutrosophic Ring and Soft Neutrosophic Field. *Neutrosophic Sets Syst.* **2014**, *3*, 53–59.
14. Ali, M.; Smarandache, F.; Shabir, M.; Vladareanu, L. Generalization of Neutrosophic Rings and Neutrosophic Fields. *Neutrosophic Sets Syst.* **2014**, *5*, 9–13.

15. Ali, M.; Shabir, M.; Smarandache, F.; Vladareanu, L. Neutrosophic LA-semigroup Rings. *Neutrosophic Sets Syst.* **2015**, *7*, 81–88.
16. Broumi, S.; Smarandache, F.; Maji, P.K. Intuitionistic Neutrosphic Soft Set over Rings. *Math. Stat.* **2014**, *2*, 120–126.
17. Smarandache, F.; Ali, M. Neutrosophic triplet group. *Neural Comput. Appl.* **2018**, *29*, 595–601. [CrossRef]
18. Vasantha, W.B.; Kandasamy, I.; Smarandache, F. *Neutrosophic Triplet Groups and Their Applications to Mathematical Modelling*; EuropaNova: Brussels, Belgium, 2017; ISBN 978-1-59973-533-7.
19. Smarandache, F.; Ali, M. Neutrosophic triplet ring and its applications. *Bull. Am. Phys. Soc.* **2017**, *62*, 7.
20. Vasantha, W.B.; Kandasamy, I.; Smarandache, F.; Zhang, X. Neutrosophic Triplets in Neutrosophic Rings. *Mathematics* **2019**, submitted.
21. Kandasamy, I. Double-valued neutrosophic sets, their minimum spanning trees, and clustering algorithm. *J. Intell. Syst.* **2018**, *27*, 163–182. [CrossRef]
22. Kandasamy, I.; Smarandache, F. Triple Refined Indeterminate Neutrosophic Sets for personality classification. In Proceedings of the 2016 IEEE Symposium Series on Computational Intelligence (SSCI), Athens, Greece, 6–9 December 2016; pp. 1–8.
23. Vasantha, W.B.; Kandasamy, I.; Smarandache, F. A Classical Group of Neutrosophic Triplet Groups Using $\{Z_{2p}, \times\}$. *Symmetry* **2018**, *10*, 194. [CrossRef]
24. Vasantha, W.B.; Kandasamy, I.; Smarandache, F. Neutrosophic Duplets of $\{Z_{pn}, \times\}$ and $\{Z_{pq}, \times\}$ and Their Properties. *Symmetry* **2018**, *10*, 345. [CrossRef]
25. Vasantha, W.B.; Kandasamy, I.; Smarandache, F. Algebraic Structure of Neutrosophic Duplets in Neutrosophic Rings $\langle Z \cup I \rangle$, $\langle Q \cup I \rangle$ and $\langle R \cup I \rangle$. *Neutrosophic Sets Syst.* **2018**, *23*, 85–95.
26. Arena, P.; Fortuna, L.; Muscato, G.; Xibilia, M.G. Multilayer Perceptrons to Approximate Quaternion Valued Functions. *Neural Netw.* **1997**, *10*, 335–342. [CrossRef]

mathematics

MDPI

Article

Neutrosophic Triplets in Neutrosophic Rings

Vasantha Kandasamy W. B. [1], Ilanthenral Kandasamy [1,*] and Florentin Smarandache [2]

[1] School of Computer Science and Engineering, VIT, Vellore 632014, India; vasantha.wb@vit.ac.in
[2] Department of Mathematics, University of New Mexico, 705 Gurley Avenue, Gallup, NM 87301, USA; smarand@unm.edu
* Correspondence: ilanthenral.k@vit.ac.in

Received: 25 May 2019; Accepted: 18 June 2019; Published: 20 June 2019

Abstract: The neutrosophic triplets in neutrosophic rings $\langle Q \cup I \rangle$ and $\langle R \cup I \rangle$ are investigated in this paper. However, non-trivial neutrosophic triplets are not found in $\langle Z \cup I \rangle$. In the neutrosophic ring of integers $Z \setminus \{0, 1\}$, no element has inverse in Z. It is proved that these rings can contain only three types of neutrosophic triplets, these collections are distinct, and these collections form a torsion free abelian group as triplets under component wise product. However, these collections are not even closed under component wise addition.

Keywords: neutrosophic ring; neutrosophic triplets; idemponents; special neutrosophic triplets

1. Introduction

Handling of indeterminacy present in real world data is introduced in [1,2] as neutrosophy. Neutralities and indeterminacies represented by Neutrosophic logic has been used in analysis of real world and engineering problems [3–5].

Neutrosophic algebraic structures such as neutrosophic rings, groups and semigroups are presented and analyzed and their application to fuzzy and neutrosophic models are developed in [6]. Subsequently, researchers have been studying in this direction by defining neutrosophic rings of Types I and II and generalization of neutrosophic rings and fields [7–12]. Neutrosophic rings [9] and other neutrosophic algebraic structures are elaborately studied in [6–8,10,13–17]. Related theories of neutrosophic triplet, duplet, and duplet set were developed by Smarandache [18]. Neutrosophic duplets and triplets have fascinated several researchers who have developed concepts such as neutrosophic triplet normed space, fields, rings and their applications; triplets cosets; quotient groups and their application to mathematical modeling; triplet groups; singleton neutrosophic triplet group and generalization; and so on [19–36]. Computational and combinatorial aspects of algebraic structures are analyzed in [37].

Neutrosophic duplet semigroup [23], classical group of neutrosophic triplet groups [27], the neutrosophic triplet group [12], and neutrosophic duplets of $\{Z_{pn}, \times\}$ and $\{Z_{pq}, \times\}$ have been analyzed [28]. Thus, Neutrosophic triplets in case of the modulo integers $Z_n (2 < n < \infty)$ have been extensively researched [27].

Neutrosophic duplets in neutrosophic rings are characterized in [29]. However, neutrosophic triplets in the case of neutrosophic rings have not yet been researched. In this paper, we for the first time completely characterize neutrosophic triplets in neutrosophic rings. In fact, we prove this collection of neutrosophic triplets using neutrosophic rings are not even closed under addition. We also prove that they form a torsion free abelian group under component wise multiplication.

2. Basic Concepts

In this section, we recall some of the basic concepts and properties associated with both neutrosophic rings and neutrosophic triplets in neutrosophic rings. We first give the following notations: I denotes the indeterminate and it is such that $I \times I = I = I^2$. I is called as the neutrosophic value. Z, Q and R denote the ring of integers, field of rationals and field of reals, respectively. $\langle Z \cup I \rangle = \{a + bI | a, b \in Z, I^2 = I\}$ is the neutrosophic ring of integers, $\langle Q \cup I \rangle = \{a + bI | a, b \in Q, I^2 = I\}$ is the neutrosophic ring of rationals and $\langle R \cup I \rangle = \{a + bI | a, b \in R, I^2 = I\}$ is the neutrosophic ring of reals with usual addition and multiplication in all the three rings.

3. Neutrosophic Triplets in $\langle Q \cup I \rangle$ and $\langle R \cup I \rangle$

In this section, we prove that the neutrosophic rings $\langle Q \cup I \rangle$ and $\langle R \cup I \rangle$ have infinite collection of neutrosophic triplets of three types. Both collections enjoy strong algebraic structures. We explore the algebraic structures enjoyed by these collections of neutrosophic triplets. Further, the neutrosophic ring of integers $\langle Z \cup I \rangle$ has no nontrivial neutrosophic triplets. An example of neutrosophic triplets in $\langle Q \cup I \rangle$ is provided before proving the related results.

Example 1. *Let $S = \langle Q \cup I \rangle, +, \times$ (or $\langle R \cup I \rangle, +, \times$) be the neutrosophic ring. If $x = a - aI \in S (a \neq 0)$, then*

$$y = \frac{1}{a} - \frac{I}{a} \in S$$

is such that

$$x \times y = (a - aI) \times \left(\frac{1}{a} - \frac{I}{a}\right) = 1 - I - I + I = 1 - I.$$

Thus, for every $x = a - aI$, of this form in S we have a unique y of the form

$$\frac{1}{a} - \frac{I}{a}$$

such that $x \times y = 1 - I$. Further, $1 - I \in S$ is such that $1 - I \times 1 - I = 1 - I + I - I = 1 - I \in S$. Thus, these triplets

$$\left\{a - aI, 1 - I, \frac{1}{a} - \frac{I}{a}\right\} \text{ and } \left\{\frac{1}{a} - \frac{I}{a}, 1 - I, a - aI\right\}$$

form neutrosophic triplets with $1 - I$ as a neutral element.

Similarly, for $aI \in S (a \neq 0)$, we have a unique

$$\frac{I}{a} \in S \text{ such that } aI \times \frac{I}{a} = I$$

and $I \times I = I$ is an idempotent. Thus,

$$\left\{aI, I, \frac{I}{a}\right\} \text{ and } \left\{\frac{I}{a}, I, aI\right\}$$

are neutrosophic triplets with I as the neutral element.

First, we prove $\langle Q \cup I \rangle$ and $\langle R \cup I \rangle$ have only I and $1 - I$ as nontrivial idempotents as invariably one idempotents serve as neutrals of neutrosophic triplets.

Theorem 1. *Let $S = \langle Q \cup I \rangle, +, \times$ (or $\{\langle R \cup I \rangle, +, \times\}$) be a neutrosophic ring. The only non-trivial idempotents in S are I and $1 - I$.*

Proof. We call 0 and $1 \in S$ as trivial idempotents. Suppose $x \in S$ is a non-trivial idempotent, then $x = aI$ or $x = a + bI \in S (a \neq 0, b \neq 0)$. Now, $x \times x = aI \times aI = a^2 I$ (as $I^2 = I$); if x is to be an idempotent, we must have $aI = a^2 I$; that is, $(a - a^2)I = 0 (I \neq 0)$, thus $a^2 = a$. However, in Q or R,

$a^2 = a$ implies $a = 0$ or $a = 1$; as $a \neq 0$, we have $a = 1$; thus, $x = I$ and x is a nontrivial idempotent in S. Now, let $y = a + bI; a \neq 0$ and $b \neq 0$ for $a = 0$ will reduce to case $y = I$ is an idempotent.

$$y^2 = (a + bI) \times (a + bI) = a^2 + b^2 I + 2abI$$

That is, $y^2 = a + bI \times a - bI = a^2 + abI + abI + b^2 I = a + bI$, equating the real and neutrosophic parts.

$$a^2 = a \text{ i.e., } a(a-1) = 0 \Rightarrow a = 1 \text{ as } a \neq 0 \text{ and } 2ab + b^2 - b = 0$$

$b(2a + b - 1) = 0; b \neq 0$, thus $2a + b - 1 = 0$; further, $a \neq 0$ as $a = 0$ will reduce to the case $I^2 = I$, thus $a = 1$. Hence, $2 + b - 1 = 0$, thus $b = -1$. Hence, $a = 1$ and $b = -1$ leading to $y = 1 - I$. Thus, only the non-trivial idempotents of S are I and $1 - I$. □

We next find the form of the triplets in S.

Theorem 2. *Let $S = \{\langle Q \cup I \rangle, +, \times\}$ (or $\langle R \cup I \rangle, +, \times$) be the neutrosophic ring. The neutrosophic triplets in S are only of the following form for $a, b \in Q$ or R.*

(i)

$$\left(a - aI, 1 - I, \frac{1}{a} - \frac{I}{a}\right) \text{ and } \left(\frac{1}{a} - \frac{I}{a}, 1 - I, a - aI\right); a \neq 0.$$

(ii)

$$\left(bI, I, \frac{I}{b}\right) \text{ and } \left(\frac{I}{b}, I, b\right); b \neq 0.$$

(iii)

$$\left(a + bI, 1, \frac{1}{a} - \frac{bI}{a(a+b)}\right); a + b \neq 0 \text{ and } \left(\frac{1}{a} - \frac{bI}{a(a+b)}, 1, a + bI\right).$$

Proof. Let S be the neutrosophic ring. Let $x = \{a + bI, e + fI, c + dI\}$ be a neutrosophic triplet in $S; a, b, c, d, e, f \in Q$ or R. We prove the neutrosophic triplets of S are in one of the forms. If x is a neutrosophic triplet, then we have

$$a + bI \times e + fI = a + bI \tag{1}$$

$$e + fI \times c + dI = c + dI \tag{2}$$

and

$$a + bI \times c + dI = e + fI \tag{3}$$

Now, solving Equation (1), we get

$$ae + (bfI + beI + afI) = a + bI$$

Equating the real and neutrosophic parts, we get

$$ae = a \tag{4}$$

$$bf + be + af = b \tag{5}$$

Expanding Equation (2), we get

$$ce + fcI + deI + fdI = c + dI.$$

Equating the real and neutrosophic parts, we get

$$ce = c \tag{6}$$

$$fc + de + fd = d. \tag{7}$$

Solving Equation (3), we get

$$ac + bcI + bdI + adI = e + fI$$

Equating the real and neutrosophic parts, we get

$$ac = e \tag{8}$$

$$bc + bd + ad = f \tag{9}$$

We find conditions so that Equations (4) and (5) are true.

Now, $ae = a$ and $bf + be + af = b$; $ae = a$ gives $a(e-1) = 0$ if $a = 0$ and $e \neq 1$ using in Equation (4), thus if $a = 0$, we get $e = 0$ and using $e = 0$ in Equation (6), we get $c = 0$. Thus, $a = c = e = 0$. This forces $b \neq 0$, $d \neq 0$ and $f \neq 0$. We solve for b, d and f using Equations (5), (7) and (9). Equations (5) and (7) gives $bf = b$ as $b \neq 0, f = 1$. Now, $fd = d$ as $f = 1; d = d$. Equation (9) gives $bd = f$ or $bd = 1$, thus

$$d = \frac{1}{b}(b \neq 0).$$

Thus, we get

$$\left(bI, I, \frac{I}{b}\right)$$

to be neutrosophic triplet then

$$\left(\frac{I}{b}, I, bI\right)$$

is also a neutrosophic triplet. Thus, we have proved (ii) of the theorem.

Assume in Equation (4) $ae = a; a \neq 0$, which forces $e = 1$. Now, using Equation (8), we get $ac = 1$, thus

$$c = \frac{1}{a}.$$

Using Equation (5), we get $bf + b + af = b$, thus $(a+b)f = 0$. If $f = 0$, then we have

$$\left(a + bI, 1, \frac{1}{a} + dI\right)$$

should be a neutrosophic triplet. That is,

$$(a + bI) \times \left(\frac{1}{a} + dI\right) = 1$$

$$1 + \frac{b}{a}I + daI + dbI = 1$$

$$\frac{b}{a} + da + db = 0$$

$$b + a^2d + abd = 0$$

$$b(ad + 1) + a^2d = 0$$

$$d(a^2 + ab) = -b.$$

$$d = \frac{-b}{a^2 + ab} = \frac{-b}{a(a+b)}$$

$a \neq 0$ and $a + b \neq 0$. $a + b \neq 0$ for if $a + b = 0$, then $b = 0$ we get $d = 0$. Thus, the trivial triplet

$$(a, 1, \frac{1}{a})$$

will be obtained. Thus, $a + b \neq 0$ and

$$\left(a + bI, 1, \frac{1}{a} - \frac{bI}{a(a+b)}\right) \text{ and } \left(\frac{1}{a} - \frac{bI}{a(a+b)}, 1, a + bI\right)$$

are neutrosophic triplets so that Condition (iii) of theorem is proved.

Now, let $f \neq 0$, thus $a + b = 0$ and $c + d = 0$. We get $a = -b$ or $b = -a$ and $d = -c$. We have already proved $c = \frac{1}{a}$. Using Equations (8) and (9) and conditions $a = -b$ and $c = -d$, we get $f = -1$.

Hence, the neutrosophic triplets are

$$\left(a - aI, 1 - I, \frac{1}{a} - \frac{I}{a}\right) \text{ and } \left(\frac{1}{a} - \frac{I}{a}, 1 - I, a - aI\right)$$

which is Condition (i) of the theorem. □

Theorem 3. *Let $S = \{\langle Q \cup I\rangle, +, \times\}$ (or $\langle R \cup I\rangle, +, \times\}$) be the neutrosophic ring.*

$$M = \left\{\left(a - aI, 1 - I, \frac{1}{a} - \frac{I}{a}\right) | a \in Q \backslash \{0\}\right\}$$

be the collection of neutrosophic triplets of S with neutral $1 - I$ is commutative group of infinite order with $(1 - I, 1 - I, 1 - I)$ as the multiplicative identity.

Proof. To prove M is a group of infinite order, we have to prove M is closed under component-wise product and has an identity with respect to which every element has an inverse.

Let

$$x = \left(a - aI, 1 - I, \frac{1}{a} - \frac{I}{a}\right) \text{ and } y = \left(c - cI, 1 - I, \frac{1}{c} - \frac{I}{c}\right) \in M$$

$$x \times y = \left(a - aI, 1 - I, \frac{1}{a} - \frac{I}{a}\right) \times \left(c - cI, 1 - I, \frac{1}{c} - \frac{I}{c}\right)$$

$$= \left(ac - acI - acI + acI, 1 - 2I + I, \frac{1}{ac} - \frac{I}{ac} - \frac{I}{ac} + \frac{I}{ac}\right)$$

$$= \left(ac - acI, 1 - I, \frac{1}{ac} - \frac{I}{ac}\right) \in M.$$

Thus, M is closed under component wise product.

We see that, when $a = 1$, we get $e = (1 - I, 1 - I, 1 - I) \in M$ is the identity of M under component wise multiplication. Clearly, $e \times x = x \times e = x$ for all $x \in M$, thus e is the identity of M. For every

$$x = \left(a - aI, 1 - I, \frac{1}{a} - \frac{I}{a}\right),$$

we have a unique

$$x^{-1} = \left(\frac{1}{a} - \frac{I}{a}, 1 - I, a - aI\right) \in M$$

such that

$$x \times x^{-1} = x^{-1} \times x = e = (1 - I, 1 - I, 1 - I)$$

$$\begin{aligned} x \times x^{-1} &= \left(a - aI, 1 - I, \frac{1}{a} - \frac{I}{a}\right) \times \left(\frac{1}{a} - \frac{I}{a}\right) - \left(\frac{1}{a} - \frac{I}{a}, 1 - I, a - aI\right) \\ &= \left(\frac{a}{a} - \frac{aI}{a} - \frac{aI}{a} + \frac{aI}{a}, 1 - 2I + I, \frac{a}{a} - \frac{aI}{a} - \frac{aI}{a} + \frac{aI}{a}\right) \\ &= (1 - I, 1 - I, 1 - I) \end{aligned}$$

as $a \neq 0$. Thus, $(M, \times)$ is a group under component wise product, which is known as the neutrosophic triplet group. □

Theorem 4. *Let* $S = \{\langle Q \cup I \rangle, +, \times\}$ *(or* $\{\langle R \cup I \rangle, +, \times\}$*) be the neutrosophic ring. The collection of neutrosophic triplets*

$$N = \left\{\left(aI, I, \frac{I}{a}\right) | a \in Q \backslash \{0\}\right\}$$

(or $R \backslash \{0\}$*) forms a commutative group of infinite order under component wise multiplication with (I, I, I) as the multiplicative identity.*

Proof. Let

$$N = \left\{\left(aI, I, \frac{I}{a}\right) | a \neq 0 \in Q \text{ or } R\right\}$$

be a collection of neutrosophic triplets. To prove N is commutative group under component wise product, let

$$x = \left(aI, I, \frac{I}{a}\right)$$

and

$$y = \left(bI, I, \frac{I}{b}\right) \in M.$$

To show $x \times y \in N$.

$$x \times y = \left(aI, I, \frac{I}{a}\right) \times \left(bI, I, \frac{I}{b}\right) = \left(abI, I, \frac{I}{ab}\right),$$

using the fact $I^2 = I$. Hence, $(N, \times)$ is a semigroup under product.

Considering $e = (I, I, I) \in N$, we see that $e \times e = x \times e = x$ for all $x \in N$.

$$e \times x = (I, I, I) \times \left(aI, I, \frac{I}{a}\right) = \left(aI, I, \frac{I}{a}\right) = x(\text{ using } I^2 = I).$$

Thus, (I, I, I) is the identity element of $(N, \times)$. For every

$$x = \left(aI, I, \frac{I}{a}\right),$$

we have a unique

$$x^{-1} = \left(\frac{I}{a}, I, a\right) \in N$$

is such that

$$x \times x^{-1} = \left(aI, I, \frac{I}{a}\right) = (I, I, I)$$

as $a \neq 0$ and $I^2 = I$.

Thus, $\{N, \times\}$ is a commutative group of infinite order.

It is interesting to note both the sets M and N are not even closed under addition.

Next, let

$$P = \left\{a + bI, 1, \frac{1}{a} - \frac{bI}{a(a+b)}; a \neq b; a + b \neq 0, a \neq 0.\right\}$$

We get

$$a + bI \times \frac{1}{a} - \frac{bI}{a(a+b)} = 1.$$

□

We call these neutrosophic triplets as special neutrosophic triplets contributed by the unity 1 of the ring which is the trivial idempotent of S; however, where it is mandatory, x and $anti(x)$ are nontrivial neutrosophic numbers with $neut(x) = 1$.

Theorem 5. *Let* $S = \langle Q \cup I \rangle, +, \times$ *(or* $\langle R \cup I \rangle, +, \times$*) be the neutrosophic ring. Let*

$$P = \left\{(a + bI, 1, \frac{1}{a} - \frac{bI}{a(a+b)}; a \neq b, \text{ where } a, b \in Q\backslash\{0\}(\text{ or } R\backslash 0) \text{ and } a + b \neq 0\right\}$$

be the collection of special neutrosophic triplets with 1 as the neutral. P is a torsion free abelian group of infinite order with $(1, 1, 1)$ *as its identity under component wise product.*

Proof. It is easily verified P is closed under the component wise product and (1, 1, 1) acts as the identity for component wise product. For every

$$x = \left(a - bI, 1, \frac{1}{a} + \frac{bI}{a(a-b)}\right) \in P,$$

we have a unique

$$y = \left(\frac{1}{a} + \frac{bI}{a(a-b)}, 1, a - bI\right) \in P$$

such that $x \times y = (1, 1, 1)$. We also see $x^n \neq (1, 1, 1)$ for any $x \in P$ and $n \neq 0 (n > 0); x \neq (1, 1, 1)$, hence P is a torsion free abelian group. □

4. Discussion and Conclusions

We show that, in the case of neutrosophic duplets in $\langle Z \cup I \rangle$, $\langle Q \cup I \rangle$ or $\langle R \cup I \rangle$, the collection of duplets $\{a - aI\}$ forms a neutrosophic subring. However, in the case of neutrosophic triplets, we show that $\langle Z \cup I \rangle$ has no nontrivial triplets and we have shown there are three distinct collection of neutrosophic triplets in $\langle R \cup I \rangle$ and $\langle Q \cup I \rangle$. We have proved there are only three types of neutrosophic triplets in these neutrosophic rings and all three of them form abelian groups that are torsion free under component wise product. For future research, we would apply these neutrosophic triplets to concepts akin to SVNS and obtain some mathematical models.

Author Contributions: Conceptualization, V.K.W.B. and F.S.; writing—original draft preparation, V.K.W.B. and I.K.; writing—review and editing, I.K.

Funding: This research received no external funding.

Acknowledgments: The authors would like to thank the reviewers for their reading of the manuscript and many insightful comments and suggestions.

Conflicts of Interest: The authors declare no conflict of interest.

References

1. Smarandache, F. *A Unifying Field in Logics: Neutrosophic Logic. Neutrosophy, Neutrosophic Set, Neutrosophic Probability and Statistics*; American Research Press: Rehoboth, DE, USA, 2005; ISBN 978-1-59973-080-6.

2. Smarandache, F. Neutrosophic set-a generalization of the intuitionistic fuzzy set. In Proceedings of the 2006 IEEE International Conference on Granular Computing, Atlanta, GA, USA, 10–12 May 2006; pp. 38–42.
3. Wang, H.; Smarandache, F.; Zhang, Y.; Sunderraman, R. Single valued neutrosophic sets. *Review* **2010**, *1*, 10–15.
4. Kandasamy, I. Double-Valued Neutrosophic Sets, their Minimum Spanning Trees, and Clustering Algorithm. *J. Intell. Syst.* **2018**, *27*, 163–182. [CrossRef]
5. Kandasamy, I.; Smarandache, F. Triple Refined Indeterminate Neutrosophic Sets for personality classification. In Proceedings of the 2016 IEEE Symposium Series on Computational Intelligence (SSCI), Athens, Greece, 6–9 December 2016; pp. 1–8. [CrossRef]
6. Vasantha, W.B.; Smarandache, F. *Basic Neutrosophic Algebraic Structures and Their Application to Fuzzy and Neutrosophic Models*; Hexis: Phoenix, AZ, USA, 2004; ISBN 978-1-931233-87-X.
7. Vasantha, W.B.; Smarandache, F. *N-Algebraic Structures and SN-Algebraic Structures*; Hexis: Phoenix, AZ, USA, 2005; ISBN 978-1-931233-05-5.
8. Vasantha, W.B.; Smarandache, F. *Some Neutrosophic Algebraic Structures and Neutrosophic N-Algebraic Structures*; Hexis: Phoenix, AZ, USA, 2006; ISBN 978-1-931233-15-2.
9. Vasantha, W.B.; Smarandache, F. *Neutrosophic Rings*; Hexis: Phoenix, AZ, USA, 2006; ISBN 978-1-931233-20-9.
10. Agboola, A.A.A.; Adeleke, E.O.; Akinleye, S.A. Neutrosophic rings II. *Int. J. Math. Comb.* **2012**, *2*, 1–12.
11. Smarandache, F. Operators on Single-Valued Neutrosophic Oversets, Neutrosophic Undersets, and Neutrosophic Offsets. *J. Math. Inf.* **2016**, *5*, 63–67. [CrossRef]
12. Smarandache, F.; Ali, M. Neutrosophic triplet group. *Neural Comput. Appl.* **2018**, *29*, 595–601. [CrossRef]
13. Agboola, A.A.A.; Akinola, A.D.; Oyebola, O.Y. Neutrosophic Rings I. *Int. J. Math. Comb.* **2011**, *4*, 115.
14. Ali, M.; Smarandache, F.; Shabir, M.; Naz, M. Soft Neutrosophic Ring and Soft Neutrosophic Field. *Neutrosophic Sets Syst.* **2014**, *3*, 53–59.
15. Ali, M.; Smarandache, F.; Shabir, M.; Vladareanu, L. Generalization of Neutrosophic Rings and Neutrosophic Fields. *Neutrosophic Sets Syst.* **2014**, *5*, 9–13.
16. Ali, M.; Shabir, M.; Smarandache, F.; Vladareanu, L. Neutrosophic LA-semigroup Rings. *Neutrosophic Sets Syst.* **2015**, *7*, 81–88.
17. Broumi, S.; Smarandache, F.; Maji, P.K. Intuitionistic Neutrosphic Soft Set over Rings. *Math. Stat.* **2014**, *2*, 120–126.
18. Smarandache, F. *Neutrosophic Perspectives: Triplets, Duplets, Multisets, Hybrid Operators, Modal Logic, Hedge Algebras and Applications*, 2nd ed.; Pons Publishing House: Brussels, Belgium, 2017; ISBN 978-1-59973-531-3.
19. Sahin, M.; Abdullah, K. Neutrosophic triplet normed space. *Open Phys.* **2017**, *15*, 697–704. [CrossRef]
20. Smarandache, F. Hybrid Neutrosophic Triplet Ring in Physical Structures. *Bull. Am. Phys. Soc.* **2017**, *62*, 17.
21. Smarandache, F.; Ali, M. Neutrosophic Triplet Field used in Physical Applications. In Proceedings of the 18th Annual Meeting of the APS Northwest Section, Pacific University, Forest Grove, OR, USA, 1–3 June 2017.
22. Smarandache, F.; Ali, M. Neutrosophic Triplet Ring and its Applications. In Proceedings of the 18th Annual Meeting of the APS Northwest Section, Pacific University, Forest Grove, OR, USA, 1–3 June 2017.
23. Zhang, X.H.; Smarandache, F.; Liang, X.L. Neutrosophic Duplet Semi-Group and Cancellable Neutrosophic Triplet Groups. *Symmetry* **2017**, *9*, 275. [CrossRef]
24. Bal, M.; Shalla, M.M.; Olgun, N. Neutrosophic Triplet Cosets and Quotient Groups. *Symmetry* **2017**, *10*, 126. [CrossRef]
25. Zhang, X.H.; Smarandache, F.; Ali, M.; Liang, X.L. Commutative neutrosophic triplet group and neutro-homomorphism basic theorem. *Ital. J. Pure Appl. Math.* **2017**. [CrossRef]
26. Vasantha, W.B.; Kandasamy, I.; Smarandache, F. *Neutrosophic Triplet Groups and Their Applications to Mathematical Modelling*; EuropaNova: Brussels, Belgium, 2017; ISBN 978-1-59973-533-7.
27. Vasantha, W.B.; Kandasamy, I.; Smarandache, F. A Classical Group of Neutrosophic Triplet Groups Using $\{Z_{2p}, \times\}$. *Symmetry* **2018**, *10*, 194. [CrossRef]
28. Vasantha, W.B.; Kandasamy, I.; Smarandache, F. Neutrosophic duplets of $\{Z_{pn}, \times\}$ and $\{Z_{pq}, \times\}$. *Symmetry* **2018**, *10*, 345. [CrossRef]
29. Vasantha, W.B.; Kandasamy, I.; Smarandache, F. Algebraic Structure of Neutrosophic Duplets in Neutrosophic Rings $\langle Z \cup I \rangle$, $\langle Q \cup I \rangle$ and $\langle R \cup I \rangle$. *Neutrosophic Sets Syst.* **2018**, *23*, 85–95.
30. Vasantha, W.B.; Kandasamy, I.; Smarandache, F. Semi-Idempotents in Neutrosophic Rings. *Mathematics* **2019**, *7*, 507. [CrossRef]

31. Smarandache, F.; Zhang, X.; Ali, M. Algebraic Structures of Neutrosophic Triplets, Neutrosophic Duplets, or Neutrosophic Multisets. *Symmetry* **2019**, *11*, 171. [CrossRef]
32. Zhang, X.H.; Wu, X.Y.; Smarandache, F.; Hu, M.H. Left (right)-quasi neutrosophic triplet loops (groups) and generalized BE-algebras. *Symmetry* **2018**, *10*, 241. [CrossRef]
33. Zhang, X.H.; Wang, X.J.; Smarandache, F.; Jaíyéolá, T.G.; Liang, X.L. Singular neutrosophic extended triplet groups and generalized groups. *Cognit. Syst. Res.* **2018**, *57*, 32–40. [CrossRef]
34. Zhang, X.H.; Wu, X.Y.; Mao, X.Y.; Smarandache, F.; Park, C. On Neutrosophic Extended Triplet Groups (Loops) and Abel-Grassmann's Groupoids (AG-Groupoids). *J. Intell. Fuzzy Syst.* **2019**. [CrossRef]
35. Zhang, X.; Hu, Q.; Smarandache, F.; An, X. On Neutrosophic Triplet Groups: Basic Properties, NT-Subgroups, and Some Notes. *Symmetry* **2018**, *10*, 289. [CrossRef]
36. Ma, Y.; Zhang, X.; Yang, X.; Zhou, X. Generalized Neutrosophic Extended Triplet Group. *Symmetry* **2019**, *11*, 327. [CrossRef]
37. Kanel-Belov, A.; Halle Rowen, L. *Computational Aspects of Polynomial Identities*; Research Notes in Mathematics; CRC Press: Boca Raton, FL, USA, 2005; ISBN 9781568811635.

mathematics

Article

Inspection Plan Based on the Process Capability Index Using the Neutrosophic Statistical Method

Muhammad Aslam * and Mohammed Albassam

Department of Statistics, Faculty of Science, King Abdulaziz University, Jeddah 21551, Saudi Arabia
* Correspondence: aslam_ravian@hotmail.com or magmuhammad@kau.edu.sa; Tel.: +966-59-3329841

Received: 25 June 2019; Accepted: 11 July 2019; Published: 16 July 2019

Abstract: The Process Capability Index (PCI) has been widely used in industry to advance the quality of a product. Neutrosophic statistics is the more generalized form of classical statistics and is applied when the data from the production process or a product lot is incomplete, incredible, and indeterminate. In this paper, we will originally propose a variable sampling plan for the PCI using neutrosophic statistics. The neutrosophic operating function will be given. The neutrosophic plan parameters will be determined using the neutrosophic optimization solution. A comparison between plans based on neutrosophic statistics and classical statistics is given. The application of the proposed neutrosophic sampling plan will be given using company data.

Keywords: acceptance number; neutrosophic approach; operating characteristics; risks; sample size

1. Introduction

Acceptance sampling is the most widely used tool for the inspection of the raw material, semi-finished product, and finished product. But, the presence of the indeterminacy in the observations or parameters may affect the performance of the sampling plan. A well-designed sampling plan used for the inspection of the product under the uncertainty and determinacy environment is needed at each stage to check that the finished product meets either the customer's upper specification limit (USL) and lower specification limit (LSL) before sending it to market. The quality of interest beyond the LSL and USL creates a non-conforming item. At the time of inspection, a random sample is taken and lot sentencing is made on the basis of this primary information about the lot. Thus, the sample information may mislead the experimenters in making the decision about the submitted product lot. There is a chance of rejecting a good lot and accepting a bad lot on the basis of the sample information. Thus, the sampling schemes are developed with the aim of reducing the cost of the inspection, non-conforming items, and minimizes the risk of the sampling. The acceptance sampling plan has two major types, known as attribute sampling plans and variable sampling plans. Attribute sampling plans are easier to apply but are more costly than the variable sampling plans. On the other hand, the variable sampling plans are more informative than attribute sampling plans [1]. A number of authors designed variable and attribute sampling plans: Jun et al. [2] studied variable sampling plans for sudden death testing; Balamurali and Jun [3] studied skip-lot sampling for the normal distribution; Fallah Nezhad et al. [4] designed a sampling plan using cumulative sums of conforming run-lengths; Pepelyshev et al. [5] applied a variable sampling plan in photovoltaic modules; Gui and Aslam [6] designed a time truncated plan for weighted exponential distribution; and Balamurali et al. [7] designed a mixed variable sampling plan.

The Process Capability Index (PCI) has been widely used in industry for quality improvement purposes and to make a relation between specification limits and process quality. Kane [8] originally proposed the PCI for classical statistics. Boyles [9] provided the bounds on the process yield for the normally distributed process. Kotz and Johnson [10] provided a detailed review of PCIs. More details on PCIs can be seen in [11]. Pearn et al. [12] discussed an effective decision method for product inspection; Montgomery [1] mentioned the applications of PCIs. Boyles [13] studied PCIs for an asymmetric tolerances case and Ebadi [14] studied a simple linear profile using PCIs. Due to the importance of the PCIs in industry, several authors focused on the development of inspection schemes using classical statistics based on PCIs for various situations including for example, and Chen et al. [15] studied PCIs for entire product inspection. Pearn et al. [12] presented an effective decision method for the inspection. Aslam et al. [16] designed various sampling plans using PCIs. Seifi and Nezhad [17] studied resubmitted sampling using PCI and Arif et al. [18] worked on a sampling plan using PCI for multiple manufacturing lines.

Fuzzy sampling plans have been widely used in the industry when the proportion of the non-conforming product is a fuzzy number [19]. Kanagawa and Ohta [20] introduced an attribute plan using fuzzy sets. Sadeghpour Gildeh et al. [19] designed a single sampling plan using fuzzy parameters. Kahraman et al. [21] designed single and double sampling plans using fuzzy approach. The PCIs using fuzzy logic can be seen in [22–24].

Smarandache [25] defined the neutrosophic logic in 1998 as the generalization of fuzzy logic. Smarandache [26] gave the idea of descriptive neutrosophic statistics. The neutrosophic statistics is the more generalized form of classical statistics and applied when the data from the production process or a product lot is incomplete, incredible, and indeterminate [26]. Chen et al. [27,28] studied the rock joint roughness coefficient using neutrosophic statistics. According to [29] "All observations and measurements of continuous variables are not precise numbers but more or less non-precise. This imprecision is different from variability and errors. Therefore also lifetime data are not precise numbers but more or less fuzzy. The best up-to-date mathematical model for this imprecision is so-called non-precise numbers".

Recently, Aslam [30] introduced the neutrosophic statistics in the area of the acceptance sampling plan. Aslam [30] proposed an acceptance sampling plan using the neutrosophic process loss function. The sampling plan for multiple manufacturing lines using the neutrosophic statistics is proposed by [31]. The sampling plan for the exponential distribution under the uncertainty is proposed by [32]. Some more details about the sampling plan using the neutrosophic plans can be seen in [33–37].

The existing sampling plans using PCIs cannot apply when the data is indeterminate or incomplete. Also, the available sampling plans using the neutrosophic statistics do not consider the PCIs for the inspection of the product. By exploring the literature and best of the author knows there is no work on the sampling plan for PCIs using the neutrosophic statistics. In this paper, we will originally propose a variable sampling plan for the PCIs using the neutrosophic statistics. The neutrosophic operating function will be given. The neutrosophic plan parameters will be determined using the neutrosophic optimization solution. A comparison between plans based on neutrosophic statistics and classical statistics is given. We expect that the proposed plan will be more effective to be applied in an uncertain environment. The application of the proposed sampling plan using neutrosophic statistics will be given using the company data.

2. Design of a Neutrosophic Plan Based on PCI

Let $n_N \epsilon \{n_L, n_U\}$ be a random sample selected from the population having some uncertain observations, where n_L and n_U are the lower and upper sample size of the indeterminacy interval, respectively. Suppose that a neutrosophic quality of interest, X_{Ni}, is expressed in the indeterminacy interval, say, $X_{Ni} \epsilon \{X_L, X_U\}; i = 1,2,3, \ldots, n_N$ having indeterminate observations follow the neutrosophic normal distribution, where X_L and X_U are the lower and the upper values, respectively, with the neutrosophic population mean $\mu_N \epsilon \{\mu_L, \mu_U\}$ and neutrosophic population standard deviation (NSD) $\sigma_N \epsilon \{\sigma_L, \sigma_U\}$ (see [26]). The neutrosophic process capability index process (NPCI), say, $\hat{c}_{N_{pk}}$, is defined as:

$$C_{N_{pk}} = Min\left\{\frac{USL - \mu_N}{3\sigma_N}, \frac{\mu_N - LSL}{3\sigma_N}\right\}; \ \mu_N \epsilon \{\mu_L, \mu_U\}, \ \sigma_N \epsilon \{\sigma_L, \sigma_U\} \quad (1)$$

where USL and LSL are the upper specification limit and lower specification limit, respectively.

Note that $C_{N_{pk}}$ reduces to PCI for classical statistics when no indeterminate observations are recorded in X_N. Usually, $\mu_N \epsilon \{\mu_L, \mu_U\}$ and $\sigma_N \epsilon \{\sigma_L, \sigma_U\}$ are unknown in practice and the best linear unbiased estimate (BLUE) of $\mu_N \epsilon \{\mu_L, \mu_U\}$ is the neutrosophic sample mean $\overline{X}_N \epsilon \{\overline{X}_L, \overline{X}_U\}$ and a BLUE of $\sigma_N \epsilon \{\sigma_L, \sigma_U\}$ is the neutrosophic sample standard deviation $s_N \epsilon \{s_L, s_U\}$ which can be used to evaluate $C_{N_{pk}}$. The $\hat{C}_{N_{pk}}$ based on sample estimate is given as by:

$$\hat{C}_{N_{pk}} = Min\left\{\frac{USL - \overline{X}_N}{3s_N}, \frac{\overline{X}_N - LSL}{3s_N}\right\}; \ \overline{X}_N \epsilon \{\overline{X}_L, \overline{X}_U\}, \ s_N = \{s_L, s_U\} \tag{2}$$

where $\overline{X}_L = \sum_{i=1}^{n} x_i^L / n_L$,

$$\overline{X}_U = \sum_{i=1}^{n} x_i^U / n_U, \ s_L = \sqrt{\sum_{i=1}^{n} \left(x_i^L - \overline{X}_L\right)^2 / n_L} \text{ and } s_U = \sqrt{\sum_{i=1}^{n} \left(x_i^U - \overline{X}_U\right)^2 / n_U}$$

To design the proposed sampling plan, it is assumed that there is uncertainty in the selection of a random sample from the submitted product lot. Thus, a random sample will be selected from a neutrosophic interval. The proposed sampling plan is stated as follows:

Step 1: Select a random sample of size $n_N \epsilon \{n_L, n_U\}$ from the product lot. Compute the statistic $\hat{C}_{N_{pk}} \epsilon Min\left\{\frac{USL - \overline{X}_N}{3s_N}, \frac{\overline{X}_N - LSL}{3s_N}\right\}; \overline{X}_N \epsilon \{\overline{X}_L, \overline{X}_U\}, s_N \epsilon \{s_L, s_U\}$.

Step 2: Accept a product lot of $\hat{C}_{N_{pk}} \geq k_N; \ k_N \epsilon \{k_{aL}, k_{aU}\}$, otherwise reject a product lot, where $k_N \epsilon \{k_{aL}, k_{aU}\}$ is the neutrosophic acceptance number. An acceptance number is also called the action number/boundary number. A product lot is rejected if the statistic $\hat{C}_{N_{pk}}$ is smaller than k_N, otherwise, the product lot is accepted.

The evaluation of the proposed sampling plan will be used on two parameters, namely $n_N = \{n_L, n_U\}$ and $k_N \epsilon \{k_{aL}, k_{aU}\}$. The neutrosophic operating characteristic (NOC) for the proposed plan is derived as follows:

$$\begin{aligned} L(p) = P\left(\hat{C}_{N_{pk}} \geq k_N\right) = P\left\{LSL + 3k_N s_N \leq \overline{X}_N \leq USL - 3k_N s_N\right\} = P\left\{\overline{X}_N + 3k_N s_N \leq USL\right\} \\ -P\left\{\overline{X}_N - 3k_N s_N \leq LSL\right\}; \ n_N \epsilon \{n_L, n_U\}, \ \overline{X}_N \epsilon \{\overline{X}_L, \overline{X}_U\} s_N \epsilon \{s_L, s_U\} \text{ and } k_N \epsilon \{k_{aL}, k_{aU}\}. \end{aligned} \tag{3}$$

Duncan [38] suggested $\overline{X}_N \pm k_N s_N; \ \overline{X}_N \epsilon \{\overline{X}_L, \overline{X}_U\}$ and $s_N = \{s_L, s_U\}$ is distributed as an approximately neutrosophic normal distribution, that is $\overline{X}_N \pm k_N s_N \sim N_N\left(\mu_N \pm c\sigma_N, \frac{\sigma_N^2}{n_N} + \frac{c^2\sigma_N^2}{2n_N}\right)$. where $N_N(.)$ shows neutrosophic normal distribution.

Suppose that quality of interest X_N beyond the USL or LSL is labeled as the defective item and this probability is defined as $p_U = P\{X_N > USL|\mu_N\}$ and $p_U = P\{X_N \langle LSL|\mu_N\}; \mu_N = \{\mu_L, \mu_U\}$. Thus, the probability of acceptance is given by the following [39]:

$$L(p) = \Phi\left\{\frac{USL - \mu_N - 3k_N\sigma_N}{(\sigma_N/n_N)\sqrt{1 + 9k_N^2/2}}\right\} - \Phi\left\{\frac{LSL - \mu_N + 3k_N\sigma_N}{(\sigma_N/n_N)\sqrt{1 + 9k_N^2/2}}\right\} \tag{4}$$

Let us define the neutrosophic standard normal random variable as:

$$Z_{N_{pU}} = \frac{USL - \mu_N}{\sigma_N} \text{ and } -Z_{N_{pL}} = \frac{LSL - \mu_N}{\sigma_N} \tag{5}$$

Now, the final form of NFOC is given by:

$$L(p) = \Phi\left\{\left(\left(Z_{Np_U} - 3k_N\right)\sqrt{\frac{n_N}{1+\left(9k_N^2/2\right)}}\right)\right\} - \Phi\left\{\left(-\left(Z_{Np_L} - 3k_N\right)\sqrt{\frac{n_N}{1+\left(9k_N^2/2\right)}}\right)\right\} \quad (6)$$

where $\Phi(.)$ is the neutrosophic cumulative standard normal distribution.

Research Methodology

To meet the given producer's risk, say, α, and the custumer's risk, say, β, the plan parameters of the proposed sampling plan will be determined in such a way that NFOC passes through the two points $(p_1, 1-\alpha)$ and (p_2, β), where p_1 is the acceptable quality limit (AQL) and p_2 is the limiting quality limit (LQL). The plan parameters of the proposed sampling plans will be determined through the following non-linear solution under the neutrosophic statistical interval method:

Minimize:

$$n_N \epsilon \{n_L, n_U\} \quad (7)$$

subject to:

$$L_N(p_1) = \Phi\left\{\left(\left(Z_{Np_{U1}} - 3k_N\right)\sqrt{\frac{n_N}{1+\left(9k_N^2/2\right)}}\right)\right\} - \Phi\left\{\left(-\left(Z_{Np_{L1}} - 3k_N\right)\sqrt{\frac{n_N}{1+\left(9k_N^2/2\right)}}\right)\right\} \geq 1-\alpha;\ k_N \epsilon \{k_{aL}, k_{aU}\};\ n_N \epsilon \{n_L, n_U\} \quad (8)$$

and:

$$L_N(p_2) = \Phi\left\{\left(\left(Z_{Np_{U2}} - 3k_N\right)\sqrt{\frac{n_N}{1+\left(9k_N^2/2\right)}}\right)\right\} - \Phi\left\{\left(-\left(Z_{Np_{L2}} - 3k_N\right)\sqrt{\frac{n_N}{1+\left(9k_N^2/2\right)}}\right)\right\} \leq \beta;\ k_N \epsilon \{k_{aL}, k_{aU}\};\ n_N \epsilon \{n_L, n_U\} \quad (9)$$

The plan parameters of the proposed plan are determined through Equations (7)–(9) using the search grid method for the various combinations of AQL and LQL. Several combinations of plan parameters in the indeterminacy interval satisfy Equations (7)–(9). The plan parameters having the smallest range in indeterminacy interval are chosen and placed in Table 1. To save the space, we present Table 1 when $\alpha = 0.05$ and $\beta = 0.10$. Similar tables for other values of α and β can be prepared. The neutrosophic lot acceptance probabilities, $L_N(p_1)$ and $L_N(p_2)$ at the consumer's risk and producer's risk are also reported in Table 1.

From Table 1, we note that, for the fixed values of all other parameters, the values of $k_N \epsilon \{k_{aL}, k_{aU}\}$; $n_N \epsilon \{n_L, n_U\}$ decrease as LQL increases. This means the indeterminacy in the sample size and acceptance number reduces. For example, under the uncertainty, when AQL = 0.001 and LQL = 0.02, the sample size will be in the interval [18,20]. This means the industrial engineers should select a sample size between 18 and 20. Furthermore, for the smaller values of AQL and LQL, larger the values of $n_N \epsilon \{n_L, n_U\}$ are required. Note here that the appropriate sample size is decided on the basis of pre-defined parameters, such as AQL, LQL, α, and β. The following algorithm is used to determine the neutrosophic plan parameters:

1. Specify the values of AQL, LQL, α and β.
2. Specify the suitable ranges for $n_N \epsilon \{n_L, n_U\}$ such that $n_L < n_U$ and $k_N \epsilon \{k_{aL}, k_{aU}\}$ such that $k_{aL} < k_{aU}$.
3. Perform the simulation by the grid search method and select those values of the neutrosophic plan parameters where $n_N \epsilon \{n_L, n_U\}$ and satisfy the conditions given in Equations (7)–(9).

Table 1. The plan parameters of the plan when $\alpha = 0.05$, $\beta = 0.10$.

p_1	p_2	n_N	k_N	$L_N(p_1)$	$L_N(p_2)$
0.001	0.002	[602, 643]	[1.093, 1.095]	[0.9500, 0.95033]	[0.0441, 0.0891]
	0.003	[218, 228]	[1.052, 1.054]	[0.9500, 0.9505]	[0.06223, 0.0898]
	0.004	[128, 133]	[1.022, 1.024]	[0.9506, 0.9513]	[0.0700, 0.0914]
	0.006	[69, 71]	[0.978, 0.980]	[0.9513, 0.9517]	[0.0807, 0.0969]
	0.008	[47, 49]	[0.946, 0.948]	[0.9506, 0.9528]	[0.0848, 0.0977]
	0.010	[36, 38]	[0.921, 0.923]	[0.9502, 0.9504]	[0.0849, 0.0958]
	0.015	[24, 28]	[0.874, 0.876]	[0.9541, 0.9675]	[0.0914, 0.0959]
	0.020	[18, 20]	[0.842, 0.844]	[0.9521, 0.9614]	[0.0761, 0.0823]
0.0025	0.030	[21, 23]	[0.793, 0.795]	[0.9529, 0.9606]	[0.0923, 0.0995]
	0.050	[13, 15]	[0.731, 0.735]	[0.9567, 0.9674]	[0.0607, 0.0754]
0.005	0.050	[19, 21]	[0.730, 0.732]	[0.9512, 0.9599]	[0.0897, 0.0967]
	0.100	[9, 11]	[0.631, 0.633]	[0.9575, 0.9740]	[0.0957, 0.0961]
0.01	0.020	[274, 290]	[0.854, 0.856]	[0.9500, 0.9504]	[0.0513, 0.0881]
	0.030	[95, 99]	[0.803, 0.805]	[0.9504, 0.9512]	[0.0696, 0.0918]
0.03	0.060	[165, 174]	[0.718, 0.720]	[0.9503, 0.9509]	[0.0581, 0.0903]
	0.090	[55, 57]	[0.659, 0.661]	[0.9505, 0.9511]	[0.0756, 0.0950]
0.05	0.100	[123, 129]	[0.647, 0.649]	[0.9502, 0.9505]	[0.0690, 0.0986]
	0.150	[41, 43]	0.584, 0.586	0.9509, 0.9530	0.0736, 0.0911

3. Comparison Study

In this section, we will compare the efficiency of the proposed plan with the sampling plan using classical statistics in terms of the sample size required for the inspection of the submitted product lot. For a fair comparison, we will consider the same values of all the specified parameters. The sample size n_N along with range ($R = n_U - n_L$) in the indeterminacy interval of the proposed plan and sample size n using classical statistics when $\alpha = 0.05$, $\beta = 0.10$ are placed in Table 2. From Table 2, it can be noted that the proposed plan provides a smaller indeterminacy interval in the sample size as compared to the plan using classical statistics. For example, when AQL = 0.001 and LQL = 0.002, the proposed plan has $n_N \in [602, 643]$ while the existing plan has $n = 1134$. Therefore, the proposed plan needs a smaller sample size and range in the indeterminacy interval for the inspection of a product lot. From this comparison, it is quite clear that the proposed plan using neutrosophic statistics is more efficient than the existing sampling plan under classical statistics in terms of sample size. In addition, the proposed plan is quite suitable, effective, and informative to be used in uncertainty than the existing plan.

Table 2. The comparison of proposed plan and the plan based on classical statistics.

p_1	p_2	Proposed Plan	Plan Based on Classical Statistics
		n_N	n
0.001	0.002	[602, 643] (R = 41)	1134 (R = 1134)
	0.003	[218, 228] (R = 10)	351 (R = 351)
	0.004	[128, 133] (R = 5)	161 (R = 161)
	0.006	[69, 71] (R = 2)	74 (R = 74)
	0.008	[47, 49] (R = 2)	47 (R = 47)

Table 2. *Cont.*

p_1	p_2	Proposed Plan	Plan Based on Classical Statistics
		n_N	n
0.01	0.020	[274, 290] (R = 16)	449 (R = 449)
	0.030	[95, 99] (R = 4)	132 (R = 132)
0.03	0.060	[165, 174] (R = 9)	240 (R = 240)
	0.090	[55, 57] (R = 2)	68 (R = 68)
0.05	0.100	[123, 129] (R = 6)	167 (R = 167)
	0.150	[41, 43] (R = 2)	46 (R = 46)

4. Application of the Proposed Plan

In this section, we will give the application of the proposed plan using the data of the amplified pressure sensor that came from industry. Viertl [29] commented that the observations obtained from the measurements are not usually precise. According to [40] "For this amplified pressure sensor process, the span is the focused characteristic". As the observations for the quality of interest are measured, some observations in the data may be indeterminate or imprecise. Under the uncertainty, the experimenter is not sure about the sample size for the inspection of a product lot when some indeterminate or imprecise observations are recorded. For this data, LSL = 1.9 V, USL 2.1. Suppose that AQL = 0.001, LQL = 0.04,$\alpha = 0.05$, and $\beta = 0.10$. The neutrosophic plan parameters from Table 1 are $n_N \epsilon \{128, 133\}$. Thus, the experimenter should select a random sample between 128 and 133. Suppose that the industrial engineers decided to select a random sample size of 128 for the inspection of a product lot. The amplified pressure sensor data of n = 128 having some indeterminate observations are reported in Table 3. Based on the given data, the neutrosophic average and standard deviation (SD) are computed as follows:

$$\overline{X}_N = \frac{\begin{matrix}[1.9422, 1.9422] + [1.9651, 1.9651] + [2.0230, 2.0230] + \ldots + [1.9994, 1.9994], \\ [1.9422, 1.9422] + [1.9651, 1.9651] + [2.0435, 2.0435] + \ldots + [2.0512, 2.0512]\end{matrix}}{128} = [1.9805, 1.9827]$$

and, similarly, $s_N = \{0.0193, 0.0225\}$.

The NPCI is computed as follows: $\hat{C}_{N_{pk}} = Min\left\{\frac{USL-\overline{X}_N}{3s_N}, \frac{\overline{X}_N-LSL}{3s_N}\right\}$, $\hat{C}_{N_{pk}} \epsilon$ $[1.7377, 2.0639]$ for $\overline{X}_N = [1.9805, 1.9827]$ and $s_N = \{0.0193, 0.0225\}$.

The proposed plan will be implemented as follows:

Step 1: Select a random sample of size $n_N = \{128, 133\}$ from a product lot. Compute the statistic $\hat{C}_{N_{pk}} \epsilon$ $[1.7377, 2.0639]$.

Step 2: Accept a product lot as $[1.7377, 2.0639] \geq \{1.022, 1.024\}$.

The application of the proposed sampling plan shows that the proposed sampling plan is quite effective, adequate, and flexible to be used under the uncertainty environment than the plan based on classical statistics which provide the determined values of the plan parameters.

Table 3. Indeterminate data of Amplified Sensors from [40].

[1.9422,1.9422]	[1.9651, 1.9651]	[2.0230, 2.0435]	[1.9712,1.9712]	[1.9975,1.9975]	[2.0164,2.0164]	[1.9927,1.9927]	[1.9566,1.9566]
[1.9738, 1.9738]	[1.9541, 1.9541]	[1.9800, 1.9800]	[1.9596, 1.9596]	[1.9811, 1.9811]	[2.0088, 2.0088]	[1.9858, 1.9858]	[1.9677, 1.9677]
[2.0001, 2.0001]	[1.9659, 1.9659]	[1.9955, 1.9955]	[1.9842, 1.9842]	[1.9909,2.0512]	[1.9829, 1.9829]	[1.9684, 1.9684]	[1.9942, 1.9942]
[1.9897, 1.9897]	[1.9836, 1.9836]	[1.9891, 1.9891]	[1.9608, 1.9608]	[2.0109, 2.0109]	[1.9912, 1.9912]	[2.0077, 2.0077]	[1.9803, 1.9803]
[2.0106, 2.0106]	[1.9885, 1.9885]	[1.9704, 1.9704]	[1.9882, 1.9882]	[1.9689, 1.9689]	[1.9553, 1.9553]	[1.9741, 1.9741]	[1.9825, 1.9825]
[1.9640, 1.9640]	[2.0187, 2.0187]	[1.9616, 1.9616]	[1.9865, 1.9865]	[1.9556, 1.9556]	[1.9817, 1.9817]	[1.9774, 1.9774]	[1.9316, 1.9316]
[1.9841, 1.9841]	[1.9919, 1.9919]	[1.9737, 1.9737]	[1.9958, 1.9958]	[2.0121, 2.0121]	[2.0021, 2.0521]	[1.9665, 1.9665]	[1.9773, 1.9773]
[1.9841, 1.9841]	[1.9570, 1.9875]	[1.9610, 1.9610]	[2.0015, 2.0015]	[1.9750, 1.9750]	[1.9825, 1.9825]	[1.9758, 1.9758]	[1.9682, 1.9682]
[1.9668, 1.9668]	[1.9696, 1.9696]	[2.0334, 2.0334]	[1.9656, 1.9656]	[1.9819, 1.9819]	[2.0116, 2.0116]	[1.9754, 1.9754]	[1.9986, 1.9986]
[2.0114, 2.0114]	[1.9861, 1.9861]	[1.9743, 1.9743]	[1.9594, 1.9594]	[1.9712,1.9914]	[1.9849, 1.9849]	[1.9711, 1.9711]	[1.9486, 1.9486]
[1.9837, 1.9837]	[1.9424, 1.9424]	[1.9744, 1.9744]	[1.9605, 1.9605]	[1.9719, 1.9719]	[1.9656, 1.9656]	[1.9549, 1.9549]	[2.0174, 2.0174]
[1.9779, 1.9779]	[2.0072, 2.0072]	[1.9875, 1.9875]	[1.9781, 1.9781]	[1.9834, 1.9834]	[1.9893, 1.9893]	[1.9276, 1.9276]	[1.9513, 1.9513]
[1.9971, 1.9971]	[1.9963, 1.9963]	[1.9375, 1.9375]	[1.9941, 1.9941]	[1.9763, 1.9763]	[2.0108, 2.0108]	[1.9687, 1.9687]	[1.9559, 1.9559]
[1.9611, 1.9611]	[1.9729, 1.9729]	[1.9992, 1.9992]	[1.9925, 1.9925]	[2.0073, 2.0073]	[1.9742, 1.9742]	[1.9557, 1.9557]	[1.9726, 1.9726]
[1.9964, 1.9964]	[1.9614, 1.9614]	[1.9768, 1.9768]	[1.9991, 1.9991]	[1.9832, 1.9832]	[1.9847, 1.9847]	[1.9849, 1.9849]	[1.9918, 1.9918]
[1.9748, 1.9748]	[1.9664, 1.9664]	[2.0035, 2.0245]	[1.9822, 1.9822]	[1.9882,1.9999]	[1.9809, 1.9809]	[1.9920, 1.9920]	[1.9994,2.0512]

5. Concluding Remarks

In this paper, we originally proposed a variable sampling plan for the PCI under the neutrosophic logic. We presented the NPCI in the paper and used it to design the sampling plan. The proposed plan is the extension of the plan using classical statistics which can be applied where data is indeterminate or unclear. The plan parameters are presented for practical use in industry. A real example from industry is also added to show the application of the proposed sampling plan. The proposed plan is designed under the assumption that the data follow the neutrosophic normal distribution which can be tested using some statistical test or graphical depictions. For non-normal data, a suitable transformation can be applied to transfer non-normal data to normal data. From the comparison study, it is concluded that the proposed plan is more efficient than the plan based on classical statistics in terms of sample size. It is recommended to use the proposed plan in the industry where the data came from the complex situation or where there is a chance of some unclear data in the sampling. The proposed sampling plan using a double sampling scheme will be considered as a future research. The proposed plan using big data can be considered as future research.

Author Contributions: Conceived and designed the experiments: M.A. (Muhammad Aslam), M.A. (Mohammed Albassam); performed the experiments: M.A. (Muhammad Aslam); analyzed the data: M.A. (Muhammad Aslam); contributed reagents/materials/analysis tools: M.A. (Muhammad Aslam); wrote the paper: M.A. (Muhammad Aslam).

Funding: This article was funded by the Deanship of Scientific Research (DSR) at King Abdulaziz University, Jeddah. The authors, therefore, acknowledge with thanks DSR technical and financial support.

Acknowledgments: The authors are deeply thankful to the editor and reviewers for their valuable suggestions to improve the quality of this manuscript.

Conflicts of Interest: The authors declare no conflict of interest regarding this paper.

References

1. Montgomery, D.C. *Introduction to statistical quality control*; John Wiley & Sons: Jefferson, MO, USA, 2007.
2. Jun, C.-H.; Balamurali, S.; Lee, S.-H. Variables sampling plans for Weibull distributed lifetimes under sudden death testing. *IEEE Trans. Reliab.* **2006**, *55*, 53–58. [CrossRef]
3. Balamurali, S.; Jun, C.H. A new system of skip-lot sampling plans having a provision for reducing normal inspection. *Appl. Stoch. Model. Bus. Ind.* **2011**, *27*, 348–363. [CrossRef]
4. Fallah Nezhad, M.S.; Akhavan Niaki, S.T.; Abooie, M.H. A new acceptance sampling plan based on cumulative sums of conforming run-lengths. *J. Ind. Syst. Eng.* **2011**, *4*, 256–264.
5. Pepelyshev, A.; Steland, A.; Avellan-Hampe, A. Acceptance sampling plans for photovoltaic modules with two-sided specification limits. *Prog. Photovolt.: Res. Appl.* **2014**, *22*, 603–611. [CrossRef]
6. Gui, W.; Aslam, M. Acceptance sampling plans based on truncated life tests for weighted exponential distribution. *Commun. Stat. Simul. Comput.* **2017**, *46*, 2138–2151. [CrossRef]
7. Balamurali, S.; Aslam, M.; Liaquat, A. Determination of a new mixed variable lot-size multiple dependent state sampling plan based on the process capability index. *Commun. Stat. Theory Methods* **2018**, *47*, 615–627. [CrossRef]
8. Kane, V.E. Process capability indices. *J. Qual. Technol.* **1986**, *18*, 41–52. [CrossRef]
9. Boyles, R.A. The Taguchi capability index. *J. Qual. Technol.* **1991**, *23*, 17–26. [CrossRef]
10. Kotz, S.; Johnson, N.L. Process capability indices—a review, 1992–2000. *J. Qual. Technol.* **2002**, *34*, 2–19. [CrossRef]
11. Wu, C.-W.; Pearn, W.; Kotz, S. An overview of theory and practice on process capability indices for quality assurance. *Int. J. Prod. Econ.* **2009**, *117*, 338–359. [CrossRef]
12. Pearn, W.L.; Wu, C.-W. An effective decision making method for product acceptance. *Omega* **2007**, *35*, 12–21. [CrossRef]
13. Boyles, R.A. Brocess capability with asymmetric tolerances. *Commun. Stat. Simul. Comput.* **1994**, *23*, 615–635. [CrossRef]

14. Ebadi, M.; Shahriari, H. A process capability index for simple linear profile. *Int. J. Adv. Manuf. Technol.* **2013**, *64*, 857–865. [CrossRef]
15. Chen, K.; Huang, M.; Li, R. Process capability analysis for an entire product. *Int. J. Prod. Res.* **2001**, *39*, 4077–4087. [CrossRef]
16. Aslam, M.; Azam, M.; Jun, C.H. Various repetitive sampling plans using process capability index of multiple quality characteristics. *Appl. Stoch. Models Bus. Ind.* **2015**, *31*, 823–835. [CrossRef]
17. Seifi, S.; Nezhad, M.S.F. Variable sampling plan for resubmitted lots based on process capability index and Bayesian approach. *Int. J. Adv. Manuf. Technol.* **2017**, *88*, 2547–2555. [CrossRef]
18. Arif, O.H.; Aslam, M.; Jun, C.-H. Acceptance sampling plan for multiple manufacturing lines using EWMA process capability index. *J. Adv. Mech. Des. Syst. Manuf.* **2017**, *11*, JAMDSM0004. [CrossRef]
19. Sadeghpour Gildeh, B.; Baloui Jamkhaneh, E.; Yari, G. Acceptance single sampling plan with fuzzy parameter. *Iran. J. Fuzzy Syst.* **2011**, *8*, 47–55.
20. Kanagawa, A.; Ohta, H. A design for single sampling attribute plan based on fuzzy sets theory. *Fuzzy Sets Syst.* **1990**, *37*, 173–181. [CrossRef]
21. Kahraman, C.; Bekar, E.T.; Senvar, O. A Fuzzy Design of Single and Double Acceptance Sampling Plans. In *Intelligent Decision Making in Quality Management*; Kahraman, C., Yanik, S., Eds.; Springer: Berlin, Germany, 2016; Volume 97, pp. 179–211.
22. Senvar, O.; Tozan, H. Process capability and six sigma methodology including fuzzy and lean approaches. In *Products and Services*; Fuerstner, I., Ed.; IntechOpen: London, UK, 2010.
23. Senvar, O.; Kahraman, C. Fuzzy Process Capability Indices Using Clements' Method for Non-Normal Processes. *J. Mult. Valued Logic Soft Comput.* **2014**, *22*, 95–121.
24. Senvar, O.; Kahraman, C. Type-2 fuzzy process capability indices for non-normal processes. *J. Intell. Fuzzy Syst.* **2014**, *27*, 769–781.
25. Smarandache, F. Neutrosophic Logic-A Generalization of the Intuitionistic Fuzzy Logic. *Multispace Multistructure. Neutrosophic Transdiscipl.* (100 Collected Papers of Science). **2010**, *4*, 396. [CrossRef]
26. Smarandache, F. Introduction to neutrosophic statistics 2014: Infinite Study. Available online: https://arxiv.org/pdf/1406.2000 (accessed on 17 June 2019).
27. Chen, J.; Ye, J.; Du, S.; Yong, R. Expressions of rock joint roughness coefficient using neutrosophic interval statistical numbers. *Symmetry* **2017**, *9*, 123. [CrossRef]
28. Chen, J.; Ye, J.; Du, S. Scale effect and anisotropy analyzed for neutrosophic numbers of rock joint roughness coefficient based on neutrosophic statistics. *Symmetry* **2017**, *9*, 208. [CrossRef]
29. Viertl, R. On reliability estimation based on fuzzy lifetime data. *J. Stat. Plan. Inference* **2009**, *139*, 1750–1755. [CrossRef]
30. Aslam, M. A New Sampling Plan Using Neutrosophic Process Loss Consideration. *Symmetry* **2018**, *10*, 132. [CrossRef]
31. Aslam, M.; Raza, M.A. (MUHAMMAD ASLAM) Design of new sampling plans for multiple manufacturing lines under uncertainty. *Int. J. Fuzzy Syst.* **2019**, *21*, 978–992. [CrossRef]
32. Aslam, M. Design of sampling plan for exponential distribution under neutrosophic statistical interval method. *IEEE Access* **2018**, *6*, 64153–64158. [CrossRef]
33. Aslam, M.; Arif, O. Testing of Grouped Product for the Weibull Distribution Using Neutrosophic Statistics. *Symmetry* **2018**, *10*, 403. [CrossRef]
34. Aslam, M.; AL-Marshadi, A. Design of Sampling Plan Using Regression Estimator under Indeterminacy. *Symmetry* **2018**, *10*, 754. [CrossRef]
35. Aslam, M. A new attribute sampling plan using neutrosophic statistical interval method. *Complex Intell. Syst.* **2019**, 1–6. [CrossRef]
36. Aslam, M. Product Acceptance Determination with Measurement Error Using the Neutrosophic Statistics. *Adv. Fuzzy Syst.* **2019**. [CrossRef]
37. Aslam, M. A New Failure-Censored Reliability Test Using Neutrosophic Statistical Interval Method. *Int. J. Fuzzy Syst.* **2019**, *21*, 1214–1220. [CrossRef]
38. Duncan, A.J. *Quality Control and Industrial Statistics*, 5th ed.; Irwin: Homewood, IL, USA, 1986.

39. Aslam, M.; Wu, C.W.; Jun, C.H.; Azam, M.; Itay, N. Developing a variables repetitive group sampling plan based on process capability index C pk with unknown mean and variance. *J. Stat. Comput. Simul.* **2013**, *83*, 1507–1517. [CrossRef]
40. Yen, C.-H.; Chang, C.-H. Designing variables sampling plans with process loss consideration. *Commun. Stat. Simul. Comput.* **2009**, *38*, 1579–1591. [CrossRef]

mathematics

MDPI

Article

Measures of Probabilistic Neutrosophic Hesitant Fuzzy Sets and the Application in Reducing Unnecessary Evaluation Processes

Songtao Shao [1,2,†,‡] **and Xiaohong Zhang** [1,3,*,‡]

1 Department of Mathematics, School of Arts and Sciences, Shaanxi University of Science & Technology, Xi'an 710021, China
2 College of Information Engineering, Shanghai Maritime University, Shanghai 201306, China
3 Department of Mathematics, College of Arts and Sciences, Shanghai Maritime University, Shanghai 201306, China
* Correspondence: zhangxiaohong@sust.edu.cn
† Current address: Xi'an Weiyang University Park, Xi'an 710016, China.
‡ These authors contributed equally to this work.

Received: 20 June 2019; Accepted: 16 July 2019; Published: 19 July 2019

Abstract: Distance measure and similarity measure have been applied to various multi-criteria decision-making environments, like talent selections, fault diagnoses and so on. Some improved distance and similarity measures have been proposed by some researchers. However, hesitancy is reflected in all aspects of life, thus the hesitant information needs to be considered in measures. Then, it can effectively avoid the loss of fuzzy information. However, regarding fuzzy information, it only reflects the subjective factor. Obviously, this is a shortcoming that will result in an inaccurate decision conclusion. Thus, based on the definition of a probabilistic neutrosophic hesitant fuzzy set (PNHFS), as an extended theory of fuzzy set, the basic definition of distance, similarity and entropy measures of PNHFS are established. Next, the interconnection among the distance, similarity and entropy measures are studied. Simultaneously, a novel measure model is established based on the PNHFSs. In addition, the new measure model is compared by some existed measures. Finally, we display their applicability concerning the investment problems, which can be utilized to avoid redundant evaluation processes.

Keywords: probabilistic neutrosophic hesitant fuzzy set; distance measure; similarity measure; entropy measure; multi-criteria decision-making (MCDM)

1. Introduction

Neutrosophic set (NS) [1,2] as a more general theory form of fuzzy sets (FS) [3] provides a simple method to describe uncertain information under the MCDM environment. Afterwards, in order to better combine with practical problems, Wang et al. proposed the single-valued neutrosophic set (SVNS) [4–6] and interval neutrosophic set (INS) [7–9] by depicting the range of different membership functions to encourage the application of FS. For instance, NS adds three independent membership functions: truth-membership function $T(x)$, indeterminacy-membership function $I(x)$ and falsity-membership function $F(x)$. In development, according to the complexity of the information in the MCDM problems, SVNS and INS have been applied to deal with some different types of problems [10–16]. When some decision makers (DMs) make a decision, some DMs may at the hesitancy among truth membership, indeterminacy membership and falsity membership. Thus, different forms of NS have been proposed, like single-valued neutrosophic hesitant FS (SVNHFS) [17–19], multi-valued NS (MVNS) [20–23], some types of linguistic NS [24–26], and other types of NS [27–32]. Some experts applied to algebraic systems [33–40], which clarified that the extended NSs are the effective tools for describing uncertainty and imprecise information, the information including imperfect, fuzzy, uncertainty and so on. Then, based on the different requirements of practical applications, the axioms of NS are investigated. The most important thing is how to minimize the loss of information when uncertain problems are resolved.

The use of truth-membership, indeterminacy-membership and falsity-membership degrees to depict the fuzziness only expresses subjective uncertainty. However, the statistical data can describe the occurrence frequency of membership degree based on objective views. The elements that decide on the accurate evaluation conclusion of MCDM include both fuzzy and statistic information. The DMs can explain the subjective information by utilizing NSs, SVNSs, SVNHSs and so on. As the amount of information increases, the impact of statistical information on decision outcomes will increase.

Xu et al. proposed hesitant probabilistic fuzzy set [41] and researched its basic operations. Next, Hao et al. [42] constructed probabilistic dual hesitant fuzzy set and applied in risk evaluation. Zhai et al. [43] took the probabilistic interval-valued intuitionistic hesitant fuzzy set and investigated its distance, similarity and entropy measures. Later, these theories have been widely studied and applied to solve MCDM problems [44–47]. However, when solving some decision problems, the decision makers will give the indeterminacy-membership hesitant degrees and corresponding probability information. In order to solve this situation, Shao et al. [48] and Peng et al. [49] established probabilistic single-valued neutrosophic hesitant fuzzy set (PSVNHFS or PNHFS) and probability multi-valued neutrosophic set (PMVN), respectively. Shao et al. investigated the basic operation laws of PNHFSs and their characteristics. Next, they established the probabilistic neutrosophic hesitant fuzzy weighted averaging (geometric) operators to fuse the uncertain information. Peng et al. presented a new QUALIFLEX method to fuse and analyze the uncertain information. The new form of expression is conductive to reducing the loss of uncertain information and improving the application in MCDM environments.

Distance measure, similarity measure and entropy measure are three effective ways to solve MCDM problems. As the key step of implicating fuzzy information explanation into MCDM, different types of distance and similarity measure for NSs [50,51], SVNSs [52,53], and SVNHFS [54,55] have been investigated. On the other hand, some ranking methods and MCDM approaches based on the measures of linguistic NSs have been established and utilized in various practical problems [56,57]. The effectiveness of similarity measure is to express the degree of similarity between factors. Additionally, the distance measure focuses on the divergence of items, which is opposite to the similarity measure. Simultaneously, similarity measure is an effective tool to express the relationship between items. Distance measure also has this characteristic.

The present notions of measures include the three independent membership degrees (truth, indeterminacy, falsity membership degrees) of fuzzy information, which can be effective to reduce the loss of information. Researchers pay attention to study the measures to improve the exactness and effectiveness in MCDM problems. According to the inner construction of present measure formulae, we establish a novel distance measure and a novel similarity. Sahin [58] proposed the Hamming distance measure of SVNHFSs as follows:

$$D_{SVNHFS} = \frac{1}{3}\sum_{x\in X}\left(\frac{1}{l}\sum_{i=1}^{l}|\alpha_{N_1}^{\mu(i)}(x) - \alpha_{N_2}^{\mu(i)}(x)| + \frac{1}{p}\sum_{i=1}^{p}|\beta_{N_1}^{\mu(i)}(x) - \beta_{N_2}^{\mu(i)}(x)| + \frac{1}{q}\sum_{i=1}^{l}|\gamma_{N_1}^{\mu(i)}(x) - \gamma_{N_2}^{\mu(i)}(x)|\right),$$

in which α, β and γ are the truth-membership, indeterminacy-membership and falsity-membership degrees of $x_i \in X$ to a situation, N_1 and N_2 are SVNHFSs. However, there are some drawbacks, for which it is necessary to be concerned. For instance, the truth-membership and falsity-membership degrees are utilized to describe DMs' determination on x to the situation A. According to DMs, there is some associated information about x to the situation A, and α and γ are given at the same time when DMs make judgements. However, β expresses the vagueness of DMs' un-known about x, and this is distinct to the α and β. Obviously, it is not logical for any DCDM problems when DMs characterize them by utilizing the same formula and equal potentiality in a measure function.

Due to the complexity of the uncertain information, the evaluation information given by the decision makers will be fused. For example, T_A, I_A and F_A describe the proportion of pros, cons, and abstentions, respectively, in the voting model. In the case of some subjective factors, the decision maker cannot be sure that it is fully or completely opposed, so some of the abstentions tend to vote in favor, expressed by TI. Similarly, IF describes the fusion information between abstention and opposition. TF describes the fusion information between approval and opposition. T_1, F_1 and I_1 represent information that is fully in favor, totally opposed, and completely abstained. Then, this type of information can be solved with neutrosophic hesitation fuzzy theory, $T_A = T_1 + TF + TI$, $F_A = F_1 + TF + IF$ and $I_A = I_1 + TI + IF$.

The whole uncertainty set is separated into vagueness, non-vagueness and hesitancy. The non-vagueness sub-domain includes truth-membership and falsity-membership regions, whereas the vagueness sub-domain is organized by the indeterminacy–membership region. The uncertainty in the non-vagueness sub-domain can be expressed as an undetermined attribute. The indeterminacy indicates that there are a variety of thoughts about x belonging to the situation A. Every thought can not be certain. Hesitancy sub-domain describes the hesitancy degrees of DMs. Thus, it is appropriate to explore and solve the uncertain information based on the vagueness, non-vagueness and hesitant degrees. The distinction among the novel measures and previous measures is distinguished.

According to the instructions above, our main aim is to accomplish the fuzzy description system based on the PNHFS. By holding more uncertainty parameters, the uncertain information is expressed. At the same time, the uncertainty information is divided more clearly. The particular introduction is related in Section 2. The second aim is to propose novel distance, similarity and entropy measures. This work is done in Section 3 exactly. We expect to take advantages of this new approach to improve the accuracy of practical MCDM results. In Section 4, the detail is described and an application case about reducing the excess re-evaluation is shown, respectively. Finally, the discussion and future research are presented followed by the Conclusions section.

2. Preliminaries

Firstly, the basic theoretical knowledge used in this paper is reviewed. For convenience, SVNHFS is simply called the neutrosophic hesitant fuzzy set (NHFS) in this work.

2.1. Several Types of NS

Definition 1. *Suppose X is a non-empty reference set. An NHFS is described by the following mathematical formula [4]:*

$$N = \{\langle x, \tilde{t}(x), \tilde{i}(x), \tilde{f}(x)\rangle | x \in X\},$$

where $\tilde{t}(x)$, $\tilde{i}(x)$ *and* $\tilde{f}(x) \in [0,1]$. $\tilde{t}$, $\tilde{i}$ *and* $\tilde{f}$ *denote three different types of degrees, respectively.* $\tilde{t} : X \to [0,1]$ *describes the truth-membership degree,* $\tilde{i} : X \to [0,1]$ *denotes the indeterminacy-membership degree,* $\tilde{f} : X \to [0,1]$ *depicts the falsity-membership degree.* $\tilde{t}(x)$, $\tilde{i}(x)$ *and* $\tilde{f}(x)$ *satisfy the following condition:* $0 \leq \tilde{t}(x) + \tilde{i}(x) + \tilde{f}(x) \leq 3$.

Definition 2. *Suppose that X is a non-empty reference set; then, an NHFS involved with X on the basis of three functions to X return three subsets of [0, 1]. Ye proposed an NHFS with the following mathematical sign [18]:*

$$N = \{\langle x, T(x), I(x), F(x)\rangle | x \in X\},$$

where $T(x)$, $I(x)$ *and* $F(x)$ *are three subsets of* $[0,1]$, *respectively. Moreover, the definition of single-valued neutrosophic hesitant fuzzy element (SVNHFE) is proposed. If* $T(x)$, $I(x)$ *and* $F(x)$ *are three finite subsets, then the SVNHFE can be expressed by*

$$\langle(\alpha_1(x), \alpha_2(x), \cdots \alpha_{L(T)}(x)), (\beta_1(x), \beta_2(x), \cdots, \beta_{L(I)}(x)), (\gamma_1(x), \gamma_2(x), \cdots, \gamma_{L(F)}(x))\rangle$$
$$= \langle T(x), I(x), F(x)\rangle,$$

in which $L(T), L(I)$ and $L(F)$ are three positive integers to describe the corresponding number of values in the $T(x), I(x)$ and $F(x)$. Simultaneously, α_a ($a \in \{1, 2, \cdots, L(T)\}$) describes the ath possible truth-membership degree, β_b ($b \in \{1, 2, \cdots, L(I)\}$) describes the bth possible indeterminacy-membership degree, and γ_c ($c \in \{1, 2, \cdots, L(F)\}$) describes the cth possible falsity-membership degree of $x \in X$ to a situation. The restrictions of SVNHFS are listed below:

$$0 \leq \alpha_a, \beta_b, \gamma_c \leq 1 \text{ and } 0 \leq \alpha^+ + \beta^+ + \gamma^+ + \leq 3, \alpha^+ = max\{\alpha_a\}, \beta^+ = max\{\beta_b\}, \gamma^+ = max\{\gamma_c\} \text{ for } x \in X.$$

After that, single-valued neutrosophic hesitant fuzzy measures and correlation coefficients, aggregation operators on SVNHFS have been investigated to solve MCDM problems, medical diagnoses and so on.

2.2. The Distance and Similarity Measures for SVNHFSs

Definition 3. *A mapping $D : NHFS(X) \times NHFS(X) \to [0, 1]$, "$\times$" is the Cartesian product. Then, D is defined to be a distance measure of NHFS, if it satisfies the following four conditions [58]: $A, B, C \in SVNHFS(X)$,*

(1) $0 \leq D(A, B) \leq 1$;
(2) $D(A, B) = 0$ *iff* $A = B$;
(3) $D(A, B) = D(B, A)$;
(4) *If* $A \subseteq B \subseteq C$, *then* $D(A, C) \geq D(A, B), D(A, C) \geq D(B, C)$.

Definition 4. *A mapping $S : NHFS(X) \times NHFS(X) \to [0, 1]$, "$\times$" is the Cartesian product. Then, S is defined a similarity measure, if S has the following four axioms [58]: $A, B, C \in NHFS(X)$,*

(1) $0 \leq S(A, B) \leq 1$;
(2) $S(A, B) = 1$ *iff* $A = B$;
(3) $S(A, B) = S(B, A)$;
(4) *If* $A \subseteq B \subseteq C$, *then* $S(A, C) \leq S(A, B), S(A, C) \leq S(B, C)$.

Definition 5. *A mapping $E : NS(X) \to [0, 1]$ is called an entropy on $NS(X)$, "$\times$" is the Cartesian product. Then, E holds the following properties [51]: $A, B \in NS(X)$,*

(1) $E(A) = 0$ *if A is a crisp set;*
(2) $E(A) = 1$ *iff* $A = \{0.5, 0.5, 0.5\}$;
(3) $E(A) \leq E(B)$ *if A is more crisper than B;*
(4) $E(A) \leq E(A^c)$, *where A^c is the complement of A.*

3. The Distance and Similarity Measures of PSVNHFS

For the content of this part, as an extended theory of FS, Shao et al. [48] first proposed the probabilistic single-valued neutrosophic hesitant fuzzy set (PSVNHFS). The PSVNHFS can better describe the uncertainty by involving objectively uncertain information and subjective uncertain information. However, the vote set was first introduced by Zhai et al. [43]. Thus, according to the division of certain opinion, indeterminacy opinion and contradictory (vagueness) opinion, inference set as a new kind of vote set is constructed and applied to the NHFS. Finally, the distance measure and similarity measure are introduced and investigated.

Definition 6. *Suppose that X is a finite reference set. A PNHFS on X is denoted by the following mathematical symbol [48]:*

$$N = \{\langle x, T(x)|P^T(x), I(x)|P^I(x), F(x)|P^F(x)\rangle | x \in X\}. \tag{1}$$

The $T(x)|P^T(x)$, $I(x)|P^I(x)$ *and* $F(x)|P^F(x)$ *are three elements of N, in which* $T(x)$, $I(x)$ *and* $F(x)$ *is defined as the possible truth-membership hesitant function, possible indeterminacy-membership hesitant function and possible falsity-membership hesitant function of x, respectively.* $P^T(x)$, $P^I(x)$ *and* $P^F(x)$ *is the probabilistic information of factors in the components* $T(x)$, $I(x)$ *and* $F(x)$, *respectively. This subjective information and objective information have the following requirements:*

$$\alpha_a, \beta_b, \gamma_c \in [0,1], 0 \le \alpha^+ + \beta^+ + \gamma^+ \le 3; P_a^T, P_b^I, P_c^F \in [0,1]; \sum_{a=1}^{L(T)} P_a^T \le 1, \sum_{b=1}^{L(I)} P_b^I \le 1, \sum_{c=1}^{L(F)} P_c^F \le 1,$$

where $\alpha_a \in T(x)$, $\beta_b \in I(x)$, $\gamma_c \in F(x)$. $\alpha^+ = max\{\alpha_a\}$, $\beta^+ = max\{\beta_b\}$, $\gamma^+ = max\{\gamma_c\}$, $P_a^T \in P^T$, $P_b^I \in P^I$, $P_c^F \in P^F$. *The symbols* $L(T)$, $L(I)$ *and* $L(F)$ *are the cardinal numbers of elements in the components* $T(x)|P^T(x)$, $I(x)|P^I(x)$ *and* $F(x)|P^F(x)$, *respectively.*

Generally, a probabilistic neutrosophic hesitant fuzzy number (PNHFN) of x is expressed by the mathematical symbol:

$$N = \langle (\alpha_1|P_1^T, \alpha_2|P_2^T, \cdots \alpha_{L(T)}|P_{L(T)}^T), (\beta_1|P_1^I, \beta_2|P_2^I, \cdots, \beta_{L(I)}|P_{L(I)}^I), (\gamma_1|P_1^F, \gamma_2|P_1^F, \cdots, \gamma_{L(F)}|P_{L(F)}^F)\rangle$$
$$= \{T|P^T, I|P^I, F|P^F\}.$$

Definition 7. *If X is a finite reference set and N is a PNHFN, then* $\tilde{N}$ *is a normalized PNHFN [49]:*

$$\tilde{N} = \{T(x)|\tilde{P}^T(x), I(x)|\tilde{P}^I(x), F(x)|\tilde{P}^F(x)\}, \tag{2}$$

where $\tilde{P}_a^T = \frac{P_a^T}{\sum P_a^T}$, $\tilde{P}_b^I = \frac{P_b^I}{\sum P_b^I}$, $\tilde{P}_c^F = \frac{P_c^F}{\sum P_c^F}$.

Example 1. *If* $X = \{x\}$ *is a reference set, an PNHFS can be denoted by*

$$N = \{x, \langle\{0.5|0.3, 0.6|0.5\}, \{0.4|0.4, 0.6|0.6\}, \{0.3|0.6\}\rangle\}.$$

For every membership function, the PNHFN $\tilde{N} = \langle\{0.5|0.3, 0.6|0.5\}, \{0.4|0.4, 0.6|0.6\}, \{0.3|0.6\}\rangle$ *independently denotes the whole uncertain area with three probabilistic membership functions, where* $\sum_{a=1}^{L(T)} P_a^T = 0.3 + 0.5 = 0.8$, $\sum_{b=1}^{L(I)} P_b^I = 0.4 + 0.6 = 1, \sum_{c=1}^{L(F)} P_c^F = 0.6$.

The PNHFS is considered a generalized theory of aforementioned various of FS, including FS, IFS, HFS, etc. Next, some special cases of normal PNHFS are introduced.

(1) If the probability values are equal for the same type of hesitant membership function, i.e.,

$$P_1^T = P_1^T = \cdots = P_{L(T)}^T, P_1^I = P_1^I = \cdots = P_{L(I)}^I, P_1^F = P_1^F = \cdots = P_{L(F)}^F.$$

Then, the normal PNHFS is reduced to the SVNHFS.

(2) If $L(T) = L(I) = L(F) = 1$ and $P_1^T = P_1^I = P_1^F = 1$, then the normal PNHFS reduces to the SVNS.

(3) If $I(x) = \varnothing$ (there is also $P^I(x) = \varnothing$), $\alpha^+ + \beta^+ \ge 1$, then the normal PNHFS reduces to the PDHFS, which can be expressed by $N = \{\langle x, T(x)|P^T(x), F(x)|P^F(x)\rangle | x \in X\}$.

(4) If the normal PNHFS satisfies the conditions in (3), and $P_1^T = P_1^T = \cdots = P_{L(T)}^T, P_1^F = P_1^F = \cdots = P_{L(F)}^F$, then the normal PNHFS reduces to the DHFS, denoted by $N = \{\langle x, T(x), F(x)\rangle | x \in X\}$

(5) If $I(x) = F(x) = \varnothing$ (there is also $P^I(x) = P^F(x) = \varnothing$), then the normal PNHFS reduces to the PHFS, the mathematical symbol is $N = \{\langle x, T(x)|P^T(x)\rangle | x \in X\}$.

(6) If the normal PNHFS satisfies the conditions in (5), and $P_1^T = P_1^T = \cdots = P_{L(T)}^T$, the normal PNHFS reduces to the HFS, denoted by $N = \{\langle x, T(x)\rangle | x \in X\}$.

(7) If $I(x) = \emptyset$ (there is also $P^I(x) = \emptyset$), $L(T) = L(F) = 1$, $P_1^T = P_1^F = 1$,$\alpha_1 + \gamma_1 \geq 1$, then the normal NHFS reduces to the IFS, denoted by $N = \{\langle x, \alpha_1, \gamma_1\rangle | x \in X\}$.

(8) If $I(x) = \emptyset$ (there is also $P^I(x) = \emptyset$), $L(T) = L(F) = 1$, $P_1^T = P_1^F = 1$, and $1 - \alpha_1 - \gamma_1 = 0$, then the normal NHFS reduces to the FS.

Definition 8. *Suppose that $X = \{x_1, x_2, \cdots, x_n\}$ is a finite reference set and N is a PNHFN, then the hesitant degree of x_i is defined by the following mathematical symbol:*

$$\chi(x_i) = 1 - \frac{1}{3}\left(\frac{1}{L(T)} + \frac{1}{L(I)} + \frac{1}{L(F)}\right); \tag{3}$$

$$\chi(N) = \frac{1}{n}\sum_{i=1}^{n} \chi(x_i), \tag{4}$$

where $L(T)$, $L(I)$ and $L(F)$ represent the total numbers of factors in the components $T(x)|\tilde{P}^T(x)$, $I(x)|\tilde{P}^I(x)$ and $F(x)|\tilde{P}^F(x)$.

The hesitant degree of x_i reflects the decision maker's degree of hesitation, the bigger $\chi(N)$, the bigger the hesitation of decision maker in making decisions. If $\chi(N) = 0$, then the decision information is completely unhesitating.

By the definition of PNHFS, we know that the information $\{\alpha_1|P_1^T, \alpha_2|P_2^T, \cdots, \alpha_{L(T)}|P_{L(T)}^T\}$ denotes the positive attitude for x to a situation A, Those data express a certain and non-vagueness component. In this case, we can not obtain effective data to denote the specific truth-membership degree. Similarly, the information elucidated by the data $\{\gamma_1|P_1^F, \gamma_2|P_2^F, \cdots, \gamma_{L(F)}|P_{L(F)}^F\}$ is like the introduction of the truth-membership hesitant degrees with probability, which denotes determinate attitude and uncertain settled data. However, the information $\{\beta_1|P_1^I, \beta_2|P_2^I, \cdots, \beta_{L(I)}|P_{L(I)}^I\}$ expresses uncertain attitude and inconclusive membership degree with probability. Thus, through the above analysis, the truth-membership hesitant degrees and false-membership hesitant degrees are considered as the components of non-vagueness subspace. The indeterminacy-membership degrees expresses the uncertain attitude. It denotes the imprecise notion of people's knowledge about x. The rest of the region denotes a contradictory (vague) attitude about whether the x belongs to an event. It represents the unexplored domain of people's knowledge about x. As people acquire more and more knowledge, the fuzzy information represented by contradictory (vague) subspace will be converted to the uncertain knowledge repressed by the information $T(x)|P^T(x)$, $I(x)|P^I(x)$ and $F(x)|P^F(x)$.

Thus, we propose a method to get all uncertain parameters and accurately describe the certain attitude subspace, indeterminate attitude subspace and contradictory (vague) subspace. Considering the certain subspace, the standpoint about the truth-membership hesitant degrees and false-membership hesitant degrees is correct. Thus, we let the truth-membership hesitant degrees have assigned positive values; the value domain is [0, 1] and then the false-membership hesitant degrees are assigned negative values; the value domain is [−1, 0]. Eventually, the value of certain attitude belongs to [−1, 1]. Obviously, by the Definition 6, the value of indeterminate attitude belongs to [0, 1]. Next, through the above analysis, we found that PNHFS is a convenient way to express fuzzy information. However, for decision makers, they prefer to get the optimal result more conveniently. However, the hesitant degree can describe the hesitation of uncertain information. Thus, we fuse the truth-membership hesitant degrees, false-membership hesitant degrees and hesitant degree into an attitude presentation. The uncertain neutrosophic space is relatively macroeconomic expressed by a certain attitude, indeterminate attitude and hesitation. The calculation process can be simplified and made more feasible for solving problems. Based on the above analysis, the definition of inference set (IS) is established as follows:

Definition 9. *Suppose that X is a finite reference set; then, a inference set (IS) is expressed by the following mathematical symbol:*

$$IS = \{\langle x, d(x), e(x), g(x)\rangle | x \in X\}, \tag{5}$$

where $IE = \langle x, d(x), e(x), g(x)\rangle$ *is defined as an inference element (IE),* $(d(x), e(x), and g(x))$ *is called an inference number (IN). The function* $d : X \to [-1, 1]$ *describes the attitude of x belonging to the situation A. It is a compositive product about the truth-membership hesitant degrees and false-membership hesitant degrees. The mapping* $e : X \to [0, 1]$ *expresses the un-vagueness opinion of x belonging to the situation A. In addition, the mapping* $g : X \to [0, 1]$ *figures the contradictory (vague) degree for people's attitudes about x belonging to the situation A. Note, when* $0 < d(x) \leq 1$, *the decision makers remain optimistic about x belonging to the situation A; when* $-1 \leq d(x) < 0$, *the decision makers are pessimistic about x belonging to the situation A. If* $d(x) = 0$, *then the decision makers' attitude is neutral.*

Example 2. *The mathematical symbol* $\langle x, 0.4, 0.7, 0.2\rangle$ *is an IE. It describes the decision maker having a 40% degree of agreement about x belonging to the situation A. However, there is a 70% degree of determination about the information on x to the situation A. In addition, there is a 20% degree of non-hesitation on the x belonging to the situation A.*

3.1. The Method of Comparing PNHFSs

In this subsection, a way to convert the PNHFE to the IE is established. Next, the PNHFS can be compared by utilizing IEs. In the entire space, the certain attitude subspace, the indeterminate attitude subspace, the contradictory (vague) attitude subspace and corresponding probabilistic values express the different meanings. The certain attitude subspace represents the degrees of agreement or disagreement about x belonging to the situation A; the indeterminate attitude subspace can be described to the lack of decision makers' information, whereas the contradictory (vague) subspace represents the contradiction of decision makers' knowledge. Additionally, the probability theory expresses uncertainty, which is shared by the certain attitude subspace, the indeterminate sub-space and contradictory (vagueness) subspace. Thus, the probability values are integrated to reduce uncertain variables. Next, in order to establish distance measure and similarity measure, a function from a PNHFS to an IS is given.

Definition 10. *Suppose that X is a finite reference set, N is a finite PNHFE, and a mapping H is defined as follows:*

$$H(N) = \{\sum_{a=1}^{L(T)} t_a P_a^T - \sum_{c=1}^{L(F)} f_c P_c^F, \sum_{b=1}^{L(I)} (1 - i_b) P_b^F, 1 - \chi(x_i)\}. \tag{6}$$

For instance, when $P_1^T = P_2^T = \cdots = P_{L(T)}^T$, $P_1^I = P_2^I = \cdots = P_{L(I)}^I$, $P_1^F = P_2^F = \cdots = P_{L(F)}^F$, the PNHFS is reduced to an NHFS. Thus, the function $H(N)$ can be transformed to an IS as

$$H(N) = \{\frac{\sum_{a=1}^{L(T)} t_a}{L(T)} - \frac{\sum_{c=1}^{L(F)} f_c}{L(F)}, \frac{\sum_{b=1}^{L(I)} (1 - i_b)}{L(I)}, 1 - \chi(x_i)\}.$$

According to Equation (6), the IS includes the probabilistic information and fuzzy information, which can be illustrated with the help of investigating the Definition 10. The formula $\sum_{a=1}^{L(T)} t_a P_a^T - \sum_{c=1}^{L(F)} f_c P_c^F$ introduces the average value of certain attitude obtained by the truth-membership subspace and the false-membership subspace. The expression $\sum_{b=1}^{L(I)} (1 - i_b) P_b^F$ explains the average degree of an un-hesitant opinion given by the indeterminate-membership subspace. Then, the formula $1 - \chi(x_i)$ illustrates the average value of the un-sloppy attitude for known information about x related to the situation A.

By Definition 6, all objective and subjective uncertain elements are considered and different types of fuzzy spaces are distinguished. However, if PNHFE is infinite, the formula 6 will change

$$H(N) = \{\int_{a=1}^{L(T)} t_a P_a^T - \int_{c=1}^{L(F)} f_c P_c^F, \int_{b=1}^{L(I)} (1 - i_b) P_b^F, 0\}. \tag{7}$$

Based on the importance of objective and subjective information, the method of comparison for IEs is defined as follows:

Definition 11. *Let X be a finite reference set,* $IE_1 = \langle d_1(x), e_1(x), g_1(x) \rangle$ *and* $IE_2 = \langle d_2(x), e_2(x), g_2(x) \rangle$ *be two IEs, then*

(1) *If* $g_1 \leq g_2$*, then* $IE_1 \leq IE_2$*;*
(2) *If* $g_1 \geq g_2$*, then* $IE_1 \geq IE_2$*;*
(3) *If* $g_1 = g_2$*, then (i) If* $e_1 \leq e_2$*, then* $IE_1 \leq IE_2$*; (ii) If* $e_1 \geq e_2$*, then* $IE_1 \geq IE_2$*;*
(4) *If* $g_1 = g_2, e_1 = e_2$*, then (i) If* $d_1 \leq d_2$*, then* $IE_1 \leq IE_2$*; (ii) If* $d_1 \geq d_2$*, then* $IE_1 \geq IE_2$*.*

The division of entire uncertain field to describe the certain, indeterminate and hesitant attitude. By Definition 9, based on the internal perspective and external perspective, the IE expresses the certain subdomain without probabilistic information. Thus, according to the degree of information obtained and the importance of experience in decision-making activities, the method of comparison for IEs is based on the rule "degree of non-hesitation, determinacy and lastly opinion".

Supposing that A and B are two PNHFEs to the finite reference set X, then the corresponding IEs can be expressed by $IE_A = \langle d_A(x), e_A(x), g_A(x) \rangle$ and $IE_B = \langle d_B(x), e_B(x), g_B(x) \rangle$, respectively. Thus, the notion of binary relation for PNHFEs can be described as follows:

Definition 12. *Suppose that A and B are two PNHFEs to the finite reference set X. Then, the binary relations for PNHFEs are given as follows:*

(1) *If* $\langle d_A(x), e_A(x), g_A(x) \rangle \geq \langle d_B(x), e_B(x), g_B(x) \rangle$*, then* $A \geq B$*;*
(2) *If* $\langle d_A(x), e_A(x), g_A(x) \rangle \leq \langle d_B(x), e_B(x), g_B(x) \rangle$*, then* $A \leq B$*;*
(3) *If* $\langle d_A(x), e_A(x), g_A(x) \rangle = \langle d_B(x), e_B(x), g_B(x) \rangle$*, then* $A = B$*.*

3.2. Distance and Similarity Measures of PNHFSs

According to the work mentioned above, the distance measure, similarity measure and entropy measure of PNHFE are established in this subsection. The inclusion between IS_A and IS_B is given. Similarity, the inclusion between $PNHFS_A$ and $PNHFS_B$ are proposed.

Suppose that X is a finite reference set, A and B are PNHFS to set X, and IS_A and IS_B are corresponding ISs of A and B, respectively.

$$A \subseteq B \text{ iff } \forall x \in X, \overline{T_A|P^{T_A}} \leq \overline{T_B|P^{T_B}}, \overline{I_A|P^{I_A}} \geq \overline{I_B|P^{I_B}}, \overline{F_A|P^{T_A}} \geq \overline{F_B|P^{F_B}} \text{ and } \chi(A) \geq \chi(B),$$

where $\overline{T_A|P^{T_A}}$ and $\overline{T_B|P^{T_B}}$ describe the average value of truth-membership hesitant degree of A and B, respectively, $\overline{I_A|P^{I_A}}$ and $\overline{I_B|P^{I_B}}$ express the average indeterminate-membership hesitant degree of A and B, respectively. Similarly, $\overline{F_A|P^{F_A}}$ and $\overline{F_B|P^{F_B}}$ represent the corresponding average false-membership hesitant degree of A and B.

Additionally, if $IS_A \subseteq IS_B$, the following conditions need to hold:

$$a_A \leq a_B, b_A \leq b_B, c_A \leq c_B.$$

Definition 13. *Suppose that X is a finite reference set, IS_A, IS_B and IS_C are three ISs in X. A function $D_{IS} : IS(X) \times IS(X) \to [0,1]$, where "$\times$" means the Cartesian production. Then, D_{IS} is called a distance measure, if D_{IS} satisfies the following three requirements:*

(1) $D_{IS}(IS_A, IS_B) = 0$ *iff* $IS_A = IS_B$;
(2) $D_{IS}(IS_A, IS_B) = D_{IS}(IS_B, IS_A)$;
(3) $D_{IS}(IS_A, IS_C) \geq D_{IS}(IS_A, IS_B)$, $D_{IS}(IS_A, IS_C) \geq D_{IS}(IS_B, IS_C)$ *when* $IS_A \subseteq IS_B \subseteq IS_C$.

Theorem 1. *Suppose that $IS_A = \{\langle d_A(x), e_A(x), g_A(x)\rangle | x \in X\}$ and $IS_B = \{\langle d_B(x), e_B(x), g_B(x)\rangle | x \in X\}$ are three ISs in X, then the function*

$$D_{IS} = AIO(MIT(MIU_1(|d_A(x) - d_B(x)|), MIU_2(|e_A(x) - e_B(x)|), MIU_3(|g_A(x) - g_B(x)|))) \quad (8)$$

is a distance measure for IS, where the mappings: $MIU_1, MIU_2, MIU_3 : [0,1] \to [0,1]$, satisfy the conditions: MIU_1, MIU_2 and MIU_3 are three monotonically increasing unary functions and $MIU_1(0) = 0$, $MIU_2(0) = 0$, $MIU_3(0) = 0$. Those functions can be the same and are not mandatory here. The mapping $MIT : [0,1]^3 \to [0,1]$ is a monotonically increasing ternary function; MIT holds the following requirements: $MIT(0,0,0) = 0$; $MIT'_1 \geq 0$, $MIT'_2 \geq 0$, and $MIT'_3 \geq 0$, MIT'_1, MIT'_2 and MIT'_1 are corresponding partial derivatives of MIU_1, MIU_2 and MIU_3, respectively. Additionally, $AIO : [0,1]^n \to [0,1]$ is an aggregation operator and the partial derivative $AIO'_i \geq 0$ $(i \in \{1, 2, \cdots, n\})$; n represses the total numbers of factors in X.

Proof. According to the conditions of MIU_1, MIU_2, MIU_3, MIT and AIO, Definition 13 (1) and (2) obviously hold. Thus, the proof process of condition (3) is listed, here. Since the restrictive conditions $IS_A \subseteq IS_B \subseteq IS_C$ hold, thus the inequalities are listed below:

$$|d_A(x) - d_C(x)| \geq |d_A(x) - d_B(x)|, |e_A(x) - e_C(x)| \geq |e_A(x) - e_B(x)|, |g_A(x) - g_C(x)| \geq |g_A(x) - g_B(x)|;$$
$$|d_A(x) - d_C(x)| \geq |d_B(x) - d_C(x)|, |e_A(x) - e_C(x)| \geq |e_B(x) - e_C(x)|, |g_A(x) - g_C(x)| \geq |g_B(x) - g_C(x)|.$$

Because functions MIU_1, MIU_2 and MIU_3 are three monotonically increasing functions, so we can get, $\forall x \in X$

$$MIU_1(|d_A(x) - d_C(x)|) \geq MIU_1(|d_A(x) - d_B(x)|), MIU_2(|e_A(x) - e_C(x)|) \geq MIU_2(|e_A(x) - e_B(x)|),$$
$$MIU_3(|g_A(x) - g_C(x)|) \geq MIU_3(|g_A(x) - g_B(x)|); MIU_1(|d_A(x) - d_C(x)|) \geq MIU_1(|d_B(x) - d_C(x)|),$$
$$MIU_2(|e_A(x) - e_C(x)|) \geq MIU_2(|e_B(x) - e_C(x)|), MIU_3(|g_A(x) - g_C(x)|) \geq MIU_3(|g_B(x) - g_C(x)|).$$

However, the partial derivatives $MIT'_1 \geq 0$, $MIT'_2 \geq 0$, and $MIT'_3 \geq 0$, thus

$$\begin{aligned}
&MIT(MIU_1(|d_A(x) - d_C(x)|), MIU_2(|e_A(x) - e_C(x)|), MIU_3(|g_A(x) - g_C(x)|)) \\
&\geq MIT(MIU_1(|d_A(x) - d_B(x)|), MIU_2(|e_A(x) - e_B(x)|), MIU_3(|g_A(x) - g_B(x)|)); \\
&MIT(MIU_1(|d_A(x) - d_C(x)|), MIU_2(|e_A(x) - e_C(x)|), MIU_3(|g_A(x) - g_C(x)|)) \\
&\geq MIT(MIU_1(|d_A(x) - d_C(x)|), MIU_2(|e_A(x) - e_C(x)|), MIU_3(|g_A(x) - g_C(x)|)).
\end{aligned}$$

According to the characteristic of function AIO, the following results are shown:

$$\begin{aligned}
&AIO(MIT(MIU_1(|d_A(x) - d_C(x)|), MIU_2(|e_A(x) - e_C(x)|), MIU_3(|g_A(x) - g_C(x)|))) \\
&\geq AIO(MIT(MIU_1(|d_A(x) - d_B(x)|), MIU_2(|e_A(x) - e_B(x)|), MIU_3(|g_A(x) - g_B(x)|))); \\
&AIO(MIT(MIU_1(|d_A(x) - d_C(x)|), MIU_2(|e_A(x) - e_C(x)|), MIU_3(|g_A(x) - g_C(x)|))) \\
&\geq AIO(MIT(MIU_1(|d_A(x) - d_C(x)|), MIU_2(|e_A(x) - e_C(x)|), MIU_3(|g_A(x) - g_C(x)|))).
\end{aligned}$$

Namely, $D_{IS}(IS_A, IS_C) = D_{IS}(IS_A, IS_B)$, $D_{IS}(IS_A, IS_C) = D_{IS}(IS_B, IS_C)$. □

Theorem 2. *Suppose that* $IS_A = \{\langle d_A(x), e_A(x), g_A(x)\rangle | x \in X\}$ *and* $IS_B = \{\langle d_B(x), e_B(x), g_B(x)\rangle | x \in X\}$ *are three ISs in X, then the function*

$$D_{IS} = AIO(MDT(MDU_1(|d_A(x) - d_B(x)|), MDU_2(|e_A(x) - e_B(x)|), MDU_3(|g_A(x) - g_B(x)|))) \quad (9)$$

is a distance measure on IS, where the mappings: $MDU_1, MDU_2, MDU_3 : [0,1] \to [0,1]$ *satisfy the conditions:* MIU_1, MIU_2 *and* MIU_3 *are three monotonically decreasing unary functions, respectively.* $MDU_1(1) = 0$, $MDU_2(1) = 0$, $MDU_3(1) = 0$. *Those functions can be the same and are not mandatory here. The mapping* $MDT : [0,1]^3 \to [0,1]$ *is a monotonically decreasing ternary function, MDT holds the following requirements:* $MDT(1,1,1) = 0$; $MDT_1' \leq 0$, $MDT_2' \leq 0$, *and* $MDT_3' \leq 0$, MDT_1', MDT_2' *and* MDT_1' *are corresponding partial derivatives of* MDU_1, MDU_2 *and* MDU_3*, respectively.* $AIO : [0,1]^n \to [0,1]$ *is an aggregation operator and the partial derivative* $AIO_i' \geq 0$ $(i \in \{1, 2, \cdots, n\})$, *n represses the total numbers of factors in X.*

Proof. Since the process of proof is similar to Theorem 1, thus the whole conditions of Definition 13 are held by Theorem 2. □

Definition 14. *Suppose that X is a finite reference set; A, B and C are three PNHFSs on X, a mapping* D_{PNHFS} : $[0,1] \times [0,1]$ *is called a distance measure on* $PNHFS(X)$*, if it holds the following three requirements: "×" is the Cartesian product,*

(1) $D_{PNHFS}(A, B) = 0$ *iff* $A = B$;
(2) $D_{PNHFS}(A, B) = D_{PNHFS}(B, A)$;
(3) *If* $A \subseteq B \subseteq C$, *then* $D_{PNHFS}(A, B) \leq D_{PNHFS}(A, C)$ *and* $D_{PNHFS}(B, C) \leq D_{PNHFS}(A, C)$.

Theorem 3. *Suppose that X is a finite reference set, A, B and C are three PNHFSs in X,* IS_A, IS_B *and* IS_C *are corresponding ISs of A, B and C, respectively. Then, a real-valued mapping:*

$$D_{PNHFS}(A, B) = MIU(D_{IS}(IS_A, IS_B))$$

is a distance measure on PNHFS(X), where $MIU : [0,1] \to [0,1]$ *is a monotonically increasing unary mapping, MIU.*

Proof. According to the conditions of Theorem 3, the mapping D_{PNHFS} holds the requirements of Definition 14 (1), (2). Thus, the requirement (3) merely needs to be proved.

Based on the explanation of $A \subseteq B \subseteq C$, $A, B, C \in PNHFS(X)$, thus, by Definition 10, the corresponding ISs of A, B, C exist in the following inclusion relation:

$$IS_A \subseteq IS_B \subseteq IS_C.$$

Obviously, the following inequalities are obtained:

$$D_{IS}(IS_A, IS_C) \geq D_{IS}(IS_A, IS_B),$$
$$D_{IS}(IS_A, IS_C) \geq D_{IS}(IS_B, IS_C).$$

Since the function MIU is a monotonically increasing unary mapping, so the following inequalities are shown:

$$MIU(D_{IS}(IS_A, IS_B)) \leq MIU(D_{IS}(IS_A, IS_C)), MIU(D_{IS}(IS_B, IS_C)) \leq MIU(D_{IS}(IS_A, IS_C)).$$

This completes the proof process. □

Example 3. *Suppose that X is a finite reference set, A, B are PNHFSs on X,* $IS_A = \{\langle d_A(x), e_A(x), g_A(x)\rangle | x \in X\}$ *and* $IS_B = \{\langle d_B(x), e_B(x), g_B(x)\rangle | x \in X\}$ *are the corresponding ISs for those two PNHFSs. Based on the Theorem 1 and Theorem 3, let* $MIU_1 = y^{\phi}$, $MIU_2 = y^{\mu}$, $MIU_3 = y^{\nu}$, $y \in [0,1], 0 \leq \phi, \mu, \nu \leq 1$. $MIT = log_4(1 + y_1 + y_2 + y_3)$, $y_1, y_2, y_3 \in [0,1]$. *Additionally, suppose* $MIU = y^{\lambda}$, *where* $y \in [0,1], 0 \leq \lambda$. *Then, we have*

$$D_1(A,B) = \frac{1}{2n} \sum_{x \in X} (log_4(1 + (\frac{|d_A(x) - d_B(x)|}{2})^{\phi} + (|e_A(x) - e_B(x)|)^{\mu} + (\frac{|g_A(x) - g_B(x)|}{2})^{\nu}))^{\lambda}. \quad (10)$$

If $\phi = \mu = \nu = \lambda = 1$, *we have*

$$D_1^{\phi=\mu=\nu=\lambda=1}(A,B) = \frac{1}{2n} \sum_{x \in X}(log_4(1 + (\frac{|d_A(x)-d_B(x)|}{2}) + (|e_A(x) - e_B(x)|) + (\frac{|g_A(x)-g_B(x)|}{2}))). \quad (11)$$

If $\phi = \mu = \nu = 2, \lambda = \frac{1}{2}$, *then*

$$D_1^{\phi=\mu=\nu=2,\lambda=\frac{1}{2}}(A,B) = \frac{1}{2n} \sum_{x \in X}((log_4(1 + \frac{|d_A(x)-d_B(x)|}{2}))^2 + (|e_A(x) - e_B(x)|)^2 + \frac{(|g_A(x)-g_B(x)|)^2}{4})^{\frac{1}{2}}. \quad (12)$$

From the formulas of $D_1(A,B)$, $D_1^{\phi=\mu=\nu=\lambda=1}(A,B)$ and $D_1^{\phi=\mu=\nu=2,\lambda=\frac{1}{2}}(A,B)$, we know that the parameters ϕ, μ, ν manage the functions of $|d_A(x) - d_B(x)|$, $|e_A(x) - e_B(x)|$ and $|g_A(x) - g_B(x)|$ to establish the internal framework of $D_1(A,B)$. However, the parameter λ is utilized to regulate the reciprocity among the $|d_A(x) - d_B(x)|$, $|e_A(x) - e_B(x)|$ and $|g_A(x) - g_B(x)|$ in the regulate area. Based on different application environments, the parameters ϕ, μ, ν are decided. Thus, for a MCDM problem, it is a tool applied to measure the distinction in their knowledge background. Thus, it is rational to decide the parameters utilized to manage the internal framework of measures based on respective importance degree. By dispatching different functions to $|d_A(x) - d_B(x)|$, $|e_A(x) - e_B(x)|$ and $|g_A(x) - g_B(x)|$, the value of adjusting the feasibility of $|d_A(x) - d_B(x)|$, $|e_A(x) - e_B(x)|$ and $|g_A(x) - g_B(x)|$ can also be solved.

Example 4. *Suppose that X, A, B,* IS_A *and* IS_B *are as mentioned above in Example 3,* $MIU_1 = ln(1 + y)$, $(y \in [0,1])$; $MIU_1 = y^{\phi}$, $(y \in [0,1]), \phi \geq 0$; $MIU_3 = y^{\mu}$, $(y \in [0,1]), \mu \geq 0$, $MIT = (y_1 \cdot y_2 \cdot y_3)^{\lambda}$, $(y_1, y_2, y_3 \in [0,1], \lambda \geq 0)$. *Additionally,* $MIU = t(ln^y)$, $(y \in [0,1], t \geq 0)$. *Then,*

$$D_2(A,B) = \sum_{x \in X} t(((ln(1 + \frac{|d_A(x) - d_B(x)|}{2}))^{\phi} |e_A(x) - e_B(x)|^{\mu} |\frac{g_A(x) - g_B(x)}{2}|^{\nu})^{\lambda})^{y}.$$

In addition, if $\phi = \mu = \nu = \lambda = t = 1$, *then*

$$D_{2,\lambda=1,y=1}^{\phi=\mu=\nu=1}(A,B) = \sum_{x \in X} ln(1 + \frac{|d_A(x) - d_B(x)|}{2})|e_A(x) - e_B(x)||\frac{g_A(x) - g_B(x)}{2}|.$$

Definition 15. *Suppose that X is a finite reference set,* IS_A, IS_B *and* IS_C *are three ISs on X,* $S_{IS} : IS(X) \times IS(X) \to [0,1]$ *is a real-valued function, where "×" is the Cartesian product. Then,* S_{IS} *is called a similarity measure on* $IS(X)$, *if it holds the following three axiomatic conditions:*

(1) $S_{IS}(IS_A, IS_B) = 1$ *iff* $IS_A = IS_B$;
(2) $S_{IS}(IS_A, IS_B) = S_{IS}(IS_B, IS_A)$;
(3) *If* $IS_A \subseteq IS_B \subseteq IS_C$, *then* $S_{IS}(IS_A, IS_B) \geq S_{IS}(IS_A, IS_C)$, $S_{IS}(IS_B, IS_C) \geq S_{IS}(IS_A, IS_C)$.

Theorem 4. *Suppose that X is a finite reference set,* $IS_A = \{\langle d_A(x), e_A(x), g_A(x)\rangle | x \in X\}$, $IS_B = \{\langle d_B(x), e_B(x), g_B(x)\rangle | x \in X\}$ *are two ISs; then, function* $S_{IS}(IS_A, IS_B)$ *is called a similarity measure, and the mathematical symbol is as follows:*

$$S_{IS}(IS_A, IS_B) = AIO(MDT(MIU_1(\frac{|d_A(x)-d_B(x)|}{2}), MIU_2(|e_A(x)-e_B(x)|), MIU_3(|g_A(x)-g_B(x)|))), \quad (13)$$

where $MIU_1, MIU_2, MIU_3 : [0,1] \to [0,1]$ hold the following conditions: MIU_1, MIU_2 and MIU_3 are three monotonically increasing unary mappings, $MIU_1(0) = MIU_2(0) = MIU_3(0) = 0$. They may be the same functions, and there are no requirements here. $MDT : [0,1]^3 \to [0,1]$ is a monotonically decreasing ternary mapping, MDT'_1, MDT'_2, MDT'_3 are three corresponding partial derivatives of MDT with respect to MIU_1, MIU_2, MIU_3, respectively. Those partial derivatives hold the following requirements: $MDT'_1 \leq 0, MDT'_2 \leq 0, MDT'_3 \leq 0$ and $MDT(0,0,0) = 1$. The mapping $AIO : [0,1]^n \to [0,1]$ is an aggregation operator, the partial derivative describes $AIO'_i \geq 0$ ($i \in \{1, 2, \cdots, n\}$); n describes the total numbers of factors in X.

Proof. The process of proof is similar to Theorem 1, thus it is unimportant here. □

Theorem 5. *Suppose that X is a finite reference set, $IS_A = \{\langle d_A(x), e_A(x), g_A(x)\rangle | x \in X\}$, $IS_B = \{\langle d_B(x), e_B(x), g_B(x)\rangle | x \in X\}$ are two ISs, then function $S_{IS}(IS_A, IS_B)$ is called a similarity measure, and the mathematical symbol is as follows:*

$$S_{IS}(IS_A, IS_B) = AIO(MIT(MDU_1(\frac{|d_A(x)-d_B(x)|}{2}), MDU_2(|e_A(x)-e_B(x)|), MDU_3(|g_A(x)-g_B(x)|))), \quad (14)$$

where $MDU_1, MDU_2, MDU_3 : [0,1] \to [0,1]$ satisfy the following requirements: MDU_1, MDU_2 and MDU_3 are three monotonically decreasing unary mappings, $MDU_1(1) = MDU_2(1) = MDU_3(1) = 0$. They may have equal functions, and there are no requirements here. $MIT : [0,1]^3 \to [0,1]$ is a monotonically increasing ternary mapping, MIT'_1, MIT'_2, MIT'_3 are three corresponding partial derivatives of MIT with respect to MIU_1, MIU_2, MIU_3, respectively. Those partial derivatives hold the following requirements: $MIT'_1 \geq 0$, $MIT'_2 \geq 0$, $MIT'_3 \geq 0$ and $MIT(0,0,0) = 0$. The mapping $AIO : [0,1]^n \to [0,1]$ is an aggregation operator and the partial derivative $AIO'_i \geq 0$ ($i \in \{1, 2, \cdots, n\}$), n describe the total numbers of factors in X.

Proof. The proof process is omitted. □

Definition 16. *Suppose that X is a finite reference set, for any three PNHFSs A, B and C on X, a function $S_{PNHFS} : PNHFS(X) \times PNHFS(X) \to [0,1]$ is called a similarity measure, if it holds the following three axiomatic conditions: "×" is the Cartesian product,*

(1) $S_{PNHFS}(A, B) = 1$ *iff* $A = B$;
(2) $S_{PNHFS}(A, B) = S_{PNHFS}(B, A)$;
(3) *If* $A \subseteq B \subseteq C$, *then* $S_{PNHFS}(A, B) \geq S_{PNHFS}(A, C)$ *and* $S_{PNHFS}(B, C) \geq S_{PNHFS}(A, C)$.

Theorem 6. *Let X be a finite reference set, and A and B be two PNHFSs on X. The IS_A and IS_B are corresponding ISs of A ,B, respectively. Then, the mapping S_{PNHFS} is called a similarity measure on PNHFS(X), and the mathematical symbol is*

$$S_{PNHFS}(A, B) = MIU(S_{IS}(IS_A, IS_B)), \quad (15)$$

where $MIU : [0,1] \to [0,1]$ is an increasing function and $MIU(0) = 0$

Proof. According to the Theorem 14, we know the proof is obvious. Thus, the process of proof is omitted. □

Example 5. *Suppose that X, A, B, IS_A, IS_B are as above mentioned, $MDU_1 = MDU_2 = MDU_3 = t^y - t$, $(0 \leq t, y \leq 1)$; $MIT = (y_1 + y_2 + y_3)^{\phi}$, $0 \leq \phi, y_1, y_2, y_3 \leq 1$. Additionally, suppose $MIU = y^{\lambda}$, $0 \leq y \leq 1, \lambda \geq 0$. The similarity measure is described as follows:*

$$S_1(A,B) = \sum_{x \in X} (t^{\frac{|d_A(x)-d_B(x)|}{2}} + t^{|e_A(x)-e_B(x)|} + t^{|g_A(x)-g_B(x)|} - 3t)^{\lambda}.$$

In addition, suppose $t = \frac{1}{3}, \phi = \lambda = 1$, thus

$$S_{1,t=\frac{1}{3}}^{\phi=\lambda=1}(A,B) = \sum_{x \in X} (\frac{1}{3})^{\frac{|d_A(x)-d_B(x)|}{2}} + (\frac{1}{3})^{|e_A(x)-e_B(x)|} + (\frac{1}{3})^{|g_A(x)-g_B(x)|} - 1.$$

Through Example 5, we know that those parameters and mappings to decide the effects of $|d_A(x) - d_B(x)|$, $|e_A(x) - e_B(x)|$ and $|g_A(x) - g_B(x)|$ to establish the internal framework of similarity measures. Those parameters and mappings' selection methods are similar to the methods of Example 3.

3.3. The Interrelations among Distance, Similarity and Entropy Measures

According to the concept of "duality", the distance and similarity measures among SVNS, IVNS were investigated. However, different knowledge backgrounds of decision makers will lead to different results. According to the interrelation among distance and similarity measures, Wang [23] first proposed the definition of entropy and across entropy of MVNS and applied them to solving MCDM problems.

In the section, the interrelations among distance, similarity and entropy measures of PNHFS are investigated. According to Subsection 3.2, the distance measure shows the difference between factors. Additionally, the similarity measure investigated the uniformity of factors. Because distance measure and similarity measure describe two opposite aspects, the relationship between these two measures is investigated based on the following theorem:

Theorem 7. *Suppose that A and B are two PNHFS on X, the distance measure $D_{PNHFS}(A,B)$ holds the conditions in Definition 14, and then $S_{PNHFS}(A,B) = FN(D_{PNHFS}(A,B))$ is a similarity measure, which holds the axiomatic conditions in Definition 16, in which $FN : [0,1] \to [0,1]$ is a fuzzy negation.*

Proof. By Definition 14 and Definition 16, the process proof is obvious, so it is omitted. □

According to the interpretation of the divisions of the neutrosophic space, to better describe stability of PNHFS, the entropy measure of a PNHFS is designed as follows:

Definition 17. *Suppose that X is a reference set, $A = \{\langle x, \{T|P^T\}, \{I|P^I\}, \{F|P^F\} \rangle | x \in X\}$ is a PNHFS in X. Then, the complement of A is expressed by the following mathematical symbol:*

$$A^c = \{\langle x, \{F|P^F\}, \{I|P^I\}, \{T|P^T\} \rangle | x \in X\}.$$

Obviously, A^c is also a PNHFS.

Definition 18. *Suppose that X is a finite reference set, A and B are two PNHFSs in X, IS_A, IS_B are corresponding ISs of A and B, respectively. Then, a function $E : PNHFS(X) \to [0,1]$ is called to be an entropy measure when it holds the following four requests:*

(1) $E(A) = 0$ *if* $A = \{\langle x, \{1|1\}, \{0|1\}, \{0|1\} \rangle | x \in X\}$ *or* $A = \{\langle x, \{0|1\}, \{0|1\}, \{1|1\} \rangle | x \in X\}$ *or* $A = \{\langle x, \{0|P_1\}, \{0|P_2\}, \{0|P_3\} \rangle | x \in X\}$;
(2) $E(A) = 1$ *if* $A = \{\langle x, \{0.5|1\}, \{0.5|1\}, \{0.5|1\} \rangle | x \in X\}$;

(3) $E(A) = E(A^c)$ *iff* $A = \{\langle x, \{T|P^T\}, \{I|P^I\}, \{F|P^F\}\rangle | x \in X\}$ *holds the requirement that* $\sum_{b=1}^{L(I)} i_b P_b^I = \sum_{c=1}^{L(F)} f_c P_c^F$, *in which* A^c *is the complement of* A.

(4) $E(B) \leq E(C)$ *when* $S_{PNHFS}(A,B) \leq S_{PNHFS}(A,C)$ *or* $D_{PNHFS}(A,B) \geq D_{PNHFS}(A,C)$, *in which* $A = \{\langle x, \{0.5|P_1\}, \{0.5|P_2\}, \{0.5|P_3\}\rangle | x \in X\}$.

Since we only are concerned with the importance of $a(x)$, $b(x)$ and $c(x)$ on the stationarity of IS, the following theorems are introduced:

Theorem 8. *Suppose that X is a finite reference set, A is a PNHFS in X, and the corresponding IS of A is described by* IS_A. *Then, the following formula:*

$$E(A) = MDT(MIU_1(|d_A(x)|), MIU_2(|2e_A(x) - 1|), MIU_3(|g_A(x)|) \quad (16)$$

is an entropy measure, in which $MIU_1, MIU_2, MIU_3 : [0,1] \to [0,1]$ *are three monotonically increasing unary mappings with* $MIU_1' \geq 0$, $MIU_2' \geq 0$, $MIU_3' \geq 0$, *and* $MIU_1(0) = MIU_2(0) = MIU_3(0) = 0$, $MIU_1(1) = MIU_2(1) = MIU_3(1) = 1$. *The function* $MDT : [0,1]^3 \to [0,1]$ *is a monotonically decreasing ternary mapping, and its partial derivatives are lower than zero with the requirements:* $MDT(0,0,0) = 1$, $MDT(1,1,1) = 0$.

Proof. The function $E(A)$ is illustrated to hold all the conditions of Definition 18.

(1) Let $A = \{\langle x, \{1|1\}, \{0|1\}, \{0|1\}\rangle | x \in X\}$, $A = \{\langle x, \{0|1\}, \{0|1\}, \{0|1\}\rangle | x \in X\}$ or $A = \{\langle x, \{0|1\}, \{0|1\}, \{1|1\}\rangle | x \in X\}$, thus the corresponding ISs of A are shown:

$$IS_A = (1,1,1) \text{ or } IS_A = (-1,1,1).$$

Next, the entropy measure of A is calculated as follows:

$$E(A) = MDT(MIU_1(1), MIU_2(1), MIU_3(1)) = MDT(1,1,1) = 0.$$

(2)

$$\begin{aligned}
& E(A) = 1 \\
& \Leftrightarrow MDT(MIU_1(|d_A(x)|), MIU_2(|2e_A(x) - 1|), MIU_3(|g_A(x)|) = 1 \\
& \Leftrightarrow MIU_1(0) = 0, MIU_2(0) = 0, MIU_3(0) = 0 \\
& \Leftrightarrow |d(x)| = 0, |2e(x) - 1| = 0, |g(x)| = 0, \\
& \Leftarrow t_a = f_c = 0.5, i_b = 0.5.\ a, b, c \in \infty.
\end{aligned}$$

(3) Let $A = \{\langle x, T_A|P^T, I_A|P^I, F_A|P^F\rangle | x \in X\}$, then the complementary of A is obtained: $A^c = \{\langle x, F_A|P^F, I_A|P^I, T_A|P^T, \rangle | x \in X\}$. By Definition 9, the following equality is obtained: $IS_A = IS_{A^c}$. Obviously, $E(A) = E(A^c)$.

(4) Suppose that B and C are two PNHFS of X, $A = \{\langle x, \{0.5|P_a^T\}, \{0.5|P_b^I\}, \{0.5|P_c^F\}\rangle | x \in X\}$. Thus, the corresponding IS of A is $IS_A = \{0,0,0\}$. By Theorem 5, the following similarity measures can be obtained:

$$S_{PNHFS}(A,B) = MIU(AIO(MIB(MDU_1(\frac{|d_B(x)|}{2}), MDU_2(|e_B(x)|), MDU_3(|g_B(x)|))));$$
$$S_{PNHFS}(A,C) = MIU(AIO((MIB(MDU_1(\frac{|d_C(x)|}{2}), MDU_2(|e_C(x)|), MDU_3(|g_C(x)|)))).$$

Since $S_{PNHFS}(A,B) \le S_{PNHFS}(A,C)$, every function is monotonous, thus, we have $|d_B(x)| \ge |d_C(x)|$, $|e_B(x)| \ge |e_C(x)|$ and $|g_B(x)| \ge |g_C(x)|$. Finally, based on the requirements of Theorem 8, $E(B) \le E(C)$.

Additionally, $D_{PNHFS}(A,B) = 1 - S_{PNHFS}(A,B)$, $D_{PNHFS}(A,C) = 1 - S_{PNHFS}(A,C)$. Thus, the process of proof based on the distance measure is omitted. □

Theorem 9. *Suppose that X is a finite reference set, A is a PNHFS on X, and IS_A is the corresponding IS about A. Then, Equation (17) is an entropy measure:*

$$E(A) = MIT(MDU_1(|d_A(x)|), MDU_2(|2e_A(x) - 1|), MDU_3(|g_A(x)|)). \tag{17}$$

$E(A)$ satisfies the following limits: $MDU_1, MDU_2, MDU_3 : [0,1] \to [0,1]$ are two monotonically decreasing unary mappings, and $MDU_1(0) = MDU_2(0) = MDU_3(0) = 1$, $MDU_1(1) = MDU_2(1) = MDU_3(1) = 0$. The mapping $MIB : [0,1]^3 \to [0,1]$ is a monotonically increasing binary function, its partial derivatives are better than 0, $MIB(0,0,0) = 0$, $MIB(1,1,1) = 1$.

Based on the Equations (16) and (17), the different entropy measures can be established.

Through the above analysis, we know that entropy measure can be depicted by the unsteadiness of a PNHFS. However, distance measure and similarity measure play a vital role. Vice versa, entropy measure can better help us to comprehend distance measurement and similarity measurement. Next, according to the distance measure and similarity measure, respectively, the entropy measure can be established.

Theorem 10. *Suppose D is a distance measure obtained according to Definition 14, $B = \{\langle x, \{0.5|P_a^T\}, \{0.5|P_b^I\}, \{0.5|P_c^F\}\rangle | x \in X\}$, then $E(A) = MDU(D_{PNHFS}(A,B))$. The $MDU : [0,1] \to [0,1]$ is a decreasing unary function, its partial derivatives are lower than 0, and $MDU(0) = 1$, $MDU(1) = 0$.*

Theorem 11. *Suppose S is a similarity measure obtained according to Definition 14, $B = \{\langle x, \{0.5|P_a^T\}, \{0.5|P_b^I\}, \{0.5|P_c^F\}\rangle | x \in X\}$, then $E(A) = MIU(S_{PNHFS}(A,B))$. The $MIU : [0,1] \to [0,1]$ is a decreasing unary function, its partial derivatives are bigger than 0, and $MDU(0) = 0$, $MDU(1) = 1$.*

The process of proof about Theorem 10 and Theorem 11 is not unfolded here. Similarity, we can also get the following theorems. The proof processes are visualized.

Theorem 12. *Supposing that D_{PNHFS} is the distance measure of PNHFS A, S_{PNHFS} is the similarity measure of PNHFS A, $B = \{\langle x, \{0.5|P_a^T\}, \{0.5|P_b^I\}, \{0.5|P_c^F\}\rangle | x \in X\}$, then $E(A) = MIB(MDU(D_{PNHFS}(A,B)), MIU(S_{PNHFS}(A,B)))$ is a entropy measure. $MIB : [0,1] \to [0,1]$ is an increasing binary function under the conditions that the partial derivatives are bigger than 0, $MIB(0,0) = 0$, $MIB(1,1) = 1$. The mappings $MDU : [0,1] \to [0,1]$ and $MIU : [0,1] \to [0,1]$ are decreasing unary function and increasing function, respectively. In addition, $MDU(0) = 1$, $MDU(1) = 0$, $MIU(0) = 0$, $MDU(1) = 1$.*

Theorem 13. *Supposing that D_{PNHFS} is the distance measure of PNHFS A, S_{PNHFS} is the similarity measure of PNHFS A, $B = \{\langle x, \{0.5|P_a^T\}, \{0.5|P_b^I\}, \{0.5|P_c^F\}\rangle | x \in X\}$, then $E(A) = MDB(MIU(D_{PNHFS}(A,B)), MDU(S_{PNHFS}(A,B)))$ is an entropy measure. $MDB : [0,1] \to [0,1]$ is a decreasing binary function under the conditions that the partial derivatives are lower than 0, $MIB(1,1) = 0$, $MIB(0,0) = 1$. The mappings $MIU : [0,1] \to [0,1]$ and $MDU : [0,1] \to [0,1]$ are increasing unary function and decreasing function, respectively. In addition, $MIU(1) = 1$, $MIU(0) = 0$, $MDU(1) = 0$, $MDU(1) = 1$.*

4. Method Analysis Based on Illustrations and Applications

4.1. Comparative Evaluations

In real life, the investment problem is a common MCDM problem, and many researchers have proposed different types of distance and similarity measures of SVNHFS to settle this problem. In this part, a famous investment selection situation is introduced. The specific evaluation and the precise data of alternatives for investment company to invest the money problem are listed in Table 1. Table 1 displays the decision matrix of four alternatives A_1, A_2, A_3, A_4 and three evaluated criteria C_1, C_2, C_3. The four alternatives are Real Estate, Oil Exploitation, Bank Financial and Western Restaurant, respectively. The three criteria are Market Prospect, Risk Assessment and Earning Cycle, respectively. The idea element $A^* = \langle 1|1, 0|0, 0|0\}$.

Table 1. Probabilistic neutrosophic hesitant fuzzy decision matrix of the investment problem.

	C_1	C_2
A_1	$\{\{0.3\|0.3, 0.4\|0.3, 0.5\|0.3\}, \{0.1\|1\}, \{0.3\|0.5, 0.4\|0.5\}\}$	$\{\{0.5\|0.5, 0.6\|0.5\}, \{0.2\|0.5, 0.3\|0.5\}, \{0.3\|0.5, 0.4\|0.5\}\}$
A_2	$\{\{0.6\|0.5, 0.7\|0.5\}, \{0.1\|0.5, 0.2\|0.5\}, \{0.2\|0.5, 0.3\|0.5\}\}$	$\{\{0.6\|0.5, 0.7\|0.5\}, \{0.1\|1\}, \{0.3\|1\}\}$
A_3	$\{\{0.5\|0.5, 0.6\|0.5\}, \{0.4\|1\}, \{0.2\|0.5, 0.3\|0.5\}\}$	$\{\{0.6\|1\}, \{0.3\|1\}, \{0.4\|1\}\}$
A_4	$\{\{0.7\|0.5, 0.8\|0.5\}, \{0.1\|1\}, \{0.1\|0.5, 0.2\|0.5\}\}$	$\{\{0.6\|0.5, 0.7\|0.5\}, \{0.1\|1\}, \{0.2\|1\}\}$
	C_3	
A_1	$\{\{0.2\|0.5, 0.3\|0.5\}, \{0.1\|0.5, 0.2\|0.5\}, \{0.5\|0.5, 0.6\|0.5\}\}$	
A_2	$\{\{0.6\|0.5, 0.7\|0.5\}, \{0.1\|0.5, 0.2\|0.5\}, \{0.1\|0.5, 0.2\|0.5\}\}$	
A_3	$\{\{0.5\|0.5, 0.6\|0.5\}, \{0.1\|1\}, \{0.3\|1\}\}$	
A_4	$\{\{0.3\|0.5, 0.5\|0.5\}, \{0.2\|1\}, \{0.1\|0.3, 0.2\|0.3, 0.3\|0.3\}\}$	

Note 1. *The data on this investment selection problem in Table 1 is in the form of PNHFNs. The PNHFS is one of the generalized from the NHFS, which we have described by item (1) after Definition 6. Thus, the definition of PNHFS can also utilized to NHFS. For instance,* $\langle\{0.5, 0.6\}, \{0.1\}, \{0.3\}\rangle$ *is an NHFE. We can describe it as* $\langle\{0.5|0.5, 0.6|0.5\}, \{0.1|1\}, \{0.3|1\}\rangle$*, which is an PNHFE.*

Note 2. *The results are listed in Table 2, and the optimal result is according to the minimum value among distance measures.*

Table 2. Results shown by Equation (10) corresponding to different parameters.

Parameter	A_1	A_2	A_3	A_4	Ranking
$D^{\phi=\mu=\nu=\lambda=1}$	0.1269	0.0635	0.0989	0.1053	$A_1 > A_4 > A_3 > A_2$
$D^{\phi=\mu=\nu=2,\lambda=\frac{1}{2}}$	0.1498	0.1063	0.1150	0.1239	$A_1 > A_4 > A_3 > A_2$
$D^{\phi=\mu=1,\nu=2,\lambda=1}$	0.0956	0.0622	0.0561	0.0792	$A_1 > A_4 > A_3 > A_2$
$D^{\phi=\nu=1,\mu=2\lambda=1}$	0.1024	0.691	0.0523	0.0733	$A_1 > A_4 > A_2 > A_3$
$D^{\phi=2,\nu=1=\mu=1\lambda=1}$	0.1871	0.1203	0.1071	0.1449	$A_1 > A_4 > A_2 > A_3$

The optimal selections are shown in Table 3. By comparing the conclusions shown by the present distance measures Xu and Xia's Method, Singh's Method, and Sahin's Method, we found that the selections calculated are the same as our method with $D^{\phi=\mu=\nu=\lambda=1}$, $D^{\phi=\mu=\nu=2,\lambda=\frac{1}{2}}$ and $D^{\phi=\mu=1,\nu=2,\lambda=1}$. However, the conclusions calculated by $D^{\phi=\nu=1,\mu=2\lambda=1}$, $D^{\phi=2,\nu=1=\mu=1\lambda=1}$ are different from the present method.

Table 3. Relationships between presenting methods and our method .

Method	Ranking	The Best Result	The Worst Result
Xu and Xia's Method	$A_1 > A_4 > A_3 > A_2$	A_1	A_2
Singh's Method	$A_1 > A_4 > A_3 > A_2$	A_1	A_2
Sahin's Method	$A_1 > A_4 > A_3 > A_2$	A_1	A_2

Thus, we deduce that the consequences may change if we change the inner frames of the distance measure formula. According to the components of $|d_A(x) - d_B(x)|$, $|e_A(x) - e_B(x)|$ and $|g_A(x) - g_B(x)|$, which describe the certain attitudes, knowledge backgrounds and hesitancy degree, respectively, we trust that the new type of distance measures are effective and significant. If the difference of the decision makers' hesitancy degree and background knowledge is relatively big, it does not have a lot of effective consult values regarding whether they have the same conclusions. However, when the difference between the decision maker's hesitation and background knowledge is not too big, analyzing the reasons for the difference in their opinions is significant. Thus, it is important for making rational decisions.

4.2. Streamlining the Talent Selection Process

In many areas of life, the existing evaluation systems are incomplete, resulting in redundancy in the evaluation processes and waste of resources. This situation results in the low efficiency of evaluation for the entire decision-making section. Through the evaluation and analysis of the existing concerned decision documents, the matter of unnecessary waste of manpower resource is extensive. For example, many companies with well-established evaluation systems are concentrated in large cities or large countries, under the context of rapid growth in information and the trend of economic globalization. In addition, the untimely exchange of information is an important reason for the waste of decision resources. In the process of multi-criteria decision-making, the final results show inaccurate features in the case of a loss of decision information. Thus, in this situation, we explain the application by taking the investment company's choice of the best investment project as an example.

ABC Investment Co., Ltd. is a large investment consulting company. The company's decision-making level is in the leading position. Thus, policymakers prefer to choose ABC Investment Co., Ltd. instead of other relatively backward companies. As a result, large investment companies are common, and small investment sectors create a waste of corporate resources. Ultimately, helping companies to share information in decision-making systems to improve decision-making processes is critical to guiding companies to choose more rational decision-making companies. Thus, when enterprises face risky decision-making problems, they should choose large decision-making departments to deal with them effectively, but not all decision-making problems blindly choose large investment departments to solve.

With regard to those decision-making issues that need to be transferred to the upper-level department for processing, the decision given by the decision-maker is a critical step. Therefore, accurate judgment, the consensus of the decision-makers at the corresponding level and the decision-making departments at higher levels provide a reference for the development of the enterprise. This can synthesize different levels of knowledge information to improve decision-making efficiency.

Combined with the above considerations, companies establish decision-making systems to improve decision-making efficiency. It is necessary for companies to have a database of their decision information. In some enterprises, decision information storage and retrieval systems have been established based on computer networks for enterprise-centric data collection and investigation. Effectively sharing decision data among decision-making departments is beneficial to the development of companies. Therefore, in reducing excessive unnecessary decisions, PNHFNs are used to express the conclusions of decision makers for the MCDM problems faced by companies.

For instance, the formula $\{\langle T|P^T, I|P^I, F|P^F\rangle\}$ is a decision maker's judgment for an MCDM problem, where T describes that the decision maker's support degrees for the problem can be solved, I indicates that the professor's indeterminacy degrees for the problem can be solved, and F expresses that the decision maker's dissentient degrees for the problem can be solved. The probabilities P^T, P^I and P^F are the corresponding statistic value of T, I and F, respectively.

Next, we introduce an illustration by utilizing the new distance and similarity measures to perfect the accurate evaluation for reducing the excessive re-evaluations. The special illustration of a talents selection problem is introduced as follows:

$C : \{A_1, A_2, A_3, A_4\}$ is a set of three investors.

$E : \{E_1, E_2\}$ is a set of two stock consultants from the higher and lower companies, respectively.

A : {RE Network Technology Company (RE); DR Biotechnology Company (DR); EV Chemical Company (EV); and FL Technology Company (FL)} is a set of stocks that the investors need to be premeditated.

Then, regarding the investment questions, the evaluation information of the two experts is described and listed in Tables 4 and 5.

Table 4. Probabilistic neutrosophic hesitant fuzzy decision matrix of E_1.

<table>
<tr><th></th><th>RE</th><th>DR</th></tr>
<tr><td>A₁</td><td>{{0.4|0.6, 0.6|0.2}, {0.4|0.6}, {0.3|0.4, 0.4|0.5}}</td><td>{{0.3|0.4, 0.6|0.4}, {0.5|0.5, 0.6|0.4}, {0.3|0.4}}</td></tr>
<tr><td>A₂</td><td>{{0.5|0.4, 0.6|0.3}, {0.4|0.2, 0.6|0.5}, {0.3|0.4}}</td><td>{{0.6|0.5}, {0.4|0.3, 0.6|0.5}, {0.4|0.6, 0.6|0.3}}</td></tr>
<tr><td>A₃</td><td>{{0.5|0.7}, {0.4|0.3, 0.5|0.4}, {0.4|0.3, 0.6|0.5}}</td><td>{{0.4|0.5, 0.6|0.5}, {0.5|0.6}, {0.4|0.4, 0.5|0.4}}</td></tr>
<tr><td>A₄</td><td>{{0.5|0.3}, {0.2|0.1, 0.4|0.5, 0.6|0.2}, {0.5|0.7}}</td><td>{{0.6|0.5}, {0.4|0.5, 0.6|0.5}, {0.5|0.3, 0.6|0.5}</td></tr>
<tr><th></th><th>EV</th><th>FL</th></tr>
<tr><td>A₁</td><td>{{0.7|0.5, 0.8|0.5}, {0.3|0.5, 0.4|0.4}, {0.5|0.6}}</td><td>{{0.5|0.4, 0.7|0.6}, {0.3|0.5, 0.5|0.4}, {0.5|0.4}}</td></tr>
<tr><td>A₂</td><td>{{0.7|0.3, 0.8|0.5}, {0.6|0.6}, {0.4|0.5, 0.6|0.4}}</td><td>{{0.6|0.4, 0.8|0.4}, {0.4|0.2, 0.6|0.5}, {0.5|0.3}}</td></tr>
<tr><td>A₃</td><td>{{0.6|0.5}, {0.4|0.5, 0.5|0.3}, {0.4|0.5, 0.6|0.4}}</td><td>{{0.6|0.5}, {0.5|0.4, 0.6|0.4}, {0.5|0.6, 0.6|0.4}}</td></tr>
<tr><td>A₄</td><td>{{0.6|0.3, 0.8|0.5}, {0.4|0.6}, {0.5|0.3, 0.6|0.5}}</td><td>{{0.6|0.5, 0.8|0.4}, {0.4|0.6}, {0.4|0.5, 0.5|0.4}</td></tr>
</table>

Table 5. Probabilistic neutrosophic hesitant fuzzy decision matrix of E_2.

<table>
<tr><th></th><th>RE</th><th>DR</th></tr>
<tr><td>A₁</td><td>{{0.6|0.5}, {0.4|0.2, 0.6|0.6}, {0.4|0.6, 0.6|0.2}}</td><td>{{0.5|0.4, 0.7|0.4}, {0.6|0.4}, {0.4|0.6, 0.5|0.4}}</td></tr>
<tr><td>A₂</td><td>{{0.3|0.4}, {0.5|0.4}, {0.2|0.2, 0.4|0.5, 0.6|0.3}}</td><td>{{0.5|0.6}, {0.6|0.4}, {0.5|0.3, 0.6|0.4}}</td></tr>
<tr><td>A₃</td><td>{{0.4|0.6, 0.6|0.2}, {0.6|0.3}, {0.5|0.4, 0.6|0.5}}</td><td>{{0.6|0.4, 0.8|0.4}, {0.5|0.3, 0.7|0.5}, {0.5|0.4}}</td></tr>
<tr><td>A₄</td><td>{{0.5|0.4, 0.6|0.4}, {0.5|0.3}, {0.3|0.4, 0.6|0.5}}</td><td>{{0.7|0.5}, {0.5|0.6, 0.6|0.3}, {0.5|0.6}</td></tr>
<tr><th></th><th>EV</th><th>FL</th></tr>
<tr><td>A₁</td><td>{{0.5|0.3, 0.6|0.5}, {0.4|0.4, 0.6|0.6}, {0.3|0.6}}</td><td>{{0.6|0.6}, {0.3|0.5}, {0.4|0.4, 0.5|0.3, 0.6|0.3}}</td></tr>
<tr><td>A₂</td><td>{{0.5|0.4, 0.6|0.3}, {0.5|0.6, 0.6|0.3}, {0.5|0.5}}</td><td>{{0.5|0.6, 0.6|0.4}, {0.4|0.5, 0.6|0.3}, {0.3|0.4}}</td></tr>
<tr><td>A₃</td><td>{{0.5|0.4, 0.6|0.5}, {0.5|0.4, 0.7|0.5}, {0.5|0.8}}</td><td>{{0.4|0.6, 0.7|0.4}, {0.3|0.4, 0.4|0.6}, {0.5|0.5}}</td></tr>
<tr><td>A₄</td><td>{{0.5|0.6}, {0.5|0.5}, {0.4|0.2, 0.6|0.5, 0.7|0.3}}</td><td>{{0.5|0.5, 0.7|0.5}, {0.5|0.4}, {0.4|0.6, 0.6|0.3}</td></tr>
</table>

First, normalize the evaluation information, since the space is limited, so the results are neglected. According to the above-mentioned explanations, the distance and similarity measures among the two reports' evaluations are calculated by utilizing the following functions:

$$D(E_1, E_2) = \begin{cases} 5log_3(1 + \frac{|d_A(x) - d_B(x)|^2}{4} + |e_A(x) - e_B(x)| + \frac{|g_A(x) - g_B(x)|}{2}), when |e_A(x) - e_B(x)| \geq 0.15; \\ 5log_3(1 + \frac{|d_A(x) - d_B(x)|}{2} + |e_A(x) - e_B(x)|^2 + \frac{|g_A(x) - g_B(x)|}{2}), when |e_A(x) - e_B(x)| \leq 0.15. \end{cases} \tag{18}$$

$$S(E_1, E_2) = \begin{cases} \frac{1}{2}((1 - \frac{2 - |d_A(x) - d_B(x)|}{2})^3 + (\frac{1}{2})^{|e_A(x) - e_B(x)|} + \frac{|g_A(x) - g_B(x)|}{2} - 0.5), when |e_A(x) - e_B(x)| \geq 0.15; \\ \frac{1}{2}((\frac{1}{2})^{3|e_A(x) - e_B(x)|} - \frac{|d_A(x) - d_A(x)|}{2} + \frac{|g_A(x) - g_B(x)|}{2} + 0.5), when |e_A(x) - e_B(x)| \leq 0.15. \end{cases} \tag{19}$$

According to the investors knowledge backgrounds, the threshold value is set to 0.15. If the difference of the stock consultant evaluation is lower than 0.15, the discussion of their evaluations is worth deeply discussing and studying, and it may be a key factor of the investment choice. Conversely, the impact of the difference in conclusions is not the most important.

Next, for the consequences of distance and similarity measures of every criterion for each investment problem, these are described by the corresponding matrices $D(E_1, E_2)$, $S(E_1, E_2)$:

$$D(E_1, E_2) = \begin{pmatrix} 0,2242, 0.0396, \widehat{0.7380}, 0.0715 \\ \widehat{0.7777}, 0.4676, 0.5701, 0.1101 \\ 0.2693, 0.3948, 0.7351, \widehat{0.7932} \\ 0.2208, 0.3892, \widehat{0.5937}, 0.2866 \end{pmatrix},$$

$$S(E_1, E_2) = \begin{pmatrix} 0.6575, 0.7257, \widehat{0.6833}, 0.7455 \\ \widehat{0.5023}, 0.6088, 0.5272, 0.6933 \\ 0.6367, 0.5848, 0.6522, \widehat{0.6589} \\ 0.6806, 0.6400, \widehat{0.5485}, 0.6628 \end{pmatrix}.$$

Based on the above conclusions, in order to confirm which criterion needs further examination, the stock consultant should discuss the threshold value of the distance value with investors. However, the similarity consequences are considered as a reference for the investor and stock consultant for the consideration of further examinations. According to this question background, 0.15 is the threshold value of distance measures for every investor (the threshold value of distance measures is determined by a third party data source, and we are not discussing this here.). On the basis of the explanation of distance measure, the threshold value of similarity measure will be determined. Next, the matrices $D(E_1, E_2)$ and $S(E_1, E_2)$ help us to understand the meaning.

Observing the matrix $D(E_1, E_2)$, investor A_1 needs to focus on EV; investor A_2 needs to focus on RE; investor A_3 needs to focus on EV and FL; and investor A_4 does not need to focus on: RE, DR and FL.

Likewise, about the matrix $S(E_1, E_2)$, for the investors A_2 and A_4, we can obtain the same conclusion as the ones explained by $D(E_1, E_2)$. However, the similarity measure of A_1 is not the smallest, and neither is the similarity measure of A_3 for $EV$and FL. Both A_1 and A_3 reflect the greater distance and similarity measures. The reason is that the context of the problem is different, and the distance and similarity measure of A_1 are investigated by the corresponding first formulas in (18) and (19); the conclusions of the A_3 are investigated by the corresponding second formula in (18) and (19). Obviously, the different knowledge background of the stock consultants caused the results of A_1, the results of A_1 are relatively less strict for rule EV. Furthermore, stock consultants need more in-depth communication to make judgments and suggestions about rules EV and FL for A_3.

However, in order to make a decision faster for A_3, the entropy measure can be utilized. For A_3, the stock consultants provide the normal probabilistic neutrosophic hesitant fuzzy information with respect to the EV listed:

$$E_1 = \langle \{0.6|1\}, \{0.4|0.625, 0.5|0.375\}, \{0.4|0.56, 0.6|0.44\} \rangle,$$
$$E_2 = \langle \{0.5|0.44, 0.6|0.56\}, \{0.5|0.44, 0.7|0.56\}, \{0.5|1\} \rangle.$$

By utilizing Equations (18) and (19), and the following entropy measures

$$E(A) = \frac{1D(A,B) + S(A,B)}{2} \tag{20}$$

to obtain the stock consultants' entropy for rule EV, in which $B = \{\langle x, \{0.5|P_1\}, \{0.5|P_2\}, \{0.5|P_3\}\rangle | x \in X\}$, we can get

$$E_1 = 0.5393; E_2 = 0.5977.$$

The bigger the entropy value, the easier it is for the stock consultant to change his/her mind. The investor should make a contract with the stock consultant E_2 first, then make a contract with E_1. Suppose the stock consultant E_2 changes his mind previously, and his opinion is closer to E_1. Then, it is not necessary for investor A_3 to make an appointment with E_1. Obviously, this method is more convenient, flexible and efficient. This method is beneficial for reducing the unnecessary selective re-examinations. In addition, the entropy measure is applied in MCDM situations, which is conducive to improving resource utilization.

It is worth noting that the evaluation information is described by PNHFS, which include the objective information and subjective degrees. The decision makers can select the optimal form of expression of PNHFS to solve practical situations.

5. Conclusions and Future Research

Based on the concept of PNHFS, the theories of NSs are enriched and its application ranges are increased. Next, the different types of fuzziness related to the uncertainty neutrosophic space are investigated. Through analysis and comparison, we know that the neutrosophic space is composed of indeterminate subspace and relatively certain subspace. These two different types of subspace should be distinguished. Simultaneously, the connections among these subspaces are investigated. According to the drawbacks of distance and similarity measures, a new method is established to describe the measures of PNHFSs. The basic axioms of measure are satisfied. Next, the connections among the novel distance, similarity and entropy measures are researched, and compared with other proposed methods. It shows that our methods are more effective. Finally, under the background of investment selection, the novel distance, similarity and entropy measures are shown for reducing the invalid evaluation processes. This is important for improving the evaluation efficiency of the entire selection system. The results have expressed that our proposed methods are meaningful and, if applied, solve the more complicated problems, like talent selections.

Furthermore, in Example 3 and Example 5, the parameters ϕ, μ, ν and λ can depict the experts' individual preferences and knowledge background. Additionally, the more information that is expressed, the more accurate the parameters will be. Thus, how to decide the parameters in measurements is a significant problem. Next, the practicality of new measures is explained by applying distance, similarity and entropy measures into the investment selection. The new distance (similarity) and entropy measures will be researched by integrating them with some related backgrounds to promote the other practical situations. Considering the privacy of information, the related situations of new measurements will help with evaluation to guide decision makers. In the future, the novel measures will be investigated and integrate some related methods in order to expand the scope of application. Based on the correlation and complexity of investors' information, the novel measures will be established. Finally, the properties of entropy measurements have not been studied in full. Thus, in the future, the axioms of the entropy measure will be given more attention. The basic operation laws of PNHFSs and IS have been omitted, so the research about this situation will be studied further.

Author Contributions: All authors have contributed equally to this paper. S.S. and X.Z. initiated the investigation and organized the draft. S.S. put forward this idea and completed the preparation of the paper. S.S. collected existing research results on PNHFS. X.Z. revised and submitted the document.

Funding: This work was supported by the National Natural Science Foundation of China (Grant No. 61573240).

Conflicts of Interest: The authors declare no conflict of interest.

References

1. Smarandache, F. *Neutrosophy: Neutrosophic Probability, Set, and Logic: Analytic Synthesis & Synthetic Analysis*; American Research Press: Santa Fe, NM, USA, 1998.
2. Smarandache, F. *A Unifying Field in Logics. Neutrosophy: Neutrosophic Probability, Set and Logic*; American Research Press: Santa Fe, NM, USA, 1999.
3. Zadeh, L.A. Fuzzy sets. *Inf. Control* **1965**, *8*, 338–353. [CrossRef]
4. Haibin, W.; Smarandache, F.; Zhang, Y.; Sunderraman, R. *Single Valued Neutrosophic Sets*; Infinite Study: Brasov, Romania, 2010.
5. Kazimieras Zavadskas, E.; Baušys, R.; Lazauskas, M. Sustainable assessment of alternative sites for the construction of a waste incineration plant by applying WASPAS method with single-valued neutrosophic set. *Sustainability* **2015**, *7*, 15923–15936. [CrossRef]
6. Şahin, R.; Küçük, A. Subsethood measure for single valued neutrosophic sets. *J. Intell. Fuzzy Syst.* **2015**, *29*, 525–530. [CrossRef]
7. Wang, H.; Smarandache, F.; Sunderraman, R.; Zhang, Y.Q. *Interval Neutrosophic Sets and Logic: Theory and Applications in Computing: Theory and Applications in Computing*; Hexis: Phoenix, AZ, USA, 2005; Volume 5.
8. Broumi, S.; Talea, M.; Smarandache, F.; Bakali, A. Decision-making method based on the interval valued neutrosophic graph. In Proceedings of the 2016 Future Technologies Conference (FTC), San Francisco, CA, USA, 6–7 December 2016; IEEE: Piscataway, NJ, USA, 2016; pp. 44–50.
9. Liu, P.; Tang, G. Some power generalized aggregation operators based on the interval neutrosophic sets and their application to decision-making. *J. Intell. Fuzzy Syst.* **2016**, *30*, 2517–2528. [CrossRef]
10. Bao, Y.L.; Yang, H.L. On single valued neutrosophic refined rough set model and its application. In *Fuzzy Multi-Criteria Decision-Making Using Neutrosophic Sets*; Springer: Cham, Switzerand, 2019; pp. 107–143.
11. Bakali, A.; Smarandache, F.; Rao, V.V. Single-Valued Neutrosophic Techniques for Analysis of WIFI Connection. In *Advanced Intelligent Systems for Sustainable Development (AI2SD'2018): Volume 5: Advanced Intelligent Systems for Computing Sciences*; Springer: Cham, Switzerand, 2019; p. 405.
12. Ashraf, S.; Abdullah, S.; Smarandache, F.; ul Amin, N. Logarithmic hybrid aggregation operators based on single valued neutrosophic sets and their applications in decision support systems. *Symmetry* **2019**, *11*, 364. [CrossRef]
13. Sun, R.; Hu, J.; Chen, X. Novel single-valued neutrosophic decision-making approaches based on prospect theory and their applications in physician selection. *Soft Comput.* **2019**, *23*, 211–225. [CrossRef]
14. Ye, J. Multiple attribute group decision-making method with single-valued neutrosophic interval number information. *Int. J. Syst. Sci.* **2019**, *50*, 152–162. [CrossRef]
15. Thong, N.T.; Dat, L.Q.; Hoa, N.D.; Ali, M.; Smarandache, F.; Son, L.H. Dynamic interval valued neutrosophic set: Modeling decision-making in dynamic environments. *Comput. Ind.* **2019**, *108*, 45–52. [CrossRef]
16. Nagarajan, D.; Lathamaheswari, M.; Broumi, S.; Kavikumar, J. A new perspective on traffic control management using triangular interval type-2 fuzzy sets and interval neutrosophic sets. *Oper. Res. Perspect.* **2019**, *6*, 100099. [CrossRef]
17. Ye, J. Multiple-attribute decision-making method under a single-valued neutrosophic hesitant fuzzy environment. *J. Intell. Syst.* **2015**, *24*, 23–36. [CrossRef]
18. Ye, J. Multiple-attribute decision-making method using similarity measures of single-valued neutrosophic hesitant fuzzy sets based on least common multiple cardinality. *J. Intell. Fuzzy Syst.* **2018**, *34*, 4203–4211. [CrossRef]
19. Li, X.; Zhang, X. Single-valued neutrosophic hesitant fuzzy Choquet aggregation operators for multi-attribute decision-making. *Symmetry* **2018**, *10*, 50. [CrossRef]
20. Peng, J.J.; Wang, J.Q.; Wu, X.H.; Wang, J.; Chen, X.H. Multi-valued neutrosophic sets and power aggregation operators with their applications in multi-criteria group decision-making problems. *Int. J. Comput. Intell. Syst.* **2015**, *8*, 345–363. [CrossRef]

21. Ji, P.; Zhang, H.-Y.; Wang, J.-Q. A projection-based TODIM method under multi-valued neutrosophic environments and its application in personnel selection. *Neural Comput. Appl.* **2018**, *29*, 221–234. [CrossRef]
22. Li, B.-L.; Wang, J.-R.; Yang, L.-H.; Li, X.-T. Multiple criteria decision-making approach with multivalued neutrosophic linguistic normalized weighted Bonferroni mean Hamacher operator. *Math. Probl. Eng.* **2018**, *2018*. [CrossRef]
23. Wang, X.; Wang, X.K.; Wang, J.Q. Cloud service reliability assessment approach based on multi-valued neutrosophic entropy and cross-entropy measures. *Filomat* **2018**, *32*, 2793–2812. [CrossRef]
24. Fang, Z.; Ye, J. Multiple attribute group decision-making method based on linguistic neutrosophic numbers. *Symmetry* **2017**, *9*, 111. [CrossRef]
25. Tian, Z.-P.; Wang, J.; Wang, J.-Q.; Zhang, H.-Y. Simplified neutrosophic linguistic multi-criteria group decision-making approach to green product development. *Group Decis. Negot.* **2017**, *26*, 597–627. [CrossRef]
26. Liu, P.; You, X. Bidirectional projection measure of linguistic neutrosophic numbers and their application to multi-criteria group decision-making. *Comput. Ind. Eng.* **2019**, *128*, 447–457. [CrossRef]
27. Broumi, S.; Smarandache, F.; Maji, P.K. Intuitionistic neutrosphic soft set over rings. *Math. Stat.* **2014**, *2*, 120–126.
28. Broumi, S.; Nagarajan, D.; Bakali, A.; Talea, M.; Smarandache, F.; Lathamaheswari, M. The shortest path problem in interval valued trapezoidal and triangular neutrosophic environment. *Complex Intell. Syst.* **2019**, 1–12. [CrossRef]
29. Broumi, S.; Bakali, A.; Talea, M.; Smarandache, F.; Singh, P.K.; Uluçay, V.; Khan, M. Bipolar complex neutrosophic sets and its application in decision making problem. In *Fuzzy Multi-Criteria Decision-Making Using Neutrosophic Sets*; Springer: Cham, Switzerland, 2019; pp. 677–710.
30. Deli, I. *npn*-Soft sets theory and their applications. *Ann. Fuzzy Math. Inf.* **2015**, *10*, 3–16.
31. Deli, I.; Broumi, S. Neutrosophic soft matrices and NSM-decision-making. *J. Intell. Fuzzy Syst.* **2015**, *28*, 2233–2241. [CrossRef]
32. Smarandache, F. n-Valued Refined Neutrosophic Logic and Its Applications to Physics; Infinite Study 2013. Available online: https://arxiv.org/abs/1407.1041 (accessed on 10 June 2019).
33. Li, L. *p*-topologicalness—A relative topologicalness in ⊤—Convergence spaces. *Mathematics* **2019**, *7*, 228. [CrossRef]
34. Li, L.; Jin, Q.; Hu, K. Lattice-valued convergence associated with CNS spaces. *Fuzzy Sets Syst.* **2019**, *317*, 91–98. [CrossRef]
35. Zhang, X.; Wu, X.; Mao, X.; Smarandache, F.; Park, C. On neutrosophic extended triplet groups (loops) and Abel-Grassmann's groupoids (AG-groupoids). *J. Intell. Fuzzy Syst.* **2019**, 1–11. [CrossRef]
36. Wu, X.; Zhang, X. The decomposition theorems of AG-neutrosophic extended triplet loops and strong AG-(l, l)-loops. *Mathematics* **2019**, *7*, 268. [CrossRef]
37. Ma, Y.; Zhang, X.; Yang, X.; Zhou, X. Generalized neutrosophic extended triplet group. *Symmetry* **2019**, *11*, 327. [CrossRef]
38. Zhang, X.; Wang, X.; Smarandache, F.; Jaíyéolá, T.G.; Lian, T. Singular neutrosophic extended triplet groups and generalized groups. *Cogn. Syst. Res.* **2019**, *57*, 32–40. [CrossRef]
39. Zhang, X.; Borzooei, R.; Jun, Y. Q-filters of quantum B-algebras and basic implication algebras. *Symmetry* **2018**, *10*, 573. [CrossRef]
40. Zhang, X.; Bo, C.; Smarandache, F.; Park, C. New operations of totally dependent-neutrosophic sets and totally dependent-neutrosophic soft sets. *Symmetry* **2018**, *10*, 187. [CrossRef]
41. Xu, Z.; Zhou, W. Consensus building with a group of decision makers under the hesitant probabilistic fuzzy environment. *Fuzzy Optim. Decis. Mak.* **2017**, *16*, 481–503. [CrossRef]
42. Hao, Z.; Xu, Z.; Zhao, H.; Su, Z. Probabilistic dual hesitant fuzzy set and its application in risk evaluation. *Knowl.-Based Syst.* **2017**, *127*, 16–28.
43. Zhai, Y.; Xu, Z.; Liao, H. Measures of probabilistic interval-valued intuitionistic hesitant fuzzy sets and the application in reducing excessive medical examinations. *IEEE Trans. Fuzzy Syst.* **2018**, *26*, 1651–1670. [CrossRef]
44. He, Y.; Xu, Z. Multi-attribute decision-making methods based on reference ideal theory with probabilistic hesitant information. *Expert Syst. Appl.* **2019**, *118*, 459–469. [CrossRef]
45. Wu, W.; Li, Y.; Ni, Z.; Jin, F.; Zhu, X. Probabilistic interval-valued hesitant fuzzy information aggregation operators and their application to multi-attribute decision-making. *Algorithms* **2018**, *11*, 120. [CrossRef]

46. Zhou, W.; Xu, Z. Group consistency and group decision-making under uncertain probabilistic hesitant fuzzy preference environment. *Inf. Sci.* **2017**, *414*, 276–288. [CrossRef]
47. Xie, W.; Ren, Z.; Xu, Z.; Wang, H. The consensus of probabilistic uncertain linguistic preference relations and the application on the virtual reality industry. *Knowl.-Based Syst.* **2018**, *162*, 14–28.
48. Shao, S.; Zhang, X.; Li, Y.; Bo, C. Probabilistic single-valued (interval) neutrosophic hesitant fuzzy set and its application in multi-attribute decision-making. *Symmetry* **2018**, *10*, 419. [CrossRef]
49. Peng, H.G.; Zhang, H.Y.; Wang, J.Q. Probability multi-valued neutrosophic sets and its application in multi-criteria group decision-making problems. *Neural Comput. Appl.* **2018**, *30*, 563–583. [CrossRef]
50. Ye, J. Similarity measures between interval neutrosophic sets and their applications in multicriteria decision-making. *J. Intell. Fuzzy Syst.* **2014**, *26*, 165–172.
51. Majumdar, P.; Samanta, S.K. On similarity and entropy of neutrosophic sets. *J. Intell. Fuzzy Syst.* **2014**, *26*, 1245–1252.
52. Ye, J. Clustering methods using distance-based similarity measures of single-valued neutrosophic sets. *J. Intell. Syst.* **2014**, *23*, 379–389. [CrossRef]
53. Garg, H. Algorithms for possibility linguistic single-valued neutrosophic decision-making based on COPRAS and aggregation operators with new information measures. *Measurement* **2019**, *138*, 278–290. [CrossRef]
54. Biswas, P.; Pramanik, S.; Giri, B.C. Some distance measures of single valued neutrosophic hesitant fuzzy sets and their applications to multiple attribute decision-making. In *New Trends in Neutrosophic Theory and Applications*; Pons Editions: Brussells, Belgium, 2016; pp. 55–63.
55. Ren, H.; Xiao, S.; Zhou, H. A Chi-square distance-based similarity measure of single-valued neutrosophic set and applications. *Int. J. Comput. Commun. Control* **2019**, *14*, 78–89. [CrossRef]
56. Cao, C.; Zeng, S.; Luo, D. A single-valued neutrosophic linguistic combined weighted distance measure and its application in multiple-attribute group decision-making. *Symmetry* **2019**, *11*, 275. [CrossRef]
57. Zhao, S.; Wang, D.; Changyong, L.; Lu, W. Induced choquet integral aggregation operators with single-valued neutrosophic uncertain linguistic numbers and their application in multiple attribute group decision-making. *Math. Probl. Eng.* **2019**, *2019*. [CrossRef]
58. Sahin, R.; Liu, P. Distance and similarity measures for multiple attribute decision making with single-valued neutrosophic hesitant fuzzy information. In *New Trends in Neutrosophic Theory and Applications*; Pons Editions: Brussells, Belgium, 2016; pp. 35–54.

mathematics

Article

Neutrosophic Quadruple Vector Spaces and Their Properties

Vasantha Kandasamy W.B. [1], Ilanthenral Kandasamy [1,*] and Florentin Smarandache [2]

1 School of Computer Science and Engineering, Vellore Institute of Technology, Vellore 632014, India
2 Department of Mathematics, University of New Mexico, 705 Gurley Avenue, Gallup, NM 87301, USA
* Correspondence: ilanthenral.k@vit.ac.in

Received: 03 July 2019; Accepted: 16 August 2019; Published: 19 August 2019

Abstract: In this paper authors for the first time introduce the concept of Neutrosophic Quadruple (NQ) vector spaces and Neutrosophic Quadruple linear algebras and study their properties. Most of the properties of vector spaces are true in case of Neutrosophic Quadruple vector spaces. Two vital observations are, all quadruple vector spaces are of dimension four, be it defined over the field of reals R or the field of complex numbers C or the finite field of characteristic p, Z_p; p a prime. Secondly all of them are distinct and none of them satisfy the classical property of finite dimensional vector spaces. So this problem is proposed as a conjecture in the final section.

Keywords: Neutrosophic Quadruple (NQ); Neutrosophic Quadruple set; NQ vector spaces; NQ linear algebras; NQ basis; NQ vector spaces; orthogonal or dual NQ vector subspaces

---

1. Introduction

In this section we just give a brief literature survey of this new field of Neutrosophic Quadruples [1]. Neutrosophic triplet groups, modal logic Hedge algebras were introduced in [2,3]. Duplet semigroup, neutrosophic homomorphism theorem and triplet loops and strong AG(1, 1) loops are defined and described in [4–6]. Neutrosophic triplet neutrosophic rings application to mathematical modelling, classical group of neutrosophic triplets on $\{Z_{2p}, \times\}$ and neutrosophic duplets in neutrosophic rings are developed and analyzed in [7–11]. Study of Algebraic structures of neutrosophic triplets and duplets, quasi neutrosophic triplet loops, extended triplet groups, AG-groupoids, NT-subgroups are carried out in [6,12–17]. Refined neutrosophic sets were developed by [18–21]. Neutrosophic algebraic structures in general were studied in [22–25]. The new notion of Neutrosophic Quadruples which assigns a known part happens to be very interesting and innovative, and was introduced by Smarandache [1,26] in 2015. Several research papers on the algebraic structure of Neutrosophic Quadruples, such as groups, monoids, ideals, BCI-algebras, BCI-positive implicative ideals, hyperstructures, BCK/BCI algebras [27–32] have been recently studied and analyzed. However in this paper authors have defined the new notion of Neutrosophic Quadruple vector spaces (NQ vector spaces) and Neutrosophic Quadruple linear algebras (NQ linear algebras) and have studied a few related properties. This work can later be used to propose neutrosophic based dynamical systems in particular in the area of hyperchoaos from cellular neural networks [33].

This paper is organized into five sections. Basic concepts needed to make this paper a self contained one is given in Section 2. NQ vector spaces are introduced in Section 3, further NQ subspaces are introduced and the notion of direct sum and NQ bases are analysed. It is shown all NQ vector spaces are of dimension 4 be it defined over R or C or Z_p, p a prime. Section 4 defines and develops the properties of NQ linear algebras. The final section proposes a conjecture which is related with the finite dimensional vector spaces, which are always isomorphic to finite direct product of fields over which the vector space is defined. Finally we give the future direction of research on this topic.

2. Basic Concepts

In this section basic concepts on vector spaces and a few of its properties and some NQ algebraic structures and their properties needed for this paper are given.

Through out this paper R denotes the field of reals, C denotes the field of complex numbers and Z_p denotes the finite field of characteristic p, p a prime. $NQ = \{(a, bT, cI, dF)$ denotes the Neutrosophic Quadruple; with a, b, c, d in R or C or Z_p, where T, I and F has the usual neutrosophic logic meaning of Truth, Indeterminate and False respectively and a denotes the known part [26].

For basic properties of vector spaces and linear algebras please refer [22].

Definition 1 ([22])**.** *A vector space or a linear space V consists of the following;*

1. *A field of R or C or Z_p of scalars.*
2. *A set V of objects called vectors.*
3. *A rule (or operation) called vector addition; which associates with each pair of vectors x, y in V; $x + y$ is in V, called sum of the vectors x and y in such a way that ;*
   (a) $x + y = y + x$ *(addition is commutative).*
   (b) $x + (y + z) = (x + y) + z$ *(addition is associative).*
   (c) *There is a unique vector 0 in V such that $x + 0 = x$ for all $x \in V$.*
   (d) *For each vector $x \in V$ there is a unique vector $-x \in V$ such that $x + -x = 0$.*
   (e) *A rule or operation called scalar multiplication that associates with each scalar $c \in R$ or C or Z_p and for a vector $x \in V$, called product denoted by '.' of c and x in such a way that for $x \in V$ and $c.x \in V$ and ;*
      i. $c.x = x.c$ *for every* $x \in V$.
      ii. $(c + d).x = c.x + d.x$
      iii. $c.(x + y) = c.x + c.y$
      iv. $c.(d.x) = (c.d)x$;

   for all $x, y \in V$ and c, d in R or C or Z_p.

We can just say $(V, +)$ is a vector space over a field R or C or Z_p if $(V, +)$ is an additive abelian group and V is compatible with the product by the scalars. If on V is defined a product such that $(V, \times)$ is a monoid and $c(x \times y) = (cx) \times y$ then V is a linear algebra over R or C or Z_p [22].

Definition 2 ([22])**.** *Let V be a vector space over R (or C or Z_p). A subspace of V is a subset W of V which is itself a vector space over R (or C or Z_p) with the operations of addition and scalar multiplication as in V.*

Definition 3. *Let V be a vector space over R (or C or Z_p). A subset B of V is said to be linearly dependent or simply dependent if there exist distinct vectors, $x_1, x_2, x_3, \ldots, x_t \in B$ and scalars $a_1, a_2, a_3, \ldots, a_t \in R$ or C or Z_p not all of which are zero such that $a_1x_1 + a_2x_2 + a_3x_3 + \ldots + a_tx_t = 0$. A set which is not linearly dependent is called independent or linearly independent. If B contains only finitely many vectors $x_1, x_2, x_3, \ldots, x_k$ we sometimes say $x_1, x_2, x_3, \ldots, x_k$ are dependent instead of saying B is dependent.*

The following facts are true [22].

1. A subset of a linearly independent set is linearly independent.
2. Any set which contains a linearly dependent subset is linearly dependent.
3. Any set which contains the zero vector (0 vector) is linearly dependent for 1.0 = 0.
4. A set B is linearly independent if and only if each finite subset of B is linearly independent; that is if and only if there exist distinct vectors $x_1, x_2, x_3, \ldots, x_k$ of B such that $a_1x_1 + a_2x_2 + a_3x_3 + \ldots + a_kx_k = 0$ implies each $a_i = 0; i = 1, 2, \ldots, k$.

For a vector space V over a field R or C or Z_p , the basis for V is a linearly independent set of vectors in V which spans the space V. We say the vector space V over R or C or Z_p is a direct sum of subspaces $W_1, W_2, \ldots, W_t$ if and only if $V = W_1 + W_2 + \ldots + W_t$ and $W_i \cap W_j$ is the zero vector for $i \neq j$ and $1 \leq i, j \leq t$.

The other properties of vector spaces are given in book [22].

Now we proceed on to recall some essential definitions and properties of Neutrosophic Quadruples [26].

Definition 4 ([26]). *The quadruple (a, bT, cI, dF) where $a, b, c, d \in R$ or C or Z_p, with T, I, F as in classical Neutrosophic logic with a the known part and (bT, cI, dF) defined as the unknown part, denoted by $NQ = \{(a, bT, cI, dF) | a, b, c, d \in R$ or C or $Z_n\}$ in called the Neutrosophic set of quadruple numbers.*

The following operations are defined on NQ, for more refer [26].

For $x = (a, bT, cI, dF)$ and $y = (e, fT, gI, hF)$ in NQ [26] have defined

$$x + y = (a, bT, cI, dF) + (e, fT, gI, hF) = (a + e, (b + f)T, (c + g)I, (d + h)F)$$

$$\text{and} \quad x - y = (a - e, (b - f)T, (c - g)I, (d - h)F)$$

are in NQ. For $x = (a, bT, cI, dF)$ in NQ and s in R or C or Z_p where s is a scalar and x is a vector in V. $s.x = s.(a, bT, cI, dF) = (sa, sbT, scI, sdF) \in V$.

If $x = 0 = (0, 0, 0, 0)$ in V usually termed as zero Neutrosophic Quadruple vector and for any scalar s in R or C or Z_p we have $s.0 = 0$.

Further $(s + t)x = sx + tx, s(tx) = (st)x, s(x + y) = sx + sy$ for all $s, t \in R$ or C or Z_p and $x, y \in NQ$. $-x = (-a, -bT, -cI, -dF)$ which is in NQ.

The main results proved in [26] and which is used in this paper are mentioned below;

Theorem 1 ([26]). *$(NQ, +)$ is an abelian group.*

Theorem 2 ([26]). *$(NQ, .)$ is a monoid which is commutative.*

We mainly use only these two results in this paper, for more literature about Neutrosophic Quadruples refer [26].

3. Neutrosophic Quadruple Vector Spaces and Their Properties

In this section we proceed on to define for the first time the new notion of Neutrosophic Quadruple vector spaces (NQ -vector spaces) their NQ vector subspaces, NQ bases and direct sum of NQ vector subspaces. All these NQ vector spaces are defined over R, the field of reals or C, the field of complex numbers and finite field of characteristic p, Z_p, p a prime. All these three NQ vector spaces are different in their properties and we prove all three NQ vector spaces defined over R or C or Z_P are of dimension 4.

We mostly use the notations from [26]. They have proved $(NQ, +) = \{(a, bT, cI, dF) | a, b, c, d \in R$ or C or Z_p, p a prime; $+\}$ is an infinite abelian group under addition.

We prove the following theorem.

Theorem 3. *$(NQ, +) = \{(a, bT, cI, dF) | a, b, c, d \in R$ or C or Z_p; p a prime, $+\}$ be the Neutrosophic quadruple group. Then $V = (NQ, +, \circ)$ is a Neutrosophic Quadruple vector space (NQ-vector space) over R or C or Z_p, where '$\circ$' is the special type of operation between V and R (or C or Z_p) defined as scalar multiplication.*

Proof. To prove V is a Neutrosophic quadruple vector space over R (or C or Z_p, p is a prime), we have to show all the conditions given in Section two (Definition 1) of this paper is satisfied. In the first place we have R or C or Z_p are field of scalars, and elements of V we call as vectors. It has been proved by [26] that $V = (NQ, +)$ is an additive abelian group, which is the basic property on V to be a vector space. Further the quadruple is defined using R or C or Z_p, p a prime, or used in the mutually exclusive sense. Now we see if $x = (a, bT, cI, dF)$ is in V and $n \in R$ (or C or Z_p) then the scalar multiplication '$\circ$' which associates with each scalar $n \in R$ and the NQ vector $x \in V$,

$n \circ x = n \circ (a, bT, cI, dF) = (n \circ a, n \circ bT, n \circ cI, n \circ dF)$ which is in V, called the product of n with x in such a way that

1. $1 \circ x = x \circ 1 \quad \forall x \in V$
2. $(nm) \circ v = n \circ (mv)$
3. $n \circ (v + w) = n \circ v + n \circ w$
4. $(m + n) \circ v = m \circ v + n \circ v$

for all $m, n \in R$ or C or Z_p and $v, w \in V$.

$0 = (0,0,0,0)$ is the zero vector of V and for 0 in R or C or Z_p; we have $0 \circ x = 0 \circ (a, bT, cI, dF) = (0,0,0,0); \forall x \in V$.

Clearly $V = (NQ, +, \circ)$ is a vector space known as the NQ vector space over R or C or Z_p. □

However we can as in case of vector spaces say in case of NQ-vector spaces also $(NQ, +)$ is a NQ vector space with special scalar multiplication $\circ$.

We now proceed on to define the concept of linear dependence, linear independence and basis of NQ vector spaces.

Definition 5. *Let $V = (NQ, +)$ be a NQ vector space over R (or C or Z_p). A subset L of V is said to be NQ linearly dependent or simply dependent, if there exists distinct vectors $a_1, a_2, \ldots, a_k \in L$ and scalars $d_1, d_2, \ldots, d_k \in R$ (or C or Z_p) not all zero such that $d_1 \circ a_1 + d_2 \circ a_2 + \ldots + d_k \circ a_k = 0$. We say the set of vectors $a_1, a_2, \ldots, a_k$ is NQ linearly independent if it is not NQ linearly dependent.*

We provide an example of this situation.

Example 4. *Let $V = (NQ, +)$ vector space over R. Let $x = (3, -4T, 5I, 2F), y = (-2, 3T, -2I, -2F)$ and $z = (-1, T, -3I, 0)$ be in V. We see $1 \circ x + 1 \circ y + 1 \circ z = (0,0,0,0)$, so x, y and z are NQ linearly dependent. Let $x = (5, 0, 0, 2F)$ and $y = (0, 5T, -3I, 0)$ be in V. We cannot find a $a, b \in R$ such that $a \circ x + b \circ y = (0,0,0,0)$. If possible $a \circ x + b \circ y = (0,0,0,0)$; this implies $a \circ 5 + b \circ 0 = 0$, forcing $a = 0$; $a \circ 0 + b \circ 5 = 0$, forcing $b = 0$; $a \circ 0 + b \circ -3 = 0$, forcing $b = 0$ and $a \circ 2 + b \circ 0 = 0$ forcing $a = 0$. Thus the equations are consistent and $a = b = 0$. So x and y are NQ linearly independent over R.*

The following properties are true in case of all vector spaces hence true in case of NQ vector spaces also.

1. A subset of a NQ linearly independent set is NQ linearly independent.
2. A set L of vectors in NQ is linearly independent if and only if for any distinct vectors $a_1, a_2, \ldots, a_k$ of L; $d_1 \circ a_1 + d_2 \circ a_2 + \ldots + d_k \circ a_k = 0$ implies each $d_i = 0$, for $i = 1, 2, \ldots, k$.

We now proceed on to define Neutrosophic Quadruple basis (NQ basis) for $V = (NQ, +)$, Neutrosophic Quadruple vector space over R or C or Z_p (or used in the mutually exclusive sense).

Definition 6. *Let $V = (NQ, +)$ vector space over R (or C or Z_p). We say a subset L of V spans V if and only if every vector in V can be got as a linear combination of elements from L and scalars from R (or C or Z_p). That is if $a_1, a_2, \ldots, a_n$ are n elements in L; then $v = d_1 \circ a_1 + d_2 \circ a_2 + \ldots + d_n \circ a_n$, is the NQ linear combination of vectors of L; where $d_1, d_2, \ldots, d_n$ are in R or C or Z_p and not all these scalars are zero.*

The Neutrosophic Quadruple basis for $V = (NQ, +)$ is a set of vectors in V which spans V. We say a set of vectors B in V is a basis of V if B is a linearly independent set and spans V over R or C or Z_p.

We say V is finite dimensional if the number of elements in basic of V is a finite set; otherwise V is infinite dimensional.

Theorem 5. *Let $V = (NQ, +)$ be the Neutrosophic Quadruple vector space over R (or C or Z_p). V is a finite dimensional NQ vector space over R (or C or Z_p) and dimension of these NQ vector spaces over R(or C or Z_p) are always four.*

Proof. Let $V = (NQ, +) = \{(a, bT, cI, dF)|a, b, c, d \in R$ (or C or Z_p), $+\}$, be the collection of all neutrosophic quadruples of the Neutrosophic Quadruple vector space over R (or C or Z_p). To prove dimension of V over R is four it is sufficient to prove that V has four linearly independent vectors which can span V, which will prove the result. Take the set $B = \{(1,0,0,0), (0,T,0,0), (0,0,I,0), (0,0,0,F)\}$ contained in V; to show B is independent and spans V it enough if we prove for any $v = (a, bT, cI, dF) \in V$, v can be represented uniquely as a linear combination of elements from B and scalars from R (or C or Z_p). Now $v = (a, bT, cI, dF) = a \circ (1,0,0,0) + b \circ (0,T,0,0) + c \circ (0,0,I,0) + d \circ (0,0,0,F)$ for the scalars $a, b, c, d \in R$ (or C or Z_p). Hence we see the elements of V are uniquely represented as a linear combination of vectors using only B, further B is a set of linearly independent elements, hence B is a basis of V and B is finite, so V is finite dimensional over R (or C or Z_p). As order of B is four, dimension of all NQ vector spaces V over R (or C or Z_p) is four. Hence the theorem. □

We call the NQ basis B as the special standard NQ basis of V.

Definition 7. *Let $V = (NQ, +)$ be a NQ vector space over R (or C or Z_p). A subset W of V is said to be Neutrosophic Quadruple vector subspace of V if W itself is a Neutrosophic Quadruple vector space over R (or C or Z_p).*

We will illustrate this situation by examples.

Example 6. *Let $V = \{NQ, +\}$ be a NQ vector space over R. $W = \{(a, bT, 0, 0)|a, b \in R\}$ is a subset of V which is a NQ vector subspace of V over R. $U = \{(0, 0, cI, dF)|c, d \in R\}$ is again a vector subspace of V and is different from W.*

We observe that the only common element between W and U is the zero quadruple vector $(0,0,0,0)$.

Further it is observed if we define the dot product or inner product on elements in V. For $x = (a, bT, cI, dF)$ and $y = (e, fT, gI, hF) \in V$, $x \bullet y$ denoted as $x \bullet y = (a \bullet e, bT \bullet fT, cI \bullet gI, dF \bullet hF)$; and $x \bullet y$ is in V. If $x \bullet y = (0,0,0,0)$ for some $x, y \in V$ then we say x is orthogonal (or dual) with y and vice versa. In fact $x \bullet y = y \bullet x$; $\forall x, y \in V$. We say two NQ vector subspaces W and U are orthogonal (or dual subspaces) if for every $x \in W$ and for every $y \in U$; $x \bullet y = (0,0,0,0)$, that is two NQ vector subspaces are orthogonal if and only if the dot product of every vector in W with every vector in U is the zero vector.

$\{(0,0,0,0)\}$ is the zero vector subspace of V. Every NQ vector subspace of V trivial or nontrivial is orthogonal with the zero vector subspace $\{(0,0,0,0)\}$ of V. V the NQ vector space is orthogonal with only the zero vector subspace of V, and with no other vector subspace of V. W orthogonal $U = W \bullet U = \{w \bullet u|w \in W$ and $u \in U\} = \{(0,0,0,0)\}$; we call the pair of NQ subspaces as orthogonal or dual NQ subspaces of V.

Definition 8. *Let $V = (NQ, +)$ be a Neutrosophic Quadruple vector space over R (or C or Z_p); $W_1, W_2, \ldots, W_n$ be n distinct NQ vector subspaces of V. We say $V = W_1 \oplus W_2 \oplus \ldots \oplus W_n$ is a direct sum of NQ vector subspaces if and only if the following conditions are true;*

1. *Every vector $v \in V$ can be written in the form $v = d_1 \circ w_1 + d_2 \circ w_2 + \ldots + d_n \circ w_n$, where $d_1, d_2, \ldots, d_n$ are in R (or C or Z_p) not all zero with $w_i \in W_i, i = 1, 2, \ldots, n$.*
2. *$W_i \bullet W_j = \{(0,0,0,0)\}$ for $i \neq j$ and true for all i, j varying in the set $\{1, 2, \ldots, n\}$.*

First we record that in case of all NQ vector spaces over R (or C or Z_p) we can have the value of n given in definition to be only four, we cannot have more than four as dimension of all NQ vector spaces are only four. Secondly the minimum of n can be two which is true in case of all vector spaces of any finite dimension. Finally we wish to prove not all NQ vector subspaces are orthogonal and there are only finitely many nontrivial NQ vector subspaces for any NQ vector space over R (or C or Z_p).

We prove as theorem a few of the properties.

Theorem 7. *Let $V = (NQ, +)$ be a NQ vector space over R (or C or Z_p). V has only finite number of NQ vector subspaces.*

Proof. We see in case of NQ vector spaces over R (or C or Z_p) the dimension is four and the special standard NQ basis for V is $B = \{(1,0,0,0),(0,T,0,0),(0,0,I,0),(0,0,0,F)\}$. So any non trivial subspace of V can be of dimension less than four; so it can be 1 or 2 or 3. Clearly there are some vector subspaces of dimension one given by, $W_1 = \langle(1,0,0,0)\rangle$, $W_2 = \langle(0,T,0,0)\rangle$, $W_3 = \langle(0,0,I,0)\rangle$, $W_4 = \langle(0,0,0,F)\rangle$, $W_5 = \langle(1,T,0,0)\rangle$, $W_6 = \langle(1,0,I,0)\rangle$, $W_7 = \langle(1,0,0,F)\rangle$, $W_8 = \langle(0,T,I,0)\rangle$, $W_9 = \langle(0,T,0,F)\rangle$, $W_{10} = \langle(0,0,I,F)\rangle$, $W_{11} = \langle(1,T,I,0)\rangle$, $W_{12} = \langle(1,T,0,F)\rangle$, $W_{13} = \langle(1,0,I,F)\rangle$, $W_{14} = \langle(0,T,I,F)\rangle$ and $W_{15} = \langle(1,T,I,F)\rangle$. Some the two dimensional vector spaces are $U_1 = \langle(1,0,0,0),(0,T,0,0)\rangle$, $U_2 = \langle(1,0,0,0),(0,0,I,0)\rangle, \ldots, U_{105} = \langle(0,T,I,F),(1,T,I,F)\rangle$;

in fact there are 105 NQ vector subspaces of dimension two. Further there are 1365 NQ vector subspaces of dimension three. Thus there are 1485 non trivial NQ vector subspaces in any NQ vector space $V = (NQ,+)$ over R (or C or Z_p). We have shown that there are four NQ vector subspaces of dimension three all of them are hyper subspaces of V, of course we are not enumerating other types of dimension three subspaces generated by vectors of the form $M_1 = \{\langle(1,T,0,0),(0,0,I,0),(0,0,0,F)\rangle\}$, or $M_2 = \{\langle(1,0,0,F),(0,0,I,0),(0,T,0,0)\rangle\}$ are spaces of dimension three which we do not take into account as hyper subspaces. □

We define the three dimensional NQ vector subspace generated only by $\{\langle(0,T,0,0),(0,0,I,0),(0,0,0,F)\rangle\}$ is defined as the special pseudo Singled Valued Neutrosophic hyper NQ vector subspace of V [22,24].

4. Neutrosophic Quadruple Linear Algebras over R or C or Z_p

In this section we take the basic concepts defined in [26] $(NQ,+)$ for the Neutrosophic Quadruple additive abelian group and $(NQ,.)$ as the commutative monoid with $(1,0,0,0)$ as the identity with respect to '.' and for any $(a,bT,cI,dF) = x$, and $y = (e,fT,gI,hF)$ in NQ [26] have defined $x.y = (ae,(af+be+bf)T,(ag+bg+ce+cf+cg)I,(ah+bh+ch+de+df+dg+dh)F)$.

Theorem 8. *$V = (NQ,+,.)$ is a Neutrosophic Quadruple linear algebra (NQ linear algebra) over R (or C or Z_p).*

Proof. To prove V is a NQ linear algebra we have to prove the following; $(NQ,+)$ is an abelian group under addition given in [26] and it is proved that $(NQ,+)$ is a vector space (Theorem 3). To prove V is a NQ linear algebra it is sufficient if we prove $(NQ,.)$ is a monoid under product '.' which is proved in [26], further $d \circ (x.y) = (d \circ x).y$ for $d \in R$ (or C or Z_p) and $x,y \in V$ which is true as $x.y$ is in V. Thus $(V,+,.)$ is a NQ linear algebra over R (or C or Z_p). □

Definition 9. *Let $V = (NQ,+,.)$ be a NQ linear algebra over R (or C or Z_p). Let W be a nonempty proper subset of V, we say W is a NQ sublinear algebra of V over R (or C or Z_p), if W itself is a linear algebra over R (or C or Z_p).*

We provide some examples of them.

Example 9. *Let $V = (NQ,+.)$ be a linear algebra over the field Z_7. $W = \{\langle(1,0,0,0)\rangle\}$ generated under $+,.$ and '$\circ$' multiplication by scalar from elements of Z_7 is a sublinear algebra and of order 7 and dimension of W over Z_7 is one. Similarly $U = \{\langle(1,t,0,0),(0,0,I,0)\rangle\}$ generated by these two vectors is a sublinear algebra of dimension two. Just we show how the product of $x = (3,4T,I,5F)$ and $y = (2,3T,4I,F)$ in V is carried out; $x.y = (6,2T,I,2F)$ which is in V.*

We can as in case of NQ vector spaces derive all properties of NQ linear algebras , further as in case of NQ vector spaces dimension of all these NQ-linear algebras is four.

We in the following section propose some open conjectures and the future work to be carried out in this direction.

5. Conclusions and Open Conjectures

In this paper for the first time we define the notion of NQ vector spaces and NQ linear algebras. All the three NQ vector spaces are of dimension four only. The NQ vector space V over R, is different from the NQ vector space W over C, and both has infinite number of vectors; but is of dimension four and U the NQ vector space over Z_p has only p^4 elements and is of dimension four.

We know the classical result on vector spaces states "A vector space V of say dimension n (n a finite integer) defined over the field F is isomorphic to $F \times F \times \ldots \times F$ n-times"; in view of this we propose the following conjectures:

1. Is the NQ vector space V defined over R isomorphic to $R \times R \times R \times R$?
2. Is the NQ vector space W defined over C isomorphic to $C \times C \times C \times C$?
3. Is the NQ vector space U defined over Z_p isomorphic to $Z_p \times Z_p \times Z_p \times Z_p$?

Finally we would be developing the new notion of NQ algebraic codes and analyse them for future research. In our opinion a new type of NQ algebraic codes can certainly be defined with appropriate modifications. Also we would develop the notion of Neutrosophic quadruples in which the unknown part would be these neutrosophic triplets or modified form of neutrosophic duplets which would be taken for further study.

Author Contributions: The contributions of authors are roughly equal.

Funding: This research received no external funding

Conflicts of Interest: The authors declare no conflict of interest.

Abbreviations

The following abbreviations are used in this manuscript:

NQ Neutrosophic Quadruple

References

1. Smarandache, F. Neutrosophic Quadruple Numbers, Refined Neutrosophic Quadruple Numbers, Absorbance Law, and the Multiplication of Neutrosophic Quadruple Numbers. *Neutrosophic Sets Syst.* **2015**, *10*, 96–98.
2. Smarandache, F.; Ali, M. Neutrosophic triplet group. *Neural Comput. Appl.* **2018**, *29*, 595–601. [CrossRef]
3. Smarandache, F. *Neutrosophic Perspectives: Triplets, Duplets, Multisets, Hybrid Operators, Modal Logic, Hedge Algebras and Applications*, 2nd ed.; Pons Publishing House: Brussels, Belgium, 2017; ISBN 978-1-59973-531-3.
4. Zhang, X.H.; Smarandache, F.; Liang, X.L. Neutrosophic Duplet Semi-Group and Cancellable Neutrosophic Triplet Groups. *Symmetry* **2017**, *9*, 275. [CrossRef]
5. Zhang, X.H.; Smarandache, F.; Ali, M.; Liang, X.L. Commutative neutrosophic triplet group and neutro-homomorphism basic theorem. *Ital. J. Pure Appl. Math.* **2017**. [CrossRef]
6. Wu, X.Y.; Zhang, X.H. The decomposition theorems of AG-neutrosophic extended triplet loops and strong AG-(l, l)-loops. *Mathematics* **2019**, *7*, 268. [CrossRef]
7. Vasantha, W.B.; Kandasamy, I.; Smarandache, F. Neutrosophic Triplets in Neutrosophic Rings. *Mathematics* **2019**, *7*, 563.
8. Vasantha, W.B.; Kandasamy, I.; Smarandache, F. *Neutrosophic Triplet Groups and Their Applications to Mathematical Modelling*; EuropaNova: Brussels, Belgium, 2017; ISBN 978-1-59973-533-7.
9. Vasantha, W.B.; Kandasamy, I.; Smarandache, F. A Classical Group of Neutrosophic Triplet Groups Using $\{Z_{2p}, \times\}$. *Symmetry* **2018**, *10*, 194. [CrossRef]
10. Vasantha, W.B.; Kandasamy, I.; Smarandache, F. Neutrosophic duplets of $\{Z_{pn}, \times\}$ and $\{Z_{pq}, \times\}$. *Symmetry* **2018**, *10*, 345. [CrossRef]
11. Vasantha, W.B.; Kandasamy, I.; Smarandache, F. Algebraic Structure of Neutrosophic Duplets in Neutrosophic Rings $\langle Z \cup I \rangle$, $\langle Q \cup I \rangle$ and $\langle R \cup I \rangle$. *Neutrosophic Sets Syst.* **2018**, *23*, 85–95.
12. Smarandache, F.; Zhang, X.; Ali, M. Algebraic Structures of Neutrosophic Triplets, Neutrosophic Duplets, or Neutrosophic Multisets. *Symmetry* **2019**, *11*, 171. [CrossRef]

13. Zhang, X.H.; Wu, X.Y.; Smarandache, F.; Hu, M.H. Left (right)-quasi neutrosophic triplet loops (groups) and generalized BE-algebras. *Symmetry* **2018**, *10*, 241. [CrossRef]
14. Zhang, X.H.; Wang, X.J.; Smarandache, F.; Jaíyéolá, T.G.; Liang, X.L. Singular neutrosophic extended triplet groups and generalized groups. *Cognit. Syst. Res.* **2018**, *57*, 32–40. [CrossRef]
15. Zhang, X.H.; Wu, X.Y.; Mao, X.Y.; Smarandache, F.; Park, C. On Neutrosophic Extended Triplet Groups (Loops) and Abel-Grassmann's Groupoids (AG-Groupoids). *J. Intell. Fuzzy Syst.* **2019**. [CrossRef]
16. Zhang, X.; Hu, Q.; Smarandache, F.; An, X. On Neutrosophic Triplet Groups: Basic Properties, NT-Subgroups, and Some Notes. *Symmetry* **2018**, *10*, 289. [CrossRef]
17. Ma, Y.; Zhang, X.; Yang, X.; Zhou, X. Generalized Neutrosophic Extended Triplet Group. *Symmetry* **2019**, *11*, 327. [CrossRef]
18. Agboola, A.A.A. On Refined Neutrosophic Algebraic Structures. *Neutrosophic Sets Syst.* **2015**, *10*, 99–101.
19. Wang, H.; Smarandache, F.; Zhang, Y.; Sunderraman, R. Single valued neutrosophic sets. *Review* **2010**, *1*, 10–15.
20. Kandasamy, I. Double-Valued Neutrosophic Sets, their Minimum Spanning Trees, and Clustering Algorithm. *J. Intell. Syst.* **2018**, *27*, 163–182. [CrossRef]
21. Kandasamy, I.; Smarandache, F. Triple Refined Indeterminate Neutrosophic Sets for personality classification. In Proceedings of the 2016 IEEE Symposium Series on Computational Intelligence (SSCI), Athens, Greece, 6–9 December 2016; pp. 1–8. [CrossRef]
22. Vasantha, W.B. *Linear Algebra and Smarandache Linear Algebra*; American Research Press: Ann Arbor, MI, USA, 2003; ISBN 978-1-931233-75-6.
23. Vasantha, W.B.; Smarandache, F. *Some Neutrosophic Algebraic Structures and Neutrosophic N-Algebraic Structures*; Hexis: Phoenix, AZ, USA, 2006; ISBN 978-1-931233-15-2.
24. Vasantha, W.B.; Smarandache, F. *Neutrosophic Rings*; Hexis: Phoenix, AZ, USA, 2006; ISBN 978-1-931233-20-9.
25. Vasantha, W.B.; Kandasamy, I.; Smarandache, F. Semi-Idempotents in Neutrosophic Rings. *Mathematics* **2019**, *7*, 507. [CrossRef]
26. Akinleye, S.A.; Smarandache, F.; Agboola, A.A.A. On neutrosophic quadruple algebraic structures. *Neutrosophic Sets Syst.* **2016**, *12*, 122–126.
27. Agboola, A.A.A.; Davvaz, B.; Smarandache, F. Neutrosophic quadruple algebraic hyperstructures. *Ann. Fuzzy Math. Inform.* **2017**, *14*, 29–42. [CrossRef]
28. Li, Q.; Ma, Y.; Zhang, X.; Zhang, J. Neutrosophic Extended Triplet Group Based on Neutrosophic Quadruple Numbers. *Symmetry* **2019**, *11*, 696. [CrossRef]
29. Jun, Y.; Song, S.Z.; Smarandache, F.; Bordbar, H. Neutrosophic quadruple BCK/BCI-algebras. *Axioms* **2018**, *7*, 41. [CrossRef]
30. Muhiuddin, G.; Al-Kenani, A.N.; Roh, E.H.; Jun, Y.B. Implicative Neutrosophic Quadruple BCK-Algebras and Ideals. *Symmetry* **2019**, *11*, 277. [CrossRef]
31. Jun, Y.B.; Song, S.-Z.; Kim, S.J. Neutrosophic Quadruple BCI-Positive Implicative Ideals. *Mathematics* **2019**, *7*, 385. [CrossRef]
32. Jun, Y.B.; Smarandache, F.; Bordbar, H. Neutrosophic N-Structures Applied to BCK/BCI-Algebras. *Information* **2017**, *8*, 128. [CrossRef]
33. Arena, P.; Baglio, S.; Fortuna, L.; Manganaro, G. Hyperchaos from cellular neural networks. *Electron. Lett.* **1995**, *31*, 250–251. [CrossRef]

mathematics

MDPI

Article

Classification of the State of Manufacturing Process under Indeterminacy

Muhammad Aslam * and Osama Hasan Arif

Department of Statistics, Faculty of Science, King Abdulaziz University, Jeddah 21589, Saudi Arabia; oarif@kau.edu.sa

* Correspondence: magmuhammad@kau.edu.sa or aslam_ravian@hotmail.com; Tel.: +966-593-329-841

Received: 27 August 2019; Accepted: 15 September 2019; Published: 19 September 2019

Abstract: In this paper, the diagnosis of the manufacturing process under the indeterminate environment is presented. The similarity measure index was used to find the probability of the in-control and the out-of-control of the process. The average run length (ARL) was also computed for various values of specified parameters. An example from the Juice Company is considered under the indeterminate environment. From this study, it is concluded that the proposed diagnosis scheme under the neutrosophic statistics is quite simple and effective for the current state of the manufacturing process under uncertainty. The use of the proposed method under the uncertainty environment in the Juice Company may eliminate the non-conforming items and alternatively increase the profit of the company.

Keywords: similarity index; diagnosis; process; indeterminacy; neutrosophic statistics

---

1. Introduction

To control the non-conforming products in the industry is an important task for industrial engineers. Their mission is to minimize the non-conforming product which can be achieved only if the problems in the manufacturing process can be tackled immediately. The control charts are essential tools in the industry to monitor the manufacturing process. These tools are used to indicate the state of the process. A timely indication about the state of the process leads to the high quality of the product. Epprecht et al. [1] and Chiu and Kuo [2] proposed a chart for monitoring one, and more than one, non-conforming product, respectively. Hsu [3] designed a variable chart using the improved sampling schemes. Ho and Quinino [4] proposed an attribute chart to control the variation in the process. Aslam et al. [5] and Aslam et al. [6] worked on a time-truncated chart for the Birnbaum-Saunders distribution and the Weibull distribution respectively. Jeyadurga et al. [7] worked on an attribute chart under truncated life tests.

To analyze the vague and fuzzy data, the fuzzy logic is applied. The fuzzy logic is applied to analyze the data when the experimenters are unsure about the exact values of the parameters. Therefore, the monitoring of the process having fuzzy data is done using the fuzzy-based control charts. Afshari and Gildeh [8] and Ercan Teksen and Anagun [9] worked on fuzzy attribute and variable charts, respectively. Fadaei and Pooya [10] worked on a fuzzy operating characteristic curve. For more details, the reader may refer to Jamkhaneh et al. [11] who discussed the rectifying fuzzy single sampling plan. Senturk and Erginel [12] studied variable control charts using fuzzy approach. Ercan Teksen and Anagun [9] worked on the fuzzy X-bar and R-charts. More details on fuzzy logic can be seen in Lee and Kim [13] and Grzegorzewski [14].

A fuzzy and imprecise data usually have indeterminate values. Fuzzy and vague data only considered the membership of the truth and false values. A neutrosophic logic deals with membership of truth, false and indeterminacy values. Therefore, the neutrosophic logic is useful to analyze the data having indeterminacy. Smarandache [15] introduced the neutrosophic statistics, which analyze the data when indeterminacy is presented. Aslam [16] and Aslam and Arif [17] introduced the neutrosophic statistics in the area of quality control. More details about the neutrosophic logic can be seen in references [18–23].

The similarity measure index (SMI) has been widely used in a variety of fields for classification purposes. In medical sciences, this index is used to classify the patients having a particular disease or not under indeterminacy, see De and Mishra [24]. By exploring the literature and to the best of the author's knowledge, there is no work on the process monitoring using SMI. In this paper, a method to classify the state of the process using SMI is introduced. The operational process of the proposed method is also given. The proposed classification method is simple in application compared to the existing method under classical statistics. It is expected that the proposed diagnosis method for the manufacturing process under the indeterminate environment will be effective, adequate and easy compared to the existing control charts under classical statistics. In Section 2, the SMI index is introduced in process control. A comparative study and application are given in Sections 3 and 4, respectively. Some concluded remarks are given in the last section.

2. The Proposed Chart Based on SMI

Suppose that $Z_N = s_N + u_N I; Z_N \in [Z_L, Z_U]$ is a neutrosophic number having a determined part s_N and an indeterminate part $u_N I$, $I \in [\text{inf}I, \text{inf}U]$ denotes the indeterminacy. Note here that $Z_N \in [Z_L, Z_U]$ is reduced to the determined number $Z_N = s_N$ when no indeterminacy is found. The practitioners cannot record observations of the variable of interest in the precise and determined form in the presence of indeterminacy. The monitoring of the data having neutrosophic numbers using classical statistics as discussed in reference [25] may mislead decision-makers regarding the state of the process. For example, the practitioners decide the process is in the control state using classical statistics, but in fact, some observations are in the indeterminacy interval. More details on this issue can be seen in reference [26]. Suppose that t_U, f_U and I_U presents the probabilities of the non-defective, defective and indeterminate. For the classification of the state of the process, let $t = 1$ and $f = 0$ show that the process is in control. Therefore, the value of SMI close to 1 indicates that the process is in control and the values away from the SMI show the process is out-of-control. The SMI from De and Mishra [24] is given by:

$$SMI = \sqrt{\left(1 - \frac{|(t_L - t_U) - (I_L - I_U) - (f_L - f_U)|}{3}\right)(1 - |(t_L - t_U) + (I_L - I_U) + (f_L - f_U)|)} \quad (1)$$

Note here $0 \le t_L, I_L, f_L \le 1$, $0 \le t_U, I_U, f_U \le 1$, $0 \le t_L + f_L \le 1$, $0 \le t_U + f_U \le 1$, $t_L + I_L + f_L \le 2$, $t_U + I_U + f_U \le 2$.

Based on SMI, the following classification procedure is proposed to diagnose the state of the manufacturing process.

Step 1: Select a random sample of size n and determine t_U, f_U and I_U.

Step 2: Compute the values of SMI. Classify the process in-control if SMI $\ge$ 0.95, otherwise, the out-of-control.

The operational process of the proposed method is also given with the help of Figure 1.

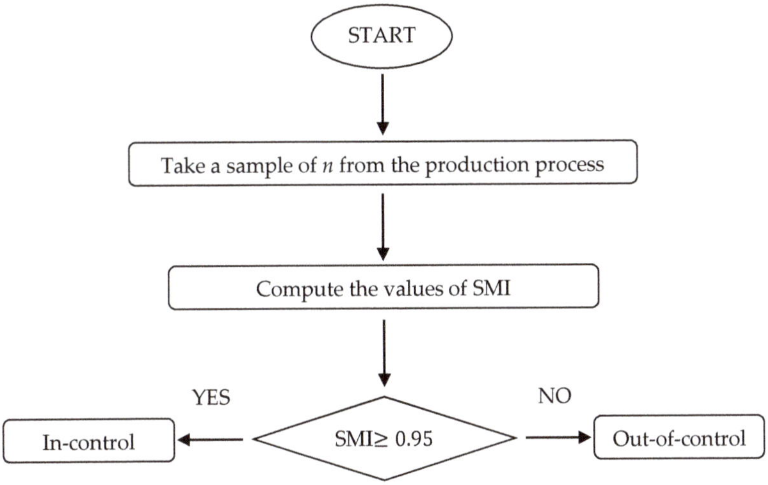

Figure 1. The operational process of the proposed method.

Note here that unlike the traditional control charts under classical statistics, the proposed chart using SMI is independent of the control limits and the control limits coefficients. The proposed chart reduces to the traditional control charts under classical statistics if no indeterminacy is found. Suppose that the probability of in-control of the process is determined from SMI. Let $SMI = P_{in}$, the P_{in} for the process is given by

$$P_{in} = \sqrt{\left(1 - \frac{|(t_L - t_U) - (I_L - I_U) - (f_L - f_U)|}{3}\right)(1 - |(t_L - t_U) + (I_L - I_U) + (f_L - f_U)|)} \quad (2)$$

The average run length (ARL) is used to see when on the average the process is expected to be out-of-control. The ARL under indeterminacy is given by:

$$ARL = \frac{1}{\left[\sqrt{\left(1 - \frac{|(t_L - t_U) - (I_L - I_U) - (f_L - f_U)|}{3}\right)(1 - |(t_L - t_U) + (I_L - I_U) + (f_L - f_U)|)}\right]} \quad (3)$$

The values of t_U, f_U and I_U for various values of n are given in Tables 1–3. Tables 1 and 2 are given when $n = 25$ and $n = 50$, respectively. Table 3 is presented for a variable sample size. In Table 4, the values of P_{in} and ARL are given for the parameters given in Tables 1–3. The classification of the state of the process based on SMI is also presented in Table 4. The process is said to be the in-control (IN) state if SMI ≥ 0.95 and the out-of-control (OOC) state if SMI < 0.95. It is noted no specific trend in ARL values. The following algorithm is used to classify the state of the process using the proposed method.

Step 1: Specify n and determine the values of t_U, f_U and I_U.
Step 2: Use the SMI to find the probability of in-control.
Step 3: Classify the process IN if SMI ≥ 0.95 and OOC if SMI < 0.95.

Table 1. Neutrosophic data when $n = 25$.

Sample No.	Sample Size	Number of Defective Units D	f_U	Number of Non-Defective Units ND	t_U	Number of Indeterminate Units I	I_U
1	25	3	0.12	21	0.84	1	0.04
2	25	4	0.16	19	0.76	2	0.08
3	25	2	0.08	23	0.92	0	0
4	25	5	0.2	20	0.8	4	0.16
5	25	2	0.08	22	0.88	1	0.04
6	25	1	0.04	22	0.88	2	0.08
7	25	0	0	20	0.8	5	0.2
8	25	4	0.16	21	0.84	0	0
9	25	6	0.24	17	0.68	2	0.08
10	25	1	0.04	23	0.92	1	0.04
11	25	2	0.08	20	0.8	3	0.12
12	25	5	0.2	18	0.72	2	0.08
13	25	4	0.16	19	0.76	2	0.08
14	25	8	0.32	16	0.64	1	0.04
15	25	3	0.12	21	0.84	1	0.04
16	25	2	0.08	21	0.84	2	0.08
17	25	5	0.2	17	0.68	3	0.12
18	25	3	0.12	19	0.76	3	0.12
19	25	7	0.28	17	0.68	1	0.04
20	25	1	0.04	23	0.92	1	0.04
21	25	0	0	23	0.92	2	0.08
22	25	2	0.08	19	0.76	4	0.16
23	25	5	0.2	17	0.68	3	0.12
24	25	7	0.28	17	0.68	1	0.04
25	25	8	0.32	17	0.68	0	0

Table 2. Neutrosophic data when $n = 50$.

Sample No.	Sample Size	Number of Defective Units D	f_U	Number of Non-Defective Units ND	t_U	Number of Indeterminate Units I	I_U
1	50	1	0.02	48	0.96	1	0.02
2	50	2	0.04	47	0.94	1	0.02
3	50	3	0.06	45	0.9	2	0.04
4	50	5	0.1	43	0.86	2	0.04
5	50	2	0.04	43	0.86	5	0.1
6	50	6	0.12	41	0.82	3	0.06
7	50	1	0.02	46	0.92	3	0.06
8	50	2	0.04	44	0.88	4	0.08
9	50	7	0.14	37	0.74	6	0.12
10	50	8	0.16	34	0.68	6	0.12
11	50	1	0.02	47	0.94	2	0.04
12	50	6	0.12	43	0.86	1	0.02
13	50	1	0.02	41	0.82	8	0.16
14	50	3	0.06	39	0.78	8	0.16
15	50	6	0.12	41	0.82	3	0.06
16	50	3	0.06	45	0.9	2	0.04
17	50	9	0.18	40	0.8	1	0.02
18	50	2	0.04	41	0.82	7	0.14
19	50	4	0.08	46	0.92	0	0
20	50	6	0.12	43	0.86	1	0.02
21	50	1	0.02	47	0.94	2	0.04
22	50	7	0.14	43	0.86	0	0
23	50	2	0.04	45	0.9	3	0.06
24	50	0	0	48	0.96	2	0.04
25	50	1	0.02	48	0.96	1	0.02

Table 3. Neutrosophic data with variable sample size.

Sample No.	Sample Size	Number of Defective units D	f_U	Number of Non-Defective Units ND	t_U	Number of Indeterminate Units I	I_U
1	100	12	0.120	78	0.780	10	0.10
2	80	8	0.100	67	0.838	5	0.06
3	80	6	0.075	69	0.863	5	0.06
4	100	9	0.090	89	0.890	2	0.02
5	110	10	0.091	99	0.900	1	0.01
6	110	12	0.109	98	0.891	0	0.00
7	100	11	0.110	85	0.850	4	0.04
8	100	16	0.160	79	0.790	5	0.05
9	90	10	0.111	66	0.733	14	0.16
10	90	6	0.067	72	0.800	12	0.13
11	110	20	0.182	89	0.809	1	0.01
12	120	15	0.125	99	0.825	6	0.05
13	120	9	0.075	108	0.900	3	0.03
14	120	8	0.067	107	0.892	5	0.04
15	110	6	0.055	95	0.864	9	0.08
16	80	8	0.100	72	0.900	0	0.00
17	80	10	0.125	69	0.863	1	0.01
18	80	7	0.088	68	0.850	5	0.06
19	90	5	0.056	78	0.867	7	0.08
20	100	8	0.080	88	0.880	4	0.04
21	100	5	0.050	88	0.880	7	0.07
22	100	8	0.080	91	0.910	1	0.01
23	100	10	0.100	88	0.880	2	0.02
24	90	6	0.067	80	0.889	4	0.04
25	90	9	0.100	80	0.889	1	0.01

Table 4. Classification of the process.

$n = 25$			$n = 50$			Variable Sample Size		
P_{in}	Classification	ARL	P_{in}	Classification	ARL	P_{in}	Classification	ARL
0.9451	OOC	18	0.9865	IN	74	0.9223	OOC	13
0.9165	OOC	12	0.9797	IN	49	0.9438	OOC	18
0.9729	IN	37	0.9660	IN	29	0.9526	IN	21
0.8265	OOC	6	0.9521	IN	21	0.9626	IN	27
0.9591	IN	24	0.9521	IN	21	0.9654	IN	29
0.9591	IN	24	0.9380	OOC	16	0.9629	IN	27
0.9309	OOC	14	0.9729	IN	37	0.9486	OOC	19
0.9451	OOC	18	0.9591	IN	24	0.9273	OOC	14
0.8869	OOC	9	0.9092	OOC	11	0.9040	OOC	10
0.9729	IN	37	0.8763	OOC	8	0.9300	OOC	14
0.9309	OOC	14	0.9797	IN	49	0.9335	OOC	15
0.9018	OOC	10	0.9521	IN	21	0.9398	OOC	17
0.9165	OOC	12	0.9380	OOC	16	0.9628	IN	27
0.8717	OOC	8	0.9237	OOC	13	0.9630	IN	27
0.9451	OOC	18	0.9380	OOC	16	0.9532	IN	21
0.9451	OOC	18	0.9660	IN	29	0.9660	IN	29
0.8869	OOC	9	0.9309	OOC	14	0.9526	IN	21
0.9165	OOC	12	0.9380	OOC	16	0.9480	OOC	19
0.8869	OOC	9	0.9729	IN	37	0.9526	IN	21
0.9729	IN	37	0.9521	IN	21	0.9591	IN	24
0.9729	IN	37	0.9797	IN	49	0.9591	IN	24
0.9165	OOC	12	0.9521	IN	21	0.9695	IN	33
0.8869	OOC	9	0.9660	IN	29	0.9591	IN	24
0.8869	OOC	9	0.9865	IN	74	0.9610	IN	26
0.8869	OOC	9	0.9865	IN	74	0.9619	IN	26

Note: IN = in-control and OOC = out-of-control.

3. Comparative Study

In this section, a comparison of the effectiveness of the proposed method is given over the control charts under classical statistics reported in reference [25]. According to Aslam et al. [26], a method which deals with indeterminacy is said to be more effective than the method which provides the determined values. The proposed method reduces to the traditional method under classical statistics if no indeterminacy is recorded. From reference [25], it is noted that the control chart under classical statistics does not consider the measure of indeterminacy which makes it limited to be used in an uncertainty environment. The performance of the existing control chart depends on the control limit coefficient which is determined through the complicated simulation process. On the other hand, the current method considered the measure of indeterminacy to evaluate the performance of the control chart. In addition, the proposed method is independent of the control limit coefficient. The proposed process can be applied easily to classify the state of the process. Note here that, the proposed method reduces to the method under classical statistics if no indeterminacy is found in the production data. The values of ARL from the proposed method and method under classical statistics discussed by Montgomery [25] are shown in Table 5 when $n = 25$ and $D = 2$. It is well-known theory that the smaller the values of ARL means more efficient the control chart process [25]. From Table 5, it can be seen that the proposed method provides the smaller values of ARL than the existing method. It means the proposed control chart has the ability to detect a shift in the process earlier than the method under classical statistics. For example, when $n = 25$ and $d = 2$, the value of ARL of the existing method from Table 5 is 37. On the other hand, the proposed method provides smaller values of ARL which are 24, 14, 18 and 12. From this comparison, it is concluded that the process is classified as IN. The industrial engineers can expect the process to be out-of-control at the 37th sample by using the existing method and on the 12th sample for sample number 22 using the proposed method. Therefore, the proposed method is efficient in detecting shifts earlier than the existing method. From this comparison, is the authors concluded that the proposed method is more effective than the existing charts as it considered the measure of indeterminacy and indicated when the process was OCC.

Table 5. The comparison of the proposed method with existing method when $n = 25$ and $D = 2$.

Sample No	ARL	Control Chart
3	37	Under classical statistics
5	24	Under neutrosophic statistics
11	14	Under neutrosophic statistics
16	18	Under neutrosophic statistics
22	12	Under neutrosophic statistics

4. Application

In this section, a discussion of the application of the proposed method in an orange juice company is given. According to Montgomery [25], "Frozen orange juice concentrate is packed in 6-oz cardboard cans. These cans are formed on a machine by spinning them from cardboard stock and attaching a metal bottom panel". By inspection, it was found that a sample of 50 juice cans was formed. Some cans were found to be leaking and some were labeled as good. For some cans, the industrial engineer is indeterminate about whether the juice product is labeled as either conforming and non-conforming. Therefore, classical statistics cannot be applied to monitor the process in the presence of indeterminacy. The data for $n = 50$ is shown in Table 2. The classification of the state of the process for the juice cans is shown in Table 4. From Table 4, it is noted that the first five subgroups show that the process is the IN control state. The 5th subgroup shows that the process is OOC and industrial engineer should take action to bring back the process in the IN state. It is noted that overall eight samples are in OOC state. From this study, it is concluded that the use of the proposed method to classify the state of the process is quite easy, effective and adequate to be applied under an uncertainty environment.

5. Conclusions and Remarks

In this paper, the diagnosis of the manufacturing process under the indeterminate environment was presented. The similarity measure index was used to find the probability of the in-control and the out-of-control of the process. The average run length (ARL) was also computed for various values of specified parameters. An industrial example was given to explain the state of the process. An industrial example under the indeterminate environment was presented. From this study, it is concluded that the proposed diagnosis scheme under the neutrosophic statistics is quite simple and effective for the current state of the manufacturing process under uncertainty. The practitioners can apply the proposed method to save time and efforts in the industry. The proposed method using non-normal measures can be considered as future research.

Author Contributions: Conceived and designed the experiments, M.A.; Performed the experiments, M.A. Analyzed the data, M.A. and O.H.A.; Contributed reagents/materials/analysis tools, M.A.; Wrote the paper, M.A. and O.H.A.

Funding: This article was funded by the Deanship of Scientific Research (DSR) at King Abdulaziz University, Jeddah. The authors, therefore, acknowledge with thanks DSR technical and financial support.

Acknowledgments: The authors are deeply thankful to editor and reviewers for their valuable suggestions to improve the quality of this manuscript.

Conflicts of Interest: The authors declare no conflict of interest.

References

1. Epprecht, E.K.; Costa, A.F.; Mendes, F.C. Adaptive Control Charts for Attributes. *IIE Trans.* **2003**, *35*, 567–582. [CrossRef]
2. Chiu, J.-E.; Kuo, T.-I. Attribute Control Chart for Multivariate Poisson Distribution. *Commun. Stat. Theory Methods* **2007**, *37*, 146–158. [CrossRef]
3. Hsu, L.-F. Note on 'Design of double-and triple-sampling X-bar control charts using genetic algorithms'. *Int. J. Prod. Res.* **2004**, *42*, 1043–1047. [CrossRef]
4. Ho, L.L.; Quinino, R.C. An attribute control chart for monitoring the variability of a process. *Int. J. Prod. Econ.* **2013**, *145*, 263–267.
5. Aslam, M.; Arif, O.H.; Jun, C.-H. An Attribute Control Chart Based on the Birnbaum-Saunders Distribution Using Repetitive Sampling. *IEEE Access* **2016**, *4*, 9350–9360. [CrossRef]
6. Aslam, M.; Arif, O.H.; Jun, C.-H. An attribute control chart for a Weibull distribution under accelerated hybrid censoring. *PLoS ONE* **2017**, *12*, e0173406. [CrossRef] [PubMed]
7. Jeyadurga, P.; Balamurali, S.; Aslam, M. Design of an attribute np control chart for process monitoring based on repetitive group sampling under truncated life tests. *Commun. Stat. Theory Methods* **2018**, *47*, 5934–5955. [CrossRef]
8. Afshari, R.; Gildeh, B.S. Construction of fuzzy multiple deferred state sampling plan. In Proceedings of the 2017 Joint 17th World Congress of International Fuzzy Systems Association and 9th International Conference on Soft Computing and Intelligent Systems (IFSA-SCIS), Otsu, Japan, 27–30 June 2017.
9. Ercan Teksen, H.; Anagun, A.S. Different methods to fuzzy X⁻-R control charts used in production: Interval type-2 fuzzy set example. *J. Enterp. Inf. Manag.* **2018**, *31*, 848–866. [CrossRef]
10. Fadaei, S.; Pooya, A. Fuzzy U control chart based on fuzzy rules and evaluating its performance using fuzzy OC curve. *TQM J.* **2018**, *30*, 232–247. [CrossRef]
11. Jamkhaneh, E.B.; Sadeghpour-Gildeh, B.; Yari, G. Important criteria of rectifying inspection for single sampling plan with fuzzy parameter. *Int. J. Contemp. Math. Sci.* **2009**, *4*, 1791–1801.
12. Senturk, S.; Erginel, N. Development of fuzzy X⁻~-R~ and X⁻~-S~ control charts using α-cuts. *Inf. Sci.* **2009**, *179*, 1542–1551. [CrossRef]
13. Lee, H.; Kim, S. Black-box classifier interpretation using decision tree and fuzzy logic-based classifier implementation. *Int. J. Fuzzy Log. Intell. Syst.* **2016**, *16*, 27–35. [CrossRef]
14. Grzegorzewski, P. On separability of fuzzy relations. *Int. J. Fuzzy Log. Intell. Syst.* **2017**, *17*, 137–144. [CrossRef]
15. Smarandache, F. *Introduction to Neutrosophic Statistics*; Infinite Study: El Segundo, CA, USA, 2014.

16. Aslam, M. Design of Sampling Plan for Exponential Distribution under Neutrosophic Statistical Interval Method. *IEEE Access* **2018**, *6*, 64153–64158. [CrossRef]
17. Aslam, M.; Arif, O.H. Testing of Grouped Product for the Weibull Distribution Using Neutrosophic Statistics. *Symmetry* **2018**, *10*, 403. [CrossRef]
18. Broumi, S.; Bakali, A.; Talea, M.; Smarandache, F.; ALi, M. Shortest Path Problem under Bipolar Neutrosphic Setting. *Appl. Mech. Mater.* **2017**, *859*, 59–66. [CrossRef]
19. Broumi, S.; Bakali, A.; Talea, M.; Smarandache, F.; Ullah, K. Bipolar Neutrosophic Minimum Spanning Tree. Available online: https://books.google.com.sa/books?id=VopuDwAAQBAJ&printsec=frontcover&hl=ar#v=onepage&q&f=false (accessed on 25 August 2019).
20. Abdel-Basset, M.; Gunasekaran, M.; Mohamed, M.; Chilamkurti, N. Three-way decisions based on neutrosophic sets and AHP-QFD framework for supplier selection problem. *Future Gener. Comput. Syst.* **2018**, *89*, 19–30. [CrossRef]
21. Abdel-Basset, M.; Gunasekaran, M.; Mohamed, M.; Smarandache, F. A novel method for solving the fully neutrosophic linear programming problems. *Neural Comput. Appl.* **2018**, *31*, 1595–1605. [CrossRef]
22. Broumi, S.; Bakali, A.; Talea, M.; Smarandache, F.; Kishore, K.K.; Şahin, R. Shortest path problem under interval valued neutrosophic setting. *J. Fundam. Appl. Sci.* **2018**, *10*, 168–174.
23. Abdel-Basset, M.; Nabeeh, N.A.; El-Ghareeb, H.A.; Aboelfetouh, A. Utilising neutrosophic theory to solve transition difficulties of IoT-based enterprises. *Enterp. Inf. Syst.* **2019**, 1–21. [CrossRef]
24. De, S.; Mishra, J. Inconsistent Data Using Neutrosophic Logic to Disease Diagnosis for Prevention. In Proceedings of the 13th International Conference on Recent Trends in Engineering Science and Management, School of Electronics and Communications Engineering, REVA, Bangalore, India, 23–24 April 2018.
25. Montgomery, D.C. *Introduction to Statistical Quality Control*; John Wiley & Sons: Jefferson City, MO, USA, 2007.
26. Aslam, M.; Khan, N.; Khan, M.Z. Monitoring the Variability in the Process Using Neutrosophic Statistical Interval Method. *Symmetry* **2018**, *10*, 562. [CrossRef]

mathematics

MDPI

Article

Time-Truncated Group Plan under a Weibull Distribution based on Neutrosophic Statistics

Muhammad Aslam [1,*], P. Jeyadurga [2], Saminathan Balamurali [2] and Ali Hussein AL-Marshadi [1]

1 Department of Statistics, Faculty of Science, King Abdulaziz University, Jeddah 21551, Saudi Arabia; aalmarshadi@kau.edu.sa
2 Department of Computer Applications, Kalasalingam Academy of Research and Education, Krishnankoil, TN 626126, India; jeyadurga@klu.ac.in (P.J.); sbmurali@rediffmail.com (S.B.)
* Correspondence: magmuhammad@kau.edu.sa or aslam_ravian@hotmail.com

Received: 26 July 2019; Accepted: 20 September 2019; Published: 27 September 2019

Abstract: The aim of reducing the inspection cost and time using acceptance sampling can be achieved by utilizing the features of allocating more than one sample item to a single tester. Therefore, group acceptance sampling plans are occupying an important place in the literature because they have the above-mentioned facility. In this paper, the designing of a group acceptance sampling plan is considered to provide assurance on the product's mean life. We design the proposed plan based on neutrosophic statistics under the assumption that the product's lifetime follows a Weibull distribution. We determine the optimal parameters using two specified points on the operating characteristic curve. The discussion on how to implement the proposed plan is provided by an illustrative example.

Keywords: time-truncated test; Weibull distribution; risk; uncertainty; neutrosophic

1. Introduction

The ambition of each producer is to globalize their business by means of marketing the products. However, few producers reach this goal since they only make sincere efforts in improving and controlling the product's quality to accomplish this target. The producer who enhances the product's quality need not concern its globalization because the continuous improvement in quality helps to increase the positive opinion of the products and to fulfill the consumer's expectations. Hence, the involvement of the producers with great efforts supports to attain the desired result and to achieve the ambition. For quality improvement and maintenance purposes, the producer uses certain statistical techniques, namely control charts and acceptance sampling (see Montgomery [1] and Schilling and Neubauer [2]). In spite of the application of control charts in quality maintenance via monitoring the manufacturing process, it is not suitable for assuring the quality of the finished products. But there is a necessity to provide quality assurance for the products before they are received by the consumer. Under this situation, the manufacturers may prefer complete inspection. However, complete inspections are not appropriate for all situations because they are costly, require quality inspectors, and are time consuming. Therefore, in most of the cases, manufacturers adopt sampling inspections to provide quality assurance. In sampling inspection, a sample of items is selected randomly from the entire lot for inspection.

Acceptance sampling is also a form of sampling inspection, in which the decision to accept or reject a lot is made based on the results of sample items taken from the concerned lot. Obviously, acceptance sampling overcomes the drawbacks of complete inspections, such as inspection cost and time consumption, since it inspects only a part of the items of the lot for making decisions. Acceptance sampling plans yield the sample size and acceptance criteria associated with the sampling rules to be implemented. For further details on acceptance sampling, one may refer to Dodge [3] and Schilling and Neubauer [2]. In the literature, several sampling plans are available for lot sentencing with different sampling procedures; however, a single-sampling plan (SSP) is the most basic, as well as the easiest, sampling plan in terms of the implementation process. In SSP, a single sample of size n is taken for lot sentencing, and the acceptance/rejection decision is made immediately by comparing the sample results with acceptance numbers determined from attribute inspections or with acceptance criteria from variables inspections. Many authors have investigated SSPs under various situations (see, for example, Loganathan et al. [4], Liu and Cui [5], Govindaraju [6], and Hu and Gui [7]).

In SSP implementation, a sample of n items is distributed to n testers, and the decision is made after consolidating the information obtained from all the testers. Obviously, it requires much time to make a decision, and the inspection cost is also high. One can overcome these drawbacks by implementing a group acceptance sampling plan (GASP) instead of using SSP. In GASP, a certain number of sample items are allocated to a single tester, and the test is conducted simultaneously on the sample items. Therefore, the testing time and inspection cost are reduced automatically under GASP when compared to SSP. It is to be mentioned that the number of testers involved in the inspection is frequently referred to as the number of groups, and the number of sample items allocated to each group is defined as the group size. For the purposes of making a decision on the lot by utilizing minimum cost and time, GASP has been used for the inspection of different quality characteristics by several authors (see, for example, Aslam and Jun [8]).

When industrial practitioners are uncertain about the parameters, the inspection cannot be done using traditional sampling plans. In this case, the use of fuzzy-based sampling plans is the best alternative to traditional sampling plans. Fuzzy-based sampling plans have been widely used for lot sentencing. Kanagawa and Ohta [9] proposed a single-attribute plan using fuzzy logic. More details on fuzzy sampling plans can be seen in Chakraborty [10], Jamkhaneh and Gildeh [11], Turanoğlu et al. [12], Jamkhaneh and Gildeh [13], Tong and Wang [14], Uma and Ramya [15], Afshari and Gildeh [16], and Khan et al. [17].

The fuzzy approach has been used to compute the degree of truth. Fuzzy logic is a special case of neutrosophic logic. The later approach computes measures of indeterminacy in addition to the first approach (see Smarandache [18]). Abdel-Basset et al. [19] discussed the application of neutrosophic logic in decision making. Abdel-Basset et al. [20] worked on linear programming using the idea of neutrosophic logic. Broumi et al. [21] provided the minimum spanning tree using neutrosophic logic. More details can be seen in [22,23]. Neutrosophic statistics is treated as an extension of classical statistics, in which set values are considered rather than crisp values. Sometimes, the data may be imprecise, incomplete, and unknown, and exact computation is not possible. Under these situations, the neutrosophic statistics concept is used (see Smarandache [24]). Broumi and Smarandache [25] discussed the correlations of sets using neutrosophic logic. More details about the use of neutrosophic logic in sets can be seen in [26–28]. But one can use a set of values (that respectively approximates these crisp numbers) for a single variable using neutrosophic statistics. Chen et al. [29,30] introduced neutrosophic numbers to solve rock engineering problems. Patro and Smarandache [31] and Alhabib et al. [32] discussed some basicsofprobablity distribution under neutrosophic numbers. Nowadays, the neutrosophic statistics concept is used for quality control purposes. When designing the control chart and sampling plans under classical statistics, it is assumed that the value which represents the quality of the product is known. But in neutrosophic statistics, such value is indeterminate or lies between an interval. Some researchers have designed the control chart and acceptance sampling plans under these statistics (see, for example, Aslam et al. [33]). Aslam [34] introduced neutrosophic statistics in the area of acceptance sampling plans. Aslam and Arif [35] proposed a sudden death testing plan under uncertainty.

As mentioned earlier, Aslam and Jun [8] designed GASP to ensure the Weibull-distributed mean life of the products under classical statistics. They determined the optimal parameters for some calculated values of failure probability; however, they did not consider the case where the failure probability is uncertain. Therefore, in this paper, we attempted to design GASP for providing Weibull-distributed mean life assurance where the values of shape parameters and failure probabilities are uncertain. That is, we considered the design of GASP under neutrosophic statistics, which is the main difference between the proposed work and the work done by Aslam and Jun [8]. We will compare the proposed plan with the existing sampling plan under classical statistics in terms of the sample size required for inspection. We expect that the proposed plan will be quite effective, adequate, and efficient compared to the existing plan in an uncertainty environment.

2. Design of the Proposed Plan using Neutrosophic Statistics

The method to design the proposed GASP for providing quality assurance of the product in terms of mean life is discussed in this section. The ratio between the true mean life and the specified mean life of the product is considered as the quality of the product. A Weibull distribution is considered as an appropriate model to express the lifetime of the product because of its flexible nature. So, we assume that the lifetime of the product $t_N \in \{t_L, t_U\}$ under study follows a neutrosophic Weibull distribution, which has the shape parameter $\delta_N \in \{\delta_L, \delta_U\}$ and scale parameter $\lambda_N \in \{\lambda_L, \lambda_U\}$. Then, the cumulative distribution function (cdf) of the Weibull distribution is obtained as follows.

$$F(t_N; \lambda_N, \delta_N) = 1 - \exp\left(-\left(\frac{t_N}{\lambda_N}\right)^{\delta_N}\right), t_N \geq 0, \lambda_N > 0, \delta_N > 0. \quad (1)$$

In this study, it is assumed that the scale parameter λ_N is unknown and the shape parameter δ_N is known. It can be seen that the cdf depends only on t_N/λ_N since the shape parameter is known. One can estimate the shape parameter from the available history of the production process when it is unknown. The true mean life of the product under the neutrosophic Weibull distribution is calculated by the following equation

$$\mu_N = \left(\frac{\lambda_N}{\delta_N}\right)\Gamma\left(\frac{1}{\delta_N}\right), \quad (2)$$

where $\Gamma(.)$ represents the complete gamma function. Then, the probability that the product will fail before it reaches the experiment time t_{N0},is denoted by p_N and is given as follows

$$p_N = 1 - \exp\left(-\left(\frac{t_{N0}}{\lambda_N}\right)^{\delta_N}\right). \quad (3)$$

As pointed out by Aslam and Jun [8], we can write t_{N0} as a constant multiple of the specified mean life μ_{N0}, such as $t_{N0} = t_0 = a\mu_0 a\mu_{N0}$ where 'a' is called the experiment termination ratio. Also, we can express the unknown scale parameter in terms of the true mean life and known shape parameters. After t_{N0}, λ_N value substitution, and possible simplification, one can obtain the probability that the product will fail before attaining the experiment time $t_{N0,}$ using the following equation.

$$p_N = 1 - \exp\left(-a^{\delta_N}\left(\frac{\mu_{N0}}{\mu_N}\right)^{\delta_N}\left(\frac{\Gamma\left(\frac{1}{\delta_N}\right)}{\delta_N}\right)^{\delta_N}\right), \quad (4)$$

With respect to the ratios between the true mean life and the specified mean life, μ_N/μ_{N0}, the acceptable quality level (AQL, i.e., p_{N1}) and limiting quality level (LQL, i.e., p_{N2}) are defined. That is, the failure probabilities obtained when the mean ratio values greater than one are taken as AQL and the same are obtained at a mean ratio equal to one are considered as LQL. The operating procedure of the proposed GASP for a time-truncated life test is described as follows:

Step 1. Take a sample of $n_N \in \{n_L, n_U\}$ items randomly from the submitted lot and distribute $r_N \in \{r_L, r_U\}$ items into $g_N \in \{g_L, g_U\}$ groups. Then, conduct the life test on the sample items for the specified time t_{N0}.

Step 2. Observe the test and count number of sample items failed in each group before reaching experiment time t_{N0} and denote it as $d_N \in \{d_L, d_U\}$.

Step 3. If at most, c_N sample items found to be failed in each of all g_N groups, then accept the lot where $c_N \in \{c_L, c_U\}$. Otherwise, reject the lot.

Two parameters used to characterize the proposed plan are number of groups g_N and the acceptance number c_N. It is to be noted that $r_N \in \{r_L, r_U\}$ denotes the number of items in each group and is called the group size. The operating procedure of the proposed GASP is represented by a flow chart and is shown in Figure 1.

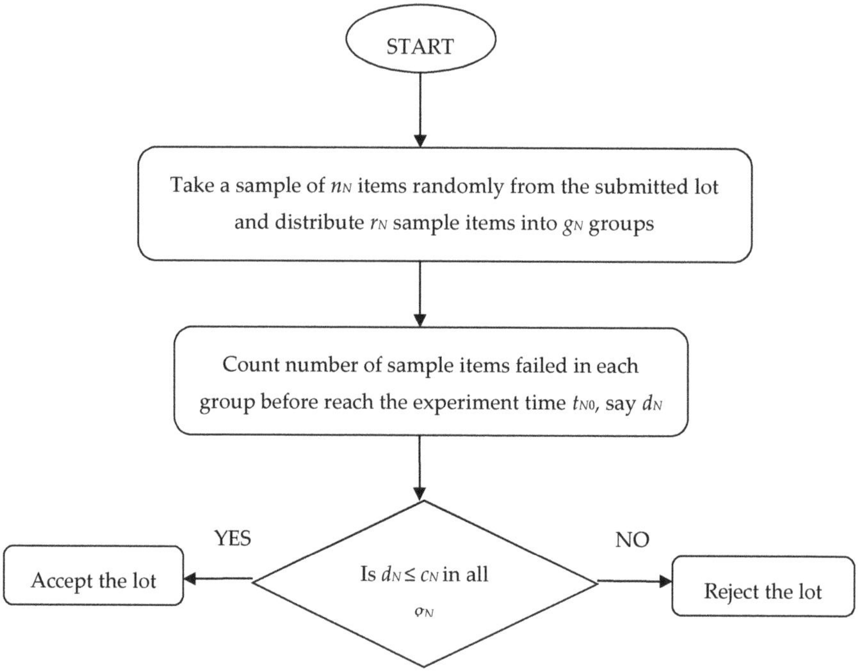

Figure 1. Operating procedure of the proposed group acceptance sampling plan (GASP) under a truncated life test.

In general, an operating characteristic (OC) function helps to investigate the performance of the sampling plan. The OC function of the proposed GASP under a Weibull model based on time-truncated test is given by

$$P_{aN}(p_N) = \left[\sum_{d_N=0}^{c_N} \binom{r_N}{d_N} p_N^{d_N} (1-p_N)^{r_N-d_N} \right]^{g_N}. \tag{5}$$

Generally, each producer wishes that the sampling plan should provide a chance greater than $(1-\alpha)$ to accept the product when the product quality is at AQL, where α is the producer's risk, whereas the consumer wants that the chance to accept the lot to be less than β when quality of the product is at LQL, where β is the consumer's risk. Obviously, the sampling plan that involves the minimum risks to both producer and consumer will be favorable. The design of the sampling plan by considering AQL and LQL, along with producer and consumer risks is known as two points on the OC curve approach and this approach is considered as the most important among others. Similarly, the sampling plan that makes its decision on the submitted lot using minimum sample size or average sample number (ASN) will be attractive. Therefore, in this study, we design GASP with the intention of assuring a Weibull-distributed mean life of the products with minimum sample size and minimum cost using two points on the OC curve approach. It should be mentioned that the ASN of the proposed plan is the product of the number of groups and group size (i.e., $n_N = g_N r_N$). For determining the optimal parameters, we use the following optimization problem.

$$\begin{gathered}\text{Minimize } g_N \\ \text{Subject to } P_a(p_{N1}) \geq 1-\alpha, \\ P_a(p_{N2}) \leq \beta, \\ g_N \geq 1, r_N > 1,\ c_N \geq 0,\end{gathered} \tag{6}$$

where p_{N1} and p_{N2} are the failure probabilities obtained from the following equations

$$p_{N1} = 1-\exp\left(-a^{\delta_{N1}}\left(\frac{\mu_{N0}}{\mu_N}\right)^{\delta_{N1}}\left(\frac{\Gamma\left(\frac{1}{\delta_{N1}}\right)}{\delta_{N1}}\right)^{\delta_{N1}}\right), \delta_{N1} \in \{\delta_{L1}, \delta_{U1}\}, \tag{7}$$

$$p_{N2} = 1-\exp\left(-a^{\delta_{N2}}\left(\frac{\mu_{N0}}{\mu_N}\right)^{\delta_{N2}}\left(\frac{\Gamma\left(\frac{1}{\delta_{N2}}\right)}{\delta_{N2}}\right)^{\delta_{N2}}\right), \delta_{N2} \in \{\delta_{L2}, \delta_{U2}\}, \tag{8}$$

$$P_{aN}(p_{N1}) = \left[\sum_{d_N=0}^{c_N}\binom{r_N}{d_N} p_{N1}^{d_N}(1-p_{N1})^{r_N-d_N}\right]^{g_N}, \tag{9}$$

$$P_{aN}(p_{N2}) = \left[\sum_{d_N=0}^{c_N}\binom{r_N}{d_N} p_{N2}^{d_N}(1-p_{N2})^{r_N-d_N}\right]^{g_N}. \tag{10}$$

In this designing, we define AQL as the failure probability corresponding to the mean ratios μ_N/μ_{N0} = 2, 4, 6, 8, 10. Similarly, the LQL is defined as the failure probability corresponding to the mean ratio $\mu_N/\mu_{N0} = 1$. The optimal parameters of the proposed GASP are determined for various combinations of group size, shape parameter, and producer's risk. We used the grid search method under neutrosophic statistics to find the optimal values of parameters $[g_L, g_U]$ and $[c_L, c_U]$. We selected those values of parameters from several combinations of parameters that satisfy the given conditions where the range between g_L and g_U is at a minimum. For this determination, we considered two sets of group sizes, such as r_N = {10, 12} and r_N = {4, 6}, and two sets of shape parameters, such as δ_N = {0.9, 1.1} and δ_N = {1.9, 2.1}. Similarly, the producer risks are assumed to be $\alpha = 0.1$ and $\alpha = 0.05$, and four values of the consumer's risk, namely β = 0.25, 0.10, 0.05, 0.01, are used. The experiment termination ratios involved in this determination are $a = 0.5$ and $a = 1$. Then, the optimal parameters are reported in Tables 1–4. We can observe the following trends from tables.

i. In most of the cases, the number of groups required for inspection decreases if the constant '*a*' increases from 0.5 to 1.
ii. For fixed values of δ_N, α, β, a, and μ_N/μ_{N0}, the number of groups increases when group size decreases. There is no particular change in the number of groups when the mean ratio increases.

Table 1. Optimal parameters of the proposed GASP under neutrosophic statistics when $r_N = [10, 12]$ and $\delta_N = [0.9, 1.1]$.

		$a = 0.5$			$a = 1.0$		
β	μ_N/μ_{N0}	g_N	c_N	$Pa_N(p_{N1})$	g_N	c_N	$Pa_N(p_{N1})$
0.25	2	[19, 44]	[6, 7]	[0.919, 0.983]	[5, 7]	[7, 8]	[0.904, 0.951]
	4	[2, 4]	[3, 4]	[0.905, 0.987]	[1, 3]	[4, 5]	[0.910, 0.957]
	6	[1, 3]	[2, 4]	[0.919, 0.999]	[1, 3]	[4, 6]	[0.974, 0.999]
	8	[1, 3]	[2, 4]	[0.956, 1.000]	[1, 3]	[4, 7]	[0.990, 1.000]
	10	[1, 3]	[2, 3]	[0.974, 0.998]	[1, 3]	[3, 7]	[0.974, 1.000]
0.1	2	*	*	*	[26, 28]	[8, 9]	[0.924, 0.969]
	4	[2, 4]	[3, 4]	[0.905, 0.987]	[1, 3]	[4, 6]	[0.910, 0.992]
	6	[2, 4]	[3, 4]	[0.969, 0.998]	[1, 3]	[4, 6]	[0.974, 0.999]
	8	[2, 4]	[2, 3]	[0.915, 0.994]	[1, 3]	[3, 5]	[0.951, 0.999]
	10	[2, 4]	[3, 4]	[0.994, 1.000]	[1, 3]	[3, 5]	[0.974, 1.000]
0.05	2	*	*	*	[34, 36]	[8, 9]	[0.902, 0.960]
	4	[7, 11]	[4, 5]	[0.935, 0.996]	[3, 5]	[5, 7]	[0.930, 0.998]
	6	[3, 6]	[3, 4]	[0.954, 0.997]	[2, 4]	[4, 5]	[0.949, 0.993]
	8	[2, 4]	[2, 3]	[0.915, 0.994]	[1, 3]	[3, 4]	[0.951, 0.991]
	10	[2, 4]	[2, 3]	[0.948, 0.998]	[1, 3]	[3, 4]	[0.974, 0.997]
0.01	2	*	*	*	*	*	*
	4	[9, 17]	[4, 5]	[0.917, 0.994]	[4, 6]	[5, 6]	[0.908, 0.984]
	6	[5, 8]	[3, 4]	[0.925, 0.996]	[2, 4]	[4, 5]	[0.949, 0.993]
	8	[5, 8]	[3, 4]	[0.968, 0.999]	[2, 4]	[3, 4]	[0.905, 0.987]
	10	[3, 5]	[2, 3]	[0.923, 0.997]	[2, 4]	[3, 6]	[0.948, 1.000]

*: Plan does not exist.

Table 2. Optimal parameters of the proposed GASP under neutrosophic statistics when $r_N = [10, 12]$ and $\delta_N = [1.9, 2.1]$.

		$a = 0.5$			$a = 1.0$		
β	μ_N/μ_{N0}	g_N	c_N	$Pa_N(p_{N1})$	g_N	c_N	$Pa_N(p_{N1})$
0.25	2	[5, 11]	[2, 3]	[0.926, 0.988]	[2, 4]	[5, 6]	[0.990, 0.995]
	4	[2, 4]	[1, 2]	[0.981, 0.999]	[1, 3]	[3, 4]	[0.998, 1.000]
	6	[2, 4]	[1, 2]	[0.996, 1.000]	[1, 3]	[2, 4]	[0.998, 1.000]
	8	[2, 4]	[1, 2]	[0.999, 1.000]	[1, 3]	[1, 5]	[0.990, 1.000]
	10	[2, 4]	[1, 2]	[0.999, 1.000]	[1, 3]	[2, 3]	[1.000, 1.000]
0.1	2	[39, 65]	[3, 4]	[0.942, 0.995]	[2, 4]	[4, 6]	[0.945, 0.995]
	4	[4, 7]	[1, 2]	[0.962, 0.999]	[1, 3]	[2, 3]	[0.985, 0.997]
	6	[3, 7]	[1, 2]	[0.994, 1.000]	[1, 3]	[2, 4]	[0.998, 1.000]
	8	[5, 7]	[1, 2]	[0.996, 1.000]	[1, 3]	[1, 3]	[0.990, 1.000]
	10	[3, 7]	[1, 2]	[0.999, 1.000]	[1, 3]	[2, 4]	[1.000, 1.000]
0.05	2	[67, 84]	[3, 4]	[0.902, 0.994]	[3, 5]	[4, 6]	[0.918, 0.994]
	4	[4, 8]	[1, 2]	[0.962, 0.998]	[1, 3]	[2, 4]	[0.985, 1.000]
	6	[5, 8]	[1, 2]	[0.989, 1.000]	[1, 3]	[2, 5]	[0.998, 1.000]
	8	[4, 8]	[1, 2]	[0.997, 1.000]	[1, 3]	[2, 5]	[1.000, 1.000]
	10	[5, 8]	[1, 2]	[0.998, 1.000]	[1, 3]	[1, 5]	[0.996, 1.000]
0.01	2	[59, 129]	[3, 4]	[0.914, 0.990]	[7, 9]	[5, 6]	[0.964, 0.989]
	4	[6, 13]	[1, 2]	[0.944, 0.997]	[2, 4]	[2, 5]	[0.970, 1.000]
	6	[9, 13]	[1, 2]	[0.981, 1.000]	[1, 3]	[1, 3]	[0.973, 1.000]
	8	[7, 13]	[1, 2]	[0.995, 1.000]	[1, 3]	[1, 2]	[0.990, 0.999]
	10	[7, 13]	[1, 2]	[0.998, 1.000]	[1, 3]	[1, 2]	[0.996, 1.000]

Table 3. Optimal parameters of the proposed GASP under neutrosophic statistics when r_N = [4, 6] and δ_N = [0.9, 1.1].

		a = 0.5			a = 1.0		
β	μ_N/μ_{N0}	g_N	c_N	$Pa_N(p_{N1})$	g_N	c_N	$Pa_N(p_{N1})$
0.25	2	*	*	*	*	*	*
	4	[6, 10]	[2, 3]	[0.932, 0.990]	[8, 10]	[3, 4]	[0.964, 0.988]
	6	[7, 10]	[2, 3]	[0.970, 0.998]	[2, 4]	[2, 3]	[0.955, 0.988]
	8	[2, 4]	[1, 2]	[0.928, 0.994]	[2, 4]	[2, 3]	[0.977, 0.996]
	10	[2, 4]	[1, 2]	[0.950, 0.997]	[1, 3]	[1, 2]	[0.923, 0.980]
0.1	2	*	*	*	*	*	*
	4	[68, 88]	[3, 4]	[0.967, 0.997]	[12, 14]	[3, 4]	[0.947, 0.983]
	6	[13, 17]	[2, 3]	[0.945, 0.997]	[3, 5]	[2, 3]	[0.933, 0.985]
	8	[13, 17]	[2, 3]	[0.973, 0.999]	[3, 5]	[2, 3]	[0.965, 0.995]
	10	[3, 5]	[1, 2]	[0.926, 0.996]	[3, 5]	[2, 3]	[0.980, 0.998]
0.05	2	*	*	*	*	*	*
	4	[103, 115]	[3, 4]	[0.951, 0.996]	[16, 18]	[3, 4]	[0.930, 0.978]
	6	[19, 22]	[2, 3]	[0.920, 0.996]	[4, 6]	[2, 3]	[0.911, 0.982]
	8	[17, 22]	[2, 3]	[0.965, 0.999]	[4, 6]	[2, 3]	[0.954, 0.994]
	10	[4, 7]	[1, 2]	[0.902, 0.994]	[4, 6]	[2, 3]	[0.973, 0.998]
0.01	2	*	*	*	*	*	*
	4	[170, 176]	[3, 4]	[0.920, 0.993]	*	*	*
	6	[23, 34]	[2, 3]	[0.904, 0.994]	[24, 26]	[3, 4]	[0.970, 0.996]
	8	[32, 34]	[2, 3]	[0.934, 0.998]	[6, 8]	[2, 3]	[0.932, 0.992]
	10	[28, 34]	[2, 3]	[0.967, 0.999]	[6, 8]	[2, 3]	[0.960, 0.997]

*: Plan does not exist.

Table 4. Optimal parameters of the proposed GASP under neutrosophic statistics when r_N = [4, 6] and δ_N = [1.9, 2.1].

		a = 0.5			a = 1.0		
β	μ_N/μ_{N0}	g_N	c_N	$Pa_N(p_{N1})$	g_N	c_N	$Pa_N(p_{N1})$
0.25	2	[156, 164]	[2, 3]	[0.902, 0.993]	[3, 5]	[2, 3]	[0.929, 0.958]
	4	[8, 23]	[1, 2]	[0.989, 1.000]	[1, 3]	[1, 3]	[0.983, 1.000]
	6	[20, 23]	[1, 2]	[0.994, 1.000]	[1, 3]	[1, 3]	[0.996, 1.000]
	8	[8, 23]	[1, 2]	[0.999, 1.000]	[1, 3]	[1, 2]	[0.999, 1.000]
	10	[9, 23]	[1, 2]	[1.000, 1.000]	[1, 3]	[1, 2]	[0.999, 1.000]
0.1	2	*	*	*	[25, 27]	[3, 4]	[0.966, 0.983]
	4	[26, 37]	[1, 2]	[0.965, 0.999]	[2, 4]	[1, 2]	[0.966, 0.995]
	6	[18, 37]	[1, 2]	[0.995, 1.000]	[2, 4]	[1, 2]	[0.992, 1.000]
	8	[18, 37]	[1, 2]	[0.998, 1.000]	[2, 4]	[1, 2]	[0.997, 1.000]
	10	[33, 37]	[1, 2]	[0.999, 1.000]	[2, 4]	[1, 2]	[0.999, 1.000]
0.05	2	*	*	*	[32, 34]	[3, 4]	[0.957, 0.978]
	4	[27, 48]	[1, 2]	[0.964, 0.999]	[3, 5]	[1, 2]	[0.949, 0.994]
	6	[23, 48]	[1, 2]	[0.993, 1.000]	[3, 5]	[1, 2]	[0.988, 0.999]
	8	[45, 48]	[1, 2]	[0.995, 1.000]	[3, 5]	[1, 2]	[0.996, 1.000]
	10	[28, 48]	[1, 2]	[0.999, 1.000]	[3, 5]	[1, 2]	[0.998, 1.000]
0.01	2	*	*	*	[49, 51]	[3, 4]	[0.935, 0.968]
	4	[27, 74]	[1, 2]	[0.964, 0.999]	[4, 6]	[1, 2]	[0.933, 0.992]
	6	[64, 74]	[1, 2]	[0.981, 1.000]	[4, 6]	[1, 2]	[0.984, 0.999]
	8	[71, 74]	[1, 2]	[0.993, 1.000]	[4, 6]	[1, 2]	[0.995, 1.000]
	10	[29, 74]	[1, 2]	[0.999, 1.000]	[4, 6]	[1, 2]	[0.998, 1.000]

*: Plan does not exist.

3. Illustrative Example

Suppose that a manufacturer wants to provide the mean life assurance for his product, and he claims that the true mean life of the product is $\mu_N = 500$ h. The quality inspector decides to check whether the manufacture's claim on the lifetime of the product is true or not and, therefore, specifies the experiment time as $t_{N0} = 500$ h. Hence, the experiment termination ratio is calculated as $a = 1.0$. The failure probability corresponding to the mean ratio $\mu_N/\mu_{N0} = 4$ is considered as AQL and the same at mean ratio 1 is taken as LQL. The consumer risk is assumed to be $\beta = 0.25$. The shape parameter of the Weibull distribution is specified as $\delta_N = 0.9$. Suppose the quality inspector wants to implement the proposed GASP under neutrosophic statistics, and he decides to allocate $r_N = 6$ items to each tester. Therefore, in order to execute the proposed plan for the above-specified conditions, we obtain the optimal parameters $g_N = [8, 10]$ and $c_N = [3, 4]$ from Table 3. The input values (or specified values) and the optimal values determined for those input values are reported in Table 5 for easy identification.

Table 5. Summary of input values and output parameters.

Input Values							Output Parameters	
μ_N	t_{N0}	a	μ_N/μ_{N0}	β	δ_N	r_N	g_N	c_N
500(h)	500(h)	1.0	4	0.25	0.9	6	[8, 10]	[3, 4]

This shows that the number of groups for the inspection lies between 8 and 10. Suppose the quality inspector chooses eight groups. Then, implementation procedure of the proposed plan is as follows.

A random sample of 48 items is chosen from the submitted lot, and 6 sample items are distributed to 8 groups. The sample items are included in the life test, and the test is conducted up to the specified time 500 h. The submitted lot is accepted if there are, at most, 3 sample items that failed before the time 500 h in each of all 8 groups. Otherwise, the lot is rejected.

4. Comparison

To show the efficiency of the proposed plan in terms of number of groups (sample size) over the existing SSP, we tabulated the optimal parameters determined for some specified values of a, r_N, and δ_N. The minimum number of groups required for inspecting the lot under neutrosophic statistics and classical statistics is shown in Table 6. We note from Table 6 that the proposed sampling plan under neutrosophic statistics has the smaller number of groups compared to the time-truncated plan under classical statistics. For example, when $a = 0.5$ and $\mu_N/\mu_{N0} = 2$, the number of groups in an indeterminate interval under classical statistics is larger than the proposed sampling plan under neutrosophic statistics. The same efficiency of the proposed plan can be observed for all other specified parameters. By comparing both sampling plans, it can be noted that time-truncated group sampling plan under neutrosophic statistics is better than the plan using classical statistics. Hence, the proposed plan is more economical than the existing plan in saving cost, time, and efforts in uncertainty environments.

Table 6. Values of the proposed GASP and single-sampling plan (SSP) under neutrosophic statistics when $r_N = [10, 12]$ and $\delta_N = [0.9, 1.1]$.

		$a = 0.5$		$a = 1.0$	
β	μ_N/μ_{N0}	GASP	SSP	GASP	SSP
		g_N	n_N	g_N	n_N
0.25	2	[19, 44]	[34, 38]	[5, 7]	[20, 23]
	4	[2, 4]	[11, 14]	[1, 3]	[7, 8]
	6	[1, 3]	[9, 10]	[1, 3]	[5, 6]
	8	[1, 3]	[6, 7]	[1, 3]	[5, 4]
	10	[1, 3]	[6, 7]	[1, 3]	[4, 4]

Table 6. *Cont.*

		$a = 0.5$		$a = 1.0$	
β	μ_N/μ_{N0}	GASP	SSP	GASP	SSP
		g_N	n_N	g_N	n_N
	2	*	[54, 59]	[26, 28]	[36, 35]
	4	[2, 4]	[17, 20]	[1, 3]	[10, 11]
0.1	6	[2, 4]	[14, 13]	[1, 3]	[9, 7]
	8	[2, 4]	[11, 13]	[1, 3]	[7, 7]
	10	[2, 4]	[8, 10]	[1, 3]	[7, 5]
	2	*	[68, 73]	[34, 36]	[48, 42]
	4	[7, 11]	[22, 27]	[3, 5]	[13, 14]
0.05	6	[3, 6]	[16, 19]	[2, 4]	[12, 10]
	8	[2, 4]	[13, 16]	[1, 3]	[10, 8]
	10	[2, 4]	[13, 16]	[1, 3]	[8, 8]
	2	*	[103, 108]	*	[68, 59]
	4	[9, 17]	[36, 36]	[4, 6]	[22, 21]
0.01	6	[5, 8]	[23, 28]	[2, 4]	[16, 15]
	8	[5, 8]	[20, 24]	[2, 4]	[12, 13]
	10	[3, 5]	[20, 20]	[2, 4]	[12, 10]

*: Plan does not exist.

5. Conclusions

In this paper, we have designed a group acceptance sampling plan for cases where the quality of the product is in determinate and vague. Therefore, neutrosophic statistics has been used in this design instead of classical statistics. The optimal parameters determined for different combinations of group sizes and shape parameters have been tabulated. It is concluded from this study that one can use the proposed plan if there is uncertainty in the product's quality. The proposed sampling plan using some other neutrosophic distributions or sampling schemes can be considered in future research. The extension of the proposed plan for big data is also a fruitful area for future research.

Author Contributions: Conceived and designed the experiments, M.A., P.J. and S.B., and A.H.A.-M. Performed the experiments, M.A. and A.H.A.-M. Analyzed the data, M.A. and A.H.A.-M. Contributed reagents/materials/analysis tools, M.A. Wrote the paper, M.A.

Funding: This article was funded by the Deanship of Scientific Research (DSR) at King Abdulaziz University, Jeddah. The authors, therefore, acknowledge and thank DSR technical and financial support.

Acknowledgments: The authors are deeply thankful to the editor and reviewers for their valuable suggestions to improve the quality of this manuscript. The author (P. Jeyadurga) would like to thank the Kalasalingam Academy of Research and Education for providing financial support through postdoctoral fellowship.

Conflicts of Interest: The authors declare no conflicts of interest regarding this paper.

Glossary

$t_N \in \{t_L, t_U\}$	lifetime of the product, where t_L is the lower value of the lifetime and t_U is the upper value of the lifetime
$\delta_N \in \{\delta_L, \delta_U\}$	shape parameter of the Weibull distribution, where δ_L is the lower value of the shape parameter and δ_U is the upper value of the shape parameter
$\lambda_N \in \{\lambda_L, \lambda_U\}$	scale parameter, where λ_L is the lower value of the scale parameter and λ_U is the upper value of the scale parameter
t_{N0}	experiment time
μ_N	true mean life
μ_{N0}	specified mean life
a	experiment termination ratio (i.e., $a = t_{N0}/\mu_{N0}$)
μ_N/μ_{N0}	ratio between the true mean life and the specified mean life

p_{N1}	acceptable quality level (AQL)
p_{N2}	limiting quality level (LQL)
$n_N \in \{n_L, n_U\}$	sample size (i.e., $n_N = g_N r_N$), where n_L is the lower value of the sample size and n_U is the upper value of the sample size
$r_N \in \{r_L, r_U\}$	group size, where r_L is the lower value of the group size and r_U is the upper value of the group size
$g_N \in \{g_L, g_U\}$	number of groups, where g_L is the lower value of the number of groups and g_U is the upper value of the number of groups
$d_N \in \{d_L, d_U\}$	number of failure items in the sample, where d_L is the lower value of the number of failure items and d_U is the upper value of the number of failure items
$c_N \in \{c_L, c_U\}$	acceptance number, where c_L is the lower value of the acceptance number and c_U is the upper value of the acceptance number
α	producer's risk
β	consumer's risk
$P_{aN}(p_N)$	Probability of acceptance at failure probability p_N
$P_{aN}(p_{N1})$	Probability of acceptance at failure probability p_{N1} or at AQL
$P_{aN}(p_{N2})$	Probability of acceptance at failure probability p_{N2} or at LQL

References

1. Montgomery, D.C. *Introduction to Statistical Quality Control*, 6th ed.; John Wiley & Sons, Inc.: New York, NY, USA, 2009.
2. Schilling, E.G.; Neubauer, D.V. *Acceptance Sampling in Quality Control*, 3rd ed.; CRC Press, Taylor and Francis Group: Boca Raton, FL, USA, 2017.
3. Dodge, H.F. Notes on the evolution of acceptance sampling plans—PartI. *J. Qual. Technol.* **1969**, *1*, 77–88. [CrossRef]
4. Loganathan, A.; Vijayaraghavan, R.; Rajagopal, K. Designing single sampling plans by variables using predictive distribution. *Econ. Qual. Control* **2010**, *25*, 301–316. [CrossRef]
5. Liu, F.; Cui, L. A new design on attribute single sampling plans. *Commun. Stat. Theory Methods* **2015**, *44*, 3350–3362. [CrossRef]
6. Govindaraju, K. Predictive design of single sampling attribute plans. *J. Stat. Comput. Simul.* **2017**, *87*, 447–456. [CrossRef]
7. Hu, M.; Gui, W. Acceptance sampling plans based on truncated life tests for Burr type X distribution. *J. Stat. Manag. Syst.* **2018**, *21*, 323–336. [CrossRef]
8. Aslam, M.; Jun, C.-H. A group acceptance sampling plan for truncated life test having Weibull distribution. *J. Appl. Stat.* **2009**, *36*, 1021–1027. [CrossRef]
9. Kanagawa, A.; Ohta, H. A design for single sampling attribute plan based on fuzzy sets theory. *Fuzzy Sets Syst.* **1990**, *37*, 173–181. [CrossRef]
10. Chakraborty, T.K. A class of single sampling plans based on fuzzy optimisation. *Opsearch* **1992**, *29*, 11–20.
11. Jamkhaneh, E.B.; Gildeh, B.S. Acceptance double sampling plan using fuzzy Poisson distribution. *World Appl. Sci. J.* **2012**, *16*, 1578–1588.
12. Turanoğlu, E.; Kaya, İ.; Kahraman, C. Fuzzy acceptance sampling and characteristic curves. *Int. J. Comput. Intell. Syst.* **2012**, *5*, 13–29. [CrossRef]
13. Jamkhaneh, E.B.; Gildeh, B.S. Chain sampling plan using fuzzy probability theory. *J. Appl. Sci.* **2011**, *11*, 3830–3838. [CrossRef]
14. Tong, X.; Wang, Z. Fuzzy acceptance sampling plans for inspection of geospatial data with ambiguity inequality characteristics. *Comput. Geosci.* **2012**, *48*, 256–266. [CrossRef]
15. Uma, G.; Ramya, K. Impact of fuzzy logic on acceptance sampling plans-A Review. *Autom. Auton. Syst.* **2015**, *7*, 181–185.
16. Afshari, R.; Gildeh, B.S. Construction of fuzzy multiple deferred state sampling plan. In Proceedings of the 2017 Joint 17th World Congress of International Fuzzy Systems Association and 9th International Conference on Soft Computing and Intelligent Systems(IFSA-SCIS), Otsu, Japan, 27–30 June 2017.
17. Khan, M.Z.; Khan, M.F.; Aslam, M.; Mughal, A.R. Design of fuzzy sampling plan using the Birnbaum-Saunders distribution. *Mathematics* **2019**, *7*, 9. [CrossRef]

18. Smarandache, F. Neutrosophic logic-A generalization of the intuitionistic fuzzy logic. *Multispace Multistruct. Neutrosophic Transdiscipl.* **2010**, *4*, 396. [CrossRef]
19. Abdel-Basset, M.; Atef, A.; Smarandache, F. A hybrid Neutrosophic multiple criteria group decision making approach for project selection. *Cogn. Syst. Res.* **2019**, *57*, 216–227. [CrossRef]
20. Abdel-Basset, M.; Gunasekaran, M.; Mohamed, M.; Smarandache, F. A novel method for solving the fully neutrosophic linear programming problems. *Neural Comput. Appl.* **2019**, *31*, 1595–1605. [CrossRef]
21. Broumi, S.; Bakali, A.; Talea, M.; Smarandache, F. Bipolar Neutrosophic Minimum Spanning Tree. InfiniteStudy, 2018. Available online: https://books.google.com.sa/books?id=VopuDwAAQBAJ&printsec=frontcover&hl=ar#v=onepage&q&f=false (accessed on 20 August 2019).
22. Abdel-Basset, M.; Manogaran, G.; Gamal, A.; Smarandache, F. A hybrid approach of neutrosophic sets and DEMATEL method for developing supplier selection criteria. *Des. Autom. Embed. Syst.* **2018**, *22*, 257–278. [CrossRef]
23. Abdel-Basset, M.; Nabeeh, N.A.; El-Ghareeb, H.A.; Aboelfetouh, A. Utilising neutrosophic theory to solve transition difficulties of IoT-based enterprises. *Enterp. Inf. Syst.* **2019**, 1–21. [CrossRef]
24. Smarandache, F. *Introduction to Neutrosophic Statistics*; InfiniteStudy: El Segundo, CA, USA, 2014.
25. Broumi, S.; Smarandache, F. Correlation coefficient of interval neutrosophic set. *Applied Mechanics and Materials* **2013**, *436*, 511–517. [CrossRef]
26. Hanafy, I.; Salama, A.; Mahfouz, K. Neutrosophic classical events and its probability. *Int. J. Math. Comput. Appl. Res. (IJMCAR)* **2013**, *3*, 171–178.
27. Smarandache, F. *Introduction to Neutrosophic Measure, Neutrosophic Integral, and Neutrosophic Probability*; InfiniteStudy: El Segundo, CA, USA, 2013.
28. Smarandache, F. *Neutrosophic Overset, Neutrosophic Underset, and Neutrosophic Offset. Similarly for Neutrosophic Over-/Under-/Off-Logic, Probability, and Statistics*; InfiniteStudy: El Segundo, CA, USA, 2016.
29. Chen, J.; Ye, J.; Du, S. Scale effect and anisotropy analyzed for neutrosophic numbers of rock joint roughness coefficient based on neutrosophic statistics. *Symmetry* **2017**, *9*, 208. [CrossRef]
30. Chen, J.; Ye, J.; Du, S.; Yong, R. Expressions of rock joint roughness coefficient using neutrosophic interval statistical numbers. *Symmetry* **2017**, *9*, 123. [CrossRef]
31. Patro, S.; Smarandache, F. *The Neutrosophic Statistical Distribution, More Problems, More Solutions*; InfiniteStudy: El Segundo, CA, USA, 2016.
32. Alhabib, R.; Ranna, M.M.; Farah, H.; Salama, A. some neutrosophic probability distributions. *Neutrosophic Sets Syst.* **2018**, *22*, 30–38.
33. Aslam, M.; Bantan, R.A.R.; Khan, N. Design of a new attribute control chart under neutrosophic statistics. *Int. J. Fuzzy Syst.* **2018**, *21*, 433–440. [CrossRef]
34. Aslam, M. A new sampling plan using neutrosophic process loss consideration. *Symmetry* **2018**, *10*, 132. [CrossRef]
35. Aslam, M.; Arif, O. Testing of grouped product for the Weibull distribution using neutrosophic statistics. *Symmetry* **2018**, *10*, 403. [CrossRef]

mathematics

MDPI

Article

A New X-Bar Control Chart for Using Neutrosophic Exponentially Weighted Moving Average

Muhammad Aslam [1,*], Ali Hussein AL-Marshadi [1] and Nasrullah Khan [2]

[1] Department of Statistics, Faculty of Science, King Abdulaziz University, Jeddah 21551, Saudi Arabia; aalmarshadi@kau.edu.sa

[2] Department of Statistics, University of Veterinary and Animal Sciences, Jhang Campus, Lahore 54000, Pakistan; nas_shan1@hotmail.com

* Correspondence: aslam_ravian@hotmail.com

Received: 22 August 2019; Accepted: 8 October 2019; Published: 12 October 2019

Abstract: The existing Shewhart X-bar control charts using the exponentially weighted moving average statistic are designed under the assumption that all observations are precise, determined, and known. In practice, it may be possible that the sample or the population observations are imprecise or fuzzy. In this paper, we present the designing of the X-bar control chart under the symmetry property of normal distribution using the neutrosophic exponentially weighted moving average statistics. We will first introduce the neutrosophic exponentially weighted moving average statistic, and then use it to design the X-bar control chart for monitoring the data under an uncertainty environment. We will determine the neutrosophic average run length using the neutrosophic Monte Carlo simulation. The efficiency of the proposed plan will be compared with existing control charts.

Keywords: neutrosophic logic; fuzzy logic; control chart; neutrosophic numbers; monitoring

---

1. Introduction

The production process may shift from the target due to a number of reasons. Therefore, to produce the product according to given specifications, it is watched to indicate any shift in the process. The control charts are popularly used in the industry to watch the production process. In the industries, usually, the Shewhart control charts are used for the monitoring of the process. Although these control charts have a simple operational procedure, they are unable to detect a small shift in the process. Therefore, the Shewhart control charts do not detect a very small shift, and cause a high non-conforming product. The applications of such charts can be seen in [1–6].

The control charts using the exponentially weighted moving average (EWMA) used the current subgroup and previous subgroup information, and were said to be more efficient in detecting a very small shift in the process. The control chart based on this statistic is more efficient than the traditional Shewhart control charts. Roberts [7] designed a control chart using this statistic first time. Haq [8] and Haq et al. [9,10] used the EWMA statistic to propose a variety of control charts. Abbasi et al. [11] and Abbasi [12] introduced its setting in normal and non-normal situations and for measurement errors, respectively. Sanusi et al. [13] presented an alternative for the EWMA-based chart when additional information about the main variable is available. References [14–17] presented such control charts. More basic information about the control charts can be seen in [18,19].

The traditional Shewhart control charts cannot be applied when uncertainty or randomness is expected in the data. The fuzzy-based control charts are the best alternative to monitor the process when observations or the parameters under study are fuzzy. As mentioned by Khademi and Amirzadeh [20], "Fuzzy data exist ubiquitously in the modern manufacturing process"; therefore, serval authors paid attention to work on such control charts, such as for example [21–26].

The traditional fuzzy logic is a special case of neutrosophic logic. The latter one has the ability to deal with the measure of indeterminacy; see Smarandache [27]. The classic statistics (CS) method is applied under the assumption that all observations in data are determined, precise, and certain. However, in the modern manufacturing process, it may not be possible to record all determined observations in the data. In this situation, the neutrosophic statistics (NS) can be applied for the analysis of the data. The NS was introduced by Smarandache [28] using neutrosophic logic, which is the generation of CS. The NS is more effective to be applied for the analysis of imprecise data than CS. Chen et al. [29,30] proved the effectiveness of the NS-based analysis. Aslam [31] introduced a new area of neutrosophic quality control (NQC). Aslam et al. [32,33] introduced NS-based attributes and variable charts. Aslam and Khan [34] proposed the X-bar chart under NS. Aslam et al. [35] designed a chart to monitor reliability under uncertainty. Aslam [36,37] proposed the attribute and variable charts using resampling under NS.

Şentürk et al. [38] proposed the EWMA control chart using the fuzzy approach, which is the special case of the control chart using the neutrosophic logic, as mentioned by Smarandache [27]. By looking into the literature of the control chart under the uncertainty environment, we did not find any work on the X-bar control chart based on the neutrosophic exponentially weighted moving average (NEWMA). In this paper, we will first introduce NEWMA. We will introduce the new Monto Carlo simulation under the neutrosophic statistical interval method (NSIM). We will determine the neutrosophic average run length (NARL) of the proposed chart to compare its performance. We hope that the proposed chart will be more sensitive in detecting a small shift in the process as compared to the traditional Shewhart X-bar chart, EWMA X-bar chart under CS [19] and X-bar chart under NS [34].

2. The Proposed NEWMA Statistics

In this section, we will introduce NEWMA statistics. Let $\overline{X}_N \epsilon \left[\frac{\sum_{i=1}^{n_L} X_i}{n_L}, \frac{\sum_{i=1}^{n_U} X_i}{n_U}\right]$; $\overline{X}_N \epsilon \left\{\overline{X}_L, \overline{X}_U\right\}$ be the neutrosophic sample average of a neutrosophic random variable (nrv) $X_{iN} \epsilon \{X_L, X_U\} = i = 1,2,3, \ldots, n_N$, where n_N is the neutrosophic sample size. Suppose that $S_N^2 = \sum_{i=1}^{n_N}\left(X_N - \overline{X}_N\right)^2 / n_N - 1$; $S_N^2 \epsilon \left\{S_L^2, S_L^2\right\}$ represents the neutrosophic sample variance. By following Smarandache [28] and Aslam [31], the neutrosophic sample average follows the neutrosophic normal distribution (NND) with a neutrosophic population mean $\mu_N = \sum_{i=1}^{N_N} X_N / N_N$; $\mu_N \epsilon \{\mu_L, \mu_U\}$ and neutrosophic population variance $\sigma_N^2 = \left[\left\{\sum_{i=1}^{n_N} (X_N - \mu_N)^2 / N_N - 1\right\} / n_N\right]$; $\sigma_N^2 \epsilon \left\{\sigma_L^2 / n_N, \sigma_L^2 / n_N\right\}$. Based on the given information, we define NEWMA statistics as follows:

$$EWMA_{N,i} = \lambda_N \overline{X}_N + (1 - \lambda_N) EWMA_{N,i-1};\ EWMA_{N,i} \epsilon \left\{EWMA_{L,i}, EWMA_{U,i}\right\} \tag{1}$$

where $\lambda_N \epsilon \{\lambda_L, \lambda_U\}$; $[0,0] \leq \lambda_N \leq [1,1]$ denotes the neutrosophic smoothing constant. Note here that $\overline{X}_N \epsilon \left\{\overline{X}_L, \overline{X}_U\right\}$ are assumed to be independent random variables with neutrosophic variance σ_N^2 / n_N ($\sigma_N^2 \epsilon \left\{\sigma_L^2 / n_N, \sigma_L^2 / n_N\right\}$ and known neutrosophic population variance, as shown in [38]. The setting of $\lambda_N \epsilon \{\lambda_L, \lambda_U\}$ is matter of personal experience. Montgomery [14] recommended that it should be selected from 0.05 to 0.25. The $EWMA_{N,i}$ follows the NND with neutrosophic mean $\mu_N \epsilon \{\mu_L, \mu_U\}$ and neutrosophic standard deviation $\frac{\sigma_N}{\sqrt{n_N}} \sqrt{\frac{\lambda_N}{2 - \lambda_N}}$.

3. The Proposed NEWMA X-Bar Control Chart

The proposed X-bar control chart using the NS is described as follows:

1. Choose a random sample of size $n_N \epsilon \{n_L, n_U\}$ and compute $EWMA_{N,i}$ statistics.

$$EWMA_{N,i} = \lambda_N \overline{X}_N + (1 - \lambda_N) EWMA_{N,i-1};\ EWMA_{N,i} \epsilon \left\{EWMA_{L,i}, EWMA_{U,i}\right\}$$

2. Declare the process is an in-control state if $LCL_N < EWMA_{N,i} < UCL_N$; otherwise, it is in an out-of-control state. Note here that $LCL_N \epsilon [LCL_L, LCL_U]$ and $LCU_N \epsilon [LCU_L, LCU_U]$ are the neutrosophic lower and upper control limits.

The proposed chart becomes a chart based on NS proposed by Aslam and Khan [34] when $\lambda_N \epsilon \{1, 1\}$. When all the observations are precise, the proposed chart becomes the traditional Shewhart chart under CS. The neutrosophic control limits are given by:

$$LCL_N = \mu_N - k_N \frac{\sigma_N}{\sqrt{n_N}} \sqrt{\frac{\lambda_N}{2-\lambda_N}};\ LCL_N \epsilon [LCL_L, LCL_U],\ k_N \in \{k_{L,}, k_U\},\ \mu_N \in \{\mu_{L,}, \mu_{U,}\} \tag{2}$$

$$UCL_N = \mu_N + k_N \frac{\sigma_N}{\sqrt{n_N}} \sqrt{\frac{\lambda_N}{2-\lambda_N}};\ LCU_N \epsilon [LCU_L, LCU_U],\ k_N \in \{k_{L,}, k_U\},\ \mu_N \in \{\mu_{L,}, \mu_{U,}\} \tag{3}$$

where $k_N \in \{k_{L,}, k_U\}$ is the neutrosophic control limits coefficient, and will be determined later. Let $\mu_{0N} \epsilon \{\mu_{0L}, \mu_{0U}\}$ be the target value for the process. According to the operational process of the proposed control, the probability that the process under the NS is an in-control state is given by:

$$P^0_{inN} = P\left(LCL_N \le \overline{X} \le UCL_N / \mu_{0N}\right);\ \mu_{0N} \epsilon \{\mu_{0L}, \mu_{0U}\} \tag{4}$$

The neutrosophic average run length (NARL) of the proposed chart is given by:

$$ARL_{0N} = \frac{1}{1 - P^0_{inN}};\ ARL_{0N} \epsilon \{ARL_{0L}, ARL_{0U}\} \tag{5}$$

Suppose now that the process has shifted to a new target at $\mu_{1N} = \mu_{0N} + d\sigma_N$; $\mu_{1N} \epsilon \{\mu_{1L}, \mu_{1U}\}$, where d is the shift constant. The neutrosophic probability of an in-control state at $\mu_{1N} \epsilon \{\mu_{1L}, \mu_{1U}\}$ is given by:

$$P^1_{in} = P\left(LCL_N \le \overline{X}_N \le UCL_N / \mu_{N1} = \mu_N + \mathrm{d}\sigma_N\right);\ \mu_{1N} \epsilon \{\mu_{1L}, \mu_{1U}\}.$$

The NARL at $\mu_{1N} \epsilon \{\mu_{1L}, \mu_{1U}\}$ is defined by:

$$ARL_{1N} = \frac{1}{1 - P^1_{in}};\ ARL_{1N} \epsilon \{ARL_{1L}, ARL_{1U}\} \tag{6}$$

4. The Proposed Neutrosophic Monte Carlo Simulation (NMCS)

As we mentioned earlier, the neutrosophic control limits coefficient $k_N \epsilon \{k_{L,}, k_U\}$ will be determined through the neutrosophic Monte Carlo Simulation (NMCS) under the given constraints. The proposed NMCS is stated as follows.

4.1. For In-Control State

Step 1: A random sample of size $n_N \epsilon \{n_L, n_U\}$ is generated from a standard normal distribution. The mean of the random sample interval of size $n_N \epsilon \{n_L, n_U\}$ is computed as $\overline{X}_N \epsilon \{\overline{X}_L, \overline{X}_U\}$ is computed. The plotting $EWMA_{N,i}$ statistic is computed as:

$$EWMA_{N,i} = \lambda_N \overline{x}_N + (1 - \lambda_N) EWMA_{N,i-1}$$

Step 2: The proposed statistic $EWMA_{N,i}$ is plotted over the $LCL_N \epsilon [LCL_L, LCL_U]$ and $LCU_N \epsilon [LCU_L, LCU_U]$ by selecting a suitable value of $k_N \epsilon \{k_{L,}, k_U\}$, and $ARL_{0N} \epsilon \{ARL_{0L}, ARL_{0U}\}$ is computed.

Step 3: The $ARL_{0N} \epsilon \{ARL_{0L}, ARL_{0U}\}$ and neutrosophic standard deviation (NSD) are computed by iterating process 10,000; only those $k_N \epsilon \{k_{L,}, k_U\}$ values along with their respective parameters are selected for which $ARL_{0N} = r_{0N}$; $ARL_{0N}\epsilon\{ARL_{0L}, ARL_{0U}\}$, where r_{0N} is the specified value of $ARL_{0N}\epsilon\{ARL_{0L}, ARL_{0U}\}$.

4.2. For Shifted Process

Step 1: For selected values of $k_N\epsilon\{k_{L,}, k_U\}$ and their corresponding parameters, $LCL_N\epsilon[LCL_L, LCL_U]$ and $LCU_N\epsilon[LCU_L, LCU_U]$ constructed.

Step 2: As per explained for the in control process in step 1, now data is generated at $\mu_{1N}\epsilon\{\mu_{1L}, \mu_{1U}\}$ and plotted on $LCL_N\epsilon[LCL_L, LCL_U]$ and $LCU_N\epsilon[LCU_L, LCU_U]$, and $ARL_{1N}\epsilon\{ARL_{1L}, ARL_{1U}\}$ is computed.

Step 3: The $ARL_{1N}\epsilon\{ARL_{1L}, ARL_{1U}\}$ is computed for a specified shift level by 10,000 iterations of the process.

Step 4: For various shifts, levels step 2 and 3 are repeated, the values $ARL_{1N}\epsilon\{ARL_{1L}, ARL_{1U}\}$ and NSD are computed at various values of d.

Note here that the proposed NMCS is the generalization of Monte Carlo simulation under CS. The values of $ARL_{1N}\epsilon\{ARL_{1L}, ARL_{1U}\}$ and NSD are determined for various values of d, $n_N\epsilon\{n_L, n_U\}$ and $\lambda_N\epsilon\{\lambda_L, \lambda_U\}$ $ARL_{0N}\epsilon\{ARL_{0L}, ARL_{0U}\}$, and are shown in Tables 1–4 for $ARL_{0N}\epsilon\{300, 300\}$ rather than $ARL_{0N}\epsilon\{370, 370\}$. The values of NARL when $n_N\epsilon[3, 5]$ and $\lambda_N\epsilon[0.08, 0.12]$ are shown in Table 1. The values of NARL when $n_N\epsilon[3, 5]$ and $\lambda_N\epsilon[0.18, 0.22]$ are shown in Table 2. The values of NARL when $n_N\epsilon[3, 5]$ and $\lambda_N\epsilon[0.28, 0.32]$ are shown in Table 3. The values of NARL when $n_N\epsilon[5, 10]$, $n_N\epsilon[5, 8]$, and $\lambda_N\epsilon[0.08, 0.12]$ are given in Table 4. From Tables 1–4, it is worth to note that when all other parameters are constant, the values of NSD are smaller for $ARL_{0N}\epsilon\{300, 300\}$ than for $ARL_{0N}\epsilon\{370, 370\}$. With the increase in $\lambda_N\epsilon\{\lambda_L, \lambda_U\}$, we note the decreasing trend in $ARL_{1N}\epsilon\{ARL_{1L}, ARL_{1U}\}$ and increasing trend in NSD. From Table 4, we observe that the indeterminacy interval in $ARL_{1N}\epsilon\{ARL_{1L}, ARL_{1U}\}$ increases as $n_N\epsilon\{n_L, n_U\}$ increases from $n_N\epsilon$ [5,8] to $n_N\epsilon$ [5,10]. On the other hand, the indeterminacy interval in NSD deceases as $n_N\epsilon\{n_L, n_U\}$ increases.

Table 1. The values neutrosophic average run length (NARL) and neutrosophic standard deviation (NSD) when $n_N\epsilon[3, 5]$ and $\lambda_N\epsilon[0.08, 0.12]$.

k_N	[2.565,2.675]		[2.655,2.765]	
d	NARL	NSD	NARL	NSD
0	[306.19,301.49]	[288.72,289.56]	[368.28,376.77]	[345.26,354.91]
0.05	[220.34,202.5]	[206.84,195.15]	[270.32,248.92]	[257.15,238.27]
0.1	[121.84,99.72]	[109.75,93.06]	[141.33,117.16]	[130.78,106.3]
0.15	[71.34,53.39]	[61.32,45.2]	[80.28,61.22]	[68.96,53.3]
0.2	[45.6,33.38]	[36.07,25.93]	[50.59,36.58]	[39.41,28.44]
0.25	[32.02,22.68]	[22.98,15.89]	[34.7,24.71]	[24.42,17.5]
0.3	[24.18,16.77]	[15.29,10.67]	[25.78,18.28]	[16.59,11.86]
0.4	[15.69,10.75]	[8.58,5.6]	[16.62,11.53]	[9.06,6.2]
0.5	[11.67,8]	[5.47,3.69]	[12.24,8.19]	[5.83,3.66]
0.6	[9.16,6.21]	[3.86,2.47]	[9.57,6.52]	[4,2.59]
0.7	[7.56,5.17]	[2.9,1.89]	[7.91,5.35]	[3.01,1.91]
0.8	[6.42,4.39]	[2.27,1.44]	[6.74,4.59]	[2.34,1.51]
0.9	[5.67,3.85]	[1.84,1.17]	[5.87,4]	[1.88,1.21]
1	[5.03,3.43]	[1.53,0.98]	[5.17,3.58]	[1.55,1.02]
1.25	[3.96,2.75]	[1.07,0.71]	[4.08,2.85]	[1.09,0.72]
1.5	[3.29,2.31]	[0.79,0.52]	[3.4,2.37]	[0.8,0.55]
1.75	[2.83,2.05]	[0.65,0.38]	[2.93,2.1]	[0.65,0.39]
2	[2.5,1.89]	[0.56,0.37]	[2.58,1.94]	[0.57,0.32]
2.5	[2.09,1.52]	[0.33,0.5]	[2.12,1.59]	[0.34,0.49]
3	[1.92,1.14]	[0.3,0.35]	[1.96,1.19]	[0.25,0.39]

Table 2. The values NARL and NSD when $n_N \epsilon [3, 5]$ and $\lambda_N \epsilon [0.18, 0.22]$.

k_N	[2.77,2.815]		[2.85,2.888]	
d	NARL	NSD	NARL	NSD
0	[306.29,304.18]	[295.94,294.05]	[368.53,367.73]	[347.63,347.95]
0.05	[248.13,232.68]	[238.69,228.36]	[303.11,279.09]	[289.54,265.81]
0.1	[155.07,128.27]	[149.33,122.61]	[187.46,150.35]	[180.34,144.18]
0.15	[93.67,71.64]	[87.13,66.2]	[110.77,82.51]	[105.67,77.1]
0.2	[60.15,42.93]	[53.64,38.28]	[69.85,47.94]	[62.94,42.65]
0.25	[40.47,27.91]	[34.74,23.37]	[45.33,31.22]	[39.11,27.02]
0.3	[29.38,19.78]	[23.75,15.61]	[32.21,21.34]	[27.03,16.82]
0.4	[17.27,11.6]	[12.42,7.85]	[18.72,12.15]	[13.34,8.17]
0.5	[11.66,7.91]	[7.53,4.58]	[12.46,8.32]	[7.82,4.77]
0.6	[8.69,5.89]	[4.75,2.88]	[9.15,6.18]	[5.16,3.08]
0.7	[6.85,4.75]	[3.42,2.09]	[7.25,4.91]	[3.63,2.17]
0.8	[5.68,3.98]	[2.55,1.6]	[5.9,4.1]	[2.66,1.64]
0.9	[4.85,3.41]	[1.99,1.26]	[5.04,3.51]	[2.04,1.28]
1	[4.24,3.01]	[1.61,1.02]	[4.38,3.11]	[1.68,1.05]
1.25	[3.25,2.38]	[1.06,0.69]	[3.36,2.43]	[1.1,0.69]
1.5	[2.67,2]	[0.77,0.52]	[2.76,2.04]	[0.79,0.52]
1.75	[2.3,1.74]	[0.59,0.49]	[2.36,1.78]	[0.6,0.49]
2	[2.04,1.51]	[0.47,0.51]	[2.08,1.56]	[0.48,0.5]
2.5	[1.7,1.13]	[0.48,0.34]	[1.76,1.17]	[0.46,0.37]
3	[1.35,1.01]	[0.48,0.12]	[1.41,1.02]	[0.49,0.14]

Table 3. The values NARL and NSD when $n_N \epsilon [3, 5]$ and $\lambda_N \epsilon [0.28, 0.32]$.

k_N	[2.85,2.865]		[2.93,2.945]	
d	NARL	NSD	NARL	NSD
0	[304.15,300.13]	[293.48,289.27]	[376.11,372.72]	[357.13,349.32]
0.05	[262.23,240.42]	[255.83,239.44]	[324.21,297.07]	[310.4,288.11]
0.1	[181.48,148.62]	[182.19,145.09]	[219.18,184.26]	[216.15,178.17]
0.15	[118.58,88.25]	[116.39,85.83]	[143.28,103.88]	[141.14,100.56]
0.2	[77.25,52.48]	[72.69,49.36]	[90.85,61.51]	[86.42,56.98]
0.25	[52.6,35.06]	[49.24,31.52]	[60.08,39.45]	[56.86,35.65]
0.3	[36.38,23.97]	[32.41,20.83]	[41.96,26.95]	[37.69,23.26]
0.4	[20.61,12.97]	[17.05,10.06]	[23.16,14.29]	[19.08,11.15]
0.5	[13.37,8.5]	[10.02,5.75]	[14.41,9.22]	[10.91,6.35]
0.6	[9.35,6.11]	[6.27,3.65]	[10.05,6.31]	[6.73,3.73]
0.7	[7.06,4.68]	[4.19,2.44]	[7.47,4.94]	[4.44,2.59]
0.8	[5.63,3.8]	[3.02,1.78]	[5.98,4]	[3.25,1.88]
0.9	[4.73,3.22]	[2.32,1.39]	[4.92,3.37]	[2.44,1.44]
1	[4.05,2.82]	[1.81,1.1]	[4.17,2.94]	[1.91,1.15]
1.25	[3.01,2.18]	[1.15,0.72]	[3.1,2.23]	[1.19,0.73]
1.5	[2.43,1.79]	[0.8,0.57]	[2.47,1.84]	[0.81,0.58]
1.75	[2.06,1.51]	[0.62,0.53]	[2.11,1.56]	[0.61,0.53]
2	[1.8,1.28]	[0.54,0.45]	[1.85,1.33]	[0.54,0.47]
2.5	[1.41,1.05]	[0.5,0.21]	[1.46,1.05]	[0.51,0.22]
3	[1.13,1]	[0.34,0.05]	[1.17,1]	[0.38,0.06]

Table 4. The values NARL and NSD when $n_N\epsilon[5,10]$, $n_N\epsilon[5,8]$,and $\lambda_N\epsilon[0.08,0.12]$.

k_N	[5,8]	[2.658,2.765]	[5,10]	[2.66,2.77]
d	NARL	NSD	NARL	NSD
0	[377.43,374.68]	[353.08,351.51]	[378.24,375.35]	[351.52,352.17]
0.05	[225.26,200.98]	[211.29,190.39]	[222.54,184.41]	[214.36,176.53]
0.1	[100.64,81.77]	[88.74,75.56]	[100.83,66.16]	[88.44,57.36]
0.15	[52.67,40.17]	[42.89,32.77]	[53.58,33.15]	[42.82,25.6]
0.2	[32.94,24.23]	[23.17,17.19]	[33.03,19.79]	[23.32,12.94]
0.25	[23.34,16.54]	[14.33,10.43]	[23.03,13.78]	[14.25,8.06]
0.3	[17.43,12.48]	[9.54,7.04]	[17.43,10.48]	[9.7,5.39]
0.4	[11.62,8.17]	[5.38,3.74]	[11.7,7.06]	[5.43,2.93]
0.5	[8.69,6.08]	[3.43,2.32]	[8.75,5.3]	[3.49,1.9]
0.6	[6.98,4.86]	[2.49,1.64]	[6.97,4.25]	[2.46,1.33]
0.7	[5.83,4.07]	[1.83,1.24]	[5.84,3.61]	[1.9,1.03]
0.8	[5.05,3.52]	[1.49,0.99]	[5.03,3.14]	[1.47,0.84]
0.9	[4.39,3.11]	[1.2,0.83]	[4.41,2.79]	[1.21,0.71]
1	[3.95,2.79]	[1.01,0.7]	[3.96,2.51]	[1.01,0.6]
1.25	[3.17,2.26]	[0.74,0.49]	[3.17,2.09]	[0.73,0.38]
1.5	[2.67,2]	[0.6,0.34]	[2.66,1.87]	[0.59,0.36]
1.75	[2.3,1.8]	[0.48,0.41]	[2.3,1.61]	[0.48,0.49]
2	[2.08,1.57]	[0.31,0.49]	[2.08,1.3]	[0.31,0.46]
2.5	[1.89,1.1]	[0.32,0.3]	[1.88,1.02]	[0.33,0.13]
3	[1.53,1]	[0.5,0.06]	[1.54,1]	[0.5,0.01]

5. Comparative Studies

In traditional control under CS, it is known that a control chart having the smaller values of average run length (ARL) and standard deviation of run length (SDRL) is said to be efficient in detecting the shift in the process. In the neutrosophic theory, according to [29,30], a method is said to be efficient if it provides the parameter in the indeterminacy interval rather than the determined values in uncertainty. As mentioned by [32], a chart under the NS is said to be more efficient if it has smaller values of NARL than the competitor's charts. We will compare the efficiency of the proposed chart in NARL with the traditional Shewhart X-bar, EWMA X-bar chart proposed by [19] and chart proposed by [34] under NS. We will compare the performance of all the charts at the same specified neutrosophic parameters. Table 5 shows the NARL values of the control charts when $n_N\epsilon[3,5]$, $ARL_{0N}\epsilon\{370,370\}$, and $\lambda_N\epsilon[0.08,0.12]$. We note that the proposed chart under the NS has smaller values of NARL as compared to the traditional Shewhart X-bar, EWMA X-bar chart [19] and charts proposed by [34]. For example, when d = 0.05, the NARL and NSD from the present chart are $ARL_{1N}\epsilon\{270.32,248.92\}$ and $NSD\epsilon$ [257.15,238.27]; from [34], it is $ARL_{1N}\epsilon[356.86,348.52]$, and from [19], they are charts 278 and 261, respectively. From this comparison, it is clear that the proposed chart has smaller values of NARL and NSD, which has the ability to detect a small shift in the process. The theoretical comparisons in NARL of the three charts show the superiority of the proposed control chart.

Table 5. The comparison between three charts.

	[19] Chart		Shewhart X-Bar Chart Under CS		Proposed Chart		[34]
	$n = 3; \lambda$	$k = 2.715$	$n = 3;\ \lambda = 1.0$	$k = 3.01$	$k_N \epsilon$ [2.655,2.765]; $n =$ [3,5]; $\lambda_N \epsilon$[0.08,0.12]		$k_N \epsilon$ [3,3.001]; $\lambda_N \epsilon$[0.08,0.12].
d	**ARL**	**SDRL**	**ARL**	**SDRL**	**NARL**	**NSD**	**NARL**
0	371.865	348.837136	370.8831	354.4075	[368.28,376.77]	[345.26,354.91]	[370.08,370.11]
0.05	278.1253	261.774943	358.9364	340.3296	[270.32,248.92]	[257.15,238.27]	[356.86,348.52]
0.1	150.7524	140.816481	328.7539	317.8932	[141.33,117.16]	[130.78,106.3]	[321.83,295.53]
0.15	86.3401	77.451166	283.8335	280.105	[80.28,61.22]	[68.96,53.3]	[275.44,233.48]
0.2	53.6373	44.328357	234.1001	229.8941	[50.59,36.58]	[39.41,28.44]	[227.54,177.61]
0.25	36.4577	28.061015	194.5368	194.1908	[34.7,24.71]	[24.42,17.5]	[184.1,133.07]
0.3	26.9744	18.701549	150.8646	152.3704	[25.78,18.28]	[16.59,11.86]	[147.43,99.48]
0.4	16.8197	9.92102	97.1724	95.70603	[16.62,11.53]	[9.06,6.2]	[93.98,56.56]
0.5	12.0976	6.090614	62.5904	60.1786	[12.24,8.19]	[5.83,3.66]	[60.65,33.38]
0.6	9.2986	4.102965	41.1474	40.18904	[9.57,6.52]	[4,2.59]	[40.01,20.55]
0.7	7.6389	3.122736	27.6095	26.86251	[7.91,5.35]	[3.01,1.91]	[27.06,13.21]
0.8	6.4074	2.341447	19.2001	18.76261	[6.74,4.59]	[2.34,1.51]	[18.78,8.85]
0.9	5.5639	1.885225	13.5113	12.74982	[5.87,4]	[1.88,1.21]	[13.37,6.18]
1	4.9515	1.570858	9.811	9.239676	[5.17,3.58]	[1.55,1.02]	[9.76,4.49]

For the summated data, we suppose that $n_N \epsilon [3, 5]$, $ARL_{0N} \epsilon \{370, 370\}$, and $\lambda_N \epsilon [0.08, 0.12]$. The 40 observations from NND are generated, having half of the data generated assuming that the process is in-control state, and next 20 observations are generated assuming that the process has shifted with $d = 0.25$. The simulated data along with $\overline{X}_N \epsilon \{\overline{X}_L, \overline{X}_U\}$ and $EWMA_{N,i}$ are shown in Table 6. From Table 1, the tabulated NARL is $ARL_{1N} \epsilon \{24.42, 17.5\}$, so it is expected that the shift should be detected between the 17th sample and the 24th sample. We constructed Figure 1 for the proposed control chart, Figure 2 for the chart proposed by [34], and Figure 3 for the traditional Shewhart X-bar chart. From Figures 1–3, it is worth noting that the proposed control chart detects the shift in the process between the 17th sample and the 24th sample. Figure 2 shows that although the process is an in-control state, some points are in an indeterminacy interval. Figure 3 shows that the process is an in-control state, and all the parameters are determined. By comparing Figures 1–3, it is concluded that the proposed control under NS is quite effective, flexible, and efficient in detecting the shift in the process as compared to the existing control charts.

Table 6. The simulated neutrosophic data.

Sr#	$\bar{X}_N$	$EWMA_N$	Sr#	$\bar{X}_N$	$EWMA_N$
1	[73.99838,73.99999]	[73.99995,74.00009]	21	[73.99971,74.00165]	[73.99984,74.00022]
2	[73.99981,73.9995]	[73.99994,74.00002]	22	[73.99993,73.99948]	[73.99985,74.00013]
3	[74.00099,74.00014]	[74.00003,74.00004]	23	[74.00076,74.00065]	[73.99992,74.00019]
4	[74.00015,73.99811]	[74.00004,73.99981]	24	[73.9993,73.99972]	[73.99987,74.00014]
5	[74.00114,73.99979]	[74.00012,73.9998]	25	[73.99958,73.99998]	[73.99985,74.00012]
6	[74.00067,73.99963]	[74.00017,73.99978]	26	[74.00036,74.00098]	[73.99989,74.00022]
7	[74.00055,74.00081]	[74.0002,73.99991]	27	[74.0003,73.99988]	[73.99992,74.00018]
8	[74.00034,73.99897]	[74.00021,73.99979]	28	[74.00039,73.99945]	[73.99996,74.00009]
9	[73.99929,73.99955]	[74.00014,73.99976]	29	[74.00027,74.00025]	[73.99998,74.00011]
10	[73.99944,73.99988]	[74.00008,73.99978]	30	[73.99993,74.00118]	[73.99998,74.00024]
11	[74.00008,74.00013]	[74.00008,73.99982]	31	[74.00062,74.00047]	[74.00003,74.00027]
12	[73.99965,74.00038]	[74.00005,73.99989]	32	[74.00077,74.00038]	[74.00009,74.00028]
13	[74.00073,73.99959]	[74.0001,73.99985]	33	[73.99993,74.00014]	[74.00008,74.00026]
14	[73.99947,73.99942]	[74.00005,73.9998]	34	[74.00065,74.00044]	[74.00012,74.00029]
15	[73.99962,73.99951]	[74.00002,73.99977]	35	[73.99977,74.00121]	[74.00009,74.0004]
16	[73.99927,74.00025]	[73.99996,73.99982]	36	[74.00068,74.00122]	[74.00014,74.00049]
17	[74.00016,74.00104]	[73.99997,73.99997]	37	[74.00078,74.00129]	[74.00019,74.00059]
18	[74.00039,74.00034]	[74.00001,74.00001]	38	[74.00079,73.99968]	[74.00024,74.00048]
19	[73.99919,74.00029]	[73.99994,74.00005]	39	[73.99973,73.99953]	[74.0002,74.00037]
20	[73.99881,73.99987]	[73.99985,74.00003]	40	[74.00126,73.99959]	[74.00028,74.00027]

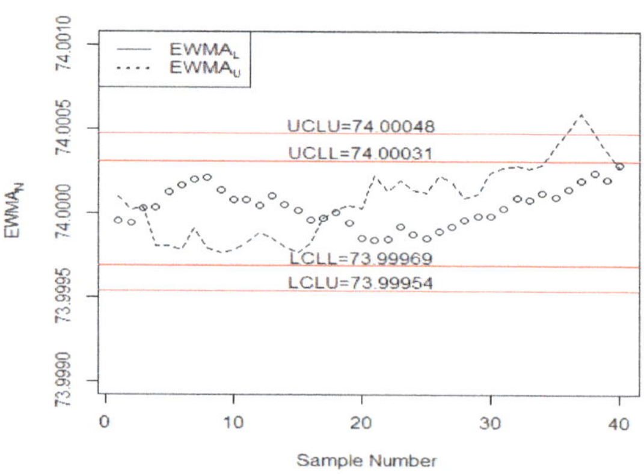

Figure 1. The proposed chart for the summated data.

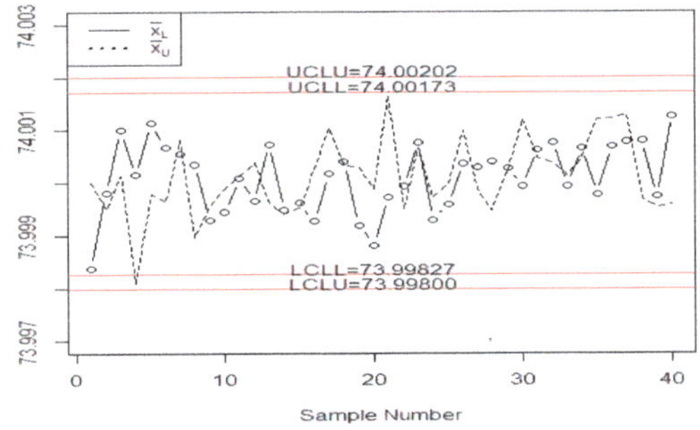

Figure 2. The Aslam and Khan (2019) chart for the summated data.

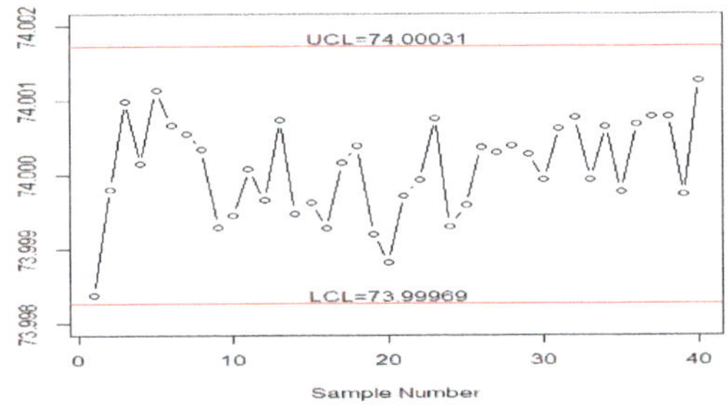

Figure 3. The X-bar chart under classic statistics (CS) for the summated data.

6. Application

A famous automobile industry situated in Saudi Arabia is interested in applying the proposed control chart under the NS for monitoring the production of engine piston rings (EPR). The EPR is an important part of the engine, which improves its efficiency by minimizing the gas or oil leakage and transforming the heat to the cylinder wall. The EPR is a continuous variable and has the possibility of imprecise, fuzzy, and in-determined values. In such a case, the use of the proposed control to monitor the production process of EPR using the proposed control chart under the NS will be more effective and informative than the use of the existing control chart. The proposed control chart will enhance the power of the monitoring of the process using the current sample and previous sample information. Furthermore, the simulation study showed the efficiency of the proposed chart over the existing chart proposed by Aslam and Khan [34]. Therefore, the use of the proposed control chart for the monitoring of ERP production in the industry will help in minimizing the non-conforming ERP product. Suppose that the automobile industry is interested in seeing the efficiency of the proposed chart when $n_N\epsilon[3,5]$, $ARL_{0N}\epsilon\{370,370\}$, and $\lambda_N\epsilon[0.12,0.12]$. The neutrosophic control limit coefficient is $k_N\epsilon\{3.001,3.002\}$. The neutrosophic data of ERP is taken from Aslam and Khan [34] and shown in Table 7 for easy reference. The neutrosophic statistic and neutrosophic control limits for monitoring the ERP data shown in Table 7 are $LCL_N\epsilon\{73.9964,73.9969\};\sigma_N\epsilon\{0.008896,0.009399\}$ and $UCL_N\epsilon\{74.0051,74.0055\};\sigma_N\epsilon\{0.008896,0.009399\}$. We constructed Figure 4 for the proposed control chart, Figure 5 for the chart proposed by Aslam and Khan [34], and Figure 6 for the traditional Shewhart X-bar chart. From Figures 4–6, it is noted that the proposed control chart shows that the process is near the neutrosophic target lines. On the other hand, the existing control chart by Aslam and Khan [34] shows much variation in the process. The traditional Shewhart has the determined values of parameters, and is not suitable in uncertainty. By comparing the three charts, it is concluded that the proposed chart has the ability to centralize EPR production process.

Table 7. The neutrosophic EPR data.

Sr#	Sample					$\bar{X}_N$	$EWMA_N$
1	[74.03,74.03]	[74.002,73.991]	[74.019,74.019]	[73.992,73.992]	[74.008,74.001]	[74.0102,74.0066]	[74.0023,74.0021]
2	[73.995,73.995]	[73.992,74.003]	[74.001,74.001]	[74.011,74.011]	[74.004,74.004]	[74.0006,74.0028]	[74.0021,74.0021]
3	[73.988,74.017]	[74.024,74.024]	[74.021,74.021]	[74.005,74.005]	[74.002,73.995]	[74.008,74.0124]	[74.0028,74.0028]
4	[74.002,74.002]	[73.996,73.996]	[73.993,73.993]	[74.015,74.015]	[74.009,74.009]	[74.003,74.003]	[74.0028,74.0028]
5	[73.992,73.992]	[74.007,74.007]	[74.015,74.015]	[73.989,73.989]	[74.014,73.998]	[74.0034,74.0002]	[74.0029,74.0013]
6	[74.009,74.009]	[73.994,74.001]	[73.997,73.997]	[73.985,73.985]	[73.993,73.993]	[73.9956,73.997]	[74.002,74.0008]
7	[73.995,73.998]	[74.006,74.006]	[73.994,73.994]	[74,74]	[74.005,74.005]	[74,74.0006]	[74.0018,74.0013]
8	[73.985,73.985]	[74.003,74.01]	[73.993,73.993]	[74.015,74.015]	[73.988,73.988]	[73.9968,73.9982]	[74.0012,74.001]
9	[74.008,74.005]	[73.995,73.995]	[74.009,74.009]	[74.005,74.005]	[74.004,74.004]	[74.0042,74.0036]	[74.0015,74.0017]
10	[73.998,73.998]	[74,74]	[73.99,73.99]	[74.007,74.007]	[73.995,73.995]	[73.998,73.998]	[74.0011,74.0013]
11	[73.994,73.998]	[73.998,73.998]	[73.994,73.994]	[73.995,73.995]	[73.99,74.001]	[73.9942,73.9972]	[74.0003,74.0009]
12	[74.004,74.004]	[74,74.002]	[74.007,74.005]	[74,74.001]	[73.996,73.996]	[74.0014,74.0016]	[74.0004,74.001]
13	[73.983,73.993]	[74.002,74.002]	[73.998,73.998]	[73.997,73.997]	[74.012,74.005]	[73.9984,73.999]	[74.0002,74.0011]
14	[74.006,74.006]	[73.967,73.985]	[73.994,73.994]	[74,74]	[73.984,73.996]	[73.9902,73.9962]	[73.999,74.0006]
15	[74.012,74.012]	[74.014,74.012]	[73.998,73.998]	[73.999,73.999]	[74.007,74.007]	[74.006,74.0056]	[73.9998,74.0019]
16	[74,74]	[73.984,73.984]	[74.005,74.005]	[73.998,73.998]	[73.996,73.996]	[73.9966,73.9966]	[73.9994,74.0013]
17	[73.994,73.994]	[74.012,74.012]	[73.986,73.986]	[74.005,74.005]	[74.007,74.007]	[74.0008,74.0008]	[73.9996,74.0014]
18	[74.006,74.006]	[74.01,74.011]	[74.018,74.018]	[74.003,74.003]	[74,74.001]	[74.0074,74.0078]	[74.0005,74.0021]
19	[73.984,73.984]	[74.002,74.002]	[74.003,74.003]	[74.005,74.005]	[73.997,73.997]	[73.9982,73.9982]	[74.0003,74.0011]
20	[74]	[74.01,74.01]	[74.013,74.009]	[74.02,74.015]	[74.003,74.003]	[74.0092,74.0074]	[74.0013,74.0018]
21	[73.982,73.982]	[74.001,74.001]	[74.015,74.015]	[74.005,74.005]	[73.996,73.996]	[73.9998,73.9998]	[74.0011,74.0012]
22	[74.004,74.004]	[73.999,73.999]	[73.99,73.99]	[74.006,74.006]	[74.009,74.002]	[74.0016,74.0002]	[74.0012,74.0011]
23	[74.01,74.01]	[73.989,73.989]	[73.99,73.99]	[74.009,74.005]	[74.014,74.011]	[74.0024,74.001]	[74.0013,74.0014]
24	[74.015,74.011]	[74.008,74.008]	[73.993,73.993]	[74,74]	[74.01,74.011]	[74.0052,74.0046]	[74.0018,74.0018]
25	[73.982,73.982]	[73.984,73.989]	[73.995,73.995]	[74.017,74.012]	[74.013,74.01]	[73.9982,73.9976]	[74.0014,74.001]

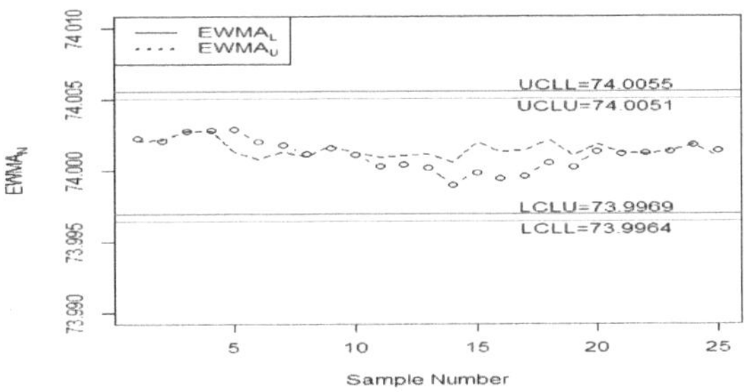

Figure 4. The proposed chart for the real data.

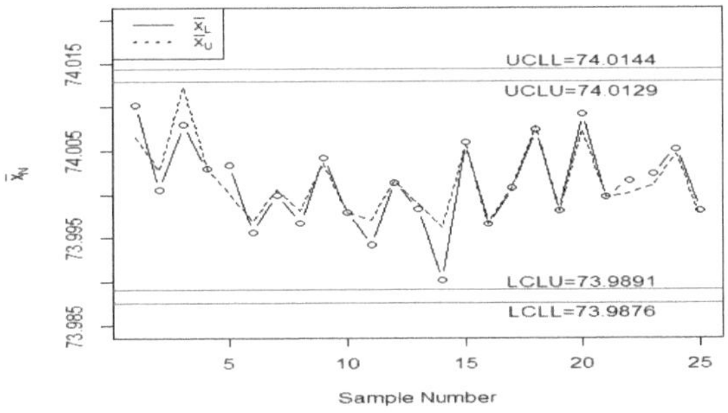

Figure 5. The Aslam and Khan (2019) chart for the real data.

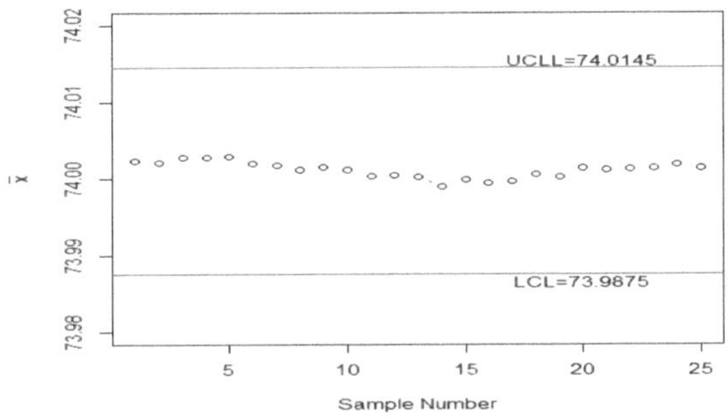

Figure 6. The X-bar chart under CS for the real data.

7. Concluding Remarks

We presented the designing of the X-bar control chart using the neutrosophic EWMA (NEWMA) statistics. The neutrosophic NEWMA and NMSC are introduced in this paper. Some tables for various neutrosophics are presented for practical use in the industry. The theoretical comparisons in the NARL and simulation study showed that the proposed chart performs better than the competitor's charts. The real example of ERP data from the automobile industry also showed the efficiency of the proposed chart. We recommend using the proposed control chart for monitoring the process in the automobile, aerospace, mobiles, water drinking, and medical instrument industries. The proposed chart can be only applied when the variable of interest follows the neutrosophic normal distribution. The proposed chart using some non-normal distributions can be considered as future research. The proposed control chart using some advanced sampling schemes will be considered as future research.

Author Contributions: M.A.; A.H.A.-M. and N.K. conceived and designed the experiments; M.A. and N.K. performed the experiments; M.A. and N.K. analyzed the data; M.A. contributed reagents/materials/analysis tools; M.A. wrote the paper.

Funding: This work was funded by the Deanship of Scientific Research (DSR), King Abdulaziz University, Jeddah. The author, therefore, gratefully acknowledges the DSR technical and financial support.

Acknowledgments: The authors are deeply thankful to the editor and reviewers for their valuable suggestions to improve the quality of this manuscript.

Conflicts of Interest: The authors declare no conflict of interest regarding this paper.

References

1. Senturk, S.; Erginel, N. Development of fuzzy X¯~-R~ and X¯~-S~ control charts using α-cuts. *Inf. Sci.* **2009**, *179*, 1542–1551. [CrossRef]
2. Hart, M.K.; Lee, K.Y.; Hart, R.F.; Robertson, J.W. Application of Attribute Control Charts to Risk-Adjusted Data for Monitoring and Improving Health Care Performance. *Qual. Manag. Healthc.* **2003**, *12*, 5–19. [CrossRef]
3. Bai, D.; Lee, K. Variable sampling interval X control charts with an improved switching rule. *Int. J. Prod. Econ.* **2002**, *76*, 189–199. [CrossRef]
4. Castagliola, P.; Celano, G.; Fichera, S.; Nunnari, V. A variable sample size S2-EWMA control chart for monitoring the process variance. *Int. J. Reliab. Qual. Saf. Eng.* **2008**, *15*, 181–201. [CrossRef]
5. Panthong, C.; Pongpullponsak, A. Non-Normality and the Fuzzy Theory for Variable Parameters Control Charts. *Thai J. Math.* **2016**, *14*, 203–213.
6. Pereira, P.; Seghatchian, J.; Caldeira, B.; Xavier, S.; de Sousa, G. Statistical methods to the control of the production of blood components: Principles and control charts for variables. *Transfus. Apher. Sci.* **2018**, *57*, 132–142. [CrossRef]
7. Roberts, S. Control chart tests based on geometric moving averages. *Technometrics* **1959**, *1*, 239–250. [CrossRef]
8. Haq, A. An improved mean deviation exponentially weighted moving average control chart to monitor process dispersion under ranked set sampling. *J. Stat. Comput. Simul.* **2014**, *84*, 2011–2024. [CrossRef]
9. Haq, A.; Brown, J.; Moltchanova, E. New exponentially weighted moving average control charts for monitoring process mean and process dispersion. *Qual. Reliab. Eng. Int.* **2015**, *31*, 877–901. [CrossRef]
10. Haq, A.; Brown, J.; Moltchanova, E.; Al-Omari, A.I. Effect of measurement error on exponentially weighted moving average control charts under ranked set sampling schemes. *J. Stat. Comput. Simul.* **2015**, *85*, 1224–1246. [CrossRef]
11. Abbasi, S.A.; Riaz, M.; Miller, A.; Ahmad, S.; Nazir, H.Z. EWMA dispersion control charts for normal and non-normal processes. *Qual. Reliab. Eng. Int.* **2015**, *31*, 1691–1704. [CrossRef]
12. Abbasi, S.A. Exponentially weighted moving average chart and two-component measurement error. *Qual. Reliab. Eng. Int.* **2016**, *32*, 499–504. [CrossRef]
13. Sanusi, R.A.; Riaz, M.; Adegoke, N.A.; Xie, M. An EWMA monitoring scheme with a single auxiliary variable for industrial processes. *Comput. Ind. Eng.* **2017**, *114*, 1–10. [CrossRef]
14. Montgomery, D.C. *Introduction to Statistical Quality Control*; John Wiley & Sons: Hoboken, NJ, USA, 2007.
15. Arshad, W.; Abbas, N.; Riaz, M.; Hussain, Z. Simultaneous use of runs rules and auxiliary information with exponentially weighted moving average control charts. *Qual. Reliab. Eng. Int.* **2017**, *33*, 323–336. [CrossRef]
16. Adeoti, O.A. A new double exponentially weighted moving average control chart using repetitive sampling. *Int. J. Qual. Reliab. Manag.* **2018**, *35*, 387–404. [CrossRef]
17. Adeoti, O.A.; Malela-Majika, J.-C. Double exponentially weighted moving average control chart with supplementary runs-rules. *Qual. Technol. Quant. Manag.* **2019**, 1–24. [CrossRef]
18. Hunter, J.S. The exponentially weighted moving average. *J. Qual. Technol.* **1986**, *18*, 203–210. [CrossRef]
19. Lucas, J.M.; Saccucci, M.S. Exponentially weighted moving average control schemes: Properties and enhancements. *Technometrics* **1990**, *32*, 1–12. [CrossRef]
20. Khademi, M.; Amirzadeh, V. Fuzzy rules for fuzzy $\overline{X}$ and R control charts. *Iran. J. Fuzzy Syst.* **2014**, *11*, 55–66.
21. Faraz, A.; Moghadam, M.B. Fuzzy control chart a better alternative for Shewhart average chart. *Qual. Quant.* **2007**, *41*, 375–385. [CrossRef]
22. Zarandi, M.F.; Alaeddini, A.; Turksen, I. A hybrid fuzzy adaptive sampling–run rules for Shewhart control charts. *Inf. Sci.* **2008**, *178*, 1152–1170. [CrossRef]

23. Faraz, A.; Kazemzadeh, R.B.; Moghadam, M.B.; Bazdar, A. Constructing a fuzzy Shewhart control chart for variables when uncertainty and randomness are combined. *Qual. Quant.* **2010**, *44*, 905–914. [CrossRef]
24. Wang, D.; Hryniewicz, O. A fuzzy nonparametric Shewhart chart based on the bootstrap approach. *Int. J. Appl. Math. Comput. Sci.* **2015**, *25*, 389–401. [CrossRef]
25. Kahraman, C.; Gülbay, M.; Boltürk, E. *Fuzzy Shewhart Control Charts, Fuzzy Statistical Decision-Making*; Springer: Berlin/Heidelberg, Germany, 2016; pp. 263–280.
26. Khan, M.Z.; Khan, M.F.; Aslam, M.; Niaki, S.T.A.; Mughal, A.R. A Fuzzy EWMA Attribute Control Chart to Monitor Process Mean. *Information* **2018**, *9*, 312. [CrossRef]
27. Smarandache, F. Neutrosophic Logic-A Generalization of the Intuitionistic Fuzzy Logic. *Multispace Multistructure Neutrosophic Transdiscipl.* **2010**, *4*, 396. [CrossRef]
28. Smarandache, F. *Introduction to Neutrosophic Statistics*; Sitech & Education: Columbus, OH, USA, 2014.
29. Chen, J.; Ye, J.; Du, S. Scale effect and anisotropy analyzed for neutrosophic numbers of rock joint roughness coefficient based on neutrosophic statistics. *Symmetry* **2017**, *9*, 208. [CrossRef]
30. Chen, J.; Ye, J.; Du, S.; Yong, R. Expressions of rock joint roughness coefficient using neutrosophic interval statistical numbers. *Symmetry* **2017**, *9*, 123. [CrossRef]
31. Aslam, M. A New Sampling Plan Using Neutrosophic Process Loss Consideration. *Symmetry* **2018**, *10*, 132. [CrossRef]
32. Aslam, M.; Bantan, R.A.; Khan, N. Design of a new attribute control chart under neutrosophic statistics. *Int. J. Fuzzy Syst.* **2019**, *21*, 433–440. [CrossRef]
33. Aslam, M.; Khan, N.; Khan, M. Monitoring the Variability in the Process Using Neutrosophic Statistical Interval Method. *Symmetry* **2018**, *10*, 562. [CrossRef]
34. Aslam, M.; Khan, N. A new variable control chart using neutrosophic interval method-an application to automobile industry. *J. Intell. Fuzzy Syst.* **2019**, *36*, 2615–2623. [CrossRef]
35. Aslam, M.; Khan, N.; Albassam, M. Control Chart for Failure-Censored Reliability Tests under Uncertainty Environment. *Symmetry* **2018**, *10*, 690. [CrossRef]
36. Aslam, M. Attribute Control Chart Using the Repetitive Sampling under Neutrosophic System. *IEEE Access* **2019**, *7*, 15367–15374. [CrossRef]
37. Aslam, M. Control Chart for Variance using Repetitive Sampling under Neutrosophic Statistical Interval System. *IEEE Access* **2019**, *7*, 25253–25262. [CrossRef]
38. Şentürk, S.; Erginel, N.; Kaya, İ.; Kahraman, C. Fuzzy exponentially weighted moving average control chart for univariate data with a real case application. *Appl. Soft Comput.* **2014**, *22*, 1–10. [CrossRef]

mathematics

MDPI

Article

Neutrosophic Portfolios of Financial Assets. Minimizing the Risk of Neutrosophic Portfolios

Marcel-Ioan Boloș [1], Ioana-Alexandra Bradea [2] and Camelia Delcea [2,*]

[1] Department of Finance and Banks, University of Oradea, 410087 Oradea, Romania; marcel.bolos@softscape.ro

[2] Department of Informatics and Cybernetics, Bucharest University of Economic Studies, 010552 Bucharest, Romania; ioana.bradea@softscape.ro

* Correspondence: camelia.delcea@csie.ase.ro; Tel.: +40-7695-432-813

Received: 16 September 2019; Accepted: 24 October 2019; Published: 3 November 2019

Abstract: This paper studies the problem of neutrosophic portfolios of financial assets as part of the modern portfolio theory. Neutrosophic portfolios comprise those categories of portfolios made up of financial assets for which the neutrosophic return, risk and covariance can be determined and which provide concomitant information regarding the probability of achieving the neutrosophic return, both at each financial asset and portfolio level and also information on the probability of manifestation of the neutrosophic risk. Neutrosophic portfolios are characterized by two fundamental performance indicators, namely: the neutrosophic portfolio return and the neutrosophic portfolio risk. Neutrosophic portfolio return is dependent on the weight of the financial assets in the total value of the portfolio but also on the specific neutrosophic return of each financial asset category that enters into the portfolio structure. The neutrosophic portfolio risk is dependent on the weight of the financial assets that enter the portfolio structure but also on the individual risk of each financial asset. Within this scientific paper was studied the minimum neutrosophic risk at the portfolio level, respectively, to establish what should be the weight that the financial assets must hold in the total value of the portfolio so that the risk is minimum. These financial assets weights, after calculations, were found to be dependent on the individual risk of each financial asset but also on the covariance between two financial assets that enter into the portfolio structure. The problem of the minimum risk that characterizes the neutrosophic portfolios is of interest for the financial market investors. Thus, the neutrosophic portfolios provide complete information about the probabilities of achieving the neutrosophic portfolio return but also of risk manifestation probability. In this context, the innovative character of the paper is determined by the use of the neutrosophic triangular fuzzy numbers and by the specific concepts of financial assets, in order to substantiating the decisions on the financial markets.

Keywords: financial assets; neutrosophicportfolio; neutrosophic portfolio return; neutrosophic portfolio risk; neutrosophic covariance

1. Introduction

The portfolios of financial assets have been the subject of numerous researches in the specialized literature, the main concern of the specialists being to identify a solution for the portfolio risk management, known being the fact that the capital market can generate huge losses if no solution is identified against the losses generated by the manifestation of the financial risk. In a first stage, to solve this sensitive problem, solutions were identified at the level of each financial asset, determining a set of three financial performance indicators that characterize the financial assets, namely: financial return, financial risk and covariance.

In order to quantify the financial asset's return, we took into consideration the profit realized by the investors, both from the price fluctuations of the financial assets and from the dividends obtained.

The methods of calculating the financial return were different; starting from the return on financial assets based on time series recorded in previous time periods, to the returns determined based on estimates for future time periods. Subsequently, the foundations of another performance indicator of financial assets known as financial risk were made, determined with the help of the statistical indicators. These statistical indicators were—the square deviation from the mean and the variance that measures the deviation of the return of financial assets from its average value. The greater this deviation, the greater the risk associated with financial assets is.

In terms of covariance, the third indicator used for evaluating the performance of financial assets, measures the intensity of the link between the return of two financial assets; simultaneously being introduced also the correlation coefficient, which, depending on the recorded value, provides information on the return of the financial assets evolution. A positive correlation coefficient indicates that the return on financial assets increases or decreases as appropriate. A negative value of the correlation coefficient indicates that the evolutions of the return of the financial assets are of opposite sign, respectively while the return of one financial asset increases, the return of the other financial asset may decrease and vice versa.

Financial performance indicators have been a step forward in evaluating the performance of financial assets but not enough. To these was added the modern portfolio theory which lays the foundation of the correlation between profitability and risk at the level of the financial assets portfolio. The mathematical model for correlating the relationship between return and risk is known as the Markowitz efficient frontier, which essentially shows that the portfolio risk of financial assets increases in proportion to the value of the portfolio's return or, on the contrary, the portfolio risk decreases in proportion to the return value; between these two variables being a direct proportionality relation. Moreover, Markowitz's frontier theory demonstrates that risk management of a financial assets portfolio is much more efficient if capital market investments are made in a diversified portfolio of financial assets [1].

Despite all the progress made by introducing the relationship between return and risk but also by diversifying the risk making investments in diversified portfolios, not enough information is provided to investors in the capital market regarding the probability of achieving the return on financial assets or financial risk. This category of information is necessary to properly substantiate the investment decision on the capital market.

To solve this problem caused by the lack of information regarding the probabilities of achieving the financial performance indicators, fuzzy neutrosophic numbers were introduced to model the performance indicators of the financial assets. The use of neutrosophic fuzzy numbers in modelling the performance indicators of financial assets brings several advantages over the existing theory so far, as:

- Modelling the financial performance indicators taking into account the probabilities of their achievement;
- Clustering, respectively, modelling the value of financial performance indicators using the linguistic values that characterize the recorded values;
- Funding the investment decisions on the capital market by selecting value ranges and probabilities of achieving the financial performance indicators desired by investors;

In order to complete the modelling of financial performance indicators with the help of fuzzy neutrosophic numbers, the present paper bases two fundamental concepts in the portfolio theory literature, namely: it introduces a new category of portfolios, respectively the neutrosophic portfolios of financial assets and bases the algorithm for minimizing the risk of the neutrosophic portfolio of financial assets. Regarding the neutrosophic portfolios of financial assets, they will provide information on the risk and return of the portfolio, together with the probabilities of their achievement, with the mention that the probability of achievement is influenced by the risk and return of each financial asset that enters into the portfolio structure.

The risk minimization algorithm of the neutrosophic portfolio of financial assets provides solutions to the investor when it seeks to minimize the risk, respectively, sets the value of the investments that the investor will make in each of the financial assets that enter the portfolio structure, so that the risk is minimal.

The paper is organized as follows: Section 2 deals with the state of the art in the area of neutrosophic theory, while stating the main characteristics and assumptions related to the structure of the financial assets. Section 3 presents the neutrosophic portfolios concept, by highlighting some of the specific notions, structure and formation related to the neutrosophic portfolios theory. Two numerical examples are provided with Section 3 in order to better explain the introduced concepts. Section 4 deals with the neutrosophic portfolio equations. Both the analytical and matrix form are discussed within this section. Sections 5 and 6 deal with the minimizing the risk of the neutrosophic portfolio consisting of two or more financial assets, while Section 7 presents the limitations of the study and draws the main concluding remarks.

2. State of the Art

2.1. The Classical Theory of Financial Asset Portfolios. A New Approach

The structure of a portfolio is based on one or more financial assets $(A_1, A_2, A_3, \ldots A_n)$ or $\left(A_i, i = \overline{1,n}\right)$. Each of the financial assets that enter into the portfolio structure is characterized by an average financial return $\left(\overline{R}_{A_i}\right)$, a financial risk of the form $\left(\sigma_{A_i}\right)$ but also of the covariance $cov\left(A_i, A_j\right)$ between the asset $\left(A_i, i = \overline{1,n}\right)$ and $\left(A_j, j = \overline{1,m}\right)$. The covariance measures the intensity of the link between the returns of the two assets. Thus, for the modern portfolio theory, the financial asset (A_i) will have the characteristic performance indicators of $A_i : \begin{cases} \overline{R}_{A_i} \\ \sigma_{A_i} \\ x_{A_i} \end{cases}$, respectively the average financial return, the financial risk and x_{A_i}, which represents the weight of a financial asset (VA_i) in the total value of the portfolio $\left(\sum_{i=1}^{n} VA_i\right)$: $x_{A_i} = \frac{VA_i}{\sum_{i=1}^{n} VA_i} \times 100[\%]$.

The calculations regarding the performance indicators of financial assets are already known in the literature. The average return of a financial asset is determined either in the form of historical yields using the arithmetic mean $\overline{R}_{A_i} = \frac{1}{N}\sum_{i=1}^{n} R_{A_i}$ or using the geometric mean according to the formula: $\overline{R}_{A_i} = \left[\prod_{i=1}^{n}\left(1 + R_{A_i}\right)\right]^{\frac{1}{n}} - 1$, either in the form of expected returns using the probabilities assigned by investors (p_i) for each evolution scenario (S) of the financial asset expected return of the form: $\overline{R}_{A_i} = \sum_{i=1}^{S} p_i \times R_{A_i}$. If the variable represented by the return on the financial asset is continuous, a normal distribution can be used, with $\overline{R}_{A_i} = 0, \sigma_{A_i} = 1$ and the distribution density function of the form: $f\left(R_{A_i}, \overline{R}_{A_i}, \sigma_{A_i}\right) = \frac{1}{\sqrt{2\pi\sigma_{A_i}}} e^{-\frac{1}{2}\frac{\left(R_{A_i} - \overline{R}_{A_i}\right)^2}{\sigma_{A_i}^2}}$, for which the probability of occurrence the expected return on the financial asset A_i will be $P\left(-1.96 \le \overline{R}_{A_i} \le +1.96\right) = 95\%$ or $\left(-1.645 \le \overline{R}_{A_i} \le +1.645\right) = 90\%$. If the variable R_{A_i} has $\overline{R}_{A_i} \neq 0$ and $\sigma_{A_i} \neq 1$, then the expected average return $\overline{R}_{A_i}$ can be transformed into a variable of the form: $\overline{R}_{A_i}(z) = \frac{R_{A_i} - \overline{R}_{A_i}}{\sigma_{A_i}}$ for which $P\left(\overline{R}_{A_i} - 1.96\sigma_{A_i} \le \overline{R}_{A_i}(z) \le \overline{R}_{A_i} + 1.96\sigma_{A_i}\right) = 95\%$ and $P\left(\overline{R}_{A_i} - 1.645\sigma_{A_i} \le \overline{R}_{A_i}(z) \le \overline{R}_{A_i} + 1.645\sigma_{A_i}\right) = 90\%$.

The financial risk, as financial assets performance indicator studied in the specialized literature, is determined with the help of the squared deviations from the mean, by using the calculation formula $\sigma_{A_i}^2 = \frac{1}{N-1}\sum_{i=1}^{N}\left(R_{A_i} - \overline{R}_{A_i}\right)^2$, as well as using the statistical indicator known as variance using the calculation formula $\sigma_{A_i} = \sqrt{\frac{1}{N-1}\sum_{i=1}^{N}\left(R_{A_i} - \overline{R}_{A_i}\right)^2}$. Regardless how the financial risk is calculated, it measures the deviation of the financial asset return R_{A_i} from its average return $\overline{R}_{A_i}$. The greater the deviation, the greater the financial asset risk is, otherwise the smaller the deviation, the smaller the financial risk, between the magnitude of the deviation and the size of the financial risk being a directly proportional relationship.

The third statistical indicator used in the portfolio theory is the covariance between two assets (A_i) and respectively $\left(A_j\right)$, established using the calculation formula: $cov\left(A_i.A_j\right) = \frac{1}{N-1}\sum_{i,j=1}^{n}\left(R_{A_i} - \overline{R}_{A_i}\right)\left(R_{A_j} - \overline{R}_{A_j}\right)$, which, as mentioned above, measures the intensity of the connections, respectively the dependency or how two assets return mutual influence each other. The correlation coefficient was introduced in the portfolio literature, as: $\rho_{i,j} = \sigma_{A_iA_j}/\sigma_{A_i}\sigma_{A_j}$ with values between $\rho_{i,j} = [-1, +1]$. If $\rho_{i,j} = -1$, the returns of the two financial assets evolve in the opposite direction, respectively when one increases the other decreases and vice versa. If $\rho_{i,j} = 0$, the returns of the two financial assets do not influence each other. If $\rho_{i,j} = +1$, the returns of the two financial assets increase or, as the case may be, they decrease simultaneously.

As mentioned previously, the performance indicators presented above, respectively the average return of the financial asset $\left(\overline{R}_{A_i}\right)$, the financial risk $\left(\sigma_{A_i}\right)$ and the covariance between two financial assets $cov\left(A_i.A_j\right)$ are specific to the financial assets which are part of a portfolio structure.

The modern theory of the financial asset's portfolio has devoted notions specific to the portfolio such as the portfolio return $\left(R_p\right)$ and the portfolio risk $\left(\sigma_P^2\right)$, in order to mathematically quantify the relationship between return and risk. The portfolio return $\left(R_p\right)$ determined by the existence of N financial assets in the portfolio is mathematically quantified as the sum of the products between the weight $\left(x_{A_i}\right)$ of each asset (A_i) in the total value of the portfolio and the average return specific to each asset $\left(\overline{R}_{A_i}\right)$, of form:

$$R_p = x_{A_1}R_{A_1} + x_{A_2}R_{A_2} + \cdots + x_{A_n}R_{A_n} = \sum_{i=1}^{N} x_{A_i}R_{A_i} \tag{1}$$

The above expression can be written in matrix form as follows:

$$R_p = \left(x_{A_1}x_{A_2}\ldots x_{A_n}\right)\begin{pmatrix} R_{A_1} \\ R_{A_2} \\ \ldots \\ R_{A_n} \end{pmatrix} = x_A^T R_A \tag{2}$$

The portfolio risk $\left(\sigma_P^2\right)$ also made up of N financial assets is determined by squared deviations from the mean and is influenced by the weight held by each financial asset in the total portfolio $\left(x_{A_i}\right)$, as well as by the individual risk of each asset entering the portfolio structure $\left(\sigma_{A_i}^2\right)$, respectively the covariance between two assets $cov\left(A_i.A_j\right)$, according to an expression of the form:

$$\begin{aligned}\sigma_P^2 = {} & x_{A_1}^2\sigma_{A_1}^2 + x_{A_2}^2\sigma_{A_2}^2 + \cdots + x_{A_n}^2\sigma_{A_n}^2 + 2x_{A_1}x_{A_2}\sigma_{A_1A_2} + 2x_{A_1}x_{A_3}\sigma_{A_1A_3} + \cdots \\ & +2x_{A_1}x_{A_n}\sigma_{A_1A_n} + 2x_{A_2}x_{A_1}\sigma_{A_2A_1} + 2x_{A_2}x_{A_3}\sigma_{A_2A_3} + \cdots \\ & +2x_{A_2}x_{A_n}\sigma_{A_2A_n} + 2x_{A_n}x_{A_1}\sigma_{A_nA_1} + 2x_{A_n}x_{A_2}\sigma_{A_nA_2} + \cdots \\ & +2x_{A_n}x_{A_{n-1}}\sigma_{A_nA_{n-1}}\end{aligned} \tag{3}$$

$$\sigma_P^2 = \sum_{i=1}^{n} x_{A_i}^2\sigma_{A_i}^2 + 2\sum_{i=1}^{n}\sum_{j=1}^{n} x_{A_i}x_{A_j}\sigma_{A_iA_j} \tag{4}$$

The portfolio risk in matrix form can be written as:

$$\sigma_P^2 = \left(x_{A_1}x_{A_2}\ldots x_{A_n}\right)\begin{pmatrix} \sigma_{A_1A_1} & \sigma_{A_1A_2} & \cdots & \sigma_{A_1A_n} \\ \sigma_{A_2A_1} & \sigma_{A_2A_2} & \cdots & \sigma_{A_2A_n} \\ \cdots & \cdots & \cdots & \cdots \\ \sigma_{A_nA_1} & \sigma_{A_nA_2} & \cdots & \sigma_{A_nA_n} \end{pmatrix}\begin{pmatrix} x_{A_1} \\ x_{A_2} \\ \cdots \\ x_{A_n} \end{pmatrix} = x_A^T\Omega x_A \tag{5}$$

In the specialized literature, starting with the modern portfolio theory, the relationship between the portfolio return and portfolio risk was established and also the concept of an optimal portfolio has been stipulated. According to this theory, financial asset portfolios are considered optimal if the portfolio return $R_p = \rho$, in which ρ has a fixed level, while the portfolio risk $\sigma_P^2 \to min$. The equations of an optimal portfolio will be of the form:

$$\begin{cases} R_p = \sum_{i=1}^{N} x_{A_i} R_{A_i} = \rho \\ \sigma_P^2 = \sum_{i=1}^{n} x_{A_i}^2 \sigma_{A_i}^2 + 2 \sum_{i=1}^{n} \sum_{j=1}^{n} x_{A_i} x_{A_j} \sigma_{A_i A_j} \to min \\ \sum_{i=1}^{n} x_{A_i} = 1 \end{cases} \quad (6)$$

The mathematical model for quantifying the relationship between the portfolio return (R_p) and its risk (σ_P^2) is known as Markowitz's frontier and has the following form:

$$\sigma_P^2 = \frac{1}{D}\left(AR_P^2 - 2BR_p + C\right) \quad (7)$$

where the coefficients A, B, C, D have the following calculation formulas (results from the literature): $A = e\Omega^{-1}e$; $B = e\Omega^{-1}R_A^T = e^T\Omega^{-1}R_A$; $C = R_A\Omega^{-1}R_A^T$. From the Markowitz's frontier relation, it emerged that there is a direct proportionality relation between risk and return, respectively, the higher the portfolio's return, the higher the risk. All the investment portfolios located on the Markowitz's frontier (the upper branch of hyperbole) are considered to be efficient portfolios. Any portfolio located for example below the Markowitz's frontier will have an equivalent portfolio located on the frontier which will have the same risk and a higher return.

Regardless the popularity of the portfolio theory and how advanced the research in the field of capital markets is, any portfolio, regardless of the number of financial assets, has a certain degree of certainty/uncertainty $(G_r(R_P))$ to realize the portfolio return and to produce the risk. This degree of portfolio returns and risk is divided into three categories:

The first category: certain degree for the portfolio return and risk $\mu(G_r(\sigma_P, R_P))$, corresponding to that situation where the portfolio return and risk have an achievement degree, estimated using professional judgment, around the value of 50%. Each portfolio constitutes a specific degree of achievement for return and risk of each portfolio.

The second category: very poor or almost null degree for the portfolio return and risk $\vartheta(G_r(\sigma_P, R_P))$, corresponding to that situation where the return and risk of a portfolio have an estimated degree of achievement of 10–20%. The causes that can lead to such situations are numerous: the assumption of a certain level of return and risk by investors, the poor ability to pay financial assets, the negative influence of national macroeconomic factors.

Third category: uncertain degree for the portfolio return and risk, noted as $\lambda(G_r(\sigma_P, R_P))$ representing the situation where the degree of return and risk is quite uncertain, estimated based on professional reasoning at 20–30%.

The introduction of these measuring degrees for the financial asset portfolios return and risk allows the creation of neutrosophic portfolios, modelled using triangular fuzzy numbers. These portfolios meet the real needs of investors on the financial market. Thus, if a portfolio will have a high degree of return and risk, the investors will have a degree of certainty that they will obtain the expected returns from the financial market. It is worth mentioning that each financial asset that constitutes the portfolio has in turn a certain return and a specific risk which will determine a certain influence on the portfolio return and risk.

The introduction of these ways of measuring the degree of portfolio return and the degree of producing the portfolio risk creates the basis for the formation of the neutrosophic portfolios of financial assets, as mentioned, modelled using the neutrosophic triangular fuzzy numbers. Neutrosophic portfolios have as performance indicators the neutrosophic portfolio return but also the neutrosophic portfolio risk.

2.2. Literature Review

Regarding the studies in the area of neutrosophic theory, it can be underline the fact that the neutrosophic theory and its derivates has been extensively applied in the last two decades various economic and social fields such as—decision making [2–13], supply chain management [14], best product selection [15], management [16] forecasting [17], sentiment analysis [18,19] and so forth.

As for the portfolio theory, there are only few studies who have tried to use the advantages of the neutrosophic theory. Islam and Ray [20] propose a multi-objective portfolio selection which is used through a neutrosophic optimization technique. The authors introduce a new objective function based on entropy and generalize the portfolio selection problem with diversification (GPSPD), stating that, as the proposed method is general, it can easily be applied to other areas in the engineering sciences or operations research. Pamucar et al. [21] propose a multicriteria decision making model in which the authors have considered the linguistic neutrosophic numbers for the purpose of eliminating the subjectivity which derives from the qualitative assessment and the assumptions made by the decision-making in complex situations.

The problem of project selection has been addressed through the use of the neutrosophic set theory by Abdel-Basset et al. [22]. The authors state in the paper the importance of a proper identification of the important criteria based on which the project selection had to be done and propose a model base on TOPSIS and DEMATEL for selection of the best project alternative. Villegas Alava et al. [23] used the single value neutrosophic numbers for project selection. In the paper, the authors present a case study for information technology project selection for proving the applicability of the proposed approach.

In the area of project management, Saleh Al-Subhi et al. [24] use the neutrosophic sets and propose a new decision making model based on neutrosophic cognitive maps and compare the proposed approach with a traditional model in order to prove its efficiency and efficacy. Perez Pupo et al. [25] use the neutrosophic theory for project management decisions, while Su et al. [26] develop a project procurement method selection model under an interval neutrosophic environment. The results gathered in the papers are compared with exiting methods and the results are encouraging.

The project risk assessment in the area of construction engineering is addressed in Reference [27] through the use of 2-tuple linguistic neutrosophic hamy mean operators. The authors provide both the theoretical background and an applicable example for better explain the proposed approach.

Regarding the identified problem within this paper, the modern portfolio theory currently quantifies with the help of Markowitz model the relationship between return and risk. The main disadvantage of this financial asset portfolio theory is that it does not provide sufficient information to investors regarding the probability of realizing the return and of producing the portfolio risk. In addition, the risk and return of the portfolio is influenced by the risk and return of each financial asset that makes up the portfolio. Under these conditions, the substantiation of the investment decision on the financial market is not based on complete information that would also include the probability of achieving the portfolio return and risk and could have as an impact a risk decrease assumed by investors.

The proposed solution in the present research paper is to use the neutrosophic triangular fuzzy numbers, that use the aforementioned categories of information, regarding the degree of achieving the return and producing the portfolio risk. At the same time, the neutrosophic triangular fuzzy numbers allow the stratification of the values recorded by each financial asset for the return and risk specific to each asset. The information resulting from neutrosophic fuzzy modelling has a much more detailed character and allows the financial market investors to more rigorously base their financial decisions. In addition, as a way of solving the problem, are proposed the concepts of neutrosophic return, neutrosophic risk and neutrosophic covariance specific to financial assets.

The innovative character of the paper is determined by the use of the neutrosophic triangular fuzzy numbers but also by the specific concepts of financial assets, namely: neutrosophic return, neutrosophic risk and/or neutrosophic covariance. The information for substantiating the decisions on the financial market is based on neutrosophic fuzzy modelling as a way to improve the decision-making process on the market.

3. The Neutrosophic Portfolios Concept. Specific Notions, Structure and Formation

The theory of neutrosophic fuzzy numbers of the form: $\widetilde{A} = \{\langle \widetilde{a}, \mu_{\widetilde{a}}, \vartheta_{\widetilde{a}}, \lambda_{\widetilde{a}} \rangle / a \in A\}$ has the characteristic that besides to the specific membership functions related to the fuzzy numbers of the form: $\mu_{\widetilde{a}} : \widetilde{A} \to [0,1]$; $\vartheta_{\widetilde{a}} : \widetilde{A} \to [0,1]$ and $\lambda_{\widetilde{a}} : \widetilde{A} \to [0,1]$, also contain the achievement degree of fuzzy numbers of the form: $(w\widetilde{A}, u\widetilde{A}, y\widetilde{A})$, with the following meanings: $w\widetilde{A}$-certainty degree for the achievement of the fuzzy number, $u\widetilde{A}$-indeterminacy degree for the achievement of the fuzzy number and $y\widetilde{A}$-falsity degree for the achievement of the fuzzy number. The membership functions for the neutrosophic fuzzy numbers of the form $\widetilde{A} = \{\langle \widetilde{a}, \mu_{\widetilde{a}}, \vartheta_{\widetilde{a}}, \lambda_{\widetilde{a}} \rangle / a \in A\}$ are determined according to their achievement degrees [28]:

The membership function for the neutrosophic numbers with truth value, the truth membership $(\mu\widetilde{A}_{(x)})$ is of the form [28]:

$$\mu\widetilde{A}_{(x)} = \begin{cases} \frac{w_{\widetilde{A}}(\widetilde{A}_x - \widetilde{A}_{a1})}{\widetilde{A}_{b1} - \widetilde{A}_{a1}} \; for\ \widetilde{A}_{a1} \le \widetilde{A}_x \le \widetilde{A}_{b1} \\ w_{\widetilde{A}} \; for\ \widetilde{A}_x = \widetilde{A}_{b1} \\ \frac{w_{\widetilde{A}}(\widetilde{A}_{c1} - \widetilde{A}_x)}{\widetilde{A}_{c1} - \widetilde{A}_{b1}} \; for\ \widetilde{A}_{b1} \le \widetilde{A}_x \le \widetilde{A}_{c1} \\ 0,\ for\ any\ other\ value\ out\ of\ range\ [\widetilde{A}_{c1}; \widetilde{A}_{a1}] \end{cases} \tag{8}$$

The membership function for the neutrosophic numbers with uncertain achievement degree, the indeterminacy membership $(\vartheta\widetilde{A}_{(x)})$ is of the form [28]:

$$\vartheta\widetilde{A}_{(x)} = \begin{cases} \frac{u_{\widetilde{A}}(\widetilde{A}_x - \widetilde{A}_{a1}) + \widetilde{A}_{b1} - \widetilde{A}_x}{\widetilde{A}_{b1} - \widetilde{A}_{a1}} \; for\ \widetilde{A}_{a1} \le \widetilde{A}_x \le \widetilde{A}_{b1} \\ u_{Ra} \; for\ \widetilde{A}_x = \widetilde{A}_{b1} \\ \frac{u_{\widetilde{A}}(\widetilde{A}_{c1} - \widetilde{A}_x) + \widetilde{A}_x - \widetilde{A}_{b1}}{\widetilde{A}_{c1} - \widetilde{A}_{b1}} \; for\ \widetilde{A}_{b1} \le \widetilde{A}_x \le \widetilde{A}_{c1} \\ 0,\ for\ any\ other\ value\ out\ of\ range\ [\widetilde{A}_{c1}; \widetilde{A}_{a1}] \end{cases} \tag{9}$$

The membership function for the neutrosophic numbers with false achievement degree, the falsity membership $(\lambda\widetilde{Ra}_{(x)})$ is of the form [28]:

$$\lambda\widetilde{A}_{(x)} = \begin{cases} \frac{y_{\widetilde{A}}(\widetilde{A}_x - \widetilde{A}_{a1}) + \widetilde{A}_{b1} - \widetilde{A}_x}{\widetilde{A}_{b1} - \widetilde{A}_{a1}} \; for\ \widetilde{A}_{a1} \le \widetilde{A}_x \le \widetilde{A}_{b1} \\ \lambda_{\widetilde{A}} \; for\ \widetilde{A}_x = \widetilde{A}_{b1} \\ \frac{y_{\widetilde{A}}(\widetilde{A}_{c1} - \widetilde{A}_x) + \widetilde{A}_x - \widetilde{A}_{b1}}{\widetilde{A}_{c1} - \widetilde{A}_{b1}} \; for\ \widetilde{A}_{b1} \le \widetilde{A}_x \le \widetilde{A}_{c1} \\ 0,\ for\ any\ other\ value\ out\ of\ range\ [\widetilde{A}_{c1}; \widetilde{A}_{a1}] \end{cases} \tag{10}$$

The neutrosophic fuzzy number theory, helps to obtain complete information about fuzzy numbers, by taking into account the achievement degrees, namely: the degree of truth, uncertainty (indeterminacy) degree or falsity degree, that are extremely useful in substantiating decisions on the capital market.

Neutrosophic portfolios can consist of two or more financial assets. Let N be the number of financial assets denoted by $(A_1, A_2, A_3, \ldots A_n)$ or $(A_i, i = \overline{1,n})$. Their characteristic is that the financial asset performance indicators, noted with A_i that enter into a neutrosophic portfolio structure are:

The neutrosophic return: $\widetilde{R_{Ai}} = \langle(\widetilde{R_{Aai}}, \widetilde{R_{Abi}}, \widetilde{R_{Aci}}); w\widetilde{R_A}, u\widetilde{R_A}, y\widetilde{R_A}\rangle$;

The neutrosophic risk: $\widetilde{\sigma}_{Ai} = \langle(\widetilde{\sigma}_{Aai}, \widetilde{\sigma}_{Abi}, \widetilde{\sigma}_{Aci}); w\widetilde{\sigma}_A, u\widetilde{\sigma}_A, y\widetilde{\sigma}_A\rangle$;

The neutrosophic covariance: $cov(\widetilde{R_{A1}}, \widetilde{R_{A2}})$;

The neutrosophic triangular fuzzy numbers that underlie the financial assets performance indicators, of the form: $\widetilde{A} = \{\langle\widetilde{a}, \mu_{\widetilde{a}}, \vartheta_{\widetilde{a}}, \lambda_{\widetilde{a}}\rangle / a \in A\}$, were defined in Bolos et al. [28] and are characterized by the membership functions of the form $\mu_{\widetilde{a}} : A \to [0,1]$; $\vartheta_{\widetilde{a}} : A \to [0,1]$ and $\lambda_{\widetilde{a}} : A \to [0,1]$ and by the achievement degree of the performance indicators, of the form: $(w\widetilde{A}, u\widetilde{A}, y\widetilde{A})$, with the following meanings: $w\widetilde{A}$-certain achievement degree for the performance indicators, $u\widetilde{A}$-indeterminate achievement degree for the performance indicators and $y\widetilde{A}$-falsity achievement degree for the performance indicators.

Definition 1. *Is defined the neutrosophic average return $\langle E_f(\widetilde{R_{Ai}}); w\widetilde{R_A}, u\widetilde{R_A}, y\widetilde{R_A}\rangle$ for the neutrosophic triangular fuzzy number $\widetilde{R_{Ai}} = \langle(\widetilde{R_{Aai}}, \widetilde{R_{Abi}}, \widetilde{R_{Aci}}); w\widetilde{R_A}, u\widetilde{R_A}, y\widetilde{R_A}\rangle$, specific for the financial asset (A_i) and component part of the neutrosophic portfolio $(\widetilde{P}; w\widetilde{P}, u\widetilde{P}, y\widetilde{P},)$ any value of the financial asset return appreciated after the achievement degree, using the following coefficients: $w\widetilde{R_A} \in [0,1]$ for certain achievement degree, $u\widetilde{R_A} \in [0,1]$ for indeterminate achievement degree and $y\widetilde{R_A} \in [0,1]$ for falsity achievement degree; determined by the calculation formula:*

$$\langle E_f(\widetilde{R_A}); w\widetilde{R_A}, u\widetilde{R_A}, y\widetilde{R_A}\rangle = \langle\left(\frac{1}{6}(\widetilde{R_{Aa1}} + \widetilde{R_{Ac1}}) + \frac{2}{3}\widetilde{R_{Ab1}}\right); w\widetilde{R_A}, u\widetilde{R_A}, y\widetilde{R_A}\rangle \tag{11}$$

Note 1: The formula for neutrosophic average return was demonstrated in Boloș et.al. [28].

Definition 2. *Is defined the neutrosophic risk $\langle \sigma f^2_{\widetilde{A_i}}; w\widetilde{\sigma}_A, u\widetilde{\sigma}_A, y\widetilde{\sigma}_A\rangle$ for the neutrosophic triangular fuzzy number $\widetilde{\sigma}_{Ai} = \langle(\widetilde{\sigma}_{Aai}, \widetilde{\sigma}_{Abi}, \widetilde{\sigma}_{Aci}); w\widetilde{\sigma}_A, u\widetilde{\sigma}_A, y\widetilde{\sigma}_A\rangle$ determined for the financial asset (A_i) and component part of the neutrosophic portfolio $(\widetilde{P}; w\widetilde{P}, u\widetilde{P}, y\widetilde{P},)$ any value of the financial asset risk appreciated after the achievement degree, using the following coefficients: $w\widetilde{\sigma}_A \in [0,1]$ for certain achievement degree, $u\widetilde{\sigma}_A \in [0,1]$ for indeterminate achievement degree and $y\widetilde{\sigma}_A \in [0,1]$ for falsity achievement degree; determined by the calculation formula:*

$$\begin{aligned}
&\langle \sigma f^2_{\widetilde{A_i}}; w\widetilde{\sigma}_A, u\widetilde{\sigma}_A, y\widetilde{\sigma}_A\rangle \\
&\quad = \langle\tfrac{1}{4}\left[(\widetilde{R_{Ab1}} - \widetilde{R_{Aa1}})^2 + (\widetilde{R_{Ac1}} - \widetilde{R_{Ab1}})^2\right]; w\widetilde{R_A}, u\widetilde{R_A}, y\widetilde{R_A}\rangle \\
&\quad + \langle\tfrac{2}{3}\left[\widetilde{R_{Aa1}}(\widetilde{R_{Ab1}} - \widetilde{R_{Aa1}}) - \widetilde{R_{Ac1}}(\widetilde{R_{Ac1}} - \widetilde{R_{Ab1}})\right]; w\widetilde{R_A}, u\widetilde{R_A}, y\widetilde{R_A}\rangle \\
&\quad + \langle\tfrac{1}{2}(\widetilde{R_{Aa1}}^2 + \widetilde{R_{Ac1}}^2); w\widetilde{R_A}, u\widetilde{R_A}, y\widetilde{R_A}\rangle - \langle\tfrac{1}{2}E^2_f(\widetilde{Ra_i}); w\widetilde{R_A}, u\widetilde{R_A}, y\widetilde{R_A}\rangle
\end{aligned} \tag{12}$$

Note 2: The formula for neutrosophic risk was demonstrated in Boloș et al. [28].

Definition 3. *Is defined the neutrosophic covariance $\langle cov(\widetilde{R_{A1}}, \widetilde{R_{A2}}); w\widetilde{R_{A1}}, u\widetilde{R_{A1}}, y\widetilde{R_{A1}}; w\widetilde{R_{A2}}, u\widetilde{R_{A2}}, y\widetilde{R_{A2}}\rangle$ for two neutrosophic triangular fuzzy numbers $\widetilde{R_{A1}} = \langle(\widetilde{R_{Aa1}}, \widetilde{R_{Ab1}}, \widetilde{R_{Ac1}}); w\widetilde{R_{A1}}, u\widetilde{R_{A1}}, y\widetilde{R_{A1}}\rangle$ and respectively $\widetilde{R_{A2}} = \langle(\widetilde{R_{Aa2}}, \widetilde{R_{Ab2}}, \widetilde{R_{Ac2}}); w\widetilde{R_{A2}}, u\widetilde{R_{A2}}, y\widetilde{R_{A2}}\rangle$ characterizing two financial assets (A_1, A_2) and component parts of the neutrosophic portfolio $(\widetilde{P}; w\widetilde{P}, u\widetilde{P}, y\widetilde{P},)$ any value of the financial asset covariance appreciated after the achievement degree, using the following coefficients: for certain achievement degree, $u\widetilde{R_{A1}}, u\widetilde{R_{A2}} \in [0,1]$ for indeterminate achievement degree and for falsity achievement degree; determined by the calculation formula:*

$$\begin{aligned}&\langle cov(\widetilde{R_{A1}},\widetilde{R_{A2}});w\widetilde{R_{A1}},u\widetilde{R_{A1}},y\widetilde{R_{A1}};\ w\widetilde{R_{A2}},u\widetilde{R_{A2}},y\widetilde{R_{A2}}\rangle =\\ &\langle\Big(\tfrac{1}{4}\big[(\widetilde{R_{Ab11}}-\widetilde{R_{Aa11}})(\widetilde{R_{Ab21}}-\widetilde{R_{Aa21}})+(\widetilde{R_{Ac11}}-\widetilde{R_{Ab11}})(\widetilde{R_{Ac21}}-\widetilde{R_{Ab21}})\big]\\ &+\tfrac{1}{3}\Big\{\big[\widetilde{R_{Aa21}}(\widetilde{R_{Ab11}}-\widetilde{R_{Aa11}})+\widetilde{R_{Aa11}}(\widetilde{R_{Ab21}}-\widetilde{R_{Aa21}})\big]\\ &-\big[\widetilde{R_{Ac11}}(\widetilde{R_{Ac21}}-\widetilde{R_{Ab21}})+\widetilde{R_{Ac21}}(\widetilde{R_{Ac11}}-\widetilde{R_{Ab11}})\big]\Big\}\\ &+\tfrac{1}{2}(\widetilde{R_{Aa11}}\widetilde{R_{Aa21}}+\widetilde{R_{Ac11}}\widetilde{R_{Ac21}})\\ &+\tfrac{1}{2}E_f(\widetilde{R_{A1}})E_f(\widetilde{R_{A2}})\Big);w\widetilde{R}_{A_1}\wedge w\widetilde{R_{A2}},u\widetilde{R_{A1}}\vee u\widetilde{R_{A2}},y\widetilde{R_{A1}}\vee y\widetilde{R_{A2}}\rangle\end{aligned} \tag{13}$$

Note 3: The formula for neutrosophic covariance was demonstrated in Boloș et al. [28]. Upon these demonstrations we will no longer return.

Definition 4. *Any portfolio P is called a neutrosophic portfolio of financial assets and is denoted* $\langle\widetilde{P};w\widetilde{P},u\widetilde{P},y\widetilde{P}\rangle$ *if it cumulativelysatisfies two conditions:*

- contains in its structure financial assets marked with $(A_i); i = \overline{2,n}$ which have as performance indicators: the neutrosophic return $\langle E_f(\widetilde{R_{Ai}}); w\widetilde{R_A}, u\widetilde{R_A}, y\widetilde{R_A}\rangle$, the neutrosophic risk $\langle \sigma f^2_{\widetilde{A_i}}; w\widetilde{\sigma}_A, u\widetilde{\sigma}_A, y\widetilde{\sigma}_A\rangle$ and the neutrosophic covariance that characterizes the intensity of the links between the neutrosophic returns of two financial assets $\langle cov(\widetilde{R_{A1}},\widetilde{R_{A2}});w\widetilde{R_{A1}},u\widetilde{R_{A1}},y\widetilde{R_{A1}};\ w\widetilde{R_{A2}},u\widetilde{R_{A2}},y\widetilde{R_{A2}}\rangle$;
- allows to calculate the return of the neutrosophic portfolio $\langle\widetilde{R}p; w\widetilde{Rp}, u\widetilde{Rp}, y\widetilde{Rp}\rangle$ and the neutrosophic portfolio risk $\langle\widetilde{\sigma_P^2}; w\widetilde{\sigma p}, u\widetilde{\sigma p}, y\widetilde{\sigma p}\rangle$ as fundamental variables that characterize any neutrosophic portfolio $\langle\widetilde{P}; w\widetilde{P}, u\widetilde{P}, y\widetilde{P}\rangle$.

Proposition 1. *The neutrosophic portfolio return* $\langle\widetilde{R}_P; w\widetilde{Rp}, u\widetilde{Rp}, y\widetilde{Rp}\rangle$ *modeled using neutrosophic triangular fuzzy numbers of the form:* $\widetilde{R_{Ai}} = \langle(\widetilde{R_{Aai}},\widetilde{R_{Abi}},\widetilde{R_{Aci}}); w\widetilde{R_A}, u\widetilde{R_A}, y\widetilde{R_A}\rangle$ is a fundamental variable that characterizes the neutrosophic portfolio and is determined by the formula:

$$\langle\widetilde{R}_P; w\widetilde{Rp}, u\widetilde{Rp}, y\widetilde{Rp}\rangle = \sum_{i=1}^{n}\langle x_{A_i}\Big(\frac{1}{6}\big(\widetilde{R}_{Aai}+\widetilde{R}_{Aci}\big)+\frac{2}{3}\widetilde{R}_{Abi}\Big); w\widetilde{R}_{A_i}, u\widetilde{R}_{A_i}, y\widetilde{R}_{A_i}\rangle \tag{14}$$

Demonstration: From the calculation relation of the neutrosophic portfolio return made up of N financial assets we know that:

$$\begin{aligned}&\langle\widetilde{R}_P; w\widetilde{Rp}, u\widetilde{Rp}, y\widetilde{Rp}\rangle\\ &= \langle x_{A_1}\widetilde{R}_{A_1}; w\widetilde{R}_{A_1}, u\widetilde{R}_{A_1}, y\widetilde{R}_{A_1}\rangle + \langle x_{A_2}\widetilde{R}_{A_2}; w\widetilde{R}_{A_2}, u\widetilde{R}_{A_2}, y\widetilde{R}_{A_2}\rangle + \cdots\\ &+\langle x_{A_n}\widetilde{R}_{A_n}; w\widetilde{R}_{A_n}, u\widetilde{R}_{A_n}, y\widetilde{R}_{A_n}\rangle\end{aligned} \tag{15}$$

The above relationship can be written as follows:

$$\langle\widetilde{R}_P; w\widetilde{Rp}, u\widetilde{Rp}, y\widetilde{Rp}\rangle = \sum_{i=1}^{n}\langle x_{A_i}\widetilde{R}_{A_i}; w\widetilde{R}_{A_i}, u\widetilde{R}_{A_i}, y\widetilde{R}_{A_i}\rangle \tag{16}$$

From the definition no.1 we know that the average neutrosophic return specific to a financial asset is of the form:

$$\langle E_f(\widetilde{R_{Ai}}); w\widetilde{R_A}, u\widetilde{R_A}, y\widetilde{R_A}\rangle = \langle\Big(\frac{1}{6}(\widetilde{R_{Aa1}}+\widetilde{R_{Ac1}})+\frac{2}{3}\widetilde{R_{Ab1}}\Big); w\widetilde{R_A}, u\widetilde{R_A}, y\widetilde{R_A}\rangle \tag{17}$$

Substituting the expression of the average neutrosophic return of a financial asset in the calculation formula of the neutrosophic portfolio return is obtained:

$$\langle\widetilde{R}_P; w\widetilde{Rp}, u\widetilde{Rp}, y\widetilde{Rp}\rangle = \sum_{i=1}^{n}\langle x_{A_i}\Big(\frac{1}{6}(\widetilde{R}_{Aai}+\widetilde{R}_{Aci})+\frac{2}{3}\widetilde{R}_{Abi}\Big); w\widetilde{R}_{A_i}, u\widetilde{R}_{A_i}, y\widetilde{R}_{A_i}\rangle \tag{18}$$

where $\widetilde{R}_{Aai}; \widetilde{R}_{Abi}; \widetilde{R}_{Aci}$ represents the financial asset return values, component part of the neutrosophic triangular fuzzy number determined according to the calculation relationships known in the specialized literature.

Example 1. *There are considered three financial assets (A_1, A_2, A_3) to which three triangular neutrosophic numbers are specified for the financial assets return, of the form:*

$$\begin{aligned} \widetilde{R}_{A1} &= \langle (0.2\ 0.3\ 0.5); 0.5,\ 0.2,\ 0.3\rangle \text{for} \widetilde{R}_A \in \\ \widetilde{R}_{A2} &= \langle (0.1\ 0.2\ 0.3); 0.6,\ 0.3,\ 0.2\rangle for \widetilde{R}_A \in [0,1; 0,3] \\ \widetilde{R}_{A3} &= \langle (0.3\ 0.4\ 0.6); 0.4,\ 0.3,\ 0.3\rangle \text{for} \widetilde{R}_A \in [0,3; 0,6] \end{aligned} \tag{19}$$

The weights held by the three financial assets in the total portfolio are determined according to the value of each financial asset and the total value of the portfolio and have the values: $x_{A_1} = 0,4$; $x_{A_2} = 0,3$ și $x_{A_3} = 0,3$. In order to establish the neutrosophic portfolio return, from proposition1 it is known that:

$$\langle \widetilde{R_P}; w\widetilde{Rp}, u\widetilde{Rp}, y\widetilde{Rp}\rangle = \sum_{i=1}^{n} \langle x_{A_i}\Big(\frac{1}{6}\big(\widetilde{R}_{Aai} + \widetilde{R}_{Aci}\big) + \frac{2}{3}\widetilde{R}_{Abi}\Big); w\widetilde{R}_{A_i}, u\widetilde{R}_{A_i}, y\widetilde{R}_{A_i}\rangle \tag{20}$$

By replacing in the above expression is obtained:

$$\begin{aligned} &\langle \widetilde{R_P}; w\widetilde{Rp}, u\widetilde{Rp}, y\widetilde{Rp}\rangle \\ &= \langle 0,4\big(\tfrac{1}{6}(0.2+0.5) + \tfrac{2}{3}\times 0.3\big); 0.5,\ 0.2,\ 0.3\rangle \\ &+\langle 0,3\big(\tfrac{1}{6}(0.1+0.3) + \tfrac{2}{3}\times 0.2\big); 0.6,\ 0.3,\ 0.2\rangle \\ &+\langle 0,3\big(\tfrac{1}{6}(0.3+0.6) + \tfrac{2}{3}\times 0.4\big); 0.4,\ 0.3,\ 0.3\rangle \end{aligned} \tag{21}$$

$$\begin{aligned} &\langle \widetilde{R_P}; w\widetilde{Rp}, u\widetilde{Rp}, y\widetilde{Rp}\rangle \\ &= \langle 0,4\big(\tfrac{1}{6}0.7 + \tfrac{2}{3}0.3\big); 0.5,\ 0.2,\ 0.3\rangle \\ &+\langle 0,3\big(\tfrac{1}{6}0.4 + \tfrac{2}{3}0.2\big); 0.6,\ 0.3,\ 0.2\rangle \\ &+\langle 0,3\big(\tfrac{1}{6}0.9 + \tfrac{2}{3}0.4\big); 0.4,\ 0.3,\ 0.3\rangle \end{aligned} \tag{22}$$

$$\begin{aligned} &\langle \widetilde{R_P}; w\widetilde{Rp}, u\widetilde{Rp}, y\widetilde{Rp}\rangle \\ &= \langle 0.4\times 0.316; 0.5,\ 0.2,\ 0.3\rangle + \langle 0.3\times 0.199; 0.5,\ 0.2,\ 0.3\rangle \\ &+\langle 0.3\times 0.416; 0.4,\ 0.3,\ 0.3\rangle \end{aligned} \tag{23}$$

$$\begin{aligned} &\langle \widetilde{R_P}; w\widetilde{Rp}, u\widetilde{Rp}, y\widetilde{Rp}\rangle \\ &= \langle 0.1264; 0.5,\ 0.2,\ 0.3\rangle + \langle 0.0597;\ 0.5,\ 0.2,\ 0.3\rangle \\ &+\langle 0.1248; 0.4,\ 0.3,\ 0.3\rangle \\ &\langle \widetilde{R_P}; w\widetilde{Rp}, u\widetilde{Rp}, y\widetilde{Rp}\rangle = \langle 0.3109; 0.5,\ 0.2,\ 0.3\rangle \end{aligned} \tag{24}$$

Result interpretation: The average neutrosophic portfolio return has a value of 31.09% with a degree of certainty of 50%, a degree of uncertainty of 20% and a degree of falsification of 30%. In order to obtain the neutrosophic portfolio return, the addition rule for two triangular neutrosophic numbers was applied according to which:

$$\widetilde{R}_{A1} + \widetilde{R}_{A2} = \langle \left(\begin{array}{c} \widetilde{R}_{Aa1} + \widetilde{R}_{Aa2}, \widetilde{R}_{Ab1} + \widetilde{R}_{Ab2}, \\ \widetilde{R}_{Ac1} + \widetilde{R}_{Ac2} \end{array} \right); wR_{A_1} \wedge w\widetilde{R}_{A2}, u\widetilde{R}_{A1} \vee u\widetilde{R}_{A2}, y\widetilde{R}_{A1} \vee y\widetilde{R}_{A2}\rangle \tag{25}$$

Proposition 2. *The neutrosophic portfolio risk noted with* $\langle \widetilde{\sigma_p^2}; w\widetilde{\sigma p}, u\widetilde{\sigma p}, y\widetilde{\sigma p}\rangle$ *modeled using the fuzzy neutrosophic numbers of the form:* $\widetilde{\sigma}_{Ai} = \langle(\widetilde{\sigma}_{Aai}, \widetilde{\sigma}_{Abi}, \widetilde{\sigma}_{Aci}); w\widetilde{\sigma}_A, u\widetilde{\sigma}_A, y\widetilde{\sigma}_A\rangle$ *is also a fundamental variable of the neutrosophic portfolio that is determined by the calculation formula:*

$$\begin{aligned}
&\langle \widetilde{\sigma_p^2}; w\widetilde{\sigma p}, u\widetilde{\sigma p}, y\widetilde{\sigma p}\rangle \\
&\quad = \sum_{i=1}^{n} x_{A_i}^2 \langle \tfrac{1}{4}\Big[(\widetilde{R}_{Abi} - \widetilde{R}_{Aai})^2 + (\widetilde{R}_{Aci} - \widetilde{R}_{Abi})^2\Big]; w\widetilde{R}_{Ai}, u\widetilde{R}_{Ai}, y\widetilde{R}_{Ai}\rangle \\
&\quad + \langle \tfrac{2}{3}\Big[\widetilde{R}_{Aai}(\widetilde{R}_{Abi} - \widetilde{R}_{Aai}) - \widetilde{R}_{Aci}(\widetilde{R}_{Aci} - \widetilde{R}_{Abi})\Big]; w\widetilde{R}_{Ai}, u\widetilde{R}_{Ai}, y\widetilde{R}_{Ai}\rangle \\
&\quad + \langle \tfrac{1}{2}(\widetilde{R_{Aai}}^2 + \widetilde{R_{Aci}}^2); w\widetilde{Ra}_i, u\widetilde{Ra}_i, y\widetilde{Ra}_i\rangle \\
&\quad - \langle \tfrac{1}{2}E_f^2(\widetilde{R}_{Ai}); w\widetilde{R}_{Ai}, u\widetilde{R}_{Ai}, y\widetilde{R}_{Ai}\rangle \\
&\quad + 2\sum_{i=1}^{n}\sum_{j=1}^{n} x_{A_i}x_{A_j} \langle\Big(\tfrac{1}{4}\Big[(\widetilde{R}_{Abi1} - \widetilde{R}_{Aai1})(\widetilde{R}_{Abj1} - \widetilde{R}_{Aaj1}) \\
&\quad + (\widetilde{R}_{Aci1} - \widetilde{R}_{Abi1})(\widetilde{R}_{Acj1} - \widetilde{R}_{Abj1})\Big] \\
&\quad + \tfrac{1}{3}\Big\{\Big[\widetilde{R}_{Aaj1}(\widetilde{R}_{Abi1} - \widetilde{R}_{Aai1}) + \widetilde{R}_{Aai1}(\widetilde{R}_{Abj1} - \widetilde{R\widetilde{R}_{A}a_{aj1}})\Big] \\
&\quad - \Big[\widetilde{R}_{Aci1}(\widetilde{R}_{Acj1} - \widetilde{R}_{Abj1}) + \widetilde{R}_{Acj1}(\widetilde{R}_{Aci1} - \widetilde{R}_{Abi1})\Big]\Big\} \\
&\quad + \tfrac{1}{2}(\widetilde{R}_{Aai1}\widetilde{R}_{Aaj1} + \widetilde{R}_{Aci1}\widetilde{R}_{Acj1}) \\
&\quad + \tfrac{1}{2}E_f(\widetilde{R}_{Ai})E_f(\widetilde{R}_{Aj})\Big); w\widetilde{R}_{A_i} \wedge w\widetilde{R}_{Aj}, u\widetilde{R}_{Ai} \vee u\widetilde{R}_{Aj}, y\widetilde{R}_{Ai} \vee y\widetilde{R}_{Aj}\rangle
\end{aligned} \tag{26}$$

Demonstration: It is known that the neutrosophic portfolio risk made up of N financial assets is of the form:

$$\begin{aligned}
&\langle \widetilde{\sigma_p^2}; w\widetilde{\sigma p}, u\widetilde{\sigma p}, y\widetilde{\sigma p}\rangle \\
&\quad = \langle x_{A_1}^2 \widetilde{\sigma}_{A_1}^2; w\widetilde{\sigma}_{A_1}, u\widetilde{\sigma}_{A_1}, y\widetilde{\sigma}_{A_1}\rangle + \langle x_{A_2}^2 \widetilde{\sigma}_{A_2}^2; w\widetilde{\sigma}_{A_2}, u\widetilde{\sigma}_{A_2}, y\widetilde{\sigma}_{A_2}\rangle + \cdots \\
&\quad + \langle x_{A_n}^2 \widetilde{\sigma}_{A_n}^2; w\widetilde{\sigma}_{A_n}, u\widetilde{\sigma}_{A_n}, y\widetilde{\sigma}_{A_n}\rangle \\
&\quad + \langle 2x_{A_1}x_{A_2}\widetilde{\sigma}_{A_1A_2}; w\widetilde{\sigma}_{A_1} \wedge w\widetilde{\sigma}_{A_2}, u\widetilde{\sigma}_{A_1} \vee u\widetilde{\sigma}_{A_2}, y\widetilde{\sigma}_{A_1} \vee y\widetilde{\sigma}_{A_2}\rangle \\
&\quad + \langle 2x_{A_1}x_{A_3}\widetilde{\sigma}_{A_1A_3}; w\widetilde{\sigma}_{A_1} \wedge w\widetilde{\sigma}_{A_3}, u\widetilde{\sigma}_{A_1} \vee u\widetilde{\sigma}_{A_3}, y\widetilde{\sigma}_{A_1} \vee y\widetilde{\sigma}_{A_3}\rangle + \cdots \\
&\quad + \langle 2x_{A_1}x_{A_n}\widetilde{\sigma}_{A_1A_n}; w\widetilde{\sigma}_{A_1} \wedge w\widetilde{\sigma}_{A_n}, u\widetilde{\sigma}_{A_1} \vee u\widetilde{\sigma}_{A_n}, y\widetilde{\sigma}_{A_1} \vee y\widetilde{\sigma}_{A_n}\rangle \\
&\quad + \langle 2x_{A_2}x_{A_1}\widetilde{\sigma}_{A_2A_1}; w\widetilde{\sigma}_{A_2} \wedge w\widetilde{\sigma}_{A_1}, u\widetilde{\sigma}_{A_2} \vee u\widetilde{\sigma}_{A_1}, y\widetilde{\sigma}_{A_2} \vee y\widetilde{\sigma}_{A_1}\rangle \\
&\quad + \langle 2x_{A_2}x_{A_3}\widetilde{\sigma}_{A_2A_3}; w\widetilde{\sigma}_{A_2} \wedge w\widetilde{\sigma}_{A_3}, u\widetilde{\sigma}_{A_2} \vee u\widetilde{\sigma}_{A_3}, y\widetilde{\sigma}_{A_2} \vee y\widetilde{\sigma}_{A_3}\rangle + \cdots \\
&\quad + \langle 2x_{A_2}x_{A_n}\widetilde{\sigma}_{A_2A_n}; w\widetilde{\sigma}_{A_2} \wedge w\widetilde{\sigma}_{A_n}, u\widetilde{\sigma}_{A_2} \vee u\widetilde{\sigma}_{A_n}, y\widetilde{\sigma}_{A_2} \vee y\widetilde{\sigma}_{A_n}\rangle \\
&\quad + \langle 2x_{A_n}x_{A_1}\widetilde{\sigma}_{A_nA_1}; w\widetilde{\sigma}_{A_n} \wedge w\widetilde{\sigma}_{A_1}, u\widetilde{\sigma}_{A_n} \vee u\widetilde{\sigma}_{A_1}, y\widetilde{\sigma}_{A_n} \vee y\widetilde{\sigma}_{A_1}\rangle \\
&\quad + \langle 2x_{A_n}x_{A_2}\widetilde{\sigma}_{A_nA_2}; w\widetilde{\sigma}_{A_n} \wedge w\widetilde{\sigma}_{A_2}, u\widetilde{\sigma}_{A_n} \vee u\widetilde{\sigma}_{A_2}, y\widetilde{\sigma}_{A_n} \vee y\widetilde{\sigma}_{A_2}\rangle + \cdots
\end{aligned} \tag{27}$$

The analytical relation above can be written as follows:

$$\begin{aligned}
&\langle \widetilde{\sigma_p^2}; w\widetilde{\sigma p}, u\widetilde{\sigma p}, y\widetilde{\sigma p}\rangle \\
&\quad = \sum_{i=1}^{n} \langle x_{A_i}^2 \widetilde{\sigma}_{A_i}^2; w\widetilde{\sigma}_{A_i}, u\widetilde{\sigma}_{A_i}, y\widetilde{\sigma}_{A_i}\rangle \\
&\quad + 2\sum_{i=1}^{n}\sum_{j=1}^{n} \langle x_{A_i}x_{A_j}\widetilde{\sigma}_{A_iA_j}; w\widetilde{\sigma}_{A_i} \wedge w\widetilde{\sigma}_{A_j}, u\widetilde{\sigma}_{A_i} \vee u\widetilde{\sigma}_{A_j}, y\widetilde{\sigma}_{A_i} \vee y\widetilde{\sigma}_{A_j}\rangle
\end{aligned} \tag{28}$$

In the neutrosophic portfolio risk relation, we substitute the expression for the determination of the mean square deviation according to the Definition 2 and the expression for the covariance according to the Definition 3, established for a financial asset and we obtain the calculation relation for determining the risk size of the portfolio according to the weight of the financial asset in the total value of the portfolio x_{A_i} but also of the individual financial asset risk $\langle \widetilde{\sigma}_{A_i}^2; w\widetilde{\sigma}_{A_i}, u\widetilde{\sigma}_{A_i}, y\widetilde{\sigma}_{A_i}\rangle$ and the covariance between two financial assets $\langle \widetilde{\sigma}_{A_iA_j}; w\widetilde{\sigma}_{A_i} \wedge w\widetilde{\sigma}_{A_j}, u\widetilde{\sigma}_{A_i} \vee u\widetilde{\sigma}_{A_j}, y\widetilde{\sigma}_{A_i} \vee y\widetilde{\sigma}_{A_j}\rangle$:

$$
\begin{aligned}
&\langle \widetilde{\sigma_p^2}; w\widetilde{\sigma p}, u\widetilde{\sigma p}, y\widetilde{\sigma p}\rangle \\
&\quad = \sum_{i=1}^{n} x_{A_i}^2 \langle \tfrac{1}{4}\Big[(\widetilde{R_{Abi}} - \widetilde{R_{Aai}})^2 + (\widetilde{R_{Aci}} - \widetilde{R_{Abi}})^2\Big]; w\widetilde{R_{Ai}}, u\widetilde{R_{Ai}}, y\widetilde{R_{Ai}}\rangle \\
&\quad + \langle \tfrac{2}{3}\Big[\widetilde{R_{Aai}}(\widetilde{R_{Abi}} - \widetilde{R_{Aai}}) - \widetilde{R_{Aci}}(\widetilde{R_{Aci}} - \widetilde{R_{Abi}})\Big]; w\widetilde{R_{Ai}}, u\widetilde{R_{Ai}}, y\widetilde{R_{Ai}}\rangle \\
&\quad + \langle \tfrac{1}{2}(\widetilde{R_{Aai}}^2 + \widetilde{R_{Aci}}^2); w\widetilde{Ra_i}, u\widetilde{Ra_i}, y\widetilde{Ra_i}\rangle \\
&\quad - \langle \tfrac{1}{2}E_f^2(\widetilde{R_{Ai}}); w\widetilde{R_{Ai}}, u\widetilde{R_{Ai}}, y\widetilde{R_{Ai}}\rangle \\
&\quad + 2\sum_{i=1}^{n}\sum_{j=1}^{n} x_{A_i}x_{A_j}\langle\Big(\tfrac{1}{4}\Big[(\widetilde{R_{Abi1}} - \widetilde{R_{Aai1}})(\widetilde{R_{Abj1}} - \widetilde{R_{Aaj1}}) \\
&\quad + (\widetilde{R_{Aci1}} - \widetilde{R_{Abi1}})(\widetilde{R_{Acj1}} - \widetilde{R_{Abj1}})\Big] \\
&\quad + \tfrac{1}{3}\Big\{\Big[\widetilde{R_{Aaj1}}(\widetilde{R_{Abi1}} - \widetilde{R_{Aai1}}) + \widetilde{R_{Aai1}}(\widetilde{R_{Abj1}} - \widetilde{R\widetilde{R_A}a_{aj1}})\Big] \\
&\quad - \Big[\widetilde{R_{Aci1}}(\widetilde{R_{Acj1}} - \widetilde{R_{Abj1}}) + \widetilde{R_{Acj1}}(\widetilde{R_{Aci1}} - \widetilde{R_{Abi1}})\Big]\Big\} \\
&\quad + \tfrac{1}{2}(\widetilde{R_{Aai1}}\widetilde{R_{Aaj1}} + \widetilde{R_{Aci1}}\widetilde{R_{Acj1}}) \\
&\quad + \tfrac{1}{2}E_f(\widetilde{R_{Ai}})E_f(\widetilde{R_{Aj}})\Big); w\widetilde{R_{A_i}} \wedge w\widetilde{R_{Aj}}, u\widetilde{R_{Ai}} \vee u\widetilde{R_{Aj}}, y\widetilde{R_{Ai}} \vee y\widetilde{R_{Aj}}\rangle
\end{aligned}
\tag{29}
$$

Example 2. *There are considered three financial assets* (A_1, A_2, A_3) *to which three triangular neutrosophic numbers are specified for the financial assets return, of the form:*

$$
\begin{aligned}
&\widetilde{R_{A1}} = \langle(0.2\ 0.3\ 0.5); 0.5,\ 0.2,\ 0.3\rangle \text{ pentru valori ale } \widetilde{R_A} \in [0,2;0,5] \\
&\widetilde{R_{A2}} = \langle(0.1\ 0.2\ 0.3); 0.6,\ 0.3,\ 0.2\rangle \text{ pentru valori ale } \widetilde{R_A} \in [0,1;0,3] \\
&\widetilde{R_{A3}} = \langle(0.3\ 0.4\ 0.6); 0.4,\ 0.3,\ 0.3\rangle \text{ pentru valori ale } \widetilde{R_A} \in [0,3;0,6]
\end{aligned}
\tag{30}
$$

The weights held by the three financial assets in the total portfolio are determined according to the value of each financial asset and the total value of the portfolio and have the values: $x_{A_1} = 0,4$; $x_{A_2} = 0,3$ și $x_{A_3} = 0,3$. In order to establish the neutrosophic portfolio risk, from proposition2 it is known that:

$$
\begin{aligned}
&\langle \widetilde{\sigma_p^2}; w\widetilde{\sigma p}, u\widetilde{\sigma p}, y\widetilde{\sigma p}\rangle \\
&\quad = \sum_{i=1}^{n} \langle x_{A_i}^2 \widetilde{\sigma}_{A_i}^2; w\widetilde{\sigma}_{A_i}, u\widetilde{\sigma}_{A_i}, y\widetilde{\sigma}_{A_i}\rangle \\
&\quad + 2\sum_{i=1}^{n}\sum_{j=1}^{n} \langle x_{A_i}x_{A_j}\widetilde{\sigma}_{A_iA_j}; w\widetilde{\sigma}_{A_i} \wedge\ w\widetilde{\sigma}_{A_j}, u\widetilde{\sigma}_{A_i} \vee u\widetilde{\sigma}_{A_j}, y\widetilde{\sigma}_{A_i} \vee y\widetilde{\sigma}_{A_j}\rangle
\end{aligned}
\tag{31}
$$

The values of the neutrosophic risk for a financial asset are determined:

$$
\begin{aligned}
\widetilde{\sigma}_{fA_1}^2 &= \langle \tfrac{1}{4}[(0.3-0.2)^2 + (0.5-0.3)^2]; 0.5,\ 0.2,\ 0.3\rangle \\
&\quad + \langle \tfrac{2}{3}(0.2(0.3-0.2) - 0.5(0.5-0.2)); 0.5,\ 0.2,\ 0.3\rangle \\
&\quad + \langle \tfrac{1}{2}(0.2^2 + 0.5^2); 0.5,\ 0.2,\ 0.3\rangle - \langle \tfrac{1}{2}(0.316)^2; 0.5,\ 0.2,\ 0.3\rangle
\end{aligned}
\tag{32}
$$

$$
\widetilde{\sigma}_{fA_1}^2 = \langle 0.0225; 0.5,\ 0.2,\ 0.3\rangle \tag{33}
$$

Proceeding in the same manner, we get the following results for $\widetilde{\sigma}_{fA_2}^2$ and $\widetilde{\sigma}_{fA_3}^2$:

$$
\widetilde{\sigma}_{fA_2}^2 = \langle 0.0180; 0.6,\ 0.3,\ 0.2\rangle \tag{34}
$$

$$
\widetilde{\sigma}_{fA_3}^2 = \langle 0.0925; 0.4,\ 0.3,\ 0.3\rangle \tag{35}
$$

We establish the covariance between financial assets according to the Definition 3 as follows:

$$\begin{aligned}\sigma_{A_1A_2} = \langle \tfrac{1}{4}[(0.3-0.2)(0.2-0.1)+(0.5-0.3)(0.3-0.2)] \\ +\tfrac{1}{3}[0.1(0.3-0.2)+0.2(0.2-0.1)] \\ -[0.5(0.3-0.2)+0.3(0.5-0.3)]+\tfrac{1}{2}(0.2*0.1+0.5*0.3) \\ +\tfrac{1}{2}0.316*0.199; 0.5\wedge 0.6,\ 0.2\vee 0.3, 0.3\vee 0.2\rangle\end{aligned}$$

$$\sigma_{A_1A_2} = \langle 0.0705; 0.6,\ 0.2,\ 0.2\rangle \tag{36}$$

In the same way, we get:

$$\sigma_{A_1A_3} = \langle 0.1914; 0.5,\ 0.2,\ 0.3\rangle \tag{37}$$

$$\sigma_{A_2A_3} = \langle 0.0805;\ 0.6,\ 0.3,\ 0.2\rangle \tag{38}$$

$$\begin{aligned}&\langle\sigma_{\tilde{p}}^2; w\widetilde{\sigma p}, u\widetilde{\sigma p}, y\widetilde{\sigma p}\rangle \\ &= \langle x_{A_1}^2\widetilde{\sigma}_{A_1}^2; w\widetilde{\sigma}_{A_1}, u\widetilde{\sigma}_{A_1}, y\widetilde{\sigma}_{A_1}\rangle + \langle x_{A_2}^2\widetilde{\sigma}_{A_2}^2; w\widetilde{\sigma}_{A_2}, u\widetilde{\sigma}_{A_2}, y\widetilde{\sigma}_{A_2}\rangle \\ &+\langle x_{A_3}^2\widetilde{\sigma}_{A_3}^2; w\widetilde{\sigma}_{A_3}, u\widetilde{\sigma}_{A_3}, y\widetilde{\sigma}_{A_3}\rangle \\ &+2x_{A_1}x_{A_2}\widetilde{\sigma}_{A_1A_2}; w\widetilde{\sigma}_{A_1}\wedge w\widetilde{\sigma}_{A_2}, u\widetilde{\sigma}_{A_1}\vee u\widetilde{\sigma}_{A_2}, y\widetilde{\sigma}_{A_1}\vee y\widetilde{\sigma}_{A_2} \\ &+\langle 2x_{A_1}x_{A_3}\widetilde{\sigma}_{A_1A_3}; w\widetilde{\sigma}_{A_1}\wedge w\widetilde{\sigma}_{A_3}, u\widetilde{\sigma}_{A_1}\vee u\widetilde{\sigma}_{A_3}, y\widetilde{\sigma}_{A_1}\vee y\widetilde{\sigma}_{A_3}\rangle \\ &+\langle 2x_{A_2}x_{A_1}\widetilde{\sigma}_{A_2A_1}; w\widetilde{\sigma}_{A_2}\wedge w\widetilde{\sigma}_{A_1}, u\widetilde{\sigma}_{A_2}\vee u\widetilde{\sigma}_{A_1}, y\widetilde{\sigma}_{A_2}\vee y\widetilde{\sigma}_{A_1}\rangle \\ &+\langle 2x_{A_2}x_{A_3}\widetilde{\sigma}_{A_2A_3}; w\widetilde{\sigma}_{A_2}\wedge w\widetilde{\sigma}_{A_3}, u\widetilde{\sigma}_{A_2}\vee u\widetilde{\sigma}_{A_3}, y\widetilde{\sigma}_{A_2}\vee y\widetilde{\sigma}_{A_3}\rangle \\ &+\langle 2x_{A_3}x_{A_1}\widetilde{\sigma}_{A_3A_1}; w\widetilde{\sigma}_{A_3}\wedge w\widetilde{\sigma}_{A_1}, u\widetilde{\sigma}_{A_3}\vee u\widetilde{\sigma}_{A_1}, y\widetilde{\sigma}_{A_3}\vee y\widetilde{\sigma}_{A_1}\rangle \\ &+\langle 2x_{A_3}x_{A_2}\widetilde{\sigma}_{A_3A_2}; w\widetilde{\sigma}_{A_3}\wedge w\widetilde{\sigma}_{A_2}, u\widetilde{\sigma}_{A_3}\vee u\widetilde{\sigma}_{A_2}, y\widetilde{\sigma}_{A_3}\vee y\widetilde{\sigma}_{A_2}\rangle\end{aligned} \tag{39}$$

$$\begin{aligned}&\langle\sigma_{\tilde{p}}^2; w\widetilde{\sigma p}, u\widetilde{\sigma p}, y\widetilde{\sigma p}\rangle \\ &= \langle 0.16\times 0.0225; 0.5,\ 0.2,\ 0.3\rangle + \langle 0.09\times 0.0180; 0.6,\ 0.3,\ 0.2\rangle \\ &+\langle 0.09\times 0.0925; 0.4,\ 0.3,\ 0.3\rangle + \langle 2\times 0.12\times 0.0705; 0.6,\ 0.2,\ 0.2\rangle \\ &+\langle 2\times 0.12\times 0.1914; 0.5,\ 0.2,\ 0.3\rangle + \langle 2\times 0.12\times 0.0705; 0.6,\ 0.2,\ 0.2\rangle \\ &+\langle 2\times 0.09\times 0.0805;\ 0.6,\ 0.3,\ 0.2\rangle \\ &+\langle 2\times 0,12\times 0.1914; 0.5,\ 0.2,\ 0.3\rangle \\ &+\langle 2\times 0.09\times 0.0805;\ 0.6,\ 0.3,\ 0.2\rangle\end{aligned} \tag{40}$$

$$\langle\sigma_p; w\widetilde{\sigma p}, u\widetilde{\sigma p}, y\widetilde{\sigma p}\rangle = \sqrt{\langle 0,168237; 0.6,\ 0.2,\ 0.2\rangle} = \langle 0,41016; 0.6,\ 0.2,\ 0.2\rangle \tag{41}$$

Interpretation: The neutrosophic portfolio return was previously determined $\langle\widetilde{R}_P; w\widetilde{Rp}, u\widetilde{Rp}, y\widetilde{Rp}\rangle = \langle 0,3109; 0.5,\ 0.2,\ 0.3\rangle$. For this neutrosophic portfolio return value corresponds a high risk $\langle\sigma_p; w\widetilde{\sigma p}, u\widetilde{\sigma p}, y\widetilde{\sigma p} = 0,41016; 0.6,\ 0.2,\ 0.2\rangle$ which confirms that between return and risk there is a directly proportional relationship. The probabilities for risk manifestation is about 60%, while the probability that the risk is certain/uncertain is 20% and the probability that the risk does not occur is quite small and has a value of 20%.

4. Neutrosophic Portfolio Equations. The Analytical and Matrix Form

Neutrosophic portfolios of the form $\langle\widetilde{P}; w\widetilde{p}, u\widetilde{p}, y\widetilde{p}\rangle$ have the characteristic that each of the financial assets they contain can be modelled using the neutrosophic performance indicators such as: neutrosophic return $\langle E_f(\widetilde{R_A}); w\widetilde{R_A}, u\widetilde{R_A}, y\widetilde{R_A}\rangle$, the neutrosophic risk $\langle\sigma f_{A_i}^2; w\widetilde{\sigma}_A, u\widetilde{\sigma}_A, y\widetilde{\sigma}_A\rangle$, the neutrosophic covariance $\langle cov(\widetilde{R_{A1}}, \widetilde{R_{A2}}); w\widetilde{R_{A1}}, u\widetilde{R_{A1}}, y\widetilde{R_{A1}};\ w\widetilde{R_{A2}}, u\widetilde{R_{A2}}, y\widetilde{R_{A2}}\rangle$.

With these neutrosophic performance indicators specific to each financial asset, are determined the two fundamental variables of the neutrosophic portfolios, namely: neutrosophic portfolio return $\langle\widetilde{R}_P; w\widetilde{Rp}, u\widetilde{Rp}, y\widetilde{Rp}\rangle$ and the neutrosophic portfolio risk $\langle\widetilde{\sigma}_p^2; w\widetilde{\sigma p}, u\widetilde{\sigma p}, y\widetilde{\sigma p}\rangle$. The neutrosophic portfolio return, according to sentence no.1 can be written in analytical form as follows:

$$\langle\widetilde{R}_P; w\widetilde{Rp}, u\widetilde{Rp}, y\widetilde{Rp}\rangle = \sum_{i=1}^{n}\langle x_{A_i}\widetilde{R}_{A_i}; w\widetilde{R}_{A_i}, u\widetilde{R}_{A_i}, y\widetilde{R}_{A_i}\rangle \tag{42}$$

The neutrosophic portfolio risk can be written in analytical form as follows:

$$\begin{aligned}&\langle\widetilde{\sigma_p^2}; w\widetilde{\sigma p}, u\widetilde{\sigma p}, y\widetilde{\sigma p}\rangle\\&\quad=\sum_{i=1}^{n}\langle x_{A_i}^2\widetilde{\sigma}_{A_i}^2; w\widetilde{\sigma}_{A_i}, u\widetilde{\sigma}_{A_i}, y\widetilde{\sigma}_{A_i}\rangle\\&\quad+2\sum_{i=1}^{n}\sum_{j=1}^{n}\langle x_{A_i}x_{A_j}\widetilde{\sigma}_{A_iA_j}; w\widetilde{\sigma}_{A_i}\wedge w\widetilde{\sigma}_{A_j}, u\widetilde{\sigma}_{A_i}\vee u\widetilde{\sigma}_{A_j}, y\widetilde{\sigma}_{A_i}\vee y\widetilde{\sigma}_{A_j}\rangle\end{aligned} \tag{43}$$

In order to form the system of equations that characterize the neutrosophic portfolios of financial assets, it should be mentioned that these portfolios are made up of financial assets whose weight in the total value of the portfolio is 100% which can be mathematically quantified by the formula:

$$\sum_{i=1}^{n} x_{A_i} = 100\% \tag{44}$$

Under these conditions, the system of equations of the neutrosophicportfolio of financial assets in analytical form will be written as follows:

$$\begin{aligned}&\langle\widetilde{R_P}; w\widetilde{Rp}, u\widetilde{Rp}, y\widetilde{Rp}\rangle=\sum_{i=1}^{n}\langle x_{A_i}\widetilde{R}_{A_i}; w\widetilde{R}_{A_i}, u\widetilde{R}_{A_i}, y\widetilde{R}_{A_i}\rangle\\&\langle\widetilde{\sigma_p^2}; w\widetilde{\sigma p}, u\widetilde{\sigma p}, y\widetilde{\sigma p}\rangle=\sum_{i=1}^{n}\langle x_{A_i}^2\widetilde{\sigma}_{A_i}^2; w\widetilde{\sigma}_{A_i}, u\widetilde{\sigma}_{A_i}, y\widetilde{\sigma}_{A_i}\rangle+\\&+2\sum_{i=1}^{n}\sum_{j=1}^{n}\langle x_{A_i}x_{A_j}\widetilde{\sigma}_{A_iA_j}; w\widetilde{\sigma}_{A_i}\wedge w\widetilde{\sigma}_{A_j}, u\widetilde{\sigma}_{A_i}\vee u\widetilde{\sigma}_{A_j}, y\widetilde{\sigma}_{A_i}\vee y\widetilde{\sigma}_{A_j}\rangle\\&\sum_{i=1}^{n} x_{A_i} = 100\%\end{aligned} \tag{45}$$

In matrix form the equations of the neutrosophicportfolio made up of Nfinancial assets will be written as follows:

$$\langle\widetilde{R_P}; w\widetilde{Rp}, u\widetilde{Rp}, y\widetilde{Rp}\rangle=(x_{A_1}x_{A_2}\dots x_{A_n})\begin{pmatrix}\langle\widetilde{R}_{A_1}; w\widetilde{R}_{A_1}, u\widetilde{R}_{A_1}, y\widetilde{R}_{A_1}\rangle\\\langle\widetilde{R}_{A_2}; w\widetilde{R}_{A_2}, u\widetilde{R}_{A_2}, y\widetilde{R}_{A_2}\rangle\\\dots\\\langle\widetilde{R}_{A_n}; w\widetilde{R}_{A_n}, u\widetilde{R}_{A_n}, y\widetilde{R}_{A_n}\rangle\end{pmatrix} \tag{46}$$

We note: $X_A^T = \left(x_{A_1}x_{A_2}\dots x_{A_n}\right)$ and

$$\langle\widetilde{R_A}; w\widetilde{R}_A, u\widetilde{R}_A, y\widetilde{R}_A\rangle=\begin{pmatrix}\langle\widetilde{R}_{A_1}; w\widetilde{R}_{A_1}, u\widetilde{R}_{A_1}, y\widetilde{R}_{A_1}\rangle\\\langle\widetilde{R}_{A_2}; w\widetilde{R}_{A_2}, u\widetilde{R}_{A_2}, y\widetilde{R}_{A_2}\rangle\\\dots\\\langle\widetilde{R}_{A_n}; w\widetilde{R}_{A_n}, u\widetilde{R}_{A_n}, y\widetilde{R}_{A_n}\rangle\end{pmatrix} \tag{47}$$

Under these conditions, the equation of the neutrosophic portfolio return will be written in matrix form as follows:

$$\langle\widetilde{R_p}; w\widetilde{Rp}, u\widetilde{Rp}, y\widetilde{Rp}\rangle = X_A^T\langle\widetilde{R_A}; w\widetilde{R}_A, u\widetilde{R}_A, y\widetilde{R}_A\rangle \tag{48}$$

The portfolio risk equation above can be written in matrix form as follows:

$$\begin{aligned}&\langle\widetilde{\sigma_p^2}; w\widetilde{\sigma p}, u\widetilde{\sigma p}, y\widetilde{\sigma p}\rangle\\&=(x_{A_1}x_{A_2}\dots x_{A_n})\begin{pmatrix}\langle\widetilde{\sigma}_{A_{11}}; w\widetilde{\sigma}_{A_{11}}, u\widetilde{\sigma}_{A_{11}}, y\widetilde{\sigma}_{A_{11}}\rangle & \dots & \langle\widetilde{\sigma}_{A_{1n}}; w\widetilde{\sigma}_{A_{1n}}, u\widetilde{\sigma}_{A_{1n}}, y\widetilde{\sigma}_{A_{1n}}\rangle\\\langle\widetilde{\sigma}_{A_{21}}; w\widetilde{\sigma}_{A_{21}}, u\widetilde{\sigma}_{A_{21}}, y\widetilde{\sigma}_{A_{21}}\rangle & \dots & \langle\widetilde{\sigma}_{A_{2n}}; w\widetilde{\sigma}_{A_{2n}}, u\widetilde{\sigma}_{A_{2n}}, y\widetilde{\sigma}_{A_{2n}}\rangle\\\dots & \dots & \dots\\\langle\widetilde{\sigma}_{A_{n1}}; w\widetilde{\sigma}_{A_{n1}}, u\widetilde{\sigma}_{A_{n1}}, y\widetilde{\sigma}_{A_{n1}}\rangle & \dots & \langle\widetilde{\sigma}_{A_{nn}}; w\widetilde{\sigma}_{A_{nn}}, u\widetilde{\sigma}_{A_{nn}}, y\widetilde{\sigma}_{A_{nn}}\rangle\end{pmatrix}\begin{pmatrix}x_{A_1}\\x_{A_2}\\\dots\\x_{A_n}\end{pmatrix}\end{aligned} \tag{49}$$

In the matrix equation of the neutrosophic portfolio risk above we note:

$$X_A^T = \left(x_{A_1} x_{A_2} \ldots x_{A_n}\right)$$

and

$$\langle \widetilde{\Omega}; w\widetilde{\sigma}_A, u\widetilde{\sigma}_A, y\widetilde{\sigma}_A \rangle = \begin{pmatrix} \langle \widetilde{\sigma}_{A_{11}}; w\widetilde{\sigma}_{A_{11}}, u\widetilde{\sigma}_{A_{11}}, y\widetilde{\sigma}_{A_{11}} \rangle & \ldots & \langle \widetilde{\sigma}_{A_{1n}}; w\widetilde{\sigma}_{A_{1n}}, u\widetilde{\sigma}_{A_{1n}}, y\widetilde{\sigma}_{A_{1n}} \rangle \\ \langle \widetilde{\sigma}_{A_{21}}; w\widetilde{\sigma}_{A_{21}}, u\widetilde{\sigma}_{A_{21}}, y\widetilde{\sigma}_{A_{21}} \rangle & \ldots & \langle \widetilde{\sigma}_{A_{2n}}; w\widetilde{\sigma}_{A_{2n}}, u\widetilde{\sigma}_{A_{2n}}, y\widetilde{\sigma}_{A_{2n}} \rangle \\ \ldots & \ldots & \ldots \\ \langle \widetilde{\sigma}_{A_{n1}}; w\widetilde{\sigma}_{A_{n1}}, u\widetilde{\sigma}_{A_{n1}}, y\widetilde{\sigma}_{A_{n1}} \rangle & \ldots & \langle \widetilde{\sigma}_{A_{nn}}; w\widetilde{\sigma}_{A_{nn}}, u\widetilde{\sigma}_{A_{nn}}, y\widetilde{\sigma}_{A_{nn}} \rangle \end{pmatrix} \quad (50)$$

Under these conditions the matrix equation of the neutrosophic portfolio risk becomes:

$$\langle \widetilde{\sigma_p^2}; w\widetilde{\sigma p}, u\widetilde{\sigma p}, y\widetilde{\sigma p} \rangle = X_A^T \langle \widetilde{\Omega}; w\widetilde{\sigma}_A, u\widetilde{\sigma}_A, y\widetilde{\sigma}_A \rangle X_A \quad (51)$$

The equation of the financial assets weight in the total value of the neutrosophic portfolio can be written as:

$$\left(x_{A_1} x_{A_2} \ldots x_{A_n}\right) \begin{pmatrix} 1 \\ 1 \\ \ldots \\ 1 \end{pmatrix} = 1 \quad (52)$$

In matrix form, the equation of weights will be written as follows: $X_A^T e = 1$.

The system of equations of the neutrosophic portfolio in matrix form will be written as follows:

$$\begin{cases} \langle \widetilde{R_p}; w\widetilde{Rp}, u\widetilde{Rp}, y\widetilde{Rp} \rangle = X_A^T \langle \widetilde{R}_A; w\widetilde{R}_A, u\widetilde{R}_A, y\widetilde{R}_A \rangle \\ \langle \widetilde{\sigma_p^2}; w\widetilde{\sigma p}, u\widetilde{\sigma p}, y\widetilde{\sigma p} \rangle = X_A^T \langle \widetilde{\Omega}; w\widetilde{\sigma}_A, u\widetilde{\sigma}_A, y\widetilde{\sigma}_A \rangle X_A \\ X_A^T e = 1 \end{cases} \quad (53)$$

Neutrosophic portfolio equations in analytical or matrix form will be used for risk minimization calculations or for determining the optimal portfolio structure depending on the needs.

5. Minimizing the Risk of the Neutrosophic Portfolio

5.1. Minimizing the Risk of the Neutrosophic Portfolio Consisting of Two Financial Assets

The purpose of this section is to determine that structure of the neutrosophic portfolio $x_{A_1} x_{A_2} \ldots x_{A_n}$ for which the risk is minimal. The financial assets that enter in the structure of the neutrosophic portfolio allow the determination of the performance indicators using the neutrosophic triangular fuzzy numbers according to theorem no.1. For this we write the equations of the neutrosophic portfolio consisting of two financial assets A_1, A_2 as follows:

$$\begin{cases} \langle \widetilde{R_p}; w\widetilde{Rp}, u\widetilde{Rp}, y\widetilde{Rp} \rangle = \langle x_{A_1} \widetilde{R}_{A_1}; w\widetilde{R}_{A1}, u\widetilde{R}_{A1}, y\widetilde{R}_{A1} \rangle + \langle x_{A_2} \widetilde{R}_{A_2}; w\widetilde{R}_{A2}, u\widetilde{R}_{A2}, y\widetilde{R}_{A2} \rangle \\ \langle \widetilde{\sigma_p^2}; w\widetilde{\sigma p}, u\widetilde{\sigma p}, y\widetilde{\sigma p} \rangle = \langle x_{A_1}^2 \widetilde{\sigma}_{A_1}^2; w\widetilde{\sigma}_{A_1}, u\widetilde{\sigma}_{A_1}, y\widetilde{\sigma}_{A_1} \rangle + \langle x_{A_2}^2 \widetilde{\sigma}_{A_2}^2; w\widetilde{\sigma}_{A_2}, u\widetilde{\sigma}_{A_2}, y\widetilde{\sigma}_{A_2} \rangle + \\ + \langle 2 x_{A_1} x_{A_2} \sigma_{A_1 A_2}; w\sigma_A, u\sigma_A, y\sigma_A \rangle \\ x_{A_1} + x_{A_2} = 1 \end{cases} \quad (54)$$

In order to establish the structure of the neutrosophic portfolio for which its risk is minimal, are imposed the minimum conditions for the portfolio risk resulted from the cancellation of the first order derivative of the neutrosophic portfolio risk:

$$\begin{cases} \frac{\partial \langle \widetilde{\sigma_p^2}; w\widetilde{\sigma p}, u\widetilde{\sigma p}, y\widetilde{\sigma p} \rangle}{\partial x_{A_1}} = 0 \\ \frac{\partial \langle \widetilde{\sigma_p^2}; w\widetilde{\sigma p}, u\widetilde{\sigma p}, y\widetilde{\sigma p} \rangle}{\partial x_{A_2}} = 0 \end{cases} \quad (55)$$

Theorem 1. *Let be two financial assets A_1, A_2 for which the neutrosophic return can be determined $\langle \widetilde{R}_{A_1}; w\widetilde{R}_{A1}, u\widetilde{R}_{A1}, y\widetilde{R}_{A1}\rangle$ and $\langle \widetilde{R}_{A_2}; w\widetilde{R}_{A2}, u\widetilde{R}_{A2}, y\widetilde{R}_{A2}\rangle$. Also, the specific neutrosophic risk can be determined for each financial asset $\langle \widetilde{\sigma}^2_{A_1}; w\widetilde{\sigma}_{A_1}, u\widetilde{\sigma}_{A_1}, y\widetilde{\sigma}_{A_1}\rangle$ and $\langle \widetilde{\sigma}^2_{A_2}; w\widetilde{\sigma}_{A_2}, u\widetilde{\sigma}_{A_2}, y\widetilde{\sigma}_{A_2}\rangle$. These two financial assets form a neutrosophic portfolio of the form: $\langle \widetilde{P}; w\widetilde{P}, u\widetilde{P}, y\widetilde{P}\rangle$. The risk of the neutrosophic portfolio has the minimum value for x_{A_1}, respectively x_{A_2} of the form:*

$$x_{A_1} = \frac{\langle \widetilde{\sigma}_{A_1A_2}; w\widetilde{\sigma}_A, u\widetilde{\sigma}_A, y\widetilde{\sigma}_A\rangle}{\langle \widetilde{\sigma}_{A_1A_2}; w\widetilde{\sigma}_A, u\widetilde{\sigma}_A, y\widetilde{\sigma}_A\rangle - \langle \widetilde{\sigma}^2_{A_1}; w\widetilde{\sigma}_{A_1}, u\widetilde{\sigma}_{A_1}, y\widetilde{\sigma}_{A_1}\rangle} \tag{56}$$

$$x_{A_2} = \frac{\langle \widetilde{\sigma}^2_{A_1}; w\widetilde{\sigma}_{A_1}, u\widetilde{\sigma}_{A_1}, y\widetilde{\sigma}_{A_1}\rangle}{\langle \widetilde{\sigma}^2_{A_1}; w\widetilde{\sigma}_{A_1}, u\widetilde{\sigma}_{A_1}, y\widetilde{\sigma}_{A_1}\rangle - \langle \widetilde{\sigma}_{A_1A_2}; w\widetilde{\sigma}_A, u\widetilde{\sigma}_A, y\widetilde{\sigma}_A\rangle} \tag{57}$$

with the condition that $\langle \widetilde{\sigma}_{A_1A_2}; w\widetilde{\sigma}_A, u\widetilde{\sigma}_A, y\widetilde{\sigma}_A\rangle \neq \langle \widetilde{\sigma}^2_{A_1}; w\widetilde{\sigma}_{A_1}, u\widetilde{\sigma}_{A_1}, y\widetilde{\sigma}_{A_1}\rangle$.

Demonstration: We know that the equations of the neutrosophic portfolio in analytical form can be written according to the above equations:

$$\begin{gathered}
\langle \widetilde{R_p}; w\widetilde{Rp}, u\widetilde{Rp}, y\widetilde{Rp}\rangle = \langle x_{A_1}\widetilde{R}_{A_1}; w\widetilde{R}_{A1}, u\widetilde{R}_{A1}, y\widetilde{R}_{A1}\rangle + \langle x_{A_2}\widetilde{R}_{A_2}; w\widetilde{R}_{A2}, u\widetilde{R}_{A2}, y\widetilde{R}_{A2}\rangle \\
\langle \widetilde{\sigma}^2_p; w\widetilde{\sigma p}, u\widetilde{\sigma p}, y\widetilde{\sigma p}\rangle = \langle x^2_{A_1}\widetilde{\sigma}^2_{A_1}; w\widetilde{\sigma}_{A_1}, u\widetilde{\sigma}_{A_1}, y\widetilde{\sigma}_{A_1}\rangle + \langle x^2_{A_2}\widetilde{\sigma}^2_{A_2}; w\widetilde{\sigma}_{A_2}, u\widetilde{\sigma}_{A_2}, y\widetilde{\sigma}_{A_2}\rangle + \\
+\langle 2x_{A_1}x_{A_2}\widetilde{\sigma}_{A_1A_2}; w\widetilde{\sigma}_A, u\widetilde{\sigma}_A, y\widetilde{\sigma}_A\rangle \\
x_{A_1} + x_{A_2} = 1
\end{gathered} \tag{58}$$

We set the minimum conditions for the neutrosophic portfolio risk and obtain:

$$\begin{cases}
\frac{\partial\langle \widetilde{\sigma}^2_p; w\widetilde{\sigma p}, u\widetilde{\sigma p}, y\widetilde{\sigma p}\rangle}{\partial x_{A_1}} = 0 \\
\frac{\partial\langle \widetilde{\sigma}^2_p; w\widetilde{\sigma p}, u\widetilde{\sigma p}, y\widetilde{\sigma p}\rangle}{\partial x_{A_2}} = 0
\end{cases} \tag{59}$$

Based on these conditions, will obtain:

$$\left\{\begin{array}{c}
\langle 2x_{A_1}\widetilde{\sigma}^2_{A_1}; w\widetilde{\sigma}_{A_1}, u\widetilde{\sigma}_{A_1}, y\widetilde{\sigma}_{A_1}\rangle + \langle 2x_{A_2}\widetilde{\sigma}_{A_1A_2}; w\widetilde{\sigma}_A, u\widetilde{\sigma}_A, y\widetilde{\sigma}_A\rangle = 0 \\
\langle 2x_{A_2}\widetilde{\sigma}^2_{A_2}; w\widetilde{\sigma}_{A_2}, u\widetilde{\sigma}_{A_2}, y\widetilde{\sigma}_{A_2}\rangle + \langle 2x_{A_1}\widetilde{\sigma}_{A_1A_2}; w\widetilde{\sigma}_A, u\widetilde{\sigma}_A, y\widetilde{\sigma}_A\rangle = 0 \\
x_{A_1} + x_{A_2} = 1
\end{array}\right. \tag{60}$$

From the last equation of the above system results that $x_{A_2} = 1 - x_{A_1}$ and replacing in the first equation we will have:

$$\langle 2x_{A_1}\widetilde{\sigma}^2_{A_1}; w\widetilde{\sigma}_{A_1}, u\widetilde{\sigma}_{A_1}, y\widetilde{\sigma}_{A_1}\rangle + \langle 2\left(1 - x_{A_1}\right)\widetilde{\sigma}_{A_1A_2}; w\widetilde{\sigma}_A, u\widetilde{\sigma}_A, y\widetilde{\sigma}_A\rangle = 0 \tag{61}$$

$$\langle 2x_{A_1}\widetilde{\sigma}^2_{A_1}; w\widetilde{\sigma}_{A_1}, u\widetilde{\sigma}_{A_1}, y\widetilde{\sigma}_{A_1}\rangle + \langle 2\widetilde{\sigma}_{A_1A_2}; w\widetilde{\sigma}_A, u\widetilde{\sigma}_A, y\widetilde{\sigma}_A\rangle - \langle 2x_{A_1}\widetilde{\sigma}_{A_1A_2}; w\widetilde{\sigma}_A, u\widetilde{\sigma}_A, y\widetilde{\sigma}_A\rangle = 0 \tag{62}$$

$$\langle x_{A_1}\widetilde{\sigma}^2_{A_1}; w\widetilde{\sigma}_{A_1}, u\widetilde{\sigma}_{A_1}, y\widetilde{\sigma}_{A_1}\rangle - \langle x_{A_1}\widetilde{\sigma}_{A_1A_2}; w\widetilde{\sigma}_A, u\widetilde{\sigma}_A, y\widetilde{\sigma}_A\rangle = -\langle 2\widetilde{\sigma}_{A_1A_2}; w\widetilde{\sigma}_A, u\widetilde{\sigma}_A, y\widetilde{\sigma}_A\rangle \tag{63}$$

$$x_{A_1}\left(\langle \widetilde{\sigma}_{A_1A_2}; w\widetilde{\sigma}_A, u\widetilde{\sigma}_A, y\widetilde{\sigma}_A\rangle - \langle \widetilde{\sigma}^2_{A_1}; w\widetilde{\sigma}_{A_1}, u\widetilde{\sigma}_{A_1}, y\widetilde{\sigma}_{A_1}\rangle\right) = \langle \widetilde{\sigma}_{A_1A_2}; w\widetilde{\sigma}_A, u\widetilde{\sigma}_A, y\widetilde{\sigma}_A\rangle \tag{64}$$

$$x_{A_1} = \frac{\langle \widetilde{\sigma}_{A_1A_2}; w\widetilde{\sigma}_A, u\widetilde{\sigma}_A, y\widetilde{\sigma}_A\rangle}{\langle \widetilde{\sigma}_{A_1A_2}; w\widetilde{\sigma}_A, u\widetilde{\sigma}_A, y\widetilde{\sigma}_A\rangle - \langle \widetilde{\sigma}^2_{A_1}; w\widetilde{\sigma}_{A_1}, u\widetilde{\sigma}_{A_1}, y\widetilde{\sigma}_{A_1}\rangle} \tag{65}$$

With the condition that: $\langle \widetilde{\sigma}_{A_1A_2}; w\widetilde{\sigma}_A, u\widetilde{\sigma}_A, y\widetilde{\sigma}_A\rangle \neq \langle \widetilde{\sigma}^2_{A_1}; w\widetilde{\sigma}_{A_1}, u\widetilde{\sigma}_{A_1}, y\widetilde{\sigma}_{A_1}\rangle$.

By substituting the formula for x_{A_1} in the expression $x_{A_2} = 1 - x_{A_1}$ it is obtained:

$$x_{A_2} = 1 - \frac{\langle \widetilde{\sigma}_{A_1A_2}; w\widetilde{\sigma}_A, u\widetilde{\sigma}_A, y\widetilde{\sigma}_A \rangle}{\langle \widetilde{\sigma}_{A_1A_2}; w\widetilde{\sigma}_A, u\widetilde{\sigma}_A, y\widetilde{\sigma}_A \rangle - \langle \widetilde{\sigma}^2_{A_1}; w\widetilde{\sigma}_{A_1}, u\widetilde{\sigma}_{A_1}, y\widetilde{\sigma}_{A_1} \rangle} \tag{66}$$

$$x_{A_2} = \frac{\langle \widetilde{\sigma}^2_{A_1}; w\widetilde{\sigma}_{A_1}, u\widetilde{\sigma}_{A_1}, y\widetilde{\sigma}_{A_1} \rangle}{\langle \widetilde{\sigma}^2_{A_1}; w\widetilde{\sigma}_{A_1}, u\widetilde{\sigma}_{A_1}, y\widetilde{\sigma}_{A_1} \rangle - \langle \widetilde{\sigma}_{A_1A_2}; w\widetilde{\sigma}_A, u\widetilde{\sigma}_A, y\widetilde{\sigma}_A \rangle} \tag{67}$$

With the same condition: $\langle \widetilde{\sigma}_{A_1A_2}; w\widetilde{\sigma}_A, u\widetilde{\sigma}_A, y\widetilde{\sigma}_A \rangle \neq \langle \widetilde{\sigma}^2_{A_1}; w\widetilde{\sigma}_{A_1}, u\widetilde{\sigma}_{A_1}, y\widetilde{\sigma}_{A_1} \rangle$.

5.2. Minimizing the Risk of the Neutrosophic Portfolio Consisting of N Financial Assets

The neutrosophic portfolio is composed of N financial assets and assumes that in the portfolio there are financial assets that allow the determination of performance indicators using neutrosophic triangular fuzzy numbers. As a result, each financial asset that is part of the portfolio allows the determination of the neutrosophic return $\langle \widetilde{R}_{A_i}; w\widetilde{R}_{Ai}, u\widetilde{R}_{Ai}, y\widetilde{R}_{Ai} \rangle$ and of the neutrosophic risk $\langle \widetilde{\sigma}^2_{A_i}; w\widetilde{\sigma}_{A_i}, u\widetilde{\sigma}_{A_i}, y\widetilde{\sigma}_{A_i} \rangle$. Each financial asset A_i holds a weight in the total value of the neutrosophic portfolio noted with x_{A_i}. According to the above mentioned, for each financial asset A_i we will have:

$$\begin{array}{l} A_1\colon x_{A_1}; \langle \widetilde{R}_{A_1}; w\widetilde{R}_{A1}, u\widetilde{R}_{A1}, y\widetilde{R}_{A1} \rangle;\ \langle \widetilde{\sigma}^2_{A_1}; w\widetilde{\sigma}_{A_1}, u\widetilde{\sigma}_{A_1}, y\widetilde{\sigma}_{A_1} \rangle \\ A_2\colon x_{A_2}; \langle \widetilde{R}_{A_2}; w\widetilde{R}_{A2}, u\widetilde{R}_{A2}, y\widetilde{R}_{A2} \rangle;\ \langle \widetilde{\sigma}^2_{A_2}; w\widetilde{\sigma}_{A_2}, u\widetilde{\sigma}_{A_2}, y\widetilde{\sigma}_{A_2} \rangle \\ \\ A_n\colon x_{A_n}; \langle \widetilde{R}_{A_n}; w\widetilde{R}_{An}, u\widetilde{R}_{An}, y\widetilde{R}_{An} \rangle;\ \langle \widetilde{\sigma}^2_{A_n}; w\widetilde{\sigma}_{A_n}, u\widetilde{\sigma}_{A_n}, y\widetilde{\sigma}_{A_n} \rangle \end{array} \tag{68}$$

The equations that describe the portfolio refer to the neutrosophic portfolio return and risk and are of the form:

$$\begin{cases} \langle \widetilde{R_p}; w\widetilde{Rp}, u\widetilde{Rp}, y\widetilde{Rp} \rangle = \sum\limits_{i=1}^{n} \langle x_{A_i}\widetilde{R}_{A_i}; w\widetilde{R}_{A_i}, u\widetilde{R}_{A_i}, y\widetilde{R}_{A_i} \rangle \\ \langle \widetilde{\sigma}^2_P; w\widetilde{\sigma p}, u\widetilde{\sigma p}, y\widetilde{\sigma p} \rangle = \\ = \sum\limits_{i=1}^{n} \langle x^2_{A_i}\widetilde{\sigma}^2_{A_i}; w\widetilde{\sigma}_{A_i}, u\widetilde{\sigma}_{A_i}, y\widetilde{\sigma}_{A_i} \rangle + 2\sum\limits_{i=1}^{n}\sum\limits_{j=1}^{n} \langle x_{A_i}x_{A_j}\widetilde{\sigma}_{A_iA_j}; w\widetilde{\sigma}_{A_i} \wedge w\widetilde{\sigma}_{A_j}, u\widetilde{\sigma}_{A_i} \vee u\widetilde{\sigma}_{A_j}, y\widetilde{\sigma}_{A_i} \vee y\widetilde{\sigma}_{A_j} \rangle \\ \sum\limits_{i=1}^{n} x_{A_i} = 100\% \end{cases} \tag{69}$$

The analytical equations for the neutrosophic portfolio return and risk are written as follows:

$$\begin{array}{l} \langle \widetilde{R_P}; w\widetilde{Rp}, u\widetilde{Rp}, y\widetilde{Rp} \rangle \\ \quad = \langle x_{A_1}\widetilde{R}_{A_1}; w\widetilde{R}_{A_1}, u\widetilde{R}_{A_1}, y\widetilde{R}_{A_1} \rangle + \langle x_{A_2}\widetilde{R}_{A_2}; w\widetilde{R}_{A_2}, u\widetilde{R}_{A_2}, y\widetilde{R}_{A_2} \rangle + \cdots \\ \quad + \langle x_{A_n}\widetilde{R}_{A_n}; w\widetilde{R}_{A_n}, u\widetilde{R}_{A_n}, y\widetilde{R}_{A_n} \rangle \end{array} \tag{70}$$

And respectively:

$$
\begin{aligned}
&\langle \widetilde{\sigma_p^2}; w\widetilde{\sigma p}, u\widetilde{\sigma p}, y\widetilde{\sigma p}\rangle \\
&\quad = \langle x_{A_1}^2 \widetilde{\sigma_{A_1}^2}; w\widetilde{\sigma}_{A_1}, u\widetilde{\sigma}_{A_1}, y\widetilde{\sigma}_{A_1}\rangle + \langle x_{A_2}^2 \widetilde{\sigma_{A_2}^2}; w\widetilde{\sigma}_{A_2}, u\widetilde{\sigma}_{A_2}, y\widetilde{\sigma}_{A_2}\rangle + \cdots \\
&\quad + \langle x_{A_n}^2 \widetilde{\sigma_{A_n}^2}; w\widetilde{\sigma}_{A_n}, u\widetilde{\sigma}_{A_n}, y\widetilde{\sigma}_{A_n}\rangle \\
&\quad + \langle 2x_{A_1}x_{A_2}\widetilde{\sigma}_{A_1A_2}; w\widetilde{\sigma}_{A_1} \wedge w\widetilde{\sigma}_{A_2}, u\widetilde{\sigma}_{A_1} \vee u\widetilde{\sigma}_{A_2}, y\widetilde{\sigma}_{A_1} \vee y\widetilde{\sigma}_{A_2}\rangle \\
&\quad + \langle 2x_{A_1}x_{A_3}\widetilde{\sigma}_{A_1A_3}; w\widetilde{\sigma}_{A_1} \wedge w\widetilde{\sigma}_{A_3}, u\widetilde{\sigma}_{A_1} \vee u\widetilde{\sigma}_{A_3}, y\widetilde{\sigma}_{A_1} \vee y\widetilde{\sigma}_{A_3}\rangle + \cdots \\
&\quad + \langle 2x_{A_1}x_{A_n}\widetilde{\sigma}_{A_1A_n}; w\widetilde{\sigma}_{A_1} \wedge w\widetilde{\sigma}_{A_n}, u\widetilde{\sigma}_{A_1} \vee u\widetilde{\sigma}_{A_n}, y\widetilde{\sigma}_{A_1} \vee y\widetilde{\sigma}_{A_n}\rangle \\
&\quad + \langle 2x_{A_2}x_{A_1}\widetilde{\sigma}_{A_2A_1}; w\widetilde{\sigma}_{A_2} \wedge w\widetilde{\sigma}_{A_1}, u\widetilde{\sigma}_{A_2} \vee u\widetilde{\sigma}_{A_1}, y\widetilde{\sigma}_{A_2} \vee y\widetilde{\sigma}_{A_1}\rangle \\
&\quad + \langle 2x_{A_2}x_{A_3}\widetilde{\sigma}_{A_2A_3}; w\widetilde{\sigma}_{A_2} \wedge w\widetilde{\sigma}_{A_3}, u\widetilde{\sigma}_{A_2} \vee u\widetilde{\sigma}_{A_3}, y\widetilde{\sigma}_{A_2} \vee y\widetilde{\sigma}_{A_3}\rangle + \cdots \\
&\quad + \langle 2x_{A_2}x_{A_n}\widetilde{\sigma}_{A_2A_n}; w\widetilde{\sigma}_{A_2} \wedge w\widetilde{\sigma}_{A_n}, u\widetilde{\sigma}_{A_2} \vee u\widetilde{\sigma}_{A_n}, y\widetilde{\sigma}_{A_2} \vee y\widetilde{\sigma}_{A_n}\rangle \\
&\quad + \langle 2x_{A_n}x_{A_1}\widetilde{\sigma}_{A_nA_1}; w\widetilde{\sigma}_{A_n} \wedge w\widetilde{\sigma}_{A_1}, u\widetilde{\sigma}_{A_n} \vee u\widetilde{\sigma}_{A_1}, y\widetilde{\sigma}_{A_n} \vee y\widetilde{\sigma}_{A_1}\rangle \\
&\quad + \langle 2x_{A_n}x_{A_2}\widetilde{\sigma}_{A_nA_2}; w\widetilde{\sigma}_{A_n} \wedge w\widetilde{\sigma}_{A_2}, u\widetilde{\sigma}_{A_n} \vee u\widetilde{\sigma}_{A_2}, y\widetilde{\sigma}_{A_n} \vee y\widetilde{\sigma}_{A_2}\rangle + \cdots \\
&\quad + \langle 2x_{A_n}x_{A_{n-1}}\widetilde{\sigma}_{A_nA_{n-1}}; w\widetilde{\sigma}_{A_n} \wedge w\widetilde{\sigma}_{A_{n-1}}, u\widetilde{\sigma}_{A_n} \vee u\widetilde{\sigma}_{A_{n-1}}, y\widetilde{\sigma}_{A_n} \vee y\widetilde{\sigma}_{A_{n-1}}\rangle
\end{aligned}
\tag{71}
$$

Theorem 2. *There are considered N financial assets $A_1, A_2, \ldots A_n$ for which the neutrosophic return can be determined $\langle \widetilde{R}_{A_1}; w\widetilde{R}_{A1}, u\widetilde{R}_{A1}, y\widetilde{R}_{A1}\rangle$, $\langle \widetilde{R}_{A_2}; w\widetilde{R}_{A2}, u\widetilde{R}_{A2}, y\widetilde{R}_{A2}\rangle$, …, $\langle \widetilde{R}_{A_n}; w\widetilde{R}_{An}, u\widetilde{R}_{An}, y\widetilde{R}_{An}\rangle$. It is also possible to determine the neutrosophic risk specific to each financial asset that is part of the portfolio $\langle \widetilde{\sigma_{A_1}^2}; w\widetilde{\sigma}_{A_1}, u\widetilde{\sigma}_{A_1}, y\widetilde{\sigma}_{A_1}\rangle$, $\langle \widetilde{\sigma_{A_2}^2}; w\widetilde{\sigma}_{A_2}, u\widetilde{\sigma}_{A_2}, y\widetilde{\sigma}_{A_2}\rangle$ and $\langle \widetilde{\sigma_{A_n}^2}; w\widetilde{\sigma}_{A_n}, u\widetilde{\sigma}_{A_n}, y\widetilde{\sigma}_{A_n}\rangle$. These N financial assets form a neutrosophic portfolio of the form: $\langle \widetilde{P}; w\widetilde{P}, u\widetilde{P}, y\widetilde{P}\rangle$. The risk of the neutrosophic portfolio has the minimum value for $x_{A_1}, x_{A_2}, \ldots, x_{A_n}$ generalized using the weight x_{A_k} of the form:*

$$
x_{A_k} = \frac{\begin{pmatrix} \langle \widetilde{\sigma}_{A_{k2}}; w\widetilde{\sigma}_{A_{k2}}, u\widetilde{\sigma}_{A_{k2}}, y\widetilde{\sigma}_{A_{k2}}\rangle \\ \langle \widetilde{\sigma}_{A_{k3}}; w\widetilde{\sigma}_{A_{k3}}, u\widetilde{\sigma}_{A_{k3}}, y\widetilde{\sigma}_{A_{k3}}\rangle \\ \cdots \\ \langle \widetilde{\sigma}_{A_{kn}}; w\widetilde{\sigma}_{A_{kn}}, u\widetilde{\sigma}_{A_{kn}}, y\widetilde{\sigma}_{A_{kn}}\rangle \end{pmatrix} - \begin{pmatrix} \langle \sigma_{A_{kk}}; w\widetilde{\sigma}_{A_{kk}}, u\widetilde{\sigma}_{A_{kk}}, y\widetilde{\sigma}_{A_{kk}}\rangle \\ \langle \sigma_{A_{kk}}; w\widetilde{\sigma}_{A_{kk}}, u\widetilde{\sigma}_{A_{kk}}, y\widetilde{\sigma}_{A_{kk}}\rangle \\ \cdots \\ \langle \sigma_{A_{kk}}; w\widetilde{\sigma}_{A_{kk}}, u\widetilde{\sigma}_{A_{kk}}, y\widetilde{\sigma}_{A_{kk}}\rangle \end{pmatrix}}{\begin{pmatrix} \langle \widetilde{\sigma}_{A_{k2}}; w\widetilde{\sigma}_{A_{k2}}, u\widetilde{\sigma}_{A_{k2}}, y\widetilde{\sigma}_{A_{k2}}\rangle \\ \langle \widetilde{\sigma}_{A_{k3}}; w\widetilde{\sigma}_{A_{k3}}, u\widetilde{\sigma}_{A_{k3}}, y\widetilde{\sigma}_{A_{k3}}\rangle \\ \cdots \\ \langle \widetilde{\sigma}_{A_{kn}}; w\widetilde{\sigma}_{A_{kn}}, u\widetilde{\sigma}_{A_{kn}}, y\widetilde{\sigma}_{A_{kn}}\rangle \end{pmatrix}}
\tag{72}
$$

with the condition that: $\begin{pmatrix} \langle \widetilde{\sigma}_{A_{k2}}; w\widetilde{\sigma}_{A_{k2}}, u\widetilde{\sigma}_{A_{k2}}, y\widetilde{\sigma}_{A_{k2}}\rangle \\ \langle \widetilde{\sigma}_{A_{k3}}; w\widetilde{\sigma}_{A_{k3}}, u\widetilde{\sigma}_{A_{k3}}, y\widetilde{\sigma}_{A_{k3}}\rangle \\ \cdots \\ \langle \widetilde{\sigma}_{A_{kn}}; w\widetilde{\sigma}_{A_{kn}}, u\widetilde{\sigma}_{A_{kn}}, y\widetilde{\sigma}_{A_{kn}}\rangle \end{pmatrix} \neq 0.$

Demonstration: The equations of the neutrosophic portfolio in analytical form have been written previously and refer to those portfolios that contain a number of N financial assets. The conditions for minimizing the risk of the neutrosophic portfolio are obtained at the points where the first order derivative of the neutrosophic portfolio risk is null, respectively from the equations:

$$
\begin{cases}
\dfrac{\partial \langle \widetilde{\sigma_p^2}; w\widetilde{\sigma p}, u\widetilde{\sigma p}, y\widetilde{\sigma p}\rangle}{\partial x_{A_1}} = 0 \\
\dfrac{\partial \langle \widetilde{\sigma_p^2}; w\widetilde{\sigma p}, u\widetilde{\sigma p}, y\widetilde{\sigma p}\rangle}{\partial x_{A_2}} = 0 \\
\cdots \\
\dfrac{\partial \langle \widetilde{\sigma_p^2}; w\widetilde{\sigma p}, u\widetilde{\sigma p}, y\widetilde{\sigma p}\rangle}{\partial x_{A_n}} = 0
\end{cases}
\tag{73}
$$

Based on the condition for minimizing the neutrosophic portfolio risk will be obtained:

$$\begin{cases} \langle 2x_{A_1}\widetilde{\sigma}^2_{A_1}; w\widetilde{\sigma}_{A_1}, u\widetilde{\sigma}_{A_1}, y\widetilde{\sigma}_{A_1}\rangle + \langle 2x_{A_2}\widetilde{\sigma}_{A_1A_2}; w\widetilde{\sigma}_{A_1} \wedge w\widetilde{\sigma}_{A_2}, u\widetilde{\sigma}_{A_1} \vee u\widetilde{\sigma}_{A_2}, y\widetilde{\sigma}_{A_1} \vee y\widetilde{\sigma}_{A_2}\rangle + \\ + \cdots + \langle 2x_{A_n}\widetilde{\sigma}_{A_1A_n}; w\widetilde{\sigma}_{A_1} \wedge w\widetilde{\sigma}_{A_n}, u\widetilde{\sigma}_{A_1} \vee u\widetilde{\sigma}_{A_n}, y\widetilde{\sigma}_{A_1} \vee y\widetilde{\sigma}_{A_n}\rangle = 0 \\ \langle 2x_{A_2}\widetilde{\sigma}^2_{A_2}; w\widetilde{\sigma}_{A_1}, u\widetilde{\sigma}_{A_1}, y\widetilde{\sigma}_{A_1}\rangle + \langle 2x_{A_1}\widetilde{\sigma}_{A_2A_1}; w\widetilde{\sigma}_{A_2} \wedge w\widetilde{\sigma}_{A_1}, u\widetilde{\sigma}_{A_2} \vee u\widetilde{\sigma}_{A_1}, y\widetilde{\sigma}_{A_2} \vee y\widetilde{\sigma}_{A_1}\rangle \\ + \cdots + \langle 2x_{A_n}\widetilde{\sigma}_{A_2A_n}; w\widetilde{\sigma}_{A_2} \wedge w\widetilde{\sigma}_{A_n}, u\widetilde{\sigma}_{A_2} \vee u\widetilde{\sigma}_{A_n}, y\widetilde{\sigma}_{A_2} \vee y\widetilde{\sigma}_{A_n}\rangle = 0 \\ \langle 2x_{A_1}\widetilde{\sigma}_{A_nA_1}; w\widetilde{\sigma}_{A_n} \wedge w\widetilde{\sigma}_{A_1}, u\widetilde{\sigma}_{A_n} \vee u\widetilde{\sigma}_{A_1}, y\widetilde{\sigma}_{A_n} \vee y\widetilde{\sigma}_{A_1}\rangle + \\ + \langle 2x_{A_n}\widetilde{\sigma}_{A_nA_1}; w\widetilde{\sigma}_{A_n} \wedge w\widetilde{\sigma}_{A_2}, u\widetilde{\sigma}_{A_n} \vee u\widetilde{\sigma}_{A_2}, y\widetilde{\sigma}_{A_n} \vee y\widetilde{\sigma}_{A_2}\rangle + \cdots + 2x_{A_n}\widetilde{\sigma}^2_{A_n}; w\widetilde{\sigma}_{A_n}, u\widetilde{\sigma}_{A_n}, y\widetilde{\sigma}_{A_n} = 0 \\ x_{A_1} + x_{A_2} + \cdots + x_{A_n} = 1 \end{cases} \tag{74}$$

In matrix form the equations above, ordered according to the neutrosophic variance are written as follows:

$$2\langle x_{A_1}\widetilde{\sigma}_{A_{11}}; w\widetilde{\sigma}_{A_{11}}, u\widetilde{\sigma}_{A_{11}}, y\widetilde{\sigma}_{A_{11}}\rangle + 2(x_{A_2} \ldots x_{A_n}) \begin{pmatrix} \langle \widetilde{\sigma}_{A_{12}}; w\widetilde{\sigma}_{A_{12}}, u\widetilde{\sigma}_{A_{12}}, y\widetilde{\sigma}_{A_{12}}\rangle \\ \langle \widetilde{\sigma}_{A_{13}}; w\widetilde{\sigma}_{A_{13}}, u\widetilde{\sigma}_{A_{13}}, y\widetilde{\sigma}_{A_{13}}\rangle \\ \ldots \\ \langle \widetilde{\sigma}_{A_{1n}}; w\widetilde{\sigma}_{A_{1n}}, u\widetilde{\sigma}_{A_{1n}}, y\widetilde{\sigma}_{A_{1n}}\rangle \end{pmatrix} = 0 \tag{75}$$

$$2\langle x_{A_2}\widetilde{\sigma}_{A_{22}}; w\widetilde{\sigma}_{A_{22}}, u\widetilde{\sigma}_{A_{22}}, y\widetilde{\sigma}_{A_{22}}\rangle + 2(x_{A_1} \ldots x_{A_n}) \begin{pmatrix} \widetilde{\sigma}_{A_{21}}; w\widetilde{\sigma}_{A_{21}}, u\widetilde{\sigma}_{A_{21}}, y\widetilde{\sigma}_{A_{21}} \\ \widetilde{\sigma}_{A_{23}}; w\widetilde{\sigma}_{A_{23}}, u\widetilde{\sigma}_{A_{23}}, y\widetilde{\sigma}_{A_{23}} \\ \ldots \\ \widetilde{\sigma}_{A_{2n}}; w\widetilde{\sigma}_{A_{2n}}, u\widetilde{\sigma}_{A_{2n}}, y\widetilde{\sigma}_{A_{2n}} \end{pmatrix} = 0 \tag{76}$$

$$\begin{aligned} &2\langle x_{A_n}\widetilde{\sigma}_{A_{nn}}; w\widetilde{\sigma}_{A_{nn}}, u\widetilde{\sigma}_{A_{nn}}, y\widetilde{\sigma}_{A_{nn}}\rangle \\ &\quad + \; + 2(x_{A_1} \ldots x_{A_{n-1}}) \begin{pmatrix} \langle \widetilde{\sigma}_{A_{n1}}; w\widetilde{\sigma}_{A_{n1}}, u\widetilde{\sigma}_{A_{n1}}, y\widetilde{\sigma}_{A_{n1}}\rangle \\ \langle \widetilde{\sigma}_{A_{n3}}; w\widetilde{\sigma}_{A_{n3}}, u\widetilde{\sigma}_{A_{n3}}, y\widetilde{\sigma}_{A_{n3}}\rangle \\ \ldots \\ \langle \widetilde{\sigma}_{A_{nn-1}}; w\widetilde{\sigma}_{A_{nn-1}}, u\widetilde{\sigma}_{A_{nn-1}}, y\widetilde{\sigma}_{A_{nn-1}}\rangle \end{pmatrix} = 0 \end{aligned} \tag{77}$$

The last neutrosophic portfolio equation $x_{A_1} + x_{A_2} + \cdots + x_{A_n} = 1$, written in the matrix form will be:

$$\left\{ x_{A_1} + \left(x_{A_2} \ldots x_{A_n}\right) \begin{pmatrix} 1 \\ \ldots \\ 1 \end{pmatrix} = 1 \right. \tag{78}$$

Resulting that:

$$x_{A_2} \ldots x_{A_n} = \left(1 - x_{A_1}\right) e^{-1} \tag{79}$$

By replacing the obtained expression for $x_{A_2} \ldots x_{A_n}$ in the first relationship, it is obtained that:

$$2\langle x_{A_1}\sigma_{A_{11}}; w\widetilde{\sigma}_{A_{11}}, u\widetilde{\sigma}_{A_{11}}, y\widetilde{\sigma}_{A_{11}}\rangle + 2(1 - x_{A_1})e^{-1} \begin{pmatrix} \langle \widetilde{\sigma}_{A_{12}}; w\widetilde{\sigma}_{A_{12}}, u\widetilde{\sigma}_{A_{12}}, y\widetilde{\sigma}_{A_{12}}\rangle \\ \langle \widetilde{\sigma}_{A_{13}}; w\widetilde{\sigma}_{A_{13}}, u\widetilde{\sigma}_{A_{13}}, y\widetilde{\sigma}_{A_{13}}\rangle \\ \ldots \\ \langle \widetilde{\sigma}_{A_{1n}}; w\widetilde{\sigma}_{A_{1n}}, u\widetilde{\sigma}_{A_{1n}}, y\widetilde{\sigma}_{A_{1n}}\rangle \end{pmatrix} = 0 \tag{80}$$

$$\begin{aligned} &x_{A_1}\left[\begin{pmatrix} \langle \widetilde{\sigma}_{A_{12}}; w\widetilde{\sigma}_{A_{12}}, u\widetilde{\sigma}_{A_{12}}, y\widetilde{\sigma}_{A_{12}}\rangle \\ \langle \widetilde{\sigma}_{A_{13}}; w\widetilde{\sigma}_{A_{13}}, u\widetilde{\sigma}_{A_{13}}, y\widetilde{\sigma}_{A_{13}}\rangle \\ \ldots \\ \langle \widetilde{\sigma}_{A_{1n}}; w\widetilde{\sigma}_{A_{1n}}, u\widetilde{\sigma}_{A_{1n}}, y\widetilde{\sigma}_{A_{1n}}\rangle \end{pmatrix} - \begin{pmatrix} \langle \sigma_{A_{11}}; w\widetilde{\sigma}_{A_{11}}, u\widetilde{\sigma}_{A_{11}}, y\widetilde{\sigma}_{A_{11}}\rangle \\ \langle \sigma_{A_{11}}; w\widetilde{\sigma}_{A_{11}}, u\widetilde{\sigma}_{A_{11}}, y\widetilde{\sigma}_{A_{11}}\rangle \\ \ldots \\ \langle \sigma_{A_{11}}; w\widetilde{\sigma}_{A_{11}}, u\widetilde{\sigma}_{A_{11}}, y\widetilde{\sigma}_{A_{11}}\rangle \end{pmatrix}\right] \\ &= \begin{pmatrix} \langle \widetilde{\sigma}_{A_{12}}; w\widetilde{\sigma}_{A_{12}}, u\widetilde{\sigma}_{A_{12}}, y\widetilde{\sigma}_{A_{12}}\rangle \\ \langle \widetilde{\sigma}_{A_{13}}; w\widetilde{\sigma}_{A_{13}}, u\widetilde{\sigma}_{A_{13}}, y\widetilde{\sigma}_{A_{13}}\rangle \\ \ldots \\ \langle \widetilde{\sigma}_{A_{1n}}; w\widetilde{\sigma}_{A_{1n}}, u\widetilde{\sigma}_{A_{1n}}, y\widetilde{\sigma}_{A_{1n}}\rangle \end{pmatrix} \end{aligned} \tag{81}$$

$$x_{A_1} = \frac{\begin{pmatrix} \langle \widetilde{\sigma}_{A_{12}}; w\widetilde{\sigma}_{A_{12}}, u\widetilde{\sigma}_{A_{12}}, y\widetilde{\sigma}_{A_{12}} \rangle \\ \langle \widetilde{\sigma}_{A_{13}}; w\widetilde{\sigma}_{A_{13}}, u\widetilde{\sigma}_{A_{13}}, y\widetilde{\sigma}_{A_{13}} \rangle \\ \cdots \\ \langle \widetilde{\sigma}_{A_{1n}}; w\widetilde{\sigma}_{A_{1n}}, u\widetilde{\sigma}_{A_{1n}}, y\widetilde{\sigma}_{A_{1n}} \rangle \end{pmatrix} - \begin{pmatrix} \langle \sigma_{A_{11}}; w\widetilde{\sigma}_{A_{11}}, u\widetilde{\sigma}_{A_{11}}, y\widetilde{\sigma}_{A_{11}} \rangle \\ \langle \sigma_{A_{11}}; w\widetilde{\sigma}_{A_{11}}, u\widetilde{\sigma}_{A_{11}}, y\widetilde{\sigma}_{A_{11}} \rangle \\ \cdots \\ \langle \sigma_{A_{11}}; w\widetilde{\sigma}_{A_{11}}, u\widetilde{\sigma}_{A_{11}}, y\widetilde{\sigma}_{A_{11}} \rangle \end{pmatrix}}{\begin{pmatrix} \langle \widetilde{\sigma}_{A_{12}}; w\widetilde{\sigma}_{A_{12}}, u\widetilde{\sigma}_{A_{12}}, y\widetilde{\sigma}_{A_{12}} \rangle \\ \langle \widetilde{\sigma}_{A_{13}}; w\widetilde{\sigma}_{A_{13}}, u\widetilde{\sigma}_{A_{13}}, y\widetilde{\sigma}_{A_{13}} \rangle \\ \cdots \\ \langle \widetilde{\sigma}_{A_{1n}}; w\widetilde{\sigma}_{A_{1n}}, u\widetilde{\sigma}_{A_{1n}}, y\widetilde{\sigma}_{A_{1n}} \rangle \end{pmatrix}} \quad (82)$$

For the weight x_{A_k} of the asset A_k it will be obtained:

$$x_{A_k} = \frac{\begin{pmatrix} \langle \widetilde{\sigma}_{A_{k2}}; w\widetilde{\sigma}_{A_{k2}}, u\widetilde{\sigma}_{A_{k2}}, y\widetilde{\sigma}_{A_{k2}} \rangle \\ \langle \widetilde{\sigma}_{A_{k3}}; w\widetilde{\sigma}_{A_{k3}}, u\widetilde{\sigma}_{A_{k3}}, y\widetilde{\sigma}_{A_{k3}} \rangle \\ \cdots \\ \langle \widetilde{\sigma}_{A_{kn}}; w\widetilde{\sigma}_{A_{kn}}, u\widetilde{\sigma}_{A_{kn}}, y\widetilde{\sigma}_{A_{kn}} \rangle \end{pmatrix} - \begin{pmatrix} \langle \sigma_{A_{kk}}; w\widetilde{\sigma}_{A_{kk}}, u\widetilde{\sigma}_{A_{kk}}, y\widetilde{\sigma}_{A_{kk}} \rangle \\ \langle \sigma_{A_{kk}}; w\widetilde{\sigma}_{A_{kk}}, u\widetilde{\sigma}_{A_{kk}}, y\widetilde{\sigma}_{A_{kk}} \rangle \\ \cdots \\ \langle \sigma_{A_{kk}}; w\widetilde{\sigma}_{A_{kk}}, u\widetilde{\sigma}_{A_{kk}}, y\widetilde{\sigma}_{A_{kk}} \rangle \end{pmatrix}}{\begin{pmatrix} \langle \widetilde{\sigma}_{A_{k2}}; w\widetilde{\sigma}_{A_{k2}}, u\widetilde{\sigma}_{A_{k2}}, y\widetilde{\sigma}_{A_{k2}} \rangle \\ \langle \widetilde{\sigma}_{A_{k3}}; w\widetilde{\sigma}_{A_{k3}}, u\widetilde{\sigma}_{A_{k3}}, y\widetilde{\sigma}_{A_{k3}} \rangle \\ \cdots \\ \langle \widetilde{\sigma}_{A_{kn}}; w\widetilde{\sigma}_{A_{kn}}, u\widetilde{\sigma}_{A_{kn}}, y\widetilde{\sigma}_{A_{kn}} \rangle \end{pmatrix}} \quad (83)$$

Respecting the condition that:

$$\begin{pmatrix} \langle \widetilde{\sigma}_{A_{k2}}; w\widetilde{\sigma}_{A_{k2}}, u\widetilde{\sigma}_{A_{k2}}, y\widetilde{\sigma}_{A_{k2}} \rangle \\ \langle \widetilde{\sigma}_{A_{k3}}; w\widetilde{\sigma}_{A_{k3}}, u\widetilde{\sigma}_{A_{k3}}, y\widetilde{\sigma}_{A_{k3}} \rangle \\ \cdots \\ \langle \widetilde{\sigma}_{A_{kn}}; w\widetilde{\sigma}_{A_{kn}}, u\widetilde{\sigma}_{A_{kn}}, y\widetilde{\sigma}_{A_{kn}} \rangle \end{pmatrix} \neq 0 \quad (84)$$

6. Numerical Applications

6.1. Numerical Application for the Case of the Neutrosophic Portfolio Consisting of Two Financial Assets

Two financial assets (A_1, A_2) are considered to which two triangular neutrosophic fuzzy numbers are specified for the financial assets return of the form:

$$\widetilde{R_{A1}} = \langle (0.2\ 0.3\ 0.5); 0.5,\ 0.2,\ 0.3 \rangle \text{ for } \widetilde{R_A} \in [0.2; 0.5]$$

$$\widetilde{R_{A2}} = \langle (0.1\ 0.2\ 0.3); 0.6,\ 0.3,\ 0.2 \rangle \text{ for } \widetilde{R_A} \in [0.1; 0.3]$$

And we aim at:

(a) determining the variance-covariance matrix;
(b) calculating the rentability and the variance for a given portfolio P formed by the two financial assets, having the structure $P = \begin{pmatrix} x_1 \\ x_2 \end{pmatrix}$ and with $x_2 = 1 - x_1$;
(c) determining the structure of a random portfolio P for which the risk is minimum (the value of the minimum risk is also required).

Starting from Example 2, the following elements are known:

- the neutrosophic risk for asset A_1 :

$$\widetilde{\sigma_f}^2 A_1 = \langle 0.0225; 0.5,\ 0.2,\ 0.3 \rangle$$

- the neutrosophic risk for asset A_2 :

$$\widetilde{\sigma}^2_{fA_2} = \langle 0.0180; 0.6,\ 0.3,\ 0.2\rangle$$

Also, from Example 2 we know that the covariance between asset A_1 and A_2 has the value:

$$\sigma_{A_1A_3} = \langle 0.1914; 0.5,\ 0.2,\ 0.3\rangle$$

(a) In this context, the variance-covariance matrix has the following form:

$$\Omega = \begin{pmatrix} \widetilde{\sigma}_{fA_1A_1} & \widetilde{\sigma}_{fA_1A_3} \\ \widetilde{\sigma}_{fA_2A_1} & \widetilde{\sigma}_{fA_2A_2} \end{pmatrix} = \begin{pmatrix} \langle 0.0225; 0.5,\ 0.2,\ 0.3\rangle & \langle 0,0705; 0.6,\ 0.2,\ 0.2\rangle \\ \langle 0.0705; 0.6,\ 0.2,\ 0.2\rangle & \langle 0.0180; 0.6,\ 0.3,\ 0.2\rangle \end{pmatrix}$$

(b) It is known that the rentability of the neutrosophic portfolio P has the following form:

$$\langle \widetilde{R_P}; w\widetilde{Rp}, u\widetilde{Rp}, y\widetilde{Rp}\rangle = \langle x_{A_1}\widetilde{R}_{A_1}; w\widetilde{R}_{A_1}, u\widetilde{R}_{A_1}, y\widetilde{R}_{A_1}\rangle + \langle x_{A_2}\widetilde{R}_{A_2}; w\widetilde{R}_{A_2}, u\widetilde{R}_{A_2}, y\widetilde{R}_{A_2}\rangle$$

Also, it is given that: $x_2 = 1 - x_1$ and that the P portfolio has the following form: $P = \begin{pmatrix} x_1 \\ x_2 \end{pmatrix}$;

By replacing these expressions in the formula for the portfolio rentability $\langle \widetilde{R_P}; w\widetilde{Rp}, u\widetilde{Rp}, y\widetilde{Rp}\rangle$ we get that:

$$\begin{aligned}\langle \widetilde{R_P}; w\widetilde{Rp}, u\widetilde{Rp}, y\widetilde{Rp}\rangle &= \langle x_{A_1}\widetilde{R}_{A_1}; w\widetilde{R}_{A_1}, u\widetilde{R}_{A_1}, y\widetilde{R}_{A_1}\rangle + \langle x_{A_2}\widetilde{R}_{A_2}; w\widetilde{R}_{A_2}, u\widetilde{R}_{A_2}, y\widetilde{R}_{A_2}\rangle \\ &= \langle x_{A_1}(0.2\ 0.3\ 0.5); 0.5,\ 0.2,\ 0.3\rangle + \langle (1 - x_{A_1})(0.1\ 0.2\ 0.3); 0.6,\ 0.3,\ 0.2\rangle\end{aligned}$$

$$\langle \widetilde{R_P}; w\widetilde{Rp}, u\widetilde{Rp}, y\widetilde{Rp}\rangle = \langle x_{A_1}(0.1\ 0.1\ 0.2); 0.6,\ 0.2,\ 0.2\rangle - (\langle 0.1\ 0.2\ 0.3); 0.6,\ 0.3,\ 0.2\rangle$$

As for the portfolio risk, it will be of the following form: $\langle \widetilde{\sigma_P^2}; w\widetilde{\sigma p}, u\widetilde{\sigma p}, y\widetilde{\sigma p}\rangle = \langle x^2_{A_1}\widetilde{\sigma}^2_{A_1}; w\widetilde{\sigma}_{A_1}, u\widetilde{\sigma}_{A_1}, y\widetilde{\sigma}_{A_1}\rangle + \langle x^2_{A_2}\widetilde{\sigma}^2_{A_2}; w\widetilde{\sigma}_{A_2}, u\widetilde{\sigma}_{A_2}, y\widetilde{\sigma}_{A_2}\rangle + \langle 2x_{A_1}x_{A_2}\sigma_{A_1A_2}; w\sigma_A, u\sigma_A, y\sigma_A\rangle$
By replacing it in the equations for x_1 and $x_2 = 1 - x_1$ we get that:

$$\begin{aligned}\langle \widetilde{\sigma_P^2}; w\widetilde{\sigma p}, u\widetilde{\sigma p}, y\widetilde{\sigma p}\rangle \\ &= \langle x^2_{A_1}\widetilde{\sigma}^2_{A_1}; w\widetilde{\sigma}_{A_1}, u\widetilde{\sigma}_{A_1}, y\widetilde{\sigma}_{A_1}\rangle + \langle (1 - x_{A_1})^2\widetilde{\sigma}^2_{A_2}; w\widetilde{\sigma}_{A_2}, u\widetilde{\sigma}_{A_2}, y\widetilde{\sigma}_{A_2}\rangle \\ &+ \langle 2x_{A_1}(1 - x_{A_1})\sigma_{A_1A_2}; w\sigma_A, u\sigma_A, y\sigma_A\rangle\end{aligned}$$

$$\begin{aligned}\langle \widetilde{\sigma_P^2}; w\widetilde{\sigma p}, u\widetilde{\sigma p}, y\widetilde{\sigma p}\rangle \\ &= \langle x^2_{A_1}0.0225; 0.5,\ 0.2,\ 0.3\rangle + \langle (1 - 2x_{A_1} + x^2_{A_1})0.0180; 0.6,\ 0.3,\ 0.2\rangle \\ &+ \langle (2x_{A_1} - 2x^2_{A_1})0.0705; 0.6,\ 0.2,\ 0.2\rangle\end{aligned}$$

$$\begin{aligned}\langle \widetilde{\sigma_P^2}; w\widetilde{\sigma p}, u\widetilde{\sigma p}, y\widetilde{\sigma p}\rangle \\ &= -\langle x^2_{A_1}0.1005; 0.6,\ 0.2,\ 0.2\rangle - \langle 0.105x_{A_1}; 0.6,\ 0.2,\ 0.2\rangle + \langle 0.141; 0.6,\ 0.2,\ 0.2\rangle\end{aligned}$$

(c) Further on, we know from Example 1 that the covariance between A_1 and A_2 has the following form:

$$\widetilde{\sigma}_{fA_1A_2} = \langle 0.0705; 0.6,\ 0.2,\ 0.2\rangle$$

By replacing the values obtained in the expression of x_{A_1} and x_{A_2} it is obtained that:

$$\widetilde{\sigma}_{fA_1A_2} = \langle 0.0705; 0.6,\ 0.2,\ 0.2\rangle$$

$$x_{A_1} = \frac{\langle \widetilde{\sigma}_{A_1A_2}; w\widetilde{\sigma}_A, u\widetilde{\sigma}_A, y\widetilde{\sigma}_A \rangle}{\langle \widetilde{\sigma}_{A_1A_2}; w\widetilde{\sigma}_A, u\widetilde{\sigma}_A, y\widetilde{\sigma}_A \rangle - \langle \widetilde{\sigma}^2_{A_1}; w\widetilde{\sigma}_{A_1}, u\widetilde{\sigma}_{A_1}, y\widetilde{\sigma}_{A_1} \rangle}$$

$$x_{A_1} = \frac{\langle 0,1914; 0.5,\ 0.2,\ 0.3 \rangle}{\langle 0,1914; 0.5,\ 0.2,\ 0.3 \rangle - \langle 0.0225; 0.5,\ 0.2,\ 0.3 \rangle} \times 100$$

$$x_{A_1} = \frac{\langle 0,1914; 0.5,\ 0.2,\ 0.3 \rangle}{\langle 0,1689; 0.5,\ 0.2,\ 0.3 \rangle} \times 100$$

$$x_{A_1} = 112.72\%$$

$$x_{A_2} = \frac{\langle \widetilde{\sigma}^2_{A_1}; w\widetilde{\sigma}_{A_1}, u\widetilde{\sigma}_{A_1}, y\widetilde{\sigma}_{A_1} \rangle}{\langle \widetilde{\sigma}^2_{A_1}; w\widetilde{\sigma}_{A_1}, u\widetilde{\sigma}_{A_1}, y\widetilde{\sigma}_{A_1} \rangle - \langle \widetilde{\sigma}_{A_1A_2}; w\widetilde{\sigma}_A, u\widetilde{\sigma}_A, y\widetilde{\sigma}_A \rangle}$$

$$x_{A_2} = \frac{\langle 0.0225; 0.5,\ 0.2,\ 0.3 \rangle}{\langle 0.0225; 0.5,\ 0.2,\ 0.3 - 0,1914; 0.5,\ 0.2,\ 0.3 \rangle} \times 100$$

$$x_{A_2} = -\frac{\langle 0.0225; 0.5,\ 0.2,\ 0.3 \rangle}{\langle 0,1689; 0.5,\ 0.2,\ 0.3 \rangle} \times 100$$

$$x_{A_2} = -13.250\%$$

The portfolio risk will be given by the relationship:

$$\begin{aligned}\langle \widetilde{\sigma}^2_p; w\widetilde{\sigma p}, u\widetilde{\sigma p}, y\widetilde{\sigma p} \rangle &= \langle x^2_{A_1} \widetilde{\sigma}^2_{A_1}; w\widetilde{\sigma}_{A_1}, u\widetilde{\sigma}_{A_1}, y\widetilde{\sigma}_{A_1} \rangle + \langle x^2_{A_2} \widetilde{\sigma}^2_{A_2}; w\widetilde{\sigma}_{A_2}, u\widetilde{\sigma}_{A_2}, y\widetilde{\sigma}_{A_2} \rangle \\ &+ \langle 2x_{A_1} x_{A_2} \sigma_{A_1A_2}; w\sigma_A, u\sigma_A, y\sigma_A \rangle\end{aligned}$$

$$\begin{aligned}\langle \widetilde{\sigma}^2_p; w\widetilde{\sigma p}, u\widetilde{\sigma p}, y\widetilde{\sigma p} \rangle &= (1.1272)^2 \langle 0.0225; 0.5,\ 0.2,\ 0.3 \rangle \\ &+ (-0.13250)^2 \langle 0.0180; 0.6,\ 0.3,\ 0.2 \rangle \\ &+ 2(1.1272)(-0.13250) \langle 0.1914; 0.5,\ 0.2,\ 0.3 \rangle\end{aligned}$$

$$\begin{aligned}\langle \widetilde{\sigma}^2_p; w\widetilde{\sigma p}, u\widetilde{\sigma p}, y\widetilde{\sigma p} \rangle &= \langle 0.0285; 0.5,\ 0.2,\ 0.3 \rangle + \langle 0.0033; 0.5,\ 0.2,\ 0.3 \rangle \\ &+ \langle 0.05717; 0.5,\ 0.2,\ 0.3 \rangle\end{aligned}$$

$$\langle \widetilde{\sigma}^2_p; w\widetilde{\sigma p}, u\widetilde{\sigma p}, y\widetilde{\sigma p} \rangle = \langle 0.08897; 0.5,\ 0.2,\ 0.3 \rangle$$

$$\langle \sigma_p; w\widetilde{\sigma p}, u\widetilde{\sigma p}, y\widetilde{\sigma p} \rangle = \sqrt{\langle \widetilde{\sigma}^2_p; w\widetilde{\sigma p}, u\widetilde{\sigma p}, y\widetilde{\sigma p} \rangle} = \sqrt{\langle 0,08897; 0.5,\ 0.2,\ 0.3 \rangle} = 0.2982$$

Conclusion: The neutrosophic portfolio risk is minimal and registers the value of 29.82%.

6.2. Numerical Application for the Case of the Neutrosophic Portfolio Consisting of N Financial Assets

For three financial assets held by three listed companies (A_1, A_2, A_3), the financial return was determined according to the information provided by the stock exchange website. The financial returns were fuzzified with the help of three triangular neutrosophic numbers and the following values were obtained:

$$\widetilde{R_{A1}} = \langle (0.3\ 0.4\ 0.6); 0.5,\ 0.2,\ 0.3 \rangle, \text{ for } \widetilde{R_A} \in [0.3; 0.6]$$

$$\widetilde{R_{A2}} = \langle (0.15\ 0.25\ 0.35); 0.6,\ 0.3,\ 0.2 \rangle, \text{ for } \widetilde{R_A} \in [0.15; 0.35]$$

$$\widetilde{R_{A3}} = \langle (0.25\ 0.45\ 0.65); 0.4,\ 0.3,\ 0.3 \rangle, \text{ for } \widetilde{R_A} \in [0.25; 0.65]$$

In order to establish:

- The variance-covariance matrix of the neutrosophic portfolio;

- The weight of financial assets $x_{A_1}, x_{A_2}, x_{A_3}$ in the total value of the neutrosophic portfolio so that the neutrosophic portfolio risk is minimal;
- The value of the neutrosophic portfolio return and risk.

The followings are undertaken:

For computing the variance-covariance matrix, the financial returns of the three assets are established in a first stage according to the calculation formula:

$$\langle \widetilde{R}_{A_i}; w\widetilde{R}_{A_i}, u\widetilde{R}_{A_i}, y\widetilde{R}_{A_i}\rangle = \langle\left(\frac{1}{6}\left(\widetilde{R}_{Aai} + \widetilde{R}_{Aci}\right) + \frac{2}{3}\widetilde{R}_{Abi}\right); w\widetilde{R}_{A_i}, u\widetilde{R}_{A_i}, y\widetilde{R}_{A_i}\rangle$$

Thus:

$$\begin{aligned}\langle \widetilde{R}_{A_1}; w\widetilde{R}_{A_1}, u\widetilde{R}_{A_1}, y\widetilde{R}_{A_1}\rangle &= \langle\left(\tfrac{1}{6}(0.3+0.6) + \tfrac{2}{3}\times 0.4\right); 0.5,\ 0.2,\ 0.3\rangle \\ &= \langle 0.166\times 0.9 + 0.666\times 0.4; 0.5,\ 0.2,\ 0.3\rangle \\ &= \langle 0.149 + 0.266; 0.5,\ 0.2,\ 0.3\rangle = \langle 0.415; 0.5,\ 0.2,\ 0.3\rangle\end{aligned}$$

Proceeding in the same way, we get the following results:

$$\langle \widetilde{R}_{A_2}; w\widetilde{R}_{A_2}, u\widetilde{R}_{A_2}, y\widetilde{R}_{A_2}\rangle = \langle 0.249; 0.6,\ 0.3,\ 0.2\rangle$$

$$\langle \widetilde{R}_{A_3}; w\widetilde{R}_{A_3}, u\widetilde{R}_{A_3}, y\widetilde{R}_{A_3}\rangle = \langle 0.448; 0.4,\ 0.3,\ 0.3\rangle$$

The following calculation formula is used to determine the variance of the three financial assets:

$$\begin{aligned}\widetilde{\sigma_f A_i} = \langle \tfrac{1}{4}\left[\left(\widetilde{Ra}_{bi} - \widetilde{Ra}_{ai}\right)^2 + \left(\widetilde{Ra}_{ci} - \widetilde{Ra}_{bi}\right)^2\right]; \widetilde{wRa}, \widetilde{uRa}, \widetilde{yRa}\rangle \\ +\langle \tfrac{2}{3}\left[\widetilde{Ra}_{ai}\left(\widetilde{Ra}_{bi} - \widetilde{Ra}_{ai}\right) - \widetilde{Ra}_{ci}\left(\widetilde{Ra}_{ci} - \widetilde{Ra}_{bi}\right)\right]; \widetilde{wRa}, \widetilde{uRa}, \widetilde{yRa}\rangle \\ +\langle \tfrac{1}{2}\left(\widetilde{Ra}^2_{ai} + \widetilde{Ra}^2_{ci}\right); \widetilde{wRa}, \widetilde{uRa}, \widetilde{yRa}\rangle - \langle \tfrac{1}{2}E^2_f\left(\widetilde{Ra}_i\right); \widetilde{wRa}, \widetilde{uRa}, \widetilde{yRa}\rangle\end{aligned}$$

By replacing the data in the above expression, we will have:

$$\begin{aligned}&\widetilde{\sigma_f a_1} = \langle \tfrac{1}{4}[(0.4-0.3)^2 + (0.6-0.4)^2]; 0.5,\ 0.2,\ 0.3\rangle + \langle \tfrac{2}{3}(0.3(0.4-0.3) - \\ &0.6(0.6-0.4)); 0.5,\ 0.2,\ 0.3\rangle + \langle \tfrac{1}{2}(0.3^2 + 0.6^2); 0.5,\ 0.2,\ 0.3\rangle - \langle \tfrac{1}{2}(0.415)^2; 0.5,\ 0.2,\ 0.3\rangle \\ &= \langle 0.0925; 0.5,\ 0.2,\ 0.3\rangle\end{aligned}$$

For the remaining of the values we will obtain:

$$\widetilde{\sigma_f a_2} = \langle 0.033; 0.6,\ 0.3,\ 0.2\rangle$$

$$\widetilde{\sigma_f a_3} = \langle 0.0910; 0.4,\ 0.3,\ 0.3\rangle$$

In order to establish the covariance between these three assets, respectively to measure the intensity of the connection between the financial assets, will be applied the following calculation formula:

$$\begin{aligned}cov(\widetilde{Ra}_i, \widetilde{Ra}_j) = \langle\Big(\tfrac{1}{4}\big[(\widetilde{Ra}_{bii} - \widetilde{Ra}_{aii})(\widetilde{Ra}_{bji} - \widetilde{Ra}_{aji}) \\ + (\widetilde{Ra}_{cii} - \widetilde{Ra}_{bii})(\widetilde{Ra}_{cji} - \widetilde{Ra}_{bji})\big] \\ + \tfrac{1}{3}\Big\{\big[\widetilde{Ra}_{aji}(\widetilde{Ra}_{bii} - \widetilde{Ra}_{aii}) + \widetilde{Ra}_{aii}(\widetilde{Ra}_{bji} - \widetilde{Ra}_{aji})\big] \\ - \big[\widetilde{Ra}_{cii}(\widetilde{Ra}_{cji} - \widetilde{Ra}_{bji}) + \widetilde{Ra}_{cji}(\widetilde{Ra}_{cii} - \widetilde{Ra}_{bii})\big]\Big\} \\ + \tfrac{1}{2}(\widetilde{Ra}_{aii}\widetilde{Ra}_{aji} + \widetilde{Ra}_{cii}\widetilde{Ra}_{cji}) \\ + \tfrac{1}{2}E_f(\widetilde{Ra}_i)E_f(\widetilde{Ra}_j)\Big); \widetilde{wRa}_i \wedge \widetilde{wRa}_j, \widetilde{uRa}_i \vee \widetilde{uRa}_j, \widetilde{yRa}_i \vee \widetilde{yRa}_j\rangle\end{aligned}$$

$$\sigma_{A_1 A_2} = \langle 0.160; 0.6,\ 0.2,\ 0.2\rangle$$

Proceeding in the same manner, the following results are obtained:

$$\sigma_{A_1A_3} = \langle 0.284; 0.6,\ 0.2,\ 0.2\rangle$$

$$\sigma_{A_2A_3} = \langle 0.171; 0.6,\ 0.2,\ 0.2\rangle$$

The variance-covariance matrix will be of the form:

$$\Omega = \begin{pmatrix} \widetilde{\sigma}_{fA11} & \widetilde{\sigma}_{fA12} & \widetilde{\sigma}_{fA13} \\ \widetilde{\sigma}_{fA21} & \widetilde{\sigma}_{fA22} & \widetilde{\sigma}_{fA23} \\ \widetilde{\sigma}_{fA31} & \widetilde{\sigma}_{fA32} & \widetilde{\sigma}_{fA33} \end{pmatrix}$$

By replacing the above values, we will have:

$$\Omega = \begin{pmatrix} \langle 0.0925; 0.5,\ 0.2,\ 0.3\rangle & \langle 0.160; 0.6,\ 0.2,\ 0.2\rangle & \langle 0.284; 0.6,\ 0.2,\ 0.2\rangle \\ \langle 0.160; 0.6,\ 0.2,\ 0.2\rangle & \langle 0.033; 0.6,\ 0.3,\ 0.2\rangle & \langle 0.171; 0.6,\ 0.2,\ 0.2\rangle \\ \langle 0.284; 0.6,\ 0.2,\ 0.2\rangle & \langle 0.171; 0.6,\ 0.2,\ 0.2\rangle & \langle 0.0910; 0.4,\ 0.3,\ 0.3\rangle \end{pmatrix}$$

According to Theorem 2 the weight of the financial asset x_{A_1} in the total value of the portfolio will be given by the relation:

$$x_{A_1} = \frac{\begin{pmatrix} \langle\widetilde{\sigma}_{A_{12}}; w\widetilde{\sigma}_{A_{12}}, u\widetilde{\sigma}_{A_{12}}, y\widetilde{\sigma}_{A_{12}}\rangle \\ \langle\widetilde{\sigma}_{A_{13}}; w\widetilde{\sigma}_{A_{13}}, u\widetilde{\sigma}_{A_{13}}, y\widetilde{\sigma}_{A_{13}}\rangle \end{pmatrix} - \begin{pmatrix} \langle\sigma_{A_{11}}; w\widetilde{\sigma}_{A_{11}}, u\widetilde{\sigma}_{A_{11}}, y\widetilde{\sigma}_{A_{11}}\rangle \\ \langle\sigma_{A_{11}}; w\widetilde{\sigma}_{A_{11}}, u\widetilde{\sigma}_{A_{11}}, y\widetilde{\sigma}_{A_{11}}\rangle \end{pmatrix}}{\begin{pmatrix} \langle\widetilde{\sigma}_{A_{12}}; w\widetilde{\sigma}_{A_{12}}, u\widetilde{\sigma}_{A_{12}}, y\widetilde{\sigma}_{A_{12}}\rangle \\ \langle\widetilde{\sigma}_{A_{13}}; w\widetilde{\sigma}_{A_{13}}, u\widetilde{\sigma}_{A_{13}}, y\widetilde{\sigma}_{A_{13}}\rangle \end{pmatrix}}$$

Through the calculations the following value is obtained:

$$x_{A_1} = 0.2376$$

For the weight of the financial asset x_{A_2} and x_{A_3} in the total value of the neutrosophic portfolio, the same calculation formula will be used and the results are the following:

$$x_{A_2} = 0.6808$$

$$x_{A_3} = 0.0816$$

Conclusion: In order to mitigate the neutrosophic portfolio risk, it is necessary to invest: in the first financial asset (A_1) a weight of $x_{A_1} = 0.2376$, in the second asset (A_2) a weight of $x_{A_2} = 0.6808$ and respectively in the third financial asset (A_3) a weight $x_{A_3} = 0.0816$.

The neutrosophic portfolio return will be:

$$\begin{aligned} \langle\widetilde{R}_p; w\widetilde{\sigma p}, u\widetilde{\sigma p}, y\widetilde{\sigma p}\rangle \\ &= \langle x_{A_1}\widetilde{R}_{A_1}; w\widetilde{\sigma}_{A_1}, u\widetilde{\sigma}_{A_1}, y\widetilde{\sigma}_{A_1}\rangle + \langle x_{A_2}\widetilde{R}_{A_2}; w\widetilde{\sigma}_{A_2}, u\widetilde{\sigma}_{A_2}, y\widetilde{\sigma}_{A_2}\rangle \\ &+ \langle x_{A_3}\widetilde{R}_{A_3}; w\widetilde{\sigma}_{A_3}, u\widetilde{\sigma}_{A_3}, y\widetilde{\sigma}_{A_3}\rangle \end{aligned}$$

By replacing in the formula, will be obtained:

$$\begin{aligned} \langle\widetilde{R}_p; w\widetilde{\sigma p}, u\widetilde{\sigma p}, y\widetilde{\sigma p}\rangle \\ &= \langle 0.2376\times 0.415; 0.5,\ 0.2,\ 0.3\rangle + \langle 0.6808\times 0.249; 0.6,\ 0.3,\ 0.2\rangle \\ &+ \langle 0.0816\times 0.448; 0.4,\ 0.3,\ 0.3\rangle \end{aligned}$$

$$\langle\widetilde{R}_p; w\widetilde{\sigma p}, u\widetilde{\sigma p}, y\widetilde{\sigma p}\rangle = \langle 0.0986; 0.5,\ 0.2,\ 0.3\rangle + \langle 0.1849; 0.6,\ 0.3,\ 0.2\rangle + \langle 0.0365; 0.4,\ 0.3,\ 0.3\rangle$$

After performing the neutrosophic calculations we obtain:

$$\langle \widetilde{R_p}; w\widetilde{\sigma p}, u\widetilde{\sigma p}, y\widetilde{\sigma p}\rangle = \langle 0.3200; 0.6,\ 0.2,\ 0.2\rangle$$

The neutrosophic portfolio risk will be as follows:

$$\begin{aligned}
&\langle \sigma_p^2; w\widetilde{\sigma p}, u\widetilde{\sigma p}, y\widetilde{\sigma p}\rangle \\
&\quad = \langle x_{A_1}^2 \widetilde{\sigma}_{A_1}^2; w\widetilde{\sigma}_{A_1}, u\widetilde{\sigma}_{A_1}, y\widetilde{\sigma}_{A_1}\rangle + \langle x_{A_2}^2 \widetilde{\sigma}_{A_2}^2; w\widetilde{\sigma}_{A_2}, u\widetilde{\sigma}_{A_2}, y\widetilde{\sigma}_{A_2}\rangle \\
&\quad + \langle x_{A_3}^2 \widetilde{\sigma}_{A_3}^2; w\widetilde{\sigma}_{A_3}, u\widetilde{\sigma}_{A_3}, y\widetilde{\sigma}_{A_3}\rangle \\
&\quad + 2x_{A_1}x_{A_2}\widetilde{\sigma}_{A_1A_2}; w\widetilde{\sigma}_{A_1} \wedge w\widetilde{\sigma}_{A_2}, u\widetilde{\sigma}_{A_1} \vee u\widetilde{\sigma}_{A_2}, y\widetilde{\sigma}_{A_1} \vee y\widetilde{\sigma}_{A_2} \\
&\quad + \langle 2x_{A_1}x_{A_3}\widetilde{\sigma}_{A_1A_3}; w\widetilde{\sigma}_{A_1} \wedge w\widetilde{\sigma}_{A_3}, u\widetilde{\sigma}_{A_1} \vee u\widetilde{\sigma}_{A_3}, y\widetilde{\sigma}_{A_1} \vee y\widetilde{\sigma}_{A_3}\rangle \\
&\quad + \langle 2x_{A_2}x_{A_1}\widetilde{\sigma}_{A_2A_1}; w\widetilde{\sigma}_{A_2} \wedge w\widetilde{\sigma}_{A_1}, u\widetilde{\sigma}_{A_2} \vee u\widetilde{\sigma}_{A_1}, y\widetilde{\sigma}_{A_2} \vee y\widetilde{\sigma}_{A_1}\rangle \\
&\quad + \langle 2x_{A_2}x_{A_3}\widetilde{\sigma}_{A_2A_3}; w\widetilde{\sigma}_{A_2} \wedge w\widetilde{\sigma}_{A_3}, u\widetilde{\sigma}_{A_2} \vee u\widetilde{\sigma}_{A_3}, y\widetilde{\sigma}_{A_2} \vee y\widetilde{\sigma}_{A_3}\rangle \\
&\quad + \langle 2x_{A_3}x_{A_1}\widetilde{\sigma}_{A_3A_1}; w\widetilde{\sigma}_{A_3} \wedge w\widetilde{\sigma}_{A_1}, u\widetilde{\sigma}_{A_3} \vee u\widetilde{\sigma}_{A_1}, y\widetilde{\sigma}_{A_3} \vee y\widetilde{\sigma}_{A_1}\rangle \\
&\quad + \langle 2x_{A_3}x_{A_2}\widetilde{\sigma}_{A_3A_2}; w\widetilde{\sigma}_{A_3} \wedge w\widetilde{\sigma}_{A_2}, u\widetilde{\sigma}_{A_3} \vee u\widetilde{\sigma}_{A_2}, y\widetilde{\sigma}_{A_3} \vee y\widetilde{\sigma}_{A_2}\rangle + \langle \sigma_p^2; w\widetilde{\sigma p}, u\widetilde{\sigma p}, y\widetilde{\sigma p}\rangle \\
&\quad = 49.42\%
\end{aligned}$$

Conclusion: In order to minimize the neutrosophic portfolio risk, the investments in financial assets must have the following weights: $x_{A_1} = 23.76\%$, $x_{A_2} = 68.08\%$ and $x_{A_3} = 8.16\%$. For these weights respective for the investments in financial assets A_1, A_2, A_3, the profitability of the neutrosophic portfolio will be of the form:

$$\langle \widetilde{R_p}; w\widetilde{\sigma p}, u\widetilde{\sigma p}, y\widetilde{\sigma p}\rangle = \langle 0.3200; 0.6,\ 0.2,\ 0.2\rangle$$

The neutrosophic portfolio risk will be minimal, respectively will have the value of $\langle \sigma_p; w\widetilde{\sigma p}, u\widetilde{\sigma p}, y\widetilde{\sigma p}\rangle = 49.42\%$. The risk was determined based on the high values of the return. The obtained results validate the risk minimization model for the neutrosophic portfolio, in the sense that for a financial return of 32%, the portfolio risk is high, reaching the value of 49.42%.

7. General Conclusions and Limitation

The neutrosophic portfolios are made up of financial assets for which it is possible to determine the financial performance indicators respectively: the neutrosophic return $\langle E_f(\widetilde{R_A}); w\widetilde{R_A}, u\widetilde{R_A}, y\widetilde{R_A}\rangle$ specific for the financial asset A_i, the neutrosophic risk $\langle \sigma f_{A_i}^2; w\widetilde{\sigma}_A, u\widetilde{\sigma}_A, y\widetilde{\sigma}_A\rangle$, specific to the same financial asset and the neutrosophic covariance $\langle cov(\widetilde{R_{A1}}, \widetilde{R_{A2}}); w\widetilde{R_{A1}}, u\widetilde{R_{A1}}, y\widetilde{R_{A1}};\ w\widetilde{R_{A2}}, u\widetilde{R_{A2}}, y\widetilde{R_{A2}}\rangle$ between two financial assets A_1 and A_2, which measures the intensity of the links between the neutrosophic returns specific to the two financial assets. Such a portfolio made up of financial assets for which financial performance indicators can be determined is called a neutrosophic portfolio.

For the neutrosophic portfolio $\langle \widetilde{P}; w\widetilde{p}, u\widetilde{p}, y\widetilde{p}\rangle$ can be determined: the neutrosophicreturn of the portfolio $\langle \widetilde{R_P}; w\widetilde{R_p}, u\widetilde{R_p}, y\widetilde{R_p}\rangle$ and the neutrosophic portfolio risk $\langle \widetilde{\sigma_p^2}; w\widetilde{\sigma p}, u\widetilde{\sigma p}, y\widetilde{\sigma p}\rangle$. These two performance indicators are fundamental indicators that characterize the neutrosophic portfolios.

Thus, the neutrosophic portfolio return is dependent on the weight held by the financial assets in the total value of the neutrosophic portfolio x_{A_i}, as well as on the neutrosophic return of each financial asset that makes up the portfolio $\langle \widetilde{R_{A_i}}; w\widetilde{R_{A_i}}, u\widetilde{R_{A_i}}, y\widetilde{R_{A_i}}\rangle$. At the same time, the neutrosophic portfolio risk is dependent on the weight held by the financial assets in the total value of the portfolio x_{A_i}, as well as on the neutrosophic risk of each financial asset of the form: $\langle \sigma f_{A_i}^2; w\widetilde{\sigma}_A, u\widetilde{\sigma}_A, y\widetilde{\sigma}_A\rangle$ and the covariance between two financial assets $\langle \widetilde{\sigma}_{A_iA_j}; w\widetilde{\sigma}_{A_i} \wedge w\widetilde{\sigma}_{A_j}, u\widetilde{\sigma}_{A_i} \vee u\widetilde{\sigma}_{A_j}, y\widetilde{\sigma}_{A_i} \vee y\widetilde{\sigma}_{A_j}\rangle$.

Also the neutrosophic portfolio risk, consisting of N financial assets admits a minimum value at the point where the first order derivative of the neutrosophic portfolio risk is zero $\frac{\partial\langle \widetilde{\sigma_p^2}; w\widetilde{\sigma p}, u\widetilde{\sigma p}, y\widetilde{\sigma p}\rangle}{\partial x_{A_i}} = 0$. Thus, it can be determined what the weight of every financial assets should be in the total value of the neutrosophic portfolio x_{A_i}, so that the portfolio risk is minimal $\langle \widetilde{\sigma_p^2}; w\widetilde{\sigma p}, u\widetilde{\sigma p}, y\widetilde{\sigma p}\rangle \to min$. From calculations but also from studying this risk category, it was found that the financial assets weight in the total value of the portfolio is dependent on the individual neutrosophic risk of each financial asset but also on the covariance between two financial assets.

The neutrosophic portfolios enables the access to complete information for the financial market investors, in order to substantiate investment decisions. This information provided by the neutrosophic portfolios refers to the probability of realizing the neutrosophic portfolio return, which in turn is influenced by the individual probabilities of achieving the desired return for each financial asset that enters the portfolio structure. Also, the neutrosophic portfolios provide information regarding the probability of producing the neutrosophic portfolio risk, which depends on the probability of producing the neutrosophic risk for each financial asset that enters the portfolio. These categories of information are stratified by means of linguistic variables, so that we will distinguish: the probability of obtaining the return and/or the production of the portfolio risk almost certainly, the probability that the return/production of the portfolio risk will not be realized and the probability that the return/production of the portfolio risk to be uncertain.

Obtaining concomitant information regarding risk and return at the level of the neutrosophic portfolio, as well as the probability of producing the risk and return for the neutrosophic portfolio as well as for the financial assets confer a strong innovative approach for this research paper. Neutrosophic portfolios also have certain limits which mainly refer to the determination of the probability of producing the risk and/or of realizing the return, both at the level of each financial asset as well as at the level of the neutrosophic portfolio as a whole.

Author Contributions: Conceptualization, M.-I.B., I.-A.B.; Data curation, C.D.; Formal analysis, M.-I.B. and I.-A.B.; Investigation, M.-I.B., I.-A.B. and C.D.; Methodology, M.-I.B. and I.-A.B.; Supervision, M.-I.B.; Validation, I.-A.B.; Visualization, C.D.; Writing—original draft, M.-I.B. and I.-A.B.; Writing—review & editing, C.D.

Funding: This research received no external funding.

Conflicts of Interest: The authors declare no conflict of interest.

References

1. Markowitz, H. Portfolio Selection. *J. Financ.* **1952**, *7*, 77.
2. Bo, C.; Zhang, X.; Shao, S. Non-dual multi-granulation neutrosophic rough set with applications. *Symmetry* **2019**, *11*, 910. [CrossRef]
3. Jamil, M.; Abdullah, S.; Yaqub Khan, M.; Smarandache, F.; Ghani, F. Application of the bipolar neutrosophic hamacher averaging aggregation operators to group decision making: An illustrative example. *Symmetry* **2019**, *11*, 698. [CrossRef]
4. Shao, S.; Zhang, X.; Zhao, Q. Multi-attribute decision making based on probabilistic neutrosophic hesitant fuzzy choquet aggregation operators. *Symmetry* **2019**, *11*, 623. [CrossRef]
5. Al-Quran, A.; Hassan, N.; Alkhazaleh, S. Fuzzy parameterized complex neutrosophic soft expert set for decision under uncertainty. *Symmetry* **2019**, *11*, 382. [CrossRef]
6. Shi, L.; Yuan, Y. Hybrid weighted arithmetic and geometric aggregation operator of neutrosophic cubic sets for MADM. *Symmetry* **2019**, *11*, 278. [CrossRef]
7. Cao, C.; Zeng, S.; Luo, D. A single-valued neutrosophic linguistic combined weighted distance measure and its application in multiple-attribute group decision-making. *Symmetry* **2019**, *11*, 275. [CrossRef]
8. Jiang, W.; Zhang, Z.; Deng, X. Multi-attribute decision making method based on aggregated neutrosophic set. *Symmetry* **2019**, *11*, 267. [CrossRef]
9. Khan, M.; Gulistan, M.; Yaqoob, N.; Khan, M.; Smarandache, F. Neutrosophic cubic einstein geometric aggregation operators with application to multi-criteria decision making method. *Symmetry* **2019**, *11*, 247. [CrossRef]
10. Abu Qamar, M.; Hassan, N. An approach toward a Q-neutrosophic soft set and its application in decision making. *Symmetry* **2019**, *11*, 139. [CrossRef]
11. Jana, C.; Pal, M. A robust single-valued neutrosophic soft aggregation operators in multi-criteria decision making. *Symmetry* **2019**, *11*, 110. [CrossRef]

12. Fan, C.; Feng, S.; Hu, K. Linguistic neutrosophic numbers einstein operator and its application in decision making. *Mathematics* **2019**, *7*, 389. [CrossRef]
13. Fan, C.; Ye, J.; Feng, S.; Fan, E.; Hu, K. Multi-criteria decision-making method using heronian mean operators under a bipolar neutrosophic environment. *Mathematics* **2019**, *7*, 97. [CrossRef]
14. Abdel-Basset, M.; Mohamed, R.; Zaied, A.E.-N.H.; Smarandache, F. A hybrid plithogenic decision-making approach with quality function deployment for selecting supply chain sustainability metrics. *Symmetry* **2019**, *11*, 903. [CrossRef]
15. Abdel-Basset, M.; Mohamed, M.; Chang, V.; Smarandache, F. IoT and its impact on the electronics market: A powerful decision support system for helping customers in choosing the best product. *Symmetry* **2019**, *11*, 611. [CrossRef]
16. Ahmad, F.; Adhami, A.Y.; Smarandache, F. Neutrosophic optimization model and computational algorithm for optimal shale gas water management under uncertainty. *Symmetry* **2019**, *11*, 544. [CrossRef]
17. Abdel-Basset, M.; Chang, V.; Mohamed, M.; Smarandche, F. A refined approach for forecasting based on neutrosophic time series. *Symmetry* **2019**, *11*, 457. [CrossRef]
18. Colhon, M.; Vlăduţescu, Ş.; Negrea, X. How objective a neutral word is? A neutrosophic approach for the objectivity degrees of neutral words. *Symmetry* **2017**, *9*, 280. [CrossRef]
19. Smarandache, F.; Colhon, M.; Vlăduţescu, Ş.; Negrea, X. Word-level neutrosophic sentiment similarity. *Appl. Soft Comput.* **2019**, *80*, 167–176. [CrossRef]
20. Islam, S.; Ray, P. Multi-objective portfolio selection model with diversification by neutrosophic optimization technique. *Neutrosophic Sets Syst.* **2018**, *21*, 74–83.
21. Pamučar, D.; Badi, I.; Sanja, K.; Obradović, R. A novel approach for the selection of power-generation technology using a linguistic neutrosophic CODAS method: A case study in libya. *Energies* **2018**, *11*, 2489. [CrossRef]
22. Abdel-Basset, M.; Atef, A.; Smarandache, F. A hybrid neutrosophic multiple criteria group decision making approach for project selection. *Cogn. Syst. Res.* **2019**, *57*, 216–227. [CrossRef]
23. Villegas Alava, M.; Delgado Figueroa, S.P.; Blum Alcivar, H.M.; Leyva Vazquez, M. Single valued neutrosophic numbers and analytic hierarchy process for project selection. *Neutrosophic Sets Syst.* **2018**, *21*, 122–130.
24. Saleh Al-Subhi, S.H.; Perez Pupo, I.; Garcia Vacacela, R.; Pinero Perez, P.Y.; Leyva Vazquez, M.Y. A new neutrosophic cognitive map with neutrosophic sets on connections, application in project management. *Neutrosophic Sets Syst.* **2018**, *22*, 63–75.
25. Perez Pupo, D.; Pinero Perez, P.Y.; Garcia Vacacela, R.; Bello, R.; Santos, O.; Leyva Vazquez, M.Y. Extensions to linguistic summaries indicators based on neutrosophic theory, applications in project management decisions. *Neutrosophic Sets Syst.* **2018**, *22*, 87–100.
26. Su, L.; Wang, T.; Wang, L.; Li, H.; Cao, Y. Project procurement method selection using a multi-criteria decision-making method with interval neutrosophic sets. *Information* **2019**, *10*, 201. [CrossRef]
27. Wu, S.; Wang, J.; Wei, G.; Wei, Y. Research on construction engineering project risk assessment with some 2-tuple linguistic neutrosophic hamy mean operators. *Sustainability* **2018**, *10*, 1536. [CrossRef]
28. Bolos, M.-I.; Bradea, I.-A.; Delcea, C. Modeling the performance indicators of financial assets with neutrosophic fuzzy numbers. *Symmetry* **2019**, *11*, 1021. [CrossRef]

mathematics

MDPI

Article

Generalized Abel-Grassmann's Neutrosophic Extended Triplet Loop

Xiaogang An [1,*], Xiaohong Zhang [1] and Yingcang Ma [2]

[1] School of Arts and Sciences, Shaanxi University of Science & Technology, Xi'an 710021, China; zhangxiaohong@sust.edu.cn

[2] School of Science, Xi'an Polytechnic University, Xi'an 710048, China; mayingcang@xpu.edu.cn

* Correspondence: anxiaogang@sust.edu.cn

Received: 5 November 2019; Accepted: 4 December 2019; Published: 9 December 2019

Abstract: A group is an algebraic system that characterizes symmetry. As a generalization of the concept of a group, semigroups and various non-associative groupoids can be considered as algebraic abstractions of generalized symmetry. In this paper, the notion of generalized Abel-Grassmann's neutrosophic extended triplet loop (GAG-NET-Loop) is proposed and some properties are discussed. In particular, the following conclusions are strictly proved: (1) an algebraic system is an AG-NET-Loop if and only if it is a strong inverse AG-groupoid; (2) an algebraic system is a GAG-NET-Loop if and only if it is a quasi strong inverse AG-groupoid; (3) an algebraic system is a weak commutative GAG-NET-Loop if and only if it is a quasi Clifford AG-groupoid; and (4) a finite interlaced AG-(l,l)-Loop is a strong AG-(l,l)-Loop.

Keywords: Abel-Grassmann's neutrosophic extended triplet loop; generalized Abel-Grassmann's neutrosophic extended triplet loop; strong inverse AG-groupoid; quasi strong inverse AG-groupoid; quasi Clifford AG-groupoid

1. Introduction

The concept of an Abel-Grassmann's groupoid (AG-groupoid) was first given by Kazim and Naseeruddin [1] in 1972 and they have called it a left almost semigroup (LA-semigroup). In [2], the same structure is called a left invertive groupoid. In [3–9], some properties and different classes of an AG-groupoid are investigated.

Smarandache proposed the new concept of neutrosophic set, which is an extension of fuzzy set and intuitionistic fuzzy set [10]. Until now, neutrosophic sets have been applied to many fields such as decision making [11–13], forecasting [14], best product selection [15], the shortest path problem [16], minimum spanning tree [17], neutrosophic portfolios of financial assets [18], etc. Some new theoretical studies are also developed [19–24]. In [25], Xiaohong Zhang introduced the concept of Abel-Grassmann's neutrosophic extended triplet loop (AG-NET-loop), and some properties and structure about AG-NET-loop are discussed. Recently, a new algebraic system, generalized neutrosophic extended triplet set, is proposed in [26].

In this paper, we combine the notions of generalized neutrosophic extended triplet set and AG-groupoid, introduce the new concept of generalized Abel-Grassmann's neutrosophic extended triplet loop (GAG-NET-loop); that is, GAG-NET-loop is both AG-groupoid and generalized neutrosophic extended triplet set. We deeply analyze the internal connecting link between GAG-NET-loop and other AG-groupoid and obtain some important results.

GAG-NET-loop is an extension of AG-NET-loop. Compared with AG-NET-loop, GAG-NET-loop relaxes the restriction on the elements in the AG-groupoid. According to our research, corresponding to the decomposition theorem of AG-NET-loop, some GAG-NET-loops can also be decomposed into smaller ones. This is also the embodiment of the research method of regular semigroups to quasi-regular semigroups in non-associative groupoid.

The paper is organized as follows. Section 2 gives the basic definitions. Some properties about finite interlaced AG-(l,l)-Loop and some structures about strong inverse AG-groupoid are discussed in Section 3. We proposed the GAG-NET-Loop and discussed its properties and structure in Section 4. Finally, the summary and future work are presented in Section 5.

2. Basic Definitions

In this section, the related research and results of the AG-NET-loop are presented. Some related notions are introduced first.

Let S be non-empty set, $*$ is a binary operation on S. If $\forall a, b \in S$, implies $a * b \in S$, then $(S, *)$ is called a groupoid. A groupoid $(S, *)$ is called an Abel-Grassmann's groupoid (AG-groupoid) [27,28] if it holds the left invertive law, that is, for all $a, b, c \in S$, $(a * b) * c = (c * b) * a$. In an AG-groupoid the medial law holds, for all $a, b, c, \in S$, $(a * b) * (c * d) = (a * c) * (b * d)$. In an AG-groupoid $(S, *)$, for all $a \in S$, $n \in Z^+$, the recursive definition of a^n is as follows: $a^1 = a, a^2 = a * a, a^3 = a^2 * a = (a * a) * a, a^4 = a^3 * a, ..., a^n = a^{n-1} * a$.

Definition 1 ([29]). *Let N be a non-empty set together with a binary operation $*$. Then, N is called a neutrosophic extended triplet set if for any $a \in N$, there exists a neutral of "a" (denoted by $neut(a)$), and an opposite of "a"(denoted by $anti(a)$), such that $neut(a) \in N$, $anti(a) \in N$ and:*

$$a * neut(a) = neut(a) * a = a,$$

$$a * anti(a) = anti(a) * a = neut(a).$$

The triplet $(a, neut(a), anti(a))$ is called a neutrosophic extended triplet.

Note that, for a neutrosophic triplet set $(N, *), a \in N$, $neut(a)$ and $anti(a)$ may not be unique. In order not to cause ambiguity, we use the following notations to distinguish: $neut(a)$ denotes any certain one of neutral of a, $\{neut(a)\}$ denotes the set of all neutral of a, $anti(a)$ denotes any certain one of opposite of a, and $\{anti(a)\}$ denotes the set of all opposite of a.

Definition 2 ([25]). *Let $(N, *)$ be a neutrosophic extended triplet set. Then, N is called a neutrosophic extended triplet loop (NET-Loop), if $(N, *)$ is well-defined, i.e., for any $a, b \in N$, one has $a * b \in N$.*

Definition 3 ([25]). *Let $(N, *)$ be a neutrosophic extended triplet loop (NET-Loop). Then, N is called an AG-NET-Loop, if $(N, *)$ is an AG-groupoid.*

An AG-NET-Loop N is called a commutative AG-NET-Loop if for all $a, b \in N, a * b = b * a$.

Theorem 1 ([25]). *Let $(N, *)$ be an AG-NET-loop. Then, for any $x, y \in \{anti(a)\}$,*

(1) $neut(a) * x = x * neut(a) = neut(a) * y$, that is, $|neut(a) * \{anti(a)\}| = 1$.
(2) $(x * neut(a)) * a = (neut(a) * x) * a = neut(a)$.
(3) $a * (x * neut(a)) = a * (neut(a) * x) = neut(a)$.
(4) $\forall a \in N, neut(a) * neut(a) = neut(a)$.

Definition 4 ([5]). *An element a of an AG-groupoid $(S, *)$ is called a regular if there exists $x \in S$ such that $a = (a * x) * a$ and S is called regular if all elements of S are regular.*

An AG-groupoid $(S, *)$ is called quasi regular if, for any $a \in S$, there exists a positive integer n such that a^n is regular.

Definition 5 ([6]). *An element a of an AG-groupoid $(S,*)$ is called a fully regular element of S if there exist some $p,q,r,s,t,u,v,w,x,y,z \in S$ ($p,q,...,z$ may be repeated) such that*

$$\begin{aligned} a &= (p*a^2)*q = (r*a)*(a*s) = (a*t)*(a*u) \\ &= (a*a)*v = w*(a*a) = (x*a)*(y*a) \\ &= (a^2*z)*a^2. \end{aligned}$$

An AG-groupoid $(S,*)$ is called fully regular if all elements of S are fully regular.

An AG-groupoid $(S,*)$ is called quasi fully regular if for any $a \in S$, there exists a positive integer n such that a^n is fully regular.

3. Strong Inverse AG-Groupoid and Finite Interlaced AG-Groupoid

Definition 6 ([30]). *An AG-groupoid $(S,*)$ is called an inverse AG-groupoid if for each element $a \in S$, there exists an element x in S such that $a = (a*x)*a$ and $x = (x*a)*x$.*

Definition 7. *An AG-groupoid $(S,*)$ is called a strong inverse AG-groupoid if for any $a \in S$, there exists a unary operation $a \rightarrow a^{-1}$ on S such that*

$$(a^{-1})^{-1} = a, \quad (a*a^{-1})*a = a*(a*a^{-1}) = a, \quad a*a^{-1} = a^{-1}*a.$$

The following example shows that an inverse AG-groupoid may not be a strong inverse AG-groupoid.

Example 1. *Let $S = \{1,2,3,4\}$, an operation $*$ on S is defined as in Table 1. Being $1 = (1*3)*1, 3 = (3*1)*3, 2 = (2*4)*2, 4 = (4*2)*4$, from Definition 6, S is an inverse AG-groupoid. Being $(1*1)*1 = 3 \neq 1, (1*2)*1 = 4 \neq 1, (1*3)*1 = 1 \neq 3 = 1*(1*3), (1*4)*1 = 2 \neq 1$, from Definition 7, S is not a strong inverse AG-groupoid.*

Table 1. The operation table of Example 1.

*	1	2	3	4
1	2	4	3	1
2	3	1	2	4
3	1	3	4	2
4	4	2	1	3

Proposition 1. *Let $(N,*)$ be an AG-NET-loop. Then, for any $a \in N$, $x \in \{anti(a)\}$,*

$$neut(neut(a)*x)*anti(neut(a)*x) = a.$$

Proof. For any $x \in \{anti(a)\}$, we have

$$\begin{aligned} (neut(a)*x)*neut(a) &= (neut(a)*x)*(a*x) \\ &= (neut(a)*a)*(x*x) \quad \textit{(applying the medial law)} \\ &= (a*neut(a))*(x*x) \\ &= (a*x)*(neut(a)*x) \quad \textit{(applying the medial law)} \\ &= neut(a)*(neut(a)*x), \end{aligned}$$

$$\begin{aligned} neut(a) * (neut(a) * x) &= (x * a) * (neut(a) * x) \\ &= (x * neut(a)) * (a * x) \quad (applying\ the\ medial\ law) \\ &= (x * neut(a)) * neut(a) \\ &= (neut(a) * neut(a)) * x \\ &= neut(a) * x, \quad (by\ Proposition\ 1(4)) \end{aligned}$$

we have $(neut(a) * x) * neut(a) = neut(a) * (neut(a) * x) = neut(a) * x$.

From Theorem 1 (2) and (3), we have

$$neut(neut(a) * x) = neut(a), a \in anti\{neut(a) * x\}.$$

From Theorem 1 (1) $neut(a) * x$ is unique, we have

$$neut(neut(a) * x) * anti(neut(a) * x) = neut(a) * a = a.$$

□

Example 2. *Let $N = \{a, b, c\}$, an operation $*$ on N is defined as in Table 2. Since $neut(a) = a, anti(a) = a, neut(b) = a, anti(b) = c, neut(c) = a, anti(c) = b$, so $(N, *)$ is an AG-NET-Loop. Being*

$$neut(neut(a) * a) * anti(neut(a) * a) = a * a = a,$$

$$neut(neut(b) * c) * anti(neut(b) * c) = neut(c) * anti(c) = b,$$

$$neut(neut(c) * b) * anti(neut(c) * b) = neut(b) * anti(b) = c,$$

that is for any $a \in N$, $x \in \{anti(a)\}$, $neut(neut(a) * x) * anti(neut(a) * x) = a$.

Table 2. An AG-NET-Loop of Example 2.

$*$	a	b	c
a	a	b	c
b	b	c	a
c	c	a	b

Theorem 2. *Let $(N, *)$ be a groupoid. Then, N is an AG-NET-Loop if and only if it is a strong inverse AG-groupoid.*

Proof. Necessity: Suppose N is an AG-NET-Loop, from Definition 3, for each $a \in N$, such that a has the neutral element and opposite element, denoted by $neut(a)$ and $anti(a)$, respectively. Set

$$a^{-1} = neut(a) * anti(a),$$

by Theorem 1 (1), $neut(a) * anti(a)$ is unique, so a^{-1} is unique. By Proposition 1, we have

$$(a^{-1})^{-1} = neut(neut(a) * anti(a)) * anti(neut(a) * anti(a)) = a.$$

Being

$$a^{-1} * a = (neut(a) * anti(a)) * a = (a * anti(a)) * neut(a) = neut(a) * neut(a) = neut(a),$$

$$\begin{aligned} a * a^{-1} &= a * (neut(a) * anti(a)) \\ &= (neut(a) * a) * (neut(a) * anti(a)) \\ &= (neut(a) * neut(a)) * (a * anti(a)) \\ &= (neut(a) * neut(a)) * neut(a) \\ &= neut(a), \end{aligned}$$

$$(a * a^{-1}) * a = neut(a) * a = a,$$

$$a * (a * a^{-1}) = a * neut(a) = a,$$

we have

$$a^{-1} * a = a * a^{-1},$$

$$(a * a^{-1}) * a = a * (a * a^{-1}) = a.$$

From Definition 7, N is a strong inverse AG-groupoid.

Sufficiency: If N is a strong inverse AG-groupoid and $a^{-1} \in N$, such that $a * a^{-1} = a^{-1} * a$ and $(a * a^{-1}) * a = a * (a * a^{-1}) = a$. Set

$$neut(a) = a * a^{-1},$$

then $neut(a) * a = (a * a^{-1}) * a = a * (a * a^{-1}) = a * neut(a) = a$, $a * (a)^{-1} = (a)^{-1} * a = neut(a)$. From Definition 3, we have that N is an AG-NET-Loop and $a^{-1} \in \{anti(a)\}$. □

Example 3. *Apply $(S, *)$ in Example 2, we know that it is an AG-NET-Loop. We show that it is a strong inverse AG-groupoid in the following.*

For b, there exists a inverse element $b^{-1} = c$, such that $(b^{-1})^{-1} = b, (b * b^{-1}) * b = b * (b * b^{-1}) = b, b * b^{-1} = b^{-1} * b = a$, so b is strong inverse. a and c are strong inverse for the same reason, so $(S, *)$ is a strong inverse AG-groupoid by Definition 7.

An AG-groupoid (S, *) is called interlaced if it satisfies $(a * a) * b = a * (a * b), a * (b * b) = (a * b) * b$ for all a, b in S. An AG-groupoid (S, *) is called locally associative if it satisfies $(a * a) * a = a * (a * a)$ for all a in S.

Theorem 3. *Let $(D, *)$ be a locally associative AG-groupoid with respect to *. If D is finite, there is an idempotent element in D. That is, $\exists a \in D, a * a = a$.*

Proof. Assume that D is a finite locally associative AG-groupoid with respect to *. Then, for any $a \in D$, $a, a * a = a^2, a * a * a = a^3, ..., a^n, ... \in D$. Since D is finite, there exists natural number m,k such that $a^m = a^{m+k}$.

Case 1: If $k = m$, then $a^m = a^{2m}$, that is, $a^m = a^m * a^m$, a^m is an idempotent element in D.

Case 2: If $k > m$, then from $a^m = a^{m+k}$ we can get

$$a^k = a^m * a^{k-m} = a^{m+k} * a^{k-m} = a^{2k} = a^k * a^k.$$

This means that a^k is an idempotent element in D.

Case 3: If $k < m$, then from $a^m = a^{m+k}$ we can get

$$a^m = a^{m+k} = a^m * a^k = a^{m+k} * a^k = a^{m+2k};$$

$$a^m = a^{m+2k} = a^m * a^{2k} = a^{m+k} * a^{2k} = a^{m+3k};$$

$$\cdots\cdots$$

$$a^m = a^{m+mk}.$$

Since m and k are natural numbers, then $mk \geq m$. Therefore, from $a^m = a^{m+mk}$, applying Case 1 or Case 2, we know that there exists an idempotent element in D. □

Definition 8 ([31]). *Let $(N,*)$ be an AG-groupoid. Then, N is called an AG-(l, l)-Loop, if for any $a \in N$, there exist two elements b and c in N that satisfy the condition: $b * a = a$, and $c * a = b$. In an AG-(l,l)-Loop, a neutral of "a" denoted by $neut_{(l,l)}(a)$.*

Definition 9 ([31]). *Let $(N,*)$ be an AG-(l, l)-Loop. Then, N is a strong AG-(l, l)-Loop if $neut_{(l,l)}(a) * neut_{(l,l)}(a) = neut_{(l,l)}(a), \forall a \in N$.*

Definition 10. *Let $(D,*)$ be an AG-(l,l)-Loop. Then, D is called an interlaced AG-(l,l)-Loop, if it satisfies $(a * a) * b = a * (a * b)$, $a * (b * b) = (a * b) * b$, for all a, b in D.*

Theorem 4. *Let $(D,*)$ be an interlaced AG-(l, l)-Loop with respect to *. If D is finite, there is an idempotent left neutral element in D. That is, $\forall a \in D, \exists s, p \in D, s * a = a, p * a = s, s * s = s$.*

Proof. Assume that D is a finite interlaced AG-(l,l)-Loop with respect to *. Then, for any $a \in D$, $\exists s, p \in D, s * a = a, p * a = s$, we have $s * a = (p * a) * a = (a * a) * p = a * (a * p) = a$,

$$\begin{aligned} a * s &= (a * (a * p)) * s \\ &= (s * (a * p)) * a \quad (by\ the\ left\ invertive\ law) \\ &= ((p * a) * (a * p)) * a \\ &= (((a * p) * a) * p) * a \quad (by\ the\ left\ invertive\ law) \\ &= (a * p) * ((a * p) * a) \quad (by\ the\ left\ invertive\ law) \\ &= ((a * p) * (a * p)) * a \quad (by\ the\ interlaced\ law) \\ &= (a * (a * p)) * (a * p) \quad (by\ the\ left\ invertive\ law) \\ &= a * (a * p) = a, \end{aligned}$$

$$s^2 * a = (s * s) * a = (a * s) * s = a,$$

$$s^3 * a = (s^2 * s) * a = (a * s) * s^2 = a * s^2 = a * (s * s) = (a * s) * s = a * s = a.$$

When $m > 3, m \equiv 0(\text{mod } 2)$, we have

$$\begin{aligned} s^m * a &= (s^{m-2} * s^2) * a \\ &= (a * s^2) * s^{m-2} \\ &= a * s^{m-2} \\ &= a * (s^{(m-2)/2} * s^{(m-2)/2}) \\ &= (a * s^{(m-2)/2}) * s^{(m-2)/2} \quad (by\ the\ interlaced\ law) \\ &= (s^{(m-2)/2} * s^{(m-2)/2}) * a \quad (by\ the\ left\ invertive\ law) \\ &= s^{m-2} * a \\ &= \ldots\ldots \\ &= s^2 * a = a. \end{aligned}$$

When $m > 3, m \equiv 1(\text{mod } 2)$, we have

$$\begin{aligned} s^m * a &= (s^{m-1} * s) * a \\ &= (a * s) * s^{m-1} \\ &= a * s^{m-1} \\ &= a * (s^{(m-1)/2} * s^{(m-1)/2}) \\ &= (a * s^{(m-1)/2}) * s^{(m-1)/2} \quad (\textit{by the interlaced law}) \\ &= (s^{(m-1)/2} * s^{(m-1)/2}) * a \\ &= s^{m-1} * a \\ &= \ldots\ldots \\ &= s^2 * a = a. \end{aligned}$$

Thus, $s, s^2, s^3 ...... s^m ......$ are all left neutral element.

Applying Theorem 3, we know that there exists an idempotent left neutral element in D. □

Theorem 5. *Assume that $(N, *)$ is a finite interlaced AG-(l,l)-Loop. Then, for all a in N, if $neut_{(l,l)}(a)$ is an idempotent element, then it is unique.*

Proof. Assume that N is a finite interlaced AG-(l,l)-Loop with respect to *. Suppose that there exist $x, y \in \{neut_{(l,l)}(a)\}, a \in N$. By Definition 8, $x * a = a, y * a = a$, and there exist $p, q \in N$ which satisfy $p * a = x, q * a = y$. If $x * x = x, y * y = y$, we have

$$x = x * x = (p * a) * x = (x * a) * p = a * p,$$

$$y = y * y = (q * a) * y = (y * a) * q = a * q,$$

$$x * y = (p * a) * y = (y * a) * p = a * p = x,$$

$$y * x = (q * a) * x = (x * a) * q = a * q = y,$$

$$x = x * y = (x * x) * y = (y * x) * x = y * x = y.$$

We know that $x = y$, $neut_{(l,l)}(a)$ is unique. □

Theorem 6. *Let $(N, *)$ be a finite interlaced AG-(l,l)-Loop. Then, N is a strong AG-(l,l)-Loop.*

Proof. For any a in N, applying Theorem 4, we have $\exists s, p \in N, s * a = a, p * a = s, s * s = s$. From this and Definition 9, we know that N is a strong AG-(l,l)-Loop. □

Example 4. *Let $S = \{1, 2, 3\}$, an operation $*$ on S is defined as in Table 3. Being $(1 * 1) * 2 = 1 * (1 * 2) = 2, 1 * (2 * 2) = (1 * 2) * 2 = 3, (1 * 1) * 3 = 1 * (1 * 3) = 3, 1 * (3 * 3) = (1 * 3) * 3 = 2, (2 * 2) * 3 = 2 * (2 * 3) = 2, 2 * (3 * 3) = (2 * 3) * 3 = 3$, and $1 * 1 = 1, 1 * 2 = 2, 3 * 2 = 1, 1 * 3 = 3, 2 * 3 = 1$, we have S is a finite interlaced AG-(l,l)-Loop by Definition 10. Being $neut_{(l,l)}(1) = neut_{(l,l)}(2) = neut_{(l,l)}(3) = 1$, 1*1=1, we have S is a strong AG-(l,l)-Loop by Definition 9.*

Table 3. A finite interlaced AG-(l,l)-Loop of Example 4.

*	1	2	3
1	1	2	3
2	2	3	1
3	3	1	2

The following example shows that a strong AG-(l,l)-Loop may not be an interlaced AG-(l,l)-Loop.

Example 5. *Let* $S = \{1, 2, 3\}$, *an operation* $*$ *on* S *is defined as in Table 4. Being* $1 * 1 = 1, 1 * 2 = 2, 2 * 2 = 1, 1 * 3 = 3, 3 * 3 = 1$, *we have* S *is a strong AG-(l,l)-Loop by Definition 9. However, it is not an interlaced AG-(l,l)-Loop because* $2 * (3 * 3) = 3 \neq 2 = (2 * 3) * 3$.

Table 4. A strong AG-(l,l)-Loop of Example 5.

$*$	**1**	**2**	**3**
1	1	2	3
2	3	1	2
3	2	3	1

4. GAG-NET-Loop

Definition 11 ([26]). *Let* N *be a non-empty set together with a binary operation* $*$. *Then,* N *is called a generalized neutrosophic extended triplet set if for any* $a \in N$, *at least exists a positive integer* n, a^n *exists neutral element (denoted by* $neut(a^n)$*) and opposite element (denoted by* $anti(a^n)$*), such that* $neut(a^n) \in N, anti(a^n) \in N$ *and*

$$a^n * neut(a^n) = neut(a^n) * a^n = a^n, a^n * anti(a^n) = anti(a^n) * a^n = neut(a^n).$$

The triplet $(a, neut(a^n), anti(a^n))$ *is called a generalized neutrosophic extended triplet with degree* n.

Definition 12. *Let* $(N, *)$ *be a generalized neutrosophic extended triplet set. Then,* N *is called a generalized Abel-Grassmann's neutrosophic extended triplet loop (GAG-NET-Loop), if the following conditions are satisfied: for all* $a, b, c \in N$, $(a * b) * c = (c * b) * a$.

A GAG-NET-Loop N *is called a commutative GAG-NET-Loop if for all* $a, b \in N, a * b = b * a$.

Example 6. *Let* $S = \{a, b, c\}$, *an operation* $*$ *on* S *is defined as in Table 5. We can see that* $(a, a, a), (a, a, b)$, *and* (a, a, c) *are neutrosophic extended triplets, but b and c do not have the neutral element and opposite element. Thus,* S *is not an AG-NET-Loop. Moreover,* $b^2 = c^2 = a$ *has the neutral element and opposite element, thus* $(S, *)$ *is a GAG-NET-Loop.* (b, a, a) *and* (c, a, a) *are generalized neutrosophic extended triplets with degree 2. We can infer that* $(S, *)$ *is a GAG-NET-Loop but not an AG-NET-Loop. Moreover it is not a commutative GAG-NET-Loop being* $b * c \neq c * b$.

Table 5. A GAG-NET-Loop of Example 6.

$*$	a	b	c
a	a	a	a
b	a	a	c
c	a	b	a

The algebraic system $(Z_n, \otimes)$, $\otimes$ is the classical mod multiplication, where $Z_n = \{[0], [1], \cdots, [n-1]\}$ and $n \in Z^+, n \geq 2$.

Example 7. *Consider* $(Z_4, \otimes)$, *an operation* $\otimes$ *on* Z_4 *is defined as in Table 6. We have:*

(1) $[0], [1]$ *and* $[3]$ *have the neutral element and opposite element.*
(2) $[2]$ *does not have the neutral element and opposite element, but we can see that* $[2]^2 = [0]$ *has the neutral element and opposite element.*

Table 6. The operation table of Z_4.

$\otimes$	[0]	[1]	[2]	[3]
[0]	[0]	[0]	[0]	[0]
[1]	[0]	[1]	[2]	[3]
[2]	[0]	[2]	[0]	[2]
[3]	[0]	[3]	[2]	[1]

Thus, Z_4 is a generalized neutrosophic extended triplet set, but it is not a neutrosophic extended triplet set. Moreover, $(Z_4, \otimes)$ is a commutative GAG-NET-Loop.

Proposition 2. *Let $(N, *)$ be a GAG-NET-Loop, $a \in N$ and $(a, neut(a^n), anti(a^n))$ is a generalized neutrosophic extended triplet with degree n. We have:*

(1) *$neut(a^n)$ is unique.*
(2) *$neut(a^n) * neut(a^n) = neut(a^n)$.*

Proof. Assume $c, d \in \{neut(a^n)\}$, so $a^n * c = c * a^n = a^n, a^n * d = d * a^n = a^n$, and there exists $x, y \in N$ such that

$$a^n * x = x * a^n = c, a^n * y = y * a^n = d.$$

We can obtain

$$c * d = (x * a^n) * d = (d * a^n) * x = a^n * x = c,$$

$$\begin{aligned} c * d &= (a^n * x) * (y * a^n) \\ &= (a^n * y) * (x * a^n) \\ &= (a^n * y) * c \\ &= (y * a^n) * c \\ &= (c * a^n) * y \\ &= a^n * y = d. \end{aligned}$$

We have $c = d = c * d$. Thus, $neut(a^n)$ is unique and $neut(a^n) * neut(a^n) = neut(a^n)$. □

Proposition 3. *Let $(N, *)$ be a GAG-NET-Loop, $a \in N$ and $(a, neut(a^n), anti(a^n))$ is a generalized neutrosophic extended triplet with degree n. Then,*

(1) *$(a^n * a^n) * a^n = a^n * (a^n * a^n)$.*
(2) *$neut(a^n) * x = neut(a^n) * y$, for any $x, y \in \{anti(a^n)\}$.*
(3) *$neut(neut(a^n)) = neut(a^n)$.*
(4) *$a^n * (x * neut(a^n)) = (x * neut(a^n)) * a^n = neut(a^n)$, for any $x \in \{anti(a^n)\}$.*
(5) *$a^n * (neut(a^n) * x) = (neut(a^n) * x) * a^n = neut(a^n)$, for any $x \in \{anti(a^n)\}$.*
(6) *$(neut(a^n) * x) * neut(a^n) = neut(a^n) * (neut(a^n) * x) = neut(a^n) * x$, for any $x \in \{anti(a^n)\}$.*
(7) *$neut(neut(a^n) * x) * anti(neut(a^n) * x) = a^n$, for any $x \in \{anti(a^n)\}$.*

Proof.

(1) For $a \in N, neut(a^n) * a^n = a^n * neut(a^n) = a^n$, we have

$$(a^n * a^n) * a^n = (a^n * a^n) * (neut(a^n) * a^n) = (a^n * neut(a^n)) * (a^n * a^n) = a^n * (a^n * a^n).$$

(2) For any $x, y \in \{anti(a^n)\}$, we have $neut(a^n) * x = (y * a^n) * x = (x * a^n) * y = neut(a^n) * y$.

(3) From Proposition 2, we have $neut(a^n)$ exists neutral element and opposite element. For any $x \in \{anti(a^n)\}$ and $y \in \{anti(neut(a^n))\}$,

$$(y * x) * a^n = (a^n * x) * y = neut(a^n) * y = neut(neut(a^n)).$$

Moreover,

$$\begin{aligned}((y * x) * a^n) * neut(a^n) &= (neut(a^n) * y) * neut(a^n) \\ &= (y * neut(a^n)) * neut(a^n) \\ &= (neut(a^n) * neut(a^n)) * y \\ &= neut(a^n) * y \\ &= neut(neut(a^n)).\end{aligned}$$

Thus, $neut(a^n) = neut(neut(a^n)) * neut(a^n) = ((y * x) * a^n) * neut(a^n) = neut(neut(a^n))$.

(4) For any $x \in \{anti(a^n)\}$, from Definition 11 and Proposition 2, we have

$$\begin{aligned}a^n * (x * neut(a^n)) &= (a^n * neut(a^n)) * (x * neut(a^n)) \\ &= (a^n * x) * (neut(a^n) * neut(a^n)) \\ &= neut(a^n) * neut(a^n) \\ &= neut(a^n),\end{aligned}$$

$$(x * neut(a^n)) * a^n = (a^n * neut(a^n)) * x = a^n * x = neut(a^n).$$

Thus, $a^n * (x * neut(a^n)) = (x * neut(a^n)) * a^n = neut(a^n)$, for any $x \in \{anti(a^n)\}$.

(5) For any $x \in \{anti(a^n)\}$, we have

$$\begin{aligned}(neut(a^n) * x) * a^n &= (neut(a^n) * x) * (neut(a^n) * a^n) \\ &= (neut(a^n) * neut(a^n)) * (x * a^n) \\ &= neut(a^n) * neut(a^n) \\ &= neut(a^n),\end{aligned}$$

$$\begin{aligned}a^n * (neut(a^n) * x) &= (neut(a^n) * a^n) * (neut(a^n) * x) \\ &= (neut(a^n) * neut(a^n)) * (a^n * x) \\ &= neut(a^n) * neut(a^n) \\ &= neut(a^n).\end{aligned}$$

Thus, $a^n * (neut(a^n) * x) = (neut(a^n) * x) * a^n = neut(a^n)$.

(6) For any $x \in \{anti(a^n)\}$, we have

$$\begin{aligned}(neut(a^n) * x) * neut(a^n) &= (neut(a^n) * x) * (a^n * x) \\ &= (neut(a^n) * a^n) * (x * x) \\ &= (a^n * neut(a^n)) * (x * x) \\ &= (a^n * x) * (neut(a^n) * x) \\ &= neut(a^n) * (neut(a^n) * x),\end{aligned}$$

$$\begin{aligned}neut(a^n) * (neut(a^n) * x) &= (x * a^n) * (neut(a^n) * x) \\ &= (x * neut(a^n)) * (a^n * x) \\ &= (x * neut(a^n)) * neut(a^n) \\ &= (neut(a^n) * neut(a^n)) * x \\ &= neut(a^n) * x.\end{aligned}$$

Thus, $(neut(a^n) * x) * neut(a^n) = neut(a^n) * (neut(a^n) * x) = neut(a^n) * x$.

(7) From (5) and (6), we have $neut(neut(a^n) * x) = neut(a^n)$, $a^n \in anti\{neut(a^n) * x\}$. From (2), $neut(a^n) * anti(a^n)$ is unique, we have

$$neut(neut(a^n) * x) * anti(neut(a^n) * x) = neut(neut(a^n) * x) * a^n = neut(a^n) * a^n = a^n.$$

□

Example 8. *Let $S = \{a, b, c, d\}$, an operation $*$ on S is defined as in Table 7. Since $neut(a) = a$, $\{anti(a)\} = \{a, b, c\}$, $neut(d) = a$, $anti(d) = d$ and $b^2 = a$, $c^2 = a$, so $(S, *)$ is a GAG-NET-Loop. We can get that (Corresponding to the results of Proposition 3):*

Table 7. A GAG-NET-Loop of Example 8.

$*$	a	b	c	d
a	a	a	a	d
b	a	a	c	d
c	a	b	a	d
d	d	d	d	a

(1) *Being $(b^2 * b^2) * b^2 = b^2 * (b^2 * b^2)$, $(d^1 * d^1) * d^1 = d^1 * (d^1 * d^1)$, that is $(a^n * a^n) * a^n = a^n * (a^n * a^n)$.*

(2) *Being $a * a = a * b = a * c$, that is for any $x, y \in \{anti(c^2)\}$, $neut(c^2) * x = neut(c^2) * y$.*

(3) *Being $neut(neut(a^1)) = neut(a^1) = a$, $neut(neut(d^1)) = neut(d^1) = a$, $neut(neut(b^2)) = neut(b^2) = a$, $neut(neut(c^2)) = neut(c^2) = a$, that is $neut(neut(a^n)) = neut(a^n)$.*

(4) *Being $c^2 * (a * neut(c^2)) = a$, $(a * neut(c^2)) * c^2 = a = neut(c^2)$, $c^2 * (b * neut(c^2)) = a$, $(b * neut(c^2)) * c^2 = a = neut(c^2)$, $c^2 * (c * neut(c^2)) = a$, $(c * neut(c^2)) * c^2 = a = neut(c^2)$, that is $c^2 * (x * neut(c^2)) = (x * neut(c^2)) * c^2 = neut(c^2)$, for any $x \in \{anti(c^2)\}$. Being $d^1 * (d * neut(d^1)) = a$, $(d * neut(d^1)) * d^1 = a = neut(d^1)$, that is $d^1 * (x * neut(d^1)) = (x * neut(d^1)) * d^1 = neut(d^1)$, for any $x \in \{anti(d^1)\}$.*

(5) *Being $c^2 * (neut(c^2) * a) = a$, $(neut(c^2) * a) * c^2 = a = neut(c^2)$, $c^2 * (neut(c^2) * b) = a$, $(neut(c^2) * b) * c^2 = a = neut(c^2)$, $c^2 * (neut(c^2) * c) = a$, $(neut(c^2) * c) * c^2 = a = neut(c^2)$, that is $c^2 * (neut(c^2) * x) = (neut(c^2) * x) * c^2 = neut(c^2)$, for any $x \in \{anti(c^2)\}$. Being $d^1 * (neut(d^1) * d) = a$, $(neut(d^1) * d) * d^1 = a = neut(d^1)$, that is $d^1 * (neut(d^1) * x) = (neut(d^1) * x) * d^1 = neut(d^1)$, for any $x \in \{anti(d^1)\}$.*

(6) *Being $neut(c^2) * a = a$, $(neut(c^2) * a) * neut(c^2) = a$, $neut(c^2) * (neut(c^2) * a) = a$; $neut(c^2) * b = a$, $(neut(c^2) * b) * neut(c^2) = a$, $neut(c^2) * (neut(c^2) * b) = a$; $neut(c^2) * c = a$, $(neut(c^2) * c) * neut(c^2) = a$, $neut(c^2) * (neut(c^2) * a) = a$; that is $(neut(c^2) * x) * neut(c^2) = neut(c^2) * (neut(c^2) * x) = neut(c^2) * x$, for any $x \in \{anti(c^2)\}$. Being $neut(d^1) * d = d$, $(neut(d^1) * d) * neut(d^1) = d$, $neut(d^1) * (neut(d^1) * d) = d$, that is $(neut(d^1) * x) * neut(d^1) = neut(d^1) * (neut(d^1) * x) = neut(d^1) * x$, for any $x \in \{anti(d^1)\}$.*

(7) *Being $neut(neut(c^2) * a) * anti(neut(c^2) * a) = a = c^2$; $neut(neut(c^2) * b) * anti(neut(c^2) * b) = a = c^2$; $neut(neut(c^2) * c) * anti(neut(c^2) * c) = a = c^2$; that is $neut(neut(c^2) * x) * anti(neut(c^2) * x) = c^2$, for any $x \in \{anti(c^2)\}$. Being $neut(neut(d^1) * d) * anti(neut(d^1) * d) = d^1$, that is $neut(neut(d^1) * x) * anti(neut(d^1) * x) = d^1$, for any $x \in \{anti(d^1)\}$.*

Proposition 4. *Let $(N, *)$ be a GAG-NET-Loop, then $\forall a, b \in N$, there are two positive integers n and m such that the following hold:*

(1) *$neut(a^n) * neut(b^m) = neut(a^n * b^m)$.*

(2) *$anti(a^n) * anti(b^m) \in \{anti(a^n * b^m)\}$.*

Proof. Being $(N,*)$ be a GAG-NET-Loop, then for $a \in N$, there is a positive integer n, such that a^n has the neutral element and opposite element, denoted by $neut(a^n)$ and $anti(a^n)$, respectively. For $b \in N$, there is a positive integer m, such that b^m has the neutral element and opposite element, denoted by $neut(b^m)$ and $anti(b^m)$, respectively. Thus,

$$\begin{aligned}(neut(a^n) * neut(b^m)) * (a^n * b^m) &= (neut(a^n) * a^n) * (neut(b^m) * b^m) \\ &= a^n * b^m.\end{aligned}$$

In the same way, we have $(a^n * b^m) * (neut(a^n) * neut(b^m)) = a^n * b^m$.
That is,

$$(a^n * b^m) * (neut(a^n) * neut(b^m)) = (neut(a^n) * neut(b^m)) * (a^n * b^m) = a^n * b^m.$$

Moreover, for any $anti(a^n) \in \{anti(a^n)\}$ and $anti(b^m) \in \{anti(b^m)\}$, we can get

$$\begin{aligned}(anti(a^n) * anti(b^m)) * (a^n * b^m) &= (anti(a^n) * a^n) * (anti(b^m) * b^m) \\ &= neut(a^n) * neut(b^m).\end{aligned}$$

Similarly, we have $(a^n * b^m) * (anti(a^n) * anti(b^m)) = neut(a^n) * neut(b^m)$. That is:

$$(a^n * b^m) * (anti(a^n) * anti(b^m)) = (anti(a^n) * anti(b^m)) * (a^n * b^m) = neut(a^n) * neut(b^m).$$

Thus, we have

$$neut(a^n) * neut(b^m) \in \{neut(a^n * b^m)\}.$$

From this, by Proposition 2, we get $neut(a^n) * neut(b^m) = neut(a^n * b^m)$. Therefore, we get $anti(a^n) * anti(b^m) \in \{anti(a^n * b^m)\}$. □

Example 9. *Apply the $(S,*)$ in Example 8, since $neut(a) = a, \{anti(a)\} = \{a,b,c\}, neut(d) = a, anti(d) = d$ and $b^2 = a, c^2 = a$, so $(S,*)$ is a GAG-NET-Loop, we can get:*

(1) Being $neut(c^2) * neut(d^1) = a$, $neut(c^2 * d^1) = a$, that is $neut(c^2) * neut(d^1) = neut(c^2 * d^1)$.
(2) Being $a * d = b * d = c * d = d$, that is $anti(c^2) * anti(d^1) \in \{anti(c^2 * d^1)\}$

Theorem 7. *Let $(N,*)$ be a GAG-NET-Loop. Then, N is a quasi regular AG-groupoid.*

Proof. For any a in N, by Definition 11 we have $(a^n * anti(a^n)) * a^n = neut(a^n) * a^n = a^n$. From this and Definition 4, we know that N is a quasi regular AG-groupoid. □

The following example shows that a quasi regular AG-groupoid may not be a GAG-NET-loop.

Example 10. *Apply the $(S,*)$ in Example 1, Being $1 = (1*3)*1, 2 = (2*4)*2, 3 = (3*1)*3, 4 = (4*2)*4$, From Definition 4, S is a quasi regular AG-groupoid. However, it is not a GAG-NET-Loop.*

Theorem 8. *Let $(N,*)$ be a GAG-NET-Loop. Then, N is a quasi fully regular AG-groupoid.*

Proof. Suppose $a \in N$ and $(a, neut(a^n), anti(a^n))$ is a generalized neutrosophic extended triplet with degree n, then there exists $m \in \{anti(a^n)\}, a^n * m = m * a^n = neut(a^n)$. Denote $p = m * neut(a^n), q = neut(a^n); r = m, s = neut(a^n); t = m, u = neut(a^n); v = m; w = m * neut(a^n); x = m, y = neut(a^n)$, then

$$
\begin{aligned}
(p * (a^n)^2) * q &= ((m * neut(a^n)) * (a^n)^2) * neut(a^n) \\
&= (((a^n)^2 * neut(a^n)) * m) * neut(a^n) \quad (by\ the\ left\ invertive\ law) \\
&= (((a^n * a^n) * neut(a^n)) * m) * neut(a^n) \\
&= (((neut(a^n) * a^n) * a^n) * m) * neut(a^n) \quad (by\ the\ left\ invertive\ law) \\
&= ((a^n * a^n) * m) * neut(a^n) \\
&= ((m * a^n) * a^n) * neut(a^n) \quad (by\ the\ left\ invertive\ law) \\
&= (neut(a^n) * a^n) * neut(a^n) \\
&= a^n * neut(a^n) = a^n,
\end{aligned}
$$

$$(r * a^n) * (a^n * s) = (m * a^n) * (a^n * neut(a^n)) = neut(a^n) * a^n = a^n,$$

$$(a^n * t) * (a^n * u) = (a^n * m) * (a^n * neut(a^n)) = neut(a^n) * a^n = a^n,$$

$$(a^n * a^n) * v = (a^n * a^n) * m = (m * a^n) * a^n = neut(a^n) * a^n = a^n,$$

$$
\begin{aligned}
w * (a^n * a^n) &= (m * neut(a^n)) * (a^n * a^n) \\
&= (m * a^n) * (neut(a^n) * a^n) \quad (by\ the\ medial\ law) \\
&= (m * a^n) * a^n \\
&= neut(a^n) * a^n = a^n,
\end{aligned}
$$

$$(x * a^n) * (y * a^n) = (m * a^n) * (neut(a^n) * a^n) = neut(a^n) * a^n = a^n.$$

Moreover, from Proposition 4, we get:

$$neut(a^n) * neut(b^m) = neut(a^n * b^m), anti(a^n) * anti(b^m) \in \{anti(a^n * b^m)\}.$$

If $b^m = a^n$, we have $neut(a^n) * neut(a^n) = neut(a^n * a^n)$, $anti(a^n) * anti(a^n) \in \{anti(a^n * a^n)\}$, there exists $k \in \{anti(a^n * a^n)\}$. Denote $z = k * m$, then

$$
\begin{aligned}
((a^n)^2 * z) * (a^n)^2 &= ((a^n * a^n) * z) * (a^n)^2 \\
&= ((z * a^n) * a^n) * (a^n)^2 \quad (applying\ the\ left\ invertive\ law) \\
&= ((a^n)^2 * a^n) * (z * a^n) \quad (applying\ the\ left\ invertive\ law) \\
&= ((a^n)^2 * a^n) * ((k * m) * a^n) \\
&= ((a^n)^2 * a^n) * ((a^n * m) * k) \quad (by\ the\ left\ invertive\ law) \\
&= ((a^n)^2 * a^n) * (neut(a^n) * k) \quad (by\ m \in \{anti(a^n)\}) \\
&= ((a^n * a^n) * (neut(a^n) * a^n)) * (neut(a^n) * k) \\
&= ((a^n * neut(a^n)) * (a^n * a^n)) * (neut(a^n) * k) \quad (applying\ the\ medial\ law) \\
&= (a^n * (a^n)^2) * (neut(a^n) * k) \\
&= (a^n * neut(a^n)) * ((a^n)^2 * k) \quad (applying\ the\ medial\ law) \\
&= a^n * neut(a^n * a^n) \quad (by\ the\ definition\ of\ k \in \{anti(a^n * a^n)\}) \\
&= a^n * (neut(a^n) * neut(a^n)) \\
&= a^n * neut(a^n) \quad (by\ Proposition\ 2(2)) \\
&= a^n.
\end{aligned}
$$

Therefore, combining above results, by Definition 5, we know that N is a quasi fully regular AG-groupoid. □

The following example shows that a quasi fully regular AG-groupoid may not be a GAG-NET-loop.

Example 11. *Applying the* $(S,*)$ *in Example 1, when* $a = 1, p = 1, q = 3, r = 4, s = 3, t = 2, u = 3, v = 2, w = 2, x = 4, y = 2, z = 3$, *we have* $a^2 = 2$, *and*

$$\begin{aligned}1 &= (1*2)*3 = (4*1)*(1*3) = (1*2)*(1*3)\\ &= (1*1)*2 = 2*(1*1) = (4*1)*(2*1)\\ &= (2*3)*2.\end{aligned}$$

When $a = 4, p = 1, q = 3, r = 4, s = 4, t = 3, u = 2, v = 3, w = 3, x = 4, y = 4, z = 2$, *we have* $a^2 = 3$, *and*

$$\begin{aligned}4 &= (1*3)*3 = (4*4)*(4*4) = (4*3)*(4*2)\\ &= (4*4)*3 = 3*(4*4) = (4*4)*(4*4)\\ &= (3*2)*3.\end{aligned}$$

Being $2^2 = 1, 3^3 = 1$, *from Definition 5, S is a quasi fully regular AG-groupoid. However, it is not a GAG-NET-Loop.*

Definition 13. *An AG-groupoid* $(S,*)$ *is called a quasi strong inverse AG-groupoid, if the following conditions are satisfied: for any* $a \in S$, *there exists a positive integer n,* $a^n \in S$, *and a unary operation* $a^n \to (a^n)^{-1}$ *on S such that*

$$((a^n)^{-1})^{-1} = a^n, \ \ (a^n * (a^n)^{-1}) * a^n = a^n * (a^n * (a^n)^{-1}) = a^n, \ \ a^n * (a^n)^{-1} = (a^n)^{-1} * a^n.$$

Theorem 9. *Let* $(N,*)$ *be a groupoid. Then, N is a GAG-NET-Loop if and only if it is a quasi strong inverse AG-groupoid.*

Proof. Necessity: Suppose N is a GAG-NET-Loop, from Definition 12, for each $a \in N$, there exists a generalized neutrosophic extended triplet with degree n denoted by $(a, neut(a^n), anti(a^n))$. Set

$$(a^n)^{-1} = neut(a^n) * anti(a^n),$$

by Proposition 3(2), $neut(a^n) * anti(a^n)$ is unique, so $(a^n)^{-1}$ is unique. By Proposition 3(7), we have

$$((a^n)^{-1})^{-1} = neut(neut(a^n) * anti(a^n)) * anti(neut(a^n) * anti(a^n)) = a^n.$$

Being

$$(a^n)^{-1} * a^n = (neut(a^n) * anti(a^n)) * a^n = (a^n * anti(a^n)) * neut(a^n) = neut(a^n) * neut(a^n) = neut(a^n),$$

$$\begin{aligned}a^n * (a^n)^{-1} &= a^n * (neut(a^n) * anti(a^n))\\ &= (neut(a^n) * a^n) * (neut(a^n) * anti(a^n))\\ &= (neut(a^n) * neut(a^n)) * (a^n * anti(a^n))\\ &= neut(a^n),\end{aligned}$$

we have

$$(a^n)^{-1} * a^n = a^n * (a^n)^{-1},$$

$$(a^n * (a^n)^{-1}) * a^n = neut(a^n) * a^n = a^n,$$

$$a^n * (a^n * (a^n)^{-1}) = a^n * neut(a^n) = a^n,$$

$$(a^n * (a^n)^{-1}) * a^n = a^n * (a^n * (a^n)^{-1}) = a^n.$$

From Definition 13, N is a quasi strong inverse AG-groupoid.

Sufficiency: If N is a quasi strong inverse AG-groupoid, and $(a^n)^{-1} \in N$, such that $a^n * (a^n)^{-1} = (a^n)^{-1} * a^n$ and $(a^n * (a^n)^{-1}) * a^n = a^n * (a^n * (a^n)^{-1}) = a^n$. Set

$$neut(a^n) = a^n * (a^n)^{-1},$$

then $neut(a^n) * a^n = (a^n * (a^n)^{-1}) * a^n = a^n * (a^n * (a^n)^{-1}) = a^n * neut(a^n) = a^n$,

$$a^n * (a^n)^{-1} = (a^n)^{-1} * a^n = neut(a^n).$$

From Definition 12, we have that N is a GAG-NET-Loop and $(a^n)^{-1} \in \{anti(a^n)\}$. □

Example 12. *Applying $(S, *)$ in Example 8, we know that it is a GAG-NET-Loop. We will show that it is a quasi strong inverse AG-groupoid in the following.*

For d, there exists an inverse element $d^{-1} = d$, such that $(d^{-1})^{-1} = d$, $(d * d^{-1}) * d = d * (d * d^{-1}) = d$, $d * d^{-1} = d^{-1} * d = a$, so d is quasi strong inverse. a is quasi strong inverse for the same reason. Moreover, being $b^2 = a, c^2 = a$, b and c are quasi strong inverse, thus $(S, *)$ is a quasi strong inverse AG-groupoid by Definition 13.

Definition 14. *Let $(N, *)$ be a GAG-NET-Loop. N is called a weak commutative GAG-NET-Loop if $\forall a, b \in N$, there exist a generalized neutrosophic extended triplet with degree n (denoted by $(a, neut(a^n), anti(a^n))$) and a generalized neutrosophic extended triplet with degree m (denoted by $(b, neut(b^m), anti(b^m))$), $n, m \in Z^+$, $a^n * neut(b^m) = neut(b^m) * a^n$.*

Example 13. *Let $S = \{1, 2, 3, 4, 5, 6, 7\}$, an operation $*$ on S is defined as in Table 8. Since $(1, 1, 1), (2, 2, 2)$ and $(6, 6, 6)$ are neutrosophic extended triplets, but 3, 4, 5, 7 do not have the neutral element and opposite element, thus S is not an AG-NET-Loop. Moreover $3^2 = 1, 4^2 = 1, 5^2 = 2, 7^2 = 6$ have the neutral element and opposite element, so $(S, *)$ is a GAG-NET-Loop. It is not a commutative GAG-NET-Loop being $3 * 1 \neq 1 * 3$. We can show that it is a weak commutative GAG-NET-Loop.*

For 1, 2, 3, 4, 5, 6 and 7, there exist positive integers 1, 1, 2, 2, 2, 1 and 2, respectively, thus $S' = \{1^1, 2^1, 3^2, 4^2, 5^2, 6^1, 7^2\} = \{1, 2, 6\}$ being $3^2 = 1, 4^2 = 1, 5^2 = 2, 7^2 = 6$. We know that $neut(1) = 1, neut(2) = 2, neut(6) = 6$, thus $\{neut(1), neut(2), neut(6)\} \subseteq S'$. In Table 8, we can get the sub algebra system $(S', *)$ of $(S, *)$ as in Table 9, and $(S', *)$ is commutative. Thus, $(S, *)$ is a weak commutative GAG-NET-Loop.

Table 8. The operation table of Example 13.

*	1	2	3	4	5	6	7
1	1	2	3	4	5	6	7
2	2	2	2	2	2	2	2
3	4	2	1	3	5	6	7
4	3	2	4	1	5	6	7
5	5	2	5	5	2	2	2
6	6	2	6	6	2	6	6
7	7	2	7	7	2	6	6

Table 9. The sub algebra system S' of S in Example 13.

*	1	2	6
1	1	2	6
2	2	2	2
6	6	2	6

Example 14. *Let $S = \{1,2,3,4\}$, an operation $*$ on S is defined as in Table 10. Being $neut(1) * 2 = 4 \neq 3 = 2 * neut(1)$, S is not a weak commutative GAG-NET-Loop. Moreover, it is not a commutative AG-NET-Loop.*

Table 10. The operation table of Example 14.

*	1	2	3	4
1	1	4	2	3
2	3	2	4	1
3	4	1	3	2
4	2	3	1	4

Proposition 5. *Let $(N,*)$ be a GAG-NET-Loop. Then, $(N,*)$ is a weak commutative GAG-NET-Loop if and only if N satisfies the following conditions: $\forall a, b \in N$, there exist a generalized neutrosophic extended triplet with degree n (denoted by $(a, neut(a^n), anti(a^n))$) and a generalized neutrosophic extended triplet with degree m (denoted by $(b, neut(b^m), anti(b^m))$), $n, m \in Z^+$, $a^n * b^m = b^m * a^n$.*

Proof. Necessity: If $(N,*)$ is a weak commutative GAG-NET-Loop, then there are two positive integers n, m, such that a^n and b^m have the neutral element and opposite element. Thus, from Definition 14, $\forall a, b \in N$, we have

$$\begin{aligned} a^n * b^m &= (neut(a^n) * a^n) * (b^m * neut(b^m)) \\ &= (neut(a^n) * b^m) * (a^n * neut(b^m)) \\ &= (b^m * neut(a^n)) * (neut(b^m) * a^n) \\ &= (b^m * neut(b^m)) * (neut(a^n) * a^n) \\ &= b^m * a^n. \end{aligned}$$

Sufficiency: If $(N,*)$ is a GAG-NET-Loop, then for $a \in N$, there is a positive integer n, such that a^n has the neutral element and opposite element, denoted by $neut(a^n)$ and $anti(a^n)$, respectively. For $b \in N$, there is a positive integer m, such that b^m has the neutral element and opposite element, denoted by $neut(b^m)$ and $anti(b^m)$, respectively. Suppose that $(N,*)$ satisfies the conditions $a^n * b^m = b^m * a^n$, From Proposition 2, we have $neut(b^m)$ exists neutral element and opposite element. We get $a^n * neut(b^m) = neut(b^m) * a^n$. From Definition 14, we know that $(N,*)$ is a weak commutative GAG-NET-Loop. □

Definition 15. *A GAG-NET-Loop $(S,*)$ is called a quasi Clifford AG-groupoid, if it is a quasi strong inverse AG-groupoid and for any $a, b \in S$, there are two positive integers n, m such that*

$$a^n * (b^m * (b^m)^{-1}) = (b^m * (b^m)^{-1}) * a^n.$$

Theorem 10. *Let $(N,*)$ be a groupoid. Then, N is a weak commutative GAG-NET-Loop if and only if it is a quasi Clifford AG-groupoid.*

Proof. Necessity: Suppose that N is a weak commutative GAG-NET-Loop. By Theorem 9, we know that N is a quasi strong inverse AG-groupoid, then $\forall a, b \in N$ there are two positive integers n, m, such that a^n and b^m have the neutral element and opposite element. Set

$$(a^n)^{-1} = neut(a^n) * anti(a^n).$$

For any $a, b \in N$, we have

$$a^n * (b^m * (b^m)^{-1}) = a^n * neut(b^m) = neut(b^m) * a^n = (b^m * (b^m)^{-1}) * a^n.$$

From Definition 15, we know that N is a quasi Clifford AG-groupoid.

Sufficiency: Assume that N is a quasi Clifford AG-groupoid, from Definition 15, it is a quasi strong inverse AG-groupoid. By Theorem 9, we know that N is a GAG-NET-Loop. Then, $\forall a, b \in N$ there are two positive integers n, m, such that a^n and b^m have the neutral element and opposite element, $(a^n)^{-1} \in N, (b^m)^{-1} \in N$. Set

$$neut(a^n) = a^n * (a^n)^{-1}, neut(b^m) = b^m * (b^m)^{-1}.$$

From Definition 15, being $a^n * (b^m * (b^m)^{-1}) = (b^m * (b^m)^{-1}) * a^n$, we have $a^n * neut(b^m) = neut(b^m) * a^n$. We can get that N is a weak commutative GAG-NET-Loop by Definition 14. □

Example 15. *Let $S = \{1,2,3,4,5,6,7,8\}$, an operation $*$ on S is defined as in Table 11. It is a weak commutative GAG-NET-Loop. We show that it is a quasi Clifford AG-groupoid. From Theorem 9, we can see that $(S,*)$ is a quasi strong inverse AG-groupoid. We just show for any $x, y \in S$, there are two positive integers n and m such that $x^n * (y^m * (y^m)^{-1}) = (y^m * (y^m)^{-1}) * x^n$.*

In Example 15, 1, 2, 3, 4, 5, 6, 7 and 8, there exist positive integers 1, 1, 2, 2, 2, 1, 2 and 2, respectively, and set $1^{-1} = 1, 2^{-1} = 2, (3^2)^{-1} = 1, (4^2)^{-1} = 1, (5^2)^{-1} = 2, 6^{-1} = 6, (7^2)^{-1} = 6, (8^2)^{-1} = 6$. For any $x, y \in \{1^1, 2^1, 3^2, 4^2, 5^2, 6^1, 7^2, 8^2\}$, without losing generality, let $x = 1, y = 2$, we can get $1^1 * (2^1 * (2^1)^{-1}) = (2^1 * (2^1)^{-1}) * 1^1 = 2$. We can verify other cases, thus $(S,*)$ is a quasi Clifford AG-groupoid.

Table 11. The operation table of Example 15.

*	1	2	3	4	5	6	7	8
1	1	2	3	4	5	6	7	8
2	2	2	2	2	2	2	2	2
3	4	2	1	3	5	6	7	8
4	3	2	4	1	5	6	7	8
5	5	2	5	5	2	2	2	2
6	6	2	6	6	2	6	6	6
7	7	2	7	7	2	6	6	6
8	8	2	8	8	2	6	6	6

Example 16. *Let $S = \{1,2,3,4,5\}$, an operation $*$ on S is defined as in Table 12. it is not a weak commutative GAG-NET-Loop. We show that there exist $x, y \in S$, for any two positive integers n and m such that $x^n * (y^m * (y^m)^{-1}) \neq (y^m * (y^m)^{-1}) * x^n$.*

In Example 16, for any $n, m \in Z^+$, $1^n = 1, 2^m = 2$ and $(1^n)^{-1} = 1, (2^m)^{-1} = 2$, but $1^n * (2^m * (2^m)^{-1}) = 4 \neq 3 = (2^m * (2^m)^{-1}) * 1^n$. That is, for $1, 2 \in S$, there are not two positive integers n, m such that $1^n * (2^m * (2^m)^{-1}) = (2^m * (2^m)^{-1}) * 1^n$. Thus, $(S,*)$ is not a quasi Clifford AG-groupoid.

Table 12. The operation table of Example 16.

*	1	2	3	4	5
1	1	4	2	3	1
2	3	2	4	1	3
3	4	1	3	2	4
4	2	3	1	4	2
5	1	4	2	3	5

5. Conclusions

We thoroughly study the GAG-NET-Loop from the perspective of the AG-groupoid theory and obtained some important results. Figures 1 and 2 give the relations of the GAG-NET-Loop and other algebraic structures.

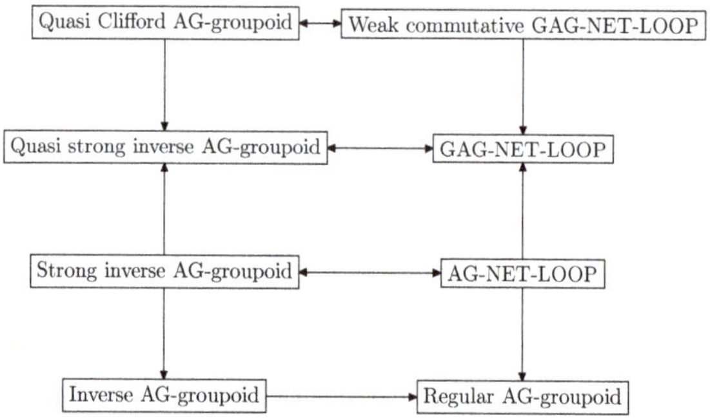

Figure 1. The relations of GAG-NET-Loop and other algebraic structures.

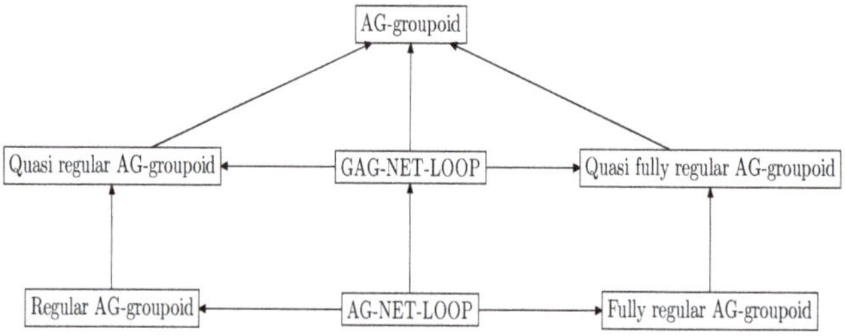

Figure 2. The relations of GAG-NET-Loop and other AG-groupoids.

As can be seen in Figure 1, we prove that the AG-NET-Loop is equal to the strong inverse AG-groupoid, the GAG-NET-Loop is equal to the quasi strong inverse AG-groupoid, and the weak commutative GAG-NET-Loop is equal to the quasi Clifford AG-groupoid.

As can be seen in Figure 2, we prove that a GAG-NET-loop is a quasi regular AG-groupoid, but a quasi regular AG-groupoid may not be a GAG-NET-loop; a GAG-NET-loop is a quasi fully regular AG-groupoid, but a quasi fully regular AG-groupoid may not be a GAG-NET-loop.

Figure 3 can be used to further express the relationships among GAG-NET-Loop and some algebraic systems. Here, as shown in Example 2, A represents a commutative AG-NET-Loop; as shown in Example 15, B represents a weak commutative GAG-NET-Loop, but it is not an AG-NET-Loop; as i s shown in Example 14, C represents a non-commutative AG-NET-Loop; D represents a GAG-NET- Loop, but it is neither an AG-NET-Loop nor a weak commutative GAG-NET-Loop; as shown in Example 10, E represents a quasi regular AG-groupoid, but it is not a GAG-NET-Loop; and as shown in Example 11, F represents a quasi fully regular AG-groupoid, but it is not a GAG-NET-Loop. A+B represents a weak commutative GAG-NET-Loop, A+C represents an AG-NET-Loop, A+B+C+D represents a

GAG-NET-Loop, A+B+C+D+E represents a quasi regular AG-groupoid, and A+B+C+D+F represents a quasi fully regular AG-groupoid.

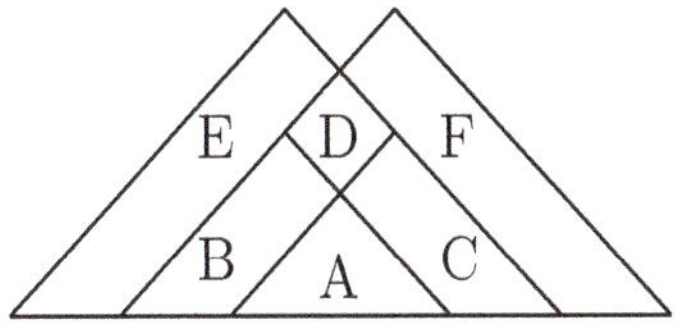

Figure 3. The relationships among some algebraic systems and GAG-NET-Loop.

All these results are interesting for the exploration of the structure characterization of GAG-NET-Loop. As the next research topics, we want to find some special GAG-NET-Loops which can be decomposed into some smaller GAG-NET-Loops, and explore the relationship between these special GAG-NET-Loops and the related AG-groupoids.

Author Contributions: Conceptualization, Funding acquisition, Writing—review and editing, X.Z.; Data curation, Software, Writing—original draft, Writing–review and editing, X.A.; Resources, Writing–review and editing, Y.M.

Funding: This research was funded by National Natural Science Foundation of China (Grant No. 61976130).

Conflicts of Interest: The authors declare no conflict of interest.

References

1. Kazim, M.; Naseeruddin, M. On almost semigroups. *Aligarh Bull. Math.* **1972**, *2*, 1–7.
2. Holgate, P. Groupoids satisfying a simple invertive law. *Math. Stud.* **1992**, *61*, 101–106.
3. Mushtaq, Q.; Iqbal, Q. Decomposition of a locally associative LA-semigroup. *Semigroup Forum* **1990**, *41*, 155–164. [CrossRef]
4. Proti, P.V.; Stevanovi, N. AG-test and some general properties of Abel-Grassmann's groupoids. *Pure Math. Appl.* **1995**, *6*, 371–383.
5. Khan, M.; Faisal; Amjad, V. On some classes of Abel-Grassmann's groupoids. *J. Adv. Res. Pure Math.* **2011**, *3*, 109–119. [CrossRef]
6. Faisal; Khan, A.; Davvaz, B. On fully regular AG-groupoids. *Afr. Mat.* **2014**, *25*, 449–459. [CrossRef]
7. Khan, M.; Anis, S. On semilattice decomposition of an Abel-Grassmann's groupoid. *Acta Math. Sin. Engl. Ser.* **2012**, *28*, 1461–1468. [CrossRef]
8. Rashad, M.; Ahmad, I.; Shah, M.; Khuhro, Z.U.A. Left transitive AG-groupoids. *Mathematics* **2014**, *46*, 547–552.
9. Iqbal, M.; Ahmad, I. Ideals in CA-AG-groupoids. *Indian J. Pure Appl. Math.* **2018**, *49*, 265–284. [CrossRef]
10. Smarandache, F. Neutrosophic set—A generalization of the intuitionistic fuzzy set. *Int. J. Pure Appl. Math.* **2005**, *24*, 287–297.
11. Peng, X.; Dai, J. A bibliometric analysis of neutrosophic set: two decades review from 1998 to 2017. *Artif. Intell. Rev.* **2018**. [CrossRef]
12. Peng, X.; Liu, C. Algorithms for neutrosophic soft decision making based on EDAS, new similarity measure and level soft set. *J. Intell. Fuzzy Syst.* **2017**, *32*, 955–968. [CrossRef]
13. Peng, X.; Dai, J. Approaches to single-valued neutrosophic MADM based on MABAC, TOPSIS and new similarity measure with score function. *Neural Comput. Appl.* **2018**, *29*, 939–954. [CrossRef]
14. Abdel-Basset, M.; Chang, V.; Mohamed, M.; Smarandche, F. A refined approach for forecasting based on neutrosophic time series. *Symmetry* **2019**, *11*, 457. [CrossRef]
15. Abdel-Basset, M.; Mohamed, M.; Chang, V.; Smarandache, F. IoT and its impact on the electronics market: a powerful decision support system for helping customers in choosing the best product. *Symmetry* **2019**, *11*, 611. [CrossRef]

16. Broumi, S.; Dey, A.; Talea, M.; Bakali, A.; Smarandache, F.; Nagarajan, D.; Lathamaheswari, M.; Kumar, R. Shortest path problem using Bellman algorithm under neutrosophic environment. *Complex Intell. Syst.* **2019**, *5*, 409–416. [CrossRef]
17. Broumi, S.; Bakali, A.; Talea, M.; Smarandache, F.; Dey, A.; Son, L.H. Spanning tree problem with neutrosophic edge weights. *Procedia Comput. Sci.* **2018**, *127*, 190–199. [CrossRef]
18. Boloş, M.I.; Bradea, I.A.; Delcea, C. Neutrosophic portfolios of financial assets. minimizing the risk of neutrosophic portfolios. *Mathematics* **2019**, *7*, 1046. [CrossRef]
19. Zhang, X.; Bo, C.; Smarandache, F.; Dai, J. New inclusion relation of neutrosophic sets with applications and related lattice structure. *Int. J. Mach. Learn. Cybern.* **2018**, *9*, 1753–1763. [CrossRef]
20. Zhang, X.; Bo, C.; Smarandache, F.; Park, C. New operations of totally dependent-neutrosophic sets and totally dependent-neutrosophic soft sets. *Symmetry* **2018**, *10*, 187. [CrossRef]
21. Zhang, X.; Smarandache, F.; Liang, X. Neutrosophic duplet semi-group and cancellable neutrosophic triplet groups. *Symmetry* **2017**, *9*, 275. [CrossRef]
22. Zhang, X.; Wang, X.; Smarandache, F.; Jaiyeola, T.G.; Lian, T. Singular neutrosophic extended triplet groups and generalized groups. *Cogn. Syst. Res.* **2019**, *57*, 32–40. [CrossRef]
23. Zhang, X.; Borzooei, R.; Jun, Y. Q-filters of quantum B-algebras and basic implication algebras. *Symmetry* **2018**, *10*, 573. [CrossRef]
24. Zhang, X.; Mao, X.; Wu, Y.; Zhai, X. Neutrosophic filters in pseudo-BCI algebras. *Int. J. Uncertain. Quan.* **2018**, *8*, 511–526. [CrossRef]
25. Zhang, X.; Wu, X.; Mao, X.; Smarandache, F.; Park, C. On neutrosophic extended triplet groups (loops) and Abel-Grassmann's groupoids (AG-groupoids). *J. Intell. Fuzzy Syst.* **2019**, *37*, 5743–5753. [CrossRef]
26. Ma, Y.; Zhang, X.; Yang, X.; Zhou, X. Generalized neutrosophic extended triplet group. *Symmetry* **2019**, *11*, 327. [CrossRef]
27. Shah, M.; Ahmad, I.; Ali, A. On quasi-cancellativity of AG-groupoids. *Int. J. Contemp. Math. Sci.* **2012**, *7*, 2065–2070.
28. Ahmad, I.; Ahmad, I.; Rashad, M. A study of anti-commutativity in AG-groupoids. *J. Math.* **2016**, *48*, 99–109.
29. Smarandache, F. *Neutrosophic Perspectives: Triplets, Duplets, Multisets, Hybrid Operators, Modal Logic, Hedge Algebras and Applications*; Pons Publishing House: Brussels, Belgium, 2017.
30. Khan, M.; Anis, S. An analogy of Clifford decomposition theorem for Abel-Grassmann groupoids. *Algebra Colloq.* **2014**, *21*, 347–353. [CrossRef]
31. Wu, X.; Zhang, X. The decomposition theorems of AG-neutrosophic extended triplet loops and strong AG-(l, l)-loops. *Mathematics* **2019**, *7*, 268. [CrossRef]

mathematics MDPI

Article

Regular CA-Groupoids and Cyclic Associative Neutrosophic Extended Triplet Groupoids (CA-NET-Groupoids) with Green Relations

Wangtao Yuan and Xiaohong Zhang *

Department of Mathematics, Shaanxi University of Science & Technology, Xi'an 710021, China; 1809007@sust.edu.cn

* Correspondence: zhangxiaohong@sust.edu.cn

Received: 18 December 2019; Accepted: 3 February 2020; Published: 6 February 2020

Abstract: Based on the theories of AG-groupoid, neutrosophic extended triplet (NET) and semigroup, the characteristics of regular cyclic associative groupoids (CA-groupoids) and cyclic associative neutrosophic extended triplet groupoids (CA-NET-groupoids) are further studied, and some important results are obtained. In particular, the following conclusions are strictly proved: (1) an algebraic system is a regular CA-groupoid if and only if it is a CA-NET-groupoid; (2) if $(S, *)$ is a regular CA-groupoid, then every element of S lies in a subgroup of S, and every $\mathcal{H}$-class in S is a group; and (3) an algebraic system is an inverse CA-groupoid if and only if it is a regular CA-groupoid and its idempotent elements are commutative. Moreover, the Green relations of CA-groupoids are investigated, and some examples are presented for studying the structure of regular CA-groupoids.

Keywords: semigroup; CA-groupoid; regular CA-groupoid; neutrosophic extended triplet (NET); Green relation

1. Introduction

The theory of group is an essential branch of algebra. The research of group has become an important trend in the theory of semigroup. Various algebraic structures are related to groups, such as regular semigroups, generalized groups, and neutrosophic extended triplet groups (see [1–6]). With the development of semigroup, the study of generalized regular semigroup has become an important topic. This paper focuses on the regularity of non-associative algebraic structures satisfying the cyclic associative law: $x(yz) = z(xy)$.

As early as 1954, Sholander [7] used the term of cyclic associative law to express the following operation law: $(ab)c = (bc)a$. Obviously, its dual form is as follows: $a(bc) = c(ab)$. At the same time, in 1954, Hosszu also used the term of cyclic associative law in the study of functional equation (see the introduction and explanation by Maksa [8]). In 1995, Kleinfeld [9] studied the rings with cyclic associative law $x(yz) = y(zx)$. Moreover, Zhan and Tan [10] introduced the notion of left weakly Novikov algebra. In many fields (such as non-associative rings and non-associative algebras [11–14]), image processing [15], and networks [16]), non-associativity has essential research significance. Since cyclic associative law is widely used in algebraic systems, we have been focusing on the basic algebraic structure of cyclic associative groupoids (CA-groupoids) and other relevant algebraic structures (see [17,18]).

Smarandache first proposed the new concept of neutrosophic set in [19]. The theory of neutrosophic set has been applied in many fields, such as applying neutrosophic soft sets in decision making, and proposing a new model of similarity in medical diagnosis and verifying its validity of l through a numerical example with practical background [20]. Later, Smarandache and colleagues extended the neutrosophic logic to the neutrosophic extended triplet group (NETG) [6]. In this paper, we analyze the structure of cyclic associative neutrosophic extended triplet groupoids (CA-NET-Groupoids).

Green's relations, first studied by Green [21] in 1951, have played a fundamental role in the development of regular semigroup theory. This has in turn completely illustrated the effectiveness of Green's method in studying semigroups, especially regular semigroups. Research on the Green relations of regular semigroups is at the core, and it involves almost all aspects of semigroup algebra theory. In 2011, Mary [22] studied the generalized inverse of semigroups by means of Green's relations. In 2017, Kufleitner and Manfred [23] considered the complexity of Green's relations when the semigroup is given by transformations on a finite set. This paper focuses on the Green's relations of CA-groupoids, in particular regular CA-groupoids. Recently, we analyzed these new results and studied them from the perspective of CA-groupoid theory. Miraculously, we obtained some unexpected results that, if S is a regular CA-groupoid, then every element of S lies in a subgroup of S, and every $\mathcal{H}$-class in S is a group.

The rest of this paper is organized as follows. In Section 2, we give the related concepts and results of the CA-groupoid. In Section 3, we give some basic concepts and examples of regular elements, strongly regular elements, inverse elements, and local associative and quasi-regular elements. In Section 4, we prove the equivalence of regular CA-groupoids and CA-NET-groupoids, and give corresponding examples. In Section 5, we discuss the Green's relations of CA-groupoids and the Green's relations of regular CA-groupoids. In Section 6, we propose a new concept of inverse CA-groupoids and prove that regular CA-groupoids, strongly regular CA-groupoids, CA-NET-groupoids, inverse CA-groupoids and commutative regular semigroups are equivalent. Finally, the summary and plans for future work are presented in Section 7.

2. Preliminaries

In this section, we give the related research and results of the CA-groupoid. Some related notions are introduced.

A groupoid is a pair $(S, \times)$ where S is a non-empty set together with a binary operation $\times$. Traditionally, the $\times$ operator is omitted without confusion.

Definition 1. *([4,5]) A groupoid $(S, \times)$ is called a neutrosophic extended triplet-groupoid NET-groupoid) if, for any $a \in S$, there exist a neutral of "a" (denoted by neut(a)), and an opposite of "a" (denoted by anti(a)), such that neut(a)$\in$S, anti(a)$\in$S, and:*

$$a \times neut(a) = neut(a) \times a = a; a \times anti(a) = anti(a) \times a = neut(a)$$

The triplet (a, neut(a), anti(a)) is called a neutrosophic extended triplet.

Let $(S, \times)$ be a groupoid. Some concepts are defined as follows:

(1) An element $a \in S$ is called idempotent if $a^2 = a$.
(2) S is called semigroup if, for any $a, b, c \in S$, $a \times (b \times c) = (a \times b) \times c$. A semigroup $(S, \times)$ is commutative if, for all $a, b \in S$, $a \times b = b \times a$.

Here, recall some basic concepts in the semigroup theory. A non-empty subset A of a semigroup $(S, \times)$ is called a left ideal if $SA \subseteq A$, a right ideal if $AS \subseteq A$, and an ideal if it is both a left and a right ideal. If a is an element of a semigroup $(S, \times)$, the smallest left ideal containing a is $Sa \cup \{a\}$, which we may conveniently write as S^1a.

An element a of a semigroup S is called regular if there exists x in S such that $a \times x \times a = a$. The semigroup S is called regular if all its elements are regular.

Among idempotents in an arbitrary semigroup, there is a natural (partial) order relation defined by the rule that $e \leq f$ if and only if $e \times f = f \times e = e$. It is easy to verify that the given relation has the properties (reflexive), (antisymmetric) that define an order relation. Certainly, it is clear that $e \leq e$, and that $e \leq f$ and $f \leq e$ together implies that $e = f$. To show transitivity, notice that, if $e \leq f$ and $f \leq g$, so that $e \times f = f \times e = e$ and $f \times g = g \times f = f$, then $e \times g = e \times f \times g = e \times f = e$ and $g \times e = g \times f \times e = f \times e = e$, and thus $e \leq g$.

Let S be a regular semigroup and let $E(S)$ denote the set of idempotents of S. For each $e \in E(S)$, let G_e be a subgroup of S with identity e. If $T(S) = \cup (G_e: e \in E(S))$ is a subsemigroup and $e, f, g \in E(S)$, $e \geq f$, and $e \geq g$ imply $f \times g = g \times f$, we term S a strongly regular semigroup [24].

An equivalent relation $\mathcal{L}$ on S is defined by the rule that $a\mathcal{L}b$ if and only if $S^1a = S^1b$; an equivalent relation $\mathcal{R}$ on S is defined by the rule that $a\mathcal{R}b$ if and only if $aS^1 = bS^1$; denote $\mathcal{H} = \mathcal{L}\cap\mathcal{R}, \mathcal{D} = \mathcal{L}\cup\mathcal{R}$, that is, $a\mathcal{H}b$ if and only if $S^1a = S^1b$ and $aS^1 = bS^1$; $a\mathcal{D}b$ if and only if $S^1a = S^1b$ or $aS^1 = bS^1$. An equivalent relation $\mathcal{J}$ on S is defined by the rule that $a\mathcal{J}b$ if and only if $S^1aS^1 = S^1bS^1$, where:

$$S^1aS^1 = SaS \cup aS \cup Sa \cup \{a\}.$$

That is, $a\mathcal{J}b$ if and only if there exists $x, y, u, v \in S^1$ for which $x \times a \times y = b$, $u \times b \times v = a$. The $\mathcal{L}$-class ($\mathcal{R}$-class, $\mathcal{H}$-class, $\mathcal{D}$-class, $\mathcal{J}$-class) containing the element a is written $\mathcal{L}a$ ($\mathcal{R}a$, $\mathcal{H}a$, $\mathcal{D}a$, $\mathcal{J}a$).

Definition 2. *([7–10,25]) Let $(S, \times)$ be a groupoid. If, for all $a, b, c \in S$,*

$$a \times (b \times c) = c \times (a \times b),$$

then $(S, \times)$ is called a cyclic associative groupoid (shortly, CA-groupoid).

Proposition 1. *([25]) Let $(S, \times)$ be a CA-groupoid. Then, for any $a, b, c, d, x, y \in S$,*

(1) $(a \times b) \times (c \times d) = (d \times a) \times (c \times b)$; and
(2) $(a \times b) \times ((c \times d) \times (x \times y)) = (d \times a) \times ((c \times b) \times (x \times y))$.

Definition 3. *([25]) A NET-groupoid $(S, \times)$ is called cyclic associative (shortly, CA-NET-groupoid) if it is cyclic associative as a groupoid. S is called a commutative CA-NET-groupoid if, for all $a, b \in N$, $a \times b = b \times a$.*

Theorem 1. *([25]) Let $(S, \times)$ be a CA-NET-groupoid. Then, for any $a, p, q \in N$ and anti(a) $\in$ {anti(a)},*

(1) $q \times$ neut(a) $\in$ {anti(a)}, for all $q \in$ {anti(a)};
(2) $p \times$ neut(a) $= q \times$ neut(a), for all $p, q \in$ {anti(a)}; and
(3) neut(p) $\times$ neut(a) = neut(a) $\times$ neut(p) = neut(a), for all $p \in$ {anti(a)}.

Remark 1. *Since there may be more than one anti-element of an element a, the symbol {anti(a)} is used to represent the set of all anti elements of a. Therefore, the meaning of $q \in$ {anti(a)} is that q is an anti-element of a.*

Theorem 2. *([25]) Let $(S, \times)$ be a CA-NET-groupoid. Denote the set of all different neutral element in S by E(S). For any $e \in E(S)$, denote $S(e) = \{a \in S \mid \text{neut}(a) = e\}$. Then, for any $e \in E(S)$, S(e) is a subgroup of S.*

3. Regular and Inverse Elements in Cyclic Associative Groupoids (CA-Groupoids)

Definition 4. *An element a of a CA-groupoid $(S, \times)$ is called regular if there exists $x \in S$ such that*

$$a = a \times (x \times a)$$

$(S, \times)$ is called a regular CA-groupoid if all its elements are regular.

Definition 5. *An element a of a CA-groupoid* $(S, \times)$ *is called strongly regular if there exists* $x \in S$ *such that*

$$a = a \times (x \times a) \text{ and } a = (a \times x) \times a$$

$(S, \times)$ *is called strongly regular CA-groupoid if all its elements are strongly regular.*

Example 1. *Denote* $S = \{a, b, c\}$ *and define operations* $\times$ *on* S*, as shown in Table 1. We can verify that* a *is strongly regular, since* $a = a \times (a \times a) = (a \times a) \times a$*;* b *is regular, since* $b = b \times (b \times b)$*. However,* b *is not strongly regular, since* $b \neq (b \times b) \times b$*, and there does not exist* $x \in S$ *such that* $b = b \times (x \times b) = (b \times x) \times b$*.*

Table 1. The operation $\times$ on S.

$\times$	**a**	**b**	**c**
a	a	a	c
b	b	a	c
c	c	c	c

Example 2. *Let* $S = \{1, 2, 3, 4\}$*. The operation* $\times$ *on* S *is defined as Table 2. We can verify that* $(S, \times)$ *is a commutative semigroup, then for any* $a, b, c \in S$*, we have* $a \times (b \times c) = (a \times b) \times c = c \times (a \times b)$*. Thus,* $(S, \times)$ *is a commutative CA-groupoid. Moreover,* $(S, \times)$ *is an AG-groupoid because* $(S, \times)$ *is a commutative CA-groupoid. In addition,* $(S, \times)$ *is a regular semigroup, because* $1 = 1 \times 1 \times 1$*,* $2 = 2 \times 2 \times 2$*,* $3 = 3 \times 1 \times 3$*,* $4 = 4 \times 2 \times 4$*.* $(S, \times)$ *is also a regular CA-groupoid, since* $1 = 1 \times (1 \times 1)$*,* $2 = 2 \times (2 \times 2)$*,* $3 = 3 \times (1 \times 3)$*,* $4 = 4 \times (2 \times 4)$*.* $(S, \times)$ *is also a regular AG-groupoid, since* $1 = (1 \times 1) \times 1$*,* $2 = (2 \times 2) \times 2$*,* $3 = (3 \times 1) \times 3$*,* $4 = (4 \times 4) \times 4$*.*

Table 2. The operation $\times$ on S.

$\times$	**1**	**2**	**3**	**4**
1	1	2	3	4
2	2	1	4	3
3	3	4	3	4
4	4	3	4	3

Example 3. *Let* $S = \{1, 2, 3, 4, 5\}$*. The operation* $\times$ *on* S *is defined as Table 3. We can verify that* $(S, \times)$ *is a strongly regular semigroup. However,* $(S, \times)$ *is not a CA-groupoid because* $3 \times (4 \times 5) \neq 5 \times (3 \times 4)$*.*

Table 3. The operation $\times$ on S.

$\times$	**1**	**2**	**3**	**4**	**5**
1	1	1	1	1	1
2	1	2	1	1	5
3	1	1	3	4	1
4	1	4	1	1	3
5	1	1	5	2	1

Example 4. *Let* $S = \{1, 2, 3, 4\}$*. The operation* $\times$ *on* S *is defined as Table 4. We can verify that* $(S, \times)$ *is a strongly regular CA-groupoid, since* $1 = 1 \times (1 \times 1) = (1 \times 1) \times 1$*,* $2 = 2 \times (4 \times 2) = (2 \times 4) \times 2$*,* $3 = 3 \times (3 \times 3) = (3 \times 3) \times 3$*,* $4 = 4 \times (2 \times 4) = (4 \times 2) \times 4$*.* $(S, \times)$ *is also a strongly regular semigroup.*

Table 4. The operation × on *S*.

×	**1**	**2**	**3**	**4**
1	1	1	1	1
2	1	4	2	3
3	1	2	3	4
4	1	3	4	2

An idea of great important in CA-groupoid theory is that of an inverse of an element.

Definition 6. *For any element a in a CA-groupoid S, we say that a^{-1} is an inverse of a if satisfied*

$$a = a \times (a^{-1} \times a), a^{-1} \times (a \times a^{-1}) = a^{-1} \tag{1}$$

Notice that an element with an inverse is necessarily regular. Less obviously, each regular element has an inverse; for if $a \times (x \times a) = a$ we need only define $a^{-1} = x \times (a \times x)$ and verify that Equation (1) are satisfied.

Theorem 3. *Let (S, ×) be a regular CA-groupoid; then, each of its elements has an inverse and the inverse is unique.*

Proof. Let x_1, x_2 be inverses of a in S. Then, we have $a = a \times (x_1 \times a)$, $x_1 = x_1 \times (a \times x_1)$ and $a = a \times (x_2 \times a)$, $x_2 = x_2 \times (a \times x_2)$,

$$x_1 = x_1 \times (a \times x_1) = x_1 \times (x_1 \times a) = x_1 \times (x_1 \times (a \times (x_2 \times a))) = x_1 \times (x_1 \times (a \times (a \times x_2)))$$

$$= x_1 \times ((a \times x_2) \times (x_1 \times a))$$

$$= x_1 \times ((a \times a) \times (x_1 \times x_2)) \text{ (Applying Proposition 1)}$$

$$= (x_1 \times x_2) \times (x_1 \times (a \times a)) = (x_1 \times x_2) \times (a \times (x_1 \times a)) = (x_1 \times x_2) \times a.$$

Similarly, we can get that $x_2 = (x_2 \times x_1) \times a$.
Then, we have

$$(x_1 \times a) \times x_2 = (x_1 \times a) \times ((x_2 \times x_1) \times a) = a \times ((x_1 \times a) \times (x_2 \times x_1)) = (x_2 \times x_1) \times (a \times (x_1 \times a))$$
$$= (x_2 \times x_1) \times a = x_2,$$

$$x_1 \times x_2 = x_1 \times ((x_2 \times x_1) \times a) = a \times (x_1 \times (x_2 \times x_1)) = (x_2 \times x_1) \times (a \times x_1)$$

$$= (x_1 \times x_2) \times (a \times x_1) \text{ (Applying Proposition 1)}$$

$$= x_1 \times ((x_1 \times x_2) \times a) = x_1 \times x_1.$$

Similarly, we can get that $x_2 \times x_1 = x_2 \times x_2$. Further, we have,

$$x_1 \times x_2 = x_1 \times ((x_1 \times a) \times x_2) = x_2 \times (x_1 \times (x_1 \times a)) = x_2 \times (a \times (x_1 \times x_1)) = (x_1 \times x_1) \times (x_2 \times a)$$
$$= (x_1 \times x_2) \times (x_2 \times a)$$

$$= (a \times x_1) \times (x_2 \times x_2) \text{ (Applying Proposition 1)}$$

$$= (a \times x_1) \times (x_2 \times x_1)$$

$$= (x_1 \times a) \times (x_2 \times x_2) \text{ (Applying Proposition 1 and } x_2 \times x_1 = x_2 \times x_2\text{)}$$

$$= x_2 \times ((x_1 \times a) \times x_2) = x_2 \times x_2.$$

Thus, $x_1 \times x_2 = x_2 \times x_1$, $x_1 = (x_1 \times x_2) \times a = (x_2 \times x_1) \times a = x_2$.

Therefore, in a regular CA-groupoid, each of its elements has an inverse and the inverse is unique.
□

Example 5. *Let S = {1, 2, 3, 4, 5, 6}. The operation × on S is defined as Table 5. We can verify that (S, ×) is a CA-groupoid; element 3 is an inverse of 3 because 3 = 3 × (3 × 3), 3 = 3 × (3 × 3), obviously element 3 is a regular; and element 5 is an inverse of 5 since 5 = 5 × (5 × 5), 5 = 5 × (5 × 5), obviously element 5 is a regular. However, elements 1, 2, 4, and 6 have no inverses because there exists no x, y, p, q∈S such that 1 = 1 × (x × 1), x = x × (1 × x); 2 = 2 × (y × 2), y = y × (2 × y); 4 = 4 × (p × 4), p = p × (4 × p); and 6 = 6 × (q × 6), q = q × (6 × q). Obviously, for any a ∈ S, if a ∉ a × S, then a has not inverse.*

Table 5. The operation × on *S*.

×	**1**	**2**	**3**	**4**	**5**	**6**
1	2	3	3	3	5	2
2	4	3	3	3	5	2
3	3	3	3	3	5	2
4	3	3	3	3	5	2
5	5	5	5	5	3	5
6	4	3	3	3	5	3

Example 6. *Let S = {1, 2, 3, 4, 5, 6}. The operation × on S is defined as Table 6. We can verify that (S, ×) is a regular CA-groupoid, since 1 = 1 × (1 × 1), 2 = 2 × (2 × 2), 3 = 3 × (3 × 3), 4 = 4 × (4 × 4), 5 = 5 × (5 × 5), 6 = 6 × (6 × 6), and the inverse is unique.*

Table 6. The operation × on *S*.

×	**1**	**2**	**3**	**4**	**5**	**6**
1	1	2	5	5	5	6
2	2	1	5	5	5	6
3	5	5	3	4	5	6
4	5	5	4	3	5	6
5	5	5	5	5	5	6
6	6	6	6	6	6	5

Definition 7. *An element a of a CA-groupoid (S, ×) is called locally associative if satisfied*

$$a \times (a \times a) = (a \times a) \times a.$$

(S, ×) is called locally associative CA-groupoid if all its elements are locally associative.

Example 7. *Let S = {1, 2, 3, 4, 5}. The operation × on S is defined as Table 7. We can verify that (S, ×) is a locally associative CA-groupoid, since 1 × (1 × 1) = (1 × 1) × 1, 2 × (2 × 2) = (2 × 2) × 2, 3 × (3 × 3) = (3 × 3) × 3, 4 × (4 × 4) = (4 × 4) × 4, and 5 × (5 × 5) = (5 × 5) × 5. However, (S, ×) is not a semigroup because (3 × 4) × 3≠ 3 × (4 × 3).*

Table 7. The operation × on *S*.

×	1	2	3	4	5
1	1	1	1	1	2
2	1	1	2	1	2
3	1	1	4	2	4
4	1	1	2	1	2
5	1	1	4	2	4

Definition 8. *An element a of a CA-groupoid (S, ×) is called quasi-regular if there exists* $x \in S$, $m \in N$ *such that*

$$a^m \times (x \times a^m) = a^m. \ (a^m \text{ is defined by } a \times a^{m-1})$$

(S, ×) is called quasi-regular CA-groupoid if all its elements are quasi-regular.

Example 8. *Let S = {1, 2, 3, 4}. The operation × on S is defined as Table 8. We can verify that (S, ×) is a quasi-regular CA-groupoid, since* $1 = 1^2 \times (3 \times 1^2)$, $2 = 2 \times (2 \times 2)$, $3 = 3 \times (3 \times 3)$, $4^2 = 4^2 \times (2 \times 4^2)$. *However, (S, ×) is not a regular CA-groupoid because there exists no x, y∈S such that* $1 = 1 \times (x \times 1)$, $4 = 4 \times (y \times 4)$. *Moreover, (S, ×) is not a semigroup because* $(4 \times 1) \times 1 \neq 4 \times (1 \times 1)$.

Table 8. The operation × on *S*.

×	1	2	3	4
1	3	2	3	2
2	2	2	2	2
3	3	2	3	2
4	4	2	2	2

Definition 9. *Let (S, ×) be a groupoid. If for all a, b, c ∈ S,*

$$a \times (b \times c) = (a \times b) \times c, \ a \times (b \times c) = c \times (a \times b),$$

then (S, ×) is called cyclic associative semigroup (shortly, CA-semigroup).

Example 9. *Suppose S = {1, 2, 3, 4} and define a binary operation × on S as shown in Table 9. We can verify that (S, ×) is a CA-groupoid, but (S, ×) is not a CA-semigroup because* $(3 \times 4) \times 3 \neq 3 \times (4 \times 3)$.

Table 9. The operation × on *S*.

×	1	2	3	4
1	1	1	1	1
2	1	1	2	1
3	1	1	4	2
4	1	1	2	1

Obviously on the CA-groupoid *S*, there is: strongly regular element ⇒ regular element ⇒ inverse element ⇒ quasi-regular element.

According to Examples 1, 2, and 5–9, we can get the relationship between CA-groupoids and related algebraic systems, which we can be expressed as Figure 1.

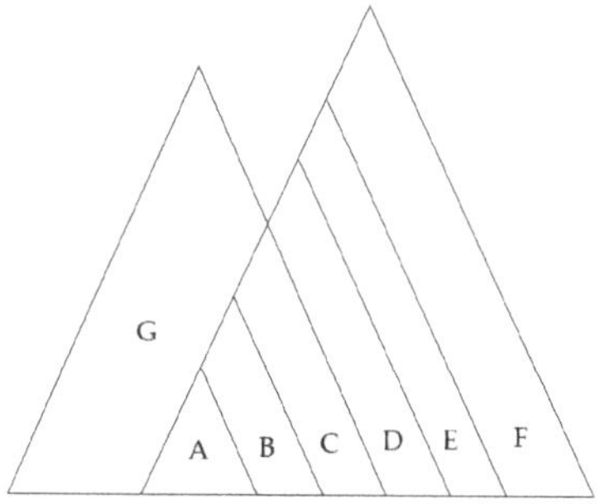

Figure 1. The relationships among some algebraic systems.

Remark 2. *In Figure 1, each letter only indicates the smallest area in which it is located. Here, A represents the set of all strongly regular CA-groupoids, and*

$A \cup B$ represents the set of all regular CA-groupoids;
$A \cup B \cup C$ represents the set of all CA-semigroups;
$A \cup B \cup C \cup D$ represents the set of all quasi-regular CA-groupoids;
$A \cup B \cup C \cup D \cup E$ represents the set of all locally associative CA-groupoids;
$A \cup B \cup C \cup D \cup E \cup F$ represents the set of all CA-groupoids; and
$A \cup B \cup C \cup G$ represents the set of all semigroups.

4. Regular Cyclic Associative Groupoids (CA-Groupoids) and Cyclic Associative Neutrosophic Extended Triplet Groupoids (CA-NET-Groupoids)

Theorem 4. *Let $(S, \times)$ be a CA-NET-groupoid. Then, its idempotents are commutative.*

Proof. Let a, b an idempotent in S; then, we have

$$(a \times b) \times (a \times b) = (b \times a) \times (a \times b) \text{ (Applying Proposition 1)}$$

$$= (b \times b) \times (a \times a) \text{ (Applying Proposition 1)} = b \times a.$$

Moreover

$$(a \times b) \times (a \times b) = (b \times (neut(b) \times a)) \times (a \times b)$$

$$= (b \times b) \times (a \times (neut(b) \times a)) \text{ (Applying Proposition 1)}$$

$$= (b \times b) \times (a \times (a \times neut(b))) = (b \times b) \times (neut(b) \times (a \times a))$$

$$= b \times (neut(b) \times a) = a \times (b \times neut(b)) = a \times b.$$

Therefore, $a \times b = b \times a$. In a CA-NET-groupoid, its idempotents are commutative. □

Corollary 1. *Every CA-NET-groupoid is commutative.*

Proof. Let $(S, \times)$ be a CA-NET-groupoid. By Theorem 4, for any $x \in S$, $neut(x)$ is idempotent. Then, for any $a, b \in S$, we have

$$neut(a) \times neut(b) = neut(b) \times neut(a),$$

Furthermore,

$$neut(a) \times b = neut(a) \times (neut(b) \times b) = b \times (neut(a) \times neut(b))$$

$$= neut(b) \times (b \times neut(a)) = (neut(b) \times neut(b)) \times (b \times neut(a))$$

$$= (neut(a) \times neut(b)) \times (b \times neut(b)) \text{ (Applying Proposition 1)}$$

$$= (neut(a) \times neut(b)) \times (neut(b) \times b)$$

$$= (b \times neut(a)) \times (neut(b) \times neut(b)) \text{ (Applying Proposition 1)}$$

$$= (b \times neut(a)) \times neut(b)$$

Further, for any $a, b \in S$, we have

$$a \times b = (neut(a) \times a) \times (neut(b) \times b)$$

$$= (b \times neut(a)) \times (neut(b) \times a) \text{ (Applying Proposition 1)}$$

$$= a \times ((b \times neut(a)) \times neut(b))$$

$$= a \times (neut(a) \times b) \text{ (by } neut(a) \times b)$$

$$= (b \times neut(a)) \times neut(b))$$

$$= b \times (a \times neut(a))$$

$$= b \times a$$

Therefore, every CA-NET-groupoid is commutative. □

Example 10. *Let S = {1, 2, 3, 4, 5}. The operation × on S is defined as Table 10. We can verify that (S, ×) is a CA-NET-groupoid, and*

$$neut(1) = 1, anti(1) = \{1, 5\}; neut(2) = 2, anti(2) = \{1, 2, 3, 4, 5\};$$

$$neut(3) = 3, anti(3) = \{3, 5\}; neut(4) = 4, anti(4) = \{1, 3, 4, 5\}; neut(5) = 5, anti(3) = 5.$$

Obviously, (S, ×) is a commutative.

Table 10. The operation × on *S*.

×	**1**	**2**	**3**	**4**	**5**
1	1	2	4	4	1
2	2	2	2	2	2
3	4	2	3	4	3
4	4	2	4	4	4
5	1	2	3	4	5

Theorem 5. *Let (S, ×) be a groupoid. Then, S is a CA-NET-groupoid if and only if it is a regular CA-groupoid.*

Proof. Assume that S is a CA-NET-groupoid. For any a in S, by Definitions 1 and 3, we have

$$a \times (anti(a) \times a) = a \times neut(a) = a.$$

From this and Definition 4, we know that element a is a regular element and S is a regular CA-groupoid.

Therefore, we prove that S is a regular CA-groupoid.

Now, we assume that S is a regular CA-groupoid. For any a in a regular CA-groupoid S, we have

$$a \times (x \times a) = a.$$

Furthermore,

$$(x \times a) \times a = (x \times a) \times (a \times (x \times a)) = (x \times a) \times ((x \times a) \times a) = a \times ((x \times a) \times (x \times a))$$
$$= a \times (a \times ((x \times a) \times x))$$
$$= a \times (x \times (a \times (x \times a)))$$
$$= a \times (x \times a) = a.$$

Therefore, there exists $(x \times a) \in S$, such that $(x \times a) \times a = a \times (x \times a) = a$.
Moreover, we have

$$(x \times a) = x \times (a \times (x \times a)) = (x \times a) \times (x \times a) = a \times ((x \times a) \times x).$$

Furthermore,

$$((x \times a) \times x) \times a = ((x \times a) \times x) \times (a \times (x \times a))$$
$$= (x \times a) \times (((x \times a) \times x) \times a) = a \times ((x \times a) \times ((x \times a) \times x))$$
$$= a \times (x \times ((x \times a) \times (x \times a)))$$
$$= a \times (x \times (x \times a)) \ (by\ (x \times a) \times (x \times a) = (x \times a))$$
$$= (x \times a) \times (a \times x)$$
$$= x \times ((x \times a) \times a) \ (by\ (x \times a) \times a = a)$$
$$= x \times a$$

Therefore, there exists $((x \times a) \times x) \in S$, such that $a \times ((x \times a) \times x) = ((x \times a) \times x) \times a = x \times a$. Then, S is a CA-NET-groupoid. □

Example 11. *Let $S = \{1, 2, 3, 4\}$. The operation $\times$ on S is defined as Table 11. We can verify that $(S, \times)$ is a CA-NET- groupoid, and $neut(1) = 1$, $anti(1) = \{1, 2, 3, 4\}$; $neut(2) = 3$, $anti(2) = 4$; $neut(3) = 3$, $anti(3) = 3$; $neut(4) = 3$, $anti(4) = 2$.*

Table 11. The operation $\times$ on S.

×	1	2	3	4
1	1	1	1	1
2	1	4	2	3
3	1	2	3	4
4	1	3	4	2

Moreover, $(S, \times)$ is a regular CA-groupoid, since $1 = 1 \times (1 \times 1)$, $2 = 2 \times (4 \times 2)$, $3 = 3 \times (3 \times 3)$, $4 = 4 \times (2 \times 4)$.

Definition 10. *Let $(S, \times)$ be a groupoid.*

(1) *If for any $a \in S$, there exist two elements b and c in S satisfying the condition $a \times b = a$ and $c \times a = b$, then S is called a CA-(r, l)-NET-groupoid.*
(2) *If for any $a \in S$, there exist two elements b and c in S satisfying the condition $a \times b = a$ and $a \times c = b$, then S is called a CA-(r, r)-NET-groupoid.*
(3) *If for any $a \in S$, there exist two elements b and c in S satisfying the condition $b \times a = a$ and $a \times c = b$, then S is called a CA-(l, r)-NET-groupoid.*
(4) *If for any $a \in S$, there exist two elements b and c in S satisfying the condition $b \times a = a$ and $c \times a = b$, then S is called a CA-(l, l)-NET-groupoid.*

Theorem 6. *Let (S, ×) be a groupoid. Then, S is a CA-(r, l)-NET-groupoid if and only if it is a regular CA-groupoid.*

Proof. Assume that S is a *CA-(r, l)-NET*-groupoid. For any a in S, by Definitions 1 and 10(1), we have

$$a \times neut(a) = a, anti(a) \times a = neut(a)$$

$$a \times (anti(a) \times a) = a \times neut(a) = a$$

From this and Definition 4, we know that element a is a regular element and S is a regular CA-groupoid. Therefore, we prove that S is a regular CA-groupoid.

Now, we assume that S is a regular CA-groupoid. For any a in a regular CA-groupoid S, we have

$$a \times (x \times a) = a.$$

Thus, there exists $(x \times a) \in S$, such that $a \times (x \times a) = a$.

Moreover, we have:

$$x \times a = (x \times a).$$

Therefore, there exists $x \in S$, such that $x \times a = (x \times a)$. Then, S is a CA-(r, l)-NET-groupoid. □

Theorem 7. *Let (S, ×) be a groupoid. Then, S is a CA-(r, r)-NET-groupoid if and only if it is a regular CA-groupoid.*

Proof. Assume that S is a CA-(r, r)-NET-groupoid. For any a in S, by Definitions 1 and 10(2), we have

$$a \times neut(a) = a, a \times anti(a) = neut(a),$$

$$a \times (anti(a) \times a) = a \times (a \times anti(a)) = a \times neut(a) = a$$

From this and Definition 4, we know that element a is a regular element and S is a regular CA-groupoid. Therefore, we prove that S is a regular CA-groupoid.

Now, we assume that S is a regular CA-groupoid, for any a in a regular CA-groupoid S, we have

$$a \times (x \times a) = a \times (a \times x) = a$$

Thus, there exists $(a \times x) \in S$, such that $a \times (a \times x) = a$.

Moreover, we have

$$a \times x = (a \times x)$$

Therefore, there exists $x \in S$, such that $a \times x = (a \times x)$. Then, S is a CA-(r, r)-NET-groupoid. □

Theorem 8. *Let (S, ×) be a groupoid. Then, S is a CA-(l, r)-NET-groupoid if and only if it is a regular CA-groupoid.*

Proof. Assume that S is a CA-(l, r)-NET-groupoid. For any a in S, by Definitions 1 and 10(3), we have

$$neut(a) \times a = a, a \times anti(a) = neut(a)$$

$$neut(a) \times a = (a \times anti(a)) \times a = (a \times anti(a)) \times (neut(a) \times a)$$
$$= (a \times a) \times (neut(a) \times anti(a)) \text{ (Applying Proposition 1)}$$
$$= (anti(a) \times a) \times (neut(a) \times a) \text{ (Applying Proposition 1)}$$
$$= (anti(a) \times a) \times a.$$

Thus, $a \times anti(a) = anti(a) \times a = neut(a)$.

Moreover, we have

$$a \times neut(a) = (neut(a) \times a) \times (anti(a) \times a)$$

$$= (a \times neut(a)) \times (anti(a) \times a) \text{ (Applying Proposition 1)}$$

$$= (a \times neut(a)) \times neut(a).$$

Thus, $a \times neut(a) = neut(a) \times a = a$.

Then,

$$(anti(a) \times a) \times a = neut(a) \times a = a \times neut(a) = a \times (anti(a) \times a) = a$$

From this and Definition 4, we know that element a is a regular element and S is a regular CA-groupoid. Therefore, we prove that S is a regular CA-groupoid.

Now, we assume that S is a regular CA-groupoid. For any a in a regular CA-groupoid S, let $a = (a \times x) \times a$. We have

$$x \times a = x \times ((a \times x) \times a) = a \times (x \times (a \times x)) = ((a \times x) \times (a \times x)),$$

$$(a \times x) \times a = (a \times x) \times ((a \times x) \times a) = a \times ((a \times x) \times (a \times x)) = a \times (x \times a) = a, a \times x = (a \times x)$$

Therefore, S is a CA-(l, r)-NET-groupoid. □

Theorem 9. *Let $(S, \times)$ be a groupoid. Then, S is a CA-(l, l)-NET-groupoid if and only if it is a regular CA-groupoid.*

Proof. Assume that S is a CA-(l, l)-NET-groupoid. For any a in S, by Definitions 1 and 10(4), we have

$$neut(a) \times a = a, anti(a) \times a = neut(a),$$

$$a \times neut(a) = (neut(a) \times a) \times (anti(a) \times a)$$

$$= (a \times neut(a)) \times (anti(a) \times a) \text{ (Applying Proposition 1)}$$

$$= (a \times neut(a)) \times neut(a)$$

Thus, $a \times neut(a) = neut(a) \times a = a$.

Then,

$$(anti(a) \times a) \times a = neut(a) \times a = a \times neut(a) = a \times (anti(a) \times a) = a.$$

From this and Definition 4, we know that element a is a regular element and S is a regular CA-groupoid. Therefore, we prove that S is a regular CA-groupoid.

Now, we assume that S is a regular CA-groupoid. For any a in a regular CA-groupoid S, let $a = (x \times a) \times a$, we have

$$x \times a = x \times ((a \times x) \times a) = a \times (x \times (a \times x)) = ((a \times x) \times (a \times x)),$$

$$(x \times a) \times a = (x \times a) \times ((x \times a) \times a) = a \times ((x \times a) \times (x \times a)) = (x \times a) \times (a \times (x \times a))$$

$$= (x \times a) \times (a \times (a \times x)) = (a \times x) \times ((x \times a) \times a) = (a \times x) \times a$$

$$= (a \times x) \times ((a \times x) \times a) \text{ (by } (a \times x) \times a = a)$$

$$= a \times ((a \times x) \times (a \times x))$$

$$= a \times (x \times a) = a.$$

Moreover, we have $x \times a = (x \times a)$. Therefore, S is a CA-(l, l)-NET-groupoid. □

Example 12. *Denote S = {1, 2, 3, 4} and define operations × on S as shown in* Table 12. *We can verify that (S, ×) is a CA-(r, l)-NET-groupoid, and,*

$$neut_{(r,\ l)}(1) = 1,\ anti_{(r,\ l)}(1) = \{1, 2, 3, 4\};\ neut_{(r,\ l)}(2) = 4,\ anti_{(r,\ l)}(2) = 2;$$

$$neut_{(r,\ l)}(3) = 3,\ anti_{(r,\ l)}(3) = 3;\ neut_{(r,\ l)}(4) = 4,\ anti_{(r,\ l)}(4) - 4$$

Table 12. The operation × on S.

×	1	2	3	4
1	1	1	1	1
2	1	4	1	2
3	1	1	3	1
4	1	2	1	4

It is easy to verify that S is also a CA-(r, r)-NET-groupoid, CA-(l, r)-NET-groupoid, CA-(l, l)-NET-groupoid. Moreover, (S, ×) is a regular CA-groupoid, since 1 = 1 × (2 × 1), 2 = 2 × (2 × 2), 3 = 3 × (3 × 3), and 4 = 4 × (4 × 4).

5. Green Relations in Cyclic Associative Groupoids (CA-Groupoids)

If a is an element of a CA-groupoid S, the smallest left ideal of S containing a is $Sa\cup\{a\}$.

Definition 11. *Let (S, ×) be a CA-groupoid, for any a, b ∈ S, define the following binary relationships:*

$$\begin{gathered} a\mathcal{L}b \Leftrightarrow Sa \cup \{a\} = Sb \cup \{b\}; \\ a\mathcal{R}b \Leftrightarrow aS \cup \{a\} = bS \cup \{b\}; \\ a\mathcal{J}b \Leftrightarrow (Sa \cup \{a\})S \cup (Sa \cup \{a\}) = (Sb \cup \{b\})S \cup (Sb \cup \{b\}); \\ \mathcal{H} = \mathcal{L} \cap \mathcal{R}. \end{gathered}$$

We call $\mathcal{L}$,$\mathcal{R}$,$\mathcal{J}$, *and* $\mathcal{H}$ the Green's relations on the CA-groupoid.

Definition 12. *Let (S, ×) be a CA-groupoid. A relation R on the set S is called left compatible (with the operation on S) if*

$$(\forall a,\ s,\ t \in S)\ (s,\ t)\ \in\ R\ \Rightarrow (a \times s\ , a \times t)\ \in R,$$

and right compatible if

$$(\forall a,\ s,\ t \in S)\ (s,\ t)\ \in\ R\ \Rightarrow (s \times a\ , t \times a)\ \in R.$$

It is called compatible if

$$(\forall s,\ t, s'\ t'\ \in S)\ [(s,\ t)\ \in\ R \text{ and } (s',\ t'\ \in R)] \Rightarrow (s \times s'\ , t \times t')\ \in R.$$

A left [right] compatible equivalence is called a left [right] congruence. A compatible equivalence relation is called a congruence.

Proposition 2. *Let a, b be elements of a CA-groupoid S. If a = b, then a*$\mathcal{L}$*a, a*$\mathcal{R}$*a. If a≠b, then a*$\mathcal{L}$*b if and only if there exists x, y in S such that x × a = b, y × b = a. In addition, a*$\mathcal{R}$ *b if and only if there exists u, v in S such that a × u = b, b × v = a.*

Another immediate property of this is as follows:

Proposition 3. $\mathcal{L}$ *is a left congruence and* $\mathcal{R}$ *is a right congruence.*

Corollary 2. *In a CA-groupoid S, $\mathcal{L}$ and $\mathcal{R}$ are not commutative. That is, as a binary relationship, $\mathcal{L}\circ\mathcal{R}\neq\mathcal{R}\circ\mathcal{L}$.*

Example 13. *Let S = {1, 2, 3, 4, 5, 6}. The operation × on S is defined as Table 13. Then, (S, ×) is a CA-groupoid.*

Table 13. The operation × on *S*.

×	1	2	3	4	5	6
1	2	3	3	3	5	2
2	4	3	3	3	5	2
3	3	3	3	3	5	4
4	3	3	3	3	5	4
5	5	5	5	5	3	5
6	3	3	3	3	5	3

$\mathcal{L} = \{<3, 5>, <5, 3>\}$, $\mathcal{R} = \{<3, 4>, <4, 3>\}$. $\mathcal{L}\circ\mathcal{R} = \{<5, 4>\} \neq \mathcal{R}\circ\mathcal{L} = \{<4, 5>\}$. Then, $\mathcal{L}$ and $\mathcal{R}$ are not commutative.

In a regular CA-groupoid *S* we have a particularly useful way of looking at the equivalences $\mathcal{L}$ and $\mathcal{R}$. First, notice that if *S* is regular then $a = a \times (x \times a) \in aS$, and similarly $a \in Sa$, $a \in SaS$. Hence, in describing the Green equivalences for a regular CA-groupoid we can drop all reference to $Sa \cup \{a\}$, and assert simply that

$$a\mathcal{L}b \Leftrightarrow Sa = Sb;$$
$$ab \Leftrightarrow aS = bS;$$
$$a\mathcal{J}b \Leftrightarrow SaS = SbS;$$
$$\mathcal{H} = \mathcal{L}\cap\mathcal{R}.$$

Definition 13. *Let (S, ×) be a regular CA-groupoid, define the following binary relationship:*

$$\mathcal{D} = \mathcal{L}\cup\mathcal{R}$$

We call $\mathcal{D}$ the Green's relations on the regular CA-groupoid.

Theorem 10. *In a regular CA-groupoid S, the relations $\mathcal{L}$ and $\mathcal{R}$ are commutative. That is, as a binary relationship, $\mathcal{L}\circ\mathcal{R} = \mathcal{R}\circ\mathcal{L}$.*

Proof. Let (*S*, ×) be a regular CA-groupoid, let $a, b \in S$, and suppose that $(a, b) \in \mathcal{L}\circ\mathcal{R}$. Then, there exists *c* in *S* such that $a\mathcal{L}c$ and $c\mathcal{R}b$. That is, there exist *x*, *y*, *u*, *v* in *S* such that

$$x \times a = c,\ c \times u = b,$$

$$y \times c = a,\ b \times v = c.$$

If we now write *d* for the element $(y \times c) \times u$ of *S*, applying Theorem 5, *S* is a CA-NET-groupoid. As such, we have

$$a \times u = (y \times c) \times u = d,$$

$$a = y \times c = y \times (b \times v) = v \times (y \times b) = b \times (v \times y) = (c \times u) \times (v \times y)$$

$$= (y \times c) \times (v \times u) \text{ (Applying Proposition 1)}$$

$$= a \times (v \times u) = u \times (a \times v) = v \times (u \times a)$$

$$= v \times (u \times (neut(a) \times a)) = v \times (a \times (u \times neut(a)))$$

$$= v \times (neut(a) \times (a \times u)) = v \times (neut(a) \times d) = d \times (v \times neut(a))$$

hence $a\mathcal{R}d$. In addition,

$$b = c \times u = (x \times a) \times u = (x \times a) \times (u \times neut(u)) = neut(u) \times ((x \times a) \times u) = neut(u) \times b$$

$$d = (y \times c) \times u = (y \times c) \times (neut(u) \times u) = u \times ((y \times c) \times neut(u)) = neut(u) \times (u \times (y \times c))$$

$$= neut(u) \times (c \times (u \times y)) = neut(u) \times (y \times (c \times u)) = neut(u) \times (y \times b) = neut(u) \times (y \times (neut(u) \times b))$$

$$= neut(u) \times (b \times (y \times neut(u))) = (y \times neut(u)) \times (neut(u) \times b) = (y \times neut(u)) \times b,$$

$$d = a \times u = (y \times c) \times u = (y \times c) \times (u \times neut(u)) = neut(u) \times ((y \times c) \times u) = neut(u) \times d,$$

$$b = c \times u = (x \times a) \times u = (x \times a) \times (neut(u) \times u) = u \times ((x \times a) \times neut(u)) = neut(u) \times (u \times (x \times a))$$

$$= neut(u) \times (u \times c) = neut(u) \times (u \times (x \times a)) = neut(u) \times (u \times (x \times (y \times c)))$$

$$= neut(u) \times ((y \times c) \times (u \times x)) = neut(u) \times (a \times (u \times x)) = neut(u) \times (x \times (a \times u)) = neut(u) \times (x \times d)$$

$$= neut(u) \times (x \times (neut(u) \times d)) = neut(u) \times (d \times (x \times neut(u))) = (x \times neut(u)) \times (neut(u) \times d)$$

$$= (x \times neut(u)) \times d,$$

thus $d\mathcal{L}b$. We deduce that $(a, b) \in \mathcal{R} \circ \mathcal{L}$. We have shown that $\mathcal{L} \circ \mathcal{R} \subseteq \mathcal{R} \circ \mathcal{L}$; the reverse inclusion follows in a similar way. □

Theorem 11. *In a regular CA-groupoid* S*,* $\mathcal{L}$ *is equivalent to* $\mathcal{R}$*. That is, as a binary relationship,* $\mathcal{L} = \mathcal{R}$*.*

Proof. By Theorem 10, we have $d\mathcal{L}\, b$. Then,

$$b = c \times u = (x \times a) \times u = (x \times a) \times (neut(u) \times u) = u \times ((x \times a) \times neut(u))$$

$$= neut(u) \times (u \times (x \times a)) = neut(u) \times (u \times c) = neut(u) \times (u \times (x \times a)) = neut(u) \times (u \times (x \times (y \times c)))$$

$$= neut(u) \times ((y \times c) \times (u \times x)) = neut(u) \times (a \times (u \times x)) = neut(u) \times (x \times (a \times u))$$

$$= neut(u) \times (x \times d) = d \times (neut(u) \times x).$$

$$d = (y \times c) \times u = (y \times c) \times (neut(u) \times u) = u \times ((y \times c) \times neut(u)) = neut(u) \times (u \times (y \times c))$$

$$= neut(u) \times (c \times (u \times y)) = neut(u) \times (y \times (c \times u)) = neut(u) \times (y \times b) = neut(u)(y \times (c \times u))$$

$$= neut(u) \times (y \times b) = b \times (neut(u) \times y).$$

Thus, $d\mathcal{R}b$.

Therefore, in a regular CA-groupoid S, $\mathcal{L}$ *is equivalent to* $\mathcal{R}$. □

Example 14. *Let* $S = \{1, 2, 3, 4, 5, 6, 7, 8\}$*. The operation* $\times$ *on* S *is defined as Table 14. Then,* $(S, \times)$ *is a regular CA-groupoid.* $\mathcal{L} = \{<1, 2>, <2, 1>, <3, 4>, <4, 3>, <5, 6>, <6, 5>, <7, 8>, <8, 7>\}$, $\mathcal{R} = \{<1, 2>, <2, 1>, <3, 4>, <4, 3>, <5, 6>, <6, 5>, <7, 8>, <8, 7>\}$, $\mathcal{L} \circ \mathcal{R} = \{<1, 1>, <2, 2>, <3, 3>, <4, 4>, <5, 5>, <6, 6>, <7, 7>, <8, 8>\} = \mathcal{R} \circ \mathcal{L} = \{<1, 1>, <2, 2>, <3, 3>, <4, 4>, <5, 5>, <6, 6>, <7, 7>, <8, 8>\}$*. Thus,* $\mathcal{L}$ *and* $\mathcal{R}$ *are commutative, and* $\mathcal{L} = \mathcal{R}$*.*

Table 14. The operation × on S.

×	1	2	3	4	5	6	7	8
1	1	2	5	5	5	6	7	8
2	2	1	5	5	5	6	7	8
3	5	5	3	4	5	6	7	8
4	5	5	4	3	5	6	7	8
5	5	5	5	5	5	6	7	8
6	6	6	6	6	6	5	7	8
7	7	7	7	7	7	7	7	8
8	8	8	8	8	8	8	8	7

Obviously on the regular CA-groupoid S, there is

$$\mathcal{H} = \mathcal{L} = \mathcal{R} = \mathcal{D} = \mathcal{J}.$$

Lemma 1. *In a regular CA-groupoid S, each $\mathcal{L}$-class contains at least one idempotent.*

Proof. For any $a \in S$, there exist $x \in S$, such that $a = a \times (x \times a)$; then,

$$(x \times a) \times (x \times a) = a \times ((x \times a) \times x) = x \times (a \times (x \times a)) = x \times a$$

Therefore, $(x \times a)$ is idempotent and $a\mathcal{L}(x \times a)$. □

Lemma 2. *Every idempotent e in a regular CA-groupoid S is a left identity for $\mathcal{L}e$.*

Proof. If $a \in \mathcal{L}e$, then $a = x \times e$. For some x in S and

$$e \times a = e \times (x \times e) = e \times (e \times x) = x \times e^2 = x \times e = a.$$

□

Proposition 4. *Let a be an element of a regular $\mathcal{L}$-class L in a regular CA-groupoid S. If $\mathcal{L}a$ contains idempotents e, then $\mathcal{L}e$ contains an inverse a^{-1} of a such that $a \times a^{-1} = a^{-1} \times a = e$.*

Proof. Since $a\mathcal{L}e$ it follows by Lemma 2 that $e \times a = a$. Again, from $a\mathcal{R}e$, it follows that there exists x in S such that $a \times x = e$. If we denote $x \times e$ by a^{-1}, we easily see that

$$a \times (a^{-1} \times a) = a \times ((x \times e) \times a) = a \times (a \times (x \times e)) = a \times (e \times (a \times x))$$

$$= a \times (e \times e) = e \times (a \times e) = e \times (e \times a) = e \times a = a,$$

$$a^{-1} \times (a \times a^{-1}) = (x \times e) \times (a \times (x \times e)) = (x \times e) \times (e \times (a \times x)) = (x \times e) \times (e \times e) = e \times ((x \times e) \times e)$$

$$= e \times (e \times (x \times e)) = e \times (e \times (e \times x)) = e \times (x \times e) = e \times (e \times x) = x \times (e \times e) = x \times e = a^{-1}.$$

Thus, a^{-1} is an inverse of a. Moreover,

$$a \times a^{-1} = a \times (x \times e) = e \times (a \times x) = e \times e = e$$

Further,

$$a \times a = (x \times e) \times a = (x \times e) \times (e \times a) = a \times ((x \times e) \times e)$$

$$= e \times (a \times (x \times e)) = e \times (e \times (a \times x)) = e \times (e \times e) = e.$$

It now follows easily that

$$a \times a^{-1} = a^{-1} \times a = e.$$

□

Theorem 12. *Let (S, ×) be a CA-groupoid. Then, the following statements are equivalent:*

(1) *S is regular;*
(2) *Every element of S lies in a subgroup of S; and*
(3) *Every $\mathcal{H}$-class in S is a group.*

Proof. (1)⇒ (2). Assume that S is a regular CA-groupoid. By Theorem 5, we know that S is a CA-NET-groupoid. By Theorem 2, we know that, in a CA-NET-groupoid S, every element of S lies in a subgroup of S. Thus, if S is a regular CA-groupoid, then every element of S lies in a subgroup of S.

(2)⇒(3). Assume that every element of S lies in a subgroup of S. Let $a \in S$; then, $a \in G$ for some subgroup G of S. Denote the identity element of G by e, and the inverse of a within G by a^{-1}. Then, from

$$e \times a = a \times e = a \text{ and } a \times a^{-1} = a^{-1} \times a = e$$

it follows that $a\mathcal{H}e$, and hence $H_a = H_e$, every $\mathcal{H}$-class in S is a group.

(3)⇒(1). Assume that every $\mathcal{H}$-class in S is a group. For each a in S, $a \in H_a$, because H_a is a group, then element a has a unique inverse a^{-1} within the group H_a. Let $x = a^{-1}$; then, it is clear that

$$a \times (x \times a) = a.$$

Therefore, S is a regular CA-groupoid. □

Example 15. *Let S = {a, b, c, d, e}. Define operation × on S as Table 15. Then, (S, ×) is a CA-groupoid.*

Table 15. The operation × on S.

×	a	b	c	d	e
a	a	b	c	d	c
b	b	a	d	c	d
c	c	d	c	d	c
d	d	c	d	c	d
e	c	d	c	d	e

(S, ×) is a regular CA-groupoid, since $a = a \times (a \times a)$, $b = b \times (b \times b)$, $c = c \times (c \times c)$, $d = d \times (d \times d)$, *and* $e = e \times (e \times e)$. *Every element of CA-groupoid S lies in a subgroup of S, because {a, b}, {c, d}, {e} is a subgroup of S. Moreover, a, b ∈ {a, b}, c, d ∈ {c, d}, and e ∈ {e}. Every* $\mathcal{H}$*-class in S is a group. Then,* H_1, H_2, H_3 *of* $\mathcal{H}$*-class in S,* $H_1 = \{a, b\}$, $H_2 = \{c, d\}$, $H_3 = \{e\}$. *Moreover,* $a \times b = b$, $b \times b = a$; $c \times d = d$, $d \times d = c$ *and* $e \times e = e$, H_1, H_2, H_3 *is a group.*

6. Relationships between Some Cyclic Associative Groupoids (CA-Groupoids)

Definition 14. *A CA-groupoid (S, ×) is called inverse CA-groupoid if there exists a unary operation* a^{-1} *on S with the properties*

$$(a^{-1})^{-1} = a, a \times (a^{-1} \times a) = a,$$

and for any x, y ∈ S,

$$(x \times x^{-1}) \times (y \times y^{-1}) = (y \times y^{-1}) \times (x \times x^{-1})$$

Theorem 13. *Let $(S, \times)$ be a CA-groupoid. Then, S is an inverse CA-groupoid if and only if it is a regular CA-groupoid and its idempotent is commutative.*

Proof. Let S be an inverse CA-groupoid, which follows if we show that every idempotent in S can be expressed in the form xx^{-1}. Let e be an idempotent in S. Then, the inverse CA-groupoid property ensures that there is an element e^{-1} in S such that $e \times (e^{-1} \times e) = e$, $(e^{-1})^{-1} = e$. Hence,

$$e^{-1} = e^{-1} \times ((e^{-1})^{-1} \times e^{-1}) = e^{-1} \times (e \times e^{-1}) = e^{-1} \times ((e \times e) \times e^{-1}) = e^{-1} \times (e^{-1} \times (e \times e)) = e^{-1} \times (e \times (e^{-1} \times e)) = e^{-1} \times e = e^{-1} \times (e \times e) = e \times (e^{-1} \times e) = e.$$

and thus $e = e^2 = e \times e = e \times e^{-1}$.

According to the definition of an inverse CA-groupoid, idempotents commute. If x, y are idempotent, then $x \times y = (x \times x^{-1}) \times (y \times y^{-1}) = (y \times y^{-1}) \times (x \times x^{-1}) = y \times x$.

Therefore, S is a regular CA-groupoid and its idempotents are commutative.

Now, we assume that S is a regular CA -groupoid and its idempotents are commutative. Then, according to regularity, for any $x \in S$, there exists $neut(x) \in S$, $anti(x) \in S$. By Theorem 1, let $anti(x) \times neut(x) = x^{-1}$; then, we have

$$(x^{-1})^{-1} = x,\ x \times (x^{-1} \times x) = x,$$

$$(x \times x^{-1}) \times (y \times y^{-1}) = neut(x) \times neut(y) = neut(y) \times neut(x) = (y \times y^{-1}) \times (x \times x^{-1})$$

Therefore, S is an inverse CA-groupoid. □

Corollary 3. *Let $(S, \times)$ be a regular CA-groupoid. Then, S is a commutative CA-groupoid.*

Proof. Let $(S, \times)$ be a regular CA-groupoid. By Theorem 5, S is a CA-NET-groupoid. By Corollary 1, S is a commutative CA-groupoid. □

Theorem 14. *Let $(S, \times)$ be a CA-groupoid. Then, the following statements are equivalent:*

(1) *S is a regular CA-groupoid;*
(2) *S is a strongly regular CA-groupoid;*
(3) *S is a CA-NET-groupoid;*
(4) *S is an inverse CA-groupoid; and*
(5) *S is a commutative regular semigroup.*

Proof. (1)⇒(2). Assume that S is a regular CA-groupoid. By Corollary 3, we know that S is a commutative CA-groupoid. Then, for any $a \in S$, there exists $x \in S$, such that $a = a \times (x \times a)$ and $a = (a \times x) \times a$. According to the definition of strongly regular CA-groupoid (Definition 5), S is a strongly regular CA-groupoid.

(2)⇒(3). Assume that S is a strongly regular CA-groupoid. By Definitions 4 and 5, S is a regular CA-groupoid. By Theorem 5, S is a CA-NET-groupoid.

(3)⇒(4). Let $(S, \times)$ be a CA-NET-groupoid. According to Theorem 4, the idempotent of S is commutative. By Theorem 5, S is a regular CA-groupoid. By Theorem 13, S is an inverse CA-groupoid.

(4)⇒(5). Let $(S, \times)$ be an inverse CA-groupoid. By Theorem 13, S is a regular CA-groupoid and its idempotent is commutative. Then, we only need proof a regular CA-groupoid is a commutative regular semigroup. By Corollary 3, S is a commutative CA-groupoid. For any $a, b, c \in S$, we have

$$a \times (b \times c) = c \times (a \times b) = (a \times b) \times c$$

and there exists $x \in S$, such that $a = a \times (x \times a) = a \times (a \times x) = (a \times x) \times a = a \times x \times a$.

Therefore, S is a commutative regular semigroup.

(5)⇒(1). Assume that $(S, \times)$ is a commutative regular semigroup. For any $a, b, c \in S$, we have

$$a \times (b \times c) = (a \times b) \times c = c \times (a \times b)$$

and there exists $x \in S$, such that $a = a \times x \times a = a \times (x \times a)$.

Therefore, S is a regular CA-groupoid. □

Example 16. *Let $S = \{1, 2, 3, 4\}$. The operation $\times$ on S is defined as Table 16. Then, $(S, \times)$ is a regular CA-groupoid, since $1 = 1 \times (1 \times 1)$, $2 = 2 \times (4 \times 2)$, $3 = 3 \times (3 \times 3)$, and $4 = 4 \times (4 \times 4)$. $(S, \times)$ is also a strongly regular CA-groupoid because $1 = 1 \times (1 \times 1)$, $1 = (1 \times 1) \times 1$; $2 = 2 \times (4 \times 2)$, $2 = (2 \times 4) \times 2$; $3 = 3 \times (3 \times 3)$, $3 = (3 \times 3) \times 3$; $4 = 4 \times (4 \times 4)$, and $4 = (4 \times 4) \times 4$. We can verify that $(S, \times)$ is a CA-NET-groupoid, and neut(1) = 1, anti(1) = 1; neut(2) = 2, anti(2) = {1, 2, 3, 4}; neut(3) = 3, anti(3) = 3; and neut(4) = 4, anti(4) = {1, 3, 4}. $(S, \times)$ is an inverse CA-groupoid, since $1 \times 2 = 2 \times 1$, $1 \times 3 = 3 \times 1$, $1 \times 4 = 4 \times 1$, $2 \times 3 = 3 \times 2$, $2 \times 4 = 4 \times 2$, and $3 \times 4 = 4 \times 3$. $(S, \times)$ is also a commutative regular semigroup because $1 = 1 \times 1 \times 1$, $2 = 2 \times 2 \times 2$, $3 = 3 \times 3 \times 3$, and $4 = 4 \times 4 \times 4$.*

Table 16. The operation $\times$ on S.

×	1	2	3	4
1	1	2	4	4
2	2	2	2	2
3	4	2	3	4
4	4	2	4	4

Corollary 4. *Let $(S, \times)$ be a strongly regular CA-groupoid. Then, S is a strongly regular semigroup.*

Proof. Let $(S, \times)$ be a strongly regular CA-groupoid. By Theorem 14 (2), (5), S is a strongly regular semigroup. □

7. Conclusions

Starting from various backgrounds (for examples, non-associative rings with $x(yz) = y(zx)$, cyclic associative Abel-Grassman groupoids, regular semigroup, and regular AG-groupoid), this paper introduces the concept of regular cyclic associative groupoid (CA-groupoid) for the first time. Furthermore, we study the relationship between regular CA-groupoids and other relevant algebraic structures. The research shows that the regular CA-groupoids, as a kind of non-associative algebraic structures, has typical representativeness and rich connotation, and is closely related to many kinds of algebraic structures. This paper concludes some important results, which are listed as follows:

(1) If an algebraic system is a regular CA-groupoid, then, each of its elements has an inverse and the inverse is unique (see Theorem 3 and Example 6).
(2) If an algebraic system is a CA-NET-groupoid, then, its idempotents are commutative (see Theorem 4).
(3) Every CA-NET-groupoid is commutative (see Corollary 1 and Example 10).
(4) An algebraic system is a regular CA-groupoid if and only if it is a CA-NET-groupoid (see Theorem 5 and Example 11).
(5) If an algebraic system is a CA-groupoid, then, $\mathcal{L}$ and $\mathcal{R}$ are not commutative. That is, as a binary relationship, $\mathcal{L}\circ\mathcal{R}\neq\mathcal{R}\circ\mathcal{L}$ (see Corollary 2 and Example 13).
(6) If an algebraic system is a regular CA-groupoid, then, the relations $\mathcal{L}$ and $\mathcal{R}$ commute. That is, as a binary relationship, $\mathcal{L}\circ\mathcal{R}=\mathcal{R}\circ\mathcal{L}$ (see Theorem 10 and Example 14).

(7) If an algebraic system is a regular CA-groupoid, then every element of S lies in a subgroup of S, and every $\mathcal{H}$-class in S is a group (see Theorem 12 and Example 15).
(8) An algebraic system is an inverse CA-groupoid if and only if it is a regular CA-groupoid and its idempotent is commutative (see Theorem 13 and Example 16).
(9) If an algebraic system is a regular CA-groupoid, then, it is a commutative CA-groupoid (see Corollary 3 and Example 16).
(10) An algebraic system is a regular CA-groupoid if and only if it is a commutative regular semigroup (see Theorem 14 and Example 16).

These results are important for exploring the structure characterizations of regular CA-groupoids and CA-NET-groupoids.

For future research, we will discuss the integration of the related topics, such as the ideals in CA-groupoids and the relationships among some algebraic structures (see [26–28]).

Author Contributions: W.Y. and X.Z. initiated the research and wrote the paper. All authors have read and agreed to the published version of the manuscript.

Funding: This research was funded by National Natural Science Foundation of China (Grant No. 61976130).

Conflicts of Interest: The authors declare no conflict of interest.

References

1. Petrich, M. The structure of completely regular semigroups. *Trans. Am. Math. Soc.* **1974**, *189*, 211–236. [CrossRef]
2. Petrich, M.; Reilly, N.R. *Completely Regular Semigroups*; John Wiley & Sons: New York, NY, USA, 1999.
3. Akinmoyewa, J.T. A study of some properties of generalized groups. *Octogon Math. Mag.* **2009**, *17*, 599–626.
4. Smarandache, F. *Neutrosophic Perspectives: Triplets, Duplets, Multisets, Hybrid Operators, Modal Logic, Hedge Algebras and Applications*; Pons Publishing House: Brussels, Belgium, 2017.
5. Smarandache, F.; Ali, M. Neutrosophic triplet group. *Neural Comput. Appl.* **2018**, *29*, 595–601. [CrossRef]
6. Zhang, X.H.; Wang, X.J.; Smarandache, F.; Jaiyeola, T.G.; Lian, T.Y. Singular neutrosophic extended triplet groups and generalized groups. *Cogn. Syst. Res.* **2019**, *57*, 32–40. [CrossRef]
7. Sholander, M. Medians, lattices, and trees. *Proc. Am. Math. Soc.* **1954**, *5*, 808–812. [CrossRef]
8. Maksa, G. CM solutions of some functional equations of associative type. *Ann. Univ. Sci. Bp. Sect. Comput.* **2004**, *24*, 125–132.
9. Kleinfeld, M. Rings with x(yz)=y(zx). *Commun. Algebra* **1995**, *23*, 5085–5093. [CrossRef]
10. Zhan, J.M.; Tan, Z.S. Left weakly Novikov algebra. *J. Math.* **2005**, *25*, 135–138.
11. Schafer, R.D. *An Introduction to Nonassociative Algebras*; Academic Press: Cambridge, MA, USA, 1966.
12. Sabinin, L.; Sbitneva, L.; Shestakov, I. *Non-Associative Algebra and Its Applications*; CRC Press: Boca Raton, FL, USA, 2006.
13. Chajda, I.; Halaš, R.; Länger, H. Operations and structures derived from non-associative MV-algebras. *Soft Comput.* **2019**, *23*, 3935–3944. [CrossRef]
14. Shah, T.; Razzaque, A.; Rehman, I.; Gondal, M.A.; Faraz, M.I.; Shum, K.P. Literature survey on non-associative rings and developments. *Eur. J. Pure Appl. Math.* **2019**, *12*, 370–408. [CrossRef]
15. Lazendic, S.; Pizurica, A.; De Bie, H. Hypercomplex algebras for dictionary learning. In Proceedings of the 7th Conference on Applied Geometric Algebras in Computer Science and Engineering–AGACSE 2018, Campinas, Brazil, 23–27 July 2018; pp. 57–64.
16. Hirsch, R.; Jackson, M.; Kowalski, T. Algebraic foundations for qualitative calculi and networks. *Theor. Comput. Sci.* **2019**, *768*, 99–116. [CrossRef]
17. Zhang, X.H.; Hu, Q.Q.; Smarandache, F.; An, X.G. On neutrosophic triplet groups: Basic properties, NT-subgroups, and some notes. *Symmetry* **2018**, *10*, 289. [CrossRef]
18. Ma, Y.C.; Zhang, X.H.; Yang, X.F.; Zhou, X. Generalized neutrosophic extended triplet group. *Symmetry* **2019**, *11*, 327. [CrossRef]
19. Smarandache, F. Neutrosophic set—A generialization of the intuituionistics fuzzy sets. *Int. J. Pure Appl. Math.* **2005**, *3*, 287–297.

20. Tehrim, S.T.; Riaz, M. A novel extension of TOPSIS to MCGDM with bipolar neutrosophic soft topology. *J. Intell. Fuzzy Syst.* **2019**, *37*, 5531–5549. [CrossRef]
21. Green, J.A. The structure of semigroups. *Ann. Math.* **1951**, *54*, 163–172. [CrossRef]
22. Mary, X. On generalized inverses and Green's relations. *Linear Algebra Appl.* **2011**, *434*, 1836–1844. [CrossRef]
23. Fleischer, L.; Kufleitner, M. Green's Relations in Finite Transformation Semigroups. *Comput. Sci. Theory Appl.* **2017**, *10304*, 112–125.
24. Warne, R.J. Orthodox congruences on strongly regular semigroups. *Comment. Math. Univ. St. Paul.* **1980**, *2*, 203–208.
25. Zhang, X.H.; Ma, Z.R.; Yuan, W.T. Cyclic associative groupoids (CA-groupoids) and cyclic associative neutrosophic extended triplet groupoids (CA-NET-groupoids). *Neutrosophic Sets Syst.* **2019**, *29*, 19–29.
26. Zhang, X.H.; Borzooei, R.A.; Jun, Y.B. Q-filters of quantum B-algebras and basic implication algebras. *Symmetry* **2018**, *10*, 573. [CrossRef]
27. Zhang, X.H.; Wu, X.Y.; Mao, X.Y.; Smarandache, F.; Park, C. On neutrosophic extended triplet groups (loops) and Abel-Grassmann's groupoids (AG-groupoids). *J. Intell. Fuzzy Syst.* **2019**, *37*, 5743–5753. [CrossRef]
28. Zhang, X.H.; Smarandache, F.; Ma, Y.C. Symmetry in hyperstructure: neutrosophic extended triplet semihypergroups and regular hypergroups. *Symmetry* **2019**, *11*, 1217. [CrossRef]

mathematics

MDPI

Article

Multi-Attribute Group Decision Making Based on Multigranulation Probabilistic Models with Interval-Valued Neutrosophic Information

Chao Zhang [1], Deyu Li [1,*], Xiangping Kang [2,3], Yudong Liang [1], Said Broumi [4] and Arun Kumar Sangaiah [5]

1 Key Laboratory of Computational Intelligence and Chinese Information Processing of Ministry of Education, School of Computer and Information Technology, Shanxi University, Taiyuan 030006, China; czhang@sxu.edu.cn (C.Z.); liangyudong@sxu.edu.cn (Y.L.)
2 Key Laboratory of Embedded System and Service Computing, Ministry of Education, Department of Computer Science and Technology, Tongji University, Shanghai 201804, China; tongji_kangxp@sina.com
3 China National Tobacco Corporation, Shanxi, Taiyuan 030000, China
4 Laboratory of Information Processing, Faculty of Science Ben M'Sik, University Hassan II, Sidi Othman, Casablanca B.P 7955, Morocco; broumisaid78@gmail.com
5 School of Computing Science and Engineering, Vellore Institute of Technology (VIT), Vellore 632014, India; arunkumarsangaiah@gmail.com
* Correspondence: lidysxu@163.com

Received: 8 January 2020; Accepted: 4 February 2020; Published: 9 February 2020

Abstract: In plenty of realistic situations, multi-attribute group decision-making (MAGDM) is ubiquitous and significant in daily activities of individuals and organizations. Among diverse tools for coping with MAGDM, granular computing-based approaches constitute a series of viable and efficient theories by means of multi-view problem solving strategies. In this paper, in order to handle MAGDM issues with interval-valued neutrosophic (IN) information, we adopt one of the granular computing (GrC)-based approaches, known as multigranulation probabilistic models, to address IN MAGDM problems. More specifically, after revisiting the related fundamental knowledge, three types of IN multigranulation probabilistic models are designed at first. Then, some key properties of the developed theoretical models are explored. Afterwards, a MAGDM algorithm for merger and acquisition target selections (M&A TSs) with IN information is summed up. Finally, a real-life case study together with several detailed discussions is investigated to present the validity of the developed models.

Keywords: multi-attribute group decision-making; granular computing; interval-valued neutrosophic information; multigranulation probabilistic models; merger and acquisition target selections

1. Introduction

1.1. A Brief Review of MAGDM

By applying decision-making issues with multiple attributes to the setting of group decision-making, multi-attribute group decision-making (MAGDM) generally provides consistent group preferences by analyzing various alternatives expressed by individual preferences [1]. To date, many granular computing (GrC)-based approaches [2–6] have been utilized to solve numerous complicated MAGDM problems, which have started a new momentum of constant development of social economy.

In the process of solving a typical MAGDM problem, it is recognized that three key challenges need to be managed reasonably, i.e., MAGDM information representation, MAGDM information fusion, MAGDM information analysis. Among the above-stated key challenges, how to express MAGDM information, especially for a complicated uncertain real-world scenario, as a standard decision matrix via alternatives and attributes is the first step to address MAGDM problems.

1.2. A Brief Review of Interval-valued Neutrosophic Information

In order to meet the demands of describing fuzzy and indeterminate information at the same time from nature and society, Smarandache [7,8] founded the notion of neutrosophic sets (NSs), which can be regarded as many generalizations of extended fuzzy sets [9] and used in plenty of meaningful areas [10–12]. An NS contains three types of membership functions (the truth, indeterminacy and falsity ones), and all of them take values in $]0^-, 1^+[$. In accordance with the mathematical formulation of NSs, using NSs directly to a range of realistic applications is relatively inconvenient because all membership functions are limited within $]0^-, 1^+[$. Thus, it is necessary to update $]0^-, 1^+[$ by virtue of standard sets and logic. Following the above-stated research route, Wang et al. [13] put forward the concept of IN sets (INSs) from the viewpoint of definitions, operational laws, and others. For each membership function in INSs, all of them take values in the power set of $[0, 1]$ instead of $]0^-, 1^+[$. Thus, INSs can tackle the first key challenge of solving typical MAGDM problems well [14–22].

1.3. A Brief Review of Multigranulation Probabilistic Models

To handle MAGDM information fusion and analysis with IN information effectively, GrC-based approaches own unique superiorities in constructing problem addressing approaches via multi-view problem solving tactics [23,24]. During the past several years, taking full advantage of GrC-based approaches, many scholars and practitioners have obtained fruitful results in merging NSs with rough sets [25–32], formal concept analysis [33,34], three-way decisions [35–37], and others [38,39]. In the current article, we plan to propose a new IN MAGDM method via multigranulation probabilistic models, which can provide a risk-based information synthesis scheme with the capability of error tolerance in light of GrC-based approaches. In particular, the notion of multigranulation rough sets (MGRSs) [40–42] and probabilistic rough sets (PRSs) [43–45] is scheduled to establish multigranulation probabilistic models, and the merits of MGRSs and PRSs can be reflected in the process of MAGDM problem addressing.

1.4. The Motivations of the Research

MGRSs play a significant role in dealing with MAGDM problems in diverse backgrounds. One one hand, some scholars have made eminent contributions to the applications of MGRSs in MAGDM problems in recent years. For instance, Zhang et al. [46,47] developed various MGRSs in the context of hesitant fuzzy and interval-valued hesitant fuzzy sets for handling person-job fit and steam turbine fault diagnosis, respectively. Sun et al. [48,49] proposed several MGRSs with linguistic and heterogeneous preference information, then they further designed corresponding MAGDM approaches. Zhan et al. [50] and Zhang et al. [51] put forward two novel covering-based MGRSs with fuzzy and intuitionistic fuzzy information for addressing MAGDM problems. On the other hand, some scholars adopted PRSs to address MAGDM problems. For instance, Liang et al. [52–54] studied novel decision-theoretic rough sets in hesitant fuzzy, incomplete, and Pythagorean information systems. Xu and Guo [55] generalized MGRSs to double-quantitative and three-way decision frameworks. Zhang et al. [56,57] combined decision-theoretic rough sets with MGRSs in Pythagorean and hesitant fuzzy linguistic information systems. In this paper, we generalize MGRSs and PRSs to IN information and apply them to M&A TSs. Specifically, the following motivations of utilizing MGRSs and PRSs in IN MAGDM problems can be summed up:

1. In order to address the challenge of MAGDM information representation, INSs take advantages of NSs and interval-valued sets at the same time. Thus, INSs play a significant role in describing indeterminate and incomplete MAGDM information.
2. In order to address the challenge of MAGDM information fusion, MGRSs excel in accelerating the information fusion procedure by virtue of processing multiple binary or fuzzy relations in parallel. In addition, classical MGRSs offer decision makers both an optimistic information fusion rule and a pessimistic counterpart. In conclusion, MGRSs play a significant role in constructing reasonable MAGDM information fusion methods [46–51].
3. In order to address the challenge of MAGDM information analysis, starting from the probability theory and Bayesian procedures, PRSs own the capability of fault tolerance. Hence, PRSs play a significant role in coping with incorrect and noisy data and they can be seen as a useful tool for robust MAGDM information analysis [52–57].

1.5. The Contributions of the Research

In this work, we aim to utilize MGRSs and PRSs in solving complicated MAGDM issues with IN information. Specifically, several comprehensive risk-based models named IN multigranulation PRSs (MG-PRSs) over two universes are looked into. Then, we further present a MAGDM approach in the setting of M&A TSs in light of the developed theoretical models that can avoid an impact of three above-mentioned challenges. Finally, a real-world example is employed to prove the validity of the established decision-making rule. In addition, it is noteworthy that plenty of interesting nonlinear modeling approaches have been proved to be successful in various applications [58–67]. For instance, Medina and Ojeda-Aciego [58] applied multi-adjoint frameworks to general t-concept lattice, and some other works on fuzzy formal contexts based on GrC-based approaches were explored in succession [64–67]. Takacs et al. [59] put forward a brand-new oft tissue model for constructing telesurgical robot systems. Gil et al. [60] studied a surrogate model based optimization of traffic lights cycles and green period ratios by means of microscopic simulation and fuzzy rule interpolation. Smarandache et al. [61] explored word-level sentiment similarities in the context of NSs, and some other meaningful works on word-level sentiment analysis were also investigated recently [62,63,68,69].

Compared with existing popular nonlinear modeling approaches, the vital contributions of the work lie in the utilization of IN information and multigranulation probabilistic models. For one thing, the above-mentioned literature on nonlinear modeling approaches can not process various practical situations with indeterminate and incomplete information effectively, thus this work shows some merits in the representation of uncertain MAGDM information. For another, the majority of nonlinear modeling approaches can not fuse and analyze multi-source information with incorrect and noisy data reasonably, thus this work shows some merits in the fusion and analysis of MAGDM problems with IN information. Moreover, several specific key contributions of the work can be further concluded below:

1. Several new IN membership degrees are put forward to handle incorrect and noisy data via the capability of fault tolerance.
2. Three types of multigranulation probabilistic models are designed according to diverse risk attitudes of decision makers, i.e., the first one in light of optimistic rules, the second one in light of pessimistic rules, and the third one in light of adjustable rules.
3. On the basis of GrC-based methods, IN MG-PRSs over two universes can address a typical IN MAGDM issue from multiple views, and integrate different individual preferences by considering risk appetites and error tolerance.

1.6. The Structure of the Research

The rest of the work is arranged below. The next section intends to review several basic knowledge on INSs, MGRSs, and PRSs. Three types of theoretical models along with their key properties are explored in Section 3. In the next section, we develop an IN MAGDM approach via multigranulation probabilistic models in the context of M&A TSs. In Section 5, a practical illustrative case study is explored to highlight the validity of the presented IN MAGDM rule. Finally, Section 6 contains several conclusive results and future study options.

2. Preliminaries

The current section plans to revisit various preliminary knowledge in terms of INSs, MGRSs, and PRSs in a brief outline.

2.1. INSs

The concept of INSs was put forward by Wang et al. [13] by updating the formulation of $]0^-, 1^+[$ from the scope of membership functions in NSs. With this update, INSs are equipped with the capability of expressing indeterminate and incomplete information simultaneously.

Definition 1 ([13]). *Suppose U is an arbitrary universe of discourse. An INS E over U is provided as the following mathematical expression:*

$$E = \{\langle x, [\mu_E^L(x), \mu_E^U(x)], [\nu_E^L(x), \nu_E^U(x)], [\omega_E^L(x), \omega_E^U(x)]\rangle \,|\, x \in U\},$$

where $\mu, \nu, \omega : U \to \text{int}[0,1]$ ($\text{int}[0,1]$ represents the set of all closed subintervals of $[0,1]$). Similar with NSs, there also exists the restriction of $0 \leq \mu_E^U(x) + \nu_E^U(x) + \omega_E^U(x) \leq 3$. Moreover, a set that includes all INSs over U is further named $IN(U)$.

For an INS E, it is noticed that E may degenerate into two special forms, i.e., an INS E is named a full INS U when $[\mu_E^L(x), \mu_E^U(x)] = [1,1]$, $[\nu_E^L(x), \nu_E^U(x)] = [0,0]$ and $[\omega_E^L(x), \omega_E^U(x)] = [0,0]$; an INS E is named an empty INS $\emptyset$ when $[\mu_E^L(x), \mu_E^U(x)] = [0,0]$, $[\nu_E^L(x), \nu_E^U(x)] = [1,1]$ and $[\omega_E^L(x), \omega_E^U(x)] = [1,1]$.

In IN MAGDM information analysis, it is common to compare the magnitude of IN numbers, thus a frequently-used method was developed in [15].

Definition 2 ([15]). *Suppose $x = \langle[\mu_E^L(x), \mu_E^U(x)], [\nu_E^L(x), \nu_E^U(x)], [\omega_E^L(x), \omega_E^U(x)]\rangle$ is an IN number. A score function with regard to x is provided as the following mathematical expression:*

$$s(x) = [\mu_E^L(x) + 1 - \nu_E^U(x) + 1 - \omega_E^U(x), \mu_E^U(x) + 1 - \nu_E^L(x) + 1 - \omega_E^L(x)],$$

for arbitrary two IN numbers x and y, $x \leq y \Leftrightarrow s(x) \leq s(y)$ is evident.

It is noted that another significant issue in IN MAGDM information analysis is viable operational laws, which are used in constructing various IN MAGDM models. In what follows, we present several IN operational laws.

Definition 3 ([13,15,22]). *Suppose E and F are arbitrary two INSs over U, then some common IN operational laws are provided as the following mathematical expressions:*

1. $E \boxplus F = \{\langle x, [1-(1-\mu_E^L(x))(1-\mu_F^L(x)), 1-(1-\mu_E^U(x))(1-\mu_F^U(x))], [\nu_E^L(x)\nu_F^L(x), \nu_E^U(x)\nu_F^U(x)], [\omega_E^L(x)\omega_F^L(x), \omega_E^U(x)\omega_F^U(x)]\rangle \,|\, x \in U\}$;
2. $E \boxtimes F = \{\langle x, [\mu_E^L(x)\mu_F^L(x), \mu_E^U(x)\mu_F^U(x)], [1-(1-\nu_E^L(x))(1-\nu_F^L(x)), 1-(1-\nu_E^U(x))(1-\nu_F^U(x))], [1-(1-\omega_E^L(x))(1-\omega_F^L(x)), 1-(1-\omega_E^U(x))(1-\omega_F^U(x))]\rangle \,|\, x \in U\}$;
3. $\lambda E = \{\langle x, [1-(1-\mu_E^L(x))^\lambda, 1-(1-\mu_E^U(x))^\lambda], [(\nu_E^L(x))^\lambda, (\nu_E^U(x))^\lambda], [(\omega_E^L(x))^\lambda, (\omega_E^U(x))^\lambda]\rangle \,|\, x \in U\}$;
4. $E^\lambda = \{\langle x, [(\mu_E^L(x))^\lambda, (\mu_E^U(x))^\lambda], [1-(1-\nu_E^L(x))^\lambda, 1-(1-\nu_E^U(x))^\lambda] [1-(1-\omega_E^L(x))^\lambda, 1-(1-\omega_E^U(x))^\lambda]\rangle \,|\, x \in U\}$;

5. $E \boxminus F = \{\langle x, [\frac{\mu_E^L(x)-\mu_F^L(x)}{1-\mu_F^L(x)}, \frac{\mu_E^U(x)-\mu_F^U(x)}{1-\mu_F^U(x)}], [\frac{\nu_E^L(x)}{\nu_F^L(x)}, \frac{\nu_E^U(x)}{\nu_F^U(x)}], [\frac{\omega_E^L(x)}{\omega_F^L(x)}, \frac{\omega_E^U(x)}{\omega_F^U(x)}]\rangle \,|x \in U\}$, *if* $E \supseteq F$, $\nu_E^L(x) \leq \frac{\nu_E^U(x)\nu_F^L(x)}{\nu_F^U(x)}$ *and* $\omega_E^L(x) \leq \frac{\omega_E^U(x)\omega_F^L(x)}{\omega_F^U(x)}$;
6. $E \boxdot F = \{\langle x, [\frac{\mu_E^L(x)}{\mu_F^L(x)}, \frac{\mu_E^U(x)}{\mu_F^U(x)}], [\frac{\nu_E^L(x)-\nu_F^L(x)}{1-\nu_F^L(x)}, \frac{\nu_E^U(x)-\nu_F^U(x)}{1-\nu_F^U(x)}], [\frac{\omega_E^L(x)-\omega_F^L(x)}{1-\omega_F^L(x)}, \frac{\omega_E^U(x)-\omega_F^U(x)}{1-\omega_F^U(x)}]\rangle \,|x \in U\}$, *if* $F \supseteq E$ *and* $\omega_E^L(x) \leq \frac{\omega_E^U(x)\omega_F^L(x)}{\omega_F^U(x)}$;
7. $E^c = \{\langle x, [\omega_E^L(x), \omega_E^U(x)], [1-\nu_E^U(x), 1-\nu_E^L(x)], [\mu_E^L(x), \mu_E^U(x)]\rangle \,|x \in U\}$;
8. $E \cup F = \{\langle x, \left[\max\left(\mu_E^L(x), \mu_F^L(x)\right), \max\left(\mu_E^U(x), \mu_F^U(x)\right)\right], \left[\min\left(\nu_E^L(x), \nu_F^L(x)\right), \min\left(\nu_E^U(x), \nu_F^U(x)\right)\right], \left[\min\left(\omega_E^L(x), \omega_F^L(x)\right), \min\left(\omega_E^U(x), \omega_F^U(x)\right)\right]\rangle \,|x \in U\}$;
9. $E \cap F = \{\langle x, \left[\min\left(\mu_E^L(x), \mu_F^L(x)\right), \min\left(\mu_E^U(x), \mu_F^U(x)\right)\right], \left[\max\left(\nu_E^L(x), \nu_F^L(x)\right), \max\left(\nu_E^U(x), \nu_F^U(x)\right)\right], \left[\max\left(\omega_E^L(x), \omega_F^L(x)\right), \max\left(\omega_E^U(x), \omega_F^U(x)\right)\right]\rangle \,|x \in U\}$;
10. $E \subseteq F \Leftrightarrow \mu_E^L(x) \leq \mu_F^L(x), \mu_E^U(x) \leq \mu_F^U(x), \nu_E^L(x) \geq \nu_F^L(x), \nu_E^U(x) \geq \nu_F^U(x), \omega_E^L(x) \geq \omega_F^L(x), \omega_E^U(x) \geq \omega_F^U(x)$;
11. $E = F \Leftrightarrow E \subseteq F, E \supseteq F$.

2.2. MGRSs

As one of the most influential generalized rough set theories, the idea of MGRSs was initially established by Qian et al. [40–42] by means of parallel computational frameworks and risk-based information fusion strategies.

Definition 4 ([40,41])**.** *Suppose $R_1, R_2, \ldots, R_m$ are m crisp binary relations. For any $X \subseteq U$, the optimistic and pessimistic multigranulation approximations of X are provided as the following mathematical expressions:*

$$\underline{\sum_{i=1}^{m} R_i}^{O}(X) = \{[x]_{R_1} \subseteq X \vee [x]_{R_1} \subseteq X \vee \ldots \vee [x]_{R_m} \,|x \in U\};$$

$$\overline{\sum_{i=1}^{m} R_i}^{O}(X) = (\underline{\sum_{i=1}^{m} R_i}^{O}(X^c))^c;$$

$$\underline{\sum_{i=1}^{m} R_i}^{P}(X) = \{[x]_{R_1} \subseteq X \wedge [x]_{R_1} \subseteq X \wedge \ldots \wedge [x]_{R_m} \,|x \in U\};$$

$$\overline{\sum_{i=1}^{m} R_i}^{P}(X) = (\underline{\sum_{i=1}^{m} R_i}^{P}(X^c))^c,$$

the pair $(\underline{\sum_{i=1}^{m} R_i}^{O}(X), \overline{\sum_{i=1}^{m} R_i}^{O}(X))$ *is named an optimistic MGRS with regard to X, whereas the pair* $(\underline{\sum_{i=1}^{m} R_i}^{P}(X), \overline{\sum_{i=1}^{m} R_i}^{P}(X))$ *is named a pessimistic MGRS with regard to X.*

2.3. PRSs

Considering that the formulation of classical rough sets is fairly strict which may affect the application range of it, hence the concept of PRSs [43–45] was developed subsequently by means of the probabilistic measure theory.

Definition 5 ([43]). *Suppose R is an equivalence relation over U, P is the probabilistic measure, then (U, R, P) is named a probabilistic approximation space. For any $0 \leq \beta < \alpha \leq 1$ and $X \subseteq U$, the lower and upper approximations of X are provided as the following mathematical expressions:*

$$\underline{R}_{\alpha}(X) = \{P(X|[x]_R) \geq \alpha | x \in U\};$$
$$\overline{R}_{\beta}(X) = \{P(X|[x]_R) > \beta | x \in U\},$$

the pair $(\underline{R}_{\alpha}(X), \overline{R}_{\beta}(X))$ is named a PRS of X with regard to (U, R, P).

3. IN MG-PRSs over Two Universes

In what follows, prior to the introduction of new theoretical models, we shall revisit the formulation of IN relations within the context of two universes [28] at first.

Definition 6 ([28]). *Suppose U and V are two an arbitrary universes of discourse. An IN relation over two universes R over $U \times V$ is provided as the following mathematical expression:*

$$R = \{\langle (x,y), [\mu_R^L(x,y), \mu_R^U(x,y)], [\nu_R^L(x,y), \nu_R^U(x,y)], [\omega_R^L(x,y), \omega_R^U(x,y)]\rangle | (x,y) \in U \times V\},$$

where $\mu, \nu, \omega : U, V \rightarrow \mathrm{int}[0,1]$. Similar with INSs, there also exists the restriction of $0 \leq \mu_R^U(x,y) + \nu_R^U(x,y) + \omega_R^U(x,y) \leq 3$. Moreover, a set that includes all IN relations over $U \times V$ is further named $INR(U \times V)$.

3.1. Optimistic IN MG-PRSs over Two Universes

It is noted that the term "optimistic" is originated from the first paper of MGRSs [40]. Within the context of MGRSs, the notion of single and multiple IN inclusion degrees is scheduled to propose at first in the current section.

Definition 7. *Suppose R_i is an IN relation over two universes over $U \times V$. For any $E \in IN(V)$, $x \in U$, $y \in V$, the single IN membership degree of x in E in terms of R_i is provided as the following mathematical expression:*

$$\eta_E^{R_i}(x) = \frac{\sum_{y \in V} R_i(x,y) E(y)}{\sum_{y \in V} R_i(x,y)},$$

based on $\eta_E^{R_i}(x)$, the multiple IN membership degrees of x in E with regard to R_i are provided as the following mathematical expressions:

$$\Psi_E^{\sum_{i=1}^{m} R_i}(x) = \min_{i=1}^{m} \eta_E^{R_i}(x);$$
$$\Omega_E^{\sum_{i=1}^{m} R_i}(x) = \max_{i=1}^{m} \eta_E^{R_i}(x),$$

$\Psi_E^{\sum_{i=1}^{m} R_i}(x)$ is named a minimal IN membership degree, whereas we call $\Omega_E^{\sum_{i=1}^{m} R_i}(x)$ a maximal IN membership degree.

In light of maximal IN membership degrees, optimistic multigranulation probabilistic models can be put forward conveniently.

Definition 8. *Suppose R_i is an IN relation over two universes over $U \times V$. For any $E \in IN(V)$, $x \in U$, $y \in V$, the two IN thresholds are represented by α and β with $\alpha > \beta$, then the lower and upper approximations of E in optimistic multigranulation probabilistic models are provided as the following mathematical expressions:*

$$\underline{\sum_{i=1}^{m} R_i}^{\Omega,\alpha}(E) = \{\Omega_E^{\sum_{i=1}^{m} R_i}(x) \geq \alpha \,|\, x \in U\};$$
$$\overline{\sum_{i=1}^{m} R_i}^{\Omega,\beta}(E) = \{\Omega_E^{\sum_{i=1}^{m} R_i}(x) > \beta \,|\, x \in U\},$$

the pair $(\underline{\sum_{i=1}^{m} R_i}^{\Omega,\alpha}(E), \overline{\sum_{i=1}^{m} R_i}^{\Omega,\beta}(E))$ *is named an optimistic IN MG-PRS over two universes of E.*

In what follows, some key properties of lower and upper approximations for optimistic multigranulation probabilistic models are explored in detail.

Proposition 1. *Suppose R_i is an IN relation over two universes over $U \times V$. For any $E, F \in IN(V)$, $x \in U$, $y \in V$, the two IN thresholds are denoted by α and β with $\alpha > \beta$, then the lower and upper approximations for optimistic multigranulation probabilistic models own the following properties:*

1. $\underline{\sum_{i=1}^{m} R_i}^{\Omega,\alpha}(\emptyset) = \emptyset$, $\overline{\sum_{i=1}^{m} R_i}^{\Omega,\beta}(V) = U$;
2. $\alpha_1 \leq \alpha_2 \Rightarrow \underline{\sum_{i=1}^{m} R_i}^{\Omega,\alpha_2}(E) \subseteq \underline{\sum_{i=1}^{m} R_i}^{\Omega,\alpha_1}(E)$, $\beta_1 \leq \beta_2 \Rightarrow \overline{\sum_{i=1}^{m} R_i}^{\Omega,\beta_2}(E) \subseteq \overline{\sum_{i=1}^{m} R_i}^{\Omega,\beta_1}(E)$;
3. $E \subseteq F \Rightarrow \underline{\sum_{i=1}^{m} R_i}^{\Omega,\alpha}(E) \subseteq \underline{\sum_{i=1}^{m} R_i}^{\Omega,\alpha}(F)$, $\overline{\sum_{i=1}^{m} R_i}^{\Omega,\beta}(E) \subseteq \overline{\sum_{i=1}^{m} R_i}^{\Omega,\beta}(F)$;
4. $\underline{\sum_{i=1}^{m} R_i}^{\Omega,\alpha}(E \cup F) \supseteq \underline{\sum_{i=1}^{m} R_i}^{\Omega,\alpha}(E) \cup \underline{\sum_{i=1}^{m} R_i}^{\Omega,\alpha}(F)$, $\overline{\sum_{i=1}^{m} R_i}^{\Omega,\beta}(E \cap F) \subseteq \overline{\sum_{i=1}^{m} R_i}^{\Omega,\beta}(E) \cap \overline{\sum_{i=1}^{m} R_i}^{\Omega,\beta}(F)$;
5. $\underline{\sum_{i=1}^{m} R_i}^{\Omega,\alpha}(E \cap F) \subseteq \underline{\sum_{i=1}^{m} R_i}^{\Omega,\alpha}(E) \cap \underline{\sum_{i=1}^{m} R_i}^{\Omega,\alpha}(F)$, $\overline{\sum_{i=1}^{m} R_i}^{\Omega,\beta}(E \cup F) \supseteq \overline{\sum_{i=1}^{m} R_i}^{\Omega,\beta}(E) \cup \overline{\sum_{i=1}^{m} R_i}^{\Omega,\beta}(F)$.

Proof.

1. $\underline{\sum_{i=1}^{m} R_i}^{\Omega,\alpha}(\emptyset) = \{\Omega_\emptyset^{\sum_{i=1}^{m} R_i}(x) \geq \alpha \,|\, x \in U\} = \{\max_{i=1}^{m} \frac{\sum_{y \in V} R_i(x,y)\emptyset}{\sum_{y \in V} R_i(x,y)} \geq \alpha \,|\, x \in U\} = \emptyset$, $\overline{\sum_{i=1}^{m} R_i}^{\Omega,\beta}(V) = \{\Omega_V^{\sum_{i=1}^{m} R_i}(x) > \beta \,|\, x \in U\} = \{\max_{i=1}^{m} \frac{\sum_{y \in V} R_i(x,y)V}{\sum_{y \in V} R_i(x,y)} > \beta \,|\, x \in U\} = U$. Thus, $\underline{\sum_{i=1}^{m} R_i}^{\Omega,\alpha}(\emptyset) = \emptyset$ and $\overline{\sum_{i=1}^{m} R_i}^{\Omega,\beta}(V) = U$ can be obtained.
2. Since $\alpha_1 \leq \alpha_2$, we have $\underline{\sum_{i=1}^{m} R_i}^{\Omega,\alpha_2}(E) = \{\Omega_E^{\sum_{i=1}^{m} R_i}(x) \geq \alpha_2 \,|\, x \in U\} \subseteq \{\Omega_E^{\sum_{i=1}^{m} R_i}(x) \geq \alpha_1 \,|\, x \in U\} = \underline{\sum_{i=1}^{m} R_i}^{\Omega,\alpha_1}(E)$. Thus, $\alpha_1 \leq \alpha_2 \Rightarrow \underline{\sum_{i=1}^{m} R_i}^{\Omega,\alpha_2}(E) \subseteq \underline{\sum_{i=1}^{m} R_i}^{\Omega,\alpha_1}(E)$ can be concluded. In a similar manner, we can also prove $\beta_1 \leq \beta_2 \Rightarrow \overline{\sum_{i=1}^{m} R_i}^{\Omega,\beta_2}(E) \subseteq \overline{\sum_{i=1}^{m} R_i}^{\Omega,\beta_1}(E)$.

3. Since $E \subseteq F$, we have $\underline{\sum_{i=1}^{m} R_i}^{\Omega,\alpha}(E) = \{\Omega_E^{\sum_{i=1}^{m} R_i}(x) \geq \alpha \,|x \in U\} = \{\max_{i=1}^{m} \frac{\sum_{y\in V} R_i(x,y)E(y)}{\sum_{y\in V} R_i(x,y)} \geq \alpha \,|x \in U\} \subseteq \{\Omega_F^{\sum_{i=1}^{m} R_i}(x) \geq \alpha \,|x \in U\} = \{\max_{i=1}^{m} \frac{\sum_{y\in V} R_i(x,y)F(y)}{\sum_{y\in V} R_i(x,y)} \geq \alpha \,|x \in U\} = \underline{\sum_{i=1}^{m} R_i}^{\Omega,\alpha}(F)$. Thus, $E \subseteq F \Rightarrow \underline{\sum_{i=1}^{m} R_i}^{\Omega,\alpha}(E) \subseteq \underline{\sum_{i=1}^{m} R_i}^{\Omega,\alpha}(F)$ can be deduced. Similarly, $E \subseteq F \Rightarrow \overline{\sum_{i=1}^{m} R_i}^{\Omega,\beta}(E) \subseteq \overline{\sum_{i=1}^{m} R_i}^{\Omega,\beta}(F)$ can also be proved.
4. $\underline{\sum_{i=1}^{m} R_i}^{\Omega,\alpha}(E \cup F) = \{\max_{i=1}^{m} \frac{\sum_{y\in V} R_i(x,y)(E\cup F)(y)}{\sum_{y\in V} R_i(x,y)} \geq \alpha \,|x \in U\} \geq \{\max_{i=1}^{m} \frac{\sum_{y\in V} R_i(x,y)E(y)}{\sum_{y\in V} R_i(x,y)} \geq \alpha \,|x \in U\}$, $\underline{\sum_{i=1}^{m} R_i}^{\Omega,\alpha}(E \cup F) = \{\max_{i=1}^{m} \frac{\sum_{y\in V} R_i(x,y)(E\cup F)(y)}{\sum_{y\in V} R_i(x,y)} \geq \alpha \,|x \in U\} \geq \{\max_{i=1}^{m} \frac{\sum_{y\in V} R_i(x,y)F(y)}{\sum_{y\in V} R_i(x,y)} \geq \alpha \,|x \in U\}$. Thus, $\underline{\sum_{i=1}^{m} R_i}^{\Omega,\alpha}(E \cup F) \supseteq \underline{\sum_{i=1}^{m} R_i}^{\Omega,\alpha}(E) \cup \underline{\sum_{i=1}^{m} R_i}^{\Omega,\alpha}(F)$ can be concluded. Similarly, we can also prove $\overline{\sum_{i=1}^{m} R_i}^{\Omega,\beta}(E \cap F) \subseteq \overline{\sum_{i=1}^{m} R_i}^{\Omega,\beta}(E) \cap \overline{\sum_{i=1}^{m} R_i}^{\Omega,\beta}(F)$.
5. According to the above conclusions, $\underline{\sum_{i=1}^{m} R_i}^{\Omega,\alpha}(E \cap F) \subseteq \underline{\sum_{i=1}^{m} R_i}^{\Omega,\alpha}(E) \cap \underline{\sum_{i=1}^{m} R_i}^{\Omega,\alpha}(F)$ and $\overline{\sum_{i=1}^{m} R_i}^{\Omega,\beta}(E \cup F) \supseteq \overline{\sum_{i=1}^{m} R_i}^{\Omega,\beta}(E) \cup \overline{\sum_{i=1}^{m} R_i}^{\Omega,\beta}(F)$ can be deduced analogously.

□

3.2. Pessimistic IN MG-PRSs over Two Universes

According to previous definitions, starting from minimal IN membership degrees, pessimistic multigranulation probabilistic models can be established in a similar way.

Definition 9. *Suppose R_i is an IN relation over two universes over $U \times V$. For any $E \in IN(V)$, $x \in U$, $y \in V$, the two IN thresholds are represented by α and β with $\alpha > \beta$, then the lower and upper approximations of E in pessimistic multigranulation probabilistic models are provided as the following mathematical expressions:*

$$\underline{\sum_{i=1}^{m} R_i}^{\Psi,\alpha}(E) = \{\Psi_E^{\sum_{i=1}^{m} R_i}(x) \geq \alpha \,|x \in U\};$$

$$\overline{\sum_{i=1}^{m} R_i}^{\Psi,\beta}(E) = \{\Psi_E^{\sum_{i=1}^{m} R_i}(x) > \beta \,|x \in U\},$$

the pair $(\underline{\sum_{i=1}^{m} R_i}^{\Psi,\alpha}(E), \overline{\sum_{i=1}^{m} R_i}^{\Psi,\beta}(E))$ is named a pessimistic IN MG-PRS over two universes of E.

Next, some key properties of lower and upper approximations for pessimistic multigranulation probabilistic models are presented, and we can prove them according to above-mentioned proofs for Proposition 1.

Proposition 2. *Suppose R_i is an IN relation over two universes over $U \times V$. For any $E, F \in IN(V)$, $x \in U$, $y \in V$, the two IN thresholds are denoted by α and β with $\alpha > \beta$, then the lower and upper approximations for pessimistic multigranulation probabilistic models own the following properties:*

1. $\underline{\sum_{i=1}^{m} R_i}^{\Psi,\alpha}(\emptyset) = \emptyset$, $\overline{\sum_{i=1}^{m} R_i}^{\Psi,\beta}(V) = U$;

2. $\alpha_1 \le \alpha_2 \Rightarrow \underline{\sum_{i=1}^{m} R_i}^{\Psi,\alpha_2}(E) \subseteq \underline{\sum_{i=1}^{m} R_i}^{\Psi,\alpha_1}(E)$, $\beta_1 \le \beta_2 \Rightarrow \overline{\sum_{i=1}^{m} R_i}^{\Psi,\beta_2}(E) \subseteq \overline{\sum_{i=1}^{m} R_i}^{\Psi,\beta_1}(E)$;
3. $E \subseteq F \Rightarrow \underline{\sum_{i=1}^{m} R_i}^{\Psi,\alpha}(E) \subseteq \underline{\sum_{i=1}^{m} R_i}^{\Psi,\alpha}(F)$, $\overline{\sum_{i=1}^{m} R_i}^{\Psi,\beta}(E) \subseteq \overline{\sum_{i=1}^{m} R_i}^{\Psi,\beta}(F)$;
4. $\underline{\sum_{i=1}^{m} R_i}^{\Psi,\alpha}(E \cup F) \supseteq \underline{\sum_{i=1}^{m} R_i}^{\Psi,\alpha}(E) \cup \underline{\sum_{i=1}^{m} R_i}^{\Psi,\alpha}(F)$, $\overline{\sum_{i=1}^{m} R_i}^{\Psi,\beta}(E \cap F) \subseteq \overline{\sum_{i=1}^{m} R_i}^{\Psi,\beta}(E) \cap \overline{\sum_{i=1}^{m} R_i}^{\Psi,\beta}(F)$;
5. $\underline{\sum_{i=1}^{m} R_i}^{\Psi,\alpha}(E \cap F) \subseteq \underline{\sum_{i=1}^{m} R_i}^{\Psi,\alpha}(E) \cap \underline{\sum_{i=1}^{m} R_i}^{\Psi,\alpha}(F)$, $\overline{\sum_{i=1}^{m} R_i}^{\Psi,\beta}(E \cup F) \supseteq \overline{\sum_{i=1}^{m} R_i}^{\Psi,\beta}(E) \cup \overline{\sum_{i=1}^{m} R_i}^{\Psi,\beta}(F)$.

3.3. Adjustable IN MG-PRSs over Two Universes

Starting from two mathematical formulations of optimistic and pessimistic multigranulation probabilistic models, the former one utilizes maximal IN membership degrees $\Omega_E^{\sum_{i=1}^{m} R_i}(x)$ to establish related IN approximations, which indicates a risk-seeking information fusion tactic in MAGDM issues. On the contrary, the latter one utilizes minimal IN membership degrees $\Psi_E^{\sum_{i=1}^{m} R_i}(x)$ to establish related IN approximations, which indicates a risk-averse information fusion tactic in MAGDM issues. However, it is noted that both optimistic and pessimistic information fusion tactics are qualitative and static, they lack the ability of expressing risks of information fusion tactics from quantitative and dynamic standpoints. In order to quantitatively and dynamically describe the risk preference of information fusion tactics, adjustable IN membership degrees should be put forward by introducing the notion of risk coefficients, then adjustable IN MG-PRSs over two universes can be developed conveniently.

In what follows, adjustable IN membership degrees are defined by means of $\Omega_E^{\sum_{i=1}^{m} R_i}(x)$ and $\Psi_E^{\sum_{i=1}^{m} R_i}(x)$.

Definition 10. *Suppose R_i is an IN relation over two universes over $U \times V$, λ ($\lambda \in [0,1]$) is a risk coefficient. For any $E \in IN(V)$, $x \in U$, $y \in V$, the adjustable IN membership degrees of x in E with regard to R_i are provided as the following mathematical expressions:*

$$\Xi_E^{\sum_{i=1}^{m} R_i}(x) = \lambda \Omega_E^{\sum_{i=1}^{m} R_i}(x) \boxplus (1-\lambda)\, \Psi_E^{\sum_{i=1}^{m} R_i}(x).$$

Next, adjustable multigranulation probabilistic models can be designed similarly.

Definition 11. *Suppose R_i is an IN relation over two universes over $U \times V$. For any $E \in IN(V)$, $x \in U$, $y \in V$, the two IN thresholds are represented by α and β with $\alpha > \beta$, then the lower and upper approximations of E in adjustable multigranulation probabilistic models are provided as the following mathematical expressions:*

$$\underline{\sum_{i=1}^{m} R_i}^{\Xi,\alpha}(E) = \{\Xi_E^{\sum_{i=1}^{m} R_i}(x) \ge \alpha \,|\, x \in U\};$$

$$\overline{\sum_{i=1}^{m} R_i}^{\Xi,\beta}(E) = \{\Xi_E^{\sum_{i=1}^{m} R_i}(x) > \beta \,|\, x \in U\},$$

the pair $\left(\underline{\sum_{i=1}^{m} R_i}^{\Xi,\alpha}(E), \overline{\sum_{i=1}^{m} R_i}^{\Xi,\beta}(E)\right)$ is named an adjustable IN MG-PRS over two universes of E.

In what follows, some key properties of lower and upper approximations for adjustable multigranulation probabilistic models are presented, and we can also prove them according to above-mentioned proofs for Proposition 1.

Proposition 3. *Suppose R_i is an IN relation over two universes over $U \times V$. For any $E, F \in IN(V)$, $x \in U$, $y \in V$, the two IN thresholds are denoted by α and β with $\alpha > \beta$, then the lower and upper approximations for adjustable multigranulation probabilistic models own the following properties:*

1. $\underline{\sum_{i=1}^{m} R_i}^{\Xi,\alpha}(\emptyset) = \emptyset$, $\overline{\sum_{i=1}^{m} R_i}^{\Xi,\beta}(V) = U$;
2. $\alpha_1 \le \alpha_2 \Rightarrow \underline{\sum_{i=1}^{m} R_i}^{\Xi,\alpha_2}(E) \subseteq \underline{\sum_{i=1}^{m} R_i}^{\Xi,\alpha_1}(E)$, $\beta_1 \le \beta_2 \Rightarrow \overline{\sum_{i=1}^{m} R_i}^{\Xi,\beta_2}(E) \subseteq \overline{\sum_{i=1}^{m} R_i}^{\Xi,\beta_1}(E)$;
3. $E \subseteq F \Rightarrow \underline{\sum_{i=1}^{m} R_i}^{\Xi,\alpha}(E) \subseteq \underline{\sum_{i=1}^{m} R_i}^{\Xi,\alpha}(F)$, $\overline{\sum_{i=1}^{m} R_i}^{\Xi,\beta}(E) \subseteq \overline{\sum_{i=1}^{m} R_i}^{\Xi,\beta}(F)$;
4. $\underline{\sum_{i=1}^{m} R_i}^{\Xi,\alpha}(E \cup F) \supseteq \underline{\sum_{i=1}^{m} R_i}^{\Xi,\alpha}(E) \cup \underline{\sum_{i=1}^{m} R_i}^{\Xi,\alpha}(F)$, $\overline{\sum_{i=1}^{m} R_i}^{\Xi,\beta}(E \cap F) \subseteq \overline{\sum_{i=1}^{m} R_i}^{\Xi,\beta}(E) \cap \overline{\sum_{i=1}^{m} R_i}^{\Xi,\beta}(F)$;
5. $\underline{\sum_{i=1}^{m} R_i}^{\Xi,\alpha}(E \cap F) \subseteq \underline{\sum_{i=1}^{m} R_i}^{\Xi,\alpha}(E) \cap \underline{\sum_{i=1}^{m} R_i}^{\Xi,\alpha}(F)$, $\overline{\sum_{i=1}^{m} R_i}^{\Xi,\beta}(E \cup F) \supseteq \overline{\sum_{i=1}^{m} R_i}^{\Xi,\beta}(E) \cup \overline{\sum_{i=1}^{m} R_i}^{\Xi,\beta}(F)$.

3.4. Relationships between Optimistic, Pessimistic, and Adjustable IN MG-PRSs over Two Universes

In previous sections, three types of multigranulation probabilistic models with IN Information are investigated in detail. The following section aims to discuss relationships between optimistic, pessimistic, and adjustable multigranulation probabilistic models.

Proposition 4. *Suppose R_i is an IN relation over two universes over $U \times V$. For any $E, F \in IN(V)$, $x \in U$, $y \in V$, the two IN thresholds are denoted by α and β with $\alpha > \beta$; then, we have:*

1. $\underline{\sum_{i=1}^{m} R_i}^{\Psi,\alpha}(E) \subseteq \underline{\sum_{i=1}^{m} R_i}^{\Xi,\alpha}(E) \subseteq \underline{\sum_{i=1}^{m} R_i}^{\Omega,\alpha}(E)$;
2. $\overline{\sum_{i=1}^{m} R_i}^{\Psi,\alpha}(E) \subseteq \overline{\sum_{i=1}^{m} R_i}^{\Xi,\alpha}(E) \subseteq \overline{\sum_{i=1}^{m} R_i}^{\Omega,\alpha}(E)$.

Proof.

1. $\underline{\sum_{i=1}^{m} R_i}^{\Omega,\alpha}(E) = \{\Omega_E^{\sum_{i=1}^{m} R_i}(x) \ge \alpha \,|\, x \in U\} = \{\max_{i=1}^{m} \eta_E^{R_i}(x) \ge \alpha \,|\, x \in U\} \ge$ $\{(\lambda \max_{i=1}^{m} \eta_E^{R_i}(x) \boxplus (1-\lambda) \min_{i=1}^{m} \eta_E^{R_i}(x)) \ge \alpha \,|\, x \in U\} = \underline{\sum_{i=1}^{m} R_i}^{\Xi,\alpha}(E) \ge$ $\{\min_{i=1}^{m} \eta_E^{R_i}(x) \ge \alpha \,|\, x \in U\} = \underline{\sum_{i=1}^{m} R_i}^{\Psi,\alpha}(E)$. Thus, $\underline{\sum_{i=1}^{m} R_i}^{\Psi,\alpha}(E) \subseteq \underline{\sum_{i=1}^{m} R_i}^{\Xi,\alpha}(E) \subseteq \underline{\sum_{i=1}^{m} R_i}^{\Omega,\alpha}(E)$ can be obtained.
2. Likewise, $\overline{\sum_{i=1}^{m} R_i}^{\Psi,\alpha}(E) \subseteq \overline{\sum_{i=1}^{m} R_i}^{\Xi,\alpha}(E) \subseteq \overline{\sum_{i=1}^{m} R_i}^{\Omega,\alpha}(E)$ can also be proved.

□

4. IN MAGDM Based on Multigranulation Probabilistic Models

In the following section, we aim to sum up a viable and effective MAGDM approach by using the newly developed theoretical models. As pointed out in Section 1, multigranulation probabilistic models can manage the three challenges of typical MAGDM situations well—to be specific, the overall study context of IN MG-PRSs over two universes is IN information, which excels in depicting indeterminate and incomplete information at the same time. In addition, the development of multigranulation probabilistic models provide decision makers with an effective strategy in MAGDM information fusion and analysis by taking superiorities of MGRSs and PRSs, and the proposed IN MG-PRSs over two universes are also equipped with the ability of describing risk preferences of information fusion quantitatively and dynamically. Hence, IN MG-PRSs over two universes play a significant role in solving MAGDM problems, and it is necessary to put forward corresponding MAGDM methods.

Next, for the sake of exploring MAGDM methods in a real-world scenario, we put the following discussions in the background of M&A TSs. We first let the universe $U = \{x_1, x_2, \ldots, x_j\}$ be a set containing selectable M&A targets, whereas universe $V = \{y_1, y_2, \ldots, y_k\}$ is a set containing assessment criteria. Then, we also let $E \in IN(V)$ be a standard set containing several needs of corporate acquirers from the aspect of assessment criteria. Afterwards, m decision makers in a group provide several relations $R_i \in INR(U \times V)$ $(i = 1, 2, \ldots, m)$ between the above-mentioned universes. Finally, an information system (U, V, R_i, E) for M&A TSs can be established as the input for the following MAGDM algorithm based on multigranulation probabilistic models.

Remark 1. *In what follows, we intend to interpret the scheme of selecting the parametric value* λ. *According to Definition 10, the adjustable IN membership degrees of* x *in* E *with regard to* R_i *are provided as* $\Xi_E^{\sum_{i=1}^m R_i}(x) = \lambda \Omega_E^{\sum_{i=1}^m R_i}(x) \boxplus (1-\lambda) \Psi_E^{\sum_{i=1}^m R_i}(x)$. *It is not difficult to see* $\Xi_E^{\sum_{i=1}^m R_i}(x) = \Omega_E^{\sum_{i=1}^m R_i}(x)$ *when* $\lambda = 1$, *whereas* $\Xi_E^{\sum_{i=1}^m R_i}(x) = \Psi_E^{\sum_{i=1}^m R_i}(x)$ *when* $\lambda = 0$. *Hence, maximal IN membership degrees* $\Omega_E^{\sum_{i=1}^m R_i}(x)$ *and minimal IN membership degrees* $\Psi_E^{\sum_{i=1}^m R_i}(x)$ *are two extreme cases of adjustable IN membership degrees* $\Xi_E^{\sum_{i=1}^m R_i}(x)$. *In light of the standpoint of risk decision-making with uncertainty from classical operational research [1],* $\Xi_E^{\sum_{i=1}^m R_i}(x) = \Omega_E^{\sum_{i=1}^m R_i}(x)$ *can be seen as the "completely risk-seeking" strategy,* $\Xi_E^{\sum_{i=1}^m R_i}(x) = \Psi_E^{\sum_{i=1}^m R_i}(x)$ *can be seen as the "completely risk-averse" strategy, and* $\Xi_E^{\sum_{i=1}^m R_i}(x) = 0.5\Omega_E^{\sum_{i=1}^m R_i}(x) \boxplus (1-0.5) \Psi_E^{\sum_{i=1}^m R_i}(x)$ *can be seen as the "risk-neutral" strategy. Moreover,* $\Xi_E^{\sum_{i=1}^m R_i}(x)$ *can be seen as the "somewhat risk-seeking" strategy when* $\lambda \in (0.5, 1)$, *and* $\Xi_E^{\sum_{i=1}^m R_i}(x)$ *can be seen as the "somewhat risk-averse" strategy when* $\lambda \in (0, 0.5)$. *According to the above-stated theoretical explanations, the parametric value* λ *represents the risk preference of different decision makers in M&A TSs. In specific, the larger of the parametric value* λ *when all decision makers are more risk-seeking, whereas the smaller of the parametric value* λ *when all decision makers are more risk-averse. In general, the parametric value* λ *is determined by all decision-makers' risk preference or the empirical studies and inherent knowledge in advance. In practical MAGDM situations, suppose there are* m *decision makers in a group, each decision maker provides his or her risk preference* λ_i $(\lambda_i \in [0,1], i = 1, 2, \ldots, m)$, *and then the final risk preference when computing* $\Xi_E^{\sum_{i=1}^m R_i}(x)$ *can be determined as* $\lambda = \frac{\sum_{i=1}^m \lambda_i}{m}$.

In what follows, an algorithm for M&A TSs by virtue of adjustable IN MG-PRSs over two universes is established.

5. An Illustrative Example

For the sake of making an efficient comparative analysis with existing similar IN MAGDM approaches, we plan to utilize the case study that was previously investigated in [28]. In what follows, we first present the general context of M&A TSs and show basic steps of obtaining the optimal M&A target by means of the newly proposed algorithm developed in Section 4.

Remark 2. *Section 4 acts as a transition part which links between the theoretical models proposed in Section 3 and the application case presented in Section 5. To be specific, we first put forward two special theoretical models named optimistic and pessimistic IN MG-PRSs over two universes. Then, we further generalize these two special theoretical models to adjustable IN MG-PRSs over two universes. All the proposed three theoretical models are foundations for addressing MAGDM problems. Next, we propose a novel algorithm for M&A TSs in light of adjustable IN MG-PRSs over two universes in Section 4. Finally, in order to show the reasonability and effectiveness of the proposed algorithm, the following section plans to conduct several quantitative and qualitative analysis via an illustrative example.*

5.1. MAGDM Procedures

In this illustrative example, we use the case study that was previously investigated in [28]. The detailed case descriptions and datasets can be found at the webpage (https://www.mdpi.com/2073-8994/9/7/126/htm).

According to Algorithm 1, we aim to obtain the optimal M&A target by means of IN MG-PRSs over two universes. First, we calculate single IN membership degrees as follows.

Algorithm 1 An algorithm for M&A TSs in light of adjustable IN MG-PRSs over two universes.

Require: An information system (U, V, R_i, E) for M&A TSs.

Ensure: The optimal alternative.

Step 1. Calculate single IN membership degrees $\eta_E^{R_i}(x) = \frac{\sum_{y \in V} R_i(x,y) E(y)}{\sum_{y \in V} R_i(x,y)}$;

Step 2. Calculate maximal IN membership degrees $\Omega_E^{\sum_{i=1}^{m} R_i}(x) = \max_{i=1}^{m} \eta_E^{R_i}(x)$ and minimal IN membership degrees $\Psi_E^{\sum_{i=1}^{m} R_i}(x) = \min_{i=1}^{m} \eta_E^{R_i}(x)$;

Step 3. Calculate the risk coefficient $\lambda = \frac{\sum_{i=1}^{m} \lambda_i}{m}$;

Step 4. Calculate adjustable IN membership degrees $\Xi_E^{\sum_{i=1}^{m} R_i}(x) = \lambda \Omega_E^{\sum_{i=1}^{m} R_i}(x) \boxplus (1-\lambda) \Psi_E^{\sum_{i=1}^{m} R_i}(x)$;

Step 5. Determine score values of $\Xi_E^{\sum_{i=1}^{m} R_i}(x)$ for all selectable M&A targets;

Step 6. The best alternative is the one with the largest score value for $\Xi_E^{\sum_{i=1}^{m} R_i}(x)$.

With regard to the relation presented in R_1, we obtain

$$\eta_E^{R_1}(x_1) = \frac{\sum_{y \in V} R_1(x_1, y) E(y)}{\sum_{y \in V} R_1(x_1, y)} = \langle [0.8477, 0.9207], [0.0378, 0.1345], [0.1771, 0.316] \rangle .$$

In a similar manner, we also obtain

$$\eta_E^{R_1}(x_2) = \langle [0.7672, 0.894], [0.0459, 0.1677], [0.3197, 0.5164] \rangle ;$$
$$\eta_E^{R_1}(x_3) = \langle [0.8212, 0.9314], [0.05, 0.1497], [0.2351, 0.3925] \rangle ;$$
$$\eta_E^{R_1}(x_4) = \langle [0.8493, 0.9465], [0.0232, 0.101], [0.2147, 0.4122] \rangle ;$$
$$\eta_E^{R_1}(x_5) = \langle [0.8774, 0.9643], [0.0466, 0.1559], [0.07, 0.1923] \rangle .$$

With regard to the relation presented in R_2, we also obtain

$$\eta_E^{R_2}(x_1) = \langle [0.8185, 0.9419], [0.033, 0.1355], [0.17, 0.3418] \rangle ;$$
$$\eta_E^{R_2}(x_2) = \langle [0.7728, 0.9177], [0.04, 0.1423], [0.285, 0.4756] \rangle ;$$
$$\eta_E^{R_2}(x_3) = \langle [0.8217, 0.9407], [0.038, 0.1294], [0.1799, 0.3357] \rangle ;$$
$$\eta_E^{R_2}(x_4) = \langle [0.8327, 0.9548], [0.022, 0.097], [0.2434, 0.4106] \rangle ;$$
$$\eta_E^{R_2}(x_5) = \langle [0.8591, 0.9645], [0.031, 0.1583], [0.096, 0.2316] \rangle .$$

With regard to the relation presented in R_3, we also obtain

$$\eta_E^{R_3}(x_1) = \langle [0.844, 0.9472], [0.03, 0.1422], [0.2104, 0.3615] \rangle ;$$
$$\eta_E^{R_3}(x_2) = \langle [0.799, 0.91], [0.0336, 0.1571], [0.3151, 0.5472] \rangle ;$$
$$\eta_E^{R_3}(x_3) = \langle [0.8388, 0.9368], [0.0424, 0.1466], [0.1803, 0.3484] \rangle ;$$
$$\eta_E^{R_3}(x_4) = \langle [0.8362, 0.9366], [0.0188, 0.1069], [0.206, 0.3834] \rangle ;$$
$$\eta_E^{R_3}(x_5) = \langle [0.8739, 0.9581], [0.029, 0.1351], [0.075, 0.2316] \rangle .$$

Next, maximal and minimal IN membership degrees can be calculated in light of Definition 7. For maximal IN membership degrees, we have

$$\Omega_E^{\sum_{i=1}^{3} R_i}(x_1) = \langle [0.8477, 0.9472], [0.0378, 0.1422], [0.2104, 0.3615] \rangle ;$$
$$\Omega_E^{\sum_{i=1}^{3} R_i}(x_2) = \langle [0.799, 0.9177], [0.0459, 0.1677], [0.3197, 0.5472] \rangle ;$$
$$\Omega_E^{\sum_{i=1}^{3} R_i}(x_3) = \langle [0.8388, 0.9407], [0.05, 0.1497], [0.2351, 0.3925] \rangle ;$$
$$\Omega_E^{\sum_{i=1}^{3} R_i}(x_4) = \langle [0.8493, 0.9548], [0.0232, 0.1069], [0.2434, 0.4122] \rangle ;$$
$$\Omega_E^{\sum_{i=1}^{3} R_i}(x_5) = \langle [0.8774, 0.9645], [0.0466, 0.1583], [0.096, 0.2316] \rangle .$$

Similarly, for minimal IN membership degrees, we also have

$$\Psi_E^{\sum_{i=1}^{3} R_i}(x_1) = \langle [0.8185, 0.9207], [0.03, 0.1345], [0.17, 0.316] \rangle ;$$
$$\Psi_E^{\sum_{i=1}^{3} R_i}(x_2) = \langle [0.7672, 0.894], [0.0336, 0.1423], [0.285, 0.4756] \rangle ;$$
$$\Psi_E^{\sum_{i=1}^{3} R_i}(x_3) = \langle [0.8212, 0.9314], [0.038, 0.1294], [0.1799, 0.3357] \rangle ;$$
$$\Psi_E^{\sum_{i=1}^{3} R_i}(x_4) = \langle [0.8327, 0.9366], [0.0188, 0.097], [0.206, 0.3834] \rangle ;$$
$$\Psi_E^{\sum_{i=1}^{3} R_i}(x_5) = \langle [0.8591, 0.9581], [0.029, 0.1351], [0.07, 0.1923] \rangle .$$

In order to make an efficient comparison with the MAGDM method proposed in [28], the risk coefficient $\lambda = 0.6$ is noted in [28]; then, we also take the risk coefficient $\lambda = 0.6$ in this case study. In what follows, adjustable IN membership degrees can be calculated. To be specific, for M&A target x_1, we have

$$\Xi_E^{\sum_{i=1}^{3} R_i}(x_1) = 0.6\Omega_E^{\sum_{i=1}^{3} R_i}(x_1) \boxplus (1-0.6)\,\Psi_E^{\sum_{i=1}^{3} R_i}(x_1) = \langle [0.8366, 0.9379], [0.0345, 0.1391], [0.1932, 0.3426] \rangle.$$

In what follows, we also have

$$\Xi_E^{\sum_{i=1}^{3} R_i}(x_2) = \langle [0.7868, 0.9089], [0.0405, 0.157], [0.3054, 0.5173] \rangle;$$
$$\Xi_E^{\sum_{i=1}^{3} R_i}(x_3) = \langle [0.832, 0.9371], [0.0448, 0.1412], [0.2112, 0.3687] \rangle;$$
$$\Xi_E^{\sum_{i=1}^{3} R_i}(x_4) = \langle [0.8429, 0.9482], [0.0213, 0.1028], [0.2277, 0.4004] \rangle;$$
$$\Xi_E^{\sum_{i=1}^{3} R_i}(x_5) = \langle [0.8705, 0.9621], [0.0385, 0.1486], [0.0846, 0.215] \rangle.$$

Finally, we calculate score values of $\Xi_E^{\sum_{i=1}^{3} R_i}(x)$ for each x_i, and the best x_i is the one with the largest score value for $\Xi_E^{\sum_{i=1}^{3} R_i}(x)$; the final ranking result shows $x_5 \succ x_1 \succ x_4 \succ x_3 \succ x_2$, i.e., the supreme alternative is x_5.

5.2. Sensitivity Analysis

In the previous section, we obtain the optimal M&A target by using adjustable multigranulation probabilistic models with the risk coefficient $\lambda = 0.6$. The following sensitivity analysis aims to investigate the influence of the risk coefficient by changing the value of λ. To be specific, supposing the value of λ is taken as 0, 0.4, 0.5, 0.6, and 1, respectively, then we can obtain the final ranking orders in Table 1 below.

Table 1. Ranking orders of M&A targets with changing values of λ.

Different λ	Information Fusion Risks	Ranking Results of M&A Targets
$\lambda = 0$	completely pessimistic (completely risk-averse)	$x_5 \succ x_1 \succ x_3 \succ x_4 \succ x_2$
$\lambda = 0.4$	somewhat pessimistic (somewhat risk-averse)	$x_5 \succ x_1 \succ x_4 \succ x_3 \succ x_2$
$\lambda = 0.5$	neutral (risk-neutral)	$x_5 \succ x_1 \succ x_4 \succ x_3 \succ x_2$
$\lambda = 0.6$	somewhat optimistic (somewhat risk-seeking)	$x_5 \succ x_1 \succ x_4 \succ x_3 \succ x_2$
$\lambda = 1$	completely optimistic (completely risk-seeking)	$x_5 \succ x_1 \succ x_4 \succ x_3 \succ x_2$

According to the final ranking orders in Table 1, it is easy to see that the best x_i is insensitive to changing values of λ, that is, all results show the best x_i is x_5. Thus, the best alternative is reliable and stable. The only difference lies in the ranking order of x_3 and x_4 when $\lambda = 0$, i.e., x_3 is superior to x_4 when $\lambda = 0$, whereas x_3 is inferior to x_4 in other situations. The cause of this phenomenon is that the changing values of λ may affect the ranking order of x_3 and x_4 when the risk preferences is completely risk-averse.

5.3. Comparative Analysis

In what follows, we aim to compare with the MAGDM method proposed in [28] to present the merits of the proposed MAGDM algorithm. In [28], the authors put forward an algorithm for M&A

TSs via IN MGRSs over two universes without the support of PRSs. The mathematical structures of optimistic and pessimistic IN MGRSs over two universes are presented below:

$$\underline{\sum_{i=1}^{m} R_i}^{O}(E) = \{\langle x, [\mu^{L}_{\underline{\sum_{i=1}^{m} R_i}^{O}(E)}(x), \mu^{U}_{\underline{\sum_{i=1}^{m} R_i}^{O}(E)}(x)], [\nu^{L}_{\underline{\sum_{i=1}^{m} R_i}^{O}(E)}(x), \nu^{U}_{\underline{\sum_{i=1}^{m} R_i}^{O}(E)}(x)], [\omega^{L}_{\underline{\sum_{i=1}^{m} R_i}^{O}(E)}(x), \omega^{U}_{\underline{\sum_{i=1}^{m} R_i}^{O}(E)}(x)]\rangle \mid x \in U\};$$

$$\overline{\sum_{i=1}^{m} R_i}^{O}(E) = \{\langle x, [\mu^{L}_{\overline{\sum_{i=1}^{m} R_i}^{O}(E)}(x), \mu^{U}_{\overline{\sum_{i=1}^{m} R_i}^{O}(E)}(x)], [\nu^{L}_{\overline{\sum_{i=1}^{m} R_i}^{O}(E)}(x), \nu^{U}_{\overline{\sum_{i=1}^{m} R_i}^{O}(E)}(x)], [\omega^{L}_{\overline{\sum_{i=1}^{m} R_i}^{O}(E)}(x), \omega^{U}_{\overline{\sum_{i=1}^{m} R_i}^{O}(E)}(x)]\rangle \mid x \in U\};$$

$$\underline{\sum_{i=1}^{m} R_i}^{P}(E) = \{\langle x, [\mu^{L}_{\underline{\sum_{i=1}^{m} R_i}^{P}(E)}(x), \mu^{U}_{\underline{\sum_{i=1}^{m} R_i}^{P}(E)}(x)], [\nu^{L}_{\underline{\sum_{i=1}^{m} R_i}^{P}(E)}(x), \nu^{U}_{\underline{\sum_{i=1}^{m} R_i}^{P}(E)}(x)], [\omega^{L}_{\underline{\sum_{i=1}^{m} R_i}^{P}(E)}(x), \omega^{U}_{\underline{\sum_{i=1}^{m} R_i}^{P}(E)}(x)]\rangle \mid x \in U\};$$

$$\overline{\sum_{i=1}^{m} R_i}^{P}(E) = \{\langle x, [\mu^{L}_{\overline{\sum_{i=1}^{m} R_i}^{P}(E)}(x), \mu^{U}_{\overline{\sum_{i=1}^{m} R_i}^{P}(E)}(x)], [\nu^{L}_{\overline{\sum_{i=1}^{m} R_i}^{P}(E)}(x), \nu^{U}_{\overline{\sum_{i=1}^{m} R_i}^{P}(E)}(x)], [\omega^{L}_{\overline{\sum_{i=1}^{m} R_i}^{P}(E)}(x), \omega^{U}_{\overline{\sum_{i=1}^{m} R_i}^{P}(E)}(x)]\rangle \mid x \in U\},$$

where

$$\mu^{L}_{\underline{\sum_{i=1}^{m} R_i}^{O}(E)}(x) = \bigvee_{i=1}^{m} \wedge_{y \in V}\{\omega^{L}_{R_i}(x,y) \vee \mu^{L}_{E}(y)\}, \ \mu^{U}_{\underline{\sum_{i=1}^{m} R_i}^{O}(E)}(x) = \bigvee_{i=1}^{m} \wedge_{y \in V}\{\omega^{U}_{R_i}(x,y) \vee \mu^{U}_{E}(y)\},$$

$$\nu^{L}_{\underline{\sum_{i=1}^{m} R_i}^{O}(E)}(x) = \bigwedge_{i=1}^{m} \vee_{y \in V}\{(1 - \nu^{U}_{R_i}(x,y)) \wedge \nu^{L}_{E}(y)\},$$

$$\nu^{U}_{\underline{\sum_{i=1}^{m} R_i}^{O}(E)}(x) = \bigwedge_{i=1}^{m} \vee_{y \in V}\{(1 - \nu^{L}_{R_i}(x,y)) \wedge \nu^{U}_{E}(y)\},$$

$$\omega^{L}_{\underline{\sum_{i=1}^{m} R_i}^{O}(E)}(x) = \bigwedge_{i=1}^{m} \vee_{y \in V}\{\mu^{L}_{R_i}(x,y) \wedge \omega^{L}_{E}(y)\}, \ \omega^{U}_{\underline{\sum_{i=1}^{m} R_i}^{O}(E)}(x) = \bigwedge_{i=1}^{m} \vee_{y \in V}\{\mu^{U}_{R_i}(x,y) \wedge \omega^{U}_{E}(y)\},$$

$$\mu^{L}_{\overline{\sum_{i=1}^{m} R_i}^{O}(E)}(x) = \bigwedge_{i=1}^{m} \vee_{y \in V}\{\mu^{L}_{R_i}(x,y) \wedge \mu^{L}_{E}(y)\}, \ \mu^{U}_{\overline{\sum_{i=1}^{m} R_i}^{O}(E)}(x) = \bigwedge_{i=1}^{m} \vee_{y \in V}\{\mu^{U}_{R_i}(x,y) \wedge \mu^{U}_{E}(y)\},$$

$$\nu^{L}_{\overline{\sum_{i=1}^{m} R_i}^{O}(E)}(x) = \bigvee_{i=1}^{m} \wedge_{y \in V}\{\nu^{L}_{R_i}(x,y) \vee \nu^{L}_{E}(y)\}, \ \nu^{U}_{\overline{\sum_{i=1}^{m} R_i}^{O}(E)}(x) = \bigvee_{i=1}^{m} \wedge_{y \in V}\{\nu^{U}_{R_i}(x,y) \vee \nu^{U}_{E}(y)\},$$

$$\omega^{L}_{\overline{\sum_{i=1}^{m} R_i}^{O}(E)}(x) = \bigvee_{i=1}^{m} \wedge_{y \in V}\{\omega^{L}_{R_i}(x,y) \vee \omega^{L}_{E}(y)\}, \ \omega^{U}_{\overline{\sum_{i=1}^{m} R_i}^{O}(E)}(x) = \bigvee_{i=1}^{m} \wedge_{y \in V}\{\omega^{U}_{R_i}(x,y) \vee \omega^{U}_{E}(y)\}.$$

$$\mu^{L}_{\underline{\sum_{i=1}^{m} R_i}^{P}(E)}(x) = \bigwedge_{i=1}^{m} \wedge_{y \in V}\{\omega^{L}_{R_i}(x,y) \vee \mu^{L}_{E}(y)\}, \ \mu^{U}_{\underline{\sum_{i=1}^{m} R_i}^{P}(E)}(x) = \bigwedge_{i=1}^{m} \wedge_{y \in V}\{\omega^{U}_{R_i}(x,y) \vee \mu^{U}_{E}(y)\},$$

$$\nu^{L}_{\underline{\sum_{i=1}^{m} R_i}^{P}(E)}(x) = \bigvee_{i=1}^{m} \vee_{y\in V}\{(1-\nu^{U}_{R_i}(x,y)) \wedge \nu^{L}_{E}(y)\},$$

$$\nu^{U}_{\underline{\sum_{i=1}^{m} R_i}^{P}(E)}(x) = \bigvee_{i=1}^{m} \vee_{y\in V}\{(1-\nu^{L}_{R_i}(x,y)) \wedge \nu^{U}_{E}(y)\},$$

$$\omega^{L}_{\underline{\sum_{i=1}^{m} R_i}^{P}(E)}(x) = \bigvee_{i=1}^{m} \vee_{y\in V}\{\mu^{L}_{R_i}(x,y) \wedge \omega^{L}_{E}(y)\}, \omega^{U}_{\underline{\sum_{i=1}^{m} R_i}^{P}(E)}(x) = \bigvee_{i=1}^{m} \vee_{y\in V}\{\mu^{U}_{R_i}(x,y) \wedge \omega^{U}_{E}(y)\},$$

$$\mu^{L}_{\overline{\sum_{i=1}^{m} R_i}^{P}(E)}(x) = \bigvee_{i=1}^{m} \vee_{y\in V}\{\mu^{L}_{R_i}(x,y) \wedge \mu^{L}_{E}(y)\}, \mu^{U}_{\overline{\sum_{i=1}^{m} R_i}^{P}(E)}(x) = \bigvee_{i=1}^{m} \vee_{y\in V}\{\mu^{U}_{R_i}(x,y) \wedge \mu^{U}_{E}(y)\},$$

$$\nu^{L}_{\overline{\sum_{i=1}^{m} R_i}^{P}(E)}(x) = \bigwedge_{i=1}^{m} \wedge_{y\in V}\{\nu^{L}_{R_i}(x,y) \vee \nu^{L}_{E}(y)\}, \nu^{U}_{\overline{\sum_{i=1}^{m} R_i}^{P}(E)}(x) = \bigwedge_{i=1}^{m} \wedge_{y\in V}\{\nu^{U}_{R_i}(x,y) \vee \nu^{U}_{E}(y)\},$$

$$\omega^{L}_{\overline{\sum_{i=1}^{m} R_i}^{P}(E)}(x) = \bigwedge_{i=1}^{m} \wedge_{y\in V}\{\omega^{L}_{R_i}(x,y) \vee \omega^{L}_{E}(y)\}, \omega^{U}_{\overline{\sum_{i=1}^{m} R_i}^{P}(E)}(x) = \bigwedge_{i=1}^{m} \wedge_{y\in V}\{\omega^{U}_{R_i}(x,y) \vee \omega^{U}_{E}(y)\}.$$

The pair $(\underline{\sum_{i=1}^{m} R_i}^{O}(E), \overline{\sum_{i=1}^{m} R_i}^{O}(E))$ is named an optimistic IN MGRS over two universes of E, whereas the pair $(\underline{\sum_{i=1}^{m} R_i}^{P}(E), \overline{\sum_{i=1}^{m} R_i}^{P}(E))$ is named a pessimistic IN MGRS over two universes of E.

More concretely, the optimistic and pessimistic IN multigranulation rough approximations in terms of an information system $(U, V, R_i, E)(i = 1,2,3)$ for M&A TSs are calculated at first, which are denoted by $\underline{\sum_{i=1}^{3} R_i}^{O}(E)$, $\overline{\sum_{i=1}^{3} R_i}^{O}(E)$, $\underline{\sum_{i=1}^{3} R_i}^{P}(E)$ and $\overline{\sum_{i=1}^{3} R_i}^{P}(E)$. Then, we further synthesize lower and upper versions of optimistic IN multigranulation rough approximations, which is denoted by $\underline{\sum_{i=1}^{3} R_i}^{O}(E) \boxplus \overline{\sum_{i=1}^{3} R_i}^{O}(E)$, whereas lower and upper versions of pessimistic counterparts can also be synthesized, which is denoted by $\underline{\sum_{i=1}^{3} R_i}^{P}(E) \boxplus \overline{\sum_{i=1}^{3} R_i}^{P}(E)$. Let the risk coefficient $\lambda = 0.6$ when integrating $\underline{\sum_{i=1}^{3} R_i}^{O}(E) \boxplus \overline{\sum_{i=1}^{3} R_i}^{O}(E)$ with $\underline{\sum_{i=1}^{3} R_i}^{P}(E) \boxplus \overline{\sum_{i=1}^{3} R_i}^{P}(E)$, i.e., the synthesized set $0.6(\underline{\sum_{i=1}^{3} R_i}^{O}(E) \boxplus \overline{\sum_{i=1}^{3} R_i}^{O}(E)) \boxplus (1-0.6)(\underline{\sum_{i=1}^{3} R_i}^{P}(E) \boxplus \overline{\sum_{i=1}^{3} R_i}^{P}(E))$ should be further obtained. The above-stated sets are obtained as follows:

$$\underline{\sum_{i=1}^{3} R_i}^{O}(E) \boxplus \overline{\sum_{i=1}^{3} R_i}^{O}(E) = \{\langle x_1, \langle [0.82,0.90], [0.15,0.24], [0.24,0.35]\rangle\rangle, \langle x_2, \langle [0.70,0.80], [0.15,0.35], [0.42,0.56]\rangle\rangle, \langle x_3, \langle [0.85,0.92], [0.08,0.18], [0.30,0.42]\rangle\rangle, \langle x_4, \langle [0.82,0.90], [0.10,0.18], [0.30,0.42]\rangle\rangle, \langle x_5, \langle [0.76,0.86], [0.15,0.24], [0.14,0.24]\rangle\rangle\};$$

$$\underline{\sum_{i=1}^{3} R_i}^{P}(E) \boxplus \overline{\sum_{i=1}^{3} R_i}^{P}(E) = \{\langle x_1, \langle [0.82,0.90], [0.10,0.21], [0.24,0.35]\rangle\rangle, \langle x_2, \langle [0.70,0.85], [0.15,0.28], [0.40,0.54]\rangle\rangle, \langle x_3, \langle [0.85,0.92], [0.08,0.18], [0.24,0.35]\rangle\rangle, \langle x_4, \langle [0.79,0.88], [0.10,0.21], [0.28,0.40]\rangle\rangle, \langle x_5, \langle [0.76,0.86], [0.15,0.28], [0.16,0.27]\rangle\rangle\};$$

$$0.6(\underline{\sum_{i=1}^{3} R_i}^{O}(E) \boxplus \overline{\sum_{i=1}^{3} R_i}^{O}(E)) \boxplus (1-0.6)(\underline{\sum_{i=1}^{3} R_i}^{P}(E) \boxplus \overline{\sum_{i=1}^{3} R_i}^{P}(E)) = \{\langle x_1, \langle [0.82,0.90], [0.13,0.23], [0.24,0.35]\rangle\rangle, \langle x_2, \langle [0.70,0.82], [0.15,0.32], [0.41,0.55]\rangle\rangle, \langle x_3, \langle [0.85,0.92], [0.08,0.18], [0.28,0.39]\rangle\rangle, \langle x_4, \langle [0.81,0.89], [0.10,0.19], [0.29,0.41]\rangle\rangle, \langle x_5, \langle [0.76,0.86], [0.15,0.25], [0.15,0.25]\rangle\rangle\}.$$

Finally, score values of $0.6(\underline{\sum_{i=1}^{3} R_i}^{O}(E) \boxplus \overline{\sum_{i=1}^{3} R_i}^{O}(E)) \boxplus (1-0.6)(\underline{\sum_{i=1}^{3} R_i}^{P}(E) \boxplus \overline{\sum_{i=1}^{3} R_i}^{P}(E))$ are calculated, and it is convenient to determine the ranking orders of five M&A targets via the above score values and obtain the optimal M&A target, which is x_3, and x_5 is ranked second. The reason for the difference with the result obtained by using the proposed method lies in IN MGRSs over two universes lacking the ability of error tolerance; the MAGDM result is sensitive to outlier values from original information for M&A TSs.

5.4. Discussion

In order to address complicated MAGDM problems effectively, three key challenges are focused on at first. Then, under the guidance of multigranulation probabilistic models, we utilize the model of INSs, MGRSs and PRSs to handle the above-mentioned challenges. Moreover, compared with existing popular nonlinear modeling approaches, such as formal concept analysis [33,34,58,64–67], control systems [59,60] and sentiment analysis [61–63,68,69], which neither effectively handle indeterminate and incomplete information in complicated MAGDM problems, nor reasonably fuse and analyze multi-source information with incorrect and noisy data, it is necessary to combine INSs, MGRSs with PRSs to develop some meaningful hybrid models along with corresponding MAGDM approaches. In light of MAGDM procedures in the current section, we sum up the merits of the proposed MAGDM algorithm below:

1. INSs act as a viable and effective tool for depicting various uncertainties in typical MAGDM situations. By dividing the notion of membership degrees into three different parts, indeterminate and incomplete MAGDM information can be described precisely.
2. In MAGDM information fusion procedures, with the support of MGRSs, the computational efficiency of information fusion can be enhanced to a large extent. Moreover, decision risks of information fusion strategies can also be modeled well.
3. Compared with [28], the proposed MAGDM algorithm excels in fusing the superiorities of PRSs in the construction of hybrid models. To be specific, IN MG-PRSs over two universes own the fault-tolerance ability when coping with incorrect and noisy data.

Hence, the developed IN MG-PRSs over two universes perform outstandingly in MAGDM information representation, information fusion, and information analysis; they provide a beneficial tool for addressing complicated MAGDM problems.

6. Conclusions

This work mainly presents a general framework for dealing with complicated IN MAGDM problems by virtue of multigranulation probabilistic models. At first, three different types of multigranulation probabilistic models are put forward, that is, the optimistic version, the pessimistic version, and the adjustable version, and both definitions and key properties are discussed in detail. Then, relationships between optimistic, pessimistic, and adjustable multigranulation probabilistic models are further explored. Afterwards, corresponding IN MAGDM approaches are proposed in the background of M&A TSs. Finally, a practical example of M&A TSs is presented with several quantitative and qualitative analysis.

In the future, it is meaningful to generalize IN MG-PRSs over two universes to more extended neutrosophic contexts such as neutrosophic duplets, triplets and multisets. Furthermore, establishing efficient IN MAGDM approaches for problems with dynamic situations, high-dimensional attributes and large-scale of alternatives are also necessary. Another future interesting study option is to apply the presented theoretical models to other areas such as clustering, feature selections, compressed sensing, image processing, etc.

Author Contributions: Conceptualization, D.L.; writing—original draft preparation, C.Z.; validation, X.K. and Y.L.; writing—review and editing, S.B. and A.K.S.; supervision, D.L. All authors have read and agreed to the published version of the manuscript.

Funding: This work was partially supported by the Key R&D program of Shanxi Province (International Cooperation, 201903D421041), National Natural Science Foundation of China (Nos. 61806116, 61672331, 61972238, 61802237, 61603278 and 61906110), Natural Science Foundation of Shanxi (Nos. 201801D221175, 201901D211176, 201901D211414), Training Program for Young Scientific Researchers of Higher Education Institutions in Shanxi, Scientific and Technological Innovation Programs of Higher Education Institutions in Shanxi (STIP) (Nos. 201802014, 2019L0066, 2019L0500), and Cultivate Scientific Research Excellence Programs of Higher Education Institutions in Shanxi (CSREP) (2019SK036).

Acknowledgments: The authors are grateful to the editor and three anonymous reviewers for their helpful and valuable comments which helped to improve the quality of the research.

Conflicts of Interest: The authors declare no conflict of interest.

References

1. Xu, Z.S. *Uncertain Multi-Attribute Decision Making: Methods and Applications*; Springer: Berlin, Germany, 2015; pp. 1–373.
2. Sun, B.Z.; Ma, W.M.; Xiao, X. Three-way group decision-making based on multigranulation fuzzy decision-theoretic rough set over two universes. *Int. J. Approx. Reason.* **2017**, *81*, 87–102. [CrossRef]
3. Liang, D.C.; Zhang, Y.R.J.; Xu, Z.S.; Darko, A.P. Pythagorean fuzzy Bonferroni mean aggregation operator and its accelerative calculating algorithm with the multithreading. *Int. J. Intell. Syst.* **2018**, *33*, 615–633. [CrossRef]
4. Zhang, C.; Li, D.Y.; Liang, J.Y. Hesitant fuzzy linguistic rough set over two universes model and its applications. *Int. J. Mach. Learn. Cyb.* **2018**, *9*, 577–588. [CrossRef]
5. Zhang, K.; Zhan, J.M.; Yao, Y.Y. TOPSIS method based on a fuzzy covering approximation space: An application to biological nano-materials selection. *Inform. Sci.* **2019**, *502*, 297–329. [CrossRef]
6. Zhang, C.; Li, D.Y.; Liang, J.Y. Interval-valued hesitant fuzzy multi-granularity three-way decisions in consensus processes with applications to multi-attribute group decision-making. *Inform. Sci.* **2020**, *511*, 192–211. [CrossRef]
7. Smarandache, F. *Neutrosophy. Neutrosophic Probability, Set, and Logic*; American Research Press: Rehoboth, DE, USA, 1998.
8. Smarandache, F. *A Unifying Field in Logics. Neutrosophy: Neutrosophic Probability, Set and Logic*; American Research Press: Rehoboth, DE, USA, 1999.
9. Peng, X.D.; Dai, J.G. A bibliometric analysis of neutrosophic set: Two decades review from 1998 to 2017. *Artif. Intell. Rev.* **2018** , 1–57. [CrossRef]
10. An, X.G.; Zhang, X.H.; Ma, Y.C. Generalized abel-grassmann's neutrosophic extended triplet loop. *Mathematics* **2019**, *7*, 1206. [CrossRef]
11. Guo, Y.H.; Sengur, A. A novel 3D skeleton algorithm based on neutrosophic cost function. *Appl. Soft. Comput.* **2015**, *36*, 210–217. [CrossRef]
12. Smarandache, F. Refined neutrosophy and lattices vs. pair structures and yinyang bipolar fuzzy set. *Mathematics* **2019**, *7*, 353. [CrossRef]
13. Wang, H.B.; Smarandache, F.; Zhang, Y.Q.; Sunderraman, R. *Interval Neutrosophic Sets and Logic: Theory and Applications in Computing*; Hexis: Phoenix, AZ, USA, 2005; pp. 1–87.
14. Broumi, S.; Smarandache, F. Correlation coefficient of interval neutrosophic set. *Appl. Mech. Mater.* **2013**, *436*, 511–517. [CrossRef]
15. Zhang, H.Y.; Wang, J.Q.; Chen, X.H. Interval neutrosophic sets and their application in multicriteria decision-making problems. *Sci. World J.* **2014**, *2014*. [CrossRef] [PubMed]
16. Fan, C.X.; Ye, J.; Feng, S.; Fan, E.; Hu, K.L. Multi-criteria decision-making method using heronian mean operators under a bipolar neutrosophic environment. *Mathematics* **2019**, *7*, 97. [CrossRef]
17. Ye, J. Multiple attribute group decision-making based on interval neutrosophic uncertain linguistic variables. *Int. J. Mach. Learn. Cyb.* **2017**, *8*, 837–848. [CrossRef]

18. Deli, I. Interval-valued neutrosophic soft sets and its decision-making. *Int. J. Mach. Learn. Cyb.* **2017**, *8*, 665–676. [CrossRef]
19. Li, X.; Zhang, X.H.; Park, C. Generalized interval neutrosophic choquet aggregation operators and their applications. *Symmetry* **2018**, *10*, 85. [CrossRef]
20. Nguyen, T.T.; Luu, Q.D.; Le, H.S.; Zhang, X.H.; Nguyen, D.H.; Ali, M.; Smarandache, F. Dynamic interval valued neutrosophic set: Modeling decision-making in dynamic environments. *Comput. Ind.* **2019**, *108*, 45–52.
21. Broumi, S.; Talea, M.; Bakali, A.; Smarandache, F.; Gai, S.Q.; Selvachandran, G. Introduction of some new results on interval-valued neutrosophic graphs. *J. Inform. Optim. Sci.* **2019**, *40*, 1475–1498. [CrossRef]
22. Rani, D.; Garg, H. Some modified results of the subtraction and division operations on interval neutrosophic sets. *J. Exp. Theor. Artif. Intell.* **2019**, *31*, 677–698. [CrossRef]
23. Xu, W.H.; Yu, J.H. A novel approach to information fusion in multi-source datasets: A granular computing viewpoint. *Inform. Sci.* **2017**, *378*, 410–423. [CrossRef]
24. Lin, B.Y.; Xu, W.H. Multi-granulation rough set for incomplete interval-valued decision information systems based on multi-threshold tolerance relation. *Symmetry* **2018**, *10*, 208. [CrossRef]
25. Broumi, S.; Smarandache, F.; Dhar, M. Rough neutrosophic sets. *Ital. J. Pure. Appl. Math.* **2014**, *32*, 493–502.
26. Broumi, S.; Smarandache, F. Soft Interval-valued neutrosophic rough sets. *Neutrosophic Sets Syst.* **2015**, *7*, 69–80. [CrossRef]
27. Zhang, C.; Zhai, Y.H.; Li, D.Y.; Mu, Y.M. Steam turbine fault diagnosis based on single-valued neutrosophic multigranulation rough sets over two universes. *J. Intell. Fuzzy. Syst.* **2016**, *31*, 2829–2837. [CrossRef]
28. Zhang, C.; Li, D.Y.; Sangaiah, A.; Broumi, S. Merger and acquisition target selection based on interval neutrosophic multigranulation rough sets over two universes. *Symmetry* **2017**, *9*, 126. [CrossRef]
29. Liu, Y.L.; Yang, H.L. Multi-granulation neutrosophic rough sets on a single domain and dual domains with applications. *Symmetry* **2017**, *33*, 1467–1478.
30. Shao, S.T.; Zhang, X.H. Measures of probabilistic neutrosophic hesitant fuzzy sets and the application in reducing unnecessary evaluation processes. *Mathematics* **2019**, *7*, 649. [CrossRef]
31. Yang, H.L.; Bao, Y.L.; Guo, Z.L. Generalized interval neutrosophic rough sets and its application in multi-attribute decision-making. *Filomat* **2018**, *32*, 11–33. [CrossRef]
32. Zhang, C.; Li, D.Y.; Kang, X.P.; Song, D.; Sangaiah, A.; Broumi, S. Neutrosophic fusion of rough set theory: An overview. *Comput. Ind.* **2020**, *115*, 103117. [CrossRef]
33. Singh, P.K. Three-way fuzzy concept lattice representation using neutrosophic set. *Int. J. Mach. Learn. Cyb.* **2017**, *8*, 69–79. [CrossRef]
34. Singh, P.K. Three-way n-valued neutrosophic concept lattice at different granulation. *Int. J. Mach. Learn. Cyb.* **2018**, *9*, 1839–1855. [CrossRef]
35. Zhang, C.; Li, D.Y.; Broumi, S.; Sangaiah, A. Medical diagnosis based on single-valued neutrosophic probabilistic rough multisets over two universes. *Symmetry* **2018**, *10*, 213. [CrossRef]
36. Abdel-Basset, M.; Manogaran, G.; Mohamed, M.; Chilamkurti, N. Three-way decisions based on neutrosophic sets and AHP-QFD framework for supplier selection problem. *Future. Gener. Comp. Sy.* **2018**, *89*, 19–30. [CrossRef]
37. Lin, J.; Yang, H.L.; Li, S.G. Three-way decision based on decision-theoretic rough sets with single-valued neutrosophic information. *Int. J. Mach. Learn. Cyb.* **2019**, 1–9. [CrossRef]
38. Abdel-Basset, M.; Mohamed, M. The role of single valued neutrosophic sets and rough sets in smart city: Imperfect and incomplete information systems. *Measurement* **2018**, *124*, 47–55. [CrossRef]
39. Thao, N.X.; Le, H.S.; Cuong, B.C.; Ali, M.; Lan, L.H. Fuzzy Equivalence on Standard and Rough Neutrosophic Sets and Applications to Clustering Analysis. In *Information Systems Design and Intelligent Applications*; Advances in Intelligent Systems and Computing, AISC; Springer: Berlin, Germany, 2018; Volume 672, pp. 834–842.
40. Qian, Y.H.; Liang, J.Y.; Yao, Y.Y.; Dang, C.Y. MGRS: A multi-granulation rough set. *Inform. Sci.* **2010**, *180*, 949–970. [CrossRef]
41. Qian, Y.H.; Li, S.Y.; Liang, J.Y.; Shi, Z.Z.; Wang, F. Mgrs: Pessimistic rough set based decisions: A multigranulation fusion strategy. *Inf. Sci.* **2014**, *264*, 196–210. [CrossRef]

42. Qian, Y.H.; Liang, X.Y.; Lin, G.P.; Guo, Q.; Liang, J.Y. Mgrs: Local multigranulation decision-theoretic rough sets. *Int. J. Approx. Reason.* **2017**, *82*, 119–137. [CrossRef]
43. Wong, S.K.M.; Ziarko, W. Comparison of the probabilistic approximate classification and the fuzzy set model. *Fuzzy Set. Syst.* **1987**, *21*, 357–362. [CrossRef]
44. Yao, Y.Y. Three-way decisions with probabilistic rough sets. *Inf. Sci.* **2010**, *180*, 341–353. [CrossRef]
45. Yao, Y.Y. The superiority of three-way decisions in probabilistic rough set models. *Inf. Sci.* **2011**, *181*, 1080–1096. [CrossRef]
46. Zhang, C.; Li, D.Y.; Mu, Y.M.; Song, D. An interval-valued hesitant fuzzy multigranulation rough set over two universes model for steam turbine fault diagnosis. *Appl. Math. Model.* **2017**, *42*, 1803–1816. [CrossRef]
47. Zhang, C.; Li, D.Y.; Zhai, Y.H.; Yang, Y.H. Multigranulation rough set model in hesitant fuzzy information systems and its application in person-job fit. *Int. J. Mach. Learn. Cyb.* **2019**, *10*, 719–729. [CrossRef]
48. Sun, B.Z.; Ma, W.M.; Li, B.J.; Li, X.N. Three-way decisions approach to multiple attribute group decision-making with linguistic information-based decision-theoretic rough fuzzy set. *Int. J. Approx. Reason.* **2018**, *93*, 424–442. [CrossRef]
49. Sun, B.Z.; Ma, W.M.; Chen, X.T.; Li, X.N. Heterogeneous multigranulation fuzzy rough set-based multiple attribute group decision-making with heterogeneous preference information. *Comput. Ind. Eng.* **2018**, *122*, 24–38. [CrossRef]
50. Zhan, J.M.; Sun, B.Z.; Alcantud, J.C.R. Covering based multigranulation (I,T)-fuzzy rough set models and applications in multi-attribute group decision-making. *Inf. Sci.* **2019**, *476*, 290–318. [CrossRef]
51. Zhang, L.; Zhan, J.M.; Xu, Z.S. Covering-based generalized IF rough sets with applications to multi-attribute decision-making. *Inf. Sci.* **2019**, *478*, 275–302. [CrossRef]
52. Liang, D.C.; Liu, D. A novel risk decision-making based on decision-theoretic rough sets under hesitant fuzzy information. *IEEE. T. Fuzzy. Syst.* **2015**, *23*, 237–247. [CrossRef]
53. Liu, D.; Liang, D.C.; Wang, C.C. A novel three-way decision model based on incomplete information system. *Knowl.-Based Syst.* **2016**, *91*, 32–45. [CrossRef]
54. Liang, D.C.; Xu, Z.S.; Liu, D.; Wu, W. Method for three-way decisions using ideal TOPSIS solutions at Pythagorean fuzzy information. *Inf. Sci.* **2018**, *435*, 282–295. [CrossRef]
55. Xu, W.H.; Guo, Y.T. Generalized multigranulation double-quantitative decision-theoretic rough set. *Knowl.-Based Syst.* **2016**, *105*, 190–205. [CrossRef]
56. Zhang, C.; Li, D.Y.; Mu, Y.M.; Song, D. A Pythagorean fuzzy multigranulation probabilistic model for mine ventilator fault diagnosis. *Complexity* **2018**, *2018*, 7125931. [CrossRef]
57. Zhang, C.; Li, D.Y.; Liang, J.Y. Multi-granularity three-way decisions with adjustable hesitant fuzzy linguistic multigranulation decision-theoretic rough sets over two universes. *Inf. Sci.* **2020**, *507*, 665–683. [CrossRef]
58. Medina, J.; Ojeda-Aciego, M. Multi-adjoint t-concept lattices. *Inf. Sci.* **2010**, *180*, 712–725. [CrossRef]
59. Takacs, A.; Kovacs, L.; Rudas, I.J.; Precup, R.E.; Haidegger, T. Models for force control in telesurgical robot systems. *Acta. Polytech. Hung.* **2015**, *12*, 95–114.
60. Gil, R.P.A.; Johanyak, Z.C.; Kovacs, T. Surrogate model based optimization of traffic lights cycles and green period ratios using microscopic simulation and fuzzy rule interpolation. *Int. J. Artif. Intell.* **2018**, *16*, 20–40.
61. Smarandache, F.; Colhon, M.; Vladutescu, S.; Negrea, X. Word-level neutrosophic sentiment similarity. *Appl. Soft. Comput.* **2019**, *80*, 167–176. [CrossRef]
62. Kandasamy, I.; Vasantha, W.B.; Obbineni, J.M.; Smarandache, F. Sentiment analysis of tweets using refined neutrosophic sets. *Comput. Ind.* **2020**, *115*, 103180. [CrossRef]
63. Liao, J.; Wang, S.G.; Li, D.Y. Identification of fact-implied implicit sentiment based on multi-level semantic fused representation. *Knowl.-Based Syst.* **2019**, *165*, 197–207. [CrossRef]
64. Zhai, Y.H.; Li, D.Y.; Zhang, J. Variable decision knowledge representation: A logical description. *J. Comput. Sci.* **2018**, *25*, 161–169. [CrossRef]
65. Zhai, Y.H.; Li, D.Y. Knowledge structure preserving fuzzy attribute reduction in fuzzy formal context. *Int. J. Approx. Reason.* **2019**, *115*, 209–220. [CrossRef]
66. Guo, Y.T.; Tsang, E.C.C.; Xu, W.H.; Chen, D.G. Local logical disjunction double-quantitative rough sets. *Inf. Sci.* **2019**, *500*, 87–112. [CrossRef]

67. Zhang, S.X.; Li, D.Y.; Zhai, Y.H.; Kang, X.P. A comparative study of decision implication, concept rule and granular rule. *Inf. Sci.* **2020**, *508*, 33–49. [CrossRef]
68. Zhao, C.J.; Wang, S.G.; Li, D.Y. Exploiting social and local contexts propagation for inducing Chinese microblog-specific sentiment lexicons. *Comput. Speech. Lang.* **2019**, *55*, 57–81. [CrossRef]
69. Zhao, C.J.; Wang, S.G.; Li, D.Y. Multi-source domain adaptation with joint learning for cross-domain sentiment classification. *Knowl.-Based Syst.* **2019**, 1–16. [CrossRef]

mathematics

MDPI

Article

An Extended TOPSIS Method with Unknown Weight Information in Dynamic Neutrosophic Environment

Nguyen Tho Thong [1,2], Luong Thi Hong Lan [3,4], Shuo-Yan Chou [5,6], Le Hoang Son [1], Do Duc Dong [2] and Tran Thi Ngan [3,*]

1 VNU Information Technology Institute, Vietnam National University, Hanoi 010000, Vietnam; nguyenthothongtt89@gmail.com or 17028003@vnu.edu.vn (N.T.T.); sonlh@vnu.edu.vn (L.H.S.)
2 VNU University of Engineering and Technology, Vietnam National University, Hanoi 010000, Vietnam; dongdoduc@gmail.com
3 Faculty of Computer Science and Engineering, Thuyloi University, 175 Tay Son, Dong Da, Hanoi 010000, Vietnam; lanlhbk@tlu.edu.vn or lanlhbk@gmail.com
4 Institute of Information Technology, Vietnam Academy of Science and Technology, Hanoi 010000, Vietnam
5 Department of Industrial Management, National Taiwan University of Science and Technology, No 43 Section 4; Keelung Rd., Taipei 10607, Taiwan; sychou@mail.ntust.edu.tw
6 Center for Cyber-Physical System Innovation, National Taiwan University of Science and Technology, Taipei 10607, Taiwan
* Correspondence: ngantt@tlu.edu.vn

Received: 9 February 2020; Accepted: 9 March 2020; Published: 11 March 2020

Abstract: Decision-making activities are prevalent in human life. Many methods have been developed to address real-world decision problems. In some practical situations, decision-makers prefer to provide their evaluations over a set of criteria and weights. However, in many real-world situations, problems include a lack of weight information for the times, criteria, and decision-makers (DMs). To remedy such discrepancies, an optimization model has been proposed to determine the weights of attributes, times, and DMs. A new concept related to the correlation measure and some distance measures for the dynamic interval-valued neutrosophic set (DIVNS) are defined in this paper. An extend Technique for Order of Preference by Similarity to Ideal Solution (TOPSIS) method in the interval-valued neutrosophic set with unknown weight information in dynamic neutrosophic environments is developed. Finally, a practical example is discussed to illustrate the feasibility and effectiveness of the proposed method.

Keywords: dynamic neutrosophic environment; dynamic interval-valued neutrosophic set; unknown weight information

1. Introduction

Multiple criteria decision-making (MCDM) problems have gained more attention to researchers in recent years. The purpose of the MCDM process is to make the best ideal choice reaching the highest standard of achievement from a set of alternatives. Existing studies of MCDM attempt to handle various kinds of multi-criteria decision-making problems. The MCDM's evaluation is decided on the basis of alternative evaluations being withdrawn from to the weights of the criteria. They are completely unknown, based upon some diverse reasons, such as time pressure, partial knowledge, incomplete attribute information, and lack of decision-makers' information, so that the overall evaluation cannot be derived. Especially, in real-world situations of group decisions, the exact appreciation of weights is important for handling MCDM problems and for making a decision. For solving such problems, several studies have attempted to develop the methods to handle the MCDM problems using various kinds of information, such as fuzzy set [1], interval fuzzy set [2,3], intuitionistic fuzzy set [4,5], hesitant fuzzy set [6], neutrosophic set [7–10], interval neutrosophic set [11–15], or single neutrosophic set [16], etc. [17–19], and various methods (e.g., maximizing deviation method, entropy, optimization method) [20–22] in which the information of criteria weights are incompletely known.

Yue et al. [23] presented a TOPSIS model to calculate the weights of the DMs under a group decision environment with individual information described as interval numbers. Sajjad Ali Khan et al. [6] introduced a study based on the combination of the maximizing deviation method and the TOPSIS method for resolving MCDM problems where the valuation information is depicted as Pythagorean hesitant fuzzy numbers and information about attribute weight is incomplete. Broumi et al. [24] proposed an extended TOPSIS method for solving multiple attribute decision-making based on two new concepts of complex neutrosophic sets. Gupta et al. [4] also extended the TOPSIS method under intuitionistic fuzzy sets and interval valued intuitionistic fuzzy sets. They considered different variations of weights of attributes depending on their subjective impression, cognitive thinking, and their psychology. Wang and Mendel [25] presented an optimization model to solve the decision-making (DM) problems on the Interval Type-2 (IT2) fuzzy set. All the DMs' information is characterized by the IT2 fuzzy set and the attribute weights' information is completely unknown. Maghrabie et al. [26] proposed a new model that used the maximizing deviation method and grey systems theory to estimate the unknown criteria weights. Peng [27] proposed a novel model for achieving unknown attribute weights and handling an IoT (Internet of Things) industry decision-making issue based on interval neutrosophic sets. Tian et al. [28] combined single-valued neutrosophic sets with completely unknown criteria weights and qualitative flexible multiple criteria method for MCDM problems. In addition, for handling multi attribute decision-making problems with interval neutrosophic information, Hong et al. [29] discussed some distance measure based on the TODIM (an acronym in Portuguese for Interactive and Multicriteria Decision Making) method.

According to above analyses, the motivations of this study are summarized as follows:

(1) Many approaches attempted to handle the MCDM problem with unknown weight information, but there is little research on discovering the weights of the DMs, the attributes, and the time in the group decision-making problems, and these methods are approximately complex.
(2) Another reason is that the TOPIS model in [13] could not work efficiently without determining the evaluation information of decision-makers and this issue was not considered in [13].
(3) In real application situations, many MCDM problems reflect a lack of weight information for the times, criteria, and decision-makers.

Therefore, we focus on the issue of multiple attribute group decision-making model based on an interval-valued neutrosophic fuzzy environment, and DMs' information is characterized by interval-valued neutrosophic fuzzy sets, and the information is completely and partially unknown. We study multiple attribute group decision-making methods with incompletely known weights of DMs, attributes, and time in the neutrosophic setting and the interval-valued neutrosophic setting.

In this paper, our aim is to propose a novel decision-making approach based on DIVNS for unknown weight information to effectively solve the above deficits. The main contributions of this paper can be summarized as follows:

- We define a new correlation measure and some distance measures for DIVNS.
- An optimization model is proposed to determine the weight information for the times, criteria, and decision-makers.
- An extend TOPSIS method under interval-valued neutrosophic set with unknown weight information in the dynamic neutrosophic environment is established.

To do that, the rest of this work is organized as follows. In Section 2, we review some basis concepts. In Section 3, we develop a TOPSIS approach to handle the MCDM problems under DIVNS in dynamic neutrosophic environments where all information of attributes, DMs, and time is completely and partially unknown. Section 4 presents the numerical results of applying our proposed method in a practical problem to demonstrate the feasibility of this method. Some comparative analyses with existing algorithms are presented in Section 5. Finally, this paper ends with some conclusions of this study in Section 6.

2. Preliminary

In this section, we review some basic knowledge, such as dynamic interval-valued neutrosophic sets and MCDM.

2.1. Dynamic Interval-Valued Neutrosophic Sets

Neutrosophic sets are characterized by truth membership (T), indeterminacy membership (I), and falsity membership (F) with the conditions as 0 <= T <= 1; 0 <= I <= 1; 0 <= F <= 1. Moreover, three membership functions have to satisfy 0 <= T + I + F <= 3. Some other concepts were designed based on neutrosophic sets such as the neutrosophic probability and neutrosophic statistics, that refer to both randomness and indeterminacy with no such contraints of memberships [30]. Herein, we extend the neutrosophic set and logic to the dynamic interval-valued neutrosophic set where each element in the new neutrosophic set is expressed by the interval-valued neutrosophic number and time sequence.

Definition 1 [31]. *Let U be a universe of discourse. A is an interval neutrosophic set expressed by:*

$$A = \left\{x, \left\langle \left[T_A^L(x), T_A^U(x)\right], \left[I_A^L(x), I_A^U(x)\right], \left[F_A^L(x), F_A^U(x)\right] \right\rangle \middle| x \in U \right\} \tag{1}$$

where $\left[T_A^L(x), T_A^U(x)\right] \subseteq [0,1]$; $\left[I_A^L(x), I_A^U(x)\right] \subseteq [0,1]$; $\left[F_A^L(x), F_A^U(x)\right] \subseteq [0,1]$ *represents truth, indeterminacy, and falsity membership functions of an element.*

Thong et al. [13] introduced the concept of a DIVNS, which is shown as follows.

Definition 2 [13]. *Let U be a universe of discourse. A is a dynamic–valued neutrosophic set (DIVNS) expressed by,*

$$A = \left\{x, \left\langle \left[T_x^L(t), T_x^U(t)\right], \left[I_x^L(t), I_x^U(t)\right], \left[F_x^L(t), F_x^U(t)\right] \right\rangle \middle| x \in U \right\} \tag{2}$$

where $t = \{t_1, t_2, \ldots, t_k\}$; $\left[T_x^L(t), T_x^U(t)\right] \subseteq [0,1]$; $\left[I_x^L(t), I_x^U(t)\right] \subseteq [0,1]$; $\left[F_x^L(t), F_x^U(t)\right] \subseteq [0,1]$ *and for convenience, we call* $\widetilde{n} = \left\langle \left[T_x^L(t), T_x^U(t)\right], \left[I_x^L(t), I_x^U(t)\right], \left[F_x^L(t), F_x^U(t)\right] \right\rangle$ *a dynamic interval–valued neutrosophic element (DIVNE).*

2.2. MCDM Problems in a Dynamic Neutrosophic Environment

Thong et al. [13] expressed MCDM problems in the dynamic neutrosophic environment as follows:

Consider a MCDM problem containing $A = \{A_1, A_2, \ldots, A_v\}$ and $C = \{C_1, C_2, \ldots, C_n\}$ and $D = \{D_1, D_2, \ldots, D_h\}$ are sets of alternatives, criteria, and decision-makers. For a decision-maker $D_q(q = 1, 2, 3, \ldots, h)$ the evaluation characteristic of an alternatives $A_m(m = 1, 2, 3, \ldots, v)$ on a criteria $C_p(p = 1, 2, 3, \ldots, n)$ in time sequence $t = \{t_1, t_2, \ldots, t_k\}$ is represented by the decision matrix where $d_{mp}^q(t) = \left\langle x_{d_{mp}}^q(t), \left(T^q\left(d_{mp}, t\right), I^q\left(d_{mp}, t\right), F^q\left(d_{mp}, t\right)\right)\right\rangle; t = \{t_1, t_2, \ldots, t_k\}$ taken by DIVNSs evaluated by decision-maker D_q.

3. An Extended TOPSIS Method for Unknown Weight Information

This section proposes the method to handle the MCDM problem that include a lack of the weight information for the times, criteria, and DMs in dynamic netrosophic environments.

3.1. Correlation Coefficient Measure for Dynamic Interval-Valued Neutrosophic Sets

We propose a novel correlation coefficient measure for DIVNSs based on the idea in [32].

Definition 3. *Let* $Y(t) = \{(x(t), \langle T^Y(x,t_l), I^Y(x,t_l), F^Y(x,t_l)\rangle), \forall t_l \in t, x \in U\}$ *and* $Z(t) = \{(x(t), \langle T^Z(x,t_l), I^Z(x,t_l), F^Z(x,t_l)\rangle), \forall t_l \in t, x \in U\}$ *be two DIVNs in* $t = \{t_1, t_2, \ldots, t_k\}$ *and* $U = (x_1, x_2, \ldots, x_n)$. *A correlation coefficient measure between A and B* is:

$$K(Y,Z) = \frac{C(Y,Z)}{max(T(Y),T(Z))} = \frac{\sum_{i=1}^{n} C(Y(x_i), Z(x_i))}{max\left(\sum_{i=1}^{n} T(Y(x_i)), \sum_{i=1}^{n} T(Z(x_i))\right)} \tag{3}$$

where $C(Y,Z)$ *is considered the correlation between two DIVNSs Y and Z;* $T(Y)$ *and* $T(Z)$ *refer to the information energies if the two DIVNSs, respectively. These components are provided by:*

$$C(Y,Z) = \frac{1}{k}\sum_{l=1}^{k}\sum_{i=1}^{n} C(Y(x_i,t_l), Z(x_i,t_l))$$
$$= \frac{1}{k}\sum_{l=1}^{k}\sum_{i=1}^{n} \frac{1}{2}\begin{bmatrix} \inf T_Y(x_i,t_l) \times \inf T_Z(x_i,t_l) + \sup T_Y(x_i,t_l) \times \sup T_Z(x_i,t_l) \\ +\inf I_Y(x_i,t_l) \times \inf I_Z(x_i,t_l) + \sup I_Y(x_i,t_l) \times \sup I_Z(x_i,t_l) \\ +\inf F_Y(x_i,t_l) \times \inf F_Z(x_i,t_l) + \sup F_Y(x_i,t_l) \times \sup F_Z(x_i,t_l) \end{bmatrix}$$
$$T(Y) = \frac{1}{k}\sum_{l=1}^{k}\sum_{i=1}^{n} T(Y(x_i,t_l))$$
$$= \frac{1}{k}\sum_{l=1}^{k}\sum_{i=1}^{n} \frac{1}{2}\begin{bmatrix} (\inf T_Y(x_i,t_l))^2 + (\sup T_Y(x_i,t_l))^2 + (\inf I_Y(x_i,t_l))^2 + (\sup I_Y(x_i,t_l))^2 \\ +(\inf F_Y(x_i,t_l))^2 + (\sup F_Y(x_i,t_l))^2 \end{bmatrix}$$
$$T(Z) = \frac{1}{k}\sum_{l=1}^{k}\sum_{i=1}^{n} T(Z(x_i,t_l))$$
$$= \frac{1}{k}\sum_{l=1}^{k}\sum_{i=1}^{n} \frac{1}{2}\begin{bmatrix} (\inf T_Z(x_i,t_l))^2 + (\sup T_Z(x_i,t_l))^2 + (\inf I_Z(x_i,t_l))^2 + (\sup I_Z(x_i,t_l))^2 \\ +(\inf F_Z(x_i,t_l))^2 + (\sup F_Z(x_i,t_l))^2 \end{bmatrix}$$

Theorem 1. *The correlation coefficient K between Y and Z satisfies the follow properties:*

(i) $0 \le K(Y,Z) \le 1$
(ii) $K(Y,Z) = K(Z,Y)$
(iii) $K(Y,Z) = 1 \Leftrightarrow Y = Z$

Proof. (i) for any $i = 1,2,3,\ldots,n$, the values of $[\inf T_Y(x_i,t_l), \sup T_Y(x_i,t_l)]$; $[\inf I_Y(x_i,t_l), \sup I_Y(x_i,t_l)]$; $[\inf F_Y(x_i,t_l), \sup F_Y(x_i,t_l)]$; $[\inf T_Z(x_i,t_l), \sup T_Z(x_i,t_l)]$; $[\inf I_Z(x_i,t_l), \sup I_Z(x_i,t_l)]$; $[\inf F_Z(x_i,t_l), \sup F_Z(x_i,t_l)] \subseteq [0,1]$ exist for any $i = 1,2,3,\ldots,n$. Thus, it is hold that $C(Y,Z) \ge 0; T(Y) \ge 0; T(Z) \ge 0$. Therefore

$$K(Y,Z) = \frac{C(Y,Z)}{max(T(Y),T(Z))} \ge 0$$

and according to the Cauchy–Schwarz inequality, it holds that:

$$K(Y,Z) = \frac{C(Y,Z)}{max(T(Y),T(Z))} \le 1$$

Therefore, $0 \le K(Y,Z) \le 1$.

(ii) It is obvious that if $Y(t) = Z(t)$, $\forall l \in \{1,2,\ldots,k\}$. We have:

$\inf T_Y(x_i,t_l) = \inf T_Z(x_i,t_l)$; $\sup T_Y(x_i,t_l) = \sup T_Z(x_i,t_l)$; $\inf I_Y(x_i,t_l) = \inf I_Z(x_i,t_l)$; $\sup I_Y(x_i,t_l) = \sup I_Z(x_i,t_l)$; $\inf F_Y(x_i,t_l) = \inf F_Z(x_i,t_l)$; $\sup F_Y(x_i,t_l) = \sup F_Z(x_i,t_l)$;

Thus, $K(Y,Z) = K(Z,Y)$. Theorem 1 is proved.

(iii) It is easily observed. □

3.2. Distance Measures for Dynamic Interval-Valued Neutrosophic Sets

In this section, we present the definitions of the Hamming and Euclidean distances between DIVNEs and distance of two dynamic interval-valued neutrosophic matrices.

Definition 4. *Let n_1 and n_2 be two DIVNEs. The dynamic interval-valued neutrosophic distance between n_1 and n_2* is determined as follows:

(i) *The Hamming distance:*

$$d_1(n_1,n_2) = \frac{1}{6\times k}\sum_{l=1}^{k}\begin{pmatrix} \left|T^L_{n_1}(t_l) - T^L_{n_2}(t_l)\right| + \left|T^U_{n_1}(t_l) - T^U_{n_2}(t_l)\right| + \left|I^L_{n_1}(t_l) - I^L_{n_2}(t_l)\right| \\ +\left|I^U_{n_1}(t_l) - I^U_{n_2}(t_l)\right| + \left|F^L_{n_1}(t_l) - F^L_{n_2}(t_l)\right| + \left|F^U_{n_1}(t_l) - F^U_{n_2}(t_l)\right| \end{pmatrix} \quad (4)$$

(ii) *The Euclidean distance:*

$$d_2(n_1,n_2) = \sqrt{\frac{1}{6\times k}\sum_{l=1}^{k}\begin{pmatrix} \left(T^L_{n_1}(t_l) - T^L_{n_2}(t_l)\right)^2 + \left(T^U_{n_1}(t_l) - T^U_{n_2}(t_l)\right)^2 + \left(I^L_{n_1}(t_l) - I^L_{n_2}(t_l)\right)^2 \\ +\left(I^L_{n_1}(t_l) - I^L_{n_2}(t_l)\right)^2 + \left(F^L_{n_1}(t_l) - F^L_{n_2}(t_l)\right)^2 + \left(F^U_{n_1}(t_l) - F^U_{n_2}(t_l)\right)^2 \end{pmatrix}} \quad (5)$$

(iii) *The geometry distance:*

$$d_3(n_1,n_2) = \left(\frac{1}{6\times k}\sum_{l=1}^{k}\begin{pmatrix} \left(T^L_{n_1}(t_l) - T^L_{n_2}(t_l)\right)^\alpha + \left(T^U_{n_1}(t_l) - T^U_{n_2}(t_l)\right)^\alpha + \left(I^L_{n_1}(t_l) - I^L_{n_2}(t_l)\right)^\alpha \\ +\left(I^L_{n_1}(t_l) - I^L_{n_2}(t_l)\right)^\alpha + \left(F^L_{n_1}(t_l) - F^L_{n_2}(t_l)\right)^\alpha + \left(F^U_{n_1}(t_l) - F^U_{n_2}(t_l)\right)^\alpha \end{pmatrix}\right)^{\frac{1}{\alpha}} \quad (6)$$

where $\alpha > 0$ and

- *If $\alpha = 1$, then equation (6) refers to the Hamming distance.*
- *If $\alpha = 2$, then equation (6) refers to the Euclidean distance.*

Therefore, the distance in Equation (6) is a generalization of distances in Equation (5) and Equation (4).

Definition 5. *Given two dynamic interval-valued neutrosophic matrices $A_1 = [\alpha(t_l)]_{h\times n}$ and $A_2 = [\beta(t_l)]_{h\times n}$, the elements of both A_1 and A_2 are described by DIVNS. After that the distance between A_1 and A_2* is defined by:

$$d(A_1,A_2) = \frac{1}{hn}\sum_{p}^{n}\sum_{q}^{h} d\left(\alpha_{qp},\beta_{qp}\right) \quad (7)$$

where $d\left(\alpha_{qp},\beta_{qp}\right)$ is the distance between two DIVNEs.

3.3. Unknown Weight Information in Dynamic Neutrosophic Environment

3.3.1. Determining the Weight of Time

It is common knowledge that the weights of time periods have an important role in MCDM problems practical application. In the followings, we present how to determine the weights of time periods in dynamic neutrosophic environments.

Definition 6. *Given a basic unit-interval monotonic (BUM) function $g : [0,1] \to [0,1]$, the time weight can be determined as follows:*

$$\lambda(t_l) = g\left(\frac{R_l}{TV}\right) - g\left(\frac{R_{l-1}}{TV}\right), \quad (8)$$

where $R_l = \sum_{j=1}^{l} V_j; TV = \sum_{i=1}^{k} V_i; V_i = 1 + T(MD_i)$; $T(MD_i)$ *denotes the support of* i^{th} *largest argument by all the other arguments:*

$$T(MD_i) = \sum_{\substack{j=1 \\ j \neq i}}^{k} Sup\left(MD_i, MD_j\right)$$

$$sup\left(MD_i, MD_j\right) = 1 - d\left(MD_i, MD_j\right)$$
$$= 1 - \frac{1}{hn}\sum_{q}^{h}\sum_{p=1}^{n}\sqrt{\frac{1}{6}\left(\begin{array}{c}\left(T^L\left(x_{pq}^i\right) - T^L\left(x_{pq}^j\right)\right)^2 + \left(T^U\left(x_{pq}^i\right) - T^U\left(x_{pq}^j\right)\right)^2 + \left(I^L\left(x_{pq}^i\right) - I^L\left(x_{pq}^j\right)\right)^2 + \\ \left(I^U\left(x_{pq}^i\right) - I^U\left(x_{pq}^j\right)\right)^2 + \left(F^L\left(x_{pq}^i\right) - F^L\left(x_{pq}^j\right)\right)^2 + \left(F^U\left(x_{pq}^i\right) - F^U\left(x_{pq}^j\right)\right)^2\end{array}\right)} \quad (9)$$

3.3.2. Determining the Weights of Decision-Makers

The weights of DMs play a critical role in MCDM problems. In this section, we present how to determine the weights of DMs in dynamic neutrosophic environment.

Definition 7. *Let* $D_1 = [\alpha(t_l)]_{v \times n}$ *and* $D_2 = [\beta(t_l)]_{v \times n}$ *be two dynamic interval-valued neutrosophic matrices, in which the elements of both* D_1 *and* D_2 *are expressed by DIVNS. Then the correlation coefficient between* D_1 *and* D_2 *is defined by:*

$$C(D_1, D_2) = \frac{1}{nv}\sum_{p}^{n}\sum_{m}^{v} K\left(\alpha_{mp}, \beta_{mp}\right) \quad (10)$$

where $K\left(\alpha_{mp}, \beta_{mp}\right)$ *is correlation coefficient measure between two DIVNEs.*

Theorem 2. *For two Dynamic interval-valued neutrosophic matrices* $D_1 = [\alpha(t_l)]_{v \times n}$ *and* $D_2 = [\beta(t_l)]_{v \times n}$ *where the elements of both* D_1 *and* D_2 *are expressed by DIVNSs,* $C(D_1, D_2)$ *satisfies the three conditions:*

(i) $0 \le C(D_1, D_2) \le 1$
(ii) $C(D_1, D_2) = C(D_2, D_1)$
(iii) $D_1 = D_2$ *if and only if* $C(D_1, D_2) = 1$

Proof. (i) According to Theorem 1, we have $0 \le K\left(\alpha_{mp}, \beta_{mp}\right) \le 1$; $m = 1, 2, 3, \ldots, v$; $p = 1, 2, 3, \ldots, n$, Thus,

$$0 \le \frac{1}{nv}\sum_{p}^{n}\sum_{m}^{v} K\left(\alpha_{mp}, \beta_{mp}\right) \le 1$$

(ii) According to Definition 3 and Theorem 1 it is easily observed.
(iii) According to Theorem 1 we obtain $C(D_1, D_2) = 1 \Leftrightarrow D_1 = D_2$
Thus, Theorem 2 is proved. □

Definition 8. *For the decision-maker* D_q*, the weights of decision-makers can be defined as follows:*

$$\omega_q = \frac{\delta_q}{\sum_{q=1}^{h} \delta_q} \quad (11)$$

where δ_q *has the form:*

$$\delta_q = \sum_{\substack{q'=1; \\ q' \neq q}}^{h} C\left(D_q, D_{q'}\right) \quad (12)$$

$C\left(D_q, D_{q'}\right)$ *is the correlation coefficient between two decision-makers* q and q'.

3.3.3. Determining the Weights of the Criteria

In real life applications, the attribute information may be completely unknown. Thus, we need to develop an integrated programming model for MCDM problems under the dynamic neutrosophic environment.

Definition 9. *Let C_p be the pth criterion and A_m be the mth alternative, the deviation value between A_m and all the other alternatives in dynamic neutrosophic environment can be calculated as:*

$$O_{mp}(w) = \sum_{\substack{k=1;\\ k\neq i}}^{v} d\left(n_{mp}, n_{kp}\right)w_p \quad (13)$$

where w_p is weight of the pth criterion. $d\left(n_{mp}, n_{kp}\right)$ is the distance between two DIVNEs.

Definition 10. *The deviation among all the alternatives to the others can be computed by the global deviation function as follows:*

$$O_p(w) = \sum_{m=1}^{v} O_{mp}(w) = \sum_{m=1}^{v} \sum_{\substack{k=1;\\ k\neq m}}^{v} d\left(n_{mp}, n_{kp}\right)w_p$$
$$\text{s.t.} \sum_{p=1}^{n} w_p = 1; w_{p\geq 0}; \quad (14)$$

By using the deviation degree between evaluations [33], the criteria weights can be calculated. Then, we construct optimization decision making model with the purpose of maximizing the decision space in the following:

$$\max O(w) = \sum_{p=1}^{n} O_p(w) = \sum_{p=1}^{n}\sum_{m=1}^{v} \sum_{\substack{k=1;\\ k\neq m}}^{v} d\left(n_{mp}, n_{kp}\right)w_p^* \rightarrow \max \quad (15)$$

where $d\left(n_{mp}, n_{kp}\right)$ is the distance between two elements. The optimization model can be solved based on the Lagrange function. Let ξ be the Lagrange multiplier. We have:

$$L\left(w_p^*, \xi\right) = O(w) - \frac{1}{2}\xi\left(\sum_{p=1}^{n} \left(w_p^*\right)^2 - 1\right)$$

$$L\left(w_p^*, \xi\right) = \sum_{p=1}^{n}\sum_{m=1}^{v} \sum_{\substack{k=1;\\ k\neq m}}^{v} d\left(n_{mp}, n_{kp}\right)w_p^* - \frac{1}{2}\xi\left(\sum_{p=1}^{n} \left(w_p^*\right)^2 - 1\right)$$

$$\begin{cases} \frac{\partial L}{\partial w_p} = \sum_{m=1}^{v} \sum_{\substack{k=1; \\ k \neq m}}^{v} d(n_{mp}, n_{kp}) - \xi w_p^* = 0 \\ \frac{\partial L}{\partial \xi} = \frac{1}{2}\left(\sum_{p=1}^{n} \left(w_p^*\right)^2 - 1 \right) = 0 \end{cases} \Rightarrow w_p^* = \frac{\sum_{m=1}^{v} \sum_{\substack{k=1; \\ k \neq m}}^{v} d(n_{mp}, n_{kp})}{\xi}$$

Since $\sum_{m=p}^{n} \left(w_p^*\right)^2 = 1$, the value of ξ can be calculated as follows:

$$\sum_{p=1}^{n} \left(\frac{\sum_{m=1}^{v} \sum_{\substack{k=1; \\ k \neq m}}^{v} d(n_{mp}, n_{kp})}{\xi} \right)^2 = 1, \text{ thus, } \xi = \sqrt{\sum_{p=1}^{n} \left[\sum_{m=1}^{v} \sum_{\substack{k=1; \\ k \neq m}}^{v} d(n_{mp}, n_{kp}) \right]^2}$$

From the above equations, a formula to calculate the criteria weights can be obtained as follows:

$$w_p^* = \frac{\sum_{m=1}^{v} \sum_{\substack{k=1; \\ k \neq m}}^{v} d(n_{mp}, n_{kp})}{\sqrt{\sum_{p=1}^{n} \left[\sum_{m=1}^{v} \sum_{\substack{k=1; \\ k \neq m}}^{v} d(n_{mp}, n_{kp}) \right]^2}} \tag{16}$$

3.4. TOPSIS Method with Unknown Weight Information in Dynamic Neutrosophic Environments

In this section, we develop a MCDM approach based on the TOPSIS model with unknown weight information in dynamic neutrosophic environments. The scheme of the proposed MCDM technique is given in Figure 1. The detailed method is constructed as follows:

Step 1. Construct the dynamic interval-valued neutrosophic decision matrix as MCDM problems expressed in Section 2.2.

Step 2. Using Equation (8) to determine the time weights $\lambda = (\lambda_1, \lambda_2, \ldots, \lambda_k)$ of k time sequence:

$$g(x) = \frac{e^{\alpha x} - 1}{e^{\alpha} - 1} \tag{17}$$

Step 3. Using Equations (10)–(12) to determine the DMs' weights $\omega = (\omega_1, \omega_2, \ldots, \omega_h)$ of h decision-makers.

Step 4. I the criteria weight information is completely unknown, we determine the criteria weights $w = (w_1, w_2, \ldots, w_n)^T$ of n criteria by using Equation (16), otherwise go to Step 5.

Step 5. Suppose $W = [\psi(t_l)]_{p\times q}; p = 1,2,3,\ldots,n; q = 1,2,3,\ldots,h; l = 1,2,3,\ldots,k$ be dynamic interval-valued neutrosophic matrix of important criteria weights. $\psi_{pq}(t_l)$ is the weight of decision-maker q^{th} to criterion p^{th} in time sequence t_l. The criteria weights $w = (w_1, w_2, \ldots, w_n)^T$ can be calculated by:

$$w_p = \left\langle \begin{array}{c} \left[\left\langle 1 - \left\{ 1 - \left(1 - \sum_{q=1}^{h} T_{pq}^{L}(\psi_{t_l}) \right)^{\frac{1}{h}} \right\}^{\frac{1}{k}} \right\rangle, \left\langle 1 - \left\{ 1 - \left(1 - \sum_{q=1}^{h} T_{pq}^{U}(\psi_{t_l}) \right)^{\frac{1}{h}} \right\}^{\frac{1}{k}} \right\rangle \right], \\ \left[\left(\sum_{q=1}^{h} I_{pq}^{L}(\psi_{t_l}) \right)^{\frac{1}{h*k}}, \left(\sum_{q=1}^{h} I_{pq}^{U}(\psi_{t_l}) \right)^{\frac{1}{h*k}} \right], \left[\left(\sum_{q=1}^{h} F_{pq}^{L}(\psi_{t_l}) \right)^{\frac{1}{h*k}}, \left(\sum_{q=1}^{h} F_{pq}^{U}(\psi_{t_l}) \right)^{\frac{1}{h*k}} \right] \end{array} \right\rangle \quad (18)$$

Step 6. The aggregate ratings of alternative m and criteria p can be estimated as:

$$x_{mp} = \frac{1}{h*k} \otimes \left\langle \begin{array}{c} \left[\left\langle 1 - \left\{ 1 - \left(1 - \sum_{q=1}^{h} T_{pmq}^{L}(x_{t_l}) \right)^{\frac{1}{h}} \right\}^{\frac{1}{k}} \right\rangle, \left\langle 1 - \left\{ 1 - \left(1 - \sum_{q=1}^{h} T_{pmq}^{U}(x_{t_l}) \right)^{\frac{1}{h}} \right\}^{\frac{1}{k}} \right\rangle \right] \\ \left[\left(\sum_{q=1}^{h} I_{pmq}^{L}(x_{t_l}) \right)^{\frac{1}{h*k}}, \left(\sum_{q=1}^{h} I_{pmq}^{U}(x_{t_l}) \right)^{\frac{1}{h*k}} \right], \overline{F_{mp}(x)} = \left[\left(\sum_{q=1}^{h} F_{pmq}^{L}(x_{t_l}) \right)^{\frac{1}{h*k}}, \left(\sum_{q=1}^{h} F_{pmq}^{U}(x_{t_l}) \right)^{\frac{1}{h*k}} \right] \end{array} \right\rangle, \quad (19)$$

Step 7: Average weighted ratings of alternatives can be calculated as follows:

Case 1: If the information about the criteria weights is known, the criteria weights is a collection of DIVNEs and the average weighted ratings of alternatives in t_l, calculated by:

$$G_m = \frac{1}{p}\sum_{p=1}^{n} \left(\left\langle \begin{array}{c} \left[T_{mp}^{L}(x) \times T_{p}^{L}(w), T_{mp}^{U}(x) \times T_{p}^{U}(w) \right], \\ \left[I_{mp}^{L}(x) + I_{p}^{L}(w) - I_{mp}^{L}(x) \times I_{p}^{L}(w), I_{mp}^{U}(x) + I_{p}^{U}(w) - I_{mp}^{U}(x) \times I_{p}^{U}(w) \right], \\ \left[F_{mp}^{L}(x) + F_{p}^{L}(w) - F_{mp}^{L}(x) \times F_{p}^{L}(w), F_{mp}^{U}(x) + F_{p}^{U}(w) - F_{mp}^{U}(x) \times F_{p}^{U}(w) \right] \end{array} \right\rangle \right); \quad (20)$$
$$m = 1,2,3,\ldots,v; p = 1,2,3,\ldots,n;$$

Case 2: If the information about the criteria weights is unknown, the criteria weights is a collection of DIVNEs and average weighted ratings of alternatives in t_l, calculated by:

$$G_m = \frac{1}{p}\sum_{p=1}^{n} \left(\left\langle \begin{array}{c} \left[1 - \left(1 - T_{mp}^{L}(x) \right)^{w_p}, 1 - \left(1 - T_{mp}^{U}(x) \right)^{w_p} \right], \\ \left[\left(I_{mp}^{L}(x) \right)^{w_p}, \left(I_{mp}^{U}(x) \right)^{w_p} \right], \left[\left(F_{mp}^{L}(x) \right)^{w_p}, \left(F_{mp}^{U}(x) \right)^{w_p} \right] \end{array} \right\rangle \right); \quad (21)$$
$$m = 1,2,3,\ldots,v; p = 1,2,3,\ldots,n;$$

Step 8: Determine the interval neutrosophic positive ideal solution (PIS, A^+) and the interval neutrosophic negative ideal solution (NIS, A^-):

$$A^+ = \{x, ([1,1],[0,0],[0,0])\} \quad (22)$$

$$A^- = \{x, ([0,0],[1,1],[1,1])\} \quad (23)$$

Step 9: Compute the distance of alternatives.

The distances of each alternative in time sequence t_l, are calculated:

$$d_m^+ = \sqrt{(G_m - A^+)^2} \quad (24)$$

$$d_m^- = \sqrt{(G_m - A^-)^2} \quad (25)$$

where d_m^+ and d_m^- represent the shortest and farthest distances of alternative A_m.

Step 10: Determine the relative closeness coefficient.

The closeness coefficient values are calculated below:

$$CC_m = \frac{d_m^-}{d_m^+ + d_m^-} \quad (26)$$

Step 11: Rank the alternatives based on the relative closeness coefficients.

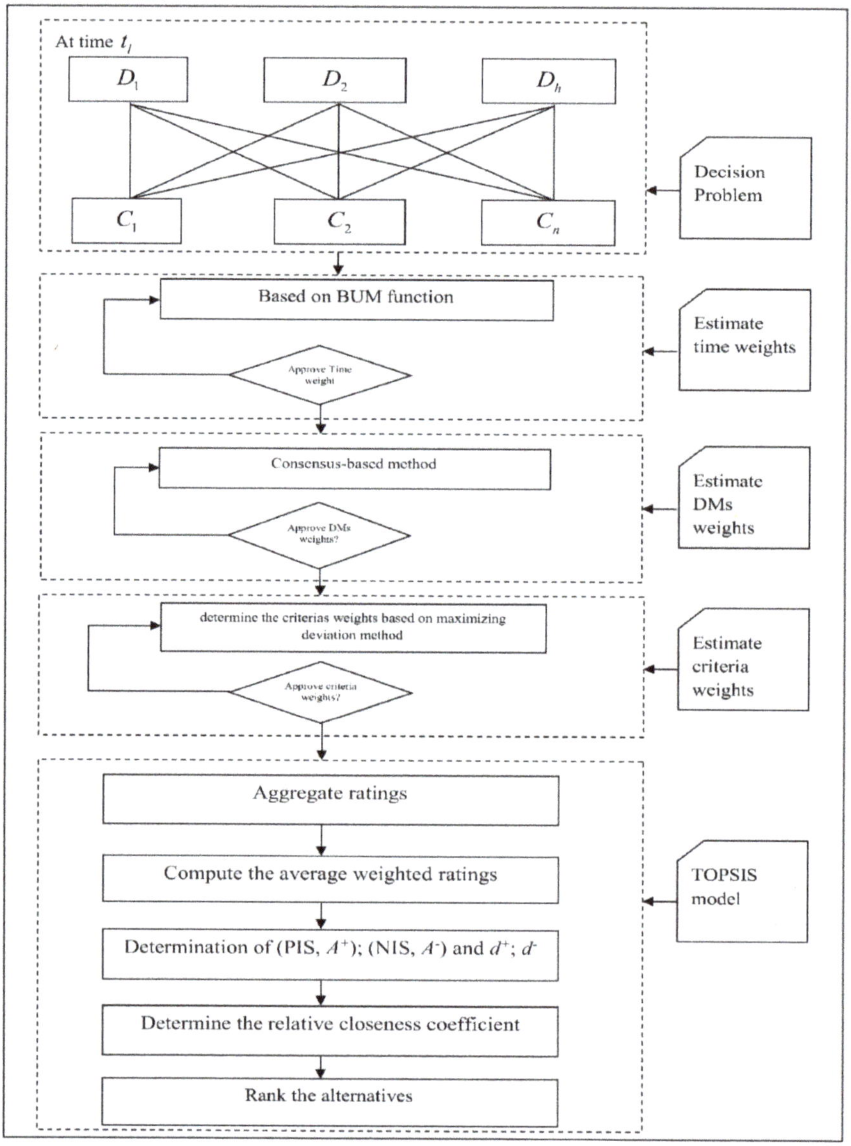

Figure 1. TOPSIS method with unknown weight information.

4. Experiments

This section applies the proposed method with dataset in [17] to evaluate lecturers' performances from ULIS, Vietnam National University, Hanoi, Vietnam. The hierarchical structure of the constructed multi-criteria decision-making problem is depicted in Figure 2 for the dataset.

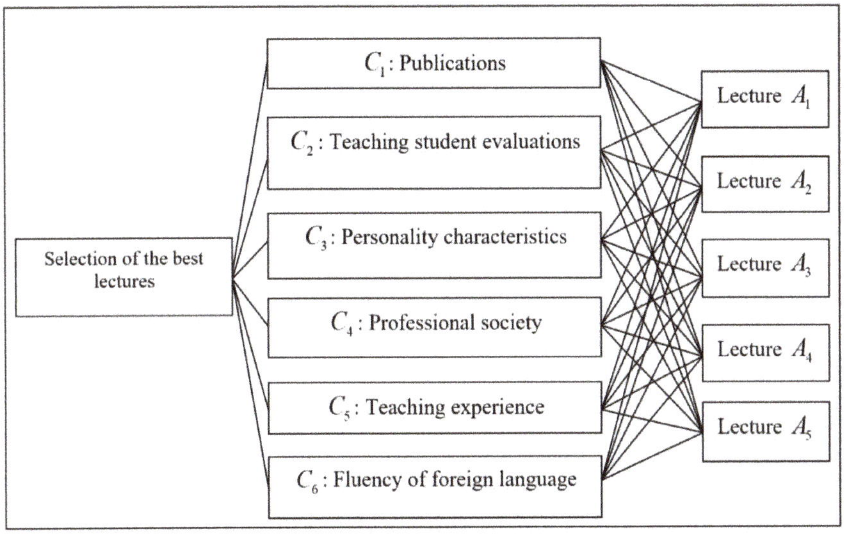

Figure 2. Evaluation lecturer's performance problem.

According to the language labels in Table 1 below, the rating of lectures through criteria sets are done by decision-makers.

Table 1. Language variables.

Language Labels	Short Labels	Values
Very-Poor	Vr	([0.1, 0.2], [0.6, 0.7], [0.7, 0.8])
Poor	Pr	([0.2, 0.3], [0.5, 0.6], [0.6, 0.7])
Medium	Mm	([0.3, 0.5], [0.4, 0.6], [0.4, 0.5])
Good	Gd	([0.5, 0.6], [0.4, 0.5], [0.3, 0.4])
Very-Good	Vd	([0.6, 0.7], [0.2, 0.3], [0.2, 0.3])

Step 1: Dynamic interval-valued neutrosophic decision matrix shown in Table 2.

Step 2: Bases on Equation (8) and BUM function in Equation (17), we receive the weights of the time periods:

$$\lambda_1 = 0.280; \lambda_2 = 0.330; \lambda_3 = 0.390$$

Step 3: Using Equations (10)–(12) to calculate weights of the DMs, we receive the weights of the DMs as follows:

$$\omega_1 = 0.330; \omega_2 = 0.337; \omega_3 = 0.333$$

Step 4: Based on the basic of maximizing deviation method and Equation (16), we receive the weights of the criteria as follows:

$$w_1 = 0.160; w_2 = 0.165; w_3 = 0.171; w_4 = 0.166; w_5 = 0.175; w_6 = 0.163$$

Step 5: Average weighted ratings are shown in Table 3.

Table 2. Dynamic interval-valued neutrosophic decision matrix.

Criteria	Lecturers	Decision Makers								
		t_1			t_2			t_3		
		D_1	D_2	D_3	D_1	D_2	D_3	D_1	D_2	D_3
C_1	A_1	Mm	Gd	Gd	Gd	Gd	Gd	Gd	Vd	Gd
	A_2	Gd	Gd	Vd	Vd	Gd	Vd	Vd	Gd	Vd
	A_3	Mm	Gd	Gd	Gd	Gd	Gd	Gd	Gd	Vd
	A_4	Go	Mm	Gd	Gd	Gd	Gd	Gd	Gd	Gd
	A_5	Mm	Gd	Mm	Go	Go	Mm	Gd	Gd	Gd
C_2	A_1	Gd	Gd	Gd	Vd	Gd	Gd	Gd	Gd	Gd
	A_2	Vd	Gd	Vd	Mm	Gd	Gd	Vd	Gd	Gd
	A_3	Vd	Gd	Gd	Gd	Mm	Gd	Gd	Mm	Gd
	A_4	Gd	Gd	Gd	Gd	Vd	Gd	Gd	Gd	Vd
	A_5	Vd	Gd	Gd	Gd	Vd	Gd	Gd	Gd	Mm
C_3	A_1	Vd	Vd	Gd	Gd	Vd	Gd	Gd	Mm	Gd
	A_2	Gd	Vd	Gd	Vd	Gd	Vd	Gd	Gd	Vd
	A_3	Gd	Vd	Vd	Gd	Gd	Gd	Gd	Vd	Gd
	A_4	Gd	Gd	Gd	Vd	Gd	Gd	Vd	Gd	Gd
	A_5	Vd	Gd	Gd	Gd	Vd	Gd	Gd	Gd	Gd
C_4	A_1	Mm	Gd	Mm	Gd	Gd	Mm	Mm	Gd	Mm
	A_2	Gd	Mm	Gd	Gd	Mm	Gd	Gd	Mm	Gd
	A_3	Gd	Gd	Gd	Gd	Gd	Mm	Gd	Gd	Vd
	A_4	Mm	Poo	Mm	Gd	Mm	Mm	Gd	Gd	Mm
	A_5	Mm	Mm	Poo	Mm	Mm	Mm	Mm	Gd	Mm
C_5	A_1	Mm	Gd	Mm	Mm	Gd	Gd	Gd	Mm	Gd
	A_2	Gd	Vd	Go	Vd	Gd	Gd	Gd	Vd	Gd
	A_3	Gd	Gd	Mm	Gd	Gd	Gd	Gd	Vd	Gd
	A_4	Vd	Gd	Gd	Vd	Gd	Gd	Vd	Gd	Gd
	A_5	Gd	Gd	Gd	Gd	Gd	Gd	Gd	Vd	Gd
C_6	A_1	Vd	Gd	Gd	Vd	Gd	Vd	Vd	Gd	Vd
	A_2	Gd	Gd	Gd	Gd	Vd	Gd	Gd	Gd	Vd
	A_3	Vd	Gd	Vd	Vd	Gd	Vd	Vd	Gd	Vd
	A_4	Gd	Vd	Gd	Gd	Vd	Gd	Gd	Gd	Gd
	A_5	Gd	Gd	Gd	Vd	Gd	Gd	Gd	Vd	Gd

Table 3. Average weighted ratings of lectures.

Lecturers	Weighted Ratings
A_1	([0.072, 0.102], [0.871, 0.906], [0.848, 0.883])
A_2	([0.083, 0.112], [0.852, 0.889], [0.833, 0.871])
A_3	([0.082, 0.110], [0.867, 0.900], [0.842, 0.878])
A_4	([0.077, 0.105], [0.867, 0.901], [0.844, 0.880])
A_5	([0.073, 0.102], [0.871, 0.907], [0.850, 0.884])

Step 6: Compute the distance of each lecture from (PIS, A^+) and (NIS, A^-). The results are shown in Table 4 below.

Table 4. The distance of each lecture.

Lecturers	d^+	d^-
A_1	0.113845	0.889443
A_2	0.128101	0.875218
A_3	0.120105	0.882727
A_4	0.117807	0.885273
A_5	0.113326	0.889768

Step 7: Calculate the closeness coefficient for lectures. Table 5 shows the values of the closeness coefficient.

Table 5. The closeness coefficient of lectures.

Lecturers	Proposed Model
A_1	0.11355
A_2	0.12778
A_3	0.11983
A_4	0.11752
A_5	0.11301

Step 8: Rank the lectures based on the values of the closeness coefficients.
Table 5 shows the ranking order is $A_2 \succ A_3 \succ A_4 \succ A_1 \succ A_5$ and A_2 is the best lecture.

5. Comparison with the Related Methods

In this section, we compare the proposed method with those in Thong et al. [17] and Peng [29] to demonstrate the advantages for unknown weight information in dynamic neutrosophic environments. Data used to prove the performance of the method are in [17]. Table 6 shows that the rankings of lectures by Thong et al. [17] as $A_2 \succ A_3 \succ A_4 \succ A_1 \succ A_5$ and Peng [29] as $A_2 \succ A_3 \succ A_1 \succ A_4 \succ A_5$. Thus, A_2 is still the best option. These results are the same as our proposed method. However, the proposed method can be solved with unknown weight information in a dynamic neutrosophic environment. Moreover, it is more generalized and flexible than Thong et al. [17]'s method with unknown weight information in a dynamic neutrosophic environment.

Table 6. A comparison study with some existing methods.

Methods	Ranking Values					Ranking Order
	A_1	A_2	A_3	A_4	A_5	
Proposed method	0.11355	0.12778	0.11983	0.11752	0.11301	$A_2 \succ A_3 \succ A_4 \succ A_1 \succ A_5$
Thong et al. [17] (Topsis Model)	0.33916	0.36694	0.35124	0.34526	0.33778	$A_2 \succ A_3 \succ A_4 \succ A_1 \succ A_5$
Peng [29] (Similarity measure)	0.92735	0.94145	0.92949	0.90850	0.89896	$A_2 \succ A_3 \succ A_1 \succ A_4 \succ A_5$

6. Conclusions

In this paper, we proposed a novel approach to solve MCDM problems in dynamic neutrosophic environments where all the information supplied by the DMs is described as interval-valued neutrosophic sets and the information about the weight of attributes, DMs, and time may be incompletely known. A new concept related to the correlation measure and some distance measures for dynamic interval-valued neutrosophic sets are defined. Then, we have proposed an extended TOPSIS method to solve MCDM problems, are is expressed with the interval-valued neutrosophic setting in dynamic neutrosophic environments. Finally, the effectiveness of the proposed method has been demonstrated with the purpose of evaluating lecturers' performance in ULIS, Vietnam National University, Hanoi, Vietnam. We considered in this situation that all the weight information about the criteria, DMs, and time is expressed with various conditions is unknown.

Since the proposed method has not demonstrated its practicality and effectiveness with more real applications and the weight information about the criteria and DMs that change over time is not mentioned in our method, in the future, we will conduct further studies to handle unknown weight information in which the criteria and DMs vary with time periods and with more real decision-making data.

Author Contributions: Data curation, L.T.H.L.; methodology, L.H.S., D.D.D. and T.T.N.; validation, N.T.T. and D.D.D.; writing—original draft, N.T.T.; writing—review & editing, S.-Y.C., L.H.S., T.T.N. and D.D.D. All authors have read and agreed to the published version of the manuscript.

Funding: This work was supported in part by the "Center for Cyber-physical System Innovation" from the Featured Areas Research Center Program within the framework of the Higher Education Sprout Project by the Ministry of Education (MOE) in Taiwan.

Acknowledgments: The first author thanks to the support from the Domestic Master/Ph.D. Scholarship Programme of the Vingroup Innovation Foundation. The authors would like to express their greatest thanks the Center for IoT Innovation (CITI), National Taiwan University of Science and Technology for their support.

Conflicts of Interest: The authors declare no conflict of interest.

References

1. Mohd, W.W.; Abdullah, L.; Yusoff, B.; Taib, C.M.; Merigo, J.M. An Integrated MCDM Model based on Pythagorean Fuzzy Sets for Green Supplier Development Program. *Malays. J. Math. Sci.* **2019**, *13*, 23–37.
2. Abdel-Basset, M.; Manogaran, G.; Gamal, A.; Smarandache, F. A group decision making framework based on neutrosophic TOPSIS approach for smart medical device selection. *J. Med. Syst.* **2019**, *43*, 38. [CrossRef]
3. Abdel-Basset, M.; Mohamed, M.; Sangaiah, A.K. Neutrosophic AHP-Delphi Group decision making model based on trapezoidal neutrosophic numbers. *J. Ambient Intell. Humaniz. Comput.* **2018**, *9*, 1427–1443. [CrossRef]
4. Basset, M.A.; Mohamed, M.; Sangaiah, A.K.; Jain, V. An integrated neutrosophic AHP and SWOT method for strategic planning methodology selection. *Benchmarking Int. J.* **2018**, *25*, 2546–2564. [CrossRef]
5. Jia, Z.; Zhang, Y. Interval-valued intuitionistic fuzzy multiple attribute group decision making with uncertain weights. *Math. Probl. Eng.* **2019**, *2019*, 5092147. [CrossRef]
6. Sajjad Ali Khan, M.; Ali, A.; Abdullah, S.; Amin, F.; Hussain, F. New extension of TOPSIS method based on Pythagorean hesitant fuzzy sets with incomplete weight information. *J. Intell. Fuzzy Syst.* **2018**, *35*, 5435–5448. [CrossRef]
7. Lourenzutti, R.; Krohling, R.A. A generalized TOPSIS method for group decision making with heterogeneous information in a dynamic environment. *Inf. Sci.* **2016**, *330*, 1–18. [CrossRef]
8. Smarandache, F. *Neutrosophy: Neutrosophic Probability, Set, and Logic: Analytic Synthesis & Synthetic Analysis*; American Research Press: Santa Fe, NM, USA, 1998.
9. Biswas, P.; Pramanik, S.; Giri, B.C. Neutrosophic TOPSIS with group decision making. In *Fuzzy Multi-Criteria Decision-Making Using Neutrosophic Sets*; Springer: Berlin, Germany, 2019; pp. 543–585.
10. Tuan, T.M.; Chuan, P.M.; Ali, M.; Ngan, T.T.; Mittal, M. Fuzzy and neutrosophic modeling for link prediction in social networks. *Evol. Syst.* **2019**, *10*, 629–634. [CrossRef]
11. Liu, P.; Wang, Y. Interval neutrosophic prioritized OWA operator and its application to multiple attribute decision making. *J. Syst. Sci. Complex.* **2016**, *29*, 681–697. [CrossRef]
12. Zhang, H.; Wang, J.; Chen, X. An outranking approach for multi-criteria decision-making problems with interval-valued neutrosophic sets. *Neural Comput. Appl.* **2016**, *27*, 615–627. [CrossRef]
13. Thong, N.T.; Dat, L.Q.; Hoa, N.D.; Ali, M.; Smarandache, F. Dynamic interval valued neutrosophic set: Modeling decision making in dynamic environments. *Comput. Ind.* **2019**, *108*, 45–52. [CrossRef]

14. Dat, L.Q.; Thong, N.T.; Ali, M.; Smarandache, F.; Abdel-Basset, M.; Long, H.V. Linguistic approaches to interval complex neutrosophic sets in decision making. *IEEE Access* **2019**, *7*, 38902–38917. [CrossRef]
15. Wang, H.; Smarandache, F.; Sunderraman, R.; Zhang, Y.Q. *Interval Neutrosophic Sets and Logic: Theory and Applications in Computing*; Hexis: Arizona, USA, 2005.
16. Broumi, S.; Son, L.H.; Bakali, A.; Talea, M.; Smarandache, F.; Selvachandran, G. Computing Operational Matrices in Neutrosophic Environments: A Matlab Toolbox. *Neutrosophic Sets Syst.* **2017**, *18*, 58–66.
17. Thong, N.T.; Giap, C.N.; Tuan, T.M.; Chuan, P.M.; Hoang, P.M. Modeling multi-criteria decision-making in dynamic neutrosophic environments bases on Choquet integral. *J. Comput. Sci. Cybern.* **2020**, *36*, 33–47.
18. Al-Quran, A.; Hashim, H.; Abdullah, L. A Hybrid Approach of Interval Neutrosophic Vague Sets and DEMATEL with New Linguistic Variable. *Symmetry* **2020**, *12*, 275. [CrossRef]
19. Xue, H.; Yang, X.; Chen, C. Possibility Neutrosophic Cubic Sets and Their Application to Multiple Attribute Decision Making. *Symmetry* **2020**, *12*, 269. [CrossRef]
20. Şahin, R.; Liu, P. Maximizing deviation method for neutrosophic multiple attribute decision making with incomplete weight information. *Neural Comput. Appl.* **2016**, *27*, 2017–2029. [CrossRef]
21. Yue, C. Entropy-based weights on decision makers in group decision-making setting with hybrid preference representations. *Appl. Soft Comput.* **2017**, *60*, 737–749. [CrossRef]
22. Wei, G.; Wang, J.; Lu, J.; Wu, J.; Wei, C.; Alsaadi, F.E.; Hayat, T. VIKOR method for multiple criteria group decision making under 2-tuple linguistic neutrosophic environment. *Econ. Res. Ekon. Istraživanja* **2019**, 1–24. [CrossRef]
23. Yue, Z. An extended TOPSIS for determining weights of decision makers with interval numbers. *Knowl. Based Syst.* **2011**, *24*, 146–153. [CrossRef]
24. Broumi, S.; Ye, J.; Smarandache, F. An extended TOPSIS method for multiple attribute decision making based on interval neutrosophic uncertain linguistic variables. *Neutrosophic Sets Syst.* **2015**, *8*, 22–31.
25. Wang, W.; Mendel, J.M. Multiple attribute group decision making with linguistic variables and complete unknown weight information. *Iran. J. Fuzzy Syst.* **2019**, *16*, 145–157. [CrossRef]
26. Maghrabie, H.F.; Beauregard, Y.; Schiffauerova, A. Multi-criteria decision-making problems with unknown weight information under uncertain evaluations. *Comput. Ind. Eng.* **2019**, *133*, 131–138. [CrossRef]
27. Peng, X. New multiparametric similarity measure and distance measure for interval neutrosophic set with IoT industry evaluation. *IEEE Access* **2019**, *7*, 28258–28280. [CrossRef]
28. Tian, C.; Zhang, W.Y.; Zhang, S.; Peng, J.J. An extended single-valued neutrosophic projection-based qualitative flexible multi-criteria decision-making method. *Mathematics* **2019**, *7*, 39. [CrossRef]
29. Hong, Y.; Xu, D.; Xiang, K.; Qiao, H.; Cui, X.; Xian, H. Multi-Attribute Decision-Making Based on Preference Perspective with Interval Neutrosophic Sets in Venture Capital. *Mathematics* **2019**, *7*, 257. [CrossRef]
30. Smarandache, F. *A Unifying Field in Logics: Neutrosophic Logic: Neutrosophy, Neutrosophic Set, Neutrosophic Probability*, 3rd ed.; American Research Press: Santa Fe, NM, USA, 2003.
31. Wang, H.; Smarandache, F.; Sunderraman, R.; Zhang, Y.Q. Interval neutrosophic sets and logic: Theory and applications in computing: Theory and applications in computing. *Infin. Study* **2005**, *5*, 1–16.
32. Xu, Z.; Chen, J.; Wu, J. Clustering algorithm for intuitionistic fuzzy sets. *Inf. Sci.* **2008**, *178*, 3775–3790. [CrossRef]
33. Yingming, W. Using the method of maximizing deviation to make decision for multiindices. *J. Syst. Eng. Electron.* **1998**, *8*, 21–26.

mathematics

MDPI

Article

Single-Valued Neutrosophic Linguistic Logarithmic Weighted Distance Measures and Their Application to Supplier Selection of Fresh Aquatic Products

Jiefeng Wang [1], Shouzhen Zeng [1,2,*] and Chonghui Zhang [3,*]

1 School of Business, Ningbo University, Ningbo 315211, China; wangjiefeng@nbu.edu.cn
2 School of Management, Fudan University, Shanghai 200433, China
3 College of Statistics and Mathematics, Zhejiang Gongshang University, Hangzhou 310018, China
* Correspondence: zengshouzhen@nbu.edu.cn (S.Z.); zhangch1988@zjgsu.edu.cn (C.Z.)

Received: 19 February 2020; Accepted: 11 March 2020; Published: 17 March 2020

Abstract: A single-valued neutrosophic linguistic set (SVNLS) is a popular fuzzy tool for describing deviation information in uncertain complex situations. The aim of this paper is to study some logarithmic distance measures and study their usefulness in multiple attribute group decision making (MAGDM) problems within single-valued neutrosophic linguistic (SVNL) environments. For achieving the purpose, SVNL weighted logarithmic averaging distance (SVNLWLAD) and SVNL ordered weighted logarithmic averaging distance (SVNLOWLAD) measures are firstly developed based on the logarithmic aggregation method. Then, the SVNL combined weighted logarithmic averaging distance (SVNLCWLAD) measure is presented by unifying the advantages of the previous SVNLWLAD and SVNLOWLAD measures. Moreover, a new MAGDM model by utilizing the SVNLCWLAD measure is presented under SVNL environments. Finally, a supplier selection for fresh aquatic products is taken as a case to illustrate the performance of the proposed framework.

Keywords: single-valued neutrosophic linguistic set; combined weighted; logarithmic distance measure; supplier selection; fresh aquatic products; MAGDM

---

1. Introduction

There are more and more vagueness and uncertainties in multiple attribute group decision making (MAGDM) problems, with the increasing complex of the evaluated objects. Therefore, researching a suitable fuzzy tool for depicting such uncertain information is a key issue in MAGDM problems. Up to now, numerous tools, such as the linguistic term set [1,2], intuitionistic fuzzy set (IFS) [3], hesitant fuzzy set [4], Pythagorean fuzzy set [5], single-valued neutrosophic set [6] and q-rung orthopair fuzzy set [7] arise at the historic moment, which greatly reduce the burden of decision makers for expressing the assessment of the attributes during the decision-making process.

Recently, Ye [8] proposed the single-valued neutrosophic linguistic set (SVNLS), which has been broadly used to handle uncertainties or vagueness under complex decision-making situations. The distinctive advantage of the SVNLS is that it combines the features of the linguistic set [2] and the single-valued neutrosophic set [5], therefore it can describe the uncertain information comprehensively and reasonably more concretely and accurately. Moreover, compared with the previous methods, such as the Pythagorean linguistic set [9] and the intuitionistic linguistic set [10], the SVNLS can overcome their defects, as it uses three elements (i.e., truth, indeterminacy and falsity) to express uncertainties of evaluated objects. So far, the SVNLS has gained increasing attention from researchers. For example, Ye [8] adapted the classic TOPSIS into SVNL environments and explored its performance in selecting suppliers. Guo and Sun [11] presented a method based on the prospect theory for decision making with SVNL information. Zhao et al. [12] introduced some SVNL induced Choquet integral aggregation operators and studied their usefulness in MAGDM. Ji et al. [13] studied the features of SVNLS and utilized it to express the uncertainties of outsourcing provider. Wang et al. [14] investigated the Maclaurin symmetric mean method for aggregating SVNL information. Chen et al. [15] presented an ordered aggregation distance measure for SVNLSs, and developed the SVNL ordered weighted averaging distance (SVNLOWAD) measure. Based on the work of Chen et al. [15], Cao et al. [16] developed a SVNL combined aggregation distance measure. Garg and Nancy [17] studied the SVNL prioritized weighted operators and used them to handle the priority relationship among attributes.

In the field of MAGDM, distance measures are often utilized to calculate the deviations between an ideal collection and the potential alternatives. Wherein the construction of distance plays a decisive role for the measurement, the weighted distance measures, including the weighted Hamming, the weighted Euclidean and the weighted Minkowski distances, are some of the most used distance measures [18]. Recently, the ordered weighted averaging distance (OWAD) measure introduced by Merigó and Gil-Lafuente [19] has become a very popular tool and gained lots of extensions, such as the linguistic OWAD [20], the induced OWAD [21,22], Heavy OWAD [23], continuous OWAD [24] intuitionistic fuzzy OWAD [25], hesitant fuzzy OWAD [26,27] and Pythagorean fuzzy OWAD measures [28,29]. More recently, Alfaro-García et al. [30] proposed a new extension of the OWAD measure, on the basis of the logarithmic aggregation method [31,32]; the result is the ordered weighted logarithmic averaging distance (OWLAD) measure. Motivated by the OWLAD, Alfaro-García et al. [33] further developed the induced OWLAD (IOWLAD) measure.

This study proposes some SVNL weighted logarithmic distance measures for highlighting the theory and application of SVNLS. Firstly, we present the SVNL weighted logarithmic averaging distance (SVNLWLAD) measure and SVNL ordered weighted logarithmic averaging distance (SVNLOWLAD) measures. Then, the SVNL logarithmic combined weighted logarithmic averaging distance (SVNLCWLAD) measure is proposed, by unifying the main advantages of the SVNLWAD and the SVNLOWLAS measures. Thus, it can weight both the SVNL deviations as well as their ordered positions, which enables its capability to overcome the limitation of the previous SVNLWLAD and SVNLOWLAD measures. The main properties and particular cases of the SVNLCWLAD are also studied. A MAGDM method based on the proposed SVNLCWLAD is formulated and its application are verified by a supplier selection problem.

The rest of this study is set out below: Section 2 reviews the backgrounds of SVNLS and the OWLAD measure. Section 3 proposes three SVNL weighted logarithmic distances, and provides some of their main properties and families. Section 4 gives a MAGDM approach based on the SVNLCWLAD measure. In Section 5, the application and merits of the presented method are discussed through a mathematical example and comparison. Finally, Section 6 summarizes the main conclusions.

2. Preliminaries

In this section, some concepts regarding the issues of the SVNLS, the OWAD and the OWALD measures are briefly reviewed.

2.1. The Single-Valued Neutrosophic Set (SVNS)

On the basis of the neutrosophic set [34], Ye [5] introduced the definition of the single-valued neutrosophic set (SVNS) for improving computational efficiency.

Definition 1 [5]. A single-valued neutrosophic set (SVNS) η in a finite set X denoted by a mathematical form:

$$\eta = \left\{ \left\langle x, T_\eta(x), I_\eta(x), F_\eta(x) \right\rangle \middle| x \in X \right\} \tag{1}$$

where $T_\eta(x)$, $I_\eta(x)$ and $F_\eta(x)$ represent the truth, the indeterminacy and the falsity-membership functions, respectively, and satisfy:

$$0 \leq T_Z(x), I_Z(x), F_Z(x) \leq 1,\ 0 \leq T_Z(x) + I_Z(x) + F_Z(x) \leq 3. \quad (2)$$

For convenience, the triplet $(T_\eta(x), I_\eta(x), F_\eta(x))$ is called the single-valued neutrosophic number (SVNN) and simply denoted as $\eta = (T_\eta, I_\eta, F_\eta)$.

2.2. The Linguistic Set

Definition 2 [2]. Let $S = \{s_\alpha | \alpha = 1, \ldots, t\}$ be a finitely ordered discrete set, where s_α denotes a linguistic term and l is an odd number. For example, taking $t = 7$, then $S = \{s_1 = extremely\ poor, s_2 = very\ poor, s_3 = poor, s_4 = fair, s_5 = good, s_6 = very\ good, s_7 = extremely\ good\}$. For actual application, we shall extend the discrete set S into a continuous set $\overline{S} = \{s_\alpha | \alpha \in R\}$ for avoiding information loss. For any linguistic terms $s_\alpha, s_\beta \in \overline{S}$, they shall satisfy following operational laws [35]:

(1) $s_\alpha \oplus s_\beta = s_{\alpha+\beta}$;
(2) $\mu s_\alpha = s_{\mu\alpha}, \mu \geq 0$;

2.3. The Single-Valued Neutrosophic Linguistic Set (SVNLS)

Definition 3 [8]. A single-valued neutrosophic linguistic set (SVNLS) ϕ in X is defined as:

$$\phi = \left\{ \left\langle x, [s_{\theta(x)}, (T_\phi(x), I_\phi(x), F_\phi(x))] \right\rangle \middle| x \in X \right\} \quad (3)$$

where $s_{\theta(x)} \in \overline{S}$, the functions $T_\phi(x)$, $I_\phi(x)$ and $F_\phi(x)$ denote the truth, indeterminacy and falsity-membership, respectively, and they have the following constraint:

$$0 \leq T_\phi(x), I_\phi(x), F_\phi(x) \leq 1,\ 0 \leq T_\phi(x) + I_\phi(x) + F_\phi(x) \leq 3. \quad (4)$$

In addition, $x = \left\langle s_{\theta(x)}, (T_x, I_x, F_x) \right\rangle$ is called the SVNL number (SVNLN) for computational convenience. Let $x_i = \left\langle s_{\theta(x_i)}, (T_{x_i}, I_{x_i}, F_{x_i}) \right\rangle (i = 1, 2)$ be two SVNLNs and $\lambda > 0$, then

(1) $x_1 \oplus x_2 = \left\langle s_{\theta(x_1)+\theta(x_2)}, (T_{x_1} + T_{x_2} - T_{x_1} * T_{x_2}, I_{x_1} * I_{x_2}, F_{x_1} * F_{x_2}) \right\rangle$;
(2) $\lambda x_1 = \left\langle s_{\lambda\theta(x_1)}, (1 - (1 - T_{x_1})^\lambda, (I_{x_1})^\lambda, (F_{x_1})^\lambda) \right\rangle$;
(3) $x_1^\lambda = \left\langle s_{\theta^\lambda(x_1)}, ((T_{x_1})^\lambda, 1 - (1 - I_{x_1})^\lambda, 1 - (1 - F_{x_1})^\lambda) \right\rangle$.

Definition 4 [8]. Let $x_i = \left\langle s_{\theta(x_i)}, (T_{x_i}, I_{x_i}, F_{x_i}) \right\rangle (i = 1, 2)$ be SVNLNs and $p > 0$, then the distance measure between x_1 and x_2 is given by the mathematical form:

$$d_{SVNL}(x_1, x_2) = \left[\left| \theta(x_1)T_{x_1} - \theta(x_2)T_{x_2} \right|^p + \left| \theta(x_1)I_{x_1} - \theta(x_2)I_{x_2} \right|^p + \left| \theta(x_1)F_{x_1} - \theta(x_2)F_{x_2} \right|^p \right]^{1/p} \quad (5)$$

On the basis of Definition 3, the SVNL weighted distance (SVNLWD) measure is formed in Equation (6), by assigning different levels of importance for the individual deviations.

$$SVNLWD((x_1, y_1), \ldots, (x_n, y_n)) = \sum_{j=1}^{n} w_j d_{SVNL}(x_j, y_j), \quad (6)$$

where the relative weight vector W satisfies $w_j \in [0, 1]$ and $\sum_{j=1}^{n} w_j = 1$.

2.4. The Ordered Weighted Logarithmic Averaging Distance (OWLAD) Measure

Motivated by the ordered weighted averaging (OWA) operator [36], Merigó and Gil-Lafuente [19] introduced the OWAD measure.

Definition 5 [19]. Let $U = \{u_1, u_2, \ldots, u_n\}$ and $V = \{v_1, v_2, \ldots, v_n\}$ be two crisp sets, $d_i = |u_i - v_i|$ be the distance between u_i and v_i, then the OWAD measure is defined as:

$$OWAD(U, V) = OWAD(d_1, d_2, \ldots, d_n) = \sum_{j=1}^{n} \omega_j d_{\sigma(j)} \tag{7}$$

where $d_{\sigma(j)} (j = 1, 2, \ldots, n)$ is the reorder values of $d_j (j = 1, 2, \ldots, n)$ such that $d_{\sigma(1)} \geq d_{\sigma(2)} \geq \cdots d_{\sigma(n)}$. The relative weight vector of the OWAD is $\omega = \left\{\omega_j \middle| \sum_{i=1}^{n} \omega_j = 1,\ \ 0 \leq \omega_j \leq 1\right\}$.

On the basis of the recent research of Zhou and Chen [31] and the OWAD measure, Alfaro-García et al. [30] introduced the OWLAD measure.

Definition 6 [30]. Let $U = \{u_1, u_2, \ldots, u_n\}$ and $V = \{v_1, v_2, \ldots, v_n\}$ be two crisp sets, $d_i = |u_i - v_i|$ be the distance between u_i and v_i, then the OWLAD measure is defined as:

$$OWLAD(U, V) = OWAD(d_1, d_2, \ldots, d_n) = \exp\left(\sum_{j=1}^{n} \omega_j \ln(d_{\sigma(j)})\right) \tag{8}$$

Alfaro-García et al. [30] studied desired properties of the OWLAD measure, such as boundedness, commutativity, idempotency and monotonicity. They also explored its different families and found that it includes many distance measures. However, the OWLAD is generally designed for aggregating crisp variables and cannot be used to handle SVNL information. What's more, it can only account for the weights of ordered deviations, but fails to consider the importance of the individual data. Therefore, we shall develop a new distance measure for overcoming the limitations of the OWLAD within SVNL environments.

3. SVNL Weighted Logarithmic Distance Measures

3.1. SVL Weighted Logarithmic Averaging Distance (SVNLWLAD) Measure

The SVNLWLAD measure is a new SVNL distance measure that utilizes the optimal logarithmic aggregation for handling SVNL deviations. It can consider the importance of the aggregated individual distances.

Definition 7. Let $d_{SVNL}(x_j, y_j)$ be the distance between two x_j, y_j $(j = 1, \ldots, n)$ defined in Equation (5), then the SVNLWLAD measure is defined as:

$$SVNLWLAD((x_1, y_1), \ldots, (x_n, y_n)) = \exp\left\{\sum_{j=1}^{n} w_j \ln\left(d_{SVNL}(x_j, y_j)\right)\right\}, \tag{9}$$

where w_j is the weight of the distance $d_{SVNL}(x_j, y_j)$ with $\sum_{j=1}^{n} w_j = 1$ and $w_j \in [0, 1]$.

Example 1. *Let* $X = (x_1, x_2, x_3, x_4, x_5) == (\langle s_2, (0.6, 0.5, 0.1)\rangle, \langle s_5, (0.6, 0.3, 0.5)\rangle, \langle s_4, (0.7, 0.2, 0.1)\rangle, \langle s_3, (0.9, 0.1, 0.6)\rangle, \langle s_4, (0.3, 0.1, 0.3)\rangle)$ *and* $Y = (y_1, y_2, y_3, y_4, y_5) = (\langle s_4, (0.2, 0.7, 0)\rangle, \langle s_6, (0.3, 0.7, 0.1)\rangle, \langle s_7, (0.6, 0.4, 0.5)\rangle, \langle s_1, (0.1, 0.7, 0.2)\rangle, \langle s_3, (0.1, 0.5, 0.6)\rangle)$ *be two SVNLSs defined in* $S =$

$\{s_1, s_2, s_3, s_4, s_5, s_6, s_7\}$. The weighting vector is supposed to be $w = (0.15, 0.25, 0.25, 0.15, 0.2)^T$. Then the computational process through the SVNLWLAD can be displayed as follows:

(1) Calculate the individual distances $d_{SVNL}(x_i, y_i)$ $(i = 1, 2, \ldots, 5)$ according to Equation (5) (let $p = 1$):

$$\begin{aligned} d_{SVNL}(x_1, y_1) &= |2\times 0.6 - 4\times 0.2| + |2\times 0.5 - 4\times 0.7| + |2\times 0.1 - 4\times 0| = 2.4, \\ d_{SVNL}(x_2, y_2) &= |5\times 0.6 - 6\times 0.3| + |5\times 0.3 - 6\times 0.7| + |5\times 0.5 - 6\times 0.1| = 5.8, \\ d_{SVNL}(x_3, y_3) &= |4\times 0.7 - 7\times 0.6| + |4\times 0.2 - 7\times 0.4| + |4\times 0.1 - 7\times 0.5| = 6.5, \\ d_{SVNL}(x_4, y_4) &= |3\times 0.9 - 1\times 0.1| + |3\times 0.1 - 1\times 0.7| + |3\times 0.6 - 1\times 0.2| = 4.2, \\ d_{SVNL}(x_5, y_5) &= |4\times 0.3 - 3\times 0.1| + |4\times 0.1 - 3\times 0.5| + |4\times 0.3 - 3\times 0.6| = 2.6. \end{aligned}$$

(2) Utilize the SVNLWLAD defined in Equation (9) to aggregate the individual distances:

$$\begin{aligned} &SVNLWLAD((x_1, y_1), \ldots, (x_5, y_5)) = \exp\left\{ \sum_{j=1}^{n} w_j \ln\left(d_{SVNL}(x_j, y_j)\right) \right\} \\ &= \exp\left\{ \sum_{j=1}^{n} (0.15\times \ln(2.4) + 0.25\times \ln(5.8) + 0.25\times \ln(6.5) + 0.15\times \ln(4.2) + 0.2\times \ln(2.6) \right\} \\ &= 4.2423 \end{aligned}$$

3.2. SVL Ordered Weighted Logarithmic Averaging Distance (SVNLOWLAD) Measure

The SVNLOWLAD operator is a useful extension of the OWLAD measure which uses SVNL information. Moreover, it can be seen as a generalization of the SVNLWLAD measure, which is characterized by its ordered mechanism of the aggregated arguments. This mechanism provides the opportunity to consider complex attitudes in the decision-making processes, as well as to handle the logarithmic deviations.

Definition 8. Let $d_{SVNL}(x_j, y_j)$ be the distance between SVNLNs x_j, y_j $(j = 1, \ldots, n)$ defined in Equation (5), then the SVNLOWLAD is defined as:

$$SVNLOWLAD((x_1, y_1), \ldots, (x_n, y_n)) = \exp\left\{ \sum_{j=1}^{n} \omega_j \ln\left(d_{SVNL}(x_{\sigma(j)}, y_{\sigma(j)})\right) \right\}, \tag{10}$$

where $d_{SVNL}(x_{\sigma(j)}, y_{\sigma(j)})(j = 1, 2, \ldots, n)$ is the reorder values of $d_{SVNL}(x_j, y_j)$ such that $d_{SVNL}(x_{\sigma(1)}, y_{\sigma(1)}) \geq \ldots. \geq d_{SVNL}(x_{\sigma(n)}, y_{\sigma(n)})$. The associated weight vector of the SVNLOWLAD is $\omega = (\omega_1, \omega_2, \ldots, \omega_n)^T$ with $\sum_{j=1}^{n} \omega_j = 1$ and $\omega_j \in [0, 1]$.

Similar to the OWLAD measure, the proposed SVNLOWLAD measure has the properties of idempotency, commutativity, monotonicity, boundedness and non-negativity. The proofs of these properties are trivial and thus omitted.

Example 2. (Continuing Example 1). Suppose the weight vector of SVNLOWLAD measure is $\omega = (0.1, 0.2, 0.25, 0.3, 0.15)^T$. Then, the computational process based on the SVNLOWLAD is displayed as follows:

(1) Compute the individual distances $d_{SVNL}(x_i, y_i)$ $(i = 1, 2, \ldots, 5)$ according to Equation (5) (obtained from example 1):

$$\begin{aligned} d_{SVNL}(x_1, y_1) = 2.4, d_{SVNL}(x_2, y_2) = 5.8, d_{SVNL}(x_3, y_3) = 6.5, \\ d_{SVNL}(x_4, y_4) = 4.2,\ d_{SVNL}(x_5, y_5) = 2.6 \end{aligned}$$

(2) Rank the $d_{SVNL}(x_i, y_i)$ $(i = 1, 2, \ldots, 5)$ in decreasing order:

$$d_{SVNL}(x_{\sigma(1)}, y_{\sigma(1)}) = d_{SVNL}(x_3, y_3) = 6.5,\ d_{SVNL}(x_{\sigma(2)}, y_{\sigma(2)}) = d_{SVNL}(x_2, y_2) = 5.8,$$
$$d_{SVNL}(x_{\sigma(3)}, y_{\sigma(3)}) = d_{SVNL}(x_4, y_4) = 4.2,\ d_{SVNL}(x_{\sigma(4)}, y_{\sigma(4)}) = d_{SVNL}(x_5, y_5) = 2.6,$$
$$d_{SVNL}(x_{\sigma(5)}, y_{\sigma(5)}) = d_{SVNL}(x_1, y_1) = 2.4.$$

(3) Utilize the SVNLOWLAD to aggregate the ordered distances:

$$SVNLOWLAD((x_1, y_1), \ldots, (x_5, y_5)) = \exp\left\{ \sum_{j=1}^{5} w_j \ln\left(d_{SVNL}(x_{\sigma(j)}, y_{\sigma(j)})\right) \right\}$$
$$= \exp\{0.1 \times \ln(6.5) + 0.2 \times \ln(5.8) + 0.25 \times \ln(4.2) + 0.3 \times \ln(2.6) + 0.15 \times \ln(2.4)\}$$
$$= 3.7266$$

3.3. SVL Combined Weighted Logarithmic Averaging Distance (SVNLCWLAD) Measure

From the previous examples, we can see that the SVNLWLAD can account for the importance of input deviations, while the SVNLOWLAD considers the weights of ordered deviations, and based on this rule, it can depict some attitudes of decision makers in decision making. However, the SVNLWLAD does not have the function of orderly aggregation, while the SVNLOWLAD cannot integrate the importance of attributes that the SVNLWLAD can. To overcome these limitations, we shall develop a new distance measure that can combine the advantages of the SVNLWLAD and the SVNLOWLAD measures.

Definition 9. Let x_j, y_j $(j = 1, \ldots, n)$ be the two collections of SVNLNs. If

$$SVNLCWLAD((x_1, y_1), \ldots, (x_n, y_n)) = \exp\left\{ \sum_{j=1}^{n} \varpi_j \ln\left(d_{SVNL}(x_{\sigma(j)}, y_{\sigma(j)})\right) \right\}, \tag{11}$$

then the SVNLCWLAD is called the SVNL combined weighted logarithmic averaging distance measure. The integrated weights ϖ_j is defined as:

$$\varpi_j = \gamma \omega_j + (1 - \gamma) w_{\sigma(j)} \tag{12}$$

where w_j is the weight of $d_{SVNL}(x_j, y_j)(j = 1, 2, \ldots, n)$ with $\sum_{j=1}^{n} w_j = 1$ and $w_j \in [0, 1]$, and the other ω_j, is the associated weight of SVNLOWLAD satisfying $\sum_{j=1}^{n} \omega_j = 1$ and $\omega_j \in [0, 1]$, parameter γ is real parameter and meeting $\gamma \in [0, 1]$.

Obviously, the SVNLCWLAD is generalized to the SVNLOWLAD and SVNLWLAD, when $\gamma = 1$ and $\lambda = 0$, respectively. Following the combined operational rules, the SVNLWLAD can be regarded as a combination of the SVNLOWLAD and SVNLWLAD measures:

$$SVNLCWLAD((x_1, y_1), \ldots, (x_n, y_n)) = \exp\left(\left\{ \gamma \sum_{j=1}^{n} \omega_j \ln\left(d_{SVNL}(x_{\sigma(j)}, y_{\sigma(j)})\right) \right\} + \left\{ (1 - \gamma) \sum_{j=1}^{n} w_j \ln\left(d_{SVNL}(x_j, y_j)\right) \right\} \right) \tag{13}$$

Example 3. (Continuing Examples 1 and 2). Let $\gamma = 0.6$ and based on the available information obtained in the examples 1 and 2, we can compute the integrated weights ϖ_j according to Equation (12):

$$\varpi_1 = 0.6 \times 0.1 + (1 - 0.6) \times 0.25 = 0.16,$$
$$\varpi_2 = 0.6 \times 0.2 + (1 - 0.6) \times 0.25 = 0.22,$$
$$\varpi_3 = 0.6 \times 0.25 + (1 - 0.6) \times 0.15 = 0.21,\ \varpi_4 = 0.6 \times 0.3 + (1 - 0.6) \times 0.2 = 0.26,$$
$$\varpi_5 = 0.6 \times 0.15 + (1 - 0.6) \times 0.15 = 0.15.$$

Perform the below aggregation, utilizing the SVNLCWLAD measure defined in Equation (11):

$$SVNLCWLAD((x_1, y_1), \ldots, (x_5, y_5)) = \exp\left\{\sum_{j=1}^{5} \varpi_j \ln\left(d_{SVNL}(x_{\sigma(j)}, y_{\sigma(j)})\right)\right\}$$
$$= \exp\{0.16 \times \ln(6.5) + 0.22 \times \ln(5.8) + 0.21 \times \ln(4.2) + 0.26 \times \ln(2.6) + 0.15 \times \ln(2.4)\}$$
$$= 3.9249$$

We can also apply the SVNLCWLAD measure given in Equation (13) to illustrate the aggregation:

$$SVNLCWLAD = \exp\left(\left\{\gamma \sum_{j=1}^{n} \omega_j \ln\left(d_{SVNL}(x_{\sigma(j)}, y_{\sigma(j)})\right)\right\} + \left\{(1-\gamma) \sum_{j=1}^{n} w_j \ln\left(d_{SVNL}(x_j, y_j)\right)\right\}\right)$$
$$= \exp(0.6 \times 1.3155 + (1 - 0.6) \times 1.4451)$$
$$= 3.9249$$

Apparently, the same results are obtained by both methods. On the other hand, following the aforementioned examples, we can see that the SVNLCWLAD combines both features of the SVNLOWLAD and the SVNLWLAD measures. Therefore, it can account for the importance of the deviations as well as highlights the ordered aggregation mechanism. Moreover, it is more convenient for application, as people can set parameters flexibly according to actual needs or their interests.

Furthermore, we can achieve some interesting SVNL distance measures, by designing the parameter γ and the weight vector in the SVNLCWLAD measure, for example:

- The SVNLOWLAD and SVNLWLAD measures are obtained when $\gamma = 1$ and $\lambda = 0$, respectively. Moreover, the more lager γ, the more importance focused on the SVNLOWLAD.
- If $w = (1, 0, 0, \ldots, 0)^T$, then max-SVNLCWLAD measure is formed.
- If $w = (0, \ldots, 0, 1)^T$, then the min-SVNLCWLAD is rendered.
- The step-SVNLCWLAD measure is obtained by designing $w_1 = \cdots = w_{k-1} = 0$, $w_k = 1$ and $w_{k+1} = \cdots = w_n = 0$.
- Based on the analysis provided in recent literature [30,33,37–40], more particular cases of the SVNLCWLAD, such as the Centered-SVNLCWLAD, Median-SVNLCWLAD and the Olympic-SVNLCWLAD measures, can be created.

According to the properties of the OWLAD measure, it is clear that the SVNLCWLAD satisfies the desirable properties of monotonicity, idempotency, boundedness and:

(1) Monotonicity: If $d_{SVNL}(x_i, y_i) \geq d_{SVNL}(x'_i, y'_i)$ for $i = 1, 2, \ldots, n$, then

$$SVNLCWLAD((x_1, y_1), \ldots, (x_n, y_n)) \geq SVNLCWLAD\left((x'_1, y'_1), \ldots, (x'_n, y'_n)\right)$$

(2) Idempotency: If $d_{SVNL}(x_i, y_i) = d$ for $i = 1, 2, \ldots, n$, then

$$SVNLCWLAD((x_1, y_1), \ldots, (x_n, y_n)) = d$$

(3) Commutativity: If $((x_1, x'_1), \ldots, (x_n, x'_n))$ is any permutation of $\left((y_1, y'_1), \ldots, (y_n, y'_n)\right)$, then

$$SVNLCWLAD((x_1, x'_1), \ldots, (x_n, x'_n)) = SVNLCWLAD\left((y_1, y'_1), \ldots, (y_n, y'_n)\right)$$

(4) Boundedness: Let $d_{\min} = \min_i\left(d\left(y_i, y'_i\right)\right)$ and $d_{\max} = \max_i\left(d\left(y_i, y'_i\right)\right)$, then

$$d_{\min} \leq SVNLCWLAD\left(\left(y_1, y'_1\right), \ldots, \left(y_n, y'_n\right)\right) \leq d_{\max}$$

In addition, we can provide a more generalized SVNL combined weighted logarithmic distance measure, by using the generalized mean method [41]; the result is the generalized SVNLCWLAD (GSVNLCWLAD) measure:

$$GSVNLCWLAD((x_1, y_1), \ldots, (x_n, y_n)) = \exp\left\{\left(\sum_{j=1}^{n} \varpi_j \ln\left(d_{SVNL}\left(x_{\sigma(j)}, y_{\sigma(j)}\right)\right)^{\lambda}\right)^{1/\lambda}\right\} \tag{14}$$

where λ is a parameter that meets $\lambda \in (-\infty, +\infty) - \{0\}$. Some representative cases of the GSVNLCWALD measure can be determined from the variation of parameter λ, for example, the SVNLCWLAD is formed when $\lambda = 1$, the SVNL combined weighted logarithmic quadratic distance (SVNLCWLQD) is obtained if $\lambda = 2$, and the SVNL combined weighted logarithmic harmonic distance (SVNLCWLHD) is rendered if $\lambda = -1$. Other more special families of the GSVNLCWLAD measure can be analyzed by using similar methods, provided in reference [41–43].

4. Application in MAGDM

The SVNLCWLAD is applicable to decision making, pattern recognition, data analysis, financial investment, social management, and many other fields. In this paper, we present its application in MAGDM problems under SVNL environments. Consider a MAGDM problem, which includes m different alternatives denoted as $B_1, B_2, \ldots, B_m$ and several experts invited to evaluate n finite attributes $A_1, A_2, \ldots, A_n$. The weight vector for these attributes is represented by $w = (w_1, w_2, \ldots, w_n)^T$ such that $w_j \in [0, 1]$ and $\sum_{j=1}^{n} w_j = 1$. Following the available information, the general procedure for MAGDM can be summarized below.

Step 1: Let each expert e_q $(q = 1, 2, \ldots, t)$(whose weight is τ_q, with $\tau_q \geq 0$ and $\sum_{q=1}^{t} \tau_q = 1$) expresses his or her assessment for different alternatives under given attributes by means of SVNLNs, thus formulate SVNL individual decision matrix $R^q = \left(r_{ij}^{(q)}\right)_{m\times n}$.

Step 2: The collective decision matrix $R = \left(r_{ij}\right)_{m\times n}$ is calculated by using the SVNL weighted average (SVNLWA) operator [8] to aggregate individual assessment, where $r_{ij} = \sum_{q=1}^{t} \tau_q r_{ij}^{(q)}$.

Step 3: Set the ideal performances for each attribute to construct the ideal scheme (Table 1).

Table 1. Ideal scheme.

	A_1	A_2	$\cdots$	A_n
I	I_1	I_2	$\ldots$	I_n

Step 4: Apply the SVNLCWLAD measure to compute the distances between the alternative $B_i (i = 1, 2, \ldots, m)$ and the ideal scheme I:

$$SVNLCWLAD(B_i, I) = \exp\left\{\sum_{j=1}^{n} \varpi_j \ln\left(d_{SVNL}\left(r_{\sigma(ij)}, I_{\sigma(j)}\right)\right)\right\} \tag{15}$$

Step 5: Sort the alternatives according to the lowest value of distance obtained in the previous step and hence, select the best one(s).

Step 6: End.

5. Numerical Example for Supplier Selection of Fresh Aquatic Products

At present, China has the largest aquatic product market in the world. With economic and social development, people's awareness for the quality and safety of aquatic products are also increasing. The most important obstacle to the further development of aquatic products has shifted from the processing field to the market circulation field. The importance and urgency of the effective maintenance of the supply chain by aquatic product processing enterprises is increasingly prominent. High-quality suppliers can provide safe and fresh raw materials and high-quality products, to help enterprises expand the market and increase competitiveness [44]. With the increasing position and role of suppliers in the production of aquatic processing enterprises, the selecting suppliers of fresh aquatic products is considered to be the most important strategic decision in the aquatic product supply chain. Thus, finding an effective method for evaluating suppliers is the key issue for buyers of fresh aquatic products. In this section, we provide uses of the proposed framework for handling this problem within SVNL environments, to highlight the theory and application of the SVNLS. Four possible fresh aquatic products suppliers $B_i(i = 1,2,3,4)$ are needed to evaluate from below attributes: A_1: quality and safety (including product safety, quality of goods, delivery performance and fulfill the full orders); A_2: costs (including material cost and transportation costs); A_3: delivery level (including delivery time, responsiveness to customers and return products time); and A_4: supply capacity (inventory amount, ability to meet delivery demand, ability to produce new raw materials and ability to receive returns products). Three experts (expert's weight $\tau = (0.37, 0.30, 0.33)$) utilize SVNL information to evaluate these alternatives under four attributes, where the linguistic term set is supposed to $S = \{s_1, s_2, s_3, s_4, s_5, s_6, s_7\}$. The results are represented by means of SVNLNs, listed in Tables 2–4.

Table 2. Single-valued neutrosophic linguistic (SVNL) decision matrix R^1.

	A_1	A_2	A_3	A_4
B_1	$\left\langle s_4^{(1)}, (0.6, 0.1, 0.2) \right\rangle$	$\left\langle s_6^{(1)}, (0.6, 0.1, 0.2) \right\rangle$	$\left\langle s_5^{(1)}, (0.7, 0.0, 0.1) \right\rangle$	$\left\langle s_3^{(1)}, (0.3, 0.1, 0.2) \right\rangle$
B_2	$\left\langle s_5^{(1)}, (0.6, 0.1, 0.2) \right\rangle$	$\left\langle s_3^{(1)}, (0.6, 0.2, 0.4) \right\rangle$	$\left\langle s_6^{(1)}, (0.6, 0.1, 0.2) \right\rangle$	$\left\langle s_4^{(1)}, (0.5, 0.2, 0.2) \right\rangle$
B_3	$\left\langle s_4^{(1)}, (0.5, 0.2, 0.3) \right\rangle$	$\left\langle s_5^{(1)}, (0.3, 0.5, 0.2) \right\rangle$	$\left\langle s_4^{(1)}, (0.3, 0.2, 0.3) \right\rangle$	$\left\langle s_3^{(1)}, (0.5, 0.3, 0.1) \right\rangle$
B_4	$\left\langle s_5^{(1)}, (0.4, 0.2, 0.3) \right\rangle$	$\left\langle s_4^{(1)}, (0.5, 0.3, 0.3) \right\rangle$	$\left\langle s_5^{(1)}, (0.4, 0.2, 0.3) \right\rangle$	$\left\langle s_3^{(1)}, (0.3, 0.2, 0.5) \right\rangle$

Table 3. SVNL decision matrix R^2.

	A_1	A_2	A_3	A_4
B_1	$\left\langle s_4^{(3)}, (0.5, 0.2, 0.2) \right\rangle$	$\left\langle s_5^{(3)}, (0.7, 0.2, 0.1) \right\rangle$	$\left\langle s_4^{(3)}, (0.6, 0.1, 0.2) \right\rangle$	$\left\langle s_3^{(3)}, (0.4, 0.1, 0.1) \right\rangle$
B_2	$\left\langle s_4^{(3)}, (0.7, 0.2, 0.2) \right\rangle$	$\left\langle s_6^{(3)}, (0.4, 0.6, 0.2) \right\rangle$	$\left\langle s_5^{(3)}, (0.5, 0.2, 0.3) \right\rangle$	$\left\langle s_5^{(3)}, (0.7, 0.2, 0.1) \right\rangle$
B_3	$\left\langle s_5^{(3)}, (0.6, 0.1, 0.3) \right\rangle$	$\left\langle s_4^{(3)}, (0.3, 0.6, 0.2) \right\rangle$	$\left\langle s_6^{(3)}, (0.5, 0.1, 0.3) \right\rangle$	$\left\langle s_4^{(3)}, (0.6, 0.2, 0.1) \right\rangle$
B_4	$\left\langle s_6^{(3)}, (0.6, 0.2, 0.4) \right\rangle$	$\left\langle s_4^{(3)}, (0.5, 0.2, 0.3) \right\rangle$	$\left\langle s_6^{(3)}, (0.5, 0.2, 0.3) \right\rangle$	$\left\langle s_5^{(3)}, (0.2, 0.1, 0.6) \right\rangle$

Table 4. SVNL decision matrix R^3.

	A_1	A_2	A_3	A_4
B_1	$\left\langle s_5^{(2)}, (0.7, 0.2, 0.3) \right\rangle$	$\left\langle s_6^{(2)}, (0.6, 0.3, 0.3) \right\rangle$	$\left\langle s_4^{(2)}, (0.8, 0.1, 0.2) \right\rangle$	$\left\langle s_4^{(2)}, (0.4, 0.2, 0.2) \right\rangle$
B_2	$\left\langle s_6^{(2)}, (0.7, 0.2, 0.3) \right\rangle$	$\left\langle s_4^{(2)}, (0.5, 0.4, 0.2) \right\rangle$	$\left\langle s_6^{(2)}, (0.7, 0.2, 0.3) \right\rangle$	$\left\langle s_5^{(2)}, (0.6, 0.2, 0.2) \right\rangle$
B_3	$\left\langle s_6^{(2)}, (0.6, 0.3, 0.4) \right\rangle$	$\left\langle s_5^{(2)}, (0.4, 0.4, 0.1) \right\rangle$	$\left\langle s_6^{(2)}, (0.4, 0.2, 0.4) \right\rangle$	$\left\langle s_4^{(2)}, (0.6, 0.1, 0.3) \right\rangle$
B_4	$\left\langle s_6^{(2)}, (0.5, 0.1, 0.2) \right\rangle$	$\left\langle s_3^{(2)}, (0.7, 0.1, 0.1) \right\rangle$	$\left\langle s_5^{(2)}, (0.4, 0.3, 0.4) \right\rangle$	$\left\langle s_5^{(2)}, (0.3, 0.1, 0.6) \right\rangle$

According to the individual opinions and weights of the experts, the collective decision matrix can be calculated by using the SVNLWA operator, shown in Table 5.

Table 5. Group SVNL decision matrix R.

	A_1	A_2	A_3	A_4
B_1	$\langle s_{4.33}, (0.611, 0.155, 0.229)\rangle$	$\langle s_{5.70}, (0.633, 0.180, 0.186)\rangle$	$\langle s_{4.37}, (0.714, 0.000, 0.155)\rangle$	$\langle s_{3.67}, (0.365, 0.128, 0.163)\rangle$
B_2	$\langle s_{4.70}, (0.666, 0.155, 0.229)\rangle$	$\langle s_{4.23}, (0.514, 0.350, 0.258)\rangle$	$\langle s_{5.70}, (0.611, 0.155, 0.258)\rangle$	$\langle s_{2.37}, (0.602, 0.200, 0.162)\rangle$
B_3	$\langle s_{4.96}, (0.566, 0.186, 0.330)\rangle$	$\langle s_{4.70}, (0.335, 0.491, 0.159)\rangle$	$\langle s_{5.26}, (0.399, 0.163, 0.330)\rangle$	$\langle s_{3.37}, (0.566, 0.185, 0.144)\rangle$
B_4	$\langle s_{5.63}, (0.450, 0.159, 0.286)\rangle$	$\langle s_{3.67}, (0.578, 0.185, 0.209)\rangle$	$\langle s_{5.30}, (0.432, 0.229, 0.330)\rangle$	$\langle s_{2.37}, (0.271, 0.129, 0.561)\rangle$

Based on the available information of the potential suppliers, the experts determine the ideal supplier that has a good performance for each attribute, shown in Table 6.

Table 6. Ideal supplier.

	A_1	A_2	A_3	A_4
I	$\langle s_7, (1, 0, 0.1)\rangle$	$\langle s_7, (0.9, 0.1, 0)\rangle$	$\langle s_6, (0.9, 0, 0)\rangle$	$\langle s_7, (0.9, 0, 0.1)\rangle$

The weighting vectors of the SVNLCWLAD measure and the attributes are considered as $\omega = (0.2, 0.3, 0.1, 0.4)^T$ and $w = (0.2, 0.3, 0.3, 0.2)^T$, respectively. Without loss of generality, let $\gamma = 0.5$, then the distances between the alternative $B_i (i = 1, 2, 3, 4)$ and the ideal scheme I are calculated by using the SVNLCWLAD as follows:

$$SVNLCWLAD(B_1, I) = 5.0778, \ SVNLCWLAD(B_2, I) = 5.7808,$$
$$SVNLCWLAD(B_3, I) = 6.7281, \ SVNLCWLAD(B_4, I) = 6.6661.$$

The smaller the value of the $SVNLVWLAD(B_i, I)$, the closer the B_i to the ideal supplier. Therefore, the alternatives are ranked as:

$$B_1 \succ B_2 \succ B_4 \succ B_3.$$

Hence, the best alternative is B_1.

Moreover, we apply two special cases of the SVNLCWLAD, i.e., the SVNLOWLAD and the SVNLWLAD measures, to calculate the distances between the alternatives and the ideal scheme. By the SVNLOWLAD measure, we have:

$$SVNLOWALD(B_1, I) = 5.1159, \ SVNLOWLAD(B_2, I =) = 5.7758,$$
$$SVNLOWLAD(B_3, I) = 6.7648, \ SVNLOWLAD(B_4, I) = 6.8483.$$

The results obtained by the SVNLWLAD measure are:

$$SVNLWALD(B_1, I) = 5.0401, \ SVNLWLAD(B_2, I =) = 5.7857,$$
$$SVNLWLAD(B_3, I) = 6.6916, \ SVNLWLAD(B_4, I) = 6.4887.$$

Thus, the ranking orders based on the SVNLOWALAD and SVNLWLAD measures are $B_1 \succ B_2 \succ B_3 \succ B_4$ and $B_1 \succ B_2 \succ B_4 \succ B_3$, respectively. Then, we obtain the same best supplier using the SVNLCWLAD, SVNLOWLAD and SVNLWLAD measures, although all the ranking orders are different. Moreover, following the analysis in the aforementioned numerical examples, the SVNLWLAD and SVNLOWLAD measures emphasize different points in aggregation process. Generally, the SVNLWLAD accounts for the importance of attributes, while the SVNLOWLAD consider the the importance of ordered deviation. However, the SVNLCWLAD measure unifies all of features of previous methods, therefore it can overcome the limitations of the previous measures and achieve a more rational aggregation result. Furthermore, the MAGDM method based on SVNLCWLAD is more flexible than the existing MAGDM approaches based on the SVNLOWAD measure [15], as decision makers can determine some desired values of γ in the SVNLCWLAD, according to their preferences or practical demands.

6. Conclusions

This paper introduces several SVNL logarithmic distance measures, including the SVNLWLAD, SVNLOWLAD and SVNLCWLAD measures. Some of their properties and particular cases are investigated. We prove that all the SVNLWLAD and SVNLOWLAD are the special cases of the SVNLCWLAD measure. Thus, the SVNLCWLAD measure combines the desired properties of SVNLWLAD and SVNLOWLAD. Moreover, it presents a more general method to handle complex situations in a more efficient and flexible way, as it can overcome the shortcomings of the existing distance measures.

Guaranteeing the quality and safety of fresh aquatic products is crucial for mankind's health and the wellbeing of fishery companies. Therefore, an appropriate supplier selection is considered as the most important strategic decision in the aquatic product supply chain. In this paper, a MAGDM approach is provided, based on the SVNLCWALD measure, and a mathematical example of selecting a fresh aquatic products problem is taken to verify its feasibility and validity. The application shows that the proposed method is effective, as the SVNLCWLAD can not only highlight the decision makers' interests through the ordered weighted mechanism, but can also integrate the importance of attributes by the weighted average function. Moreover, it provides a possibility for decision makers to flexibly select the parameter, based on the demands for the specific problem or actual interests. In addition, this study also presents an effective guideline for selecting suppliers in other industries.

In subsequent work, we will consider the application of the proposed method in other fields, such as pattern recognition, innovation management and investment selection [45–50]. We also develop some new extensions of the proposed distance measures in complex fuzzy situations.

Author Contributions: Writing—original draft preparation, S.Z.; Writing—review and editing, C.Z.; formal analysis, J.W. All authors have read and agreed to the published version of the manuscript.

Funding: This paper was supported by Major Humanities and Social Sciences Research Projects in Zhejiang Universities (no. 2018QN058), China Postdoctoral Science Foundation (no. 2019M651403), Zhejiang Province Natural Science Foundation (No. LY18G010007; No. LQ20G010001) and Ningbo Science and Technology Project (no. 2019C50008).

Conflicts of Interest: The authors declare that there is no conflict of interest regarding the publication of this paper.

References

1. Herrera, F.; Herrera-Viedma, E. Linguistic decision analysis: Steps for solving decision problems under linguistic information. *Fuzzy Sets Syst.* **2000**, *115*, 67–82. [CrossRef]

2. Zadeh, L.A. Fuzzy sets. *Inf. Control* **1965**, *18*, 338–353. [CrossRef]
3. Atanassov, K. Intuitionistic fuzzy sets. *Fuzzy Sets Syst.* **1986**, *20*, 87–96. [CrossRef]
4. Torra, V. Hesitant fuzzy sets. *Int. J. Intell. Syst.* **2010**, *25*, 529–539. [CrossRef]
5. Yager, R.R. Pythagorean membership grades in multi-criteria decision making. *IEEE Trans. Fuzzy Syst.* **2014**, *22*, 958–965. [CrossRef]
6. Ye, J. Multicriteria decision-making method using the correlation coefficient under single-valued neutrosophic environment. *Int. J. Gen. Syst.* **2013**, *42*, 386–394. [CrossRef]
7. Yager, R.R. Generalized orthopair fuzzy sets. *IEEE Trans. Fuzzy Syst.* **2017**, *25*, 1222–1230. [CrossRef]
8. Ye, J. An extended TOPSIS method for multiple attribute group decision making based on single valued neutrosophic linguistic numbers. *J. Intell. Fuzzy Syst.* **2015**, *28*, 247–255. [CrossRef]
9. Jin, F.F.; Pei, L.D.; Chen, H.Y.; Langari, R.; Liu, J.P. A novel decision-making model with pythagorean fuzzy linguistic information measures and its application to a sustainable blockchain product assessment problem. *Sustainability* **2019**, *11*, 5630. [CrossRef]
10. Liu, P.D.; Wang, Y.M. Multiple attribute group decision making methods based on intuitionistic linguistic power generalized aggregation operator. *Appl. Soft Comput.* **2014**, *17*, 90–104. [CrossRef]
11. Guo, Z.X.; Sun, F.F. Multi-attribute decision making method based on single-valued neutrosophic linguistic variables and prospect theory. *J. Intell. Fuzzy Syst.* **2019**, *37*, 5351–5362. [CrossRef]
12. Zhao, S.P.; Wang, D.; Liang, C.Y.; Lu, W.X. Induced Choquet Integral aggregation operators with single-valued neutrosophic uncertain linguistic numbers and their application in multiple attribute group decision-making. *Math. Probl. Eng.* **2019**. [CrossRef]
13. Ji, P.; Zhang, H.Y.; Wang, J.Q. Selecting an outsourcing provider based on the combined MABAC-ELECTRE method using single-valuedneutrosophic linguistic sets. *Comput. Ind. Eng.* **2018**, *120*, 429–441. [CrossRef]
14. Wang, J.Q.; Yang, Y.; Li, L. Multi-criteria decision-making method based on single-valued neutrosophic linguistic Maclaurin symmetric mean operators. *Neural Comput. Appl.* **2018**, *30*, 1529–1547. [CrossRef]
15. Chen, J.; Zeng, S.Z.; Zhang, C.H. An OWA distance-based, single-valued neutrosophic linguistic topsis approach for green supplier evaluation and selection in low-carbon supply chains. *Int. J. Environ. Res. Public Health* **2018**, *15*, 1439. [CrossRef]
16. Cao, C.D.; Zeng, S.Z.; Luo, D.D. A single-valued neutrosophic linguistic combined weighted distance measure and its application in multiple-attribute group decision-making. *Symmetry* **2019**, *11*, 275. [CrossRef]
17. Garg, H.; Nancy. Linguistic single-valued neutrosophic prioritized aggregation operators and their applications to multiple-attribute group decision-making. *J. Ambient Intell. Humaniz. Comput.* **2018**, *9*, 1975–1997. [CrossRef]
18. Merigó, J.M.; Palacios-Marqués, D.; Soto-Acosta, P. Distance measures, weighted averages, OWA operators and Bonferroni means. *Appl. Soft Comput.* **2017**, *50*, 356–366. [CrossRef]
19. Merigó, J.M.; Gil-Lafuente, A.M. New decision-making techniques and their application in the selection of financial products. *Inf. Sci.* **2010**, *180*, 2085–2094. [CrossRef]
20. Merigó, J.M.; Casanovas, M. Decision making with distance measures and linguistic aggregation operators. *Int. J. Fuzzy Syst.* **2010**, *12*, 190–198.
21. Merigó, J.M.; Casanovas, M. Decision making with distance measures and induced aggregation operators. *Comput. Ind. Eng.* **2011**, *60*, 66–76. [CrossRef]
22. Xian, S.D.; Sun, W.J.; Xu, S.H.; Gao, Y.Y. Fuzzy linguistic induced OWA Minkowski distance operator and its application in group decision making. *Pattern Anal. Appl.* **2016**, *19*, 325–335. [CrossRef]
23. Merigó, J.M.; Casanovas, M.; Zeng, S.Z. Distance measures with heavy aggregation operators. *Appl. Math. Model.* **2014**, *38*, 3142–3153. [CrossRef]
24. Zhou, L.G.; Xu, J.X.; Chen, H.Y. Linguistic continuous ordered weighted distance measure and its application to multiple attributes group decision making. *Appl. Soft Comput.* **2014**, *25*, 266–276. [CrossRef]
25. Zeng, S.Z.; Su, W.H. Intuitionistic fuzzy ordered weighted distance operator. *Knowl. Based Syst.* **2011**, *24*, 1224–1232. [CrossRef]

26. Zeng, S.Z.; Xiao, Y. A method based on TOPSIS and distance measures for hesitant fuzzy multiple attribute decision making. *Technol. Econ. Dev. Econ.* **2018**, *24*, 969–983. [CrossRef]
27. Xu, Z.S.; Xia, M.M. Distance and similarity measures for hesitant fuzzy sets. *Inf. Sci.* **2011**, *181*, 2128–2138. [CrossRef]
28. Qin, Y.; Liu, Y.; Hong, Z.Y. Multicriteria decision making method based on generalized Pythagorean fuzzy ordered weighted distance measures. *J. Intell. Fuzzy Syst.* **2017**, *33*, 3665–3675. [CrossRef]
29. Zeng, S.Z.; Chen, J.P.; Li, X.S. A hybrid method for pythagorean fuzzy multiple-criteria decision making. *Int. J. Inf. Technol. Decis. Mak.* **2016**, *15*, 403–422. [CrossRef]
30. Alfaro-García, V.G.; Merigó, J.M.; Gil-Lafuente, A.M.; Kacprzyk, J. Logarithmic aggregation operators and distance measures. *Int. J. Intell. Syst.* **2018**, *33*, 1488–1506. [CrossRef]
31. Zhou, L.; Chen, H.; Liu, J. Generalized logarithmic proportional averaging operators and their applications to group decision making. *Knowl. Based Syst.* **2012**, *36*, 268–279. [CrossRef]
32. Zhou, L.; Tao, Z.; Chen, H.Y.; Liu, J. Generalized ordered weighted logarithmic harmonic averaging operators and their applications to group decision making. *Soft Comput.* **2014**, *19*, 715–730. [CrossRef]
33. Alfaro-Garcia, V.G.; Merigo, J.M.; Plata-Perez, L.; Alfaro-Calderon, G.G.; Gil-Lafuente, A.M. Induced and logarithmic distances with multi-region aggregation operators. *Technol. Econ. Dev. Econ.* **2019**, *25*, 664–692. [CrossRef]
34. Smarandache, F. *Neutrosophy, Neutrosophic Probability, Set, and Logic. Proquest Information & Learning*; American Research Press: Ann Arbor, MI, USA, 1998.
35. Xu, Z.S. A note on linguistic hybrid arithmetic averaging operator in multiple attribute group decision making with linguistic information. *Group Decis. Negot.* **2006**, *15*, 593–604. [CrossRef]
36. Yager, R.R. On ordered weighted averaging aggregation operators in multi-criteria decision making. *IEEE Trans. Syst. Man Cybern. B* **1988**, *18*, 183–190. [CrossRef]
37. Zeng, S.Z.; Su, W.H.; Zhang, C.H. Intuitionistic fuzzy generalized probabilistic ordered weighted averaging operator and its application to group decision making. *Technol. Econ. Dev. Econ.* **2016**, *22*, 177–193. [CrossRef]
38. Yu, L.P.; Zeng, S.Z.; Merigo, J.M.; Zhang, C.H. A new distance measure based on the weighted induced method and its application to Pythagorean fuzzy multiple attribute group decision making. *Int. J. Intell. Syst.* **2019**, *34*, 1440–1454. [CrossRef]
39. Zeng, S.Z.; Peng, X.M.; Baležentis, T.; Streimikiene, D. Prioritization of low-carbon suppliers based on Pythagorean fuzzy group decision making with self-confidence level. *Econ. Res. Ekon. Istraživanja* **2019**, *32*, 1073–1087. [CrossRef]
40. Zeng, S.Z. Pythagorean fuzzy multiattribute group decision making with probabilistic information and OWA approach. *Int. J. Intell. Syst.* **2017**, *32*, 1136–1150. [CrossRef]
41. Merigó, J.M.; Yager, R.R. Generalized moving averages, distance measures and OWA operators. *Int. J. Uncertain. Fuzziness Knowl. Based Syst.* **2013**, *21*, 533–559. [CrossRef]
42. Garg, H.; Rani, D. Exponential, logarithmic and compensative generalized aggregation operators under complex intuitionistic fuzzy environment. *Group Decis. Negot.* **2019**, *28*, 991–1050. [CrossRef]
43. Rahman, K.; Abdullah, S. Some induced generalized interval-valued Pythagorean fuzzy Einstein geometric aggregation operators and their application to group decision-making. *Comput. Appl. Math.* **2019**, *38*, 139–154. [CrossRef]
44. Li, J.R.; Lu, H.X.; Zhu, J.L.; Wang, Y.B.; Li, X.P. Aquatic products processing industry in China: Challenges and outlook. *Trends Food Sci. Technol.* **2009**, *20*, 73–77. [CrossRef]
45. Zeng, S.Z.; Chen, S.M.; Kuo, L.W. Multiattribute decision making based on novel score function of intuitionistic fuzzy values and modified VIKOR method. *Inf. Sci.* **2019**, *488*, 76–92. [CrossRef]
46. Zhang, C.H.; Wang, Q.; Zeng, S.Z.; Balezentis, T.; Streimikiene, D.; Alisauskaite-Seskiene, I.; Chen, X. Probabilistic multi-criteria assessment of renewable micro-generation technologies in households. *J. Clean. Prod.* **2019**, *212*, 582–592. [CrossRef]
47. Balezentis, T.; Streimikiene, D.; Melnikienė, R.; Zeng, S.Z. Prospects of green growth in the electricity sector in Baltic States: Pinch analysis based on ecological footprint. *Resour. Conserv. Recycl.* **2019**, *142*, 37–48. [CrossRef]
48. Luo, D.D.; Zeng, S.Z.; Chen, J. A probabilistic linguistic multiple attribute decision making based on a new correlation coefficient method and its application in hospital assessment. *Mathematics* **2020**, *8*, 340. [CrossRef]

49. Zhang, C.H.; Chen, C.; Streimikiene, D.; Balezentis, T. Intuitionistic fuzzy multimoora approach for multi-criteria assessment of the energy storage technologies. *Appl. Soft Comput.* **2019**, *79*, 410–423. [CrossRef]
50. Jin, H.H.; Ashraf, S.; Abdullah, S.; Qiyas, M.; Bano, M.; Zeng, S.Z. Linguistic spherical fuzzy aggregation operators and their applications in multi-attribute decision making problems. *Mathematics* **2019**, *7*, 413. [CrossRef]

MDPI
St. Alban-Anlage 66
4052 Basel
Switzerland
Tel. +41 61 683 77 34
Fax +41 61 302 89 18
www.mdpi.com

Mathematics Editorial Office
E-mail: mathematics@mdpi.com
www.mdpi.com/journal/mathematics

www.ingramcontent.com/pod-product-compliance
Lightning Source LLC
LaVergne TN
LVHW070225100526
838202LV00015B/2093

* 9 7 8 3 0 3 9 4 3 2 8 8 2 *